Concepts in
Biology
Twelfth Edition

Eldon D. Enger

Frederick C. Ross

David B. Bailey

Delta College

McGraw Hill **Higher Education**

Boston   Burr Ridge, IL   Dubuque, IA   Madison, WI   New York   San Francisco   St. Louis
Bangkok   Bogotá   Caracas   Kuala Lumpur   Lisbon   London   Madrid   Mexico City
Milan   Montreal   New Delhi   Santiago   Seoul   Singapore   Sydney   Taipei   Toronto

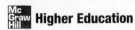

 Higher Education

CONCEPTS IN BIOLOGY, TWELFTH EDITION

2 3 4 5 6 7 8 9 0 WCK/WCK 0 9 8 7

ISBN 978–0–07–298657–0
MHID 0–07–298657–3

Publisher: *Janice Roerig-Blong*
Sponsoring Editor: *Thomas C. Lyon*
Developmental Editor: *Darlene M. Schueller*
Marketing Manager: *Tamara Maury*
Senior Project Manager: *Mary E. Powers*
Senior Production Supervisor: *Laura Fuller*
Senior Media Project Manager: *Jodi K. Banowetz*
Media Producer: *Eric A. Weber*
Senior Designer: *David W. Hash*
Cover/Interior Designer: *Rokusek Design*
(USE) Cover Image: Bobcat Lynx rufus portrait rocky mountains, © *Oxford Scientific/Daniel Cox*
Senior Photo Research Coordinator: *Lori Hancock*
Photo Research: *LouAnn K. Wilson*
Supplement Producer: *Melissa M. Leick*
Compositor: *Carlisle Communications, Ltd.*
Typeface: *10/12 Sabon*
Printer: *Quebecor World Versailles, KY*

The credits section for this book begins on page 642 and is considered an extension of the copyright page.

Library of Congress Cataloging-in-Publication Data

Enger, Eldon D.
  Concepts in biology / Eldon D. Enger, Frederick C. Ross, David B. Bailey. — 12th ed.
    p. cm.
  Includes index.
  ISBN 978–0–07–298657–0—ISBN 0–07–298657–3 (hard copy : alk. paper)
  1. Biology. I. Ross, Frederick C. II. Bailey, David B. III. Title.

QH308.2.E54   2007
570—dc22   2006002607
                                                     CIP

www.mhhe.com

Meet the Authors

Eldon D. Enger (Center)

Eldon D. Enger is a professor emeritus of biology at Delta College, a community college near Saginaw, Michigan. He received his B.A. and M.S. degrees from the University of Michigan. Professor Enger has over 30 years of teaching experience, during which he taught biology, zoology, environmental science, and several other courses, and he was very active in curriculum and course development. Professor Enger is an advocate for variety in teaching methodology. He feels that, if students are provided with varied experiences, they are more likely to learn. In addition to the standard textbook assignments, lectures, and laboratory activities, his classes were likely to include writing assignments, student presentation of lecture material, debates by students on controversial issues, field experiences, individual student projects, and discussions of local examples and relevant current events.

Professor Enger has been a Fulbright Exchange Teacher to Australia and Scotland, received the Bergstein Award for Teaching Excellence and the Scholarly Achievement Award from Delta College.

Professor Enger is married, has two sons, and enjoys a variety of outdoor pursuits, such as cross-country skiing, hiking, hunting, fishing, camping, and gardening. Other interests include reading a wide variety of periodicals, beekeeping, singing in a church choir, and preserving garden produce.

Frederick C. Ross (Right)

Fred Ross is a professor emeritus of biology at Delta College, a community college near Saginaw, Michigan. He received his B.S. and M.S. from Wayne State University, Detroit, Michigan, and has attended several other universities and institutions. Professor Ross has over 30 years' teaching experience, including junior and senior high school. He has been very active in curriculum development and has developed the courses "Infection Control and Microbiology" and "AIDS and Infectious Diseases," a PBS ScienceLine course. He has also been actively involved in the National Task Force of Two Year College Biologists (American Institute of Biological Sciences); N.S.F. College Science Improvement Program (COSIP); Evaluator for Science and Engineering Fairs; Michigan Community College Biologists (MCCB); Judge for the Michigan Science Olympiad and the Science Bowl; and a member of the Topic Outlines in Introductory Microbiology Study Group of the American Society for Microbiology.

Professor Ross involved his students in a variety of learning techniques and was a prime advocate of writing-to-learn. Besides writing, his students were typically engaged in active learning techniques, including use of inquiry based learning, the Internet, e-mail communications, field experiences, classroom presentation, and lab work. The goal of his classroom presentations was to actively engage the minds of his students in understanding the material, not just memorization of "scientific facts."

David B. Bailey (Left)

David B. Bailey is an associate professor of biology at Delta College, a community college near Saginaw, Michigan. He received his B.A. from Hiram College, Hiram, Ohio, and his Ph. D. from Case Western Reserve University in Cleveland, Ohio. Dr. Bailey has been teaching in classrooms and labs for 10 years in both community colleges and 4-year institutions. He has taught general biology, introductory zoology, cell biology, molecular biology, biotechnology, genetics, and microbiology. Dr. Bailey is currently directing Delta's General Education Program.

Dr. Bailey strives to emphasize critical thinking skills so that students can learn from each other. Practicing the scientific method and participating in discussions of literature, religion, and movies, students are able to learn how to practice appropriate use of different critical thinking styles. Comparing different methods of critical thinking for each of these areas, students develop a much more rounded perspective on their world.

Dr. Bailey's community involvement includes participating with the Michigan Science Olympiad. In his spare time, he enjoys camping, swimming, beekeeping, and wine-making.

Brief Contents

Contents

PART III

Molecular Biology, Cell Division, and Genetics   143

14   The Formation of Species and Evolutionary Change   281

15   Ecosystem Dynamics: The Flow of Energy and Matter   303

16   Community Interactions   323

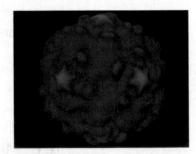

PART V

The Origin and Classification of Life 401

PART VI
Physiological Processes   519

Table of Boxes

Preface

The origin of this book remains deeply rooted in our concern for the education of college students in the field of biology. We believe that large, thick books intimidate introductory-level students who are already anxious about taking science courses. With each edition, we have worked hard to provide a book that is useful, interesting, and engaging to students while introducing them to the core concepts and current state of the science.

The Twelfth Edition

There are several things about the twelfth edition of *Concepts in Biology* that we find exciting. This revision, as with previous editions, is very much a collaborative effort. When we approach a revision, we carefully consider comments and criticisms of reviewers and discuss how to address their suggestions and concerns. As we proceed through the revision process, we solicit input from one another and we critique each other's work.

Overview of Changes to *Concepts in Biology*, Twelfth Edition

Opening Chapter Vignette
A new opening chapter vignette sets the scene for the student by demonstrating the relevance of that chapter's content to their daily lives.

Enhanced Visuals and Page Layout
The art program has been a prime focus for this edition. We have added and updated a large percentage of line drawings and photographs. Page design has been updated to provide students with a more open, inviting, and useable text.

Increased Student Accessibility
The readability of the text has been improved by the use of additional subheadings and a more direct writing style.

See the list of topical additions and updates made to the twelfth edition.

Biochemical Pathways
In previous editions, biochemical pathways were presented in one chapter (chapter 6). We decided to split this chapter into two chapters: Biochemical Pathways—Cellular Respiration and Biochemical Pathways—Photosynthesis. The complexity of these metabolic pathways has long been difficult for students to grasp. To help with this problem, and with the encouragement of reviewers, we moved the material on ATP, electron transport, and proton pumps to the previous chapter—Enzymes, Coenzymes, and Energy. These changes should allow students to more easily sort out the details of each pathway and understand the information.

New Chapter—Applications of Biotechnology
In previous editions, biotechnology was addressed in the same chapter as DNA replication and protein synthesis. This new chapter presents biotechnology as the processes of making comparisons of DNA or transferring DNA from one cell to another. This chapter also provides students with fundamental information on current topics, such as why biotechnology works, gene sequencing, the Human Genome Project, biotechnology ethics, DNA comparisons, stem cells, the polymerase chain reaction, and gene cloning.

Cell Division Chapter
The mitosis and meiosis chapters have been combined to stress comparisons between these two cellular processes and their implications. Treating both topics as one chapter will encourage students to find similarities in the names for the different stages. At the same time, students will also be more likely to explore the significant differences between these two types of cell division.

New End-of-Chapter Basic Review Questions
New end-of-chapter basic review questions have been added. These questions will provide students with more help in mastering chapter concepts.

Part I—Introduction

Chapter 1—What Is Biology?

The major changes to this chapter involve updating the figures and adding subheadings to help the reader recognize the organization of the material in the chapter. There are nine new illustrations and several of the original figures have been modified.

Part II—Cornerstones: Chemistry, Cells, and Metabolism

Chapter 2—The Basics of Life: Chemistry

This chapter has been restructured to make an otherwise challenging topic more manageable. Many figures have been changed in order to enhance learning. There is also a new Outlooks box on how buffers work.

Chapter 3—Organic Molecules—the Molecules of Life

Additional subheadings, examples, and figures allow students to more easily follow the flow of ideas. New information is presented on reducing LDL levels and controlling cholesterol. There is a new Thinking Critically box that focuses on the use of alcoholic beverages and societal issues concerning its use.

Chapter 4—Cell Structure and Function

This chapter was reorganized with extensive revision of the sections on cell size, cellular membranes, and diffusion and osmosis. There is a new How Science Works box on the development of the fluid-mosaic model. Most of the illustrations are new or revised.

Chapter 5—Enzymes, Coenzymes, and Energy

New material on Gaucher disease has been added, along with many new illustrations. In addition, the reorganization of the chapter dealing with biochemistry resulted in the addition of a new section on enzymatic reactions used in processing energy and matter. This section presents information on biochemical pathways, generating ATP, electron transport, and proton pumps.

Chapter 6—Biochemical Pathways—Cellular Respiration

The original chapter on biochemical pathways has been divided into two chapters. This chapter focuses on the process of cellular respiration. The first section presents material on how organisms handle energy to meet their needs. There are two new Outlooks boxes: one on the processes of spoiling and souring and one on lipid metabolism and ketoacidosis. The new Thinking Critically feature challenges students to trace the biochemical pathways they would follow if they were a hydrogen atom as they moved through the process of aerobic cellular respiration.

Chapter 7—Biochemical Pathways—Photosynthesis

Photosynthesis is now in its own chapter, with many new illustrations.

Part III—Molecular Biology, Cell Division, and Genetics

Chapters in this part have been updated to reflect recent advances in fields of molecular biology, cellular biology, and biotechnology. These four chapters are revised to emphasize the genetic implications of cell division, Mendelian genetics, and biotechnology for the cell's protein synthesis activities.

Chapter 8—DNA and RNA: The Molecular Basis of Heredity

This chapter presents the traditional sequence of DNA replication, followed by protein synthesis and the effects of mutation on protein synthesis. The illustration updates for this edition more clearly present transcription, translation, and DNA replication. New illustrations address gene structure, splicing, and the effects of mutations.

Chapter 9—Cell Division—Replication and Reproduction

Chapter 9 presents both mitosis and meiosis in one chapter to stress their common cellular processes and contrast their differences. Cancer is used to provide the student with insight into the regulatory mechanisms used to control cell division and how cancer treatments frequently attempt to use cellular processes to the advantage of the patient.

Chapter 10—Patterns of Inheritance

Two new tools for teaching genetics problems are presented. The Gene Key helps students organize allele interactions by directing the students to identify and examine phenotype and genotype information for each gene. The Solution Pathway helps students identify and organize the information in a genetics problem by placing the information from each problem into a single, generalized process that stresses key concepts, such as gamete formation, independent assortment, segregation, and allelic interactions.

Chapter 11—Applications of Biotechnology

This is a new chapter for *Concepts in Biology*. In previous editions, biotechnology was addressed in the same chapter as DNA replication and protein synthesis. This chapter presents biotechnology as a process of making comparisions of DNA or transferring DNA from one cell to another. This chapter emphasizes the end results of biotechnology by examining its applications. This will help the student more easily identify the timeliness and relevancy of this topic. Techniques common to biotechnology, such as gel electrophoresis, DNA sequencing, and PCR, are treated in boxes separate from the chapter's conceptual development. This allows the instructor the choice to focus on the uses and implications of biotechnology or to include a deeper discussion of how biotechnology is actually accomplished in a laboratory setting.

Part IV—Evolution and Ecology

Chapter 12—Diversity Within Species and Population Genetics
The sequence of presentation within this chapter has been changed. Many figures are either modified or new.

Chapter 13—Evolution and Natural Selection
Information on the development of evolutionary thought has been updated and moved from a later chapter. There are many new figures.

Chapter 14—The Formation of Species and Evolutionary Change
New examples have been included throughout, along with many new illustrations. Two new How Science Works boxes have been added: one on evolution and the domesticated dog and one on the taxonomic status of Neandertals. There is also a new section on homologous and analogous structures, and material on human evolution has been updated.

Chapter 15—Ecosystem Dynamics: The Flow of Energy and Matter

Chapter 16—Community Interactions
Chapters 15 and 16 have been completely reorganized and over 80% of the figures are new or revised. Chapter 15 now focuses on ecosystem-level concepts, such as energy flow, trophic relationships, and nutrient cycles. Chapter 16 deals with concepts related to interactions among organisms within communities. The concepts of niche, habitat, kinds of organism interactions, biomes, and succession are some of the topics covered in chapter 16. It also has a new Outlooks box on biodiversity hotspots.

Chapter 17—Population Ecology
There is a new section on population distribution and a new Outlooks box on the lesser snow goose population problem. Two-thirds of the figures are new or revised.

Chapter 18—Evolutionary and Ecological Aspects of Behavior
There has been extensive reorganization of the material in the chapter, and the sections on anthropomorphism, territorial behavior, and social behavior were extensively revised. A new section on communication was added and many of the figures are new or revised.

Part V—The Origin and Classification of Life

Chapter 19—The Origin of Life and the Evolution of Cells
This chapter was structurally reorganized and many new heads were added to help the reader make sense of this complex topic. The section on the origin of eukaryotic cells was rewritten. About half of the figures are new or revised.

Chapter 20—The Classification and Evolution of Organisms
Most of the changes to this chapter are organizational. The How Science Works box on cladistics was greatly modified and the table on human classification was greatly simplified. Over half of the figures are new or revised.

Chapter 21—The Nature of Microorganisms

Chapter 22—The Plant Kingdom

Chapter 23—The Animal Kingdom
Chapters 21–23 have been completely rewritten. Each chapter is now clearly organized on a phylogenetic basis and discusses the major structural and functional features of each taxon. In addition, the ecological significance of each kind of organism is discussed. Most of the figures in these chapters are new or revised. Several new boxed readings discuss the microbial ecology of a cow, fairy rings, the terminology used in classifying plants, the major food crops of the world, and the discovery of coelacanths.

Part VI—Physiological Processes

Chapter 24—Materials Exchange in the Body
New material has been added on white blood cells, the nature of plasma, and blood proteins. A new section has been added that presents the lymphatic system, including material on lymph organs and the role this system plays.

Chapter 25—Nutrition: Food and Diet
New to this chapter is a How Science Works box on estimating basal metabolic rate. There is a new section on Dietary Reference Intakes. There is updated material on the Food Guide Pyramid, diet and weight control, and osteoporosis. The concept of fitness is also introduced.

Chapter 26—The Body's Control Mechanisms and Immunity
A section covering the body's defense mechanisms has been added, along with many new illustrations. This section presents the topics of nonspecific defenses, specific defenses, allergic reactions, and autoimmune and immunodeficiency diseases. Also new are materials on homeostasis, negative and positive feedback, and somatic and autonomic nervous systems, as well as an Outlooks box on organ transplants.

Chapter 27—Human Reproduction, Sex, and Sexuality
There is a new section presenting the spectrum of human sexuality. Updated information is included on Klinefelter's syndrome, Barr bodies, the SRY gene, and erectile dysfunction. There is a new Outlooks box on menstruation and endometriosis and two new How Science Works boxes: one on assisted reproductive technology and one on the history of pregnancy testing. A new table presents the common causes of infertility.

Features

Opening Vignette   The vignette is designed to pique students' interest and help them recognize the application and relevance of the topics presented in each chapter.

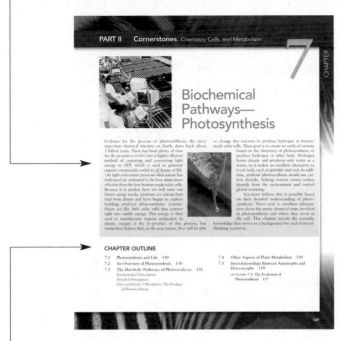

Chapter Outline   At the opening of each chapter, the outline lists the major headings in the chapter, as well as the boxed readings.

Quality Visuals   The line drawings and photographs illustrate concepts or associate new concepts with previously mastered information. Every illustration emphasizes a point or helps teach a concept.

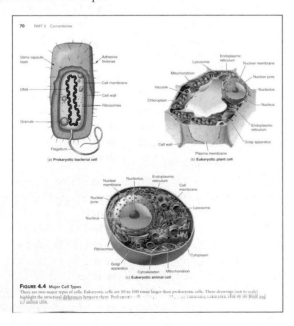

Topical Headings   Throughout each chapter, headings subdivide the material into meaningful sections that help readers recognize and organize information.

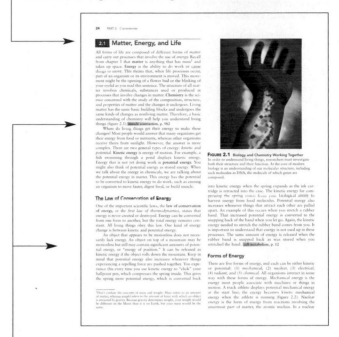

Cross-References   Highlighted cross-references call attention to related or background material located elsewhere in the text. These guide students to review basic coverage as they study new topics and to see connections between concepts.

How Science Works and Outlooks   Each of these boxed readings was designed to catch readers' interest by providing alternative views, historical perspectives, or interesting snippets of information related to the content of the chapter.

Chapter Summary   The summary at the end of each chapter clearly reviews the concepts presented.

Summary

Enzymes are protein catalysts that speed up the rate of chemical reactions without any significant increase in the temperature. They do this by lowering activation energy. Enzymes have a very specific structure, which matches the structure of particular substrate molecules. The substrate molecule comes in contact with only a specific part of the enzyme molecule—the attachment site. The active site of the enzyme is the place where the substrate molecule is changed. The enzyme-substrate complex reacts to form the end product. The protein nature of enzymes makes them sensitive to environmental conditions, such as temperature and pH, that change the structure of proteins. The number and kinds of enzymes are ultimately controlled by the genetic information of the cell. Other kinds of molecules, such as coenzymes, inhibitors, and competing enzymes, can influence specific enzymes. Changing conditions within the cell shift its enzymatic priorities by influencing the turnover number.

Enzymes are also used to speed and link chemical reactions into biochemical pathways. The energy currency of the cell, ATP, is produced by enzymatic pathways known as electron transport and proton pumping. The four concepts of biochemical pathways, ATP production, electron transport, and the proton pump are all interrelated.

Thinking Critically   This feature gives students an opportunity to think through problems logically and arrive at conclusions based on the concepts presented in the chapters. Guidelines to assist students in thinking about these questions are found on the Assessment Review Instruction System (ARIS) site.

Thinking Critically

*For guidelines to answer this question, visit the **Concepts in Biology, 12/e website.***

The scientific method is central to all work that a scientist does. Can this method be applied to the ordinary activities of life? How might a scientific approach change how you choose your clothing, your recreational activities, or a car? Can these choices be analyzed scientifically? Should they be analyzed scientifically? Is there anything wrong with looking at these decisions from a scientific point of view?

Page-Referenced Key Terms   A list of page-referenced key terms in each chapter helps students identify the vocabulary they need to understand the concepts and ideas presented in the chapter. Definitions are found in the glossary at the end of the text. Students can practice learning key terms with interactive flash cards on the Assessment Review Instruction System (ARIS) site.

Basic Review and Concept Review Questions   Students can assess their knowledge by answering the basic review questions. The answers to the basic review questions are given at the end of the question set, so students can get immediate feedback. The concept review questions are designed as a "writing-to-learn" experience. Students are asked to address the concept questions by writing a few sentences, making a list, or composing a paragraph. Concept review questions are answered on the student site of the Assessment Review Instruction System (ARIS).

Teaching Supplements for the Instructor

McGraw-Hill offers a variety of tools and technology products to support the twelfth edition of *Concepts in Biology.*

Laboratory Manual   The laboratory manual features 30 carefully designed, class-tested learning activities. Each exercise contains an introduction to the material, step-by-step procedures, ample space to record and graph data, and review questions. The activities give students an opportunity to go beyond reading and studying to actually participate in the process of science.

Digital Content Manager This collection of multimedia resources provides tools for rich visual support of your lectures. You can utilize artwork from the text in multiple formats to create customized classroom presentations, visually based tests and quizzes, dynamic course website content, or attractive printed support materials. The following assets are grouped by chapter on a cross-platform CD-ROM or DVD:

Art—Full-color digital files of all illustrations in the book, plus the same art saved in an unlabeled version, can be readily incorporated into lecture presentations, exams, or custom-made classroom materials.

Photos—All photos from the text are available in color. A separate folder contains hundreds of additional photos relative to the study of biology.

Tables—Every table that appears in the text is provided in electronic format.

Animations—Full-color presentations of key biological processes have been brought to life via animation. These animations offer flexibility for instructors and were designed to be used in lecture.

PowerPoint Lecture Outlines—A ready-made presentation that combines lecture notes and art is written for each chapter. The outlines can be used as they are, or you can customize them to preferred lecture topics and sequences.

Active Art—Illustrations depicting key processes have been converted to a format that allows the artwork to be edited inside PowerPoint. Each piece can be broken down to its core elements, grouped or ungrouped, and edited to create customized illustrations.

Instructor's Testing and Resource CD This cross-platform CD-ROM provides a wealth of resources. Among the supplements featured on this CD-ROM is a computerized test bank that uses testing software to quickly create customized exams. Word files of the test bank questions are provided for those instructors who prefer to work outside of the test-generator software.

Other assets on the Instructor's Testing and Resource CD include an instructor's manual and the Laboratory Resource Guide. Both of these assets can also be accessed via the Assessment Review Instruction System (ARIS) site at *www.mhhe.com/enger12*.

For each chapter, the instructor's manual contains a brief narrative summarizing the chapter and goals/objectives.

The Laboratory Resource Guide provides information on acquiring, organizing, and preparing laboratory equipment and supplies. Estimates of the time required for students to complete individual laboratory experiences are provided, as well as answers to questions in the laboratory manual.

Overhead Transparencies A set of 250 full-color transparencies is available free to adopters of the twelfth edition of *Concepts in Biology*. This set includes figures and some tables from the text.

Assessment Review Instruction System (ARIS)

www.mhhe.com/enger12

McGraw-Hill's ARIS for *Concepts in Biology* is a complete electronic homework and course management system, designed for greater ease of use than any other system available. Free upon adoption of *Concepts in Biology*, you can create and share course materials and assignments with colleagues with a few clicks of the mouse. You can edit questions, import your own content, and create announcements and due dates for assignments. ARIS has automatic grading and reporting of homework, quizzing, and testing. Once a student is registered in the course, all student activity within McGraw-Hill's ARIS is automatically recorded and available to you through a fully integrated grade book, which can be downloaded to Excel.

eInstruction Classroom Performance System (CPS) Wireless technology brings interactivity into the classroom or lecture hall. The Classroom Performance System (CPS) is an interactive system that allows you to administer questions electronically during class. Instructors and students receive immediate feedback through wireless response pads, which are easy to use and engage students. eInstruction can be used to

- Take attendance
- Administer quizzes and tests
- Create a lecture with intermittent questions
- Manage lectures and student comprehension through use of the CPS grade book
- Integrate interactivity into PowerPoint presentations

Course Delivery Systems With help from our partners, WebCT, Blackboard, eCollege, and other course management systems, you can take complete control over your course content. These course cartridges also provide online testing and powerful student tracking features.

McGraw-Hill: *Biology Digitized Video Clips on DVD* McGraw-Hill is pleased to offer adopting instructors a new presentation tool—digitized biology video clips on DVD. Licensed from some of the highest-quality science video producers in the world, these brief segments range from about 5 seconds to just under 3 minutes in length and cover all areas of general biology from cells to ecosystems. Engaging and informative, McGraw-Hill's digi-

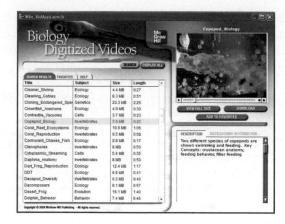

tized biology videos will help capture students' interest while illustrating key biological concepts and processes such as mitosis, how cilia and flagella work, and how some plants have evolved into carnivores.

Learning Supplements for the Student

Student Study Guide

This student study guide provides an overview for each chapter along with several learning activities. There is a focus on vocabulary terms in a section entitled "What do you mean by that?". There are several activities that help students master the material. These include: questions with short answers, label/diagram/explain activities, and multiple choice questions. A new section entitled "Asking the right question" provides open-ended questions and an opportunity for students to develop questions that they can ask their instructors. Answers to all of the questions are provided in the study guide.

Assessment Review Instruction System (ARIS)

The ARIS site at *www.mhhe.com/enger12* offers a vast array of online content to fortify the learning experience. This site features quizzes, interactive activities, and study tools tailored to coincide with each chapter of the text.

Acknowledgments

A large number of people have helped us write this text. Our families continued to give understanding and support as we worked on this revision. We acknowledge the thousands of students in our classes who have given us feedback over the years concerning the material and its relevancy. They were the best possible sources of criticism.

We would like to thank Susan Rouse from Southern Wesleyan University, who prepared the PowerPoint lecture outlines.

We gratefully acknowledge the invaluable assistance of the following reviewers throughout the development of the manuscript:

Consuelo Alvarez, *Longwood University*
Dianne L. Anderson, *San Diego City College*
Estry Z. Ang, *University of Pittsburgh at Greensburg*
Bert Atsma, *Union County College*
Gretchen S. Bernard, *Moraine Valley Community College*
Charles L. Biles, *East Central University*
Dennis Bogyo, *Valdosta State University*
James L. Botsford, *New Mexico State University*
Richard R. Bounds, *Mount Olive College*
Robert Boyd, *Auburn University*
Clarence Branch Jr., *St. Augustine's College*
James M. Britton, *Arkansas State University–Beebe*
Dale F. Burnside, *Lenoir-Rhyne College*
Michelle Cawthorn, *Georgia Southern University*
Bane W. Cheek, *Polk Community College*
Nicholas J. Cheper, *East Central University*
Andrew R. Dyer, *University of South Carolina Aiken*
Ana Esther Escandon, *Los Angeles Harbor College*
Mark L. Fink, *Longwood University*
Edison R. Fowlks, *Hampton University*
Andrew Goliszek, *North Carolina A&T State University*
Stanley Z. Guffey, *University of Tennessee*
Nicholas D. Hackett, *Moraine Valley Community College*
Richard Hanke, *Rose State College*
James A. Hewlett, *Finger Lakes Community College*
Mario Hollomon, *Texas Southern University*
R. A. Karpie, *Brevard Community College, Melbourne Campus*
Lee Kavaljian, *California State University, Sacramento*
Mary Lehman, *Longwood University*
N. J. LoCascio, *Marshall University*
Craig Longtine, *North Hennepin Community College*
Blase Maffia, *University of Miami*
Karen Benn Marshall, *Montgomery College–Takoma Park Campus*
David McFadden, *California State University, Fresno*
Dalivan Melendrez, *Harold Washington College*
Mark A. Melton, *St. Augustine's College*
Timothy D. Metz, *Campbell University*
Hao Nguyen, *California State University, Sacramento*
Steven J. Oliver, *Worcester State College*
Brian K. Paulson, *California University of Pennsylvania*
Kathleen Pelkki, *Saginaw Valley State University*
Nancy A. Penncavage, *Suffolk County Community College*
Susan Phillips, *Brevard Community College*
Robert L. Pope, *Miami-Dade College, Kendall Campus*
James V. Price, *Utah Valley State College*
Lisa Rapp, *Springfield Technical Community College*
Timothy K. Revell, *Mount San Antonio College*

Penelope ReVelle, *Community College of Baltimore County–Essex*
Todd A. Rimkus, *Marymount University*
Bill Rogers, *Ball State University*
Steve Rothenberger, *University of Nebraska–Kearney*
Mark W. Salata, *Gordon College*
Leba Sarkis, *Aims Community College*
Roy D. Schodtler, *Lake Land College*
Cara Shillington, *Eastern Michigan University*
Kerri M. Skinner, *University of Nebraska–Kearney*
Del William Smith, *Modesto Junior College*
Minou Djawdan Spradley, *San Diego City College*
Carol St. Angelo, *Hofstra University*
Erika Stephens, *Arkansas State University*
Robert J. Sullivan, *Marist College*
Willetta Toole-Simms, *Azusa Pacific University*
Randall L. Tracy, *Worcester State College*
Dirk Vanderklein, *Montclair State University*
Charles A. Wade, *C. S. Mott Community College*
D. Alexander Wait, *Southwest Missouri State University*
James A. Wallis II, *St. Petersburg College*
Jennifer M. Warner, *University of North Carolina at Charlotte*

Craig Weaver, *Brevard Community College*
Wayne H. Whaley, *Utah Valley State College*
Mary L. Wilson, *Gordon College*
Heather Wilson-Ashworth, *Utah Valley State College*
Edwin M. Wong, *Western Connecticut State University*
James R. Yates, *University of South Carolina Aiken*
Calvin Young, *Fullerton College*
Danette D. Young, *Virginia State University*
Brenda Zink, *Northeastern Junior College*
Michelle Zurawski, *Moraine Valley Community College*

We also want to express our appreciation to the entire McGraw-Hill book team for their wonderful work in putting together this edition. Tom Lyon, sponsoring editor, has continued to support this project over the years with enthusiasm and creative ideas. Darlene Schueller oversaw the many facets of the developmental stages, Mary Powers kept everything running smoothly through the production process, Lori Hancock assisted us with the photos, David Hash provided us with a beautiful design, and Tamara Maury who promoted the text and educated the sales reps on its message.

1

What Is Biology?

How does your life compare with that of someone who lived 300 years ago? People did not know what caused disease, there were few machines, and travel involved walking or using horse-drawn vehicles. Communication was by word-of-mouth or hand-carried letter. Many scientific discoveries led to improvements in the way people lived. However, they did not always accept scientific advances or new technology. For example, people resisted using vaccines to control disease and substituting machines for the animals that pulled wagons.

Do we still resist scientific advances today? Researchers' measurements confirm that, since machines first began to replace the use of animals for power, the amount of carbon dioxide in the atmosphere has risen steadily. Furthermore, there is a strong correlation between increasing carbon dioxide and a warming climate. How have you been affected by this new scientific knowledge?

Scientific studies show that smoking causes lung cancer, emphysema, and heart disease in many people who smoke, yet cigarettes are still legal and companies are allowed to advertise them to the public. Why do people ignore this information?

The study of how diet and exercise affect health is based on many kinds of scientific investigations. However, most people obtain their information about nutrition from advertising and highly marketed diet books, rather than scientifically reputable sources. Few of us exercise regularly and 50% of Americans are obese or overweight.

This chapter introduces you to the process of science. It will help you see how science works and appreciate that science involves a way of thinking that requires the careful evaluation and testing of ideas. Science cannot make decisions for you, but it can present you with information that can help you make better decisions throughout your life.

CHAPTER OUTLINE

1.1 The Significance of Biology in Your Life

Many students question the need for science courses, such as biology, especially when their area of study is not science-related. However, it is becoming increasingly important for everyone to be able to recognize the power and limitations of science. We must all understand how scientists think and how the actions of societies change the world in which we and other organisms live. Consider how the following issues could influence your future:

- Does electromagnetic radiation from electric power lines, computer monitors, cell phones, and microwave ovens affect living things?
- Is DNA testing reliable enough to serve as evidence in court cases?
- Can we control a person's weight with a pill?
- Can scientists manipulate our genes to control disease conditions we have inherited?
- Will the thinning of the ozone layer of the upper atmosphere increase the incidence of skin cancer?
- Will a vaccine for AIDS be developed in the next 10 years?
- Will new, inexpensive, socially acceptable methods of birth control be developed that can slow world population growth?
- Are human activities causing the world to get warmer?
- Does the extinction of one species really affect the balance of life in an area?

As an informed citizen, you can have a great deal to say about how these problems are analyzed and what actions provide appropriate solutions. In a democracy, it is assumed that the public has gathered enough information to make intelligent decisions (figure 1.1). This is why an understanding of the nature of science and fundamental biological concepts is so important for any person, regardless of his or her occupation. *Concepts in Biology* was written with this philosophy in mind. This book presents core concepts selected to help you become more aware of how biology influences nearly every aspect of your life.

Most of the important questions of today can be considered from philosophical, scientific, and social points of view. However, none of these approaches individually answers those questions. For example, it is a fact that the human population of the world is growing rapidly. Philosophically, we may all agree that the rate of population growth should be slowed. Science can provide information about why populations grow and which actions will be the most effective in slowing population growth. Science can also develop effective methods of birth control. Social leaders can suggest strategies for population control that are acceptable within a society. It is important to recognize that science does not have the answers to all of our problems. In this situation, society must make the fundamental philosophical decisions about reproductive rights and the morality of various control methods.

FIGURE 1.1 Biology in Everyday Life
These news headlines reflect a few of the biologically based issues that face us every day. Although articles such as these seldom propose solutions, they do inform the general public, so that people can begin to explore possibilities and make intelligent decisions that lead to solutions.

1.2 Science and the Scientific Method

Most textbooks define **biology** as the science that deals with life. This definition seems clear until you begin to think about what the words *science* and *life* mean.

Science is actually a *process* used to solve problems or develop an understanding of natural events that involves testing possible answers. The process has become known as the *scientific method*. The **scientific method** is a way of gaining information (facts) about the world by forming possible solutions to questions, followed by rigorous testing to determine if the proposed solutions are supported.

Basic Assumptions in Science

When using the scientific method, scientists make some basic assumptions:

- There are specific causes for events observed in the natural world.
- The causes can be identified.
- There are general rules or patterns that can be used to describe what happens in nature.
- An event that occurs repeatedly probably has the same cause.

- What one person observes can be observed by others.
- The same fundamental rules of nature apply, regardless of where and when they occur.

For example, we have all observed lightning with thunderstorms. According to the assumptions that have just been stated, we should expect that there is a cause of all cases of lightning, regardless of where or when they occur, and that all people could make the same observations. We know from scientific observations and experiments that (1) lightning is caused by a difference in electrical charge, (2) the behavior of lightning follows the same general rules as those for static electricity, and (3) all lightning that has been measured has had the same cause wherever and whenever it has occurred.

Cause-and-Effect Relationships

Scientists distinguish between situations that are merely correlated (happen together) and those that are correlated and show *cause-and-effect relationships*. Many events are correlated, but not all correlations show cause-and-effect. When an event occurs as a direct result of a previous event, a cause-and-effect relationship exists. For example, lightning and thunder are correlated and have a cause-and-effect relationship. Lightning causes thunder.

The relationship between autumn and trees dropping their leaves is more difficult to sort out. Because autumn brings colder temperatures, many people assume that the cold temperatures cause the leaves to turn color and fall. Cold temperatures are correlated with falling leaves. However, there is no cause-and-effect relationship. The cause of the change in trees is actually the shortening of days that occurs in the autumn.

Experiments have shown that artificially shortening the length of days in a greenhouse will cause trees to drop their leaves, with no change in temperature. Knowing that a cause-and-effect relationship exists enables us to predict what will happen, should the same set of circumstances occur in the future.

The Scientific Method

The scientific method involves an orderly, careful search for information. It also involves a continual checking and rechecking to see if previous conclusions are still supported by new information. If new evidence is not supportive, scientists discard or change their original ideas. Thus, scientific ideas undergo constant reevaluation, criticism, and modification as new discoveries are made. The scientific method has several important, identifiable components:

- Careful observation
- The construction and testing of hypotheses
- An openness to new information and ideas
- A willingness to submit one's ideas to the scrutiny of others

However, the scientific method is not an inflexible series of steps that must be followed in a specific order. Figure 1.2 shows how these steps may be linked, and table 1.1 gives an example of how scientific investigation proceeds from an initial question to the development of theories and laws.

Observation

Scientific inquiry often begins with an observation. We make an **observation** when we use our senses (i.e., smell, sight, hearing, taste, touch) or an extension of our senses (e.g., microscope,

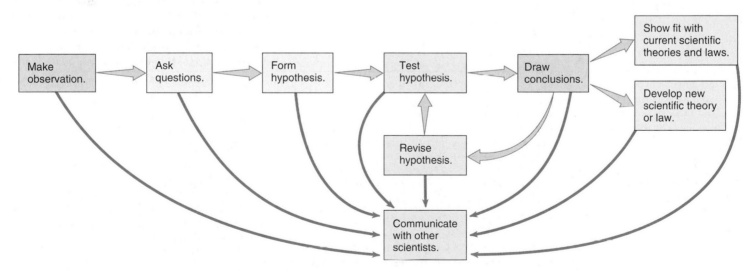

FIGURE 1.2 The Scientific Method
The scientific method is a way of thinking that involves making hypotheses about observations and testing the validity of the hypotheses. When hypotheses are disproved, they can be revised and tested in their new form. Throughout the scientific process, people communicate about their ideas. Scientific theories and laws develop as a result of people recognizing broad areas of agreement about how the world works. These laws and theories help people develop their approaches to scientific questions.

TABLE 1.1

The Nature of the Scientific Method

Component of Science Process	Description of Process	Example of the Process in Action
Make observations.	Recognize that something has happened and that it occurs repeatedly. (Empirical evidence is gained from experience or observation.)	Doctors observe that many of their patients who are suffering from tuberculosis fail to be cured by the use of the medicines (antibiotics) traditionally used to treat the disease.
Ask questions.	Ask questions about the observation, evaluate the questions, and keep the ones that will be answerable.	Have the drug companies modified the antibiotics? Are the patients failing to take the antibiotics as prescribed? Has the bacterium that causes tuberculosis changed?
Explore other sources of information.	Go to the library. Talk to others who are interested in the same problem. Communicate with other researchers to help determine if your question is a good one or if others have already answered it.	Read medical journals. Contact the Centers for Disease Control and Prevention. Consult experts in tuberculosis. Attend medical conventions. Contact drug companies and ask if their antibiotic formulation has been changed.
Form a hypothesis.	Pose a possible answer to your question. Be sure that it is testable and that it accounts for all the known information. Recognize that your hypothesis may be wrong.	Hypothesis: Tuberculosis patients who fail to be cured by standard antibiotics have tuberculosis caused by antibiotic-resistant populations of the bacterium *Mycobacterium tuberculosis*.
Test the hypothesis (experimentation).	Set up an experiment that will allow you to test your hypothesis using a control group and an experimental group. Be sure to collect and analyze the data carefully.	Set up an experiment in which samples of tuberculosis bacteria are collected from two groups of patients: those who are responding to antibiotic therapy and those who are not responding to antibiotic therapy. Grow the bacteria in the lab and subject them to the antibiotics normally used to see if the bacteria from these two groups of patients respond differently. Experiments consistently show that the patients who are not recovering have strains of bacteria that are resistant to the antibiotic being used.
Find agreement with existing scientific laws and theories or construct new laws or theories.	If your findings are seen to fit with current information, the scientific community will recognize them as being consistent with current scientific laws and theories. In rare instances, a new theory or law may develop as a result of research.	Your results are consistent with the following laws and theories: • Mendel's laws of heredity state that characteristics are passed from parent to offspring. • The theory of natural selection predicts that, when populations of *Mycobacterium tuberculosis* are subjected to antibiotics, the bacteria that survive will pass on their ability to survive exposure to antibiotics to the next generation and that the next generation will have a higher incidence of these characteristics.
Form a conclusion and communicate it.	You arrive at a conclusion. Throughout the process, communicate with other scientists by both informal conversation and formal publications.	You conclude that the antibiotics are ineffective because the bacteria are resistant to the antibiotics. You write a scientific article describing the experiment and your conclusions.

tape recorder, X-ray machine, thermometer) to record an event. There is a difference between a scientific observation and simple awareness. For example, you may hear a sound or see an image without really observing it. Do you know what music was being played the last time you were in a shopping mall? You certainly heard it but, if you are unable to tell someone else what it was, you didn't "observe" it. If you had prepared yourself to observe the music being played, you would now be able to identify it. When scientists talk about their observations, they are referring to careful, thoughtful recognition of an event—not just casual notice. Scientists train themselves to improve their observational skills, because careful observation is important in all parts of the scientific method (figure 1.3).

Questioning and Exploration

As scientists make observations, they begin to develop *questions*. How does this happen? What causes it to occur? When will it take place again? Can I control the event to my benefit? Forming questions is not as simple as it might seem, because the way you ask questions determines how you answer them. A question that is too broad or too complex may be impossible to answer; therefore, a great deal of effort is put into asking the question in the right way. In some situations, this is the most time-consuming part of the scientific method; asking the right question is critical to how you look for answers.

Do cats hunt only when they are hungry?

If given a choice between mice and cat food, which would a cat choose?

FIGURE 1.4 Questioning
The scientific method involves forming questions about what you observe.

FIGURE 1.3 Observation
Careful observation is an important part of the scientific method. This technician is making observations on the characteristics of the soil and recording the results.

Let's say that you have observed a cat catch, kill, and eat a mouse. You could ask several kinds of questions:

1a. Did the cat like the taste of the mouse?
1b. If given a choice between mice and canned cat food, which would a cat choose?
2a. What motivates a cat to hunt?
2b. Do cats hunt only when they are hungry?

Although the two sets of questions attempt to obtain similar information, 1b and 2b are much easier to answer than 1a and 2a (figure 1.4).

Once a decision has been made about what question to ask, scientists *explore other sources of knowledge* to gain more information. Perhaps a question has already been answered by someone else. Perhaps several possible answers have already been rejected. Knowing what others have already done can save time and energy. This process usually involves reading appropriate science publications, exploring information on the Internet, and contacting fellow scientists interested in the same field of study. After exploring these sources of information, a decision is made about whether to continue to consider the question. If the scientist is still intrigued by the question, he or she constructs a formal hypothesis and continues the process of inquiry at a different level.

Constructing Hypotheses

A **hypothesis** is a statement that provides a possible answer to a question or an explanation for an observation that can be tested. A good hypothesis must have the following characteristics. It must be logical. It must account for all the relevant information currently available. It must allow one to predict future events relating to the question being asked. It must be testable. Furthermore, if one has a choice of several competing hypotheses, one should use the simplest hypothesis with the fewest assumptions. Just as deciding which questions to ask is often difficult, forming a hypothesis requires much critical thought and mental exploration.

Testing Hypotheses

Scientists test a hypothesis to see if it is supported or disproved. If they disprove the hypothesis, they reject it and must construct a new hypothesis. However, if they cannot disprove a hypothesis, they are more confident in the validity of the hypothesis, even though they have not proven it true in all cases and for all time. Science always allows for the questioning of ideas and the substitution of new explanations as new information is obtained. As new information is obtained, an alternative hypothesis may become apparent and may explain the situation better than the original hypothesis. It is also possible, however, that the scientists have not made the appropriate observations to indicate that the hypothesis is wrong.

The test of a hypothesis can take several forms. First, testing may simply involve collecting relevant information that already exists. For example, suppose you visited a cemetery and observed, from reading the tombstones, that an unusually large number of people of various ages died in the same year. You could hypothesize that an epidemic of disease or a natural disaster caused the deaths. To test this hypothesis, you could consult historical newspaper accounts for that year.

Second, in some cases, additional observations may be all that is necessary to test a hypothesis. For example, suppose you hypoth-

esized that a certain species of bird uses cavities in trees as places to build nests. You could observe several birds of the species and record the kinds of nests they build and where they build them.

Third, a common method for testing a hypothesis involves devising an experiment. An **experiment** is a re-creation of an event or occurrence in a way that enables a scientist to support or disprove a hypothesis. This can be difficult, because a particular event may involve many separate factors, called **variables.** For example when a bird sings many activities of its nervous and muscular systems are involved. It is also stimulated by a wide variety of environmental factors. Understanding the factors involved in birdsong production might seem an impossible task. To help unclutter such a situation, scientists break it up into a series of simple questions and use a *controlled experiment* to answer each question.

A **controlled experiment** allows scientists to construct a situation so that only one variable is present. A typical controlled experiment includes two groups: one group in which the variable is manipulated in a particular way and one group in which there is no manipulation. The group in which there is no manipulation of the variable is called the **control group;** the other group is called the **experimental group.**

The situation involving birdsong production would have to be broken down into a large number of simple questions, such as the following: Do both males and females sing? Do they sing during all parts of the year? Is the song the same in all cases? Do some individuals sing more than others (figure 1.5)? What parts of their body are used to produce the song? What situations cause birds to start or stop singing? Each question would provide the basis for the construction of a hypothesis, which could be tested by an experiment. Each experiment would provide information about a small part of the total process of birdsong production. For example, in order to test the hypothesis that male sex hormones produced by the testes are involved in stimulating male birds to sing, an experiment could be performed in which

FIGURE 1.5 Testing Hypotheses
In order to determine the significance of birdsong production to this golden cheeked warbler, hypotheses must be tested.

one group of male birds had their testes removed (the experimental group) but the control group was allowed to develop normally.

The presence or absence of testes would be manipulated by the scientist in the experiment and would be the **independent variable.** The singing behavior of the males would be the **dependent variable,** because, if sex hormones are important, the singing behavior observed will change, depending on whether the males have testes or not (the independent variable). In an experiment, there should be only one independent variable, and the dependent variable is expected to change as a direct result of the manipulation of the independent variable. After the experiment, the new data (facts) gathered would be analyzed. If there were no differences in singing between the two groups, scientists could conclude that the independent variable evidently did not have a cause-and-effect relationship with the dependent variable (singing). However, if there were a difference, it would be likely that the independent variable was responsible for the difference between the control and experimental groups. In the case of songbirds, removal of the testes does change their singing behavior.

Scientists draw their most reliable conclusions from multiple experiments. This is because random events having nothing to do with the experiment may have altered one set of results and suggest a cause-and-effect relationship when none actually exists. For example, if the experimental group of birds became ill with bird flu, they would not sing. Scientists use several strategies to avoid the effects of random events in their experiments, including using large numbers of animals in experiments and having other scientists repeat their experiments at other locations. With these strategies, it is less likely that random events will lead to false conclusions.

Scientists must make sure that an additional variable is not accidentally introduced into experiments. For example, the operation necessary to remove the testes of male birds might cause illness or discomfort in some birds, resulting in less singing. A way to overcome this difficulty would be to subject all the birds to the same surgery but to remove the testes of only half of them. (The control birds would still have their testes.) The results of an experiment are only scientifically convincing when there is just one variable, when it has been repeated many times, and when the results are consistent.

During experimentation, scientists learn new information and formulate new questions, which can lead to even more experiments. One good experiment can result in many new questions and experiments. For example, the discovery of the structure of the DNA molecule by James D. Watson and Francis W. Crick, resulted in thousands of experiments and stimulated the development of the entire field of molecular biology (figure 1.6).

As the processes of questioning and experimentation continue, it often happens that new evidence continually and consistently supports the original hypothesis and other closely related hypotheses. When this happens, the scientific community begins to see how these hypotheses and facts fit together into a broad pattern, and a theory comes into existence.

FIGURE 1.6 One Discovery Leads to Others
In 1953, James D. Watson (left) and Francis W. Crick (right) determined the structure of the DNA molecule, which contains the genetic information of a cell. This photograph shows the model of DNA they constructed. The discovery of the structure of the DNA molecule was followed by much research into how the molecule codes information, how it makes copies of itself, and how the information is put into action.

The Development of Theories and Laws

A **theory** is a widely accepted, plausible, general statement about fundamental concepts in science that explain *why* things happen. An example of a biological theory is the germ theory of disease. This theory states that certain diseases, called *infectious* diseases, are caused by living microorganisms that are capable of being transmitted from one person to another. When these microorganisms reproduce within a person and the populations of microorganisms increase, they cause disease. As you can see, this is a very broad statement, which is the result of years of observation, questioning, experimentation, and data analysis. The germ theory of disease provides a broad overview of the nature of infectious diseases and methods for their control. However, we also recognize that each kind of microorganism has particular characteristics, which determine the kind of disease condition it causes and the appropriate methods of treatment. Furthermore, we recognize that there are many diseases that are not caused by microorganisms.

Because we are so confident that the germ theory explains why some kinds of diseases spread from one person to another, we use extreme care to protect people from infectious microorganisms. Common ways in which we prevent the spread of germs is by treating drinking water, sterilizing surgical instruments, and protecting persons with weakened immune systems from sources of infection.

Theories are different from hypotheses. A hypothesis provides a possible explanation for a specific question, a theory is a broad concept that shapes how scientists look at the world and how they frame their hypotheses. For example, when a new disease is encountered, one of the first questions asked is "What causes this disease?" A hypothesis might be constructed, which states, "The disease is caused by a microorganism." This is a logical hypothesis, because it is consistent with the general theory that many kinds of diseases are caused by microorganisms (the germ theory of disease). Because theories are broad, unifying statements, there are few of them. However, just because theories exist does not mean that testing stops. As scientists continue to gain new information, they may find exceptions to a theory or, rarely, disprove a theory.

A **scientific law** is a uniform or constant fact of nature that describes *what* happens in nature. An example of a biological law is the biogenetic law, which states that all living things come from preexisting living things. Although laws describe what happens and theories describe why things happen, there is one way in which laws and theories are similar. Both laws and theories have been examined repeatedly and are regarded as excellent predictors of how nature behaves.

In the process of sorting out the way the world works, scientists use generalizations to help them organize information. However, the generalizations must be backed up with facts. The relationship between facts and generalizations is a two-way street. Often as observations are made and hypotheses are tested, a pattern emerges, which leads to a general conclusion, principle, or theory. This process of developing general principles from the examination of many sets of specific facts is called **inductive reasoning** or **induction**. For example, when people examine hundreds of species of birds, they observe that all kinds lay eggs. From these observations, they may develop the principle that laying eggs is a fundamental characteristic of birds, without examining every species of bird.

Once a rule, principle, or theory is established, it can be used to predict additional observations in nature. The process of using general principles to predict the specific facts of a situation is called **deductive reasoning** or **deduction**. For example, after the general principle that birds lay eggs is established, one might deduce that a newly discovered species of bird also lays eggs. In the process of science, both induction and deduction are important thinking processes used to increase our understanding of the nature of our world.

Communication

One central characteristic of the scientific method is the importance of communication among colleagues. For the most part, science is conducted out in the open, under the critical eyes of others who are interested in the same kinds of questions. An important part of the communication process involves the publication of articles in scientific journals about one's research, thoughts, and opinions. This communication can occur at any point during the process of scientific discovery.

Scientists may ask questions about unusual observations. They may publish preliminary results of incomplete experiments. They may publish reports that summarize large bodies of material. And they may publish strongly held opinions that are not supportable with current data. This provides other

FIGURE 1.7 Communication
One important way that scientists communicate is through publications in scientific journals.

scientists with an opportunity to criticize, make suggestions, or agree (figure 1.7). Scientists attend conferences, where they can engage in dialog with colleagues. They also interact in informal ways by phone, e-mail, and the Internet. The result is that most of science is subjected to examination by many minds as it is discovered, discussed, and refined.

1.3 Science, Nonscience, and Pseudoscience

Fundamental Attitudes in Science

As you can see from our discussion of the scientific method, a scientific approach to the world requires a certain way of thinking. A scientist is a healthy skeptic who separates facts

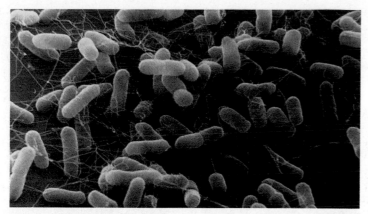

(a)

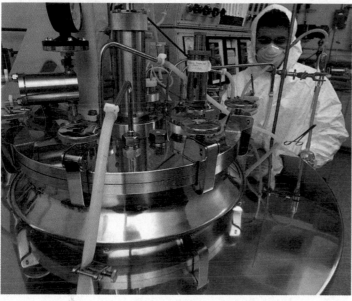

(b)

FIGURE 1.8 Genetic Engineering

Genetic engineers have modified the genetic code of bacteria, such as *Escherichia coli*, commonly found in the colon *(a)* to produce useful products, such as vitamins, protein, and antibiotics. The bacteria can be cultured in vats, where the genetically modified bacteria manufacture their products *(b)*. The products can be extracted from the mixture in the vat.

from opinions. Ideas are accepted because there is much supporting evidence from numerous studies, not because influential people have strongly held opinions.

Careful attention to detail is also important. Because scientists publish their findings and their colleagues examine their work, they have a strong desire to produce careful work that can be easily defended. This does not mean that scientists do not speculate and state opinions. When they do, however, they take great care to clearly distinguish fact from opinion.

There is also a strong ethic of honesty. Scientists are not saints, but the fact that science is conducted openly in front of one's peers tends to reduce the incidence of dishonesty. In addition, the scientific community strongly condemns and severely penalizes those who steal the ideas of others, perform shoddy science, or falsify data. Any of these infractions can lead to the loss of one's job and reputation.

Theoretical and Applied Science

The scientific method has helped us understand and control many aspects of our natural world. Some information is extremely important in understanding the structure and functioning of things in nature but at first glance appears to have little practical value. For example, the discovery of the structure of deoxyribonucleic acid (DNA) answered a fundamental question about the nature of genetic material. Initially, this information had little practical value. However, as people began to use this new knowledge, they developed many practical applications based on an understanding of the nature of DNA. For example, scientists known as *genetic engineers* have altered the chemical code system of microorganisms, in order to produce many new drugs, such as antibiotics, hormones, and enzymes. However, to produce these complex chemicals so easily, genetic engineers needed information from the basic, theoretical sciences of microbiology, molecular biology, and genetics (figure 1.8).

Another example of how fundamental research can lead to practical application is the work of Louis Pasteur (1822–1895), a French chemist and microbiologist. Pasteur was interested in the theoretical problem of whether life could be generated from nonliving material. Much of his theoretical work led to practical applications in disease control. His theory that microorganisms cause diseases and decay led to the development of vaccinations against rabies and the development of pasteurization for the preservation of foods (figure 1.9).

FIGURE 1.9 Louis Pasteur and Pasteurized Milk

Louis Pasteur (1822–1895) performed many experiments while he studied the question of the origin of life, one of which led directly to the food-preservation method now known as pasteurization.

Science and Nonscience

Both scientists and nonscientists seek to gain information and improve understanding in their fields of study. The differences between science and nonscience are based on the assumptions and methods used to gather and organize information and, most important, the way the assumptions are tested. The difference between a scientist and a nonscientist is that a scientist continually challenges and tests principles and assumptions to determine cause-and-effect relationships. A nonscientist may not be able to do so or may not believe that this is important. For example, a historian may have the opinion that, if President Lincoln had not appointed Ulysses S. Grant to be a general in the Union Army, the Confederate States of America would have won the Civil War. Although there can be considerable argument about the topic, there is no way that it can be tested. Therefore, such speculation about historical events is not scientific. This does not mean that history is not a respectable field of study, only that it is not science. Historians simply use the standards of critical thinking that are appropriate to their field of study and that can provide insights into the role military leadership plays in the outcome of conflicts.

Once you understand the scientific method, you won't have any trouble identifying astronomy, chemistry, physics, geology, and biology as sciences. But what about economics, sociology, anthropology, history, philosophy, and literature? All of these fields may make use of certain central ideas that are derived in a logical way, but they are also nonscientific in some ways. Some things are beyond science and cannot be approached using the scientific method. Art, literature, theology, and philosophy are rarely thought of as sciences. They are concerned with beauty, human emotion, and speculative thought, rather than with facts and verifiable laws.

Many fields of study have both scientific and nonscientific aspects. For example, the styles of clothing people wear are often shaped by the artistic creativity of designers and shrewd marketing by retailers. Originally, animal hides, wool, cotton, and flax were the only materials available, and the color choices were limited to the natural colors of the material or dyes extracted from nature. Scientific discoveries led to the development of synthetic fabrics and dyes, machines to construct clothing, and new kinds of fasteners which allowed for new styles and colors (figure 1.10). Similarly, economists use mathematical models and established economic laws to make predictions about future economic conditions. However, the reliability of predictions is a central criterion of science, so the regular occurrence of unpredicted economic changes indicates that economics is far from scientific. Many aspects of anthropology and sociology are scientific, but they cannot be considered true sciences, because many of the generalizations in these fields cannot be tested by repeated experimentation. They also do not show a significantly high degree of cause-and-effect, or they have poor predictive value.

(a)　　　　　　　　　　　　　(b)

FIGURE 1.10 Science and Culture

Although the design of clothing is not a scientific enterprise, scientific discoveries have altered the choices available.
(a) Originally, clothing could be made only from natural materials with simple construction methods. *(b)* The discovery of synthetic fabrics and dyes and the invention of specialized fasteners resulted in increased variety and specialization of clothing.

Pseudoscience

Pseudoscience (*pseudo* = false) is a deceptive practice that uses the appearance or language of science to convince, confuse, or mislead people into thinking that something has scientific validity. When pseudoscientific claims are closely examined, they are not found to be supported by unbiased tests.

For example, nutrition is a respectable scientific field; however, many individuals and organizations make unfounded claims about their products and diets (figure 1.11). Because of nutritional research, we all know that we must obtain certain nutrients, such as amino acids, vitamins, and minerals, from the food we eat or we may become ill. However, in most cases, it has not been demonstrated that the nutritional supplements so vigorously advertised are as useful or desirable as claimed. Rather, the advertisements select bits of scientific information about the fact that amino acids, vitamins, and minerals are essential to good health and then use this information to create the feeling that nutritional supplements are necessary or can improve health. In reality, the average person eating a varied diet can obtain all these nutrients in adequate amounts.

Another related example involves the labeling of products as organic or natural. Marketers imply that organic or natural products have greater nutritive value because they are organically grown (grown without pesticides or synthetic fertilizers) or because they come from nature. Although there are

FIGURE 1.11 Pseudoscience—"Nine out of 10 Doctors Surveyed Recommend Brand X"
Pseudoscience is designed to mislead. There are several ways in which this image and the statement can be misleading. You can ask yourself two questions. First, is the person in the white coat a physician? Second, how many doctors were asked for a recommendation and how were they selected? If only 10 doctors were asked, the sample size was too small. Perhaps the doctors who participated were selected to obtain the desired outcome. Finally, the doctors could have been surveyed in such a way as to obtain the desired answer, such as "Would you recommend Brand X over Dr. Pete's snake oil?"

questions about the health effects of trace amounts of pesticides in foods, no scientific study has shown that a diet of natural or organic products has any benefit over other diets. The poisons curare, strychnine, and nicotine are all organic molecules that are produced in nature by plants that can be grown organically, but we wouldn't want to include them in our diet.

The Limitations of Science

Science is a way of thinking that involves testing possible answers to questions. Therefore, the scientific method can be applied only to questions that have factual bases. Questions about morals, value judgments, social issues, and attitudes cannot be answered using the scientific method. What makes a painting great? What is the best type of music? Which wine is best? What color should I paint my house? These questions are related to values, beliefs, and tastes; therefore, the scientific method cannot be used to answer them.

Science is also limited by the ability of people to figure out how the natural world works. People are fallible and do not always come to the right conclusions because they lack information or misinterpret it. However science is self-correcting and,

as new information is gathered, old, incorrect ways of thinking are changed or discarded. For example, at one time scientists were sure that the Sun went around the Earth. They observed that the Sun rose in the east and traveled across the sky to set in the west. Because scientists could not feel the Earth moving, it seemed perfectly logical that the Sun traveled around the Earth. Once they understood that the Earth rotated on its axis, they began to realize that the rising and setting of the Sun could be explained in other ways. A completely new concept of the relationship between the Sun and the Earth developed (figure 1.12). Although this kind of study seems rather primitive to us today, this change in thinking about the relationship between the Sun and the Earth was a very important step forward in our understanding of the universe.

People need to understand that science cannot answer all the problems of our time. Although science is a powerful tool, there are many questions it cannot answer and many problems it cannot solve. Most of the problems societies face are

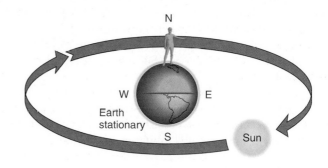

Scientists thought that the Sun revolved around the Earth.

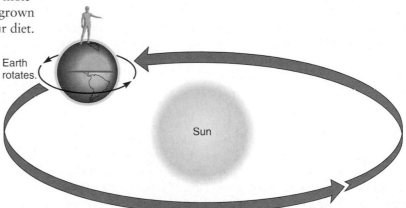

We now know that the Earth rotates on its axis and revolves around the Sun.

FIGURE 1.12 Science Is Willing to Challenge Previous Beliefs
Science must always be aware that new discoveries may force a reinterpretation of previously held beliefs. Early scientists thought that the Sun revolved around the Earth. This was certainly a reasonable theory at the time. People saw the sun rise in the east and set in the west, and it looked as if the sun moved through the sky. Today, we know that the Earth revolves around the Sun and that the apparent motion of the Sun in the sky is caused by the Earth rotating on its axis.

generated by the behavior and desires of people. Famine, drug abuse, war, and pollution are human-caused and must be resolved by humans. Science provides some important tools for social planners, politicians, and ethical thinkers. However, science does not have, nor does it attempt to provide, all the answers to the problems of the human race. Science is merely one of the tools at our disposal.

1.4   The Science of Biology

The science of biology is, broadly speaking, the study of living things. However, there are many specialty areas of biology, depending on the kind of organism studied or the goals a person has. Some biological studies are theoretical, such as establishing an evolutionary tree of life, understanding the significance of certain animal behaviors, or determining the biochemical steps involved in photosynthesis. Other fields of biology are practical—for example, medicine, crop science, plant breeding, and wildlife management. There is also just plain fun biology—fly-fishing for trout or scuba diving on a coral reef.

At the beginning of the chapter, we defined *biology* as the science that deals with life. But what distinguishes living things from those that are not alive? You would think that a biology textbook could answer this question easily. However, this is more than just a theoretical question. In recent years, it has become necessary to construct legal definitions of *life*, especially of when it begins and ends. The legal definition of *death* is important, too, because it may determine whether a person will receive life insurance benefits or if body parts may be used in transplants. In the case of a heart transplant, the person donating the heart may be legally "dead" but the heart certainly isn't. It is removed while it is still alive, even though the person is not. In other words, there are different kinds of death. There is death of the whole living unit and death of each cell within the living unit. A person actually "dies" before every cell has died. Death, then, is the absence of life, but that still doesn't tell us what life is.

What Makes Something Alive?

Living things have abilities and structures not typically found in things that were never living. The ability to manipulate energy and matter is unique to living things. **Energy** is the ability to do work or cause things to move. **Matter** is anything that has mass and takes up space. Developing an understanding of how living things modify matter and use energy will help you appreciate how living things differ from nonliving objects. Living things show five characteristics that nonliving things do not: (1) metabolic processes, (2) generative processes, (3) responsive processes, (4) control processes, and (5) a unique structural organization. It is important to recognize that, although these characteristics are typical of all living things, they may not all be present in each organism at every point in time. For example, some individuals may reproduce or grow only at certain times. This section briefly

FIGURE 1.13 Metabolism
The metabolic processes of this hummingbird include the intake of nutrients in the form of nectar from flowers.

introduces these basic characteristics of living things, which will be expanded on in the rest of the text.

Metabolic Processes

Metabolic processes are all the chemical reactions and associated energy changes that take place within an organism (figure 1.13). This set of reactions is often simply referred to as **metabolism.** Energy is necessary for movement, growth, and many other activities. The energy that organisms use is stored in the chemical bonds of complex molecules. The chemical reactions used to provide energy and raw materials to organisms occur in sequence and are controlled. Metabolic processes consist of three kinds of activities: *nutrient uptake, nutrient processing,* and *waste elimination.* Nutrient uptake occurs when living things expend energy to take in nutrients (raw materials) from their environment. Many animals take in these materials by eating other organisms. Microorganisms and plants absorb raw materials into their cells to maintain their lives. In all organisms, once inside cells, nutrients enter a network of chemical reactions. These reactions process the nutrients to manufacture new parts, make repairs, reproduce, and provide energy for essential activities. However, not all materials entering a living thing are valuable to it. Some portions of nutrients are useless or even harmful, and organisms eliminate these portions as waste. Metabolic processes also produce unusable heat energy, which can be considered a waste product.

Generative Processes

Generative processes are activities that result in an increase in the size of an organism—*growth*—or an increase in the number of individuals in a population—*reproduction* (figure 1.14).

FIGURE 1.14 Generative Processes
Reproduction is a generative process.

Growth and reproduction are directly related to metabolism, because neither can occur without gaining and processing nutrients.

During growth, living things add to their structure, repair parts, and store nutrients for later use. In large organisms, growth usually involves an increase in the number of cells present.

Reproduction is also an essential characteristic of living things. Because all organisms eventually die, life would cease to exist without reproduction. Organisms can reproduce in two basic ways. Some reproduce by *sexual reproduction*, in which two individuals each contribute sex cells, which leads to the creation of a new, unique, organism. *Asexual reproduction* occurs when an organism makes identical copies of itself.

Summer coat   Winter coat

FIGURE 1.15 Responsive Processes
The change in coat color of this varying hare is a response to changing environmental conditions.

Responsive Processes

Responsive processes allow organisms to respond to changes in their surroundings in a meaningful way. There are three categories of responsive processes: *irritability, individual adaptation,* and *evolution,* which is also known as *adaptation of populations.*

Irritability is an individual's ability to recognize a stimulus and respond rapidly to it, such as your response to a loud noise, beautiful sunset, or noxious odor. The response occurs only in the individual receiving the stimulus, and the reaction is rapid, because there are structures and processes already in place that receive the stimulus and cause the response. One-celled organisms, such as protozoa and bacteria, can sense and orient to light. Many plants orient their leaves to follow the sun. Animals use sense organs, nerves, and muscles to monitor and respond to changes in their environment.

Individual adaptation also results from an individual's reaction to a stimulus, but it is slower than an irritability response, because it requires growth or some other fundamental change in an organism. For example, during the summer the varying hare has brown fur. However, the shortening days of autumn cause the genes responsible for the production of brown pigment to be "turned off" and new, white hair grows (figure 1.15). Plants also show individual adaptation to changing day length. Lengthening days stimulate the production of flowers and shortening days result in falling leaves. Similarly, your body will respond to lower oxygen levels by producing more oxygen-carrying red blood cells. Many athletes like to train at high elevations because the increased number of red blood cells resulting from exposure to low oxygen levels delivers more oxygen to their muscles.

Evolution involves changes in the characteristics displayed within a population. It is a slow change in the genetic makeup of a *population* of organisms over generations. Evolution enables a species (a population of a specific kind of organism) to adapt to long-term changes in its environment. For example, between about 1.8 million and 11,000 years ago, the climate was cold and large continental glaciers covered northern Europe and North America. The plants and animals were adapted to these conditions. As the climate slowly warmed over the last 11,000 years, many of these species went extinct, whereas others adapted and continue in a modified form. For example, mammoths and mastodons were unable to adapt to the changing environment and became extinct, but some species, such as moose, elk, and wolves, were able to adapt to a warming environment and still exist today. Similarly, the development of the human brain and its ability to reason allowed our prehuman ancestors to craft and use tools. Their use of tools allowed them to survive and succeed in a great variety of environmental conditions.

Control Processes

Control processes are mechanisms that ensure an organism will carry out all metabolic activities in the proper sequence (*coordination*) and at the proper rate (*regulation*).

Coordination occurs within an organism at several levels. At the metabolic level, all the chemical reactions of an organism are coordinated and linked together in specific pathways. The control of all the reactions ensures efficient, stepwise handling of the nutrients needed to maintain life. The molecules responsible for coordinating these metabolic reactions are known as *enzymes*. **Enzymes** are molecules, produced by organisms, that are able to control the rate at which life's chemical reactions occur. Enzymes also regulate the amount of nutrients processed into other forms. Enzymes will be discussed in detail in chapter 5.

Coordination also occurs at the organism level. When an insect walks, the muscles of its six legs are coordinated, so that orderly movement results. In plants, regulatory chemicals assure the proper sequence of events that result in growth in the spring and early summer, followed by flowering and the development of fruit later in the year.

Regulation involves altering the rate of processes. Many of the internal activities of an organism are interrelated and regulated, so that a constant internal environment is maintained. The process of maintaining a constant internal environment is called **homeostasis.** For example, when we begin to exercise we use up oxygen more rapidly, so the amount of oxygen in the blood falls. In order to maintain a constant internal environment, the body must obtain more oxygen. This requires more rapid contractions of the muscles that cause breathing and a more rapid and forceful pumping of the heart to get blood to the lungs. These activities must occur together at the right time and at the correct rate; when they do, the level of oxygen in the blood will remain normal while supporting the additional muscular activity (figure 1.16).

Unique Structural Organization

The **unique structural organization** of living things can be seen at the cell and organism levels. **Cells** are the fundamental structural units of all living things. Cells have an outer limiting membrane and several kinds of internal structures. Each structure has specific functions. Some living things, such as people, consist of trillions of cells, whereas others, such as bacteria and yeasts, consist of only one cell. Nonliving materials, such as rocks, water, and gases, do not have a cellular structure.

An **organism** is any living thing that is capable of functioning independently, whether it consists of a single cell or a complex group of interacting cells (figure 1.17). Each kind of organism has specific structural characteristics, which it shares with all other organisms of the same kind. You recognize an African elephant, a redwood tree, or a sunflower as having certain characteristics, although other organisms may not be as easy to distinguish.

Figure 1.18 summarizes the characteristics of living things.

The Levels of Biological Organization

Biologists and other scientists like to organize vast amounts of information into conceptual chunks that are easy to relate to one another. One important concept in biology is that all living

FIGURE 1.16 Control Processes
Running involves changes in heart rate, breathing rate, and muscular activity in a controlled manner.

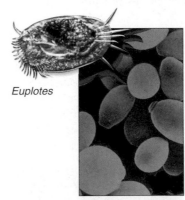

Euplotes

Yeast

Orchid

Humans

FIGURE 1.17 Structural Organization
Each organism, whether it is simple or complex, independently carries on metabolic, generative, responsive, and control processes. Some organisms, such as yeast or the protozoan *Euplotes,* consist of single cells, whereas others, such as orchids and humans, consist of many cells organized into complex structures.

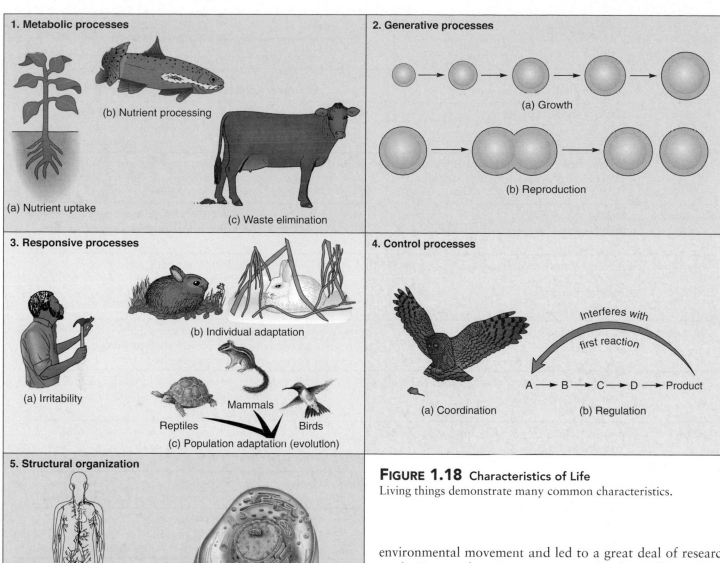

FIGURE 1.18 Characteristics of Life

Living things demonstrate many common characteristics.

things share the structural and functional characteristics we have just discussed. Another important concept is that living things can be studied at several levels (table 1.2). When biologists seek solutions to a particular problem, they may attack it at several levels at the same time. They must understand the molecules that make up living things; how the molecules are incorporated into cells; how tissues, organs, and systems function within an organism; and how changes in individual organisms affect populations and ecological relationships.

For example, in 1962, biologist Rachel Carson wrote a book entitled *Silent Spring,* in which she linked the use of certain kinds of persistent pesticides with changes in the populations of animals. This controversial book launched the modern

environmental movement and led to a great deal of research on the impact of persistent organic molecules on living things. The pesticide DDT presents a good case study of how biologists must be aware of the various levels of organization when studying any problem. DDT is an organic molecule that dissolves readily in fats and oils. It also breaks down slowly, so once present it continues to have its effects for years. Since DDT dissolves in oils, it is often concentrated in the fatty portions of animals. When a carnivore (meat-eater) eats an animal with DDT in its fat, the carnivore receives an increased dose of the toxin. Birds are particularly affected by DDT, because it interferes with the ability of many kinds of birds to synthesize egg shells. As a result, the fragile shells are easily broken, the birds' ability to reproduce falls sharply, and their populations decline. Carnivorous birds, such as eagles, are particularly vulnerable to increased levels of DDT in their bodies, because carnivores consume fats from their prey. biomagnification, p. 349

Thus, to determine why the populations of certain birds were declining, biologists had to study what was happening at several levels of organization, including (1) the nature of the molecules involved, (2) how different kinds of animals fit into food chains, (3) where DDT was found in the bodies of

TABLE 1.2

Levels of Organization for Living Things

Category	Characteristics/Explanation	Example/Application
Biosphere	The worldwide ecosystem	Human activity affects the climate of the Earth. Global climate change and the hole in ozone layer are examples of human impacts on the biosphere.
Ecosystem	Communities (groups of populations) that interact with the physical world in a particular place	The Everglades ecosystem involves many kinds of organisms, the climate, and the flow of water to south Florida.
Community	Populations of different kinds of organisms that interact with one another in a particular place	The populations of trees, insects, birds, mammals, fungi, bacteria, and many other organisms interact in any location.
Population	A group of individual organisms of a particular kind	The human population currently consists of over 6 billion individuals. The current population of the California condor is about 220 individuals.
Organism	An independent living unit	Some organisms consist of many cells—you, a morel mushroom, a rose bush. Others are single cells—yeast, pneumonia bacterium, *Amoeba*.
Organ system	A group of organs that work together to perform a particular function	The circulatory system consists of a heart, arteries, veins, and capillaries, all of which are involved in moving blood from place to place.
Organ	A group of tissues that work together to perform a particular function	An eye contains nervous tissue, connective tissue, blood vessels, and pigmented tissues, all of which are involved in sight.
Tissue	Groups of cells that work together to perform particular functions	Blood, muscle cells, and the layers of the skin are all groups of cells and each performs a specific function.
Cell	The smallest unit that displays the characteristics of life	Some organisms are single cells. Within multicellular organisms are several kinds of cells—heart muscle cells, nerve cells, white blood cells.
Molecules	Specific arrangements of atoms	Living things consist of special kinds of molecules, such as proteins, carbohydrates, and DNA, as well as common molecules, such as water.
Atoms	The fundamental units of matter	There are about 100 different kinds of atoms such as hydrogen, oxygen and nitrogen.

animals, (4) which organs DDT affected, and ultimately (5) how DDT impaired the ability of specialized cells to produce egg shells.

The Significance of Biology

To a great extent, we owe our high standard of living to biological advances in two areas: food production and disease control. Plant and animal breeders have modified organisms to yield greater amounts of food than did older varieties. A good example is the changes that have occurred in corn. Corn, a kind of grass, produces its seeds on a cob. The original corn plant had very small cobs, which, were perhaps only 3 or 4 centimeters long. Selective breeding has produced vari-

eties of corn with much larger cobs and more seeds per cob, increasing the yield greatly. In addition, plant breeders have created varieties, such as sweet corn and popcorn, with special characteristics. Similar improvements have occurred in wheat, rice, oats, and other cereal grains. The improvements in the plants, along with better farming practices (also brought about through biological experimentation), have greatly increased food production.

Animal breeders also have had great successes. The pig, chicken, and cow of today are much different animals from those available even 100 years ago. Chickens lay more eggs, beef cattle grow faster, and dairy cows give more milk (figure 1.19). All these improvements increase the amount of food available and raise our standard of living.

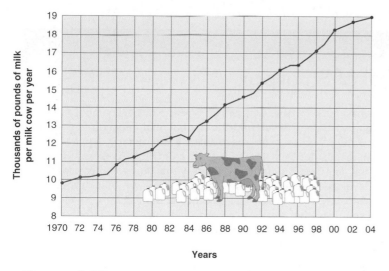

FIGURE 1.19 Biological Research Improves Food Production

This graph illustrates a steady increase in milk yield, largely because of changing farming practices and selective breeding programs.

Biological research has also improved food production by developing controls for the disease organisms, pests, and weeds that reduce yields. Biologists must understand the nature of these harmful organisms to develop effective control methods.

There also has been fantastic progress in the area of human health. An understanding that diseases such as cholera, typhoid fever, and dysentery spread from one person to another through the water supply led to the development of treatment plants for sewage and drinking water. Recognizing that diseases such as botulism and salmonella spread through food led to guidelines for food preservation and preparation that greatly reduced the incidence of these diseases. Many other diseases, such as polio, whooping cough, measles, and mumps, can be prevented by vaccinations (How Science Works 1.1). Unfortunately, the vaccines have worked so well that some people no longer bother to get them, and some of these diseases, such as diphtheria, are reappearing. They have not been eliminated, and people who are not protected by vaccinations are still susceptible to them. By helping us understand how the human body works, biological research has led to the development of treatments that can control chronic diseases, such as diabetes, high blood pressure, and even some kinds of cancer. Unfortunately, all these advances in health contribute to another major biological problem: the increasing size of the human population.

The Consequences of Not Understanding Biological Principles

A lack of understanding of biological principles also has consequences. At one time, it was thought that the protection of specific land areas would preserve endangered ecosystems.

However, it is now recognized that many activities outside park and preserve boundaries are also important. For example, although Everglades National Park in Florida has been well managed by the National Park Service, this ecosystem is experiencing significant destruction. Commercial and agricultural development adjacent to the park has caused groundwater levels in the Everglades to drop so low that the park's very existence is threatened. Fertilizer has entered the park from surrounding farmland and has encouraged the growth of plants that change the nature of the ecosystem. In 2000, Congress authorized the expenditure of $1.4 billion to begin to implement a plan that will address the problems of water flow and pollution. The major goals are to reduce the amount of nutrients entering from farms and to increase the flow of water to the Everglades from Lake Okeechobee to the north.

In North America, the introduction of exotic (foreign) species of plants and animals has had disastrous consequences in a number of cases (figure 1.20). Both the

FIGURE 1.20 Exotic Animals

Exotic organisms such as starlings and zebra mussels have altered natural ecosystems by replacing native species.

HOW SCIENCE WORKS 1.1

Edward Jenner and the Control of Smallpox

Edward Jenner (1749–1823) was born in Berkeley, Gloucestershire, in western England. Since he wanted to become a doctor, he became an apprentice to a local doctor. This was the typical training for physicians at that time. After his apprenticeship, he went to London and studied with an eminent surgeon. In 1773, he returned to Berkeley and practiced medicine there for the rest of his life.

At that time in Europe and Asia, smallpox was a common disease, which nearly everyone developed, usually early in life. Many children died of it, and many who survived were disfigured by scars. It was known that people who had had smallpox once were protected from future infection. If children were deliberately exposed to smallpox when they were otherwise healthy, a mild form of the disease often developed, and they were protected from future smallpox infections. Indeed, in the Middle East, people were deliberately infected by scratching material from the pocks of an infected person into their skin. This practice was introduced to England in 1717 by Lady Mary Wortley Montagu, the wife of the ambassador to Turkey. She had observed the practice of deliberate infection in Turkey and had had her own children inoculated. This practice had become common in England by the early 1700s, and Jenner carried out such deliberate inoculations as part of his practice. He also frequently came into contact with individuals who had smallpox, as well as people infected with cowpox—a mild disease similar to smallpox.

In 1796, Jenner introduced a safer way to protect against smallpox as a result of his 26-year study of cowpox and smallpox. Jenner had made two important *observations*. First, many milkmaids and other farmworkers developed a mild illness, with pocklike sores, after milking cows that had cowpox sores on their teats. Second, very few of those who had been infected with cowpox became sick with smallpox. He asked the *question* "Why don't people who have had cowpox get smallpox?" He developed the *hypothesis* that the mild disease caused by cowpox somehow protected them from the often fatal smallpox. This led him to perform an *experiment*. In his first experiment, he took puslike material from a sore on the hand of a milkmaid named Sarah Nelmes and rubbed it into

small cuts on the arm of an 8-year-old boy named James Phipps. James developed the normal mild infection typical of cowpox and completely recovered. Subsequently, Jenner inoculated James with material from a smallpox patient. (Recall that this was a normal practice at the time.) James did not develop any disease. Jenner's *conclusion* was that deliberate exposure to cowpox had protected James from smallpox. Eventually the word *vaccination* was used to describe the process. It was derived from the Latin words for *cow (vacca)* and *cowpox disease (vaccinae)* (box figure).

When these results became known, public reaction was mixed. Some people thought that vaccination was the work of the devil. However, many European rulers supported Jenner by encouraging their subjects to be vaccinated. Napoleon and the empress of Russia were very influential and, in the United States, Thomas Jefferson had some members of his family vaccinated. Many years later, following the development of the *germ theory of disease,* it was discovered that cowpox and smallpox are caused by viruses that are similar in structure. Exposure to the cowpox virus allows the body to develop immunity against both the cowpox virus and the smallpox virus. In the mid 1900s a slightly different virus was used to develop a vaccine against smallpox, which was used worldwide. In 1979, almost 200 years after Jenner developed his vaccination, the Centers for Disease Control and Prevention (CDC) in the United States and the World Health Organization (WHO) of the United Nations declared that smallpox had been eradicated.

Today, vaccinations (immunizations) are used to control many diseases that used to be common. Many of them were known as childhood diseases, because essentially all children got them. Today, they are rare in populations that are vaccinated. The following chart shows the schedule of immunizations recommended by the Advisory Committee on Immunization Practices of the American Academy of Pediatrics and American Academy of Family Physicians.

The painting depicts Edward Jenner vaccinating James Phipps.

HOW SCIENCE WORKS 1.1 (*continued*)

Recommended Childhood and Adolescent Immunization Schedule United States, 2005

AGE VACCINE	Birth	1 month	2 months	4 months	6 months	12 months	15 months	18 months	24 months	4—6 years	11—12 years	13—18 years
Hepatitis B		First	Second			Third				Hep B series if missed (3 doses)		
		Three-dose series—certain infants										
DPT: diphtheria, tetanus, pertussis (whooping cough)			First	Second	Third		Fourth			Fifth	Tetanus and diphtheria	Tetanus and diphtheria if dose missed
Haemophilus influenzae type B influenza			First	Second	Third	Fourth						
Inactivated poliovirus			First	Second	Third					Fourth		
Pneumococcal conjugate (pneumonia)			First	Second	Third	Fourth			Additional vaccinations for high-risk groups			
MMR: measles, mumps, rubella (German measles)						First				Second	Second if missed	
Varicella (chicken pox)						First			Two doses if first missed			
Hepatitis A									Children in certain parts of country (2 doses)			
Influenza					Yearly				Yearly for high-risk groups			

Source: Advisory Committee on Immunization Practices, American Academy of Pediatrics and American Academy of Family Physicians.

American chestnut and the American elm have been nearly eliminated by diseases that were introduced by accident. Another accidental introduction, the zebra mussel, has greatly altered freshwater lakes and rivers in the central and eastern parts of the United States. They filter tiny organisms from the water and deprive native organisms of this food source. In addition, they attach themselves to native mussels, often causing their death.

Other organisms have been introduced on purpose because of shortsightedness or a lack of understanding about biology. The European starling and the English (house) sparrow were both introduced into this country by people who thought they were doing good. Both of these birds have multiplied greatly and have displaced some native birds. Many people want to have exotic animals as pets. When these animals escape or are intentionally released, they can become established in local ecosystems and endanger native organisms. For example, Burmese pythons are commonly kept as pets. Today, they are common in the Everglades and kill and eat native species. Large pythons have even been observed attacking alligators. The introduction of exotic plants has also caused problems. At one time, people were encouraged to plant a shrub known as autumn olive as a wildlife food. The plant produces many small fruits, which are readily eaten by many kinds of birds and mammals. However, because the animals eat the fruits and defecate the seeds everywhere, autumn olive spreads rapidly. Today, it is recognized as an invasive plant needing to be controlled.

Advances in technology and our understanding of human biology have presented us with difficult ethical issues, which we have not been able to resolve satisfactorily. Major advances in health care have prolonged the lives of people who would have died if they had lived a generation earlier. Many of the techniques and machines that allow us to preserve and extend life are extremely expensive and are therefore unavailable to most citizens of the world. Many people lack even the most basic health care, while the rich nations of the world spend millions of dollars to have cosmetic surgery and to keep comatose patients alive with the assistance of machines.

Future Directions in Biology

Where do we go from here? Although the science of biology has made major advances, many problems remain to be solved. For example, scientists are seeking major advances in the control of the human population, and there is a continued interest in the development of more efficient methods of producing food.

One area that will receive much attention in the next few years is the relationship between genetic information and such diseases as Alzheimer's disease, stroke, arthritis, and cancer.

These and many other diseases are caused by abnormal body chemistry, which is the result of hereditary characteristics. Curing hereditary diseases is a big job. It requires a thorough understanding of genetics and the manipulation of hereditary information in all of the trillions of cells of the organism.

Another area that will receive much attention in the next few years is ecology. Climate change, pollution, and the destruction of natural ecosystems to feed a rapidly increasing human population are all still severe problems. We face two tasks. The first is to improve technology and our understanding about how things work in our biological world. The second, and probably the more difficult, is to educate and remind people that their actions determine the kind of world in which the next generation will live.

It is the intent of science to learn what is going on by gathering facts objectively and identifying the most logical courses of action. It is also the role of science to identify cause-and-effect relationships and note their predictive value in ways that will improve the environment for all forms of life—including us. Scientists should also make suggestions to politicians and other policy makers about which courses of action are the most logical from a scientific point of view.

Summary

The science of biology is the study of living things and how they interact with their surroundings. Science can be distinguished from nonscience by the kinds of laws and rules that are constructed to unify the body of knowledge. Science involves the continuous testing of rules and principles by the collection of new facts. In science, these rules are usually arrived at by using the scientific method— observation, questioning, the exploration of resources, hypothesis formation, and the testing of hypotheses. When general patterns are recognized, theories and laws are formulated. If a rule is not testable, or if no rule is used, it is not science. Pseudoscience uses scientific appearances to mislead.

Living things show the characteristics of (1) metabolic processes, (2) generative processes, (3) responsive processes, (4) control processes, and (5) a unique structural organization. Biology has been responsible for major advances in food production and health. The incorrect application of biological principles has sometimes led to the destruction of useful organisms and the introduction of harmful ones. Many biological advances have led to ethical dilemmas, which have not been resolved. In the future, biologists will study many things. Two areas that are certain to receive attention are ecology and the relationship between heredity and disease.

Key Terms

Use the interactive flash cards on the **Concepts in Biology,** *12/e website to help you learn the meaning of these terms.*

atoms 16
biology 2
biosphere 16
cells 14
community 16
control group 6
control processes 13
controlled experiment 6
deductive reasoning
   (deduction) 8
dependent variable 7
ecosystem 16
energy 12
enzymes 14
experiment 6
experimental group 6
generative processes 12
homeostasis 14
hypothesis 6
independent variable 7

inductive reasoning
   (induction) 8
matter 12
metabolic processes 12
metabolism 12
molecules 16
observation 3
organ 16
organ system 16
organism 14
population 16
pseudoscience 10
responsive processes 13
science 2
scientific law 8
scientific method 2
theory 7
tissue 16
variables 6

Basic Review

1. Which one of the following distinguishes science from nonscience?

   a. the collection of information

   b. the testing of a hypothesis

   c. the acceptance of the advice of experts

   d. information that never changes

2. A hypothesis must account for all available information, be logical, and be _____.

3. A scientific theory is

   a. a guess as to why things occur.

   b. always correct.

   c. a broad statement that ties together many facts.

   d. easily changed.

4. Pseudoscience is the use of the appearance of science to _____.

5. Economics is not considered a science because

   a. it does not have theories.

   b. it does not use facts.

   c. many economic predictions do not come true.

   d. economists do not form hypotheses.

6. Reproduction is

   a. a generative process.

   b. a responsive process.

   c. a control process.

   d. a metabolic process.

7. The smallest independent living unit is the _____.

8. The smallest unit that displays characteristics of life is the _____.

9. An understanding of the principles of biology will prevent policy makers from making mistakes. (T/F)

10. Three important advances in the control of infectious diseases are safe drinking water, safe food, and _____.

Answers

1. b   2. testable   3. c   4. mislead   5. c   6. a
7. organism   8. cell   9. F   10. vaccination

Concept Review

Answers to Concept Review questions can be found on the **Concepts in Biology,** *12/e website.*

1.1   The Significance of Biology in Your Life

1. List five issues that biological research may help us solve in the near future.

1.2   Science and the Scientific Method

2. What is the difference between simple correlation and a cause-and-effect relationship?

3. How does a hypothesis differ from a scientific theory or a scientific law?

4. List three objects or processes you use daily that are the result of scientific investigation.

5. The scientific method cannot be used to deny or prove the existence of God. Why?

6. What are controlled experiments? Why are they necessary to support a hypothesis?

7. List the parts of the scientific method.

1.3 Science, Nonscience, and Pseudoscience

8. What is the difference between science and nonscience?
9. How can you identify pseudoscience?
10. Why is political science not a science?

1.4 The Science of Biology

11. List three advances that have occurred as a result of biology.
12. List three mistakes that could have been avoided had we known more about living things.
13. What is biology?
14. List five characteristics of living things.
15. What is the difference between regulation and coordination?

Thinking Critically

For guidelines to answer this question, visit the **Concepts in Biology,** *12/e website.*

The scientific method is central to all work that a scientist does. Can this method be applied to the ordinary activities of life? How might a scientific approach change how you choose your clothing, your recreational activities, or a car? Can these choices be analyzed scientifically? Should they be analyzed scientifically? Is there anything wrong with looking at these decisions from a scientific point of view?

2 CHAPTER

The Basics of Life
Chemistry

For thousands of years, people have noticed similarities among many kinds of organisms. As time passed, scientists realized that all living things consist of the same basic elements. Scientists recognized that the same principles that govern the interactions of nonliving chemicals in a test tube also control organisms. By studying the behavior of matter, we can explore the principles that underlie the structure and function of all organisms.

To some degree, we all need to understand a little chemistry to deal with topics that, at first glance, might seem unrelated. Consider the tsunami that struck Asia in December 2004, killing an estimated 176,000 people. It was feared that this number could double because of infectious diseases. Therefore, it became crucial to prevent the spread of disease. This involved the use of disinfectants and sanitizing agents. Disinfectants and sanitizers are chemicals, and their ability to control the spread of infectious diseases is based on chemical principles. For example, the chemical chlorine attacks and destroys proteins, which are important components of all organisms including microorganisms. This makes chlorine capable of killing many disease-causing microbes, such as bacteria, viruses, and fungi. The disinfecting properties of chlorine were discovered as a result of studying its chemistry. In turn this has led to its being used by rescue workers to help control the spread of diseases after disasters such as the tsunami. Chlorine and other disinfectants enable humans to combat many diseases ranging from the deadly (such as cholera, and hepatitis) to the merely annoying (such as the common cold).

CHAPTER OUTLINE

2.1 Matter, Energy, and Life

All forms of life are composed of different forms of *matter* and carry out processes that involve the use of *energy*. Recall from chapter 1 that **matter** is anything that has mass[1] and takes up space. **Energy** is the ability to do work or cause things to move. This means that, when life processes occur, part of an organism or its environment is moved. This movement might be the opening of a flower bud or the blinking of your eyelid as you read this sentence. The structure of all matter involves chemicals, substances used or produced in processes that involve changes in matter. **Chemistry** is the science concerned with the study of the composition, structure, and properties of matter and the changes it undergoes. Living matter has the same basic building blocks and undergoes the same kinds of changes as nonliving matter. Therefore, a basic understanding of chemistry will help you understand living things (figure 2.1). muscle contraction, p. 582

Where do living things get their energy to make these changes? Most people would answer that many organisms get their energy from food or nutrients, whereas other organisms receive theirs from sunlight. However, the answer is more complex. There are two general types of energy: *kinetic* and *potential*. **Kinetic energy** is energy of motion. For example, a fish swimming through a pond displays kinetic energy. Energy that is not yet doing work is **potential energy**. You might also think of potential energy as stored energy. When we talk about the energy in chemicals, we are talking about the potential energy in matter. This energy has the potential to be converted to kinetic energy to do work, such as causing an organism to move faster, digest food, or build muscle.

The Law of Conservation of Energy

One of the important scientific laws, the **law of conservation of energy**, or the *first law of thermodynamics*, states that energy is never created or destroyed. Energy can be converted from one form to another, but the *total* energy remains constant. All living things obey this law. One kind of energy change is between kinetic and potential energy.

An object that appears to be motionless does not necessarily lack energy. An object on top of a mountain may be motionless but still may contain significant amounts of potential energy, or "energy of position." It can be released as kinetic energy if the object rolls down the mountain. Keep in mind that potential energy also increases whenever things experiencing a repelling force are pushed together. You experience this every time you use kinetic energy to "click" your ballpoint pen, which compresses the spring inside. This gives the spring more potential energy, which is converted back

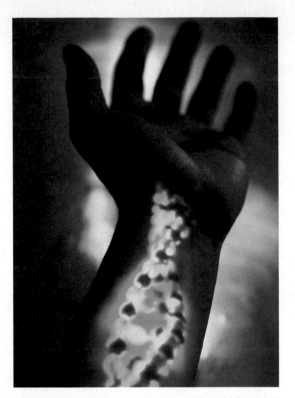

FIGURE 2.1 Biology and Chemistry Working Together
In order to understand living things, researchers must investigate both their structure and their function. At the core of modern biology is an understanding of our molecular structure, including such molecules as DNA, the molecule of which genes are composed.

into kinetic energy when the spring expands as the ink cartridge is retracted into the case. The kinetic energy for compressing the spring comes from your biological ability to harvest energy from food molecules. Potential energy also increases whenever things that attract each other are pulled apart. An example of this occurs when you stretch a rubber band. That increased potential energy is converted to the snapping back of the band when you let go. Again, the kinetic energy needed to stretch the rubber band comes from you. It is important to understand that energy is not used up in these processes. The same amount of energy is released when the rubber band is snapped back as was stored when you stretched the band. cell metabolism, p. 12

Forms of Energy

There are five forms of energy, and each can be either kinetic or potential: (1) *mechanical*, (2) *nuclear*, (3) *electrical*, (4) *radiant*, and (5) *chemical*. All organisms interact in some way with these forms of energy. *Mechanical energy* is the energy most people associate with machines or things in motion. A track athlete displays potential mechanical energy at the start line; the energy becomes kinetic mechanical energy when the athlete is running (figure 2.2). *Nuclear energy* is the form of energy from reactions involving the innermost part of matter, the atomic nucleus. In a nuclear

---

[1]Don't confuse the concepts of mass and weight. *Mass* refers to an amount of matter, whereas *weight* refers to the amount of force with which an object is attracted by gravity. Because gravity determines weight, your weight would be different on the Moon than it is on Earth, but your mass would be the same.

FIGURE 2.2 Potential Mechanical Energy

The runners in *(a)* have the potential energy needed to make the run. Converting this form of energy to kinetic energy takes place during the race *(b)*.

power plant, nuclear energy is used to generate electrical energy. *Electrical energy,* or electricity, is the flow of charged particles. All organisms use charged particles as a part of their metabolism. *Radiant energy* is most familiar as heat and visible light, but there are other forms as well, such as X-radiation and microwaves. *Chemical energy* is a kind of internal potential energy that is stored in matter and can be released as kinetic energy when chemicals are changed from one form to another. For example, the burning of wood involves converting the chemical energy of wood into heat and light. A slower, controlled burning process, called *cellular respiration,* releases energy from food in living systems.

2.2 The Nature of Matter

The idea that substances are composed of very small particles goes back to early Greek philosophers. During the fifth century B.C., Democritus wrote that matter was empty space filled with tremendous numbers of tiny, indivisible particles called *atoms.* (The word *atom* is from the Greek word meaning *uncuttable.*)

Structure of the Atom

Recall from chapter 1 that atoms are the smallest units of matter that can exist alone. **Elements** are fundamental chemical substances made up of collections of only one kind of atom. For example, hydrogen, helium, lead, gold, potassium, and iron are all elements. There are over 100 elements. Each is composed of collections of a particular atom. To understand how the atoms of various elements differ from each other, we need to look at the structure of atoms (How Science Works 2.1).

Atoms are constructed of three major subatomic particles: *neutrons, protons,* and *electrons.* A **neutron** is a heavy subatomic particle that does not have a charge; it is located in the central core of each atom. The central core is called the **atomic nucleus.** The mass of the atom is concentrated in the atomic nucleus. A **proton** is a heavy subatomic particle that has a positive charge; it is also located in the atomic nucleus. An **electron** is a light subatomic particle with a negative electrical charge that moves about outside the atomic nucleus in regions known as *energy levels* (figure 2.3). An **energy level** is a region of space surrounding the atomic nucleus that contains electrons with certain amounts of energy. The number of electrons an atom has determines the space, or volume, it takes up.

All the atoms of an element have the same number of protons. The number of protons determines the element's identity. For example, carbon always has 6 protons; no other element has that number. Oxygen always has 8 protons. The **atomic number** of an element is the number of protons in an atom of that element; therefore, each element has a unique atomic number. Because oxygen has 8 protons, its atomic number is 8. The mass of a proton is 1.67×10^{-24} grams. Because this is an extremely small mass and is awkward to express, 1 proton is said to have a mass of 1 **atomic mass unit** (abbreviated as AMU) (table 2.1).

TABLE 2.1			
Comparison of Atomic Particles			
	Protons	**Neutrons**	**Electrons**
Location	Nucleus	Nucleus	Outside nucleus
Charge	Positive (+)	None (neutral)	Negative (−)
Number present	Identical to atomic number	Atomic weight minus atomic number	Equal to number of protons
Mass	1 AMU	1 AMU	1/1,836 AMU

HOW SCIENCE WORKS 2.1

The Periodic Table of the Elements

Traditionally, the elements have been represented in a shorthand form by letters called *chemical symbols*. The table that displays these symbols is called *periodic* because the properties of the elements recur periodically (at regular intervals) when the elements are listed in order of their size. The table has horizontal rows of elements called *periods*. The vertical columns are called *families*. The periods and families consist of squares, with each element having its own square in a specific location. This arrangement has a meaning, both about atomic structure and about chemical functions. The periods are numbered from 1 to 7 on the left side. Period 1, for example, has only two elements: H (hydrogen) and He (helium). Period 2 starts with Li (lithium) and ends with Ne (neon). The two rows at the bottom of the table are actually part of periods 6 and 7 (between atomic numbers 57 and 72, 89 and 104). They are moved so that the table is not so wide.

Families are identified with Roman numerals and letters at the top of each column. Family IIA, for example, begins with Be (beryllium) at the top and has Ra (radium) at the bottom. The A families are in sequence from left to right. The B families are not in sequence, and one group contains more elements than the others. The elements in vertical columns have the same configurations, and that structure is responsible for the chemical properties of an element. Don't worry—you will not have to memorize the entire table. The 11 main elements comprising living things have the chemical symbols C, H, O, P, K, I, N, S, Ca, Fe, and Mg. (A mnemonic trick to help you remember them is CHOPKINS CaFé, Mighty good!).

diet and nutrition, p. 542

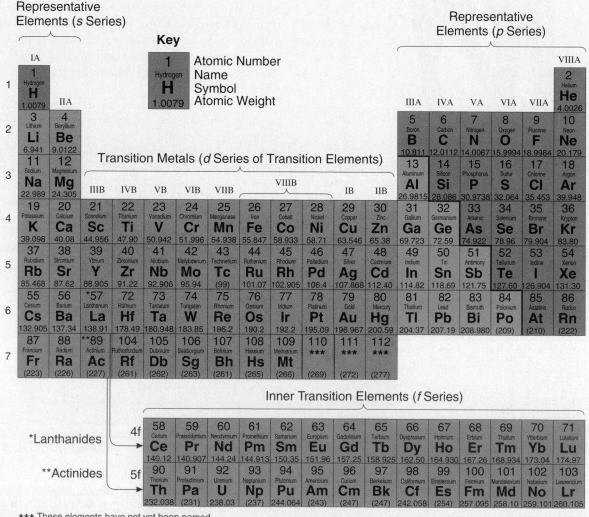

Periodic Table of the Elements

The table provides information about all the known elements. Notice that the atomic weights of the elements increase as you read left to right along the periods. Reading top to bottom in a family gives you a glimpse of a group of elements that have similar chemical properties.

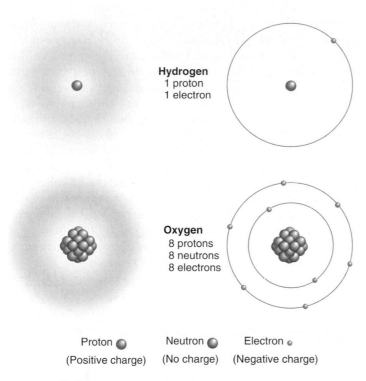

Hydrogen
1 proton
1 electron

Oxygen
8 protons
8 neutrons
8 electrons

Proton ● (Positive charge)   Neutron ● (No charge)   Electron ● (Negative charge)

FIGURE 2.3 Atomic Structure
While these two are shown as being the same size, a hydrogen atom is actually smaller than an oxygen atom.

Elements May Vary in Neutrons but Not Protons

Although all atoms of the same element have the same number of protons and electrons, they do not always have the same number of neutrons. In the case of oxygen, over 99% of the atoms have 8 neutrons, but others have more or fewer neutrons. Each atom of the same element with a different number of neutrons is called an **isotope** of that element. Since neutrons have a mass very similar to that of protons, isotopes that have more neutrons have a greater mass than those that have fewer neutrons.

Elements occur in nature as a mixture of isotopes. The **atomic weight** of an element is an average of all the isotopes present in a mixture in their normal proportions. For example, of all the hydrogen isotopes on Earth, 99.985% occur as an isotope without a neutron and 0.015% as an isotope with 1 neutron. There is a third isotope with 2 neutrons, and is even more rare. When the math is done to account for the relative amounts of these three isotopes of hydrogen, the atomic weight turns out to be 1.0079 AMU.

The sum of the number of protons and neutrons in the nucleus of an atom is called the **mass number.** Mass numbers are used to identify isotopes. The most common isotope of hydrogen has 1 proton and no neutrons. Thus, its mass number is 1 (1 proton + 0 neutrons = 1). A hydrogen atom with 1 proton and 1 neutron has a mass number of 1 + 1, or 2, and is referred to as hydrogen-2, also called deuterium. A hydrogen atom with 1 proton and 2 neutrons has a mass

number of 1 + 2, or 3, and is referred to as hydrogen-3, also called tritium (figure 2.4).

Subatomic Particles and Electrical Charge

Subatomic particles were named to reflect their electrical charge. Protons have a positive (+) electrical charge. Neutrons, the second type of particle in the atomic nucleus, are neutral, since they lack an electrical charge (0). Electrons have a negative (−) electrical charge. Because positive and negative particles are attracted to one another, electrons are held near the nucleus. However, their kinetic energy (motion) keeps them from combining with the nucleus. The overall electrical charge of an atom is neutral (0) because the number of protons (positively charged) equals the number of electrons (negatively charged). For instance, hydrogen, with 1 proton, has 1 electron; carbon, with 6 protons, has 6 electrons. You can determine the number of either of these two particles in an atom if you know the number of the other particle.

Scientists' understanding of the structure of an atom has changed since the concept was first introduced. At one time,

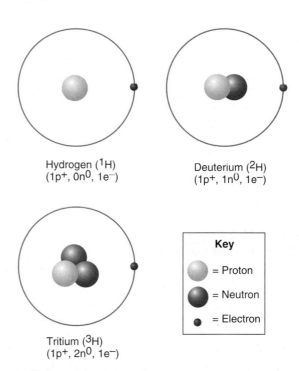

Hydrogen (^{1}H)
(1p$^+$, 0n^0, 1e$^-$)

Deuterium (^{2}H)
(1p$^+$, 1n^0, 1e$^-$)

Tritium (^{3}H)
(1p$^+$, 2n^0, 1e$^-$)

Key
● = Proton
● = Neutron
● = Electron

FIGURE 2.4 Isotopes of Hydrogen
The most common form of hydrogen is the isotope that is composed of 1 proton and no neutrons. It has the mass number 1 AMU. The isotope *deuterium* is 2 AMU and has 1 proton and 1 neutron. *Tritium*, 3 AMU, has 2 neutrons and 1 proton. Each of these isotopes of hydrogen also has 1 electron but, because the mass of an electron is so small, the electrons do not contribute significantly to the mass as measured in AMU. All three isotopes of hydrogen are found on Earth, but the most frequently occurring has 1 AMU and is commonly called *hydrogen*. Most scientists use the term *hydrogen* in a generic sense (i.e., the term is not specific but might refer to any or all of these isotopes).

people thought of atoms as miniature solar systems, with the nucleus in the center and electrons in orbits, like satellites, around the nucleus. However, as more experimental data were gathered and interpreted, a new model was formulated.

The Position of Electrons

In contrast to the "solar system" model, electrons are now believed to occupy certain areas around the nucleus, the energy levels. Each energy level contains electrons moving at approximately the same speed; therefore, electrons of a given level have about the same amount of kinetic energy. Each energy level is numbered in increasing order, with energy level 1 containing electrons closest to the nucleus, with the lowest amount of energy. The electrons in energy level 2 have more energy and are farther from the nucleus than those found in energy level 1. Electrons in energy level 3 having electrons with even more energy are still farther from the nucleus than those in level 2 and so forth.

Electrons do not encircle the atomic nucleus in two-dimensional paths. Some move around the atomic nucleus in a three-dimensional region that is spherical, forming cloud-like or fuzzy layers about the nucleus. Others move in a manner that resembles the figure 8, forming fuzzy regions that look like dumbbells or hourglasses (figure 2.5).

The first energy level is full when it has 2 electrons. The second energy level is full when it has 8 electrons; the third energy level, 8; and so forth (table 2.2). Also note in table 2.2 that, for some of the atoms (He, Ne, Ar), the outermost energy level contains the maximum number of electrons it can hold. Elements such as He and Ne, with filled outer energy levels, are particularly stable.

All atoms have a tendency to seek such a stable, filled outer energy level arrangement, a tendency referred to as the *octet (8) rule*. (Hydrogen and helium are exceptions to this rule and have a filled outer energy level when they have 2 electrons.) The rule states that atoms attempt to acquire an outermost energy level with 8 electrons through chemical reactions. Because elements such as He and Ne have full outermost energy levels under ordinary circumstances, they do not normally undergo chemical reactions and are therefore referred to as *noble* (implying that they are too special to interact with other elements) or *inert*. Atoms of other elements have outer energy levels that are not full; for example, H, C, Mg, and Ca will undergo reactions to fill their outermost energy level in order to become stable. It is important for chemists and biologists to focus on electrons in the outermost energy level, because it is these electrons that are involved in the chemical activities of all life.

2.3 The Kinetic Molecular Theory and Molecules

Greek philosopher Aristotle (384–322 B.C.) rejected the idea of atoms. He believed that matter was continuous and made up of only four parts: earth, air, fire, and water. Aristotle's belief about matter predominated through the 1600s. Galileo and Newton, however, believed the ideas about matter being composed of tiny particles, or atoms, because this theory seemed to explain matter's behavior. Widespread acceptance of the atomic model did not occur, however, until strong evidence was developed through the science of chemistry in the late 1700s and early 1800s. The experiments finally led to a collection of assump-

TABLE 2.2

Number of Electrons in Energy Level

Element	Symbol	Atomic Number	Energy Level 1	Energy Level 2	Energy Level 3	Energy Level 4
Hydrogen	H	1	1			
Helium	He	2	2			
Carbon	C	6	2	4		
Nitrogen	N	7	2	5		
Oxygen	O	8	2	6		
Neon	Ne	10	2	8		
Sodium	Na	11	2	8	1	
Magnesium	Mg	12	2	8	2	
Phosphorus	P	15	2	8	5	
Sulfur	S	16	2	8	6	
Chlorine	Cl	17	2	8	7	
Argon	Ar	18	2	8	8	
Potassium	K	19	2	8	8	1
Calcium	Ca	20	2	8	8	2

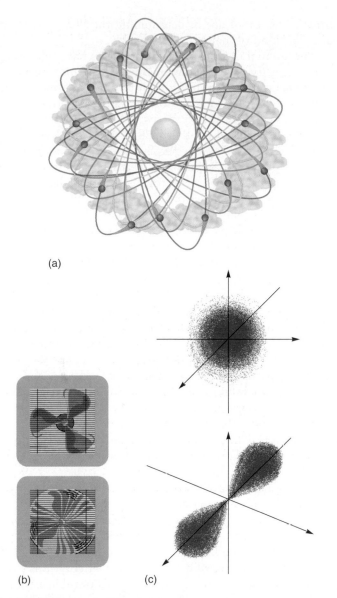

(a)

(b)   (c)

FIGURE 2.5 The Electron Cloud

Electrons are moving around the nucleus so fast that they can be thought of as forming a cloud around it, rather than an orbit or a single track. *(a)* You might think of the electron cloud as hundreds of photographs of an atom. Each photograph shows where an electron was at the time the picture was taken. However, when the next picture is taken, the electron is somewhere else. In effect, an electron appears to be everyplace in its energy level at the same time, just as the fan blade of a window fan is everywhere at once when it is running. *(b)* No matter where you stick your finger in the fan, you will be touched by the moving blade. Although we are able to determine where an electron is at a given time, we do not know the exact path it uses to go from one place to another.

(c) This is a better way to represent the positions of electrons in spherical and hourglass configurations.

tions about the small particles of matter and the space around them; these assumptions came to be known as the kinetic molecular theory. The **kinetic molecular theory** states that all matter is made up of tiny particles, which are in constant motion.

The Formation of Molecules

Because atoms tend to fill their outer energy levels, they often interact with other atoms. Recall from chapter 1 that a **molecule** is the smallest particle of a chemical compound. Some atoms, such as oxygen, hydrogen, and nitrogen, bond to form *diatomic (di = two)* molecules. In our atmosphere, these elements are found as the gases H_2, O_2, and N_2. The subscript indicates the number of atoms of an element in a single molecule of a substance. Other elements are not normally diatomic but exist as single, or *monatomic (mon = one)*, units—for example, the gases helium (He) and neon (Ne). These chemical symbols, or initials, indicate a single atom of that element.

When two different kinds of atoms combine, they form compounds. A **compound** is a chemical substance made up of atoms of two or more elements combined in a specific ratio and arrangement. The attractive forces that hold the atoms of a molecule together are called **chemical bonds.** Molecules can consist of two or more atoms of the same element (such as O_2 or N_2) or of specific numbers of atoms of different elements.

The **formula** of a compound describes what elements it contains (as indicated by a chemical symbol) and in what proportions they occur (as indicated by the subscript number). For example, pure water is composed of two atoms of hydrogen and one atom of oxygen. It is represented by the chemical formula H_2O. The subscript "2" indicates two atoms of the element hydrogen, and the symbol for oxygen without a subscript indicates that there is only 1 atom of oxygen present in this molecule.

2.4   Molecules and Kinetic Energy

Common experience shows that all matter has a certain amount of kinetic energy. For instance, if you were to open a bottle of perfume in a closed room with no air movement, it wouldn't take long for the aroma to move throughout the room. The kinetic molecular theory explains this by saying that the molecules *diffuse*, or spread, throughout the room because they are in constant, random motion. This theory also predicts that the rate at which they diffuse depends on the temperature of the room—the higher the air temperature, the greater the kinetic energy of the molecules and the more rapid the diffusion of the perfume.

Temperature is a measure of the average kinetic energy of the molecules making up a substance. The two most common numerical scales used to measure temperature are the Fahrenheit scale and the Celsius scale. When people comment on the temperature of something, they usually are making a comparison. For example, they may say that the air temperature today is "colder" or "hotter" than it was yesterday. They may also refer to a scale for comparison, such as "the temperature is 20°C [68°F]."

Heat is the total internal kinetic energy of molecules. Heat is measured in units called *calories.* A **calorie** is the amount of heat necessary to raise the temperature of 1 gram of water 1 degree Celsius (°C). The concept of heat is not the same as the concept of temperature. Heat is a *quantity* of energy. Temperature deals with the comparative hotness or coldness of things. The heat, or internal kinetic energy, of molecules can change as a result of interactions with the environment. This is what happens when you rub your hands together. Friction results in increased temperatures because molecules on one moving surface catch on another surface, stretching the molecular forces that are holding them. They are pulled back to their original position with a "snap," resulting in an increase of vibrational kinetic energy. *Heat (measured in calories) and temperature (measured in Celsius or Fahrenheit) are not the same thing but are related to one another.* The heat that an object possesses cannot be measured with a thermometer. What a thermometer measures is the temperature of an object. The temperature is really a measure of how fast the molecules of the substance are moving and how often they bump into other molecules, a measure of their kinetic energy. If heat energy is added to an object, the molecules vibrate faster. Consequently, the temperature rises, because the added heat energy results in a speeding up of the movement of the molecules. Although there is a relationship between heat and temperature, the amount of heat, in calories, that an object has depends on the size of the object and its particular properties, such as its density, volume, and pressure. temperature regulation in animals, p. 493

Why do we take a person's body temperature? The body's size and composition usually do not change in a short time, so any change in temperature means that the body has either gained or lost heat. If the temperature is high, the body has usually gained heat as a result of increased metabolism. This increase in temperature is a symptom of abnormality, as is low a body temperature.

2.5 Physical Changes—Phases of Matter

There are implications to the kinetic molecular theory. First, the amount of kinetic energy that particles contain can change. Molecules can gain or lose energy from their surroundings, resulting in changes in their behavior. Second, molecules have an attraction for one another. This force of

attraction is important in determining the phase in which a particular kind of matter exists.

The amount of kinetic energy molecules have, the strength of the attractive forces between molecules, and the kind of arrangements they form result in three **phases of matter:** solid, liquid, and gas (figure 2.6). A **solid** (e.g., bone) consists of molecules with strong attractive forces and low kinetic energy. The molecules are packed tightly together. With the least amount of kinetic energy of all the phases of matter, these molecules vibrate in place and are at fixed distances from one another. Powerful forces bind them together. Solids have definite shapes and volumes under ordinary temperature and pressure conditions. The hardness of a solid is its resistance to forces that tend to push its molecules farther apart. There is less kinetic energy in a solid than in a liquid of the same material.

A **liquid** (e.g., the watery component of blood and lymph) has molecules with enough kinetic energy to overcome the attractive forces that hold molecules together. Thus, although the molecules are still strongly attracted to each other, they are slightly farther apart than in a solid. Because they are moving more rapidly, and the attractive forces can be overcome, they sometimes slide past each other as they move. Although liquids can change their shape under ordinary conditions, they maintain a fixed volume under ordinary temperature and pressure conditions—that is, a liquid of a certain volume will take the shape of the container into which it is poured, but it will take up the same amount of space regardless of its shape. This gives liquids the ability to *flow,* so they are called *fluids.*

A **gas** (e.g., air and the components of air that are present in the blood) is made of molecules that have a great deal of kinetic energy. The attraction the gas molecules have for each other is overcome by the speed with which the individual molecules move. Because gas molecules are moving faster than the molecules of solids or liquids, their collisions tend to push them farther apart, so a gas expands to fill its container. The shape of the container and the pressure determine the shape and volume of the gas. The term *vapor* is used to describe the gaseous form of a substance, that is normally in the liquid phase. Water vapor, for example, is the gaseous form of liquid water. nitrogen cycle, p. 313

2.6 Chemical Changes—Forming New Kinds of Matter

Atoms interact with other atoms to fill their outermost energy level with electrons to become more stable. When this happens we call the interaction a *chemical reaction*. A **chemical reaction** is a change in matter in which different chemical substances are created by forming or breaking chemical bonds. When a chemical reaction occurs the interacting atoms may become attached, or bonded, to one another by a chemical bond. Two types of bonds are (1) ionic bonds and (2) covalent bonds.

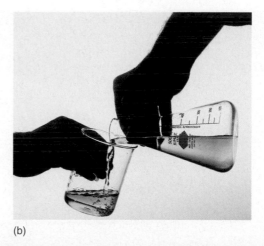

(b)

(a)   (c)

FIGURE 2.6 Phases of Matter

(a) In a solid, such as this rock, molecules vibrate around a fixed position and are held in place by strong molecular forces. *(b)* In a liquid, molecules can rotate and roll over each other, because the kinetic energy of the molecules is able to overcome the molecular forces. *(c)* Inside the bubble, gas molecules move rapidly in random, free paths.

Ionic Bonds and Ions

Any positively or negatively charged atom or molecule is called an **ion. Ionic bonds** are formed after atoms *transfer* electrons to achieve a full outermost energy level. Electrons are donated or received in the transfer, forming a positive and a negative ion, a process called *ionization*. The force of attraction between oppositely charged ions forms ionic bonds, and *ionic compounds* are the result. Ionic compounds are formed when an element from the left side of the periodic table (those eager to gain electrons) reacts with an element from the right side (those eager to donate electrons). This results in the formation of a stable group, which has an orderly arrangement and is a crystalline solid (figure 2.7).

Ions and ionic compounds are very important in living systems. For example, sodium chloride is a crystal solid known as table salt. A positively charged sodium ion is formed when a sodium atom loses 1 electron. This results in a stable, outermost energy level with 8 electrons. When an atom of chlorine receives an electron to stabilize its outermost energy level, it becomes a negative ion. All positively charged ions are called *cations* and all negative charged ions are called *anions* (figure 2.8). When these oppositely charged ions are close to one another, the attractive force between them forms an ionic bond. The dots in the following diagram represent the electrons in the outermost energy levels of each atom. This kind of diagram is called an electron dot formula. nerve impulse, p. 569

$$Na\cdot + \cdot \ddot{\underset{\cdot\cdot}{Cl}}\cdot \longrightarrow Na^{+}Cl^{-}$$

FIGURE 2.7 Crystals
A crystal is composed of ions that are bonded together and form a three-dimensional structure. Crystals grow with the addition of atoms to their surface.

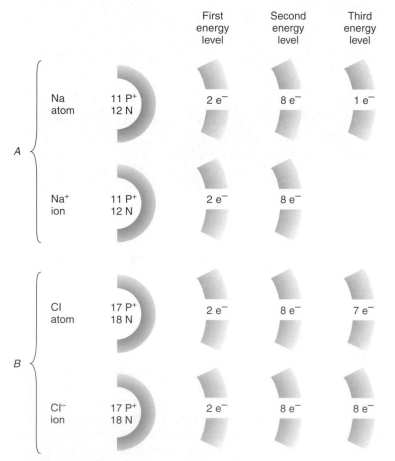

FIGURE 2.8 Ion Formation
A sodium atom *(A)* has 2 electrons in the first energy level, 8 in the second energy level, and 1 in the third level. When it loses its 1 outer electron, it becomes a sodium cation. An atom of chlorine *(B)* has 2 electrons in the first energy level, 8 in the second, and only 7 in the third energy level. To become stable, chlorine picks up an extra electron from an electron donor to fill its outermost energy level, thus satisfying the octet rule. It becomes a chlorine anion.

When many ionic compounds are dissolved in water, the ionic bonds are broken and the ions separate, or *dissociate,* from one another. For example, solid sodium chloride dissociates in water to become ions in solution:

$$NaCl \rightarrow Na^+ + Cl^-$$

Any substance that dissociates into ions in water and allows the conduction of electric current is called an *electrolyte.*

Covalent Bonds

Most substances do not have the properties of ionic compounds, because they are not composed of ions. Most substances are composed of electrically neutral groups of atoms that are tightly bound together. As noted earlier, many gases are diatomic, occurring naturally as two of the same kinds of atoms bound together as an electrically neutral molecule. Hydrogen, for example, occurs as molecules of H_2 and no ions are involved. The hydrogen atoms are held together by a **covalent bond,** a chemical bond formed by the sharing of a pair of electrons. In the diatomic hydrogen molecule, each hydrogen atom contributes a single electron to the shared pair. Hydrogen atoms both share one pair of electrons, but other elements might share more than one pair. <mark>structure of carbohydrates, p. 000</mark>

Consider how the covalent bond forms between two hydrogen atoms by imagining two hydrogen atoms moving toward one another. Each atom has a single electron. As the atoms move closer and closer together, their outer energy levels begin to overlap. Each electron is attracted to the oppositely charged nucleus of the other atom and the overlap tightens. Then, the repulsive forces from the like-charged nuclei stop the merger. A state of stability is reached between the 2 nuclei and 2 electrons, because the outermost energy level is full and an H_2 molecule has been formed. The electron pair is now shared by both atoms, and the attraction of each nucleus for the electron of the other holds the atoms together (figure 2.9).

Dots can be used to represent the electrons in the outer energy levels of atoms. If each atom shares one of its electrons with the other, the two dots represent the bonding pair of electrons shared by the two atoms. Bonding pairs of electrons are often represented by a simple line between two atoms, as in the following example:

H : H is shown as H – H

and

:O: is shown as
H H

A covalent bond in which a single pair of electrons is shared by two atoms is called a *single covalent bond* or, simply, a single bond. Some atoms can share more than one electron pair. A double bond is a covalent bond formed when *two pairs*

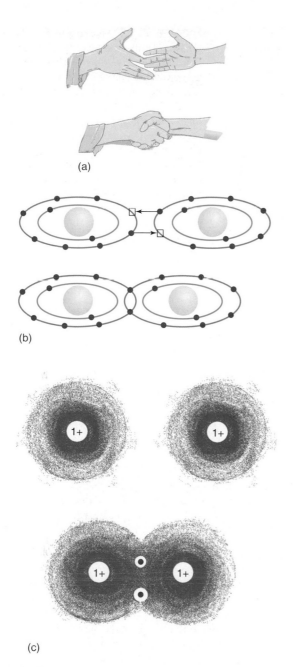

FIGURE 2.10 Ethylene and the Ripening Process
The ancient Chinese knew from observation that fruit would ripen faster if placed in a container of burning incense, but they did not realize the incense released ethylene. We now know that ethylene stimulates the ripening process; it is used commercially to ripen fruits that are picked green.

FIGURE 2.9 Covalent Bond Between Hydrogen Atoms
(a) When two people shake hands, they need to be close enough to each other that their hands can overlap to form a handshake. *(b)* When two fluorine atoms come so close to each other that the locations of the outermost electrons overlap, an electron from each one can be shared to "fill" the outermost energy levels. *(c)* When two hydrogen atoms bond, a new electron distribution pattern forms around the entire molecule, and both electrons share the outermost molecular energy level.

A triple bond is a covalent bond formed when *three pairs* of electrons are shared by two atoms. Triple bonds occur mostly in compounds with atoms of the elements C and N. Atmospheric nitrogen gas, for example, forms a triple covalent bond:

of electrons are shared by two atoms. This happens mostly in compounds involving atoms of the elements C, N, O, and S. For example, ethylene, a gas given off by ripening fruit, has a double bond between the two carbons (figure 2.10). The electron dot formula for ethylene is

2.7 Water: The Essence of Life

Water seems to be a simple molecule, but it has several special properties that make it particularly important for living things. A water molecule is composed of two atoms of hydrogen and one atom of oxygen joined by covalent bonds.

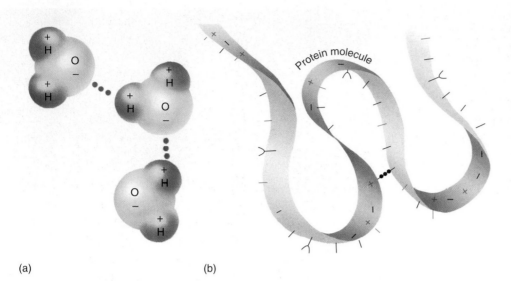

(a)

(b)

FIGURE 2.11 Hydrogen Bonds
(a) Water molecules arrange themselves so that their positive portions are near the negative portions of other water molecules. The attraction force of a single hydrogen bond is indicated as three dots in a row. It is this kind of intermolecular bonding that accounts for water's unique chemical and physical properties. Without such bonds, life as we know it on Earth would be impossible. *(b)* The large protein molecule here also has polar areas. When the molecule is folded so that the partially positive areas are near the partially negative areas, a slight attraction forms and tends to keep it folded.

However, the electrons in these covalent bonds are not shared equally. Oxygen, with 8 protons, has a greater attraction for the shared electron than does hydrogen, with its single proton. Therefore, the shared electrons spend more time around the oxygen part of the molecule than they do around the hydrogen. As a result, the oxygen end of the molecule is more negative than the hydrogen end.

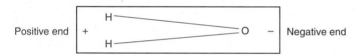

Positive end + — Negative end

When the electrons in a covalent bond are not equally shared, the molecule is said to be *polar* and the covalent bonds are called *polar covalent bonds*.

When the negative end of a polar molecule is attracted to the positive end of another polar molecule, the hydrogen is located between the two molecules. Because in polar molecules the positive hydrogen end of one molecule is attracted to the negative end of another molecule, these attractive forces are often called *hydrogen bonds*. **Hydrogen bonds** can be intermolecular (between molecules) or intramolecular (within molecules) forces of attraction. They occur only between hydrogen and oxygen or hydrogen and nitrogen. As intramolecular forces, *hydrogen bonds hold molecules together. Because they do not bond atoms together, they are not considered true chemical bonds*. This attraction is usually represented as three dots between the attracted regions. This weak force of attraction is not responsible for forming molecules, but it is important in determining the three-dimensional shape of a molecule. For example, when a very large molecule, such as a protein, has some regions that are slightly positive and others that are slightly negative, these areas attract each other and result in the coiling or folding of these threadlike molecules (figure 2.11). Because water is a polar covalent compound (it has slightly + and – ends), it

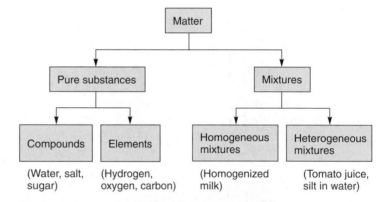

FIGURE 2.12 How Do Mixtures Compare?
Matter can be a pure substance or a mixture. The term *homogeneous* means "the same throughout." Homogenized milk has the same composition throughout the container. Before milk was homogenized (i.e., vigorously shaken to break fat into small globules), it would "separate." The cream (which floats to the top) could be skimmed off the milk below, leaving skimmed milk. A heterogeneous mixture does not have the same composition throughout.

has several significant physical and biological properties (Outlooks 2.1).

Mixtures and Solutions

A **mixture** is matter that contains two or more substances that are not in set proportions (figure 2.12). A **solution** is a liquid mixture of ions or molecules of two or more substances. For example, salt water can be composed of varying amounts of NaCl and H_2O. If the components of the mixture are distributed equally throughout, the mixture is homogenous. The process of making a solution is called *dissolving*. The amounts of the component parts of a solution

OUTLOOKS 2.1

Water and Life

1. **Water has a high surface tension.** Because water molecules are polar, hydrogen bonds form between water molecules, and they stick more to one another than to air molecules. Thus, water tends to pull together to form a smooth surface where water meets air. This layer can be surprisingly strong. For instance, some insects can walk on the surface of a pond. The tendency of water molecules to stick to each other and to some other materials explains why water can make things wet. It also explains why water climbs through narrow tubes, called capillary tubes. This *capillary action* also helps water move through soil, up the vessels in plants' stems, and through the capillaries (tiny blood vessels) in animals.

2. **Water has unusually high heats of vaporization and fusion.** Because polar water molecules stick to one another, an unusually large amount of heat energy is required to separate them. Water resists changes in temperature. It takes 540 calories of heat energy to convert 1 gram of liquid water to its gaseous state, water vapor. This means that large bodies of water, such as lakes and rivers, must absorb enormous amounts of energy before they will evaporate and leave the life within them high and dry. This also means that humans can get rid of excess body heat by sweating because, when the water evaporates, it removes heat from the skin. On the other hand, a high heat of fusion means that this large amount of heat energy must be removed from liquid water before it changes from a liquid to its solid state, ice. Therefore, water can remain liquid and a suitable home for countless organisms long after the atmospheric temperature has reached the freezing point, 0°C (32°F).

3. **Water has unusual density characteristics.** Water is most dense at 4°C. As heat energy is lost from a body of water and its temperature falls below 4°C, its density decreases and this less dense, colder water is left on top. As the surface water reaches the freezing point and changes from its liquid to its solid phase, the molecules form new arrangements, which resemble a honeycomb. The spaces between the water molecules make the solid phase, ice, less dense than the water beneath and the ice floats. It is the surface water that freezes to a solid, covering the denser, liquid water and the living things in it.

4. **Water's specific gravity is also an important property.** Water has a density of 1 gram/cubic centimeter at 4°C. Anything with a higher density sinks in water, and anything with a lower density floats. *Specific gravity* is the ratio of the density of a substance to the density of water. Therefore, the specific gravity of water is 1.00. Any substance with a specific gravity less than 1.00 floats. If you mix water and gasoline, the gasoline (specific gravity of 0.75) floats to the top. People also vary in the specific gravity of their bodies. Some persons find it very easy to float in water, whereas others find it impossible. This is directly related to each person's specific gravity, which is a measure of the person's ratio of body fat to muscle and bone.

5. **Water is considered the universal solvent,** because most other chemicals can be dissolved in water. This means that, wherever water goes—through the ground, in the air, or through an organism—it carries chemicals. Water in its purest form is even capable of acting as a solvent for oils.

6. **Water comprises 50–60% of the bodies of most living things.** This is important, because the chemical reactions of all living things occur in water.

7. **Water vapor in the atmosphere is known as humidity, which changes with environmental conditions.** The ratio of how much water vapor is in the air to how much water vapor could be in the air at a certain temperature is called *relative humidity*. Relative humidity is closely associated with your comfort. When the relative humidity and temperature are high, it is difficult to evaporate water from your skin, so it is more difficult to cool yourself and you are uncomfortably warm.

8. **Water's specific gravity changes with its physical phase. Ice is also more likely to change from a solid to a liquid (melt) as conditions warm.** If the specific gravity of water did not decrease when it freezes, then the ice would likely sink and never thaw. Our life-giving water would be trapped in ocean-sized icebergs. Ice also provides a protective layer for the life under the ice sheet.

are identified by the terms *solvent* and *solute*. The **solvent** is the component present in the larger amount. The **solute** is the component that dissolves in the solvent. Many combinations of solutes and solvents are posssible. If one of the components of a solution is a liquid, it is usually identified as the solvent. An *aqueous solution* is a solution of a solid, liquid, or gas in water. When sugar dissolves in water, sugar molecules separate from one another. The molecules become uniformly dispersed throughout the molecules of water. In an aqueous salt solution, however, the salt dissociates into sodium and chlorine ions.

The relative amounts of solute and solvent are described by the *concentration* of a solution. In general, a solution with a large amount of solute is "concentrated," and a solution with much less solute is "dilute," although these are somewhat arbitrary terms.

2.8 Chemical Reactions

When compounds are broken or formed, new materials with new properties are produced. This kind of a change in matter is called a *chemical change,* and the process is called a *chemical reaction*. In a chemical reaction, the elements stay the same but the compounds and their properties change when the elements are bonded in new combinations. All living things can use energy and matter. In other words, they are constantly performing chemical reactions.

Chemical reactions produce new chemical substances with greater or smaller amounts of potential energy. Energy is absorbed to produce new chemical substances with more potential energy. Energy is released when new chemical substances are produced with less potential energy. For example, new chemical substances are produced in green plants through the process of *photosynthesis*. A green plant uses radiant energy (sunlight), carbon dioxide, and water to produce new chemical materials and oxygen. These new chemical materials, the stuff that makes up leaves, roots, and wood, contain more chemical energy than the carbon dioxide and water from which they were formed.

A **chemical equation** is a way of describing what happens in a chemical reaction. For example, the chemical reaction of photosynthesis is described by the equation

$$\underset{\substack{\text{energy}\\\text{(sunlight)}}}{} + \underset{\substack{\text{carbon}\\\text{dioxide}\\\text{molecules}}}{} + \underset{\substack{\text{water}\\\text{molecules}}}{} \rightarrow \underset{\substack{\text{plant}\\\text{material}\\\text{molecules}\\\text{(ex., sugar)}}}{} + \underset{\substack{\text{oxygen}\\\text{molecules}}}{}$$

$$6\,CO_2 + 6\,H_2O \rightarrow C_6H_{12}O_6 + 6\,O_2$$

In chemical reactions, **reactants** are the substances that are changed (in photosynthesis, the carbon dioxide molecules and water molecules); they appear on the left side of the equation. The equation also indicates that energy is absorbed; the term *energy* appears on the left side. The arrow indicates the direction in which the chemical reaction is occurring; it means "yields." The new chemical substances are on the right side

and are called **products.** Reading the photosynthesis reaction as a sentence, you would say, "Carbon dioxide and water use energy to react, yielding plant materials and oxygen."

Notice in the photosynthesis reaction that there are numbers preceding some of the chemical formulas and subscripts within each chemical formula. The number preceding each of the chemical formulas indicates the number of each kind of molecule involved in the reaction. The subscripts indicate the number of each kind of element in a single molecule of that compound. Chemical reactions always take place in whole number ratios. That is, only whole molecules can be involved in a chemical reaction. It is not possible to have half a molecule of water serve as a reactant or become a product. Half a molecule of water is not water. Furthermore, the numbers of atoms of each element on the reactant side must equal the numbers on the product side. Because the preceding equation has equal numbers of each element (C, H, O) on both sides, the equation is said to be "balanced."

Five of the most important chemical reactions that occur in organisms are (1) oxidation-reduction, (2) dehydration synthesis, (3) hydrolysis, (4) phosphorylation, and (5) acid-base reactions.

Oxidation-Reduction Reactions

An **oxidation-reduction reaction** is a chemical change in which electrons are transferred from one atom to another and, with it, the energy contained in its electrons. As implied by the name, such a reaction has two parts and each part tells what happens to the electrons. *Oxidation* is the part of an oxidation-reduction reaction in which an atom loses an electron. *Reduction* is the part of an oxidation-reduction reaction in which an atom gains an electron. When the term *oxidation* was first used, it specifically meant reactions involving the combination of oxygen with other atoms. But fluorine, chlorine, and other elements were soon recognized to participate in similar reactions, so the definition was changed to describe the shifts of electrons in the reaction. The name also implies that, in any reaction in which oxidation occurs, reduction must also take place. One cannot take place without the other. Cellular respiration is an oxidation-reduction reaction that occurs in all cells:

$$\underset{\text{sugar}}{C_6H_{12}O_6} + \underset{\text{oxygen}}{6\,O_2} \rightarrow \underset{\text{water}}{6\,H_2O} + \underset{\substack{\text{carbon}\\\text{dioxide}}}{6\,CO_2} + \text{energy}$$

In this cellular respiration reaction, sugar is being oxidized (losing its electrons) and oxygen is being reduced (gaining the electrons from sugar). The high chemical potential energy in the sugar molecule is released, and the organism uses some of this energy to perform work. In the previously mentioned photosynthesis reaction, water is oxidized (loses its electrons) and carbon dioxide is reduced (gains the electrons from water). The energy required to carry out this reaction comes from the sunlight and is stored in the product, sugar.

Dehydration Synthesis Reactions

Dehydration synthesis reactions are chemical changes in which water is released and a larger, more complex molecule is made (synthesized) from smaller, less complex parts. The water is a product formed from its component parts (H and OH), which are removed from the reactants. Proteins, for example, consist of a large number of amino acid subunits joined together by dehydration synthesis:

$$NH_2CH_2CO—OH + H—NH\ CH_2CO—OH$$
amino acid 1   +   amino acid 2

$$NH_2CH_2CO—NH\ CH_2CO—OH + H—OH$$
Protein   + water (H_2O)

The building blocks of protein (amino acids) are bonded to one another to synthesize larger, more complex product molecules (i.e., protein). In dehydration synthesis reactions, water is produced as smaller reactants become chemically bonded to one another, forming fewer but larger product molecules.

Hydrolysis Reactions

Hydrolysis reactions are the opposite of dehydration synthesis reactions. In a hydrolysis reaction, water is used to break the reactants into smaller, less complex products:

$$NH_2CH_2CO—NH\ CH_2CO—OH + H—OH$$
Protein   + water (H_2O)
$$NH_2CH_2CO—OH + H—NH\ CH_2CO—OH$$
amino acid 1   +   amino acid 2

A more familiar name for this chemical reaction is *digestion*. This is the kind of chemical reaction that occurs when a protein food, such as meat, is digested. Notice in the previous example that the H and OH component parts of the reactant water become parts of the building block products. In hydrolysis reactions, water is used as a reactant, and larger molecules are broken down into smaller units.

Phosphorylation Reactions

A **phosphorylation reaction** takes place when a cluster of atoms known as a *phosphate group*

is added to another molecule. This cluster is abbreviated in many chemical formulas in a shorthand form as P, and only the P is shown when a phosphate is transferred from one molecule to another. This is a very important reaction, because the bond between a phosphate group and another atom con-

FIGURE 2.13 Phosphorylation and Muscle Contractions
When the phosphate group is transferred between molecules, energy is released which powers muscle contractions.

tains the potential energy that is used by all cells to power numerous activities. Phosphorylation reactions result in the transfer of their potential energy to other molecules to power the activities of all organisms (figure 2.13).

high potential energy	low potential energy	low potential energy	high potential energy

$$Q–P \quad + \quad Z \quad \rightarrow \quad Q \quad + \quad Z–P$$

This type of reaction is commonly involved in providing the kinetic energy needed by all organisms. It can also take place in reverse. When this occurs, energy must be added from the environment (sunlight or another phosphorylated molecule) and is stored in the phosphorylated molecule.

Acid-Base Reactions

Acid-base reactions take place when the ions of an acid interact with the ions of a base, forming a salt and water (see section 2.9). An aqueous solution containing dissolved acid is a solution containing hydrogen ions. If a solution containing a second ionic basic compound is added, a mixture of ions results. While they are mixed together, a reaction can take place—for example,

$$H^+ Cl^- + Na^+ OH^- \rightarrow Na^+ Cl^- + HOH$$
hydrochloric + sodium   → sodium   + water
acid   hydroxide   chloride   (H_2O)

In an acid-base reaction, the H from the acid becomes chemically bonded to the OH of the base. This type of reaction frequently occurs in organisms and their environment. Because acids and bases can be very harmful, reactions in which they neutralize one another protect organisms from damage.

2.9 Acids, Bases, and Salts

Acids, bases, and salts are three classes of biologically important compounds (table 2.3). Their characteristics are determined by the nature of their chemical bonds. **Acids** are ionic compounds that release hydrogen ions in solution. A hydrogen atom without its electron is a proton. You can think of an acid, then, as a substance able to donate a proton to a solution. Acids have a sour taste, such as that of citrus fruits. However, tasting chemicals to see if they are acids can be very hazardous, because many are highly corrosive. An example of a common acid is the phosphoric acid—H_3PO_4— in cola soft drinks. It is a dilute solution of this acid that gives cola drinks their typical flavor. Hydrochloric acid is another example:

A **base** is the opposite of an acid, in that it is an ionic compound, which, when dissolved in water, removes hydrogen ions from solution. Bases, or *alkaline* substances, have a slippery feel on the skin. They have a caustic action on living tissue by converting the fats in living tissue into a water-soluble substance. A similar reaction is used to make soap by mixing a strong base with fat. This chemical reaction gives soap its slippery feeling. Bases are also used in alkaline batteries. Weak bases have a bitter taste—for example, the taste of broccoli, turnip, and cabbage. Many kinds of bases release a group of hydrogen ions known as a **hydroxide ions,** or an OH^- group. This group is composed of an oxygen atom and a hydrogen atom bonded together, but with an additional electron. The hydroxide ion is negatively charged; therefore, it will remove positively charged hydrogen ions from solution. A very strong base used in oven cleaners is sodium hydroxide, NaOH. Notice that ions that free in solution are always written with the type and number of their electrical charge as a superscript.

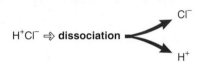

Acids are iconically bonded molecules which when placed in water dissociate, releasing hydrogen (H^+) ions.

Basic (alkaline) substances are ionically bonded molecules, which when placed in water dissociate, releasing hydroxide (OH^-) ions.

TABLE 2.3

Some Common Acids, Bases, and Salts

Acids

Acetic acid	CH_3COOH	Weak acid found in vinegar
Carbonic acid	H_2CO_3	Weak acid of carbonated beverages that provides bubbles or fizz
Lactic acid	$CH_3CHOHCOOH$	Weak acid found in sour milk, sauerkraut, and pickles
Phosphoric acid	H_3PO_4	Weak acid used in cleaning solutions, added to carbonated cola beverages for taste
Sulfuric acid	H_2SO_4	Strong acid used in batteries

Bases

Sodium hydroxide	NaOH	Strong base also called lye or caustic soda; used in oven cleaners
Potassium hydroxide	KOH	Strong base also known as caustic potash; used in drain cleaners
Magnesium hydroxide	$Mg(OH)_2$	Weak base also known as milk of magnesia; used in antacids and laxatives

Salts

Alum	$Al(SO_4)_2$	Found in medicine, canning, and baking powder
Baking soda	$NaHCO_3$	Used in fire extinguishers, antacids, baking powder, and sodium bicarbonate
Chalk	$CaCO_3$	Used in antacid tablets
Epsom salts	$MgSO_4 \cdot H_2O$	Used in laxatives and skin care
Trisodium phosphate (TSP)	Na_3PO_4	Used in water softeners, fertilizers, and cleaning agents

Acids and bases are also spoken of as being strong or weak (Outlooks 2.2). Strong acids (e.g., hydrochloric acid) are those that dissociate nearly all of their hydrogens when in solution. Weak acids (e.g., phosphoric acid) dissociate only a small percentage of their hydrogens. Strong bases dissociate nearly all of their hydroxides (NaOH); weak bases, only a small percentage. The weak base sodium bicarbonate, $NaHCO_3$, will react with acids in the following manner:

$$NaHCO_3 + HCl \rightarrow NaCl + CO_2 + H_2O$$

Notice that sodium bicarbonate does not contain a hydroxide ion but it is still a base, because it removes hydrogen ions from solution.

The degree to which a solution is acidic or basic is represented by a quantity known as **pH**. The pH scale is a measure of hydrogen ion concentration (figure 2.14). A pH of 7 indicates that the solution is neutral and has an equal number of H^+ ions and OH^- ions to balance each other. As the pH number gets smaller, the number of hydrogen ions in the solution increases. A number higher than 7 indicates that the solution has more OH^- than H^+. Pure water has a pH of 7. As the pH number gets larger, the number of hydroxide ions increases. It is important to note that the pH scale is logarithmic—that is, a change in one pH number is actually a 10-fold change in real numbers of OH^- or H^+. For example, there is 10 times more H^+ in a solution of pH 5 than in a solution of pH 6 and 100 times more H^+ in a solution of pH 4 than in a solution of pH 6. enzymes and acid, p. 100

Ionically Bonded Molecule (water) $H^+OH^- \Rightarrow$ **dissociation** $\longrightarrow$ OH^- $\longrightarrow$ H^+

When water dissociates, it releases both hydrogen (H^+) and hydroxide (OH^-) ions. It is neither a base nor an acid. Its pH is 7, neutral.

Salts are ionic compounds that do not release either H^+ or OH^- when dissolved in water; thus, they are neither acids nor bases. However, they are generally the result of the reaction between an acid and a base in a solution. For example, when an acid, such as HCl, is mixed with NaOH in water, the H^+ and the OH^- combine with each other to form pure water, H_2O. The remaining ions (Na^+ and Cl^-) join to form the salt NaCl:

$$HCl + NaOH \rightarrow Na^+ + Cl^- + H^+ + OH^- \rightarrow$$
$$NaCl + H_2O$$

The chemical reaction that occurs when acids and bases react with each other is called *neutralization*. The acid no longer acts as an acid (it has been neutralized) and the base no longer acts as a base.

As you can see from figure 2.14, not all acids or bases produce the same pH. Some compounds release hydrogen ions very easily, cause low pHs, and are called strong acids.

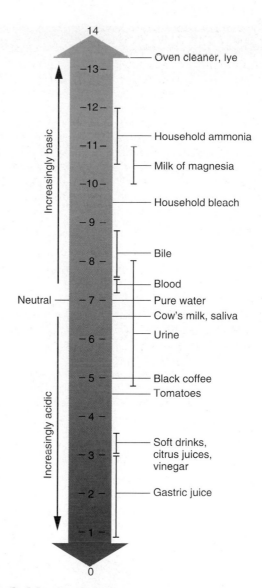

FIGURE 2.14 The pH Scale
The concentration of acid (proton donor or electron acceptor) is greatest when the pH number is lowest. As the pH number increases, the concentration of base (proton acceptor or electron donor) increases. At a pH of 7, the concentrations of H^+ and OH^- are equal. As the pH number gets smaller, the solution becomes more acidic. As the pH number gets larger, the solution becomes more basic, or alkaline.

Hydrochloric acid (HCl) and sulfuric acid (H_2SO_4) are strong acids (figure 2.15a). Many other compounds give up their hydrogen ions grudgingly and therefore do not change pH very much. They are known as weak acids. Carbonic acid (H_2CO_3) and many organic acids found in living things are weak acids. Similarly, there are strong bases, such as sodium hydroxide (NaOH) and weak bases, such as sodium bicarbonate—$Na^+(HCO_3)^-$. evolution of life, p. 110

OUTLOOKS 2.2

Maintaining Your pH—How Buffers Work

Acids, bases, and salts are called *electrolytes,* because, when these compounds are dissolved in water, the solution of ions allows an electrical current to pass through it. Salts provide a variety of ions essential to the human body. Small changes in the levels of some ions can have major effects on the functioning of the body. The respiratory system and kidneys regulate many of the body's ions. Because many kinds of chemical activities are sensitive to changes in the pH of the surroundings, it is important to regulate the pH of the blood and other body fluids within very narrow ranges. Normal blood pH is about 7.4. Although the respiratory system and kidneys are involved in regulating the pH of the blood, there are several systems in the blood that prevent wide fluctuations in pH.

Buffers are mixtures of *weak acids* and the *salts of weak acids* that tend to maintain constant pH, because the mixture can either accept or release hydrogen ions (H^+). The weak acid can release hydrogen ions (H^+) if a base is added to the solution, and the negatively charged ion of the salt can accept hydrogen ions (H^+) if an acid is added to the solution.

One example of a buffer system in the body is a phosphate buffer system, which consists of the weak acid dihydrogen phosphate ($H_2PO_4^-$) and the salt of the weak acid monohydrogen phosphate ($HPO_4^=$).

$$H_2PO_4^- \rightleftharpoons H^+ + HPO_4^=$$

weak acid   hydrogen +   salt of
               ion   weak acid

(The two arrows indicate that this is in balance, with equal reactions in both directions.) The addition of an acid to the mixture causes the equilibrium to shift to the left.

$$H_2PO_4^- \rightleftharpoons H^+ + HPO_4^= + \text{added } H^+$$

Notice that the arrow pointing to the right is shorter than the arrow pointing to the left. This indicates that H^+ is combining with HPO_4^- and additional $H_2PO_4^-$ is being formed. This removes the additional hydrogen ions from solution and ties them up in the $H_2PO_4^-$, so that the amount of free hydrogen ions in the solution remains constant.

Similarly, if a base is added to the mixture, the equilibrium shifts to the right, additional hydrogen ions are released to tie up the hydroxyl ions, and the pH remains unchanged.

$$H_2PO_4^- + \text{added } OH^- \rightleftharpoons H^+HPO_4^= + HOH$$

Seawater is a buffer solution that maintains a pH of about 8.2. Buffers are also added to medicines and to foods. Many lemon-lime carbonated beverages, for example, contain citric acid and sodium citrate (salt of the weak acid), which forms a buffer in the acid range. The beverage label may say that these chemicals are to impart and regulate "tartness." In this case, the tart taste comes from the citric acid, and the addition of sodium citrate makes it a buffered solution.

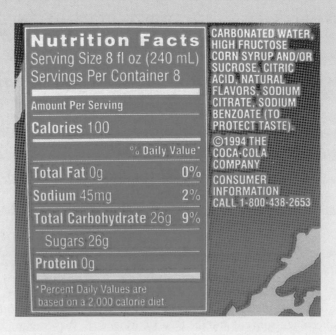

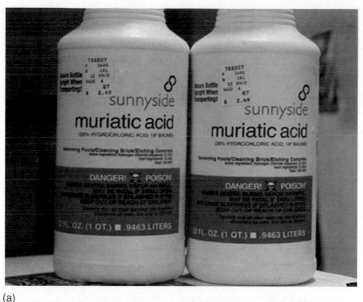

(a)   (b)

FIGURE 2.15 Strong Acid and Strong Base

(a) Hydrochloric acid (HCL) has the common name of muriatic acid. It is a strong acid used in low concentrations to clean swimming pools and brick surfaces. It is important that you wear protective equipment when working with a solution of muriatic acid. *(b)* Liquid-Plumr ® is a good example of a drain cleaner with a strong base. The active ingredient is NaOH.

Summary

The study of life involves learning about the structure and function of organisms. All organisms display the chemical and physical properties typical of all matter and energy. The two kinds of energy used by organisms are potential and kinetic. The kinetic molecular theory states that all matter is made up of tiny particles, which are in constant motion.

Energy can be neither created nor destroyed, but it can be converted from one form to another. Potential energy and kinetic energy can be interconverted. The amount of kinetic energy that the molecules of various substances contain determines whether they are solids, liquids, or gases. Temperature is a measure of the average kinetic energy of the molecules making up a substance. Heat is the total internal kinetic energy of molecules. The random motion of molecules, which is due to their kinetic energy, results in their being distributed throughout available space, forming mixtures.

There are many kinds of atoms, whose symbols and traits are described by the periodic table of the elements. These atoms differ from one another by the number of protons and electrons they contain. Each is given an atomic number, based on the number of protons in the nucleus, and an atomic weight, an average of all the isotopes of a particular element. The mass number is the sum of the number of protons and neutrons in the nucleus of an atom.

All matter is composed of atoms, which are composed of an atomic nucleus and electrons. The atomic nucleus can contain protons and neutrons, whereas the electrons encircle the nucleus at different energy levels. Atoms tend to seek their most stable configuration and follow the octet rule, which states that they all seek a filled outermost energy level.

Atoms may be combined by chemical reactions into larger units called molecules. There are many kinds of molecules. Two kinds of chemical bonds allow molecules to form—ionic bonds and covalent bonds. A third bond, the hydrogen bond, is a weaker bond that holds molecules together and may help large molecules maintain a specific shape. Molecules are described by their chemical formulas, which state the number and kinds of components of which they are composed.

An ion is an atom that is electrically unbalanced. Ions interact to form ionic compounds, such as acids, bases, and salts. Compounds that release hydrogen ions when mixed in water are called acids; those that remove hydrogen ions are called bases. A measure of the hydrogen ions present in a solution is the pH of the solution.

Water is one of the most important compounds required by all organisms. This polar molecule has many unique properties, which allow organisms to survive and reproduce. Without water, life as we know it on Earth would not be possible.

How atoms achieve stability is the nature of chemical reactions. Five of the most important chemical reactions that occur in organisms are (1) oxidation-reduction, (2) dehydration synthesis, (3) hydrolysis, (4) phosphorylation, and (5) acid-base reactions.

Acids, bases, and salts are three classes of biologically important molecules. The hydrogen ion releasing or acquiring properties of acids and bases make them valuable in all organisms.

Salts are a source of many essential ions. Although acids and bases may be potentially harmful, buffer systems help in maintain pH levels.

Key Terms

*Use the interactive flash cards on the **Concepts in Biology**, 12/e website to help you learn the meaning of these terms.*

acid-base reactions 37
acids 38
atomic mass unit 25
atomic nucleus 25
atomic number 25
atomic weight 27
bases 38
calorie 29
chemical bonds 29
chemical equation 36
chemical reaction 30
chemistry 24
compound 29
covalent bond 32
dehydration synthesis reactions 37
electron 25
elements 25
energy 24
energy level 25
formula 29
gas 30
heat 29
hydrogen bonds 34
hydrolysis reactions 37
hydroxide ions 38
ion 31
ionic bonds 31

isotope 27
kinetic energy 24
kinetic molecular theory 29
law of conservation of energy 24
liquid 30
mass number 27
matter 24
mixture 34
molecule 29
neutron 25
oxidation-reduction reaction 36
pH 39
phases of matter 30
phosphorylation reaction 37
potential energy 24
products 36
proton 25
reactants 36
salts 39
solid 30
solute 36
solution 34
solvent 36
temperature 29

Basic Review

1. _____ is the total internal kinetic energy of molecules.
2. The atomic weight of the element sodium is
   a. 22.989.
   b. 11.
   c. 10.252.
   d. 11+22.989.

3. Which is *not* a pure substance?
   a. the compound sugar
   b. the element oxygen
   c. a mixture of milk and honey
   d. the compound table salt
4. When a covalent bond forms between two kinds of atoms that are the same, the result is known as a
   a. a mixture.
   b. a crystal.
   c. a dehydration chemical reaction.
   d. diatomic molecule.
5. In this kind of chemical reaction, two molecules interact, resulting in the formation of a molecule of water and a new, larger end product.
   a. hydrolysis
   b. dehydration synthesis
   c. phosphorylation
   d. acid-base reaction
6. Which of the following is an acid?
   a. HCl
   b. NaOH
   c. KOH
   d. $CaCO_3$
7. Salts are compounds that do not release either _____ or _____ ions when dissolved in water.
8. This intramolecular force under the right conditions can result in a molecule that is coiled or twisted into a complex, three-dimensional shape.
   a. covalent bond
   b. ionic bond
   c. hydrogen bond
   d. cement bond
9. A triple covalent bond is represented by which of the following?
   a. a single, fat, straight line
   b. a single, thin, straight line
   c. three separate, thin lines
   d. three thin, curved lines
10. Electron clouds, or routes, traveled by electrons are sometimes drawn as spherical or _____ shapes.

Answers

1. Heat   2. a   3. c   4. d   5. b   6. a   7. H^+, OH^- 8. c
9. c   10. hourglass

Concept Review

*Answers to the Concept Review questions can be found on the **Concepts in Biology,** 12/e website.*

2.1   Matter, Energy, and Life

1. What is chemistry?
2. What does chemistry have to do with biology?

2.2   The Nature of Matter

3. List the five forms of energy.
4. What is the difference between potential and kinetic energy?
5. Define the term *work*.

2.3   The Kinetic Molecular Theory and Molecules

6. What is the difference between an atom and an element?
7. What is the difference between a molecule and a compound?
8. On what bases are solids, liquids, and gases differentiated?
9. What relationship does kinetic energy have to the three phases of matter?

2.4   Molecules and Kinetic Energy

10. What is the difference between temperature and heat?
11. What is a calorie?

2.5   Physical Changes—Phases of Matter

12. How many protons, electrons, and neutrons are in a neutral atom of potassium having an atomic weight of 39?
13. How would two isotopes of oxygen differ?
14. Define the terms *AMU* and *atomic number*.

2.6   Chemical Changes—Forming New Kinds of Matter

15. Why are the outermost electrons of an atom important?
16. Name two kinds of chemical bonds that hold atoms together. How do these bonds differ from one another?

2.7   Water: The Essence of Life

17. What is the difference between a polar and a nonpolar molecule?
18. What is different about a hydrogen bond in comparison with covalent and ionic bonds?
19. What is the difference between a solute and a solvent?
20. What relationship does kinetic energy have to homogeneous solutions?

2.8   Chemical Reactions

21. Give an example of an ion exchange reaction.
22. What happens during an oxidation-reduction reaction?
23. Describe the difference between a reactant and a product.

2.9   Acids, Bases, and Salts

24. What does it mean if a solution has a pH of 3, 12, 2, 7, or 9?
25. If the pH of a solution changes from 8 to 9, what happens to the hydroxide ion concentration?

Thinking Critically

*For guidelines to answer this question, visit the **Concepts in Biology,** 12/e website.*

Sodium bicarbonate ($NaHCO_3$) is a common household chemical known as baking soda, bicarbonate of soda, or bicarb. It has many uses and is a component of many products, including toothpaste and antacids, swimming pool chemicals, and headache remedies. When baking soda comes in contact with hydrochloric acid, the following reaction occurs:

$$HCl + NaHCO_3 \rightarrow NaCl + CO_2 + H_2O$$

What happens to the atoms in this reaction? In your description, include changes in chemical bonds, pH, and kinetic energy. Why is baking soda such an effective chemical in the previously mentioned products? Try this at home: Place a pinch of sodium bicarbonate ($NaHCO_3$) on a plate. Add two drops of vinegar. Observe the reaction. Based on the previous reaction, can you explain chemically what has happened?

CHAPTER

3

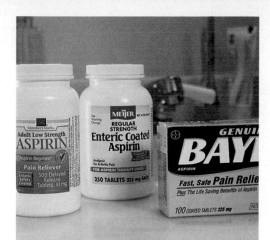

Organic Molecules—
the Molecules of Life

When you first look at the material in this chapter, you may think that learning organic chemistry will be a real "headache." You might even look for a pain reliever—aspirin. Consider, though, that aspirin is an organic molecule. An estimated 80 million tablets are consumed in the United States daily by children and adults. Aspirin is made by several companies and sold under many brand names. All aspirin tablets contain the same chemical.

Aspirin's chemical name is acetylsalicylic acid. The name aspirin is an acronym for *a*cetyl (a portion of the chemical), *spir* from *spir*ea (an ornamental plant), and *in* as in salic*in* (the compound from which aspirin is derived). In the past few decades, several other pain-relieving drugs have been introduced. However, at this time, no other drug has been as successful or as safe.

Aspirin went on the market as a prescription drug in 1899 after being discovered only 2 years earlier by a chemist working for the German company Bayer. Even primitive humans were familiar with its value as a pain reliever. At that time, the bark of the white willow tree was known for its "magical," pain-relieving power. Its pain-relief properties stem from its ability to inhibit the body's production of organic chemicals known as prostaglandins, which promote pain and inflammation. People in many cultures chewed the bark for its medicinal value. Famous Greek philosopher/scientist and father of medicine Hippocrates (2000 B.C.) knew that chewing this bark relieved pain, inflammation, and fever. We now know that it can also help many people prevent heart attacks and stroke if a low dose (81 mg) is taken every other day. If taken immediately after a heart attack, aspirin can also prevent stroke and thousands of deaths because of its anti-clotting properties.

CHAPTER OUTLINE

3.1 Molecules Containing Carbon

The principles and concepts discussed in chapter 2 apply to all types of matter—nonliving as well as living. Living systems are composed of various types of molecules. Most of the chemicals described in chapter 2 do not contain carbon atoms and, so, are classified as **inorganic molecules.** This chapter is mainly concerned with more complex structures, **organic molecules,** which contain carbon atoms arranged in rings or chains. The words *organic, organism, organ,* and *organize* are all related. Organized objects have parts that fit together in a meaningful way. Organisms are living things that are organized. Animals, for example, have within their body's organ systems, and their organs are composed of unique kinds of molecules that are *organic.* origin of life, p. 406

The original meanings of the terms *inorganic* and *organic* came from the fact that organic materials were thought either to be alive or to be produced only by living things. A very strong link exists between organic chemistry and the chemistry of living things, which is called **biochemistry** or biological chemistry. Modern chemistry has considerably altered the original meanings of the terms *organic* and *inorganic,* because it is now possible to manufacture unique organic molecules that cannot be produced by living things. Many of the materials we use daily are the result of the organic chemist's art. Nylon, aspirin, polyurethane varnish, silicones, Plexiglas, food wrap, Teflon, and insecticides are just a few of the unique synthetic molecules that have been invented by organic chemists (figure 3.1).

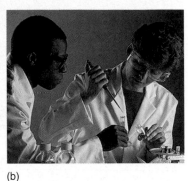

(a)   (b)

FIGURE 3.2 Natural and Synthetic Organic Compounds
(a) The photograph shows the collection of latex from the rubber tree *(Castilla elastica).* After processing, this naturally occurring organic material will be converted into products such as gloves, condoms, and tubing. *(b)* Shows organic chemists testing a new synthetic rubber.

Many organic chemists have taken their lead from living organisms and have been able to produce organic molecules more efficiently, or in forms that are slightly different from the original natural molecules. Some examples of these are rubber, penicillin, certain vitamins, insulin, and alcohol (figure 3.2). Another example is the insecticide Pyrethrin. It is based on a natural insecticide and is widely used for agricultural and domestic purposes; it is from the chrysanthemum plant, *Pyrethrum cinerariaefolium.*

Carbon: The Central Atom

All organic molecules, whether natural or synthetic, have certain common characteristics. The carbon atom, which is the central atom in all organic molecules, has some unusual properties. Carbon is unique in that it can combine with other carbon atoms to form long chains. In many cases, the ends of these chains may join together to form ring structures (figure 3.3). Only a few other atoms have this ability.

Also unusual is that these bonding sites are all located at equal distances from one another. If you were to stick four nails into a rubber ball so that the nails were equally distributed around the ball, you would have a good idea of the three-dimensional arrangement of the bonds (figure 3.4). These bonding sites are arranged this way because, in the carbon atom, there are 4 electrons in the outermost energy level. However, these 4 electrons do not stay in the standard positions described in chapter 2. They distribute themselves differently—that is, into four propeller-shaped electron paths. This allows them to be as far away from each other as possible. Carbon atoms are usually involved in covalent bonds.

FIGURE 3.1 Some Common Synthetic Organic Materials
These are only a few examples of useful organic compounds invented and manufactured by chemists.

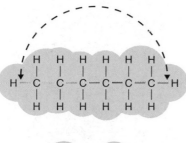

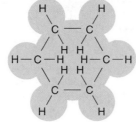

FIGURE 3.3 Chain and Ring Structures
The ring structure shown on the bottom is formed by joining the two ends of a chain of carbon atoms.

Because carbon has four places it can bond, the carbon atom can combine with four other atoms by forming four separate, *single* covalent bonds with other atoms. This is the case with the methane molecule, which has four hydrogen atoms attached to a single carbon atom (review figure 3.4). Pure methane is a colorless, odorless gas that makes up 95% of natural gas. The aroma of natural gas is the result of mercaptan and trimethyl disulfide added for safety to let consumers know when a leak occurs.

Some atoms may be bonded to a single atom more than once. This results in a slightly different arrangement of bonds around the carbon atom. An example of this type of bonding occurs when oxygen is attracted to a carbon. An atom of oxygen has 2 electrons in its outermost energy level. If it shares 1 of these with a carbon and then shares the other with the same carbon, it forms a double bond. A **double bond** is two

covalent bonds formed between two atoms that share two pairs of electrons. Oxygen is not the only atom that can form double bonds, but double bonds are common between oxygen and carbon. The double bond is denoted by two lines between the two atoms:

$$-C=O$$

Two carbon atoms might form double bonds between each other and then bond to other atoms at the remaining bonding sites. Figure 3.5 shows several compounds that contain double bonds.

Some organic molecules contain *triple covalent bonds*; the flammable gas acetylene, HC≡CH, is one example. Others—such as hydrogen cyanid HC≡N—have biological significance. This molecule inhibits the production of energy and can cause death.

Isomers

Although many kinds of atoms can be part of an organic molecule, only a few are commonly found. Hydrogen (H) and oxygen (O) are almost always present. Nitrogen (N), sulfur (S), and phosphorus (P) are also very important in specific types of organic molecules.

An enormous variety of organic molecules is possible, because carbon is able to (1) bond at four different places, (2) form long chains, and (3) combine with many other kinds of atoms. The types of atoms in the molecule are important in determining the properties of the molecule. The three-dimensional arrangement of the atoms within the molecule is also important. Because most inorganic molecules are small and involve few atoms, a group of atoms can be usually arranged in only one way to form a molecule. There is only one arrangement for a single oxygen atom and two hydrogen

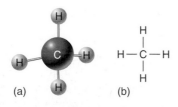

(a)   (b)

FIGURE 3.4 A Methane Molecule
(a) The arrangement of bonding sites around the carbon is similar to a ball with four equally spaced nails in it. A methane molecule is composed of one carbon atom bonded with four hydrogen atoms. These bonds are formed at the four bonding sites of the carbon. For the sake of simplicity, all future diagrams of molecules will be two-dimensional drawings, although in reality they are three-dimensional molecules. *(b)* Each line in the diagram represents a covalent bond between the two atoms where a pair of electrons is being shared. This is a shorthand way of drawing the pair of shared electrons.

FIGURE 3.5 Double Bonds
These diagrams show several molecules that contain double bonds. A double bond is formed when two atoms share two pairs of electrons with each other.

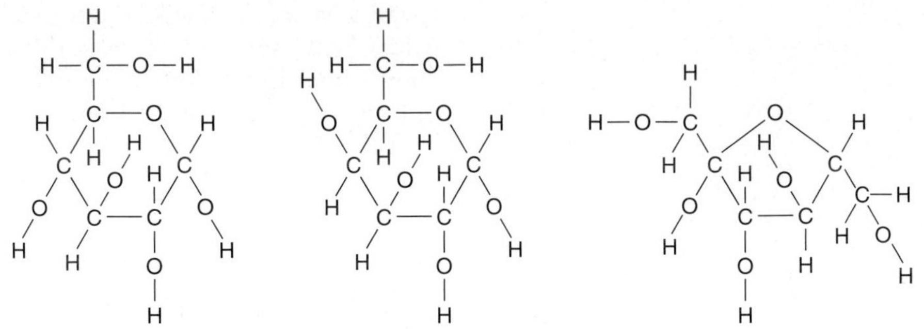

FIGURE 3.6 Structural Formulas for Several Hexoses

Three 6-carbon sugars—hexoses (*hex* = 6; *-ose* = sugar)—are represented here. All have the same empirical formula ($C_6H_{12}O_6$), but each has a different structural formula. These three are called *structural isomers*. Structural isomers may act quite differently than each other.

atoms in a molecule of water. In a molecule of sulfuric acid, there is only one arrangement for the sulfur atom, the two hydrogen atoms, and the four oxygen atoms.

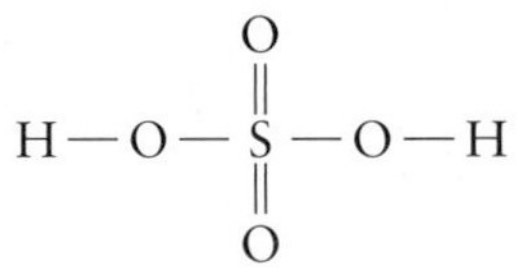

Sulfuric (battery) acid

However, consider these two organic molecules:

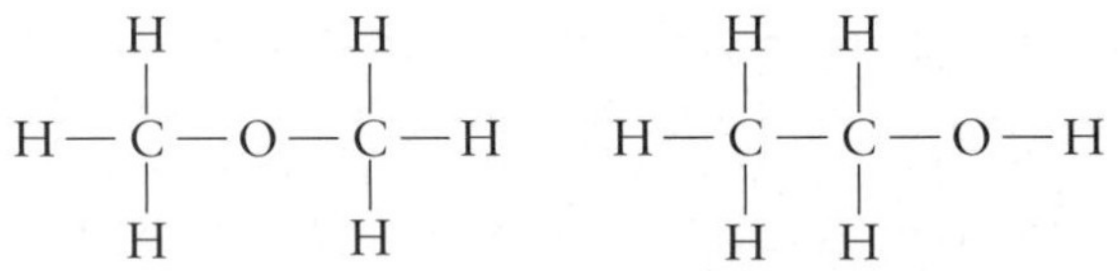

Dimethyl ether   Ethyl alcohol
(as found in alcoholic beverages)

Both the dimethyl ether and the ethyl alcohol contain two carbon atoms, six hydrogen atoms, and one oxygen atom, but they are quite different in their arrangement of atoms and in the chemical properties of the molecules. The first is an ether; the second is an alcohol. Because the ether and the alcohol have the same number and kinds of atoms, they are said to have the same *empirical formula,* which in this case is written C_2H_6O. An empirical formula simply indicates the number of each kind of atom within the molecule. The arrangement of the atoms and their bonding within the molecule are indicated in a *structural formula.* Figure 3.6 shows several structural formulas for the empirical formula $C_6H_{12}O_6$. Molecules that have the same empirical formula but different structural formulas are called **isomers** (How Science Works 3.1).

The Carbon Skeleton and Functional Groups

At the core of all organic molecules is a **carbon skeleton,** which is composed of rings or chains (sometimes branched) of carbon. It is this carbon skeleton that determines the overall shape of the molecule. The differences among various kinds of organic molecules are determined by three factors: (1) the length and arrangement of the carbon skeleton, (2) the kinds and location of the atoms attached to it, and (3) the way these attached atoms are combined. These specific combinations of atoms, called **functional groups,** are frequently found on organic molecules. The kind of functional groups attached to a carbon skeleton determine the specific chemical properties of that molecule. By learning to recognize some of the functional groups, you can identify an organic molecule and predict something about its activity. Figure 3.7 shows some of the functional groups that are important in biological activity. Remember that a functional group does not exist by itself; it is part of an organic molecule. Outlooks 3.1 explains how chemists and biologists diagram the kinds of bonds formed in organic molecules.

Macromolecules of Life

Macromolecules (*macro* = large) are very large organic molecules. We will look at four important kinds of macromolecules: carbohydrates, proteins, nucleic acids, and lipids. Carbohydrates, proteins, and nucleic acids are all *polymers* (*poly* = many; *mer* = segments). **Polymers** are combinations of many smaller, similar building blocks called *monomers* (*mono* = single) bonded together.

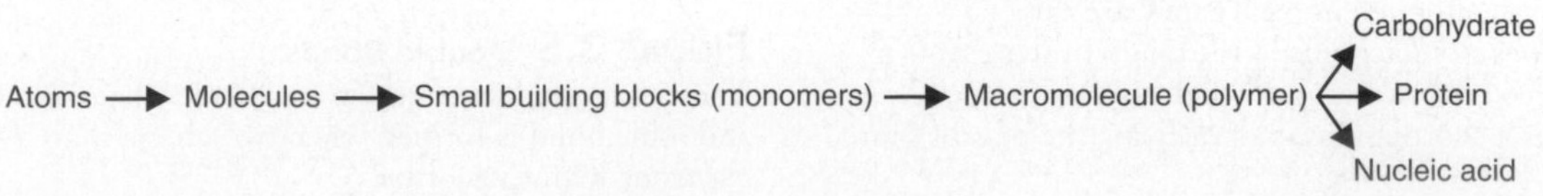

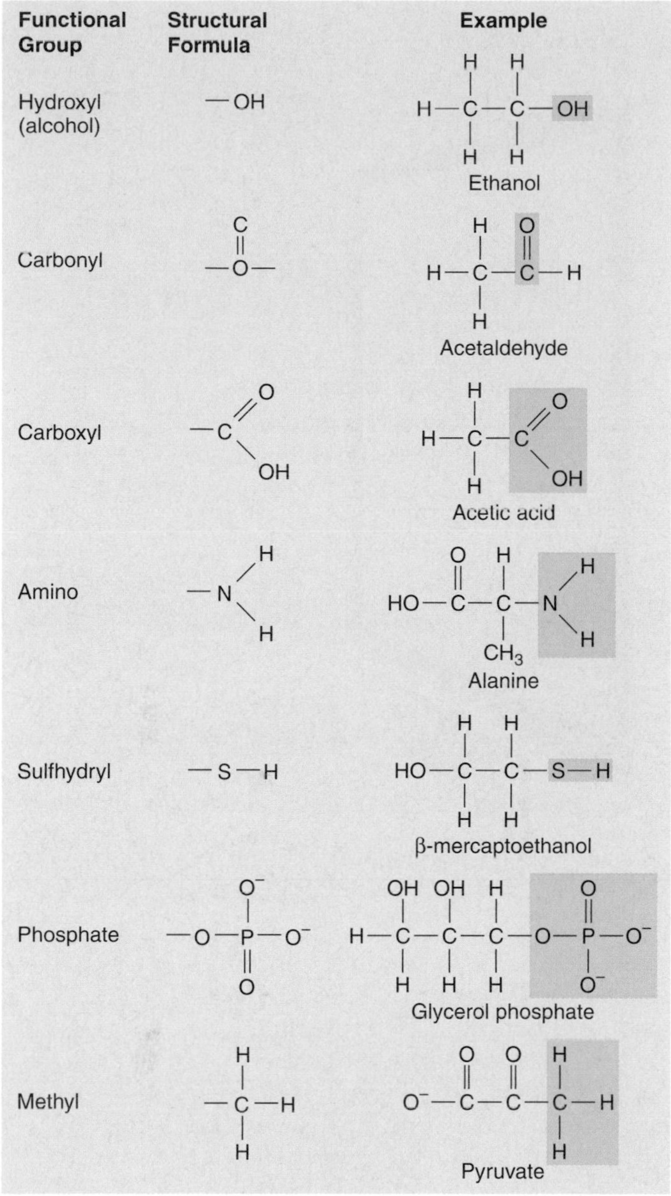

Functional Group	Structural Formula	Example
Hydroxyl (alcohol)	—OH	Ethanol
Carbonyl		Acetaldehyde
Carboxyl		Acetic acid
Amino		Alanine
Sulfhydryl	—S—H	β-mercaptoethanol
Phosphate		Glycerol phosphate
Methyl		Pyruvate

FIGURE 3.7 Functional Groups
These are some of the groups of atoms that frequently attach to a carbon skeleton. The nature of the organic compound changes as the nature of the functional group changes from one molecule to another.

Although lipids are macromolecules, they are not polymers. A polymer is similar to a pearl necklace or a boat's anchor chain. All polymers are constructed of similar segments (pearls or links) hooked together to form one large product (necklace or anchor chain).

The monomers in a polymer are usually combined by a dehydration synthesis reaction (*de* = remove; *hydro* = water; *synthesis* = combine). This reaction occurs when two smaller molecules come close enough to have an —OH removed from one and an —H removed from the other. These are combined to form a molecule of water (H_2O), and the remaining two seg

ments are combined to form the macromolecule. Figure 3.8a shows the removal of water from between two monomers. Notice that, in this case, the structural formulas are used to help identify where this is occurring. The chemical equation also indicates the removal of water. You can easily recognize a dehydration synthesis reaction, because the reactant side of the equation shows numerous, small molecules, whereas the product side lists fewer, larger products and water.

The reverse of a dehydration synthesis reaction is known as hydrolysis (*hydro* = water; *lyse* = to split or break). Hydrolysis is the process of splitting a larger organic molecule into two or more component parts by adding water (figure 3.8b). The digestion of food molecules in the stomach is an example of hydrolysis.

3.2 Carbohydrates

Carbohydrates are composed of carbon, hydrogen, and oxygen atoms linked together to form monomers called *simple sugars* or *monosaccharides* (*mono* = single; *saccharine* = sweet, sugar). Carbohydrates play a number of roles in living things. They are an immediate source of energy (sugars), provide shape to certain cells (cellulose in plant cell walls), and are the components of many antibiotics and coenzymes. They are also an essential part of genes, the nucleic acids.

Simple Sugars

The empirical formula for a simple sugar is easy to recognize, because there are equal numbers of carbons and oxygens and twice as many hydrogens—for example, $C_3H_6O_3$ or $C_5H_{10}O_5$. The ending *-ose* indicates that you are dealing with a carbohydrate. Simple sugars are usually described by the number of carbons in the molecule. A tri*ose* has 3 carbons, a pent*ose* has 5, and a hex*ose* has 6. If you remember that the number of carbons equals the number of oxygen atoms and that the number of hydrogens is double that number, these names tell you the empirical formula for the simple sugar.

Simple sugars, such as glucose, fructose, and galactose, provide the chemical energy necessary to keep organisms alive. *Glucose*, $C_6H_{12}O_6$, is the most abundant carbohydrate; it serves as a food and a basic building block for other carbohydrates. Glucose (also called *dextrose*) is found in the sap of plants; in the human bloodstream, it is called *blood sugar*. Corn syrup, which is often used as a sweetener, is mostly glucose. Fructose, as its name implies, is the sugar that occurs in fruits, and it is sometimes called *fruit sugar*. Glucose and fructose have the same emperical formula but have different structural formulas—that is, they are isomers (figure 3.9). A mixture of glucose and fructose is found in honey. This mixture is also formed when table sugar (sucrose) is reacted with water in the presence of an acid, a reaction that takes place in the preparation of canned fruit and candies. The mixture of glucose and fructose is called *invert sugar*. Thanks to fructose, invert sugar is about twice as sweet to the taste as the same

Generic Drugs and Mirror Image Isomers

Isomers that are mirror images of each other are called *mirror image isomers*, *stereo isomers*, *enantomers*, or *chiral compounds*. The difference among stereo isomers is demonstrated by shining polarized light through the two types of sugar. The light coming out the other side of a test tube containing D-glucose will be turned to the right—that is, *dextro*rotated. When polarized light is shown through a solution of L-glucose, the light coming out the other side will be rotated to the left—that is, *levo*rotated.

The results of this basic research have been used in the pharmaceutical and health care industries. When drugs are synthesized in large batches in the lab, many contain 50% "D" and 50% "L" enantomers. Various so-called generic drugs are less expensive because they are a mixture of the two enantomers and have not undergone the more thorough and more expensive chemical processes involved in isolating only the "D" or "L" form of the drug. Generic drugs that are mixtures may be (1) just as effective as the pure form, (2) slower acting, (3) or not effective if prescribed to an individual who is a "nonresponder" to that medication. It is for these last two reasons that physicians must be consulted when there is a choice between a generic and a nongeneric medication.

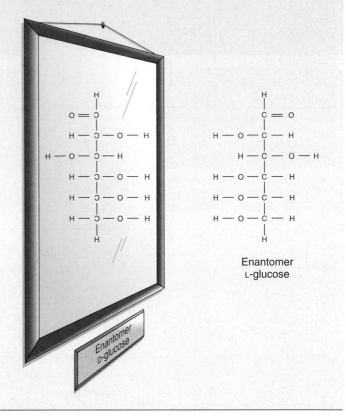

Enantomer
L-glucose

Enantomer
D-glucose

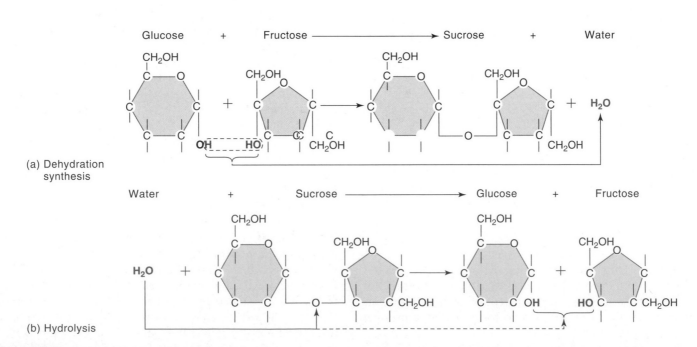

(a) Dehydration synthesis

(b) Hydrolysis

FIGURE 3.8 Polymer Formation and Breakdown

(a) In the dehydration synthesis reaction illustrated here, the two —OH groups line up next to each other, so that an —OH group can be broken from one of the molecules and an —H can be removed from the other. The H —and the —OH are then combined to form water, and the oxygen that remains acts as an attachment site between the two sugar molecules. *(b)* A hydrolysis reaction is the opposite of a dehydration synthesis reaction. Carefully compare the two. Many structural formulas appear to be complex at first glance, but, if you look for the points where subunits are attached and dissect each subunit, they become much simpler to deal with.

OUTLOOKS 3.1

Chemical Shorthand

You have probably noticed that sketching the entire structural formula of a large organic molecule takes a great deal of time. If you know the structure of the major functional groups, you can use several shortcuts to more quickly describe chemical structures. When multiple carbons with 2 hydrogens are bonded to each other in a chain, it is sometimes written as follows:

$$-C-C-C-C-C-C-C-C-C-C-C-C-$$

It can also be written:

$$-CH_2-CH_2-CH_2-CH_2-CH_2-CH_2-CH_2-CH_2-CH_2-CH_2-CH_2-CH_2-$$

More simply, it can be written $(-CH_2-)_{12}$. If the 12 carbons were in a pair of two rings, we probably would not label the carbons or hydrogens unless we wished to focus on a particular group or point. We would probably draw the two 6-carbon rings with only hydrogen attached as follows:

$$(-CH_2-)_{12}$$

Don't let these shortcuts throw you. You will soon find that you will be putting a —H group onto a carbon skeleton and neglecting to show the bond between the oxygen and hydrogen. Structural formulas are regularly included in the package insert information of most medications.

or

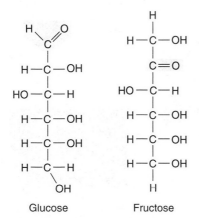

Glucose   Fructose

FIGURE 3.9 Molecular Formulas of Glucose and Fructose
Glucose and fructose have the same empirical formula ($C_6H_{12}O_6$) but different structural formulas. Because they have different structural formulas, they are handled differently by the body when they are metabolized.

amount of sucrose (table 3.1). Invert sugar also attracts water (is hygroscopic). Brown sugar feels moister than white, granulated sugar because it contains more invert sugar. Therefore, baked goods made with brown sugar are moist and chewy.

Cells can use simple sugars as building blocks of other more complex molecules. Sugar molecules are a part of other, larger molecules such as the genetic material, DNA, and the important energy transfer molecule, ATP. ATP has a simple sugar (ribose) as part of its structural makeup.

Complex Carbohydrates

Simple sugars can be combined with each other to form **complex carbohydrates** (figure 3.10). When two simple sugars bond to each other, a *disaccharide* (*di* = two) is formed; when three bond together, a *trisaccharide* (*tri* =

TABLE 3.1

Relative Sweetness of Various Sugars and Sugar Substitutes

Type of Sugar or Artificial Sweetener	Relative Sweetness
Lactose (milk sugar)	0.16
Maltose (malt sugar)	0.33
Glucose	0.75
Sucrose (table sugar)	**1.00**
Fructose (fruit sugar)	1.75
Cyclamate	30.00
Aspartame	150.00
Saccharin	350.00
Sucralose	600.00

The most common disaccharide is *sucrose,* ordinary table sugar. Sucrose occurs in high concentrations in sugarcane and sugar beets. It is extracted by crushing the plant materials, then dissolving the sucrose from the materials with water. The water is evaporated and the crystallized sugar is decolorized with charcoal to produce white sugar. Other common disaccharides are *lactose* (milk sugar) and *maltose* (malt sugar). All three of these disaccharides have similar properties, but maltose tastes only about one-third as sweet as sucrose. Lactose tastes only about one-sixth as sweet as sucrose. No matter which disaccharide is consumed (sucrose, lactose, or maltose), it is converted into glucose and transported by the bloodstream for use by the body.

All the complex carbohydrates are polysaccharides and formed by dehydration synthesis reactions. Some common

three) is formed. Generally, a complex carbohydrate that is larger than this is called a *polysaccharide* (*poly* = many). For example, when glucose and fructose are joined together, they form a disaccharide, with the loss of a water molecule (review figure 3.8).

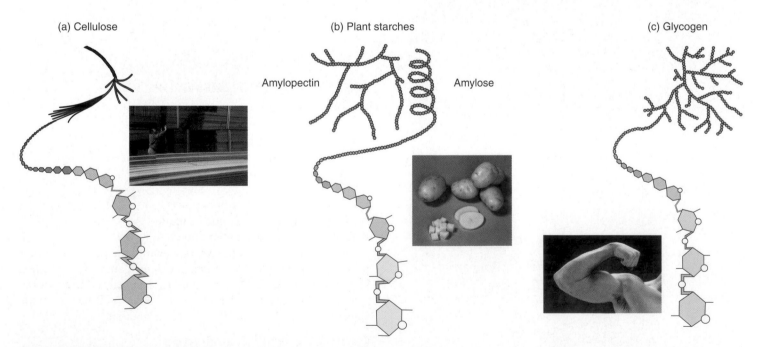

(a) Cellulose   (b) Plant starches   (c) Glycogen

Amylopectin   Amylose

FIGURE 3.10   Complex Carbohydrates
Three common complex carbohydrates are *(a)* cellulose (wood fibers), *(b)* plant starches (amylose and amylopectin), and *(c)* glycogen (sometimes called animal starch). Glycogen is found in muscle cells. Notice how all are similar in that they are all polymers of simple sugars, but they differ in how they are joined together. Although many organisms are capable of digesting (hydrolyzing) the bonds that are found in glycogen and plant starch molecules, few are able to break those that link the monosaccharides of cellulose.

examples of polysaccharides are starch and glycogen. Cellulose is an important polysaccharide used in constructing the cell walls of plant cells. Humans cannot digest (hydrolyze) this complex carbohydrate, so we are not able to use it as an energy source. On the other hand, animals known as ruminants (e.g., cows and sheep) and termites have microorganisms within their digestive tracts that digest cellulose, making it an energy source for them. Plant cell walls add bulk or fiber to our diet, but no calories. Fiber is an important addition to the diet, because it helps control weight and reduces the risk of colon cancer. It also controls constipation and diarrhea, because these large, water-holding molecules make these conditions less of a problem.

3.3 Proteins

Proteins are polymers made up of monomers known as *amino acids*. An **amino acid** is a short carbon skeleton that contains an amino functional group (nitrogen and two hydrogens) attached on one end of the skeleton and a carboxylic acid group at the other end. In addition, the carbon skeleton may have one of several different "side chains" on it (figure 3.11). There are about 20 naturally occurring amino acids (Outlooks 3.2).

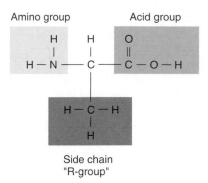

FIGURE 3.11 **The Structure of an Amino Acid**
An amino acid is composed of a short carbon skeleton with three functional groups attached: an amino group, a carboxylic acid group (acid group), and an additional variable group (a side chain, or "R-group"). It is the side chain that determines which specific amino acid is constructed.

OUTLOOKS 3.2

Interesting Amino Acids

Humans require nine amino acids in their diet: threonine, tryptophan, methionine, lysine, phenylalanine, isoleucine, valine, histidine, and leucine. They are called *essential amino acids* because the body is not able to manufacture them. The body uses these essential amino acids in the synthesis of the proteins required for good health. For example, the sulfur-containing amino acid methionine is essential for the absorption and transportation of the elements selenium and potassium. It also prevents excess fat buildup in the liver, and it traps heavy metals, such as lead, cadmium, and mercury, bonding with them so that they can be excreted from the body. Because essential amino acids are not readily available in most plant proteins, they are most easily acquired through meat, fish, and dairy products.

- Lysine is found in foods such as yogurt, fish, chicken, brewer's yeast, cheese, wheat germ, pork, and other meats. It improves calcium uptake; in concentrations higher than the amino acid arginine, it helps control cold sores (herpes virus infection).
- Tryptophan is found in turkey, dairy products, eggs, fish, and nuts. It is required for the manufacture of hormones, such as serotonin, prolactin, and growth hormone; it has been shown to be of value in controlling depression, premenstrual syndrome (PMS), insomnia, migraine headaches, and immune function disorders.

Other amino acids also play interesting roles in human metabolism:

- Glutamic acid is found in animal and vegetable proteins. It is used in monosodium glutamate (MSG), a flavor-enhancing salt. It is required for the synthesis of folic acid. In some people, folic acid can accumulate in brain cells and cause brain damage following stroke. Folic acid is necessary during pregnancy to decrease the fetus's chance of developing spina bifida.
- Asparagine is found in asparagus. Most people have genes that cause asparagine to be converted to very smelly compounds (methyl thioacrylate and methyl 3-(methylthio) thiopropionate), which are excreted in their urine.

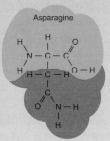

FIGURE 3.12 A Peptide Bond
Part of this protein polypeptide is made up of the amino acids glycine, alanine, and leucine. The bond that results from a dehydration synthesis reaction between two amino acids is called a peptide bond. This bond forms as a result of the removal of the hydrogen and hydroxide groups. In the formation of this bond, the nitrogen is bonded directly to the carbon.

The Structure of Proteins

Amino acids can bond together by dehydration synthesis reactions. When two amino acids undergo dehydration synthesis, the nitrogen of the amino group of one is bonded to the carbon of the acid group of another. This covalent bond is termed a *peptide bond* (figure 3.12).

You can imagine that, by using 20 different amino acids as building blocks, you can construct millions of combinations. Each of these combinations is termed a **polypeptide** chain. A specific polypeptide is composed of a specific sequence of amino acids bonded end to end. Protein molecules are composed of individual polypeptide chains or groups of chains forming a particular configuration. There are four levels, or degrees, of protein structure: primary, secondary, tertiary, and quaternary structure.

Primary Structure

A listing of the amino acids in their proper order within a particular polypeptide is its *primary structure*. The specific sequence of amino acids in a polypeptide is controlled by the genetic information of an organism. *Genes are specific messages that tell the cell to link particular amino acids in a specific order; that is, they determine a polypeptide's primary structure. The kinds of side chains on these amino acids influence the shape that the polypeptide forms, as well as its function.* Many polypeptides fold into globular shapes after they have been made as the molecule bends. Some of the amino acids in the chain can form bonds with their neighbors.

Secondary Structure

Some sequences of amino acids in a polypeptide are likely to twist, whereas other sequences remain straight. These twisted forms are referred to as the *secondary structure* of polypeptides. For example, at this secondary level some proteins (e.g., hair) take the form of an *alpha helix:* a shape like that of a twisted ladder. Like most forms of secondary structure, the shape of the alpha helix is maintained by hydrogen bonds

formed between different amino acid side chains at different locations in the polypeptide. Remember from chapter 2 that these forces of attraction do not form molecules but result in the orientation of one part of a molecule to another part within the same molecule. Other polypeptides form hydrogen bonds that cause them to make several flat folds that resemble a pleated skirt. This is called a *beta pleated sheet.*

Tertiary Structure

It is possible for a single polypeptide to contain one or more coils and pleated sheets along its length. As a result, these different portions of the molecule can interact to form an even more complex globular structure. This occurs when the coils and pleated sheets twist and combine with each other. The complex, three-dimensional structure formed in this manner is the polypeptide's *tertiary* (third-degree) *structure.* A good example of tertiary structure can be seen when a coiled phone cord becomes so twisted that it folds around and back on itself in several places. The oxygen-holding protein found in muscle cells, myoglobin, displays tertiary structure. It is composed of a single (153 amino acids) helical molecule folded back and bonded to itself in several places.

Quaternary Structure

Frequently, several different polypeptides, each with its own tertiary structure, twist around each other and chemically combine. The larger, globular structure formed by these interacting polypeptides is referred to as the protein's *quaternary* (fourth-degree) *structure.* The individual polypeptide chains are bonded to each other by the interactions of certain side chains, which can form disulfide covalent bonds (figure 3.13). Quaternary structure is displayed by the protein molecules called *immunoglobulins,* or *antibodies,* which are involved in fighting infectious diseases, such as the flu, the mumps, and chicken pox. immunity, p. 586

The Form and Function of Proteins

If a protein is to do its job effectively, it must have a particular three-dimensional shape. The protein's shape can be altered by changing the order of the amino acids, which causes different cross-linkages to form. Figure 3.14 shows the importance of the protein's three-dimensional shape.

For example, normal hemoglobin found in red blood cells consists of two kinds of polypeptide chains, called the alpha and beta chains. The beta chain is 146 amino acids long. If just one of these amino acids is replaced by a different one, the hemoglobin molecule may not function properly. A classic example of this results in a condition known as *sickle-cell anemia.* In this case, the sixth amino acid in the beta chain, which is normally glutamic acid, is replaced by valine. What might seem like a minor change causes the hemoglobin to fold differently. The red blood cells that contain this altered hemoglobin assume a sickle shape when the body is deprived of an adequate supply of oxygen. sickle-cell gene frequency, p. 158

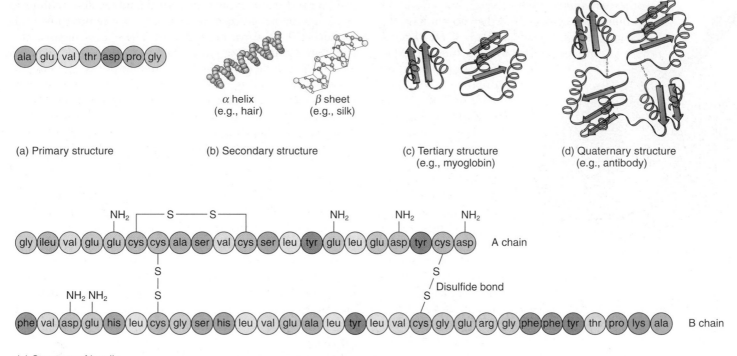

(a) Primary structure

(b) Secondary structure

α helix
(e.g., hair)

β sheet
(e.g., silk)

(c) Tertiary structure
(e.g., myoglobin)

(d) Quaternary structure
(e.g., antibody)

(e) Structure of insulin

FIGURE 3.13 Levels of Protein Structure

(a) The primary structure of a molecule is simply a list of its component amino acids in the order in which they occur. *(b)* This figure illustrates the secondary structure of protein molecules or how one part of the molecule is initially attached to another part of the same molecule. *(c)* If already folded parts of a single molecule attach at other places, the molecule is said to display tertiary (third-degree) structure. *(d)* Quaternary (fourth-degree) structure is displayed by molecules that are the result of two separate molecules (each with its own tertiary structure) combining into one large macromolecule. *(e)* The protein insulin is composed of two polypeptide chains bonded together at specific points by reactions between the side chains of particular sulfur-containing amino acids. The side chains of one interact with the side chains of the other and form a unique three-dimensional shape. The bonds that form between the polypeptide chains are called disulfide bonds.

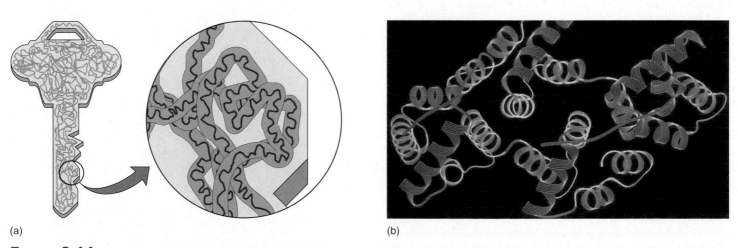

(a)

(b)

FIGURE 3.14 The Three-Dimensional Shape of Proteins

(a) The specific arrangement of amino acids in a polypeptide allows the amino acid side chains to bond with other amino acids. These stabilizing interactions result in a protein with a specific surface geometry. This three-dimensional shape can be compared to the three-dimensional shape of a specific key. *(b)* This is the structure of the protein annexin as determined by X-ray diffraction. This molecule is located just inside cell membranes and is involved in the transport of materials through the membrane. The surface of this molecule has the shape required to attach to the molecule being transported.

In other situations, two proteins may have the same amino acid sequence but they do not have the same three-dimensional form. The difference in shape affects how they function. Mad cow disease (bovine spongiform encephalopathy—BSE) and Creutzfeldt-Jakob disease (CJD) are caused by rogue proteins called *prions*. The prions that cause these diseases have an amino acid sequence identical to a normal brain protein but are folded differently. The normal brain protein contains helical segments, whereas the corresponding segments of the prion protein are pleated sheets. When these malformed proteins enter the body, they cause normal proteins to fold differently. This causes the death of brain cells which causes loss of brain function and eventually death.

Changing environmental conditions also influence the shape of proteins. Energy in the form of heat or light may break the hydrogen bonds within protein molecules. When this occurs, the chemical and physical properties of the protein are changed and the protein is said to be **denatured.** (Keep in mind that a protein is a molecule, not a living thing, and therefore cannot be "killed.") A common example of this occurs when the gelatinous, clear portion of an egg is cooked and the protein changes to a white solid. Some medications, such as insulin, are proteins and must be protected from denaturation so as not to lose their effectiveness. For protection, such medications may be stored in brown bottles to protect them from light or may be kept under refrigeration to protect them from heat.

What Do Proteins Do?

There are thousands of kinds of proteins in living things, and they can be placed into three categories based on the functions they serve. **Structural proteins** are important for maintaining the shape of cells and organisms. The proteins that make up cell membranes, muscle cells, tendons, and blood cells are examples of structural proteins. The protein collagen, found throughout the human body, gives tissues shape, support, and strength. **Regulator proteins,** the second category of proteins, help determine what activities will occur in the organism. Regulator proteins include enzymes and some hormones. These molecules help control the chemical activities of cells and organisms. Enzymes are important, and they are dealt with in detail in chapter 5. Some examples of enzymes are the digestive enzymes in the intestinal tract. Three hormones that are regulator proteins are insulin, glucagon, and oxytocin. Insulin and glucagon, produced by different cells of the pancreas, control the amount of glucose in the blood. If insulin production is too low, or if the molecules are improperly constructed, glucose molecules are not removed from the bloodstream at a fast enough rate. The excess sugar is then eliminated in the urine. Other symptoms of excess "sugar" in the blood include excessive thirst and even loss of consciousness. When blood sugar is low, glucagon is released from the pancreas to stimulate the breakdown of glycogen. The disease caused by improperly functioning insulin is known as *diabetes.* Oxytocin, a third protein hormone, stimulates the contraction of the uterus during childbirth. It is an organic molecule that has been produced artificially (e.g., Pitocin), and used by physicians to induce labor.

Carrier proteins are the third category. These pick up and deliver molecules at one place and transport them to another. For example, proteins regularly attach to cholesterol entering the system from the diet, forming molecules called *lipoproteins,* which are transported through the circulatory system. The cholesterol is released at a distance from the digestive tract, and the proteins return to pick up more entering dietary cholesterol.

3.4 | Nucleic Acids

Nucleic acids are complex organic polymers that store and transfer genetic information within a cell. There are two types of nucleic acids: deoxyrybonucleic acid (DNA) and ribonucleic acid (RNA). DNA serves as genetic material, whereas RNA plays a vital role in using genetic information to manufacture proteins. All nucleic acids are constructed of monomers known as **nucleotides.** Nucleotides also provide an immediate source of energy for cellular reactions. Each nucleotide is composed of three parts: (1) a 5-carbon simple sugar molecule, which may be ribose or deoxyribose, (2) a phosphate group, and (3) a nitrogenous base. The nitrogenous base may be one of five types. Two of the types are the larger, double-ring molecules Adenine and Guanine. The smaller bases are the single-ring bases Thymine, Cytosine, and Uracil (i.e., A, G, T, C, and U) (figure 3.15). Nucleotides (monomers) are linked together in long sequences (polymers), so that the sugar and phosphate sequence forms a "backbone" and the nitrogenous bases stick out to the side. DNA has deoxyribose sugar and the bases A, T, G, and C, whereas RNA has ribose sugar and the bases A, U, G, and C (figure 3.16).

DNA

Deoxyribonucleic acid (DNA) is composed of two strands, which form a ladderlike structure thousands of nucleotides long. The two strands are attached by hydrogen bonds between their bases according to the base pair rule. *The base paring rule states that Adenine on one strand always pairs with thymine on the other strand, A with T (in the case of RNA, adenine always pairs with uracil—A with U) and Guanine always pairs with cytosine—G with C.*

A T (or A U) and G C

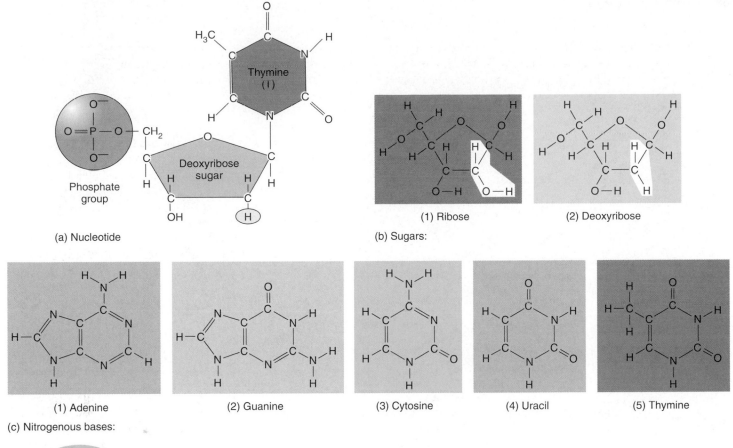

(a) Nucleotide

(b) Sugars: (1) Ribose (2) Deoxyribose

(c) Nitrogenous bases: (1) Adenine (2) Guanine (3) Cytosine (4) Uracil (5) Thymine

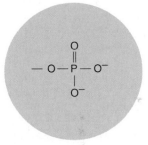

(d) A phosphate group

FIGURE 3.15 The Building Blocks of Nucleic Acids

(a) A complete DNA nucleotide composed of a sugar, phosphate, and a nitrogenous base. *(b)* The two possible sugars used in nucleic acids, ribose and deoxyribose. *(c)* The five possible nitrogenous bases: adenine (A), guanine (G), cytosine (C), uracil (U), and thymine (T). *(d)* A phosphate group.

A meaningful genetic message, a gene, is written using the nitrogenous bases as letters along a section of a strand of DNA, such as the base sequence CATTAGACT (figure 3.17). The strand that contains this message is called the *coding strand,* from which comes the term *genetic code.* To make a protein, the cell reads the coding strand and uses sets of 3 bases. In the example sequence, sets of three bases are CAT, TAG, and ACT. This system is the basis of the genetic code for all organisms. Directly opposite the coding strand is a sequence of nitrogenous bases that are called *non-coding,* because the sequence of letters make no "sense," but this strand protects the coding strand from chemical and physical damage. Both strands are twisted into a helix—that is, a molecule turned around a tubular space, like a twisted ladder.

The information carried by DNA can be compared to the information in a textbook. Books are composed of words (constructed from individual letters) in particular combina-

tions, organized into chapters. In the same way, DNA is composed of tens or thousands of nucleotides (letters) in specific three letter sequences (words) organized into genes (chapters). Each chapter gene carries the information for producing a protein, just as the chapter of a book carries information relating to one idea. The order of nucleotides in a gene is directly related to the order of amino acids in the protein for which it codes. Just as chapters in a book are identified by beginning and ending statements, different genes along a DNA strand have beginning and ending signals. They tell when to start and when to stop reading the gene. Human body cells contain 46 strands (books) of helical DNA, each containing many genes (chapters). These strands are called **chromosomes** when they become supercoiled in preparation for cellular reproduction. Before cellwar reproduction, the DNA makes copies of the coding and non-coding strands, ensuring that the offspring, or *daughter cells,* will receive a

(a) DNA single strand

(b) RNA

A single nucleotide

Backbone

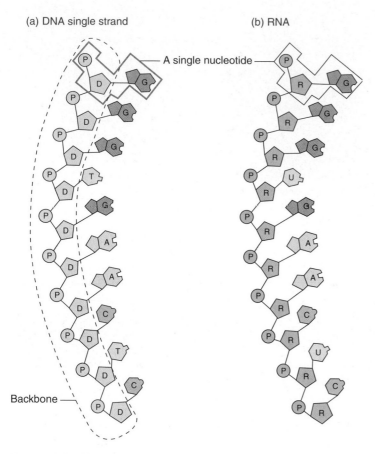

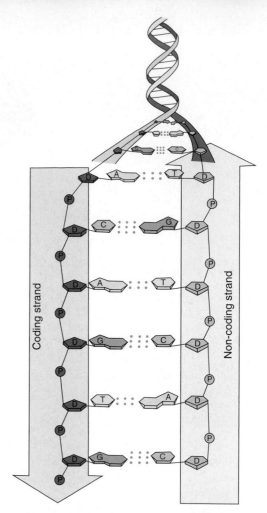

FIGURE 3.16 DNA and RNA

(a) A single strand of DNA is a polymer composed of nucleotides. Each nucleotide consists of deoxyribose sugar, phosphate, and one of four nitrogenous bases: A, T, G, or C. Notice the backbone of sugar and phosphate. *(b)* RNA is also a polymer, but each nucleotide is composed of ribose sugar, phosphate, and one of four nitrogenous bases: A, U, G, or C.

FIGURE 3.17 DNA

The genetic material is really double-stranded DNA molecules comprised of sequences of nucleotides that spell out an organism's genetic code. The coding strand of the double molecule is the side that can be translated by the cell into meaningful information. The genetic code has the information for telling the cell what proteins to make, which in turn become the major structural and functional components of the cell. The non-coding strand is unable to code for such proteins.

full complement of the genes required for their survival (figure 3.18). A gene is a segment of DNA that is able to (1) replicate by directing the manufacture of copies of itself; (2) mutate, or chemically change, and transmit these changes to future generations; (3) store information that determines the characteristics of cells and organisms; and (4) use this information to direct the synthesis of structural, carrier, and regulator proteins.

RNA

Ribonucleic acid (RNA) is found in three forms. **Messenger RNA (mRNA)** is a single-strand copy of a portion of the coding strand of DNA for a specific gene. When mRNA is formed on the surface of the DNA, the base pair rule applies. However, because RNA does not contain thymine, it pairs U with A instead of T with A. After mRNA is formed and

peeled off, it links with a cellular structure called the *ribosome,* where the genetic message can be translated into a protein molecule. Ribosomes contain another type of RNA, **ribosomal RNA (rRNA)**. rRNA is also an RNA copy of DNA, but after being formed it becomes twisted and covered in protein to form a ribosome. The third form of RNA, **transfer RNA (tRNA)**, is also a copy of different segments of DNA, but when peeled off the surface each segment takes the form of a cloverleaf. tRNA molecules are responsible for transferring or carrying specific amino acids to the ribosome, where all three forms of RNA come together and cooperate in the manufacture of protein molecules (figure 3.19).

Whereas the specific sequence of nitrogenous bases correlates with the coding of genetic information, the energy transfer function of nucleic acids is correlated with the number of phosphates each contains. A nucleotide with 3 phosphates has more energy than a nucleotide with only 1 or 2 phosphates.

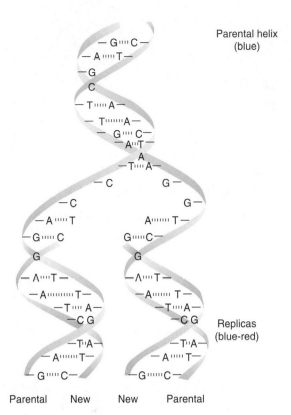

Figure 3.18 **Passing on Information to the Next Generation**

This is a generalized illustration of DNA replication. Each daughter cell receives a copy of the double helix. The helices are identical to each other and identical to the original double strands of the parent cell.

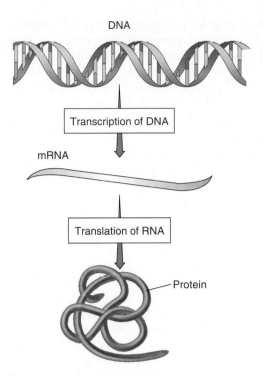

Figure 3.19 **The Role of RNA**

The entire process of protein synthesis begins with DNA. All forms of RNA (messenger, transfer, and ribosomal) are copies of different sequences of coding strand DNA and each plays a different role in protein synthesis. When the protein synthesis process is complete, the RNA can be reused to make more of the same protein coded for by the mRNA. This is similar to replaying a videotape in a VHS machine. The tape is like the mRNA, and the tape machine is like the ribosome. Eventually, the tape and machine will wear out and must be replaced. In a cell, this involves the synthesis of new RNA molecules from food.

All of the different nucleotides are involved in transferring energy in phosphorylation reactions. One of the most important, ATP (*a*denosine *tri*phosphate) and its role in metabolism will be discussed in chapter 6.

3.5 Lipids

There are three types of **lipids**: *true fats* (e.g., olive oil), *phospholipids* (the primary component of cell membranes), and *steroids* (some hormones). In general, lipids are large, nonpolar (do not have a positive end and a negative ends), organic molecules that do not dissolve easily in polar solvents, such as water. For example, nonpolar vegetable oil molecules do not dissolve in polar water molecules; they separate. Molecules in this group are generally called **fats**. They are not polymers, as are carbohydrates, proteins, and nucleic acids. Fats are soluble in nonpolar substances, such as ether or acetone. Just like carbohydrates, lipids are composed of carbon, hydrogen, and oxygen. They do not, however, have the same ratio of carbon, hydrogen, and oxygen in their empirical formulas. Lipids gen-

erally have very small amounts of oxygen, compared with the amounts of carbon and hydrogen. *Simple lipids* are not able to be broken down into smaller, similar subunits. *Complex lipids* can be hydrolyzed into smaller, similar units.

True (Neutral) Fats

True (neutral) fats are important, complex organic molecules that are used to provide energy, among other things. The building blocks of a fat are a *glycerol* molecule and *fatty acids*. **Glycerol** is a carbon skeleton that has three alcohol groups attached to it. Its chemical formula is $C_3H_5(OH)_3$. At room temperature, glycerol looks like clear, lightweight oil. It is used under the name *glycerin* as an additive to many cosmetics to make them smooth and easy to spread.

$$H - \underset{\underset{H}{|}}{\overset{\overset{OH}{|}}{C}} - \underset{\underset{H}{|}}{\overset{\overset{OH}{|}}{C}} - \underset{\underset{H}{|}}{\overset{\overset{OH}{|}}{C}} - H$$

Glycerol

A **fatty acid** is a long-chain carbon skeleton that has a carboxyl functional group. If the carbon skeleton has as much hydrogen bonded to it as possible, it is called **saturated.** The saturated fatty acid shown in figure 3.20a is stearic acid, a component of solid meat fats, such as mutton tallow. Notice that, at every point in this structure, the carbon has as much hydrogen as it can hold. Saturated fats are generally found in animal tissues—they tend to be solids at room temperatures. Some other examples of saturated fats are butter, whale blubber, suet, lard, and fats associated with such meats as steak and pork chops.

A fatty acid is said to be **unsaturated** if the carbons are double-bonded to each other at one or more points. The occurrence of a double bond in a fatty acid is indicated by the Greek letter ω (omega), followed by a number indicating the location of the first double bond in the molecule. Counting begins from the omega end, that is the end farthest from the carboxylic acid functional group. Oleic acid, one of the fatty acids found in olive oil, is comprised of 18 carbons with a single double bond between carbons 9 and 10. Therefore, it is chemically designated C18:1ω9 and is a monounsaturated fatty acid. This fatty acid is commonly referred to as an omega-9 fatty acid. The unsaturated fatty acid in figure 3.20b is linoleic acid, a component of sunflower and safflower oils. Notice that there are two double bonds between the carbons and fewer hydrogens than in the saturated fatty acid. Linoleic acid is chemically a polyunsaturated fatty acid with two double bonds and is designated C18:2ω6, an omega-6 fatty acid. This indicates that the first double bond of this 18-carbon molecule is between carbons 6 and 7. Because the human body cannot make this fatty acid and must be taken in as a part of the diet, it is called an *essential fatty acid*. The other essential fatty acid, linolenic acid (figure 3.20c), is C18:3ω3; it has three double bonds. This fatty acids is commonly referred to as an omega-3 fatty acid. One key function of these essential fatty acids is the synthesis of the prostaglandin hormones that are necessary in controlling cell growth and specialization.

Sources of Omega-3 Fatty Acids	Sources of Omega-6 Fatty Acids
Certain fish oils (salmon, sardines, herring)	Corn oil
Flaxseed oil	Peanut oil
Soybeans	Cottonseed oil
Soybean oil	Soybean oil
Walnuts	Sesame oil
Canola oil	Sunflower oil
Green, leafy vegetables	Safflower oil

Many unsaturated fats are plant fats or oils—they are usually liquids at room temperatures. Peanut, corn, and olive oils are mixtures of true fats and are considered unsaturated because they have double bonds between the carbons of the carbon skeleton. A polyunsaturated fatty acid is one that has several double bonds in the carbon skeleton. When glycerol and 3 fatty acids are combined by three dehydration synthe-

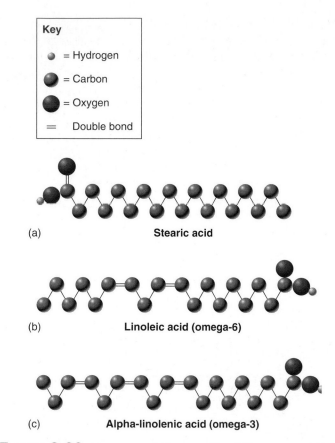

Key

- = Hydrogen
- = Carbon
- = Oxygen
= Double bond

(a) **Stearic acid**

(b) **Linoleic acid (omega-6)**

(c) **Alpha-linolenic acid (omega-3)**

FIGURE 3.20 Structure of Saturated and Unsaturated Fatty Acids

(a) Stearic acid is an example of a saturated fatty acid. *(b)* Linoleic acid is an example of an unsaturated fatty acid. It is technically an omega-6 fatty acid, because the first double bond occurs at carbon number 6. *(c)* An omega-3 fatty acid, linolenic acid. Both linoleic and linolenic acids are essential fatty acids for humans.

sis reactions, a fat is formed. That dehydration synthesis is almost exactly the same as the reaction that causes simple sugars to bond.

In nature, most unsaturated fatty acids have hydrogen atoms that are on the same side of the double-bonded carbons. These are called *cis* fatty acids. If the hydrogens are on opposite sides of the double bonds, they are called *trans* fatty acids.

cis double bond: oleic acid

trans double bond: elaidic acid

Trans fatty acids are found in grazing animals, such as cattle, sheep, and horses. Therefore, humans acquire them in their diets in the form of meat and dairy products. French fries, donuts, cookies, and crackers are foods high in *trans* fatty acids. *Trans* fatty acids are also formed during the *hydrogenation* of either vegetable or fish oils. The hydrogenation process breaks the double bonds in the fatty acid chain and adds more hydrogen atoms. This can change the liquid to a solid. Many product labels list the term *hydrogenated*. This process extends shelf life and allows producers to convert oils to other solids, such as margarine.

Clinical studies have shown that *trans* fatty acids tend to raise total blood cholesterol levels, but less than the more saturated fatty acids. Dietary *trans* fatty acids also tend to raise the so-called bad fats (low-density lipoproteins, LDLs) and lower the so-called good fats (high-density lipoproteins, HDLs) when consumed instead of *cis* fatty acids. It is suspected that this increases the risk for heart disease. Because of the importance of *trans* fatty acids in cardiovascular health, the U.S. Department of Health and Human Services (HHS) requires that the amount of *trans* fatty acids in foods be stated under the listed amount of saturated fat. The HHS suggests that a person eat no more than 20 grams of saturated fat a day (about 10% of total calories), including *trans* fatty acids.

Fats are important molecules for storing energy. There is more than twice as much energy in a gram of fat as in a gram of sugar—9 Calories versus 4 Calories. This is important to an organism, because fats can be stored in a relatively small space yet yield a high amount of energy. Fats in animals also provide protection from heat loss; some animals have an insulating layer of fat under the skin. The thick layer of blubber in whales, walruses, and seals prevents the loss of internal body heat to the cold, watery environment in which they live. The same layer of fat and the fat deposits around some internal organs (such as the kidneys and heart) cushion the organs from physical damage. If a fat is formed from a glycerol molecule and 3 attached fatty acids, it is called a *triglyceride*; if 2, a *diglyceride*; if 1, a *monoglyceride* (figure 3.21). Triglycerides account for about 95% of the fat stored in human tissue.

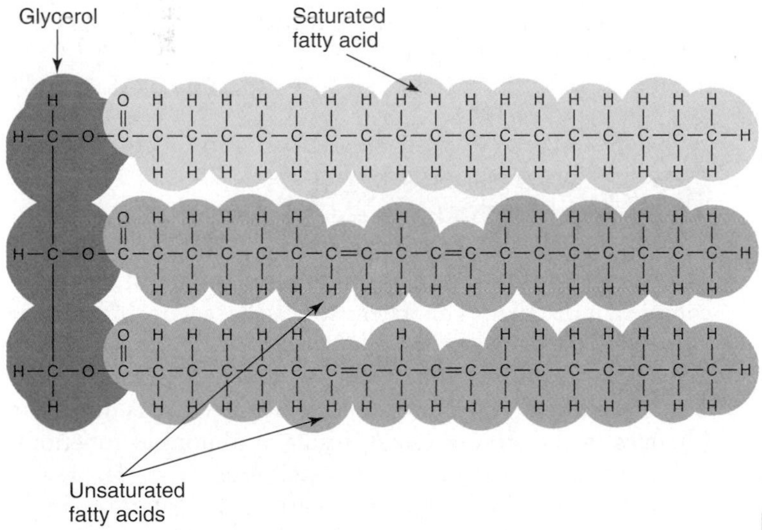

Phospholipids

Phospholipids are a class of complex, water-insoluble organic molecules that resemble neutral fats but contain a charged phosphate group (PO_4) in their structure (figure 3.22).

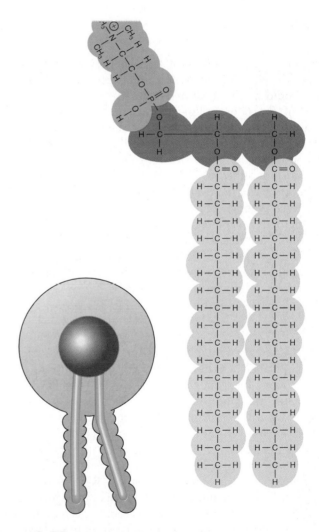

FIGURE 3.22 A Phospholipid Molecule
This molecule is similar to a fat but has a phosphate group (purple) in its structure. The phosphate group and 2 fatty acids are bonded to the glycerol by a dehydration synthesis reaction. Molecules such as these are also known as lecithins. Phospholipid molecules may be shown as a balloon with two strings. The balloon portion is the glycerol and phosphate group, which are soluble in water. The strings are the fatty acid segments of the molecule, which are not water-soluble.

Glycerol

Saturated fatty acid

Unsaturated fatty acids

FIGURE 3.21 A Fat Molecule
The arrangement of the 3 fatty acids (yellow) attached to a glycerol molecule (red) is typical of the formation of a fat. The structural formula of the fat appears to be very cluttered until you dissect the fatty acids from the glycerol; then, it becomes much more manageable. This example of a triglyceride contains a glycerol molecule, 2 unsaturated fatty acids (linoleic acid), and a third saturated fatty acid (stearic acid).

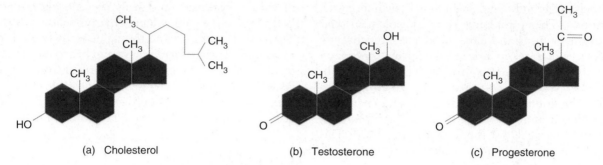

FIGURE 3.23 Steroids

(a) Cholesterol is produced by the human body and is found in cell membranes. *(b)* Testosterone increases during puberty, causing the male sex organs to mature. *(c)* Progesterone is a female sex hormone produced by the ovaries and placenta. Notice the slight structural differences among these molecules.

Phospholipids are important because they are a major component of cell membranes. Without these lipids, the cell contents would not be separated from the exterior environment. Some of the phospholipids are better known as *lecithins*. Found in cell membranes, lecithins help in the emulsification of fats—that is, they help separate large portions of fat into smaller units. This allows the fat to mix with other materials. Lecithins are added to many types of food for this purpose (chocolate bars, for example). Some people take lecithin as nutritional supplements because they believe it leads to healthier hair and better reasoning ability. But once inside the intestines, lecithins are destroyed by enzymes, just as any other phospholipid is (Outlooks 3.3). Because phospholipids are essential components of the membranes of all cells, they will also be examined in chapter 4.

Steroids

Steroids, another group of lipid molecules, are characterized by their arrangement of interlocking rings of carbon. We have already mentioned one steroid molecule: cholesterol. Serum cholesterol (the kind found in the blood and associated with lipoproteins) has been implicated in many cases of atherosclerosis. However, the body makes this steroid for use as a component of cell membranes. The body also uses it to make bile acids. These products of the liver are channeled into the intestine to emulsify fats. Cholesterol is necessary for the manufacture of vitamin D, which assists in the proper development of bones and teeth. Cholesterol molecules in the skin react with ultraviolet light to produce vitamin D. Figure 3.23 illustrates some of the steroid compounds, such as testosterone and progesterone, that are typically manufactured by organisms. conception control, p. 612

Cholesterol plays both positive and negative roles in metabolism. Regulating the amount of cholesterol in the body to prevent its negative effects can be difficult, because the body makes it and it is consumed in the diet. Recall that saturated fats cause the body to produce more cholesterol, increasing the risk for diseases such as atherosclerosis. By watching your diet, you can reduce the amount of cholesterol in your blood serum by about 20%, **as much as taking a drug.** Therefore, it is best to eat foods that are low in cholesterol. Because many foods that claim to be low- or no-cholesterol have high levels of saturated fats, they should also be avoided in order to control serum cholesterol levels.

Many steroid molecules are sex hormones. Some of them regulate reproductive processes, such as egg and sperm production (see chapter 21); others regulate such things as salt concentration in the blood.

Summary

The chemistry of living things involves a variety of large and complex molecules. This chemistry is based on the carbon atom and the fact that carbon atoms can connect to form long chains or rings. This results in a vast array of molecules. The structure of each molecule is related to its function. Changes in the structure may result in abnormal functions, called disease. Some of the most common types of organic molecules found in living things are carbohydrates, lipids, proteins, and nucleic acids. Table 3.2 summarizes the major types of biologically important organic molecules and how they function in living things.

OUTLOOKS 3.3

Fat and Your Diet

When triglycerides are eaten in fat-containing foods, digestive enzymes hydrolyze them into glycerol and fatty acids. These molecules are absorbed by the intestinal tract and coated with protein to form lipoprotein, as shown in the accompanying diagram. The combination allows the fat to dissolve better in the blood, so that it can move throughout the body in the circulatory system.

Five types of lipoproteins found in the body are

1. Chylomicrons
2. Very-low-density lipoproteins (VLDLs)
3. Low-density lipoproteins (LDLs)
4. High-density lipoproteins (HDLs)
5. Lipoprotein a [Lp(a)]

Chylomicrons are very large particles formed in the intestine; they are 80–95% triglycerides. As the chylomicrons circulate through the body, cells remove the triglycerides in order to make sex hormones, store energy, and build new cell parts. When most of the triglycerides have been removed, the remaining portions of the chylomicrons are harmlessly destroyed.

The VLDLs and LDLs are formed in the liver. VLDLs contain all types of lipid, protein, and 10–15% cholesterol, whereas the LDLs are about 50% cholesterol. As with the chylomicrons, the body uses these molecules for the fats they contain. However, in some people, high levels of LDL and lipoprotein a [Lp(a)] in the blood are associated with atherosclerosis, stroke, and heart attack. It appears that saturated fat disrupts the clearance of LDLs from the bloodstream. Thus, while in the blood, LDLs may stick to the insides of the vessels, forming deposits, which restrict blood flow and contribute to high blood pressure, strokes, and heart attacks. Even though they are 30% cholesterol, a high level of HDLs (made in the intestine), compared with LDLs and [Lp(a)], is associated with a lower risk for atherosclerosis. One way to reduce the risk of this disease is to lower your intake of LDLs and [Lp(a)]. This can be done by reducing your consumption of saturated fats. An easy way to remember the association between LDLs and HDLs is "L = Lethal" and "H = Healthy" or "Low = Bad" and "High = Good." The federal government's cholesterol guidelines recommend that all adults get a full lipoprotein profile (total cholesterol, HDL, LDL, and triglycerides) once every five years.

They also recommend a sliding scale for desirable LDL levels; however, recent studies suggest that one's LDL level should be as low as possible.

One way to reduce LDL levels is through a process called *LDL apheresis*. This is similar to kidney dialysis but is designed to remove only particles with LDL, VLDL, and [Lp(a)]. As the blood is removed and passed through an apheresis machine, it is separated into cells and plasma. The plasma is then passed over a column containing a material that grabs onto the LDL particles and removes them, leaving the HDL cholesterol and other important blood components. The plasma is then returned to the patient.

Experimental methods currently being explored that result in raising the relative amounts of protective HDL levels include

1. The use of drugs (e.g., Torcetrapib) that stimulate HDL production
2. The injection of an altered form of HDL that is better at helping protect against heart disease by removing plaque from the bloodstream
3. The insertion of an omega-3 fatty acid–producing gene into cells that lack the ability to produce such fatty acids

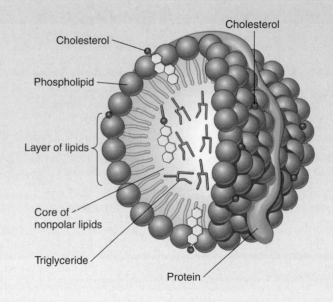

Normal HDL Values
Men: 40–70 mg/dL
Women: 40–85 mg/dL
Children: 30–65 mg/dL
Minimum desirable: 40 mg/dL

For Total Cholesterol Levels
Desirable: Below 200 mg/dL

Normal LDL Values
Men: 91–100 mg/dL
Women: 69–100 mg/dL

Heart attack risk level/acceptable LDL levels:
Low risk: no higher than 70 mg/dL
Moderate risk: between 70 and 160 mg/dL
High risk: at or above 160 mg/dL

Borderline: 200–240 mg/dL

Normal VLDL Values
Men: 0–40 mg/dL
Women: 0–40 mg/dL

Desirable: below 40 mg/dL

Undesirable: Above 240 mg/dL

TABLE 3.2

Types of Organic Molecules Found in Living Things

Type of Organic Molecule	Basic Subunit	Function	Examples
Carbohydrates	Simple sugar/monosaccharides	Provide energy	Glucose, fructose
	Complex carbohydrates/ polysaccharides	Provide support	Cellulose, starch, glycogen
Proteins	Amino acid	Maintain the shape of cells and parts of organisms	Cell membrane Hair Antibodies Clotting factors Enzymes Muscle
		As enzymes, regulate the rate of cell reactions	Ptyalin in the mouth
		As hormones, effect physiological activity, such as growth or metabolism	Insulin
		Serve as molecules that carry other molecules to distant places in the body	Lipoproteins, hemoglobin
Nucleic acids	Nucleotide	Store and transfer genetic information that controls the cell	DNA
		Are involved in protein synthesis	RNA
Lipids 1. Fats	Glycerol and fatty acids	Provide energy Provide insulation Serve as shock absorbers	Lard Olive oil Linseed oil Tallow
2. Phospholipids	Glycerol, fatty acids, and phosphate group	Form a major component of the structure of the cell membrane	Cell membrane
3. Steroids and prostaglandins	Structure of interlocking carbon rings	Often serve as hormones that control the body processes	Testosterone Vitamin D Cholesterol

Key Terms

Use the interactive flash cards on the **Concepts in Biology,** *12/e website to help you learn the meaning of these terms.*

Basic Review

1. A(n) _____ formula indicates the number of each kind of atom within a molecule.

2. The name of this functional, $-NH_2$,

   a. amino.

   b. alcohol.

   c. carboxylic acid.

   d. aldehyde.

3. Molecules that have the same empirical formula but different structural formulas are called

   a. ions.

   b. isomers.

   c. icons.

   d. radicals.

4. Which is *not* a macromolecule?

   a. carbohydrate

   b. protein

   c. sulfuric acid

   d. steroid

5. Which is *not* a polymer?

   a. insulin

   b. DNA

   c. fatty acid

   d. RNA

6. The monomer of a complex carbohydrate is

   a. an amino acid.

   b. a monosaccharide.

   c. a nucleotide.

   d. a fatty acid.

7. When blood sugar is low, this protein hormone is released from the pancreas to stimulate the breakdown of glycogen.

   a. glucagon

   b. estrogen

   c. oxytocin

   d glycine

8. Mad cow disease is caused by a

   a. virus.

   b. bacteria.

   c. prion.

   d. hormone.

9. _____ occurs when the shape of a macromolecule altered as a result of exposure to excess heat or light.

10. By watching your diet it is possible to reduce the amount of cholesterol in your blood serum by about _____ %.

Answers

1. emperical 2. a 3. b 4. c 5. c 6. b 7. a 8. c 9. Denaturation 10. 20%

Concept Review

*Answers to the Concept Review questions can be found on the **Concepts in Biology**, 12/e website.*

3.1 Molecules Containing Carbon

1. What is the difference between inorganic and organic molecules?

2. What two characteristics of the carbon molecule make it unique?

3. Diagram an example of each of the following: amino acid, simple sugar, glycerol, fatty acid.

4. Describe five functional groups.

5. List three monomers and the polymers that can be constructed from them.

3.2 Carbohydrates

6. Give two examples of simple sugars and two examples of complex sugars.

7. What are the primary characteristics used to identify a compound as a carbohydrate?

3.3 Proteins

8. How do the primary, secondary, tertiary, and quaternary structures of proteins differ?

9. List the three categories of proteins and describe their functions.

3.4 Nucleic Acids

10. Describe how DNA differs from and is similar to RNA both structurally and functionally.

11. List the nitrogenous bases that base pair in DNA, in RNA.

3.5 Lipids

12. Describe three kinds of lipids.

13. What is meant by HDL, LDL, and VLDL? Where are they found? How do they relate to disease?

Thinking Critically

*For guidelines to answer this question, visit the **Concepts in Biology**, 12/e website.*

Archeologists, anthropologists, chemists, biologists, and health care professionals agree that the drinking of alcohol dates back thousands of years. Evidence also exists that this practice has occurred in most cultures around the world. Use the Internet to search out answers to the following questions:

1. What is the earliest date for which there is evidence for the production of ethyl alcohol?
2. In which culture did alcohol drinking first occur?
3. What is the molecular formula and structure of ethanol?
4. Do alcohol and water mix?
5. How much ethanol is consumed in the form of beverages in the United States each year?
6. What is the legal limit to be considered intoxicated in your state?
7. How is the legal limit in your state measured?
8. Why is there a tax on alcoholic beverages?
9. How do the negative effects of drinking alcohol compare for men and women?
10. Have researchers demonstrated any beneficial effects of drinking alcohol?

Compare what you thought you knew to what archeologists, anthropologists, chemists, biologists, can now support with scientific evidence.

CHAPTER

4

Cell Structure and Function

Have you ever thought of your body as a machine made of many specialized parts? Like a machine, you will not function properly unless all of your parts are in good working order. Large organisms, such as humans, have many millions of cells, which are specialized for particular functions. For example, red blood cells are specialized to carry oxygen, muscle cells to contract, nerve cells to carry messages, and pancreas cells to produce insulin. When any of these specialized cells do not function properly, the entire body suffers from a disease.

Nearly all drugs work by changing how certain cells function. Organ transplants are possible because antirejection drugs prevent the cells of the immune system from destroying transplanted hearts, kidneys, or livers. Pain can

be controlled because painkillers, such as aspirin make nerve cells less sensitive to painful stimuli. Many kinds of infectious disease can be controlled because antibiotics kill or prevent the reproduction of bacteria, but the antibiotics have little effect on human cells.

The development of drugs requires an intimate understanding of the structure and function of cells. In particular, drug researchers must know how each kind of cell differs from the others. Once it is known how a drug affects cells, it is then possible to adjust the dose, so that the drug's effects can do good, rather than cause harm. This chapter provides background about different kinds of cells, the structures they contain, and how they work.

CHAPTER OUTLINE

4.1 The Development of the Cell Theory

The **cell theory** states that all living things are made of cells. The **cell** is the basic structural and functional unit of living things and is the smallest unit that displays the characteristics of life. However, the concept of a cell did not emerge all at once but, rather, was developed and modified over several centuries. It is still being modified today. The ideas of hundreds of people were important in the development of the cell theory, but certain key people can be identified.

Some History

The first person to use the term *cell* was Robert Hooke (1635–1703) of England. He used a simple kind of microscope to study thin slices of cork from the bark of a cork oak tree. He saw many cubicles fitting neatly together, which reminded him of the barren rooms (cells) in a monastery (figure 4.1). He used the term *cell* when he described his observations in 1666 in the publication *Micrographia*. The tiny boxes Hooke saw were, in fact, only the cell walls that surrounded the once living portions of plant cells.

We now know that the **cell wall** of a plant cell is produced on the outside of the cell and is composed of the complex carbohydrate called cellulose. It provides strength and protection to the living contents of the cell. Although the cell wall appears to be a rigid, solid layer of material, it is actually composed of many interwoven strands of cellulose molecules. Thus, most kinds of molecules pass easily through it.

Anton van Leeuwenhoek (1632–1723), a Dutch merchant who sold cloth, was one of the first individuals to carefully study magnified cells. He apparently saw a copy of Hooke's *Micrographia* and began to make his own microscopes, so that he could study biological specimens. He was

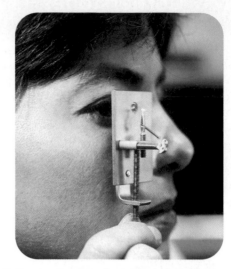

FIGURE 4.2 Anton van Leeuwenhoek's Microscope
Although van Leeuwenhoek's microscope had only one lens, the lens quality was so good that he was able to see cells clearly. This replica of his microscope shows that it is a small, simple apparatus.

interested in magnifying glasses, because magnifiers were used to count the number of threads in cloth. He used a very simple kind of microscope that had only one lens. Basically, it was a very powerful magnifying glass (figure 4.2). What made his microscope better than others of the time was his ability to grind very high-quality lenses. He used his skill at lens grinding to make about 400 lenses during his lifetime. One of his lenses was able to magnify 270 times. He used his keen eyesight and lens-making skills to pursue his intense curiosity about tiny living things. He made thousands of observations of many kinds of microscopic objects. He also made very detailed sketches of the things he viewed with his simple microscopes and communicated his findings to Robert Hooke and the Royal Society of London. His work stimulated further investigation of magnification techniques and descriptions of cell structures.

When van Leeuwenhoek discovered that he could see things moving in pond water using his microscope, his curiosity stimulated him to look at a variety of other things. He studied many things such as blood, semen, feces, pepper, and tartar, for example. He was the first to see individual cells and recognize them as living units, but he did not call them cells. The name he gave to the "little animals" he saw moving around in the pond water was *animalcules*.

Although Hooke, van Leeuwenhoek, and others continued to make observations, nearly 200 years passed before it was generally recognized that all living things are made of cells and that these cells can reproduce themselves. In 1838, Mathias Jakob Schleiden of Germany stated that all plants

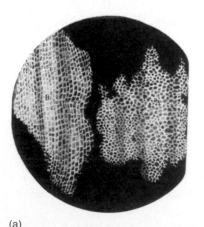

(a)

(b)

FIGURE 4.1 Hooke's Observations
The concept of a cell has changed considerably over the past 300 years. Robert Hooke's idea of a cell *(a)* was based on his observation of slices of cork (cell walls of the bark of the cork oak tree). *(b)* Hooke constructed his own simple microscope to be able to make these observations.

are made up of smaller cellular units. In 1839, Theodor Schwann, another German, published the idea that all animals are composed of cells.

Soon after the term *cell* caught on, it was recognized that the cell wall was essentially lifeless and that it was really the contents of the cell that had "life." This living material was termed **protoplasm,** which means *first-formed substance.* Scientists used the term *protoplasm* to distinguish between the living portion of the cell and the nonliving cell wall. As better microscopes were developed, people began to distinguish two different regions of protoplasm. One region, called the *nucleus,* appeared as a central body within a more fluid material surrounding it. Today, we know the **nucleus** is the part of a cell that contains the genetic information. **Cytoplasm** was the name given to the fluid portion of the protoplasm surrounding the nucleus. Although the term *protoplasm* is seldom used today, the term *cytoplasm* is still commonly used.

The development of special staining techniques, better light microscopes, and ultimately powerful electron microscopes revealed that the cytoplasm contains many structures, called **organelles** *(little organs)* (figure 4.3). Further research has shown that each kind of organelle has certain functions related to its structure.

Prokaryotic and Eukaryotic Cells

All living things are cells or composed of cells, which have an outer membrane, cytoplasm, and genetic material. In addition, the cells of some organisms have a cell wall surrounding these materials. Today, biologists recognize two very different kinds of cells. The differences are found in the details of their structure.

Prokaryotic cells are structurally simple cells that lack a nucleus and most other cellular organelles. The fossil record shows that prokaryotic cells were the first kind of cells to develop about 3.5 billion years ago. Today, we recognize that the organisms commonly called bacteria are prokaryotic cells. **Eukaryotic cells** are much more structurally complex cells that have a nucleus and many kinds of organelles. The first eukaryotic cells show up in the fossil record about 1.8 billion years ago. The cells of plants, animals, fungi, protozoa, and algae are eukaryotic (figure 4.4).

4.2 | Cell Size

Cells of different kinds vary greatly in size (figure 4.5). In general, the cells of prokaryotic organisms are much smaller than those of eukaryotic organisms. Prokaryotic cells are typically 1–2 micrometers in diameter, whereas eukaryotic cells are typically 10–100 times larger.

Some basic physical principles determine how large a cell can be. A cell must transport all of its nutrients and all of its wastes through its outer membrane to stay alive. Cells are limited in size because, as a cell becomes larger, adequate

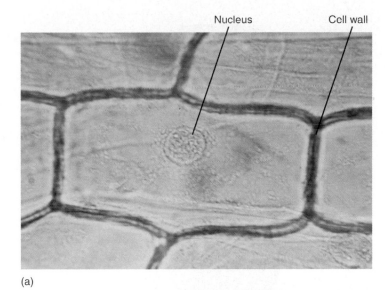

(a)

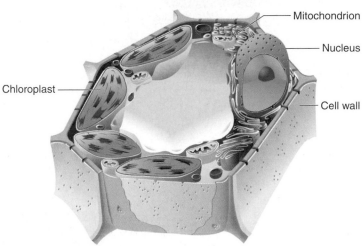

(b)

FIGURE 4.3 Cellular Organelles
(a) One of the first things that people were able to distinguish in cells was the nucleus. *(b)* Today, with the use of more powerful microscopes, it is possible to identify a variety of organelles within a cell.

transport of materials through the membrane becomes more difficult. The difficulty arises because, as the size of a cell increases, the amount of living material (the cell's volume) increases more quickly than the size of the outer membrane (the cell's surface area). As cells grow, the amount of surface area increases by the square (X^2) but volume increases by the cube (X^3). This mathematical relationship between the surface area and volume is called the *surface area-to-volume ratio* and is shown for a cube in figure 4.6. Notice that, as the cell becomes larger, both surface area and volume increase. Most important, volume increases *more* quickly than surface area, causing the surface area-to-volume ratio to decrease. As the cell's volume increases, the cell's metabolic requirements increase but its ability to satisfy those requirements are limited

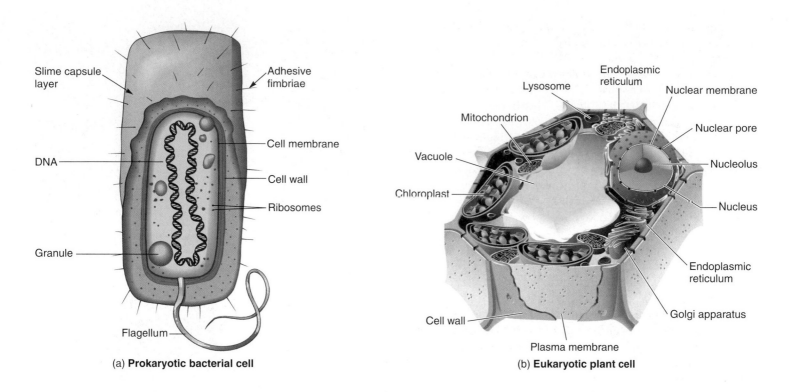

Slime capsule layer

Adhesive fimbriae

DNA

Cell membrane

Cell wall

Ribosomes

Granule

Flagellum

(a) **Prokaryotic bacterial cell**

Lysosome

Endoplasmic reticulum

Nuclear membrane

Mitochondrion

Nuclear pore

Vacuole

Nucleolus

Chloroplast

Nucleus

Endoplasmic reticulum

Golgi apparatus

Cell wall

Plasma membrane

(b) **Eukaryotic plant cell**

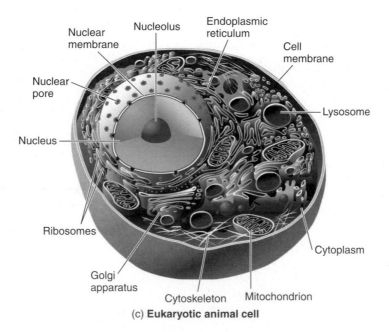

Nuclear membrane

Nucleolus

Endoplasmic reticulum

Cell membrane

Nuclear pore

Nucleus

Lysosome

Ribosomes

Cytoplasm

Golgi apparatus

Cytoskeleton

Mitochondrion

(c) **Eukaryotic animal cell**

FIGURE 4.4 Major Cell Types

There are two major types of cells. Eukaryotic cells are 10 to 100 times larger than prokaryotic cells. These drawings (not to scale) highlight the structural differences between them. Prokaryotic cells are represented by a *(a)* bacterium; eukaryotic cells by *(b)* plant and *(c)* animal cells.

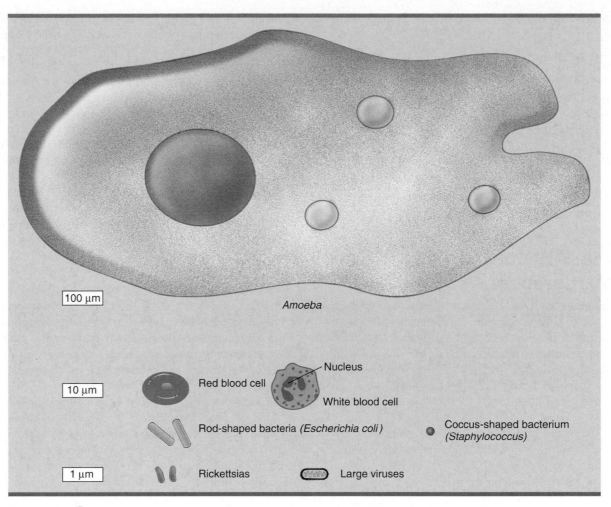

FIGURE 4.5 Comparing Cell Sizes

Most cells are too small to be seen with the naked eye. Prokaryotic cells are generally about 1–2 micrometers in diameter. Eukaryotic cells are generally much larger and generally range between 10 and 100 micrometers. A micrometer is 1/1,000 of a millimeter. A sheet of paper is about 1/10 of a millimeter thick, which is about 100 micrometers. Therefore, some of the largest eukaryotic cells are just visible to the naked eye.

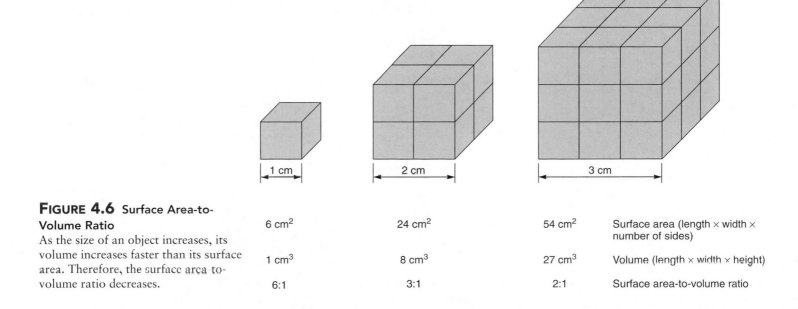

FIGURE 4.6 Surface Area-to-Volume Ratio

As the size of an object increases, its volume increases faster than its surface area. Therefore, the surface area-to-volume ratio decreases.

	1 cm	2 cm	3 cm	
	6 cm^2	24 cm^2	54 cm^2	Surface area (length × width × number of sides)
	1 cm^3	8 cm^3	27 cm^3	Volume (length × width × height)
	6:1	3:1	2:1	Surface area-to-volume ratio

by the surface area through which the needed materials must pass. Consequently, most cells are very small.

There are a few exceptions to this general rule, but they are easily explained. For example, what we call the yolk of a chicken's egg cell is a single cell. However, the only part of an egg cell that is metabolically active is a small spot on its surface. The largest portion of the egg cell is simply inactive stored food called *yolk*. Similarly, some plant cells are very large but consist of a large, centrally located region filled with water. Again, the metabolically active portion of the cell is at the surface, where exchange of materials with the surroundings is possible.

4.3 The Structure of Cellular Membranes

One feature common to all cells is the presence of **cellular membranes,** thin sheets composed primarily of phospholipids and proteins. The current model of how cellular membranes are constructed is known as the *fluid-mosaic model.* The **fluid-mosaic model,** considers cellular membranes to consist of two layers of phospholipid molecules and that the individual phospholipid molecules are able to move about within the structure of the membrane (How Science Works 4.1). Many kinds of proteins and some other molecules are found among the phospholipid molecules within the membrane and on the membrane surface. The individual molecules of the membrane remain associated with one another

because of the physical interaction of its molecules with its surroundings. The phospholipid molecules of the membrane have two ends, which differ chemically. One end, which contains phosphate, is soluble in water and is therefore called **hydrophilic** (*hydro* = water; *phile* = loving). The other end of the phospholipid molecule consists of fatty acids, which are not soluble in water, and is called **hydrophobic** (*phobia* = fear).

In diagrams, phospholipid molecules are commonly represented as a balloon with two strings (figure 4.7). The balloon represents the water-soluble phosphate portion of the molecule and the two strings represent the 2 fatty acids. Consequently, when phospholipid molecules are placed in water, they form a double-layered sheet, with the water-soluble (hydrophilic) portions of the molecules facing away from each other. This is commonly referred to as a *phospholipid bilayer* (figure 4.8). If phospholipid molecules are shaken in a glass of water, the molecules automatically form double-layered membranes. It is important to understand that the membranes formed are not rigid but, rather, resemble a heavy olive oil in consistency. The component phospholipid molecules are in constant motion as they move with the surrounding water molecules and slide past one another. Other molecules found in cell membranes are cholesterol, proteins, and carbohydrates.

Because cholesterol is not water-soluble, it is found in the middle of the membrane, in the hydrophobic region. It appears to play a role in stabilizing the membrane and keeping it flexible. There are many different proteins associated with the membrane. Some are found on the surface, some are partially submerged in the membrane, and others traverse the

FIGURE 4.7 A Phospholipid Molecule

Phospholipids have a hydrophobic (water-insoluble) portion and a hydrophilic (water-soluble) portion. The hydrophilic portion contains phosphate and is represented as a balloon in many diagrams. The fatty acids are represented as two strings on the balloon.

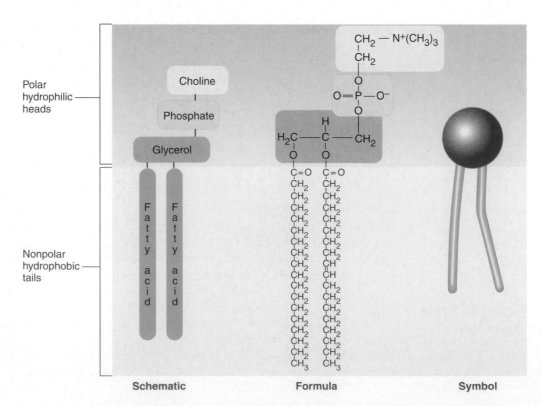

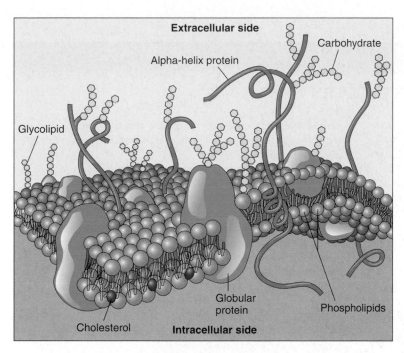

Extracellular side

Carbohydrate

Alpha-helix protein

Glycolipid

Globular protein

Cholesterol

Intracellular side

Phospholipids

FIGURE 4.8 The Nature of Cellular Membranes

The membranes in all cells are composed primarily of protein and phospholipids. Two layers of phospholipid are oriented so that the hydrophobic fatty acid ends extend toward each other and the hydrophilic phosphate-containing portions are on the outside. Proteins are found buried within the phospholipid layer and are found on both surfaces of the membrane. Cholesterol molecules are also found among the phospholipid molecules. Carbohydrates are often attached to one surface of the membrane.

membrane and protrude from both surfaces. These proteins serve a variety of functions, including helping transport molecules across the membrane, serving as attachment points for other molecules, and serving as identity tags for cells. Carbohydrates are typically attached to the membranes on the outside of cells. They appear to play a role in cell-to-cell interactions and are involved in binding with regulatory molecules.

4.4 Organelles Composed of Membranes

Although both prokaryotic and eukaryotic cells have membranes, eukaryotic cells have many more specialized organelles composed of membranes than do prokaryotic cells.

Plasma Membrane

The outer limiting boundary of both prokaryotic and eukaryotic cells is known as the **plasma membrane,** or **cell membrane.** It is composed of a single phospholipid bilayer and serves as a barrier between the cell contents and the external environment. However, it is not just a physical barrier. It has many different functions. In many ways, it functions in a

manner analogous to a border between countries, which delimits countries but also allows for some restricted movement across the border. The plasma membrane has several important features.

Metabolic Activities

Because the plasma membrane is part of a living unit, it is metabolically active. Many important chemical reactions take place within the membrane or on its inside or outside surface. Many of these chemical reactions involve transport of molecules.

Movement of Molecules Across the Membrane

Because cells must continuously receive nutrients and rid themselves of waste products, there is a constant traffic of molecules across the membrane. See section 4.7 for a detailed discussion of the many ways by which molecules enter and leave cells. Many of the proteins that are associated with the plasma membrane are involved in moving molecules across the membrane. Some proteins are capable of moving from one side of the plasma membrane to the other and shuttle certain molecules across the membrane. Others extend through the membrane and form channels through which substances can travel. Some of these channels operate like border checkpoints, which open and close when circumstances dictate. Some molecules pass through the membrane passively, whereas others are assisted by metabolic activities within the membrane.

Inside and Outside

The inside of the plasma membrane is different from its outside. The presence of specific proteins or carbohydrates is important in establishing this difference. The carbohydrates that are associated with the plasma membrane are usually found on the outside of the membrane, where they are bound to proteins or lipids. Many important activities take place on only one of the surfaces of the plasma membrane because of the way the two sides differ.

Identification

The outside surface of the plasma membrane has many proteins, which act as recognition molecules. Each organism has a unique combination of these molecules. Thus, the presence of these molecules enables one cell or one organism to recognize cells that are like it and those that are different. For example, if a disease organism enters your body, the cells of your immune system use the proteins on the invader's surface to identify it as being foreign. Immune system cells can then destroy the invader.

In humans, there is a group of such protein molecules, collectively known as *histocompatibility antigens* (*histo* = tissue). Each person has a specific combination of these proteins. It is the presence of these antigens that is responsible for the rejection of transplanted tissues or organs from donors that are "incompatible." In large part, a person's pattern of histocompatibility antigens is hereditary; for instance, in

HOW SCIENCE WORKS 4.1

Developing the Fluid-Mosaic Model

The fluid-mosaic model describes the current understanding of how cellular membranes are organized and function. As is typical during the development of most scientific understandings, the fluid-mosaic model was formed as a result of the analysis of data from many experiments. We will look at three characteristics of cellular membranes and how certain experiments and observations about these characteristics led scientists to develop the fluid-mosaic model.

1. *What is the chemical nature of cellular membranes and how do they provide a barrier between the contents of the cell and the cell's environment?*

    In 1915, scientists isolated cellular membranes from other cellular materials and chemically determined that they consisted primarily of lipids and proteins. The scientists recognized that, because lipids do not mix with water, a layer of lipid could serve as a barrier between the watery contents of a cell and its watery surroundings.

2. *How are the molecules arranged within the membrane?*

    Nearly 10 years after it became known that cellular membranes consist of lipids and proteins, two scientists reasoned from the chemical properties of lipids and proteins that cellular membranes probably consist of two layers of lipid. This arrangement became known as a bilayer. They were able to make this deduction because they understood the chemical nature of lipids and how they behave in water. But this model did not account for the proteins, which were known to be an important part of cellular membranes because proteins were usually isolated from cellular membranes along with lipids. Also, artificial cellular membranes—made only of lipids—did not have the same chemical properties as living cellular membranes.

    The first model to incorporate proteins into the cellular membrane was incorrect. It was called the sandwich model, because it placed the lipid layers of the cellular membrane between two layers of protein, which were exposed to the cell's watery environment and cytoplasm. Although incorrect, the sandwich model was very popular

into the 1960s, because it was supported by images from electron microscopes, which showed two dark lines, with a lighter area between them.

One of the biggest problems with the sandwich model was that the kinds of proteins isolated from the cellular membrane were strongly hydrophobic. A sandwich model with the proteins on the outside required these hydrophobic proteins to be exposed to water, which would have been an unstable arrangement.

In 1972, two scientists proposed that the hydrophobic proteins are actually made stable because they are submerged in the hydrophobic portion of the lipid bilayer. This hypothesis was supported by an experimental technique called freeze-fracture.

Freeze-fracture experiments split a frozen lipid bilayer, so that the surface between the two lipid layers could be examined by electron microscopy. These experiments showed large objects (proteins) sitting in a smooth background (phospholipids), similar to the way paving stones sit in a sidewalk. These experiments supported the hypothesis that the proteins are not on the surface but, rather, are incorporated into the lipid bilayer.

3. *How do these protein and lipid molecules interact with one another within the cellular membrane?*

    The answer to this question was provided by a series of hybrid-cell experiments. In these experiments, proteins in a mouse cell and proteins in a human cell were labeled differently. The two cells were fused, so that their cellular membranes were connected. At first, one-half of the new hybrid cell contained all mouse proteins. The other half of the new hybrid cell contained all human proteins. However, over several hours, the labeled proteins were seen to mix until the mouse and human proteins were evenly dispersed. Seeing this dispersion demonstrated that molecules in cellular membranes move. Cellular membranes consist of a mosaic of protein and lipid molecules, which move about in a fluid manner.

---

identical twins, the cells of both individuals have very similar proteins. Therefore, in transplant situations, the cells of the immune system would see the cells of the donor twin to be the same as those on the cell surfaces of the recipient twin. When closely related donors are not available, physicians try to find donors whose histocompatability antigens are as similar as possible to those of recipients.

Attachment Sites

Some molecules on the outside surface of the plasma membrane serve as attachment sites for specific chemicals, bacteria, protozoa, white blood cells, and viruses. Many

dangerous agents cannot stick to the surface of cells and therefore do not cause harm. For this reason, cell biologists explore the exact structure and function of these cell surface molecules. They are also attempting to identify molecules that can interfere with the binding of viruses and bacteria to cells in the hope of controlling infections. For example, human immunodeficiency virus (HIV) attaches to specific molecules on the surface of certain immune system cells and nerve cells. If these attachment sites could be masked, the virus would not be able to attach to the cells and cause disease. Drugs that function this way are called "blockers."

Signal Transduction

Another way in which attachment sites are important is in signal transduction. **Signal transduction** is the process by which cells detect specific signals and transmit these signals to the cell's interior. These signals can be physical (electrical or heat) or chemical. Some chemicals are capable of passing directly through the membrane of specific target cells. Once inside, they can pass on their message to regulator proteins. These proteins then enter into chemical reactions, which result in a change in the cell's behavior. For example, estrogen produced in one part of the body travels through the bloodstream and passes through the tissue to make direct contact with specific target cells. Once the hormone passes through the plasma membrane of the target cells, the message is communicated to begin the process of female sex organ development. This is like a person smelling the cologne of his or her date through a curtain. The aroma molecules pass through the curtain to the person's nose and stimulate a response.

However, most signal molecules are not capable of entering cells in such a direct manner. Most signal molecules remain outside their target cells. When they arrive at the cell, they attach to a receptor site molecule embedded in the membrane. The signal molecule is often called the *primary messenger*. The receptor/signal molecule combination initiates a sequence of events within the membrane that transmits information through the membrane to the interior, generating internal signal molecules, called *secondary messengers*. In many other cases, the receptor molecule is a protein that spans the cell membrane. This protein is capable of binding the signal molecule outside of the cell and then generates a secondary messenger inside the cell.

The secondary messengers are molecules or ions that begin a cascade of chemical reactions that causes the target cell to change how it functions. This is like your mother sending your little brother to tell you it is time for dinner. Your mother provides the primary message, your little brother provides the secondary message, and you respond by going to the dinner table. In a cell, such signal transduction results in a change in the cell's chemical activity. Often, this is accomplished by turning genes on or off. For example, when a signal molecule called *epidermal growth factor (EGF)* attaches to the receptor protein of skin cells, it triggers a chain of events inside the plasma membrane of the cells. These changes within the plasma membrane produce secondary messengers, ultimately leading to gene action, which in turn causes cell growth and division.

Endoplasmic Reticulum

There are many other organelles in addition to the plasma membrane, that are composed of membranes. Each of these membranous organelles has a unique shape or structure associated with its particular functions (figure 4.9). One of the most common organelles found in cells, the **endoplasmic reticulum (ER)**, is a set of folded membranes and tubes throughout the cell. This system of membranes provides a large surface on which chemical activities take place. Because the ER has an enormous surface area, many chemical reactions can be carried out in an extremely small space. Picture the vast surface area of a piece of newspaper crumpled into a tight little ball. The surface contains hundreds of thousands of tidbits of information in an orderly arrangement, yet it is packed into a very small volume.

Proteins on the surface of the ER are actively involved in controlling and encouraging chemical activities—whether they are reactions involving cell growth and development or reactions resulting in the accumulation of molecules from the environment. The arrangement of the proteins allows them to control the sequences of metabolic activities, so that chemical reactions can be carried out very rapidly and accurately.

On close examination with an electron microscope, it is apparent that there are two types of ER—rough and smooth. The rough ER appears rough because it has *ribosomes* attached to its surface. **Ribosomes** are nonmembranous organelles that are associated with the synthesis of proteins from amino acids. They are "protein-manufacturing machines." Therefore, cells with an extensive amount of rough ER—for example, human pancreas cells—are capable of synthesizing large quantities of proteins. Smooth ER lacks attached ribosomes but is the site of many other important cellular chemical activities, including fat metabolism and detoxification reactions involved in the destruction of toxic substances, such as alcohol and drugs. Human liver cells are responsible for detoxification reactions and contain extensive smooth ER.

In addition, the spaces between the folded membranes serve as canals for the movement of molecules within the cell. This system of membranes allows for the rapid distribution of molecules within a cell.

Golgi Apparatus

Another organelle composed of membrane is the **Golgi apparatus.** Animal cells contain several such structures and plant cells contain hundreds. The typical Golgi apparatus consists of 5 to 20 flattened, smooth, membranous sacs, which resemble a stack of flattened balloons (figure 4.10). The Golgi apparatus has several functions. It modifies molecules shipped to it from elsewhere in the cell, it manufactures some polysaccharides and lipids, and it packages molecules within sacs.

There is a constant traffic of molecules through the Golgi apparatus. Tiny, membranous sacs called *vesicles* deliver molecules to one surface of the Golgi apparatus. Many of these vesicles are formed by the endoplasmic reticulum and contain proteins. These vesicles combine with the sacs of the Golgi apparatus and release their contents into it. Many kinds of chemical reactions take place within the Golgi apparatus. Ultimately, new sacs, containing "finished products," are produced from the other surface of the Golgi apparatus. The Golgi apparatus produces many kinds of vesicles. Each has a different function. Some are transported within the cell and combine with other membrane structures, such as the endoplasmic

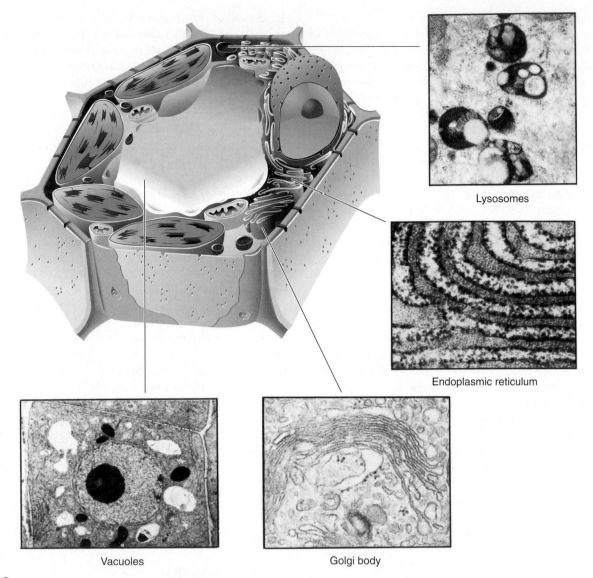

Lysosomes

Endoplasmic reticulum

Vacuoles

Golgi body

FIGURE 4.9 Membranous Cytoplasmic Organelles

Certain structures in the cytoplasm are constructed of membranes. Membranes are composed of protein and phospholipids. The four structures here—vacuoles, a Golgi body, endoplasmic reticulum, and lysosomes—are constructed of single-layer membranes.

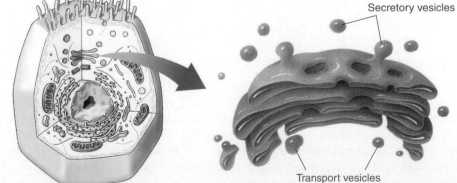

FIGURE 4.10 Golgi Apparatus

The Golgi apparatus is a series of membranous sacs that accept packages of materials and produce vesicles containing materials that are transported throughout the cell and to the exterior of the cell.

Secretory vesicles

Transport vesicles

reticulum. Some migrate to the plasma membrane and combine with it. These vesicles release molecules such as mucus, cellulose, glycoproteins, insulin, and enzymes to the outside of the cell. In plant cells, cellulose-containing vesicles are involved in producing new cell wall material. Finally, some of the vesicles produced by the Golgi apparatus contain enzymes that can break down the various molecules of the cell, causing its destruction. These vesicles are known as *lysosomes.*

Lysosomes

Lysosomes are tiny vesicles that contain enzymes capable of digesting carbohydrates, nucleic acids, proteins, and lipids. Because cells are composed of these molecules, these enzymes must be controlled in order to prevent the destruction of the cell. This control is accomplished very simply. The enzymes of lysosomes function best at a pH of about 5. The membrane, which is the outer covering of the lysosome, transports hydrogen ions into the lysosome and creates the acidic conditions these enzymes need. Since the pH of a cell is generally about 7, these enzymes will not function if released into the cell cytoplasm.

The functions of lysosomes are basically digestion and destruction. For example, in many kinds of protozoa, such as *Paramecium* and *Amoeba,* food is taken into the cell in the form of a membrane-enclosed food vacuole. Lysosomes combine with food vacuoles and break down the food particles into smaller and smaller molecular units, which the cell can use.

In a similar fashion, lysosomes destroy disease-causing microorganisms, such as bacteria, viruses, and fungi. The microorganisms become surrounded by membranes from the endoplasmic reticulum. Lysosomes combine with the membranes surrounding these invaders and destroy them. This kind of activity is common in white blood cells that engulf and destroy disease-causing organisms. Lysosomes are also involved in the breakdown of worn-out cell organelles by fusing with them and destroying them (figure 4.11).

Peroxisomes

Another organelle that consists of many kinds of enzymes surrounded by a membrane is the *peroxisome.* **Peroxisomes** were first identified by the presence of an enzyme, catalase, that breaks down hydrogen peroxide (H_2O_2). Peroxisomes differ from lysosomes in that peroxisomes are not formed by the Golgi apparatus and they contain different enzymes. It appears that peroxisomes are formed by the synthesis of a sac of membrane into which the enzymes are imported. The enzymes of peroxisomes have been shown to be important in many kinds of chemical reactions, including the breakdown of long-chain fatty acids to shorter molecules, the synthesis of cholesterol and the bile acids produced from cholesterol, and the synthesis of specific lipid molecules present in the plasma membranes of specialized cells, such as nerve cells.

Vacuoles and Vesicles

There are many kinds of membrane enclosed containers in cells known as *vacuoles* and *vesicles.* **Vacuoles** are the larger structures and **vesicles** are the smaller ones. They are frequently described by their function. Specialized water vacuoles called

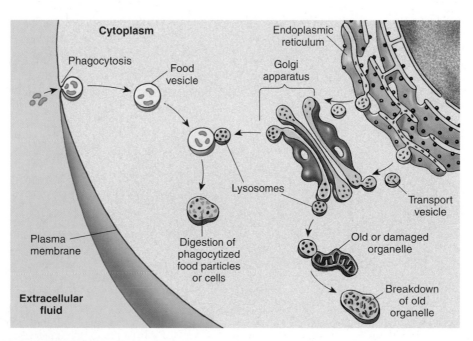

FIGURE 4.11 Lysosome Function
Lysosomes contain enzymes that are capable of digesting many kinds of materials. They are involved in the digestion of food vacuoles, harmful organisms, and damaged organelles.

contractile vacuoles are able to forcefully expel excess water that has accumulated in the cytoplasm. The contractile vacuole is a necessary organelle in cells that live in freshwater because water constantly diffuses into the cell. In most plants, there is one huge, centrally located, water-filled vacuole. Animal cells typically have many small vacuoles and vesicles throughout the cytoplasm.

Nuclear Membrane

Just as a room is a place created by walls, a floor, and a ceiling, a cell's nucleus is a place created by the **nuclear membrane.** If the nuclear membrane were not formed around the cell's genetic material, the organelle called the nucleus would not exist. This membrane separates the genetic material (DNA) from the cytoplasm. Because they are separated, the cytoplasm and the nuclear contents can maintain different chemical compositions. The nuclear membrane is composed of two layers of membrane. It has large openings through this double-membrane structure, called *nuclear pore complexes* (figure 4.12). The nuclear pore complexes consist of proteins, which collectively form barrel-shaped pores through the membrane. These pores allow relatively large molecules, such as RNA, to pass through the nuclear membrane. Thousands of molecules move in and out through these pores each second.

Interconversion of Membranes

It is important to remember that all membranous structures in cells are composed of two layers of phospholipid with associated proteins and other molecules. Furthermore, all of these membranous organelles can be converted from one form to another (figure 4.13). For example, the plasma membrane is continuous with the endoplasmic reticulum; as a cell becomes larger, some of the endoplasmic reticulum moves to the surface to become plasma membrane. Similarly, the nuclear membrane is connected to the endoplasmic reticulum. Remember also that the Golgi apparatus receives membrane-enclosed packages from the endoplasmic reticulum and produces lysosomes that combine with other membrane-enclosed structures and secretory vesicles that fuse with the plasma membrane. Thus, this entire set of membrane material is in a constant state of flux.

Energy Converters—Mitochondria and Chloroplasts

Two other organelles composed of membranes are *mitochondria* and *chloroplasts*. Both types of organelles are associated with energy conversion reactions in the cell. Mitochondria and chloroplasts are different from other kinds of membranous structures in four ways. First, their membranes are chemically different from those of other membranous organelles; second, they are composed of double layers of membrane—an inner and an outer membrane; third, both of these structures have ribosomes and DNA that are similar to those of bacteria; fourth, these two structures have a certain degree of independence from the rest of the cell—they have a limited ability to reproduce themselves but must rely on DNA from the cell nucleus for assistance. It is important to understand that cells cannot make mitochondria or chloroplasts by themselves. The DNA of the organelle is necessary for their reproduction.

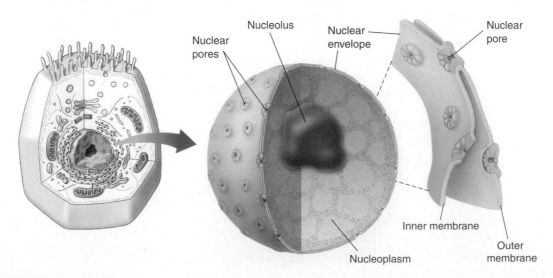

FIGURE 4.12 Nuclear Membrane
The nuclear membrane is a double membrane separating the nuclear contents from the cytoplasm. Pores in the nuclear membrane allow molecules as large as proteins to pass through.

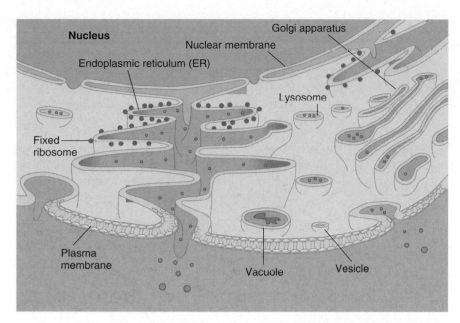

FIGURE 4.13 Interconversion of Membranous Organelles

Eukaryotic cells contain a variety of organelles composed of phospholipids and proteins. Each has a unique shape and function. Many of these organelles are interconverted from one to another as they perform their essential functions. Plasma membrane can become vacuolar membrane or endoplasmic reticulum, which can become vesicular membrane, which in turn can become Golgi or nuclear membrane.

Mitochondrion

The **mitochondrion** is an organelle that contains the enzymes responsible for aerobic cellular respiration. It consists of an outer membrane and an inner folded membrane. The individual folds of the inner membrane are known as **cristae** (figure 4.14a). **Aerobic cellular respiration** is the series of enzyme-controlled reactions involved in the release of energy from food molecules and requires the participation of oxygen molecules. Some of the enzymes are dissolved in the fluid inside the mitochondrion. Others are incorporated into the structure of the membranes and are arranged in an orderly sequence. The number of mitochondria per cell varies from less than ten to over 1,000 depending on the kind of cell. Cells involved in activities that require large amounts of energy, such as muscle cells, contain the most mitochondria. When properly stained, they can be seen with a compound light microscope. When cells are functioning aerobically, the mitochondria swell with activity. When this activity diminishes, though, they shrink and appear as threadlike structures. The details of the reactions involved in aerobic cellular respiration and their relationship to the structure of mitochondria will be discussed in chapter 6.

Chloroplast

The **chloroplast** is a membranous saclike organelle responsible for the process of photosynthesis. Chloroplasts contain the green pigment, **chlorophyll,** and are found in cells of plants and other eukaryotic organisms that carry out photosynthesis. **Photosynthesis** is a metabolic process in which light energy is converted to chemical bond energy. Chemical-bond energy is found in food molecules. The cells

of some organisms contain one large chloroplast; others contain hundreds of smaller chloroplasts. A study of the ultrastructure—that is, the structures seen with an electron microscope—of a chloroplast shows that the entire organelle is enclosed by a membrane. Inside are other membranes throughout the chloroplast, forming networks and structures of folded membrane. As shown in figure 4.14b, in some areas, these membranes are stacked up or folded back on themselves. Chlorophyll molecules are attached to these membranes. These areas of concentrated chlorophyll are called **thylakoid** membranes and are stacked up to form the **grana** of the chloroplast. The space between the grana, which has no chlorophyll, is known as the **stroma.** The details of how photosynthesis occurs and how this process is associated with the structure of the chloroplast will be discussed in chapter 7.

4.5 Nonmembranous Organelles

Suspended in the cytoplasm and associated with the membranous organelles are various kinds of structures that are not composed of phospholipids and proteins arranged in sheets.

Ribosomes

Ribosomes are nonmembranous organelles responsible for the synthesis of proteins from amino acids. They are composed of RNA and protein. Each ribosome is composed of

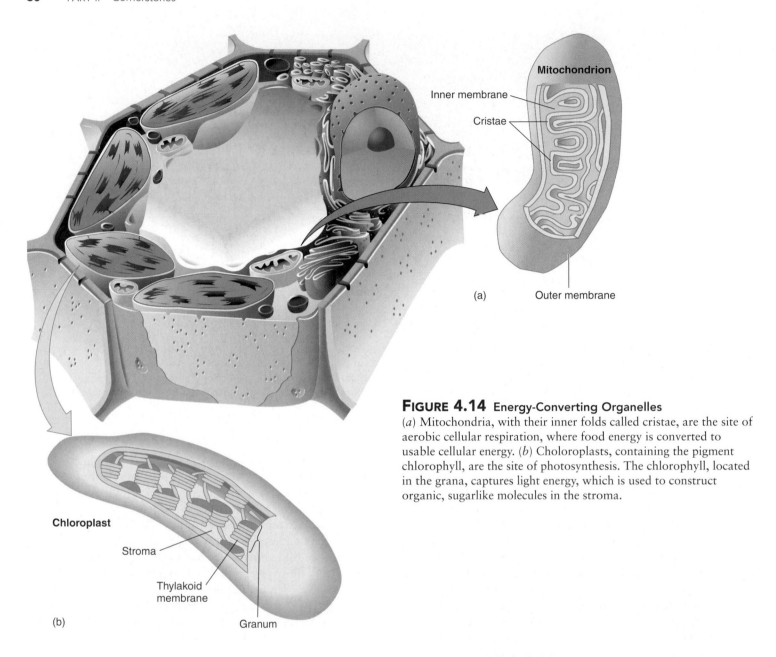

Mitochondrion

Inner membrane

Cristae

(a)

Outer membrane

Chloroplast

Stroma

Thylakoid
membrane

(b)

Granum

FIGURE 4.14 **Energy-Converting Organelles**
(*a*) Mitochondria, with their inner folds called cristae, are the site of aerobic cellular respiration, where food energy is converted to usable cellular energy. (*b*) Choloroplasts, containing the pigment chlorophyll, are the site of photosynthesis. The chlorophyll, located in the grana, captures light energy, which is used to construct organic, sugarlike molecules in the stroma.

two subunits—a large one and a small one. Ribosomes assist in the process of joining amino acids together to form proteins. Many ribosomes are attached to the endoplasmic reticulum. Because ER that has attached ribosomes appear rough when viewed through an electron microscope it is called rough ER. Areas of rough ER are active sites of protein production (figure 4.15). Many ribosomes are also found floating freely in the cytoplasm wherever proteins are being assembled. Cells that are actively producing protein (e.g., liver cells) have great numbers of free and attached ribosomes. The details of how ribosomes function in protein synthesis will be discussed in chapter 8.

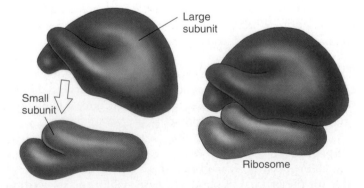

Large
subunit

Small
subunit

Ribosome

FIGURE 4.15 Ribosomes
Each ribosome is constructed of two subunits. Each of the subunits is composed of protein and RNA. These globular organelles are associated with the construction of protein molecules from individually amino acids.

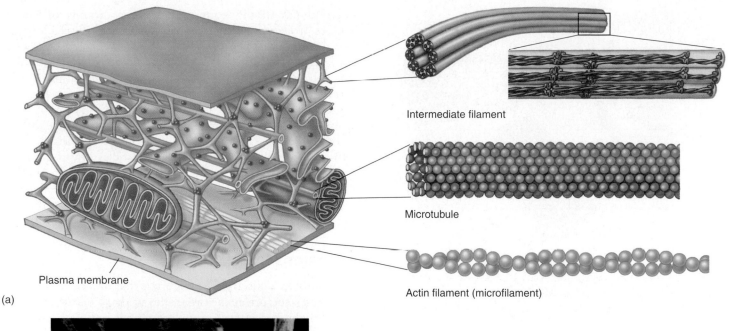

Intermediate filament

Microtubule

Actin filament (microfilament)

Plasma membrane

(a)

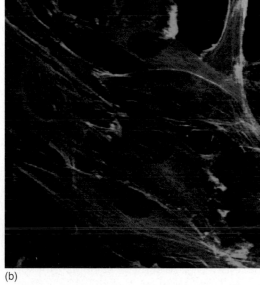

(b)

FIGURE 4.16 The Cytoskeleton

Microtubules, actin filaments, and intermediate filaments are all interconnected within the cytoplasm of the cell. *(a)* These structures, along with connections to other cellular organelles, form a cytoskeleton for the cell. The cellular skeleton is not a rigid, fixed-in-place structure but, rather, changes as the actin and intermediate filaments and microtubule component parts are assembled and disassembled. *(b)* The elements of the cytoskeleton have been labeled with a fluorescent dye to make them visible. The microtubules have fluorescent red dye, and actin filaments are green.

Microtubules, Microfilaments, and Intermediate Filaments

The interior of a cell is not simply filled with liquid cytoplasm. Among the many types of nonmembranous organelles found there are elongated protein structures known as **microtubules, microfilaments,** and **intermediate filaments.** All three types of organelles interconnect and some are attached to the inside of the plasma membrane, forming the **cytoskeleton** of the cell (figure 4.16). These cellular components provide the cell with shape, support, and the ability to move.

Think of the cytoskeleton components as the internal supports and cables required to construct a circus tent. The shape of the flexible canvas cover (i.e., the plasma membrane) is determined by the location of internal tent poles (i.e., micro-

tubules) and the tension placed on them by attached wire or rope cables (i.e., intermediate filaments and actin filaments).

Just as in the tent analogy, when one of the actin filaments or intermediate filaments is adjusted, the shape of the entire cell changes. For example, when a cell is placed on a surface to which it cannot stick, the internal tensions created by the cytoskeleton components can pull together and cause the cell to form a sphere. A cell's cytoskeleton also changes shape dramatically when the cell divides. It constricts, pulling the cell together in the middle and allowing the membrane to be sealed between the two new daughter cells.

Actin filaments and microtubules of the cytoskeleton also transport organelles from place to place within the cytoplasm. In addition, information can be transported through the cytoskeleton. Enzymes attached to the cytoskeleton are activated when the cell is touched. Some of these events even affect gene activity.

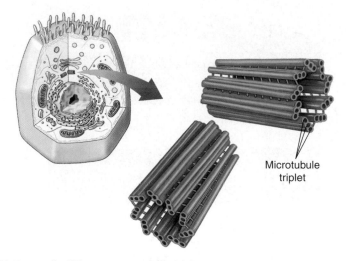

Microtubule triplet

FIGURE 4.17 The Centriole

These two sets of short microtubules are located just outside the nuclear membrane in many types of cells.

Centrioles

An arrangement of two sets of microtubules at right angles to each other makes up a structure known as a **centriole.** The centrioles of many cells are located in a region called the *centrosome.* The centrosome is often referred to as the microtubule organizing center and is usually located close to the nuclear membrane. Centrioles operate by organizing microtubules into a complex of strings called *spindle fibers.* The *spindle* is the structure on which chromosomes are attached, so that they can be separated properly during cell division. Each set of microtubules is composed of nine groups of short microtubules arranged in a cylinder (figure 4.17). The functions of centrioles and spindle fibers in cell division will be referred to again in chapter 9. One curious fact about centrioles is that they are present in most animal cells but not in many types of plant cells, although plant cells do have a centrosome. Other structures, called *basal bodies*, resemble centrioles and are located at the base of cilia and flagella. mitosis, p. 164

Cilia and Flagella

Many cells have microscopic, hairlike structures projecting from their surfaces: **cilia** and **flagella** (figure 4.18). In general, flagella are long and few in number; cilia are short and more numerous. They are similar in structure and each functions to move the cell through its environment or to move the environment past the cell. Flagella have an undulating, whiplike motion; cilia move back and forth like oars on a boat. They are constructed of a cylinder of nine sets of microtubules similar to those in the centriole, but they have an additional two

microtubules in the center. These long strands of microtubules project from the cell surface and are covered by plasma membrane. When cilia and flagella are sliced crosswise, their cut ends show a *9 + 2 arrangement* of microtubules. The cell can control the action of these microtubular structures, enabling them to be moved in a variety of ways. The protozoan *Paramecium* is covered with thousands of cilia, which move in a coordinated, rhythmic way to move the cell through the water. Similarly, the cilia on the cells that line the human trachea move mucus and particles trapped in the mucus from deep within the lungs.

Some kinds of prokaryotic cells also have flagella. However, their structure and the way they function are quite different from those of eukaryotic cells.

Inclusions

Inclusions are collections of materials that do not have as well defined a structure as the organelles we have discussed so far. They might be concentrations of stored materials, such as starch grains, sulfur, or oil droplets, or they might be a collection of miscellaneous materials known as **granules.** Unlike organelles, which are essential to the survival of a cell, inclusions are generally only temporary sites for the storage of nutrients and wastes.

Some inclusion materials are harmful to other cells. For example, rhubarb leaf cells contain an inclusion composed of oxalic acid, an organic acid. Needle-shaped crystals of calcium oxalate can cause injury to the kidneys of an organism that eats rhubarb leaves. The sour taste of this compound aids in the rhubarb plant's survival by discouraging animals from eating it. Similarly, certain bacteria store, in their inclusions, crystals of a substance known to be harmful to insects. Spraying plants with these bacteria is a biological method of controlling the insect pest population while not interfering with the plant or with humans.

In the past, cell structures such as ribosomes, mitochondria, and chloroplasts were also called *granules* because their structure and function were not clearly known. As scientists learn more about inclusions and other unidentified particles in the cells, they, too, will be named and more fully described.

4.6 Nuclear Components

As stated at the beginning of this chapter, one of the first structures to be identified in cells was the nucleus. If the nucleus is removed from a cell or the cell loses its nucleus, the cell can live only a short time. For example, human red blood cells begin life in bone marrow, where they have nuclei. Before they are released into the bloodstream to carry oxygen and carbon dioxide, they lose their nuclei. As a consequence, red blood cells are able to function only for about 120 days before they disintegrate. A double layer of membrane sur-

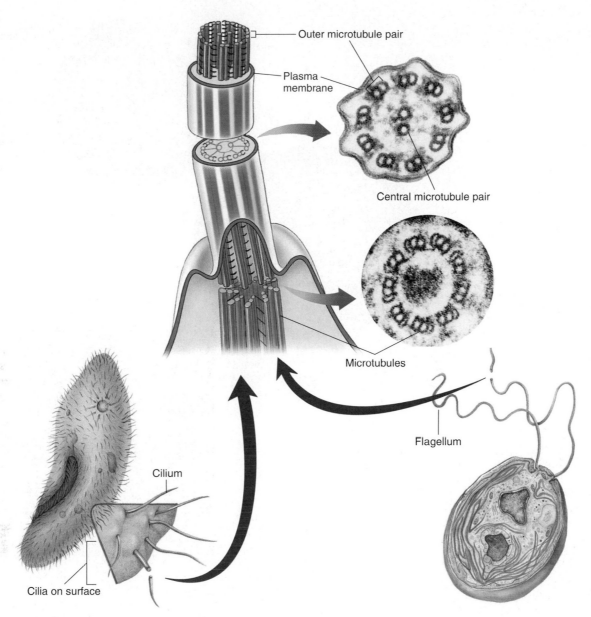

Outer microtubule pair

Plasma membrane

Central microtubule pair

Microtubules

Flagellum

Cilium

Cilia on surface

FIGURE 4.18 Cilia and Flagella

Cilia and flagella function like oars or propellers that move the cell through its environment or move the environment past the cell. Cilia and flagella are constructed of groups of microtubules, as in the ciliated protozoan shown on the left and the flagellated alga on the right. Flagella are usually less numerous and longer than cilia.

rounds the nucleus, and the membrane has pores. The nucleus contains DNA. Because DNA has the directions for building the proteins a cell needs to function, it is easy to understand why a cell lacking a nucleus dies.

When nuclear structures were first identified, it was noted that certain dyes stained some parts of the nuclear contents more than others. The parts that stained more heavily were called **chromatin,** which means colored material. Chromatin is composed of long molecules of DNA, along with proteins. Most of the time, the chromatin is arranged as a long, tangled mass of threads in the nucleus. However, during cell division, the chro-

matin becomes tightly coiled into short, dense structures called **chromosomes** (*chromo* = color; *some* = body). Chromatin and chromosomes are really the same molecules, but they differ in structural arrangement. In addition to chromosomes, the nucleus may also contain one, two, or several *nucleoli*. A **nucleolus** is the site of ribosome manufacture. Specific parts of the cell's DNA are involved in the manufacture of ribosomes. These DNA regions become organized at a particular place within the nucleus and produce ribosomes. The nucleolus is composed of this DNA, specific granules and fibers used in the manufacture of ribosomes, and partially completed ribosomes. The final

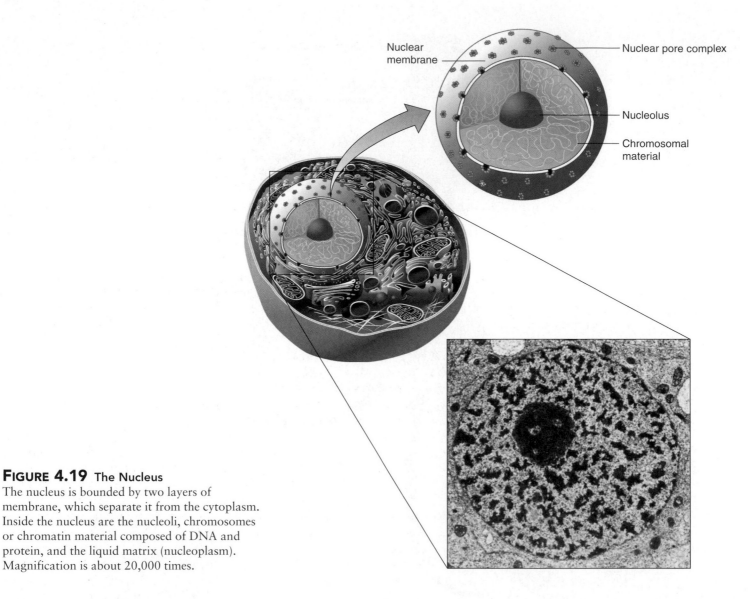

Nuclear membrane

Nuclear pore complex

Nucleolus

Chromosomal material

FIGURE 4.19 The Nucleus
The nucleus is bounded by two layers of membrane, which separate it from the cytoplasm. Inside the nucleus are the nucleoli, chromosomes or chromatin material composed of DNA and protein, and the liquid matrix (nucleoplasm). Magnification is about 20,000 times.

component of the nucleus is its liquid matrix, called the **nucleoplasm.** It is a colloidal mixture composed of water, nucleic acids, the molecules used in the construction of ribosomes, and other nuclear material (figure 4.19).

4.7   Exchange Through Membranes

If a cell is to stay alive, it must be able to exchange materials with its surroundings. Because all cells are surrounded by a plasma membrane, the nature of the membrane influences what materials can pass through it. There are six ways in which materials enter and leave cells: diffusion, osmosis, facilitated diffusion, active transport, endocytosis, and exocytosis. The same mechanisms are involved in the movement of materials across the membranes of the various cellular organelles.

Diffusion

Molecules are in a constant state of motion. Although in solids molecules tend to vibrate in place, in liquids and gases they are able to move past one another. Because the motion of molecules is random, there is a natural tendency in gases and liquids for molecules of different types to mix completely with each other.

Consider a bottle of ammonia. When you open the bottle, the ammonia molecules and air molecules begin to mix and you smell the ammonia. Ammonia molecules leave the bottle and enter the bottle. Molecules from the air enter and leave the bottle. However, more ammonia molecules leave the bottle than enter it. This overall movement is termed **net movement,** the movement in one direction minus the movement in the opposite direction. The direction in which the greatest number of molecules of a particular kind moves (net movement) is determined by the difference in concentration of the molecules

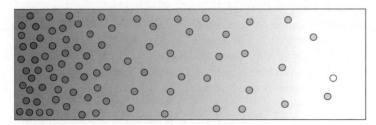

FIGURE 4.20 Concentration Gradient
The difference in concentrations of molecules over a distance is called a concentration gradient. This bar shows a concentration gradient in which the molecules (represented by circles) are more concentrated on the left than on the right.

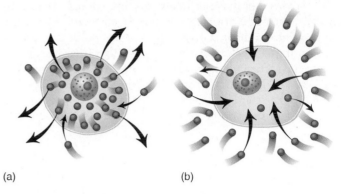

(a)   (b)

FIGURE 4.21 Diffusion
As a result of molecular motion, molecules move from areas where they are concentrated to areas where they are less concentrated. These figures show molecules leaving and entering animal cells by diffusion. The net direction of movement is controlled by concentration (always high to low concentration), and the energy necessary is supplied by the kinetic energy of the molecules themselves. Part *(a)* shows net diffusion out of the cell, whereas *(b)* shows net diffusion into the cell.

in different places. **Diffusion** is the net movement of a kind of molecule from a place where that molecule is in higher concentration to a place where that molecule is less concentrated. The difference in concentration of the molecules over a distance is known as a **concentration gradient** or **diffusion gradient** (figure 4.20). When no concentration gradient exists, the movement of molecules is equal in all directions, and the system has reached a state of **dynamic equilibrium**. There is an equilibrium because there is no longer a net movement (diffusion), because the movement in one direction equals the movement in the other. It is dynamic, however, because the system still has energy, and the molecules are still moving.

The rate at which diffusion takes place is determined by several factors. Diffusion occurs faster if the molecules are small, if they are moving rapidly, and if there is a large concentration gradient.

Diffusion can take place only as long as there are no barriers to the free movement of molecules. In the case of a cell, the plasma membrane surrounds the cell and serves as a partial barrier to the movement of molecules through it. However, the membrane is composed of phospholipid and protein molecules that are in constant motion and form temporary openings that allow some small molecules to cross from one side of the membrane to the other. Other molecules are unable to pass through the membrane by diffusion. If a molecule is able to pass through the membrane, the membrane is **permeable** to the molecules. Because the plasma membrane allows only certain molecules to pass through it, it is **selectively permeable.** A molecule's ability to pass through the membrane depends on its size, electrical charge, and solubility in the phospholipid membrane. In certain cases, the membrane differentiates on the basis of molecular size; that is, the membrane allows small molecules, such as water, to pass through but prevents the passage of larger molecules. The membrane may also regulate the passage of ions. If a particular portion of the membrane has a large number of positive ions on its surface, positively charged ions in the environment will be repelled and prevented from crossing the membrane. Molecules that are able to dissolve in phospholipids, such as vitamins A and D, can pass through the membrane rather easily; however, many molecules cannot pass through at all.

The cell has no control over the rate or direction of diffusion. The direction of diffusion is determined by the relative concentration of specific molecules on the two sides of the membrane, and the energy that causes diffusion to occur is supplied by the kinetic energy of the molecules themselves (figure 4.21). Diffusion is a passive process, which does not require any energy expenditure on the part of the cell.

Diffusion is an important means by which materials are exchanged between a cell and its environment. For example, cells constantly use oxygen in various chemical reactions. Consequently, the oxygen concentration in cells always remains low. The cells, then, contain a lower concentration of oxygen than does the environment outside the cells. This creates a concentration gradient, and the oxygen molecules diffuse from the outside of the cell to the inside.

In large animals, many cells are buried deep within the body; if it were not for the animals' circulatory systems, cells would have little opportunity to exchange gases directly with their surroundings. Oxygen can diffuse into blood through the membranes of the lungs, gills, or other moist surfaces of an animal's body. The circulatory system then transports the oxygen-rich blood throughout the body, and the oxygen automatically diffuses into cells. This occurs because the concentration of oxygen inside cells is lower than that of the blood. The opposite is true of carbon dioxide. Animal cells constantly produce carbon dioxide as a waste product, so there is always a high concentration of it within the cells. These molecules diffuse from the cells into the blood, where the concentration of carbon dioxide is kept constantly low, because the blood is pumped to the moist surfaces (e.g., gills, lungs) and the carbon dioxide again diffuses into the surrounding environment. In a similar manner, many other types of molecules constantly enter and leave cells.

In persons who have difficulty getting enough oxygen to their cells, their condition can be improved by increasing the concentration gradient. Oxygen makes up about 20 percent of the air. If this concentration is artificially raised by supplying a special source of oxygen, diffusion from the lungs to the blood will take place more rapidly. This will help assure that oxygen reaches the body cells that need it, and some of the person's symptoms can be controlled.

Osmosis

Water molecules easily diffuse through cell membranes. **Osmosis** is the net movement (diffusion) of water molecules through a selectively permeable membrane. Although osmosis is important in living things, it will take place in any situation in which there is a selectively permeable membrane and a difference in water concentration in the solutions on opposite sides of the membrane. For example, consider a solution of 90% water and 10% sugar separated by a selectively permeable membrane from a sugar solution of 60% water and 40% sugar (figure 4.22). The membrane allows water molecules to pass freely but prevents the larger sugar molecules from crossing. There is a higher concentration of water molecules in one solution, compared with the concentration of water molecules in the other, so more of the water molecules move from the solution with 90% water to the other solution, with 60% water. Be sure that you recognize (1) that osmosis is really diffusion in which the diffusing substance is water and (2) that the regions of different concentrations are separated by a membrane that is more permeable to water than the substance dissolved in the water.

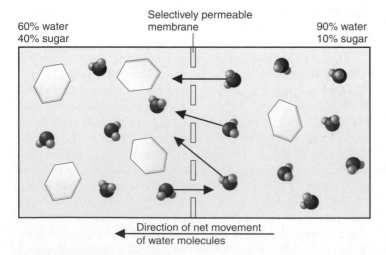

60% water
40% sugar

Selectively permeable
membrane

90% water
10% sugar

Direction of net movement
of water molecules

FIGURE 4.22 Osmosis
When two solutions with different percentages of water are separated by a selectively permeable membrane, there will be a net movement of water from the solution with the highest percentage of water to the one with the lowest percentage of water.

It is important to understand that, when one adds something to a water solution, the percentage of the water in the solution declines. For example, pure water is 100% water. If you add salt to the water, the solution contains both water and salt and the percentage of water is less than 100%. Thus, the more material you add to the solution, the lower the percentage of water.

A proper amount of water is required if a cell is to function efficiently. Too much water in a cell may dilute the cell contents and interfere with the chemical reactions necessary to keep the cell alive. Too little water in the cell may result in a buildup of poisonous waste products. As with the diffusion of other molecules, osmosis is a passive process, because the cell has no control over the diffusion of water molecules. This means that the cell can remain in balance with an environment only if that environment does not cause the cell to lose or gain too much water.

If cells contain a concentration of water and dissolved materials equal to that of their surroundings, the cells are said to be **isotonic** to their surroundings. For example, the ocean contains many kinds of dissolved salts. Organisms such as sponges, jellyfishes, and protozoa are isotonic to the ocean, because the amount of material dissolved in their cellular water is equal to the amount of salt dissolved in the ocean's water.

If an organism is to survive in an environment that has a different concentration of water than does its cells, it must expend energy to maintain this difference. Organisms that live in freshwater have a lower concentration of water (a higher concentration of dissolved materials) than their surroundings and tend to gain water by osmosis very rapidly. They are said to be **hypertonic** to their surroundings, and the surroundings are **hypotonic,** compared with the cells. These two terms are always used to compare two different solutions. The hypertonic solution is the one with more dissolved material and less water; the hypotonic solution has less dissolved material and more water.

Organisms whose cells gain water by osmosis must expend energy to eliminate any excess if they are to keep from swelling and bursting. Similarly, organisms that are hypotonic to their surroundings (have a higher concentration of water than their surroundings) must drink water or their cells will shrink. Most ocean fish are in this situation. They lose water by osmosis to their salty surroundings and must drink seawater to keep their cells from shrinking. Because they are taking in additional salt with the seawater they drink, they must expend energy to excrete this excess salt.

The concept of osmosis is important in surgical and other medical situations. Often, people are given materials by intravenous injections. However, the solutions added must have the right balance between water and dissolved substances, or red blood cells may be injured (figure 4.23).

Under normal conditions, when we drink small amounts of water, the cells of the brain swell a little, and signals are sent to the kidneys to rid the body of excess water. By contrast, persons who are dehydrated, such as marathon runners, may drink large quantities of water in a very short time following a race. This rapid addition of water to the body may

(a) Hypertonic solution

(b) Isotonic solution

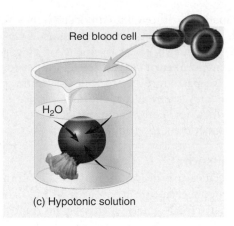
Red blood cell

H_2O

(c) Hypotonic solution

FIGURE 4.23 Osmotic Influences on Cells

The cells in these three diagrams were subjected to three different environments. (*a*) The cell is in a hypertonic solution. Water has diffused from the cell to the environment, because a high concentration of water was in the cell and the cell has shrunk. (*b*) The cell is isotonic to its surroundings. The water concentration inside the red blood cell and the water concentration in the environment are in balance with each other, so the movement of water into the cell equals the movement of water out of the cell, and the cell has its normal shape. (*c*) The cell is in a hypotonic solution. The cell has accumulated water from the environment, because there is a higher concentration of water outside the cell than inside.

TABLE 4.1

Effects of Osmosis on Various Cell Types

Cell Type	What Happens When Cell Is Placed in Hypotonic Solution	What Happens When Cell Is Placed in Hypertonic Solution
With cell wall (e.g., bacteria, fungi, plants)	Water enters the cell, causing it to swell and generate pressure. However, the cell does not burst because the presence of an inelastic cell wall on the outside of the plasma membrane prevents the membrane from stretching and rupturing.	Water leaves the cell and the cell shrinks. The plasma membrane pulls away from inside the cell wall; the cell contents form a small mass.
Without cell wall (e.g., human red blood cells)	Water enters the cell and it swells, causing the plasma membrane to stretch and rupture.	Water leaves the cell and it shrinks into a compact mass.

cause abnormal swelling of brain cells, because the excess water cannot be gotten rid of rapidly enough. If this happens, the person may lose consciousness or even die because the brain cells have swollen too much.

Plant cells also experience osmosis. If the water concentration outside the plant cell is higher than the water concentration inside, more water molecules enter the cell than leave. This creates internal pressure within the cell. But plant cells do not burst, because they are surrounded by a strong cell wall. Lettuce cells that are crisp are ones that have gained water so that there is high internal pressure. Wilted lettuce has lost some of its water to its surroundings, so that it has only slight internal pressure. Osmosis occurs when you put salad dressing on a salad. Because the dressing has a very low water concentration, water from the lettuce diffuses from the cells into the surroundings. Salad that has been "dressed" too long becomes limp and unappetizing (table 4.1).

Controlled Methods of Transporting Molecules

So far, we have considered only situations in which cells have no control over the movement of molecules. Cells cannot rely solely on diffusion and osmosis, however, because many of the molecules they require either cannot pass through the plasma membrane or occur in relatively low concentrations in the cell's surroundings.

Facilitated Diffusion

Some molecules move across the membrane by combining with specific carrier proteins. When the rate of diffusion of a substance is increased in the presence of a carrier, it is called **facilitated diffusion.** Because this movement is still diffusion, the net direction of movement is in accordance with the concentration gradient. Therefore, this is considered a *passive* *transport* method, although it can occur only in living organisms with the necessary carrier proteins. One example of facilitated diffusion is the movement of glucose molecules across the membranes of certain cells. In order for the glucose molecules to pass into these cells, specific proteins are required to carry them across the membrane. The action of the carrier does not require an input of energy other than the molecules' kinetic energy (figure 4.24).

Active Transport

When molecules are moved across the membrane from an area of *low* concentration to an area of *high* concentration, the cell must expend energy. This is the opposite direction molecules move in osmosis and diffusion. The process of using a carrier protein to move molecules up a concentration gradient is called **active transport** (figure 4.25). Active transport is very specific: Only certain molecules or ions can be moved in this way, and they must be carried by specific proteins in the membrane. The action of the carrier requires an input of energy other than the molecules' kinetic energy; therefore, this process is termed *active* transport. For example, some ions, such as sodium and potassium, are actively pumped across plasma membranes. Sodium ions are pumped out of cells up a concentration gradient. Potassium ions are pumped into cells up a concentration gradient. nervous system function, p. 569

Endocytosis and Exocytosis

Larger particles or collections of materials can be transported across the plasma membrane by being wrapped in membrane, rather than passing through the membrane molecule by molecule. When materials enter a cell in this manner, it is called **endocytosis.** When materials are transported out of cells in membrane-wrapped packages, it is known as **exocytosis** (figure 4.26). Endocytosis can be divided into three sorts of activities: phagocytosis, pinocytosis, and receptor mediated endocytosis. **Phagocytosis** is the process of engulfing large particles, such as cells. For example, protozoa engulf food and white blood cells engulf bacteria by wrapping them with membrane and taking them into the cell. Because of this, white blood cells often are called *phagocytes.* When phagocytosis occurs, the material to be engulfed touches the surface of the cell and causes a portion of the outer plasma membrane to be indented. The indented plasma membrane is pinched off inside the cell to form a sac containing the engulfed material. Recall that this sac, composed of a single membrane, is called a vacuole. Once inside the cell, the membrane of the vacuole fuses with the membrane of lysosomes, and the enzymes of the lysosomes break down the contents of the vacuole.

    Pinocytosis is the process of engulfing liquids and the materials dissolved in the liquids. In this form of endocytosis, the sacs that are formed are very small, compared with those formed during phagocytosis. Because of their small size they

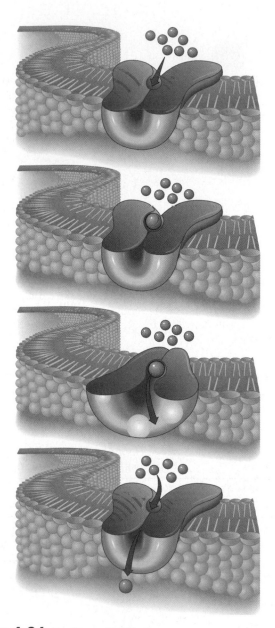

FIGURE 4.24 Facilitated Diffusion

This method of transporting materials across membranes is a diffusion process (i.e., a net movement of molecules from a high to a low concentration). However, the process is helped (facilitated) by a particular membrane protein. The molecules being moved through the membrane attach to a specific transport carrier protein in the membrane. This causes a change in its shape of the protein, which propels the molecule or ion from inside to outside or from outside to inside.

are called vesicles. In fact, an electron microscope is needed in order to see vesicles.

    Receptor mediated endocytosis is the process in which molecules from the cell's surroundings bind to receptor mol-

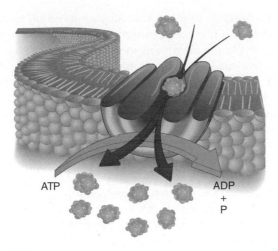

FIGURE 4.25 Active Transport
The action of the carrier protein requires an input of energy (the compound ATP) other than the kinetic energy of the molecules; therefore, this process is termed *active* transport. Active transport mechanisms can transport molecules or ions up a concentration gradient from a low concentration to a higher concentration.

ecules on the plasma membrane. The membrane then folds in and engulfs these molecules. Because receptor molecules are involved, the cell can gather specific necessary molecules from its surroundings and take the molecules into the cell.

Exocytosis occurs in the same manner as endocytosis. Membranous sacs containing materials from the cell migrate to the plasma membrane and fuse with it. This results in the sac contents' being released from the cell. Many materials, such as mucus, digestive enzymes, and molecules produced by nerve cells, are released in this manner.

4.8 Prokaryotic and Eukaryotic Cells Revisited

Now that you have an idea of how cells are constructed, we can look at the great diversity of the kinds of cells that exist. You already know that there are significant differences between prokaryotic and eukaryotic cells.

Because prokaryotic and eukaryotic cells are so different and prokaryotic cells show up in the fossil records much earlier, the differences between the two kinds of cells are

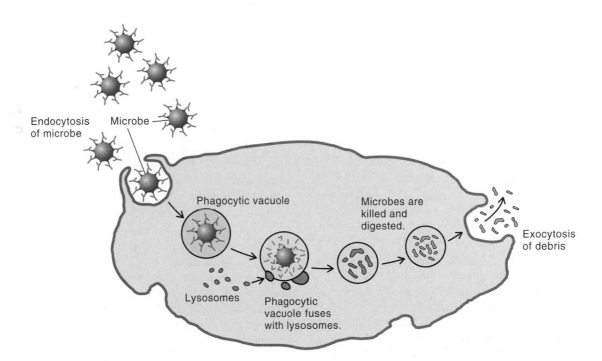

FIGURE 4.26 Endocytosis and Exocytosis
The sequence illustrates a white blood cell engulfing a microbe and surrounding it with a membrane via endocytosis. Once encased in a portion of the plasma membrane (now called a phagocytic vacuole), lysosomes add their digestive enzymes to it, which speeds the breakdown of the contents of the vacuole. Finally, the digested material moves from the vacuole to the inner surface of the plasma membrane, where the contents are discharged by exocytosis.

used to classify organisms. In addition, there are two kinds of organisms that have prokaryotic cells. Thus, biologists have classified organisms into three large categories, called **domains.** The following diagram illustrates how living things are classified:

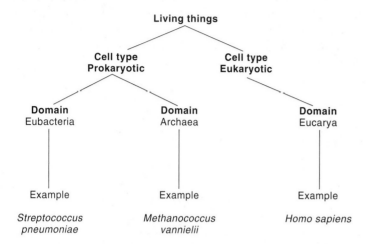

The Domain Eubacteria contains most of the organisms commonly referred to as *bacteria.* The Domain Archaea contains many kinds of organisms that have significant biochemical differences from the Eubacteria. Many of the Archaea have special metabolic abilities and live in extreme environments of high temperature or extreme saltiness. All other living things are based on the eukaryotic cell plan. All the members of the kingdoms Protista (algae and protozoa), Fungi, Plantae (plants), and Animalia (animals) are comprised of eukaryotic cells.

Prokaryotic Cell Structure

Prokaryotic cells, the Eubacteria and Archaea, do not have a typical nucleus bound by a nuclear membrane, nor do they contain mitochondria, chloroplasts, Golgi, or extensive networks of ER. However, prokaryotic cells contain DNA and enzymes and are able to reproduce and engage in metabolism. They perform all of the basic functions of living things with fewer and simpler organelles. Although some Eubacteria have a type of green photosynthetic pigment and carry on photosynthesis, they do so without chloroplasts and use somewhat different chemical reactions. microbes, p. 442

Most Eubacteria are surrounded by a *capsule,* or slime layer, which is composed of a variety of compounds. In certain bacteria, this layer is responsible for their ability to stick to surfaces (including host cells) and to resist phagocytosis. Many bacteria also have fimbriae, hairlike protein structures, which help the cell stick to objects. Those with flagella are capable of propelling themselves through the environment. Below the capsule is the rigid cell wall, comprised of a unique protein/carbohydrate complex called peptidoglycan. This gives the cell the strength to resist osmotic pressure changes and gives it shape. Just beneath the wall is the plasma membrane. Thinner

and with a slightly different chemical composition from that of eukaryotes, the plasma membrane carries out the same functions as the plasma membrane in eukaryotes. Most bacteria are either rod-shaped (bacilli), spherical (cocci), corkscrew-shaped (spirilla), or comma-shaped (vibrio). The genetic material within the cytoplasm is DNA in the form of a loop.

The Archaea share many characteristics with the Eubacteria. Many have a rod or spherical shape, although some are square or triangular. Some have flagella and have cell walls, but the cell walls are made of a different material than that of Eubacteria.

One significant difference between the cells of Eubacteria and Archaea is in the chemical makeup of their ribosomes. The ribosomes of Eubacteria contain different proteins from those found in the cells of Eucarya or Archaea. Bacterial ribosomes are also smaller. This discovery was important to medicine, because many cellular forms of life that cause common diseases are bacterial. As soon as differences in the ribosomes were noted, researchers began to look for ways in which to interfere with the bacterial ribosome's function, but *not* interfere with the ribosomes of eukaryotic cells. **Antibiotics,** such as streptomycin, are the result of this research. This drug combines with bacterial ribosomes and causes bacteria to die because it prevents production of the proteins essential to survival of bacteria. Because eukaryotic ribosomes differ from bacterial ribosomes, streptomycin does not interfere with the normal function of the ribosomes in human cells.

Eukaryotic Cell Structure

Eukaryotic cells contain a true nucleus and most of the membranous organelles described earlier. Eukaryotic organisms can be further divided into several categories, based on the specific combination of organelles they contain. The cells of plants, fungi, protozoa and algae, and animals are all eukaryotic. The most obvious characteristic that sets plants and algae apart from other organisms is their green color, which indicates that the cells contain chlorophyll in chloroplasts. Chlorophyll is necessary for photosynthesis—the conversion of light energy into chemical-bond energy in food molecules. Another distinguishing characteristic of plants and algae is the presence of cellulose in their cell walls (table 4.2).

The fungi are a distinct group of organisms that lack chloroplasts but have a cell wall. However, the cell wall is made from a polysaccharide, called chitin, rather than cellulose. Organisms that belong in this category of eukaryotic cells include yeasts, molds, mushrooms, and the fungi that cause such human diseases as athlete's foot, jungle rot, and ringworm.

Eukaryotic organisms that lack cell walls and chloroplasts are placed in separate groups. Organisms that consist of only one cell are called protozoans—examples are *Amoeba* and *Paramecium.* They have all the cellular organelles described in this chapter except the chloroplast; therefore, protozoans must consume food as do fungi and multicellular animals.

TABLE 4.2

Comparison of Various Kinds of Cells

Prokaryotic Cells		Eukaryotic Cells			
Cells are smaller than eukaryotic cells. DNA is not separated from the cytoplasm by a membrane. Cells have few membranous organelles.		Cells are generally much larger than prokaryotic cells. DNA is found within a nucleus, which is separated from the cytoplasm by a membrane. Cells contain many complex organelles.			
Domain Eubacteria	**Domain Archaea**		**Domain Eucarya**		
		Kingdom Protista	Kingdom Fungi	Kingdom Plantae	Kingdom Animalia
Kingdoms not specified	Kingdoms Euryarchaeota, Korarchaeota, Krenarchaeota				
1. Single-celled organisms 2. Some cause disease. 3. Most are ecologically important. 4. Cyanobacteria are able to perform a kind of photosynthesis.	1. Single-celled organisms 2. They typically generate their own food. 3. Most live in extreme environments.	1. Single-celled organisms commonly called algae and protozoa 2. Some form colonies of cells. 3. Some have cell walls and chloroplasts.	1. Multicellular organisms 2. Cell wall contains chitin. 3. None have chloroplasts. 4. Many kinds of decay organisms and parasites are fungi.	1. Multicellular organisms 2. Cell wall contains cellulose. 3. Chloroplasts are present.	1. Multicellular organisms 2. They do not have a cell wall. 3. They lack chloroplasts.
Examples: *Steptococcus pneumonia* and *Escherichia coli*	Examples: *Methanococcus* and *Thermococcus*	Examples: *Amoeba* and *Spirogyra*	Examples: yeast, molds, and mushrooms	Examples: moss, ferns, cone-bearing trees, and flowering plants	Examples: worms, insects, starfish, frogs reptiles, birds, and mammals

Note: Viruses are not included in this classification system, because viruses are not composed of the basic cellular structural components. They are composed of a core of nucleic acid (DNA or RNA, never both) and a surrounding coat, or capsid, composed of protein. For this reason, viruses are called acellular or noncellular.

Although the differences in these groups of organisms may seem to set them worlds apart, their similarity in cellular structure is one of the central themes unifying the field of biology. One can obtain a better understanding of how cells operate in general by studying specific examples. Because the organelles have the same general structure and function, regardless of the kind of cell in which they are found, we can learn more about how mitochondria function in plants by studying how mitochondria function in animals. There is a commonality among all living things with regard to their cellular structure and function.

Summary

The concept of the cell has developed over a number of years. Initially, only two regions, the cytoplasm and the nucleus, could be identified. At present, numerous organelles are recognized as essential components of both prokaryotic and eukaryotic cell types. The structure and function of some of these organelles are compared in table 4.3. This table also indicates whether the organelle is unique to prokaryotic or eukaryotic cells or is found in both.

The cell is the common unit of life. Individual cells and their structures are studied to discover how they function as individual living organisms and as parts of many-celled beings. Knowing how prokaryotic and eukaryotic cell types resemble each other and differ from each other helps physicians control some organisms dangerous to humans.

There are several ways in which materials enter or leave cells. These include diffusion and osmosis, which involve the net movement of molecules from an area of high to low concentration. In addition, there are several processes that involve activities on the part of the cell to move things across the membrane. These include facilitated diffusion, which uses carrier molecules to diffuse across the membrane; active transport, which uses energy from the cell to move materials from low to high concentration; and endocytosis and exocytosis, in which membrane-enclosed packets are formed.

Key Terms

*Use the interactive flash cards on the **Concepts in Biology**, 12/e website to help you learn the meaning of these terms.*

active transport 88
aerobic cellular respiration 79
antibiotics 90
cell 68
cell theory 68
cell wall 68
cellular membranes 72
centriole 82
chlorophyll 79
chloroplast 79
chromatin 83
chromosome 83
cilia 82
concentration gradient (diffusion gradient) 85
cristae 79
cytoplasm 69
cytoskeleton 81
diffusion 85
domain 90
dynamic equilibrium 85
endocytosis 88
endoplasmic reticulum (ER) 75
eukaryotic cells 69
exocytosis 88
facilitated diffusion 87
flagella 82
fluid-mosaic model 72
Golgi apparatus 75
grana 79
granules 82
hydrophilic 72
hydrophobic 72
hypertonic 86
hypotonic 86
inclusions 82
intermediate filaments 81
isotonic 86
lysosomes 77
microfilaments 81
microtubules 81
mitochondrion 79
net movement 84
nuclear membrane 78
nucleolus 83
nucleoplasm 84
nucleus 69
organelles 69
osmosis 86
peroxisomes 77
phagocytosis 88
photosynthesis 79
pinocytosis 88
plasma membrane (cell membrane) 73
prokaryotic cells 69
protoplasm 69
receptor mediated endocytosis 88
ribosomes 75
selectively permeable 85
signal transduction 75
stroma 79
thylakoid 79
vacuoles 77
vesicles 77

Basic Review

1. The first structure to be distinguished within a cell was the _____.
2. Membranous structures in cells are composed of
   a. phosopholipid.
   b. cellulose.
   c. ribosomes.
   d. chromatin.
3. The Golgi apparatus produces
   a. ribosomes.
   b. DNA.
   c. lysosomes.
   d. endoplasmic reticulum.
4. If a cell has chloroplasts, it is able to carry on photosynthesis. (T/F)

TABLE 4.3

Summary of the Structure and Function of the Cellular Organelles

Organelle	Type of Cell in Which Located	Structure	Function
Plasma membrane	Prokaryotic and eukaryotic	Membranous; typical membrane structure; phospholipid and protein present	Controls passage of some materials to and from the environment of the cell
Inclusions (granules)	Prokaryotic and eukaryotic	Nonmembranous; variable	May have a variety of functions
Chromatin material	Prokaryotic and eukaryotic	Nonmembranous; composed of DNA and proteins	Contains the hereditary information the cell uses in its day-to-day life and passes it on to the next generation of cells
Ribosomes	Prokaryotic and eukaryotic	Nonmembranous; protein and RNA structure	Are the site of protein synthesis
Microtubules, microfilaments, and intermediate filaments	Eukaryotic	Nonmembranous; strands composed of protein	Provide structural support and allow for movement
Nuclear membrane	Eukaryotic	Membranous; double membrane formed into a single container of nucleoplasm and nucleic acids	Separates the nucleus from the cytoplasm
Nucleolus	Eukaryotic	Nonmembranous; group of RNA molecules and DNA located in the nucleus	Is the site of ribosome manufacture and storage
Endoplasmic reticulum	Eukaryotic	Membranous; folds of membrane forming sheets and canals	Is a surface for chemical reactions and intracellular transport system
Golgi apparatus	Eukaryotic	Membranous; stack of single membrane sacs	Is associated with the production of secretions and enzyme activation
Vacuoles and vesicles	Eukaryotic	Membranous; microscopic single membranous sacs	Contain materials
Peroxisomes	Eukaryotic	Membranous; submicroscopic membrane-enclosed vesicle	Contain enzymes to break down hydrogen peroxide and perform other functions
Lysosomes	Eukaryotic	Membranous; submicroscopic membrane-enclosed vesicle	Separate certain enzymes from cell contents
Mitochondria	Eukaryotic	Membranous; double membranous organelle: large membrane folded inside a smaller membrane	Are the site of aerobic cellular respiration associated with the release of energy from food
Chloroplasts	Eukaryotic	Membranous; double membranous organelle: inner membrane contains chlorophyll	Are the site of photosynthesis associated with the capture of light energy and the synthesis of carbohydrate molecules
Centriole	Eukaryotic	Two clusters of nine microtubules	Is associated with cell division
Contractile vacuole	Eukaryotic	Membranous; single-membrane container	Expels excess water
Cilia and flagella	Eukaryotic and prokaryotic	Nonmembranous; prokaryotes composed of a single type of protein arranged in a fiber that is anchored into the cell wall and membrane; eukaryotes consist of tubules in a 9 + 2 arrangement	Cause movement

5. The nucleolus is

   a. where the DNA of the cell is located.

   b. found only in prokaryotic cells.

   c. found in the cytoplasm.

   d. where ribosomes are made and stored.

6. Diffusion occurs

   a. if molecules are evenly distributed.

   b. because of molecular motion.

   c. only in cells.

   d. when cells need it.

7. Prokaryotic cells are larger than eukaryotic cells. (T/F)

8. Osmosis involves the diffusion of _____ through a selectively permeable membrane.

9. The structure of the plasma membrane contains proteins. (T/F)

10. Which one of the following have cell walls made of cellulose?

    a. animals

    b. protozoa

    c. fungi

    d. plants

Answers

1. nucleus   2. a   3. c   4. T   5. d   6. b   7. F   8. water   9. T   10. d

Concept Review

Answers to the Concept Review questions can be found on the Concepts in Biology, 12/e website.

4.1   The Development of the Cell Theory

1. Describe how the concept of the cell has changed over the past 200 years.

2. Define *cytoplasm*.

4.2   Cell Size

3. On the basis of surface area-to-volume ratio, why do cells tend to remain small?

4. What is the advantage of having numerous folds in a cell's outer membrane?

4.3   The Structure of Cellular Membranes

5. What are the differences between the cell wall and the cellular membrane?

6. Describe the structure of cellular membranes based on the fluid-mosaic model.

4.4   Organelles Composed of Membranes

7. List the membranous organelles of a eukaryotic cell and describe the function of each.

8. Define the following terms: *stroma, grana, cristae*.

9. Describe the functions of the plasma membrane.

4.5   Nonmembranous Organelles

10. List the nonmembranous organelles of the cell and describe their functions.

4.6   Nuclear Components

11. Define the following terms: *chromosome, chromatin*.

12. What is a nucleolus?

13. Describe the structure of a chromosome.

4.7   Exchange Through Membranes

14. Describe three methods that allow the exchange of molecules between cells and their surroundings?

15. How do diffusion, facilitated diffusion, osmosis, and active transport differ?

16. Why does putting salt on meat preserve it from spoilage by bacteria?

4.8   Prokaryotic and Eukaryotic Cells Revisited

17. List five differences in structure between prokaryotic and eukaryotic cells.

18. What two types of organisms have prokaryotic cell structure?

Thinking Critically

For guidelines to answer this question, visit the Concepts in Biology, 12/e website.

A primitive type of cell consists of a membrane and a few other cell organelles. This first kind of cell lives in a sea that contains three major kinds of molecules with the following characteristics:

X
Inorganic
High concentration outside cell
Essential to life of cell
Small and can pass through the membrane

Y
Organic
High concentration inside cell
Essential to life of cell
Large and cannot pass through the membrane

Z
Organic
High concentration inside cell
Poisonous to the cell
Small and can pass through the membrane

With this information and your background in cell structure and function, osmosis, diffusion, and active transport, decide whether this kind of cell will continue to live in this sea, and explain why or why not.

Enzymes, Coenzymes, and Energy

Medical professionals primarily concerned with the effect of alcohol on humans assumed that only the liver is able to neutralize alcohol. However, a recent article in the *New England Journal of Medicine* reported the discovery of a stomach enzyme that also neutralizes alcohol. Normally, this enzyme reduces the alcohol level in the stomach by 20%. This, in turn, reduces the amount of alcohol entering the bloodstream and being transported to the brain. However, large amounts of alcohol interfere with the enzyme. In the reported study, the male alcoholics lost some of their ability to digest alcohol in the stomach, and the females experienced a complete loss of the enzyme. The report stated, "For them [the women], drinking alcohol has the same effect as injecting it intravenously."

CHAPTER OUTLINE

5.1   How Cells Use Enzymes

All living things require energy and building materials in order to grow and reproduce. Energy may be in the form of visible light, or it may be in energy-containing covalent bonds found in nutrients. **Nutrients** are molecules required by organisms for growth, reproduction, or repair—they are a source of energy and molecular building materials. The formation, breakdown, and rearrangement of molecules to provide organisms with essential energy and building blocks are known as *biochemical reactions*. Most reactions require an input of energy to get them started. This energy is referred to as **activation energy.** This energy is used to make reactants unstable and more likely to react (figure 5.1).

If organisms are to survive, they must obtain sizable amounts of energy and building materials in a very short time. Experience tells us that the sucrose in candy bars contains the potential energy needed to keep us active, as well as building materials to help us grow (sometimes to excess!). Yet, random chemical processes alone could take millions of years to break down a candy bar, releasing its energy and building materials. Of course, living things cannot wait that long. To sustain life, biochemical reactions must occur at extremely rapid rates. One way to increase the rate of any chemical reaction and make its energy and component parts

available to a cell is to increase the temperature of the reactants. In general, the hotter the reactants, the faster they will react. However, this method of increasing reaction rates has a major drawback when it comes to living things: Organisms die because cellular proteins are denatured before the temperature reaches the point required to sustain the biochemical reactions necessary for life. This is of practical concern to people who are experiencing a fever. Should the fever stay too high for too long, major disruptions of cellular biochemical processes could be fatal.

There is a way of increasing the rate of chemical reactions without increasing the temperature. This involves using a **catalyst,** a chemical that speeds the reaction but is not used up in the reaction. It can be recovered unchanged when the reaction is complete. Catalysts lower the amount of activation energy needed to start the reaction. A cell manufactures specific proteins that act as catalysts. An **enzyme** is a protein molecule that acts as a catalyst to speed the rate of a reaction. Enzymes can be used over and over again until they are worn out or broken. The production of these protein catalysts is under the direct control of an organism's genetic material (DNA). The instructions for the manufacture of all enzymes are found in the genes of the cell. Organisms make their own enzymes. How the genetic information is used to direct the synthesis of these specific protein molecules will be discussed in chapter 7.

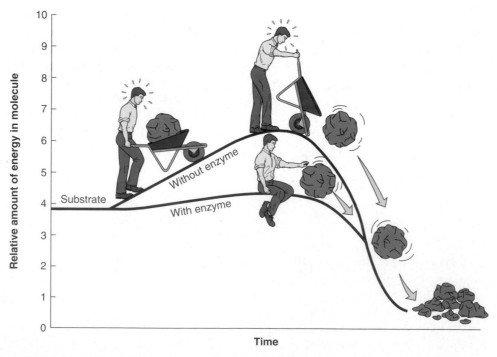

FIGURE 5.1 The Lowering of Activation Energy

Enzymes operate by lowering the amount of energy needed to get a reaction going—the activation energy. When this energy is lowered, the nature of the bonds is changed, so they are more easily broken. Although the figure shows the breakdown of a single reactant into many end products (as in a hydrolysis reaction), the lowering of activation energy can also result in bonds being broken so that new bonds may be formed in the construction of a single, larger end product from several reactants (as in a synthesis reaction).

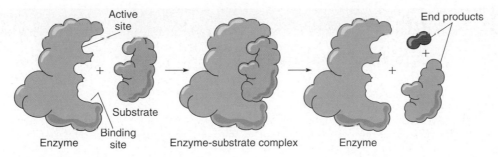

FIGURE 5.2 Enzyme-Substrate Complex Formation

During an enzyme-controlled reaction, the enzyme and substrate come together to form a new molecule—the enzyme-substrate complex molecule. This molecule exists for only a very short time. During that time, the activation energy is lowered and bonds are changed. The result is the formation of a new molecule or molecules, called the end products of the reaction. Notice that the enzyme comes out of the reaction intact and ready to be used again.

5.2 How Enzymes Speed Chemical Reaction Rates

As the instructions for the production of an enzyme are read from the genetic material, a specific sequence of amino acids is linked together at the ribosomes. Once bonded, the chain of amino acids folds and twists to form a globular molecule with a particular three-dimensional shape.

Enzymes Bind to Substrates

It is the nature of its three-dimensional shape that allows an enzyme to combine with a reactant and lower the activation energy. Each enzyme has a specific size and three-dimensional shape, which in turn is specific to the kind of reactant with which it can combine. The enzyme physically fits with the reactant. The molecule to which the enzyme attaches itself (the reactant) is known as the **substrate.** When the enzyme attaches itself to the substrate molecule, a new, temporary molecule—the **enzyme-substrate complex**—is formed (figure 5.2). When the substrate is combined with the enzyme, its chemical bonds are less stable and more likely to be altered and form new bonds. The enzyme is specific because it has a particular shape, which can combine only with specific parts of certain substrate molecules (Outlooks 5.1).

You can think of an enzyme as a tool that makes a job easier and faster. For example, the use of an open-end crescent wrench can make the job of removing or attaching a nut and bolt go much faster than doing the same job by hand. To do this job, the proper wrench must be used. Just any old tool (screwdriver or hammer) won't work! The enzyme must also physically attach itself to the substrate; therefore, there is a specific **binding site,** or **attachment site,** on the enzyme surface. Figure 5.3 illustrates the specificity of both wrench and enzyme. Note that the wrench and enzyme are recovered unchanged after they have been used. This means that the enzyme and wrench can be used again. Eventually, like wrenches, enzymes wear out and have to be replaced by syn-

thesizing new ones using the instructions provided by the cell's genes. Generally, only very small quantities of enzymes are necessary, because they work so fast and can be reused.

Both enzymes and wrenches are specific in that they have a particular surface geometry, or shape, which matches the geometry of their respective substrates. Note that both the enzyme and the wrench are flexible. The enzyme can bend or fold to fit the substrate, just as the wrench can be adjusted to fit the nut. This is called the *induced fit hypothesis.* The fit is induced because the presence of the substrate causes the enzyme to mold or adjust itself to the substrate as the two come together.

The **active site** is the place on the enzyme that causes a specific part of the substrate to change. It is the place where chemical bonds are formed or broken. (Note in the case illustrated in figure 5.3 that the active site is the same as the binding site. This is typical of many enzymes.) This site is where the activation energy is lowered and the electrons are shifted to change the bonds. The active site may enable a positively charged surface to combine with the negative portion of a reactant. Although the active site molds itself to a substrate, enzymes cannot fit all substrates. Enzymes are specific to certain substrates or a group of very similar substrate molecules. One enzyme cannot speed the rate of all types of biochemical reactions. Rather, a special enzyme is required to control the rate of each type of reaction occurring in an organism.

Naming Enzymes

Because an enzyme is specific to both the substrate to which it can attach and the reaction it can encourage, a unique name can be given to each enzyme. The first part of an enzyme's name is the name of the molecule to which it can become attached. The second part of the name indicates the type of reaction it facilitates. The third part of the name is "-ase," the ending that indicates it is an enzyme. For example, *DNA polymerase* is the name of the enzyme that attaches to the molecule DNA and is responsible for increasing its length through a polymerization reaction. A few enzymes (e.g., pepsin and

OUTLOOKS 5.1

Enzymes and Stonewashed "Genes"

The popularity of stonewashed jeans grew dramatically in the late 1960s. To get the stonewashed effect, the denim was washed with stones; the stones rubbed against the denim, wearing the blue dye off the surface of the material. But the stones also damaged the cotton fibers, which shortened the life of the fabric, a feature that many consumers found unac-

ceptable. Now, to create the stonewashed look and still maintain strong cotton fibers, enzymes are used to "digest," or hydrolyze, the blue dye on the surface of the fabric. Because the enzyme is substrate or dye specific, the cotton fibers are not harmed.

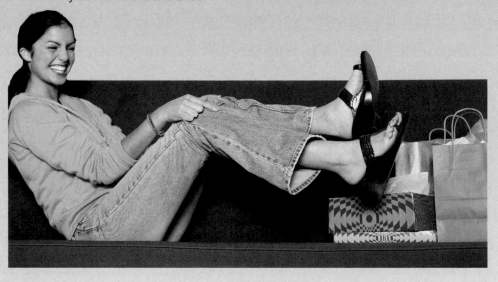

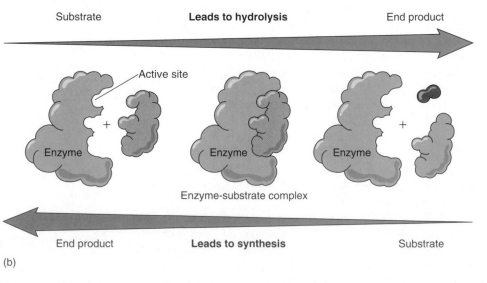

(a)

FIGURE 5.3 It Fits, It's Fast, and It Works

(a) Although removing the wheel from this bicycle could be done by hand, using an open-end crescent wrench is more efficient. The wrench is adjusted and attached, temporarily forming a nut-bolt-wrench complex. Turning the wrench loosens the bonds holding the nut to the bolt and the two are separated. Using the wrench makes the task much easier. *(b)* An enzyme will "adjust itself" as it attaches to its substrate, forming a temporary enzyme-substrate complex. The presence and position of the enzyme in relation to the substrate lowers the activation energy required to alter the bonds.

trypsin) are still referred to by their original names. The enzyme responsible for the dehydration synthesis reactions among several glucose molecules to form glycogen is known as *glycogen synthetase*. The enzyme responsible for breaking the bond that attaches the amino group to the amino acid arginine is known as *arginine aminase*. When an enzyme is very common, its formal name is shortened: The salivary enzyme involved in the digestion of starch is *amylose* (starch) *hydrolase*; it is generally known as *amylase*. Other enzymes associated with the human digestive system are noted in table 25.2.

5.3 Cofactors, Coenzymes, and Vitamins

Certain enzymes need an additional molecule to help them function. **Cofactors** are inorganic ions or organic molecules that serve as enzyme helpers. Ions such as zinc, iron, and magnesium assist enzymes in their performance as catalysts. These ions chemically combine with the enzyme. A **coenzyme** is an organic molecule that functions as a cofactor. Organic cofactors are made from molecules such as certain amino

acids, nitrogenous bases, and vitamins. **Vitamins** are a group of unrelated organic molecules used in the making of certain coenzymes; they also play a role in regulating gene action. Vitamins are either water-soluble (e.g., vitamin B complex) or fat-soluble (e.g., vitamin A). For example, the vitamin riboflavin (B_2) is metabolized by cells and becomes **flavin adenine dinucleotide (FAD)**. The vitamin niacin is metabolized by cells and becomes **nicotinamide adenine dinucleotide (NAD$^+$)**. Coenzymes such as NAD$^+$ and FAD are used to carry electrons to and from many kinds of oxidation-reduction reactions. NAD$^+$, FAD, and other coenzymes are bonded only temporarily to their enzymes and therefore can assist various enzymes in their reaction. Coenzymes and vitamins are required because cells are not able to manufacture these molecules (figure 5.4).

Another vitamin, pantothenic acid, becomes coenzyme A (CoA), a molecule used to carry a specific 2-carbon molecule, acetyl, generated in one reaction to another reaction. Like enzymes, the cell uses inorganic cofactors, coenzymes, and vitamins repeatedly until these molecules are worn out and destroyed. Coenzymes play vital roles in metabolism. Without them, most cellular reactions would come to an end and the cell would die. electron transport, p. 115

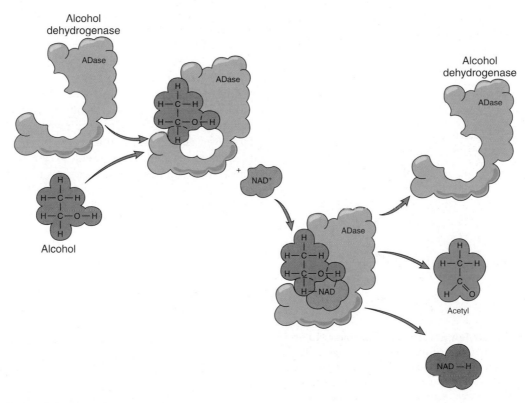

FIGURE 5.4 The Role of Coenzymes
NAD is a coenzyme that works with the enzyme alcohol dehydrogenase (ADase) during the breakdown of alcohol. The coenzyme carries the hydrogen from the alcohol molecule after it is removed by the enzyme. Notice that the hydrogen on the alcohol is picked up by the NAD. The use of the coenzyme NAD makes the enzyme function more efficiently, because one of the end products of this reaction (hydrogen) is removed from the reaction site. Because the hydrogen is no longer close to the reacting molecules, the overall direction of the reaction is toward the formation of acetyl. This encourages more alcohol to be broken down.

5.4 How the Environment Affects Enzyme Action

An enzyme forms a complex with one substrate molecule, encourages a reaction to occur, detaches itself, and then forms a complex with another molecule of the same substrate. The number of molecules of substrate with which a single enzyme molecule can react in a given time (e.g., reactions per minute) is called the **turnover number.**

Sometimes, the number of jobs an enzyme can perform during a particular time period is incredibly large—ranging between a thousand (10^3) and 10 thousand trillion (10^{16}) times faster per minute than uncatalyzed reactions. Without the enzyme, perhaps only 50 or 100 substrate molecules might be altered in the same time. With this in mind, let's identify the ideal conditions for an enzyme and consider how these conditions influence the turnover number.

Temperature

An important environmental condition affecting enzyme-controlled reactions is temperature (figure 5.5), which has two effects on enzymes: (1) It can change the rate of molecular motion, and (2) it can cause changes in the shape of an enzyme. An increase in temperature increases molecular motion. Therefore, as the temperature of an enzyme-substrate system increases, the amount of product molecules formed increases, up to a point. The temperature at which the rate of formation of enzyme-substrate complex is fastest is termed the *optimum temperature. Optimum* means the best or most

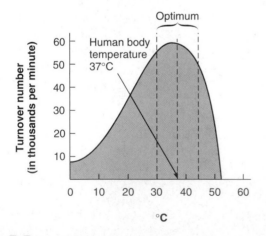

FIGURE 5.5 The Effect of Temperature on Turnover Number As the temperature increases, the turnover number increases. The increasing temperature increases molecular motion and may increase the number of times an enzyme contacts and combines with a substrate molecule. Temperature may also influence the shape of the enzyme molecule, making it fit better with the substrate. At high temperatures, the enzyme molecule is irreversibly changed, so that it can no longer function as an enzyme. At that point, it has been denatured. Notice that the enzyme represented in this graph has an optimum (best) temperature range of between 30°C and 45°C.

productive quantity or condition. In this case, the optimum temperature is the temperature at which the product is formed most rapidly. As the temperature decreases below the optimum, molecular motion slows, and the rate at which the enzyme-substrate complexes form decreases. Even though the enzyme is still able to operate, it does so very slowly. Foods can be preserved by storing them in freezers or refrigerators because the enzyme-controlled reactions of the food and spoilage organisms are slowed at lower temperatures.

When the temperature is raised above the optimum, some of the enzyme molecules are changed in such a way that they can no longer form the enzyme-substrate complex; thus, the reaction slows. If the temperature continues to increase, more and more of the enzyme molecules become inactive. If the temperature is high enough, it causes permanent changes in the three-dimensional shape of the molecules. The surface geometry of the enzyme molecule is not recovered, even when the temperature is reduced. Recall the wrench analogy. When a wrench is heated above a certain temperature, the metal begins to change shape. The shape of the wrench is changed permanently, so that, even if the temperature is reduced, the surface geometry of the end of the wrench is permanently lost. When this happens to an enzyme, it has been *denatured*. A denatured enzyme is one whose protein structure has been permanently changed, so that it has lost its original biochemical properties. Because enzymes are molecules and are not alive, they are not killed but, rather, denatured. For example, although egg white is not an enzyme, it is a protein and provides a common example of what happens when denaturation occurs as a result of heating. As heat is applied to the egg white, it is permanently changed from a runny substance to a rubbery solid (denatured).

Many people have heard that fevers cause brain damage in children. Brain damage from a fever can result from the denaturation of proteins if the fever is over 42°C (107.6°F). However, denaturation and brain damage from fevers is rare, because untreated fevers seldom go over 40.5°C (105°F) unless the child is overdressed or trapped in a hot place. Generally, the brain's thermostat will stop the fever from going above (41.1°C) 106°F .

pH

Another environmental condition that influences enzyme action is pH. The three-dimensional structure of a protein leaves certain side chains exposed. These side chains may attract ions from the environment. Under the right conditions, a group of positively charged hydrogen ions may accumulate on certain parts of an enzyme. In an environment that lacks these hydrogen ions, this would not happen. Thus, variation in the enzyme's shape could be caused by a change in the number of hydrogen ions present in the solution. Because the environmental pH is so important in determining the shapes of protein molecules, there is an optimum pH for each specific enzyme. The enzyme will fit with the substrate only when it has the proper shape, and it has the proper shape only when it is at the right pH. Many enzymes function best at a

FIGURE 5.6 The Effect of pH on the Turnover Number

As the pH changes, the turnover number changes. The ions in solution alter the environment of the enzyme's active site and the overall shape of the enzyme. The enzymes illustrated here are human amylase, pepsin, and trypsin. Amylase is found in saliva and is responsible for hydrolyzing starch to glucose. Pepsin is found in the stomach and hydrolyzes protein. Trypsin is produced in the pancreas and enters the small intestine, where it also hydrolyzes protein. Notice that each enzyme has its own pH range of activity, the optimum (shown in the color bars) being different for each.

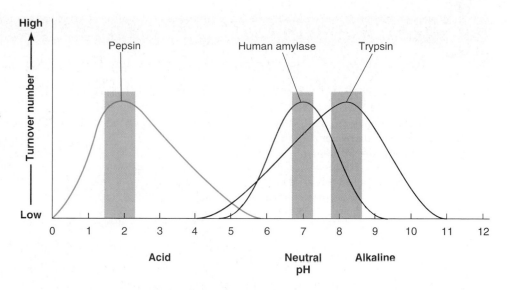

pH close to neutral (7). However, a number of enzymes perform best at pHs quite different from 7. Pepsin, an enzyme found in the stomach, works well at an acid pH of 1.5 to 2.2, whereas arginase, an enzyme in the liver, works well at a basic pH of 9.5 to 9.9 (figure 5.6).

Enzyme-Substrate Concentration

In addition to temperature and pH, the concentration of enzymes, substrates, and products influences the rates of enzymatic reactions. Although the enzyme and the substrate are in contact with one another for only a short time, when there are huge numbers of substrate molecules it may happen that all the enzymes present are always occupied by substrate molecules. When this occurs, the rate of product formation cannot be increased unless the number of enzymes is increased. Cells can do this by synthesizing more enzymes. However, just because there are more enzyme molecules does not mean that any one enzyme molecule will work any faster. The turnover number for each enzyme stays the same. As the enzyme concentration increases, the amount of product formed increases in a specified time. A greater number of enzymes are turning over substrates; they are not turning over substrates faster. Similarly, if enzyme numbers are decreased, the amount of product formed declines.

We can also look at this from the point of view of the substrate. If substrate is in short supply, enzymes may have to wait for a substrate molecule to become available. Under these conditions, as the amount of substrate increases, the amount of product formed increases. The increase in product is the result of more substrates' being available to be changed. When there is a very large amount of substrate, all the enzymes will be occupied all the time. However, if given enough time, even a small amount of enzyme can eventually change all the substrate to product; it just takes longer. To see how abnormal enzyme activity may result in a metabolic disorder, see How Science Works 5.1.

5.5 Cellular Control Processes and Enzymes

In any cell, there are thousands of kinds of enzymes. Each controls specific chemical reactions and is sensitive to changing environmental conditions, such as pH and temperature. For a cell to stay alive in an ever-changing environment, its countless chemical reactions must be controlled. Recall from chapter 1 that control processes are mechanisms that ensure that an organism will carry out all metabolic activities in the proper sequence (coordination) and at the proper rate (regulation). The *coordination* of enzymatic activities in a cell results when specific reactions occur in a given sequence—for example, $A \rightarrow B \rightarrow C \rightarrow D \rightarrow E$. This ensures that a particular nutrient will be converted to a particular end product necessary to the survival of the cell. Should a cell be unable to coordinate its reactions, essential products might be produced at the wrong time or never be produced at all, and the cell would die. The *Regulation* of biochemical reactions is the way a cell controls the amount of chemical product produced. The expression "having too much of a good thing" applies to this situation. For example, if a cell manufactures too much lipid, the presence of those molecules could interfere with other life-sustaining reactions, resulting in the cell's death. On the other hand, if a cell does not produce enough of an essential molecule, such as a hydrolytic (digestive) enzyme, it might also die. The cellular-control process involves both enzymes and genes.

Enzymatic Competition for Substrates

Enzymatic competition results whenever there are several kinds of enzymes available to combine with the same kind of substrate molecule. Although all these different enzymes may combine with the same substrate, they do not have the same chemical effect on the substrate, because each converts the substrate to different end products. For example, acetyl is a

HOW SCIENCE WORKS 5.1

Metabolic Disorders Related to Enzyme Activity—Fabray's Disease and Gaucher Disease

Fabray's disease is a fat-storage disorder caused by a deficiency of an enzyme known as ceramidetrihexosidase, also called alpha-galactosidase A. This enzyme is involved in the breakdown of lipids. The gene for the production of this enzyme is located on the X chromosome. Normally, a woman has 2 X chromosomes; if 1 of these chromosomes contains this abnormal form of the gene, she is considered to be a "carrier" of this trait. Some carriers show cloudiness of the cornea of their eyes. Normally, males have 1 X and 1 Y chromosome. Therefore, if their mother is a carrier, they have a 50:50 chance of inheriting this trait from their mother. Males with this abnormality have burning sensations in their hands and feet, which become worse when they exercise and in hot weather. Most have small, raised, reddish-purple blemishes on their skin. As they grow older, they are at risk for strokes, heart attacks, and kidney damage. Some affected people develop gastrointestinal problems. They have frequent bowel movements shortly after eating. It is hoped that enzyme replacement and eventually gene therapy will allow patients to control, if not eliminate, the symptoms of Fabray's disease.

Gaucher disease is an inherited, enzyme deficiency disorder. People with this disease have a deficiency in the enzyme glucocerebrosidase, which is necessary for the breakdown of the fatty acid glucocerebroside. People with Gaucher disease cannot break down this fatty acid as they should; instead, it becomes abnormally stored in certain cells of the bone marrow, spleen, and liver. People may experience enlargement of the liver and spleen and bone pain, degeneration, and fractures. They may also show symptoms of anemia, fatigue, easy bruising, and a tendency to bleed. Gaucher disease is diagnosed through DNA testing, which identifies certain mutations in the glucocerebrosidase gene on chromosome 1. In the past, the treatment for Gaucher disease has relied on periodic blood transfusions, partial or total spleen removal, and pain relievers. More recently, however, enzyme replacement therapy has been used. This treatment relies on a chemically modified form of the enzyme glucocerebrosidase that has been specifically targeted to bone cells.

---

substrate that can be acted upon by three different enzymes: citrate synthetase, fatty acid synthetase, and malate synthetase (figure 5.7). Which enzyme has the greatest success depends on the number of each type of enzyme available and the suitability of the environment for the enzyme's operation. The enzyme that is present in the greatest number or is best suited to the job in the environment of the cell wins, and the amount of its end product becomes greatest.

Gene Regulation

The number and kind of enzymes produced are regulated by the cell's genes. It is the job of chemical messengers to inform the genes as to whether specific enzyme-producing genes should be turned on or off or whether they should have their protein-producing activities increased or decreased. **Gene-regulator proteins** are chemical messengers that inform the genes of the cell's need for enzymes. Gene-regulator proteins that decrease protein production are called *gene-repressor*

proteins, whereas those that increase protein production are *gene-activator proteins*. Look again at figure 5.7. If the cell were in need of protein, gene-regulator proteins could increase the amount of malate synthetase. This would result in an increase in the amount of acetyl being converted to malate. The additional malate would then be modified into one of the amino acids needed to produce the needed protein. On the other hand, if the cell required energy, an increase in the amount of citrate synthetase would cause more acetyl to be metabolized to release this energy. When the enzyme fatty acid synthetase is produced in greater amounts, it outcompetes the other two; the acetyl is used in fat production and storage.

Inhibition

An **inhibitor** is a molecule that attaches itself to an enzyme and interferes with that enzyme's ability to form an enzyme-substrate complex. For example, one of the early kinds of pesti-

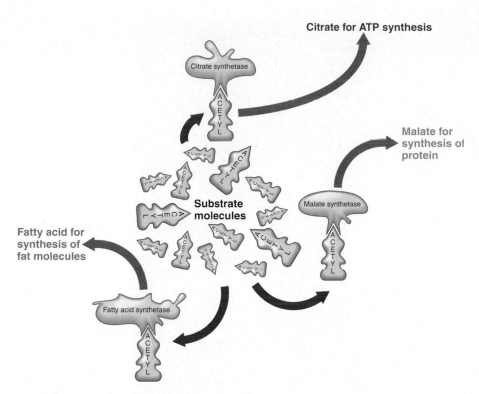

FIGURE 5.7 Enzymatic Competition

Acetyl can serve as a substrate for a number of reactions. Three such reactions are shown here. Whether it becomes a fatty acid, malate, or citrate is determined by the enzymes present. Each of the three enzymes can be thought of as being in competition for the same substrate— the acetyl molecule. The cell can partially control which end product will be produced in the greatest quantity by producing greater numbers of one kind of enzyme and fewer of the other kinds. If citrate synthetase is present in the highest quantity, more of the acetyl substrate will be acted upon by that enzyme and converted to citrate, rather than to the other two end products, malate and fatty acids.

cides used to spray fruit trees contained arsenic. The arsenic attached itself to insect enzymes and inhibited the normal growth and reproduction of insects. Organophosphates are pesticides that inhibit several enzymes necessary for the operation of the nervous system. When they are incorporated into nerve cells, they disrupt normal nerve transmission and cause the death of the affected organisms (figure 5.8). In humans, death that is due to pesticides is usually caused by uncontrolled muscle contractions, resulting in breathing failure.

Competitive Inhibition

Some inhibitors have a shape that closely resembles the normal substrate of the enzyme. The enzyme is unable to distinguish the inhibitor from the normal substrate, so it combines with either or both. As long as the inhibitor is combined with an enzyme, the enzyme is ineffective in its normal role. Some of these enzyme-inhibitor complexes are permanent. An inhibitor removes a specific enzyme as a functioning part of the cell: The reaction that enzyme catalyzes no longer occurs, and none of the product is formed. This is termed **competitive inhibition** because the inhibitor molecule competes with the normal substrate for the active site of the enzyme (figure 5.9).

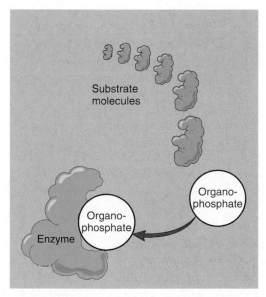

No end product is released

FIGURE 5.8 Inhibition of Enzyme at Active Site

Organophosphate pesticides are capable of attaching to the enzyme acetylcholinesterase, preventing it from forming an enzyme substrate complex with its regular substrate.

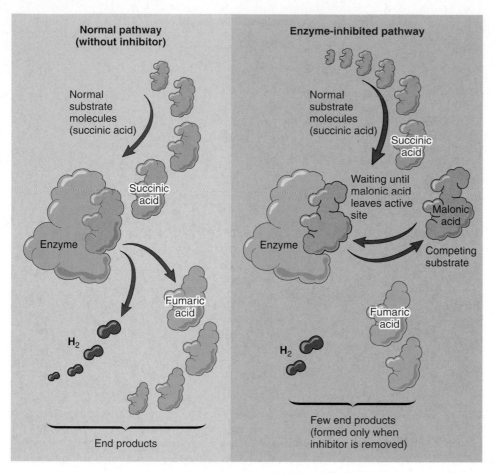

FIGURE 5.9 Competitive Inhibition
The left-hand side of the illustration shows the normal functioning of the enzyme. On the right-hand side, the enzyme is unable to attach to succinic acid. This is because an inhibitor, malonic acid, is attached to the enzyme and prevents the enzyme from forming the normal complex with succinic acid. As long as malonic acid stays attached in the active site, the enzyme will be unable to produce fumaric acid. If the malonic acid is removed, the enzyme will begin to produce fumaric acid again. Its attachment to the enzyme in this case is not permanent but, rather, reduces the number of product molecules formed per unit of time, its turnover number.

Scientists use their understanding of enzyme inhibition to control disease. For instance, sulfa drugs are used to control a variety of bacteria, such as the bacterium *Streptococcus pyogenes,* the cause of strep throat and scarlet fever. The drug resembles one of the bacterium's necessary substrates and, when it attaches to the active site, it prevents some of the cell's enzymes from producing an essential cell component. As a result, the bacterial cell dies, because its normal metabolism is not maintained. Because people do not normally produce this enzyme, they are not harmed by sulfa drugs.

Negative-Feedback Inhibition

Negative-feedback inhibition is another method of controlling the synthesis of many molecules within a cell. This control process occurs within an enzyme-controlled reaction sequence. As the number of end products increases, some product molecules *feed back* to one of the previous reactions and have a *negative effect* on the enzyme controlling that reaction; that is, they *inhibit*, or prevent, that enzyme from performing at its best.

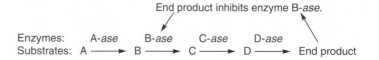

If the enzyme is inhibited, the end product can no longer be produced at the same rapid rate, and its concentration falls. When there are too few end product molecules to have a negative effect on the enzyme, the enzyme is no longer inhibited. The enzyme resumes its previous optimum rate of operation, and the end product concentration begins to increase. With this kind of regulation, the amount of the product rises and falls within a certain range and never becomes too large or small.

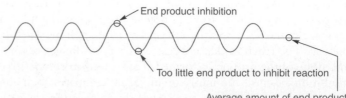

5.6 | Enzymatic Reactions Used in Processing Energy and Matter

All living organisms require a constant supply of energy to sustain life. They obtain this energy through enzyme-controlled chemical reactions, which release the internal potential energy stored in the chemical bonds of molecules (figure 5.10). Burning wood is a chemical reaction that results in the release of energy by breaking chemical bonds. The chemical bonds of cellulose are broken, and smaller end prod-ucts of carbon dioxide (CO_2) and water (H_2O) are produced. There is less potential energy in the chemical bonds of carbon dioxide and water than in the complex organic cellulose molecules, and the excess energy is released as light and heat.

Biochemical Pathways

In living things, energy is also released but it is released in a series of small steps and each controlled by a specific enzyme. Each step begins with a substrate, which is converted to a product, which in turn becomes the substrate for a different enzyme. Such a series of enzyme-controlled reactions is called a **biochemical pathway,** or a **metabolic pathway.** The processes of photosynthesis, respiration, protein synthesis, and many other cellular activities consist of a series of bio-chemical pathways. Biochemical pathways that result in the breakdown of compounds are generally referred to as **catab-olism.** Biochemical pathways that result in the synthesis of new, larger compounds are known as **anabolism.** Figure 5.11 illustrates the nature of biochemical pathways. enzymes, p. 97

One of the amazing facts of nature is that most organisms use the same basic biochemical pathways. Thus, the reactions in an elephant are essentially the same as those in a shark, a petunia, and an earthworm. However, because the kinds of enzymes an organism is able to produce depends on its genes, some variation occurs in the details of the biochemical path-ways of different organisms. The fact that so many organisms use essentially the same biochemical processes is a strong argument for the idea of evolution from a common ancestor. Once a successful biochemical strategy evolved, the genes and the pathway were retained (conserved) by evolutionary descendents, with slight modifications of the scheme.

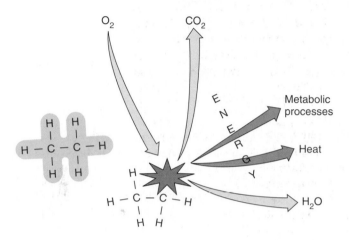

FIGURE 5.10 Life's Energy: Chemical Bonds
All living things use the energy contained in chemical bonds. As organisms break down molecules, they can use the energy released for metabolic processes, such as movement, growth, and reproduction. In all cases, there is a certain amount of heat released when chemical bonds are broken.

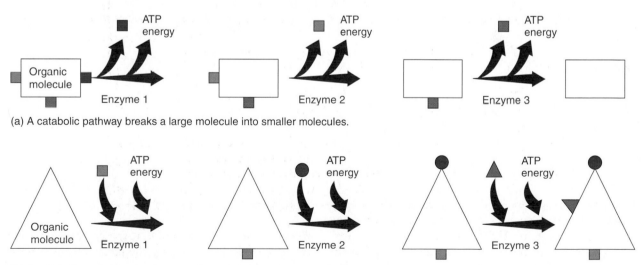

(a) A catabolic pathway breaks a large molecule into smaller molecules.

(b) An anabolic pathway combines smaller molecules to form a larger molecule.

FIGURE 5.11 Biochemical Pathways
Biochemical pathways are the result of a series of enzyme-controlled reactions. In each step, a substrate is acted upon by an enzyme to produce a product. The product then becomes the substrate for the next enzyme in the chain of reactions. Such pathways can be used to break down molecules, build up molecules, release energy, and perform many other actions.

Generating Energy in a Useful Form: ATP

The transfer of chemical energy within living things is handled by a nucleotide known as **adenosine triphosphate (ATP)**. Chemical energy is stored when ATP is made and is released when it is broken apart. An ATP molecule is composed of a molecule of adenine (a nitrogenous base), ribose (a sugar), and 3 phosphate groups (figure 5.12). If only 1 phosphate is present, the molecule is known as adenosine monophosphate (AMP), an RNA nucleotide.

When a second phosphate group is added to the AMP, a molecule of adenosine **di**phosphate (ADP) is formed. The ADP, with the addition of even more energy, is able to bond to a third phosphate group and form ATP. (Recall from chapter 3 that the addition of phosphate to a molecule is called a *phosphorylation reaction.*) The bonds holding the last 2 phosphates to the molecule are easily broken to release energy for cellular processes that require energy. Because the bond between these phosphates is so easy for a cell to use, it is called a **high-energy phosphate bond**. These bonds are often shown as solid, curved lines in diagrams. Both ADP and ATP, because they contain high-energy bonds, are very unstable molecules and readily lose their phosphates. When this occurs, the energy held in the phosphate's high-energy bonds can be transferred to a lower-energy molecule or released to the environment. Within a cell, specific enzymes (phosphorylases) speed this release of energy as ATP is broken down to ADP and P (phosphate). When the bond holding the third phosphate of an ATP molecule is broken, energy is released for use in other activities. phosphorylation reactions, p. 37

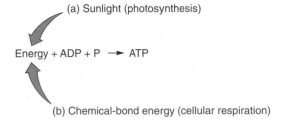

FIGURE 5.12 Adenosine Triphosphate (ATP)
An ATP molecule is an energy carrier. A molecule of ATP consists of several subunits: a molecule of adenine, a molecule of ribose, and 3 phosphate groups. The 2 end phosphate groups are bonded together by high-energy bonds. These bonds are broken easily, so they release a great amount of energy. Because they are high-energy bonds, they are represented by curved, solid lines.

When energy is being harvested from a chemical reaction or another energy source, such as sunlight, it is stored when a phosphate is attached to an ADP to form ATP.

An analogy that might be helpful is to think of each ATP molecule used in the cell as a rechargeable battery. When the power has been drained, it can be recharged numerous times before it must be recycled (figure 5.13).

Electron Transport

Another important concept that can be applied to many different biochemical pathways is the mechanism of *electron transport*. Because the electrons of an atom are on its exterior, the electrons in the outer energy level can be lost more easily to the surroundings, particularly if they receive additional energy and

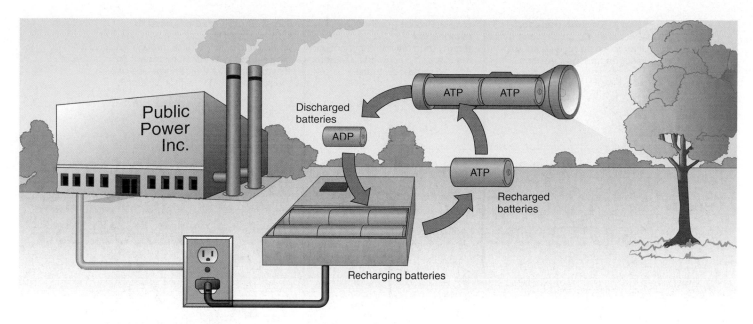

FIGURE 5.13 ATP: The Power Supply for Cells

When rechargeable batteries in a flashlight have been drained of their power, they can be recharged by placing them in a specially designed battery charger. This enables the right amount of power from a power plant to be packed into the batteries for reuse. Cells operate in much the same manner. When the cell's "batteries," ATP, are drained while powering a job, such as muscle contraction, the discharged "batteries," ADP, can be recharged back to full ATP power.

move to a higher energy level. When they fall back to their original position, they give up that energy. This activity takes place whenever electrons gain or lose energy. In living things, such energy changes are harnessed by special molecules that capture such "excited" electrons which can be transferred to other chemicals. These electron-transfer reactions are commonly called *oxidation-reduction reactions.* In oxidation-reduction (redox) reactions, the molecules losing electrons become oxidized and those gaining electrons become reduced. The molecule that loses the electron loses energy; the molecule that gains the electron gains energy. redox, p. 36

There are many different electron acceptors or carriers in cells. However, the three most important are the coenzymes: *n*icotinamide *a*denine *d*inucleotide (NAD$^+$), *n*icotinamide *a*denine *d*inucleotide *p*hosphate (NADP$^+$), and *f*lavin *a*denine *d*inucleotide (FAD). Recall that niacin is needed to make NAD$^+$ and NADP$^+$ and the riboflavin is needed to make FAD. Because NAD$^+$, NADP$^+$, FAD, and similar molecules accept and release electrons, they are often involved in oxidationreduction reactions. When NAD$^+$, NADP$^+$, and FAD accept electrons, they become negatively charged. Thus, they readily pick up hydrogen ions (H$^+$), so when they become reduced they are shown as NADH, NADPH, and FADH$_2$. Therefore, it is also possible to think of these molecules as *hydrogen carriers.* In many biochemical pathways, there is a series of enzyme controlled oxidation-reduction reactions (electron-transport reactions) in which each step results in the transfer of a small amount of energy from a higher-energy molecule to a lower-energy molecule (figure 5.14). Thus, electron transport is often tied to the formation of ATP.

Proton Pump

In many of the oxidation-reduction reactions that take place in cells, the electrons that are transferred come from hydrogen atoms. A hydrogen nucleus (proton) is formed whenever electrons are stripped from hydrogen atoms. When these higher-energy electrons are transferred to lower-energy states, often protons are pumped across membranes. This creates a region with a high concentration of protons on one side of the membrane. Therefore, this process is referred to as a *proton pump.* The "pressure" created by this high concentration of protons is released when protons flow through pores in the membrane back to the side from which they were pumped. As they pass through the pores, an enzyme, ATPase (a phosphorylase), uses their energy to speed the formation of an ATP molecule by bonding a phosphate to an ADP molecule. Thus, making a proton gradient is an important step in the production of much of the ATP produced in cells (review figure 5.14).

The four concepts of biochemical pathways, ATP production, electron transport, and the proton pump—are all interrelated. We will use these concepts to examine particular aspects of photosynthesis and respiration in chapters 6 and 7.

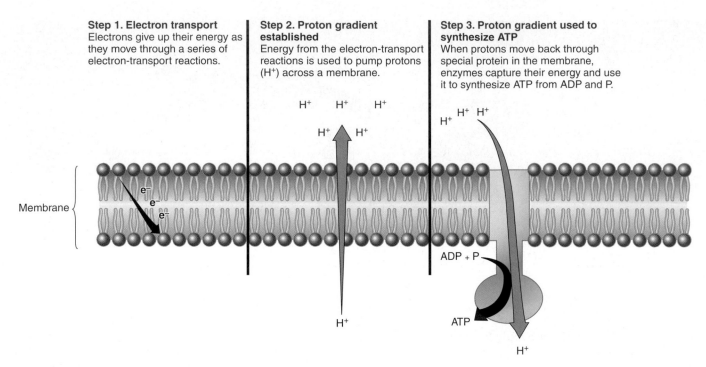

Step 1. Electron transport
Electrons give up their energy as they move through a series of electron-transport reactions.

Step 2. Proton gradient established
Energy from the electron-transport reactions is used to pump protons (H^+) across a membrane.

Step 3. Proton gradient used to synthesize ATP
When protons move back through special protein in the membrane, enzymes capture their energy and use it to synthesize ATP from ADP and P.

FIGURE 5.14 Electron Transport and Proton Gradient
The transport of high-energy electrons through a series of electron carriers can allow the energy to be released in discrete, manageable packets. In some cases, the energy given up is used to move or pump protons (H^+) from one side of a membrane to the other and a proton concentration gradient is established. When the protons flow back through the membrane, enzymes in the membrane can capture energy and form ATP.

Summary

Enzymes are protein catalysts that speed up the rate of chemical reactions without any significant increase in the temperature. They do this by lowering activation energy. Enzymes have a very specific structure, which matches the structure of particular substrate molecules. The substrate molecule comes in contact with only a specific part of the enzyme molecule—the attachment site. The active site of the enzyme is the place where the substrate molecule is changed. The enzyme-substrate complex reacts to form the end product. The protein nature of enzymes makes them sensitive to environmental conditions, such as temperature and pH, that change the structure of proteins. The number and kinds of enzymes are ultimately controlled by the genetic information of the cell. Other kinds of molecules, such as coenzymes, inhibitors, and competing enzymes, can influence specific enzymes. Changing conditions within the cell shift its enzymatic priorities by influencing the turnover number.

Enzymes are also used to speed and link chemical reactions into biochemical pathways. The energy currency of the cell, ATP, is produced by enzymatic pathways known as electron transport and proton pumping. The four concepts of biochemical pathways, ATP production, electron transport, and the proton pump are all interrelated.

Key Terms

*Use the interactive flash cards on the **Concepts in Biology, 12/e** website to help you learn the meaning of these terms.*

activation energy 96
active site 97
adenosine triphosphate (ATP) 106
anabolism 105
binding site (attachment site) 97
biochemical pathway (metabolic pathway) 105
catabolism 105
catalyst 96
coenzyme 99
cofactors 99
competitive inhibition 103
enzymatic competition 101
enzyme 96
enzyme-substrate complex 97

flavin adenine dinucleotide (FAD) 99
gene-regulator proteins 102
high-energy phosphate bond 106
inhibitor 102
negative-feedback inhibition 104
nicotinamide adenine dinucleotide (NAD$^+$) 99
nicotinamide adenine dinucleotide phosphate (NADP$^+$) 107
nutrients 96
substrate 97
turnover number 100
vitamins 99

Basic Review

1. Something that speeds the rate of a chemical reaction but is not used up in that reaction is called a
   a. catalyst.
   b. catabolic molecule.
   c. coenzyme.
   d. ATP.

2. The amount of energy it takes to a get chemical reaction going is known as
   a. starting energy.
   b. ATP.
   c. activation energy.
   d. denaturation.
   e. Q.

3. A molecule that is acted upon by an enzyme is a
   a. cofactor.
   b. binding site.
   c. vitamin.
   d. substrate.

4. Your cells require _____ to manufacture certain coenzymes.

5. When a protein's three-dimensional structure has been altered to the extent that it no longer functions, it has been
   a. denatured.
   b. killed.
   c. anabolized.
   d. competitively inhibited.

6. Whenever there are several different enzymes available to combine with a given substrate, _____ results.

7. In _____, a form of enzyme control, the end product inhibits one step of its formation when its concentration becomes high enough.

8. Which of the following contains the greatest amount of potential chemical-bond energy?
   a. AMP
   b. ADP
   c. ATP
   d. ARP

9. Electron-transfer reactions are commonly called _____ reactions.

10. As electrons pass through the pores, an enzyme, _____ (a phosphorylase), uses electrons energy to speed the formation of an ATP molecule by bonding a phosphate to an ADP molecule.

Answers

1. a   2. c   3. d   4. vitamins   5. a
6. enzymatic competition   7. negative feedback   8. c
9. oxidation-reduction   10. ATPase

Concept Review

Answers to the Concept Review questions can be found on the **Concepts in Biology,** *12/e website.*

5.1   How Cells Use Enzymes

1. What is the difference between a catalyst and an enzyme?
2. Describe the sequence of events in an enzyme-controlled reaction.
3. Would you expect a fat and a sugar molecule to be acted upon by the same enzyme? Why or why not?
4. Where in a cell would you look for enzymes?

5.2   How Enzymes Speed Chemical Reaction Rates

5. What is turnover number? Why is it important?
6. What is meant by the term *binding site? Active site?*

5.3   Enzymes, Coenzymes, and Vitamins

5.4   How the Environment Affects Enzyme Action

7. How do these three types of molecules relate to one another? enzymes, coenzymes, and vitamins
8. Why must a vitamin be a part of the diet?
9. How does changing temperature affect the rate of an enzyme-controlled reaction?
10. What factors in a cell can speed up or slow down enzyme reactions?
11. What is the relationship between vitamins and coenzymes?
12. What effect might a change in pH have on enzyme activity?

5.5   Cellular-Controlling Processes and Enzymes

13. What is enzyme competition, and why is it important to all cells?
14. Describe the nature and action of an enzyme inhibitor.

5.6   Enzymatic Reactions Used in Processing Energy and Matter

15. What is a biochemical pathway, and what does it have to do with enzymes?
16. Describe what happens during electron transport and what it has to do with a proton pump.

Thinking Critically

*For guidelines to answer this question, visit the **Concepts in Biology**, 12/e website.*

The following data were obtained by a number of Nobel Prize–winning scientists from Lower Slobovia. As a member of the group, interpret the data with respect to the following:

1. Enzyme activities
2. Movement of substrates into and out of the cell
3. Competition among various enzymes for the same substrate
4. Cell structure

Data
  a. A lowering of the atmospheric temperature from 22°C to 18°C causes organisms to form a thick, protective coat.
  b. Below 18°C, no additional coat material is produced.
  c. If the cell is heated to 35°C and then cooled to 18°C, no coat is produced.
  d. The coat consists of a complex carbohydrate.
  e. The coat will form even if there is a low concentration of simple sugars in the surroundings.
  f. If the cell needs energy for growth, no cell coats are produced at any temperature.

CHAPTER 6

Biochemical Pathways—Cellular Respiration

There are over 100 species of the fungus *Penicillium,* and each characteristically produces reproductive spores that line up and look like a hairbrush. The Latin word *penicillus* means little brush. These fungi do more than just produce the antibiotic penicillin. Many people are familiar with the blue, cottony growth on citrus fruits. It appears to be blue because of the pigment produced in the mold's spores. The blue cheeses, such as Cambozola, Stilton, Gorgonzola, and the original Roquefort, all have that color. Each of these cheeses is "aged" with *Penicillium roquefortii* to produce the characteristic color, texture, and flavor. The differences in the blue cheeses are determined by the kinds of milk used and the conditions under which the aging occurs. True Roquefort cheese is made from sheep's milk and aged in caves at Roquefort, France. American blue cheese is made from cow's milk and aged at various places in the United States. The blue color has become a very important feature of those cheeses. However, a team of researchers found a mutant species of *P. roquefortii* that produces spores having no blue color. The cheese made from that mold is "white" blue cheese. The flavor is exactly the same as blue cheese, but the cheese is commercially worthless, because people want the blue color.

How cheeses become different from one another in texture, flavor, and aroma is only one example of a vital metabolic process known as cellular respiration. Whether *Penicillium* or a person, all organisms must carry out cellular respiration; should they stop, they would die. Although the exact details of how each organism goes about performing this kind of metabolism differ, the basic concepts are the same. This chapter will present the basics of this process.

CHAPTER OUTLINE

6.1 Energy and Organisms

There are hundreds of different chemical reactions taking place within the cells of organisms. Many of these reactions are involved in providing energy for the cells. Organisms are classified into groups based on the kind of energy they use. Organisms that are able to use basic energy sources, such as sunlight, to make energy-containing organic molecules from inorganic raw materials are called **autotrophs** (*auto* = self; *troph* = feeding). There are also prokaryotic organisms that use inorganic chemical reactions as a source of energy to make larger, organic molecules. This process is known as **chemosynthesis.** Therefore, there are at least two kinds of autotrophs: Those that use light are called *photosynthetic* autotrophs and those that use inorganic chemical reactions are called *chemosynthetic* autotrophs. All other organisms require organic molecules as food and are called **heterotrophs** (*hetero* = other; *troph* = feeding). Heterotrophs get their energy from the chemical bonds of food molecules, such as carbohydrates, fats, and proteins, which they must obtain from their surroundings.

Within eukaryotic cells, certain biochemical processes are carried out in specific organelles. Chloroplasts are the sites of photosynthesis, and mitochondria are the sites of most of the reactions of cellular respiration (figure 6.1). Because prokaryotic cells lack mitochondria and chloroplasts, they carry out photosynthesis and cellular respiration within the cytoplasm or on the inner surfaces of the cell membrane or on other special membranes. Table 6.1 provides a summary of the concepts just discussed and how they are related to one another. organelles, p. 73

This chapter will focus on the reactions involved in the processes of cellular respiration. In **cellular respiration,** organisms control the release of chemical-bond energy from large, organic molecules and use the energy for the many activities necessary to sustain life. All organisms, whether autotrophic or heterotrophic, must carry out cellular respiration if they are to survive. Because nearly all organisms use organic molecules as a source of energy, they must obtain these molecules from their environment or manufacture these organic molecules, which they will later break down. Thus, photosynthetic organisms produce food molecules, such as carbohydrates, for themselves as well as for all the other organisms that feed on them. There are many variations of cellular respiration. Some organisms require the presence of oxygen for these processes, called *aerobic* processes. Other organisms carry out a form of respiration that does not require oxygen; these processes are called *anaerobic.*

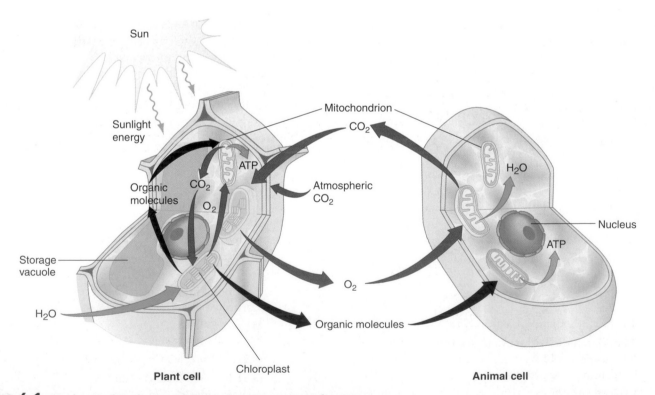

FIGURE 6.1 Biochemical Pathways That Involve Energy Transformation
Photosynthesis and cellular respiration both involve a series of chemical reactions that control the flow of energy. Organisms that contain photosynthetic machinery are capable of using light, water, and carbon dioxide to produce organic molecules, such as sugars, proteins, lipids, and nucleic acids. Oxygen is also released as a result of photosynthesis. In aerobic cellular respiration, organic molecules and oxygen are used to provide the energy to sustain life. Carbon dioxide and water are also released during aerobic respiration.

TABLE 6.1

Summary of Biochemical Pathways, Energy Sources, and Kinds of Organisms

Autotroph or Heterotroph	Biochemical Pathways	Energy Source	Kinds of Organisms	Notes
Autotroph	Photosynthesis	Light	Prokaryotic—certain bacteria	Prokaryotic photosynthesis is somewhat different from eukaryotic photosynthesis and does not take place in chloroplasts.
			Eukaryotic—plants and algae	Eukaryotic photosynthesis takes place in chloroplasts.
Autotroph	Chemosynthesis	Inorganic chemical reactions	Prokaryotic—certain bacteria and archaea	There are many types of chemosynthesis.
Autotroph and heterotroph	Cellular respiration	Oxidation of large organic molecules	Prokaryotic—bacteria and archaea	There are many forms of respiration. Some organisms use aerobic cellular respiration; others use anaerobic cellular respiration.
			Eukaryotic—plants, animals, fungi, algae, protozoa	Most respiration in eukaryotic organisms takes place in mitochondria and is aerobic.

6.2 An Overview of Aerobic Cellular Respiration

Aerobic cellular respiration is a specific series of enzyme-controlled chemical reactions in which oxygen is involved in the breakdown of glucose into carbon dioxide and water and the chemical-bond energy from glucose is released to the cell in the form of ATP. Although the actual process of aerobic cellular respiration involves many enzyme-controlled steps, the net result is that a reaction between sugar and oxygen results in the formation of carbon dioxide and water with the release of energy. The following equation summarizes this process:

$$\text{glucose} + \text{oxygen} \rightarrow \begin{array}{c}\text{carbon}\\\text{dioxide}\end{array} + \text{water} + \text{energy}$$
$$C_6H_{12}O_6 + 6\,O_2 \rightarrow 6\,CO_2 + 6\,H_2O + \text{energy}$$
$$(\text{ATP} + \text{heat})$$

Covalent bonds are formed by atoms sharing pairs of fast-moving, energetic electrons. Therefore, the covalent bonds in the sugar glucose contain chemical potential energy. Of all the covalent bonds in glucose (O—H, C—H, C—C), those easiest to get at are the C—H and O—H bonds on the outside of the molecule. When these bonds are broken, two things happen:

1. The energy of the electrons can be used to phosphorylate ADP molecules to produce higher-energy ATP molecules.

2. Hydrogen ions (protons) are released. The ATP is used to power the metabolic activities of the cell. The chemical activities that remove electrons from glucose result in the glucose being *oxidized.*

These high-energy electrons must be controlled. If they were allowed to fly about at random, they would quickly combine with other molecules, causing cell death. Electron-transfer molecules, such as NAD^+ and FAD, temporarily hold the electrons and transfer them to other electron-transfer molecules. ATP is formed when these transfers take place (see chapter 5). Once energy has been removed from electrons for ATP production, the electrons must be placed in a safe location. In *aerobic* cellular respiration, these electrons are ultimately attached to oxygen. Oxygen serves as the final resting place of the less energetic electrons. When the electrons are added to oxygen, it becomes a negatively charged ion, $O^=$.

Because the oxygen has gained electrons, it has been *reduced. Thus, in the aerobic cellular respiration of glucose, glucose is oxidized and oxygen is reduced.* If something is oxidized (loses electrons), something else must be reduced (gains electrons). A molecule cannot simply lose its electrons—they have to go someplace! Eventually, the positively charged hydrogen ions that were released from the glucose molecule combine with the negatively charged oxygen ion to form water.

Once all the hydrogens have been stripped off the glucose molecule, the remaining carbon and oxygen atoms are

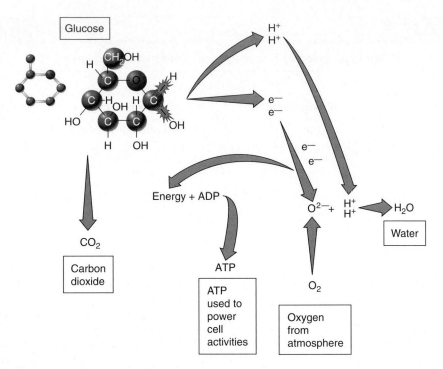

FIGURE 6.2 **Aerobic Cellular Respiration and Oxidation-Reduction Reaction**
During aerobic cellular respiration, a series of oxidation-reduction reactions takes place. When the electrons are removed (oxidation) from sugar, it is unable to stay together and breaks into smaller units. The reduction part of the reaction occurs when these electrons are attached to another molecule. In aerobic cellular respiration, the electrons are eventually picked up by oxygen and the negatively charged oxygen attracts two positively charged hydrogen ions (H^+) to form water.

rearranged to form individual molecules of CO_2. All the hydrogen originally a part of the glucose has been moved to the oxygen to form water. All the remaining carbon and oxygen atoms of the original glucose are now in the form of CO_2. The energy released from this process is used to generate ATP (figure 6.2).

In cells, these reactions take place in a particular order and in particular places within the cell. In eukaryotic cells, the process of releasing energy from food molecules begins in the cytoplasm and is completed in the mitochondrion. There are three distinct enzymatic pathways involved (figure 6.3): glycolysis, the Krebs cycle, and the electron-transport system.

Glycolysis

Glycolysis (*glyco* = sugar; *lysis* = to split) is a series of enzyme-controlled reactions that takes place in the cytoplasm of cells, which results in the breakdown of glucose with the release of electrons and the formation of ATP. During glycolysis, the 6-carbon sugar glucose is split into two smaller, 3-carbon molecules, which undergo further modification to form pyruvic acid or pyruvate.[1] Enough energy is released to produce two ATP molecules. Some of the bonds holding hydrogen atoms to the glucose molecule are broken, and the electrons are picked up by electron carrier molecules (NAD^+)

and transferred to a series of electron-transfer reactions known as the electron-transport system (ETS).

The Krebs Cycle

The **Krebs cycle** is a series of enzyme-controlled reactions that takes place inside the mitochondrion, which completes the breakdown of pyruvic acid with the release of carbon dioxide, electrons, and ATP. During the Krebs cycle, the pyruvic acid molecules produced from glycolysis are further broken down. During these reactions, the remaining hydrogens are removed from the pyruvic acid, and their electrons are picked up by the electron carriers NAD^+ and FAD. These electrons are sent to the electron-transport system. A small amount of ATP is also formed during the Krebs cycle. The carbon and oxygen atoms that are the remains of the pyruvic acid molecules are released as carbon dioxide (CO_2).

---

[1]Several different ways of naming organic compounds have been used over the years. For our purposes, pyruvic acid and pyruvate are really the same basic molecule although technically, pyruvate is what is left when pyruvic acid has lost its hydrogen ion: pyruvic acid → H^+ + pyruvate. You also will see terms such as lactic acid and lactate and citric acid and citrate and many others used in a similar way.

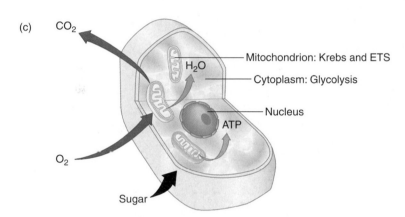

(a) Glucose + specific sequence of reactions controlled by enzymes + O_2 ⟶ H_2O + CO_2 + ATP

(b)

NADH

NADH

FADH$_2$

O_2

Glycolysis — e^-

Krebs cycle

Electron-transport system

e^-

e^-

Glucose (6-carbon) → Pyruvic acid (3 carbons) → Acetyl-CoA (2 carbons) → H_2O

2 ATP — 2 ATP — 32 ATP → 36 ATP

CO_2 — CO_2 — CO_2

(c)

CO_2

H_2O — Mitochondrion: Krebs and ETS

Cytoplasm: Glycolysis

Nucleus

ATP

O_2

Sugar

FIGURE 6.3 Aerobic Cellular Respiration: Overview
(a) This sequence of reactions in the aerobic oxidation of glucose is an overview of the energy-yielding reactions of a cell. *(b)* Glycolysis, the Krebs cycle, and the electron-transport system (ETS) are each a series of enzyme-controlled reactions that extract energy from the chemical bonds in a glucose molecule. During glycolysis, glucose is split into pyruvic acid and ATP and electrons are released. During the Krebs cycle, pyruvic acid is further broken down to carbon dioxide with the release of ATP and the release of electrons. During the electron-transport system, oxygen is used to accept electrons, and water and ATP are produced. *(c)* Glycolysis takes place in the cytoplasm of the cell. Pyruvic acid enters mitochondria, where the Krebs cycle and electron-transport system (ETS) take place.

The Electron-Transport System (ETS)

The **electron-transport system (ETS)** is a series of enzyme-controlled reactions that converts the kinetic energy of hydrogen electrons to ATP. The electrons are carried to the electron-transport system from glycolysis and the Krebs cycle by NADH and FADH$_2$. The electrons are transferred through a series of oxidation-reduction reactions involving enzymes until eventually the electrons are accepted by oxygen atoms to form oxygen ions ($O^=$). During this process, a great deal of ATP is produced. The ATP is formed as a result of a proton gradient established when the energy of electrons is used to pump protons across a membrane. The subsequent movement of protons back across the membrane results in ATP formation. The negatively charged oxygen atoms attract two positively charged hydrogen ions to form water (H_2O).

Aerobic respiration can be summarized as follows. *Glucose* enters glycolysis and is broken down to pyruvic acid, which enters the Krebs cycle, where the pyruvic acid molecules are further dismantled. The remains of the pyruvic acid molecules are released as *carbon dioxide*. The electrons and

hydrogen ions released from glycolysis and the Krebs cycle are transferred by NADH and FADH$_2$ to the electron-transport system, where the electrons are transferred to *oxygen* available from the atmosphere. When hydrogen ions attach to oxygen ions, *water* is formed. *ATP* is formed during all three stages of aerobic cellular respiration, but most comes from the electron-transfer system.

6.3 The Metabolic Pathways of Aerobic Cellular Respiration

This discussion of aerobic cellular respiration is divided into two levels: a fundamental description and a detailed description. It is a good idea to begin with the simplest description and add layers of understanding as you go to additional levels. Ask your instructor which level is required for your course of study.

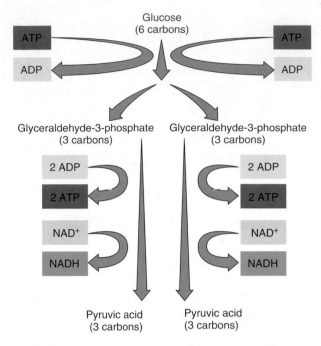

FIGURE 6.4 Glycolysis: Fundamental Description
Glycolysis is the biochemical pathway many organisms use to oxidize glucose. During this sequence of chemical reactions, the 6-carbon molecule of glucose is oxidized. As a result, pyruvic acid is produced, electrons are picked up by NAD^+, and ATP is produced.

Fundamental Description

Glycolysis

Glycolysis is a series of enzyme-controlled reactions that takes place in the cytoplasm. During glycolysis, a 6-carbon sugar molecule (glucose) has energy added to it from two ATP molecules. Adding this energy makes some of the bonds of the glucose molecule unstable, and the glucose molecule is more easily broken down. After passing through several more enzyme-controlled reactions, the 6-carbon glucose is broken down to two 3-carbon molecules known as glyceraldehyde-3-phosphate (also known as phosphoglyceraldehyde[2]), which undergo additional reactions to form pyruvic acid ($CH_3COCOOH$).

Enough energy is released by this series of reactions to produce four ATP molecules. Because two ATP molecules were used to start the reaction and four were produced, there is a *net gain* of two ATPs from the glycolytic pathway (figure 6.4). During the process of glycolysis, some hydrogens and their electrons are removed from the organic molecules being processed and picked up by the electron-transfer molecule

---

[2]As with many things in science, the system for naming organic chemical compounds has changed. In the past, the term *phosphoglyceraldehyde* was commonly used for this compound and was used in previous editions of this text. However, today the most commonly used term is *glyceraldehyde-3-phosphate*. In order to reflect current usage more accurately, the term *glyceraldehyde-3-phosphate* is used in this edition.

NAD^+ to form NADH. Enough hydrogens are released during glycolysis to form 2 NADHs. The NADH with its extra electrons contains a large amount of potential energy, which can be used to make ATP in the electron-transport system. The job of the coenzyme NAD^+ is to transport these energy-containing electrons and protons safely to the electron-transport system. Once they have dropped off their electrons, the oxidized NAD^+s are available to pick up more electrons and repeat the job. The following is a generalized reaction that summarizes the events of glycolysis:

$$\text{glucose} + 2 \text{ ATP} + 2 \text{ NAD}^+ \rightarrow$$
$$4 \text{ ATP} + 2 \text{ NADH} + 2 \text{ pyruvic acid}$$

The Krebs Cycle

The series of reactions known as the Krebs cycle takes place within the mitochondria of cells. It gets its name from its discoverer, Hans Krebs, and the fact that the series of reactions

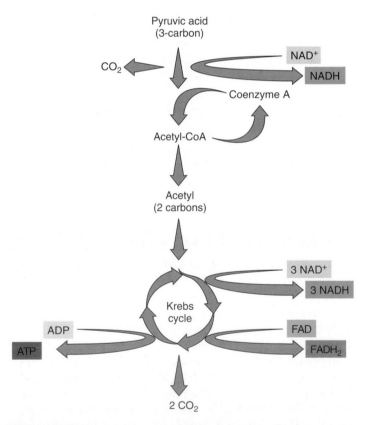

FIGURE 6.5 Krebs Cycle: Fundamental Description
The Krebs cycle takes place in the mitochondria of cells to complete the oxidation of glucose. During this sequence of chemical reactions, a pyruvic acid molecule produced from glycolysis is stripped of its hydrogens. The hydrogens are picked up by NAD^+ and FAD for transport to the ETS. The remaining atoms are reorganized into molecules of carbon dioxide. Enough energy is released during the Krebs cycle to form 2 ATPs. Because 2 pyruvic acid molecules were produced from glycolysis, the Krebs cycle must be run twice in order to complete their oxidation (once for each pyruvic acid).

begins and ends with the same molecule. The Krebs cycle is also known as the citric acid cycle and the TriCarboxylic Acid cycle (TCA). The 3-carbon pyruvic acid molecules released from glycolysis enter the mitochondria, are acted upon by specific enzymes, and are converted to 2-carbon **acetyl** molecules. At the time the acetyl is produced, 2 hydrogens are attached to NAD^+ to form NADH. The carbon atom that was removed is released as carbon dioxide. The acetyl molecule is attached to *coenzyme A (CoA)* and proceeds through the Krebs cycle. During the Krebs cycle (figure 6.5), the acetyl is completely oxidized.

The remaining hydrogens and their electrons are removed. Most of the electrons are picked up by NAD^+ to form NADH, but at one point in the process FAD picks up electrons to form $FADH_2$. Regardless of which electron carrier is being used, the electrons are sent to the electron-transport system. The remaining carbon and oxygen atoms are combined to form CO_2. As in glycolysis, enough energy is released to generate 2 ATP molecules. At the end of the Krebs cycle, the acetyl has been completely broken down (oxidized) to CO_2. The energy in the molecule has been transferred to ATP, NADH, or $FADH_2$. Also, some of the energy has been released as heat. For each of the acetyl molecules that enters the Krebs cycle, 1 ATP, 3 NADHs, and 1 $FADH_2$. If we count the NADH produced during glycolysis, when acetyl was formed, there are a total of 4 NADHs for each pyruvic acid that enters a mitochondrion. The following is a generalized equation that summarizes those reactions:

$$\text{pyruvic acid} + \text{ADP} + 4\,\text{NAD}^+ + \text{FAD} \longrightarrow$$
$$\longrightarrow 3\,\text{CO}_2 + 4\,\text{NADH} + \text{FADH}_2 + \text{ATP}$$

The Electron-Transport System

Of the three steps of aerobic cellular respiration, (glycolysis, Krebs cycle, and electron-transport system) cells generate the greatest amount of ATP from the electron-transport system (figure 6.6). During this stepwise sequence of oxidation-reduction reactions, the energy from the NADH and $FADH_2$ molecules generated in glycolysis and the Krebs cycle is used to produce ATP. Iron-containing *cytochrome* (*cyto* = cell; *chrom* = color) enzyme molecules are located on the membranes of the mitochondrion. The energy-rich electrons are passed (*transported*) from one cytochrome to another, and the energy is used to pump protons (hydrogen ions) from one side of the membrane to the other. The result of this is a higher concentration of hydrogen ions on one side of the membrane. As the concentration of hydrogen ions increases on one side, a proton gradient builds up. Because of this concentration gradient, when a membrane channel is opened, the protons flow back to the side from which they were pumped. As they pass through the channels, a phosphorylase enzyme (ATPase) speeds the formation of an ATP molecule by bonding a phosphate to an ADP molecule (phosphorylation). When all the electrons and hydrogen ions are accounted for, a total of 32 ATPs are formed from the electrons and hydro-

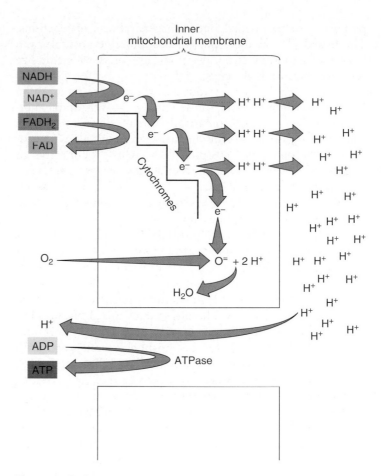

FIGURE 6.6 The Electron-Transport System: Fundamental Description

The electron-transport system (ETS) is also known as the cytochrome system. With the help of enzymes, the electrons are passed through a series of oxidation-reduction reactions. The energy the electrons give up is used to pump protons (H^+) across a membrane in the mitochondrion. When protons flow back through the membrane, enzymes in the membrane cause the formation of ATP. The protons eventually combine with the oxygen that has gained electrons, and water is produced.

gens removed from the original glucose molecule. The hydrogens are then bonded to oxygen to form water. The following is a generalized reaction that summarizes the events of the electron-transport system:

$$6\,\text{O}_2 + 8\,\text{NADH} + 4\,\text{FADH}_2 + 32\,\text{ADP} \longrightarrow$$
$$\longrightarrow 8\,\text{NAD}^+ + 4\,\text{FAD} + 32\,\text{ATP} + 12\,\text{H}_2\text{O}$$

Detailed Description

Glycolysis

The first stage of the cellular respiration process takes place in the cytoplasm. This first step, known as glycolysis, consists of the enzymatic breakdown of a glucose molecule without the use of molecular oxygen. Because no oxygen is required, glycolysis is called an anaerobic process. The glycolysis pathway can be divided into two general sets of reactions. The first reactions make the glucose molecule unstable, and later oxidation-reduction reactions are used to synthesize ATP and capture hydrogens.

Some energy must be added to the glucose molecule in order to start glycolysis, because glucose is a very stable molecule and will not automatically break down to release energy. In glycolysis, the initial glucose molecule gains a phosphate to become glucose-6-phosphate, which is converted to fructose-6-phosphate. When a second phosphate is added, fructose-1,6-bisphosphate (P—C_6—P) is formed. This 6-carbon molecule is unstable and breaks apart to form two 3-carbon, glyceraldehyde-3-phosphate molecules.

Each of the two glyceraldehyde-3-phosphate molecules acquires a second phosphate from a phosphate supply normally found in the cytoplasm. Each molecule now has 2 phosphates attached, 1, 3 isphosphoglycerate (P—C_3—P). A series of reactions follows, in which energy is released by breaking chemical bonds that hold the phosphates to 1,3 bisphosphoglycerate. The energy and the phosphates are used to produce ATP. Since there are 2 1,3 bisphosphoglycerate each with 2 phosphates, a total of 4 ATPs are produced. Because 2 ATPs were used to start the process, a net yield of 2 ATPs results. In addition, 4 hydrogen atoms detach from the carbon skeleton and their electrons are transferred to NAD^+ to form NADH, which transfers the electrons to the electron-transport system. The 3-carbon pyruvic acid molecules that remain are the raw material for the Krebs cycle. Because glycolysis occurs in the cytoplasm and the Krebs cycle takes place inside mitochondria, the pyruvic acid must enter the mitochondrion before it can be broken down further.

In summary, the process of glycolysis takes place in the cytoplasm of a cell, where glucose ($C_6H_{12}O_6$) enters a series of reactions that

1. Requires the use of 2 ATPs
2. Ultimately results in the formation of 4 ATPs
3. Results in the formation of 2 NADHs
4. Results in the formation of 2 molecules of pyruvic acid $CH_3COCOOH$)

Because 2 molecules of ATP are used to start the process and a total of 4 ATPs are generated, each glucose molecule that undergoes glycolysis produces a net yield of 2 ATPs (Figure 6.7).

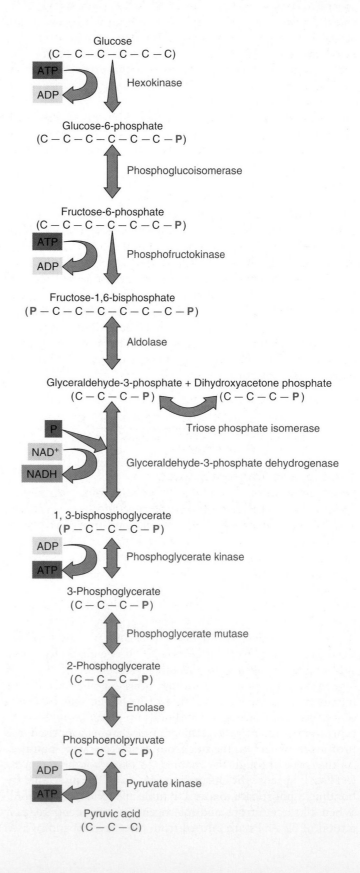

FIGURE 6.7 Glycolysis: Detailed Description

Glycolysis is a process that takes place in the cytoplasm of cells. It does not require the use of oxygen, so it is an anaerobic process. During the first few steps, phosphates are added from ATP and ultimately the 6-carbon sugar is split into two 3-carbon compounds. During the final steps in the process, NAD^+ accepts electrons and hydrogen to form NADH and ATP is produced. Two ATPs form for each of the 3-carbon molecules that are processed in glycolysis. Because there are two 3-carbon compounds, a total of 4 ATPs are formed. However, because 2 ATPs were used to start the process, there is a net gain of 2 ATPs. Pyruvic acid (pyruvate) is left at the end of glycolysis.

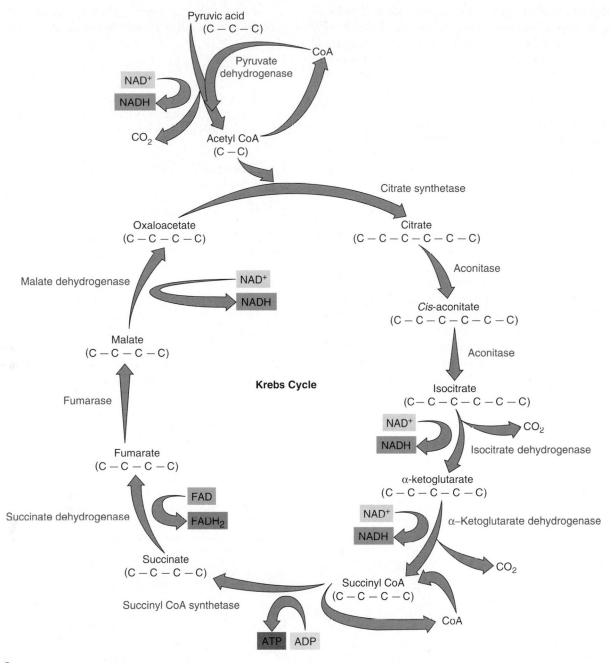

FIGURE 6.8 Krebs Cycle: Detailed Descriptions

The Krebs cycle occurs within the mitochondrion. Pyruvate enters the mitochondrion from glycolysis and is converted to a 2-carbon compound, acetyl. With the help CoA, the 2-carbon acetyl combines with 4-carbon oxaloacetate to form a 6-carbon citrate molecule. Through a series of reactions in the Krebs cycle, electrons are removed and picked up by NAD^+ and FAD to form NADH and $FADH_2$, which will be shuttled to the electron-transport system. Carbons are removed as carbon dioxide. Enough energy is released that 1 ATP is formed for each acetyl that enters the cycle.

The Krebs Cycle

After pyruvate (pyruvic acid) enters the mitochondrion, it is first acted upon by an enzyme, along with a molecule known as *coenzyme* A (*CoA*) (figure 6.8). This results in three significant products. Hydrogen atoms are removed and NADH is formed, a carbon is removed and carbon dioxide is formed, and a 2-carbon acetyl molecule is formed, which temporarily attaches to coenzyme A to produce acetyl-coenzyme A.

(These and subsequent reactions of the Krebs cycle take place in the fluid between the membranes of the mitochondrion.) The acetyl coenzyme A enters the series of reactions known as the Krebs cycle. During the Krebs cycle, the acetyl is systematically dismantled. Its hydrogen atoms are removed and the remaining carbons are released as carbon dioxide.

The first step in this process involves the acetyl coenzyme A. The acetyl portion of the complex is transferred to a 4-carbon compound called *oxaloacetate* (*oxaloacetic acid*) and a new 6-carbon citrate molecule (citric acid) is formed. The coenzyme A is released to participate in another reaction with pyruvic acid. This newly formed citrate is broken down in a series of reactions, which ultimately produces oxaloacetate, which was used in the first step of the cycle (hence, the names Krebs cycle, citric acid cycle, and tricarboxylic acid cycle). The compounds formed during this cycle are called *keto acids*.

In the process, electrons are removed and, along with protons, become attached to the coenzymes NAD^+ and FAD. Most become attached to NAD^+ but some become attached to FAD. As the molecules move through the Krebs cycle, enough energy is released to allow the synthesis of 1 ATP molecule for each acetyl that enters the cycle. The ATP is formed from ADP and a phosphate already present in the mitochondria. For each pyruvate molecule that enters a mitochondrion and is processed through the Krebs cycle, 3 carbons are released as 3 carbon dioxide molecules, 5 pairs of hydrogen atoms are removed and become attached to NAD^+ or FAD, and 1 ATP molecule is generated. When both pyruvate molecules have been processed through the Krebs cycle, (1) all the original carbons from the glucose have been released into the atmosphere as 6 carbon dioxide molecules; (2) all the hydrogen originally found on the glucose has been transferred to either NAD^+ or FAD to form NADH or $FADH_2$; and (3) 2 ATPs have been formed from the addition of phosphates to ADPs (review figure 6.8).

In summary, the Krebs cycle takes place within the mitochondria. For each pyruvate molecule that enters the Krebs cycle:

1. The three carbons of the pyruvate are released as carbon dioxide (CO_2).
2. Five pairs of hydrogens become attached to hydrogen carriers to become 4 NADHs and 1 $FADH_2$.
3. One ATP is generated.

The Electron-Transport System

The series of reactions in which energy is transferred from the electrons and protons carried by NADH and $FADH_2$ is known as the electron-transport system (ETS) (figure 6.9). This is the final stage of aerobic cellular respiration and is dedicated to generating ATP. The reactions that make up the electron-transport system are a series of oxidation-reduction reactions in which the electrons are passed from one electron carrier molecule to another until ultimately they are accepted by oxygen atoms. The negatively charged oxygen combines with the hydrogen ions to form water. It is this step that makes the process aerobic. Keep in mind that potential energy increases whenever things experiencing a repelling force are pushed together, such as adding the third phosphate to an ADP molecule. Potential energy also increases whenever things that attract each other are pulled apart, as in the separation of the protons from the electrons.

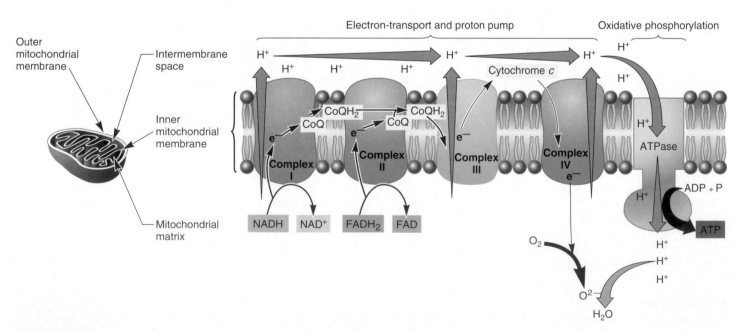

FIGURE 6.9 The Electron-Transport System: Detailed Description
Most of the ATP produced by aerobic cellular respiration comes from the ETS. NADH and $FADH_2$ deliver electrons to the enzymes responsible for the ETS. There are several protein complexes in the inner membrane of the mitochondrion, each of which is responsible for a portion of the reactions that yield ATP. The energy of electrons is given up in small amounts and used to pump protons into the intermembrane space. When these protons flow back through pores in the membrane, ATPase produces ATP. The electrons eventually are transferred to oxygen and the negatively charged oxygen ions accept protons to form water.

Let's now look in just a bit more detail at what happens to the electrons and protons that are carried to the electron-transport systems by NADH and FADH₂ and how these activities are used to produce ATP. The mitochondrion consists of two membranes—an outer, enclosing membrane and an inner, folded membrane. The reactions of the ETS are associated with this inner membrane. Within the structure of the membrane are several *enzyme complexes,* which perform particular parts of the ETS reactions (review figure 6.9). The production of ATPs involves two separate but connected processes. Electrons carried by NADH enter reactions in enzyme complex I, where they lose some energy and are eventually picked up by a coenzyme (coenzyme Q). Electrons from FADH₂ enter enzyme complex II and also are eventually transferred to coenzyme Q. Coenzyme Q transfers the electrons to enzyme complex III. In complex III, the electrons lose additional energy and are transferred to cytochrome c, which transfers electrons to enzyme complex IV. In complex IV, the electrons are eventually transferred to oxygen. As the electrons lose energy in complex I, complex III, and complex IV, additional protons are pumped into the intermembrane space. When these protons flow down the concentration gradient through channels in the membrane, phosphorylase enzymes (ATPase) in the membrane are able to use the energy to generate ATP.

A total of 12 pairs of electrons and hydrogens are transported to the ETS from glycolysis and the Krebs cycle for each glucose that enters the process. In eukaryotic organisms, the pairs of electrons can be accounted for as follows: 4 pairs are carried by NADH and were generated during glycolysis outside the mitochondrion, 8 pairs are carried as NADH and were generated within the mitochondrion, and 2 pairs are carried by FADH₂ and were generated within the mitochondrion.

- For each of the 8 NADHs generated within the mitochondrion, enough energy is released to produce 3 ATP molecules. Therefore, 24 ATPs are released from these electrons carried by NADH.
- In eukaryotic cells, the electrons released during glycolysis are carried as NADH and converted to FADH₂ in order to shuttle them into the mitochondria. Once they are inside the mitochondria, they follow the same pathway as the other FADH₂s.

The electrons carried by FADH₂ are lower in energy. When these electrons go through the series of oxidation-reduction reactions, they release enough energy to produce a total of 8 ATPs. Therefore, a total of 32 ATPs are produced from the hydrogen electrons that enter the ETS.

Finally, a complete accounting of all the ATPs produced during all three parts of aerobic cellular respiration results in a total of 36 ATPs: 32 from the ETS, 2 from glycolysis, and 2 from the Krebs cycle.

In summary, the electron-transport system takes place within the mitochondrion, where

1. Oxygen is used up as the oxygen atoms receive the hydrogens from NADH and FADH₂ to form water (H_2O).
2. NAD^+ and FAD are released, to be used over again.
3. Thirty-two ATPs are produced.

6.4   Aerobic Cellular Respiration in Prokaryotes

The discussion so far in this chapter has dealt with the process of aerobic cellular respiration in eukaryotic organisms. However, some prokaryotes also use aerobic cellular respiration. Because prokaryotes do not have mitochondria, there are some differences between what they do and what eukaryotes do. The primary difference involves the electrons carried from glycolysis to the electron-transport system. In eukaryotes, the electrons released during glycolysis are carried by NADH and transferred to FAD to form FADH₂ in order to get the electrons across the outer membrane of the mitochondrion. Because FADH₂ results in the production of fewer ATPs than NADH, there is a cost to the eukaryotic cell of getting the electrons into the mitochondrion. This transfer is not necessary in prokaryotes, so they are able to produce a theoretical 38 ATPs for each glucose metabolized, rather than the 36 ATPs produced by eukaryotes (table 6.2).

TABLE 6.2		
Aerobic ATP Production: Prokaryotic vs. Eukaryotic Cells		
Cellular Respiration Stage	**Prokaryotic Cells ATP Theoretically Generated**	**Eukaryotic Cells ATP Theoretically Generated**
Glycolysis	Net gain 2 ATP	Net gain 2 ATP
Krebs cycle	2 ATP	2 ATP
ETS	34 ATP	32 ATP
Total	38 ATP	36 ATP

6.5 Anaerobic Cellular Respiration

Although aerobic cellular respiration is the fundamental process by which most organisms generate ATP, some organisms do not have the necessary enzymes to carry out the Krebs cycle and ETS. Most of these are prokaryotic organisms, but there are certain eukaryotic organisms, such as yeasts, that can live in the absence of oxygen and do not use the Krebs cycle and ETS. Even within organisms, there are differences in the metabolic activities of cells. Some of their cells are unable to perform aerobic respiration, whereas others are able to survive for periods of time without it. However, all of these cells still need a constant supply of ATP. An organism that does not require O_2 as its final electron acceptor is called *anaerobic* (*an* = without; *aerob* = air) and performs anaerobic cellular respiration. Although they do not use oxygen, some anaerobic organisms are capable of using other inorganic or organic molecules as their final electron acceptors. The acceptor molecule might be sulfur, nitrogen, or other inorganic atoms or ions. It might also be an organic molecule, such as pyruvic acid ($CH_3COCOOH$).

Fermentation is the word used to describe anaerobic pathways that oxidize glucose to generate ATP energy by using an organic molecule as the ultimate hydrogen acceptor. Anaerobic respiration is the incomplete oxidation of glucose and results in the production of smaller hydrogen-containing organic molecules and energy in the form of ATP and heat. Many organisms that perform anaerobic cellular respiration use the glycolysis pathway to obtain energy from sugar molecules. Typically, glucose proceeds through the glycolysis pathway, producing pyruvic acid. The pyruvic acid then undergoes one of several alternative changes, depending on the kind of organism and the specific enzymes it possesses. Some organisms are capable of returning the electrons removed from sugar in the earlier stages of glycolysis to the pyruvic acid formed at the end of glycolysis. When this occurs, the pyruvic acid is converted into lactic acid, ethyl alcohol, acetone, or other organic molecules (figure 6.10). The organisms that produce ethyl alcohol have the enzymes necessary to convert pyruvic acid to ethyl alcohol (ethanol) and carbon dioxide. The formation of molecules such as alcohol and lactic acid is necessary to regenerate the NAD^+ needed for use in glycolysis. It must be done here, because it is not being regenerated by an ETS, as happens in aerobic respiration. Although many products can be formed from pyruvic acid, we will look at only two anaerobic pathways in detail.

Alcoholic Fermentation

Alcoholic fermentation is the anaerobic respiration pathway that yeast cells follow when oxygen is lacking in their environment. In this pathway, the pyruvic acid is converted to ethanol (a 2-carbon alcohol, C_2H_5OH) and carbon dioxide. Yeast cells then are able to generate only 4 ATPs from glycolysis. The cost for glycolysis is still 2 ATPs; thus, for each glucose a yeast cell oxidizes, it profits by 2 ATPs.

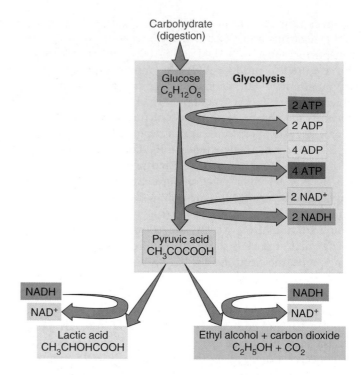

Fermentation Product	Possible Source	Importance
Lactic acid	Bacteria: *Lactobacillus bulgaricus*	Aids in changing milk to yogurt
	Homo sapiens Muscle cells	Produced when O_2 is limited; results in pain and muscle inaction
Ethyl alcohol +CO_2	Yeast: *Saccharomyces cerevisiae*	Brewing and baking

FIGURE 6.10 Fermentations

The upper portion of this figure is a simplified version of glycolysis. Many organisms can carry out the process of glycolysis and derive energy from it. The ultimate end product is determined by the kinds of enzymes the specific organism can produce. The synthesis of these various molecules is the organism's way of oxidizing NADH to regenerate NAD^+ and reducing pyruvic acid to a new end product.

Although during alcoholic fermentation yeasts get ATP and discard the waste products ethanol and carbon dioxide, these waste products are useful to humans. In making bread, the carbon dioxide is the important end product; it becomes trapped in the bread dough and makes it rise—the bread is *leavened*. Dough that has not undergone this process is called *unleavened*. The alcohol produced by the yeast evaporates

OUTLOOKS 6.1

Souring vs. Spoilage

The fermentation of carbohydrates to organic acid products, such as lactic acid, is commonly called *souring*. Cultured sour cream, cheese, and yogurt are produced by the action of fermenting bacteria. Lactic-acid bacteria of the genus *Lactobacillus* are used in the fermentation process. While growing in the milk, the bacteria convert lactose to lactic acid, which causes the milk to change from a liquid to a solid curd. The higher acid level also inhibits the growth of spoilage microorganisms.

Spoilage, or putrefaction, is the anaerobic respiration of proteins with the release of nitrogen and sulfur-containing organic compounds as products. Protein fermentation by the bacterium *Clostridium* produces foul-smelling chemicals such as putrescine, cadaverine, hydrogen sulfide, and methyl mercaptan. *Clostridium perfringens* and *C. sporogenes* are the two anaerobic bacteria associated with the disease gas gangrene. A gangrenous wound is a foul-smelling infection resulting from the fermentation activities of those two bacteria.

during the baking process. In the brewing industry, ethanol is the desirable product produced by yeast cells. Champagne, other sparkling wines, and beer are products that contain both carbon dioxide and alcohol. The alcohol accumulates, and the carbon dioxide in the bottle makes them sparkling (bubbly) beverages. In the manufacture of many wines, the carbon dioxide is allowed to escape, so these wines are not sparkling; they are called "still" wines.

Lactic Acid Fermentation

In **lactic acid fermentation,** the pyruvic acid ($CH_3COCOOH$) that results from glycolysis is converted to lactic acid ($CH_3CHOHCOOH$) by the transfer of electrons that had been removed from the original glucose. In this case, the net profit is again only 2 ATPs per glucose. The buildup of the waste product, lactic acid, eventually interferes with normal metabolic functions and the bacteria die. The lactic acid waste product from these types of anaerobic bacteria are used to make yogurt, cultured sour cream, cheeses, and other fermented dairy products. The lactic acid makes the milk protein coagulate and become puddinglike or solid. It also gives the products their tart flavor, texture, and aroma (Outlooks 6.1).

In the human body, different cells have different metabolic capabilities. Nerve cells must have a constant supply of oxygen to conduct aerobic cellular respiration. Red blood cells lack mitochondria and must rely on the anaerobic process of lactic acid fermentation to provide themselves with energy. Muscle cells can do either. As long as oxygen is available to skeletal muscle cells, they function aerobically. However, when oxygen is unavailable—because of long periods of exercise or heart or lung problems that prevent oxygen from getting to the skeletal muscle cells—the cells make a valiant effort to meet energy demands by functioning anaerobically.

While skeletal muscle cells are functioning anaerobically, they accumulate lactic acid. This lactic acid must ultimately be metabolized, which requires oxygen. Therefore, the accumulation of lactic acid represents an *oxygen debt*, which must be repaid in the future. It is the lactic acid buildup that makes muscles tired when we exercise. When the lactic acid concentration becomes great enough, lactic acid fatigue results. As a person cools down after a period of exercise, breathing and heart rate stay high until the oxygen debt is repaid and the level of oxygen in the muscle cells returns to normal. During

this period, the lactic acid that has accumulated is converted back into pyruvic acid. The pyruvic acid can then continue through the Krebs cycle and the ETS as oxygen becomes available. In addition to what is happening in the muscles, much of the lactic acid is transported by the bloodstream to the liver, where about 20% is metabolized through the Krebs cycle and 80% is resynthesized into glucose.

6.6 Metabolic Processing of Molecules Other Than Carbohydrates

Up to this point, we have discussed only the methods and pathways that allow organisms to release the energy tied up in carbohydrates (sugars). Frequently, cells lack sufficient carbohydrates for their energetic needs but have other materials from which energy can be removed. Fats and proteins, in addition to carbohydrates, make up the diet of many organisms. These three foods provide the building blocks for the cells, and all can provide energy. Carbohydrates can be digested to simple sugars, proteins can be digested to amino acids, and fats can be digested to glycerol and fatty acids. The basic pathways organisms use to extract energy from fat and protein are the same as for carbohydrates: glycolysis, the Krebs cycle, and the electron-transport system. However, there are some additional steps necessary to get fats and proteins ready to enter these pathways and several points in glycolysis and the Krebs cycle where fats and proteins enter to be respired.

Fat Respiration

A triglyceride (also known as a neutral fat) is a large molecule that consists of a molecule of glycerol with 3 fatty acids attached to it. Before these fats can be broken down to release energy, they must be converted to smaller units by digestive processes. Several enzymes are involved in these steps. The first step is to break the bonds between the glycerol and the fatty acids. Glycerol is a 3-carbon molecule that is converted into glyceraldehyde-3-phosphate. Because glyceraldehyde-3-phosphate is involved in one of the steps in glycolysis, it can enter the glycolysis pathway (figure 6.11). The remaining fatty acids are often long molecules (typically 14 to 20 carbons long), which also must be processed before they can be further metabolized. First, they need to enter the mitochondrion, where subsequent reactions take place. Once inside the mitochondrion, each long chain of carbons that makes up the carbon skeleton is hydrolyzed (split by the addition of a water molecule) into 2-carbon fragments. Next, each of the 2-carbon fragments is converted into acetyl. The acetyl molecules are carried into the Krebs cycle by coenzyme A molecules. Once in the Krebs cycle, they proceed through the Krebs cycle just like the acetyls from glucose (Outlooks 6.2).

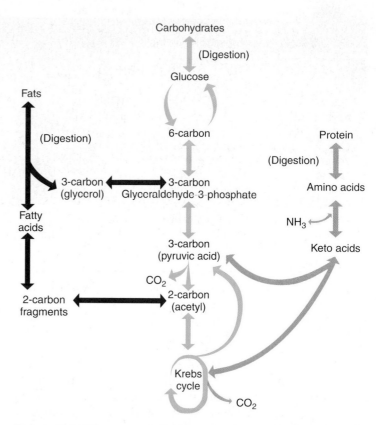

FIGURE 6.11 The Interconversion of Fats, Carbohydrates, and Proteins

Cells do not necessarily use all food as energy. One type of food can be changed into another type to be used as raw materials for the construction of needed molecules or for storage. Notice that many of the reaction arrows have two heads (i.e., these reactions can go in either direction). For example, glycerol can be converted into glyceraldehyde-3-phosphate and glyceraldehyde-3-phosphate can become glycerol.

By following the glycerol and each 2-carbon fragment through the cycle, you can see that each molecule of fat has the potential to release several times as much ATP as does a molecule of glucose. Each glucose molecule has 6 pairs of hydrogen, whereas a typical molecule of fat has up to 10 times that number. This is why fat makes such a good long-term energy storage material. It is also why it takes so long for people on a weight-reducing diet to remove fat. It takes time to use all the energy contained in the fatty acids. On a weight basis, there are twice as many calories in a gram of fat as there are in a gram of carbohydrate.

In summary, fats are an excellent source of energy and the storage of fat is an important process. Furthermore, other kinds of molecules can be converted to fat. You already know that people can get fat from eating sugar. Notice in figure 6.12 that both carbohydrates and fats can enter the Krebs cycle and release energy. Although people require both fats

OUTLOOKS 6.2

Lipid Metabolism and Ketoacidosis

In starvation and severe diabetes mellitus, the body does not metabolize sugars properly, and it shifts to using fats as its main source of energy. When this occurs, the Krebs cycle is unable to perform as efficiently and the acetyl CoA does not move into the mitochondria. It accumulates in the blood. To handle this problem, the liver converts the acetyl CoA to *ketone bodies*.

Ketone bodies include compounds such as acetoacetic acid, β-hydroxybutryic acid, and acetone. As ketone bodies accumulate in the blood, the pH decreases and the person experiences *ketosis*, or *ketoacidosis*, with symptoms such as an increased breathing rate; in untreated cases, it can lead to depression of the central nervous system, coma, and death.

and carbohydrates in their diets, they need not be in precise ratios; the body can make some interconversions. This means that people who eat excessive amounts of carbohydrates will deposit body fat. It also means that people who starve can generate glucose by breaking down fats and using the glycerol to synthesize glucose. nutrition, p. 542

Protein Respiration

Proteins can be catabolized and interconverted just as fats and carbohydrates are (review figure 6.12). The first step in using protein for energy is to digest the protein into individual amino acids. Each amino acid then needs to have the amino group (—NH$_2$) removed. The remaining carbon skeleton, a keto acid, enters the respiratory cycle as acetyl, pyruvic acid, or one of the other types of molecules found in the Krebs cycle. As the acids progress through the Krebs cycle, the electrons are removed and sent to the ETS, where their energy is converted into the chemical-bond energy of ATP. The amino group that was removed from the amino acid is converted into ammonia. Some organisms excrete ammonia directly; others convert ammonia into other nitrogen-containing compounds, such as urea or uric acid. All of these molecules are toxic and must be eliminated. They are transported in the blood to the kidneys, where they are eliminated. In the case of a high-protein diet, increasing fluid intake will allow the kidneys to remove the urea or uric acid efficiently. kidneys, p. 535

When proteins are eaten, they are digested into their component amino acids. These amino acids are then available to be used to construct other proteins. Proteins cannot be stored; if they or their component amino acids are not needed immediately, they will be converted into fat or carbohydrates or will be metabolized to provide energy. This presents a problem for individuals who do not have ready access to a continuous source of amino acids in their diet (e.g., individuals on a low-protein diet). If they do not have a source of dietary protein, they must dismantle proteins from important cellular components to supply the amino acids they need. This is why proteins and amino acids are considered an important daily food requirement.

One of the most important concepts is that carbohydrates, fats, and proteins can all be used to provide energy. The fate of any type of nutrient in a cell depends on the cell's momentary needs. An organism whose daily food-energy intake exceeds its daily energy expenditure will convert only the necessary amount of food into energy. The excess food will be interconverted according to the enzymes present and the organism's needs at that time. In fact, glycolysis and the Krebs cycle allow molecules of the three major food types (carbohydrates, fats, and proteins) to be interchanged.

As long as a person's diet has a certain minimum of each of the three major types of molecules, a cell's metabolic machinery can manipulate molecules to satisfy its needs. If a person is on a starvation diet, the cells will use stored carbohydrates first. Once the carbohydrates are gone (after about 2 days), the cells will begin to metabolize stored fat. When the fat is gone (after few days to weeks), the proteins will be used. A person in this condition is likely to die (How Science Works 6.1).

HOW SCIENCE WORKS 6.1

Applying Knowledge of Biochemical Pathways

As scientists have developed a better understanding of the processes of aerobic cellular respiration and anaerobic cellular respiration, several practical applications of this knowledge have developed:

- Although for centuries people have fermented beverages such as beer and wine, they were often plagued by sour products that were undrinkable. Once people understood that there were yeasts that produced alcohol under *anaerobic* conditions and bacteria that converted alcohol to acetic acid under *aerobic* conditions, it was a simple task to prevent acetic acid production by preventing oxygen from getting to the fermenting mixture.

- When it was discovered that the bacterium that causes gas gangrene is anaerobic and is, in fact, poisoned by the presence of oxygen, various oxygen therapies were developed to help cure patients with gangrene. Some persons with gangrene are placed in *hyperbaric chambers,* with high oxygen levels under pressure. In other patients, only the affected part of the body is enclosed. Under such conditions, the gangrene-causing bacteria die or are inhibited.

- Because many disease-causing organisms are prokaryotic and have somewhat different pathways and enzymes than do eukaryotic organisms, it is possible to develop molecules, antibiotics, that selectively interfere with the enzymes of prokaryotes without affecting eukaryotes, such as humans.

- When physicians recognized that the breakdown of fats releases ketone bodies, they were able to diagnose diseases such as diabetes and anorexia more easily, because people with these illnesses have bad breath.

High-Efficiency Anaerobic Bioreactors

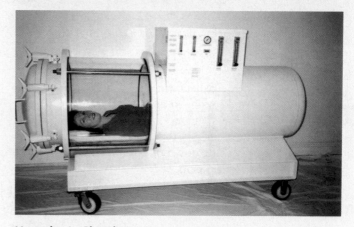

Hyperbaric Chamber

Summary

In aerobic cellular respiration, organisms convert foods into energy (ATP) and waste materials (carbon dioxide and water). Three distinct metabolic pathways are involved in aerobic cellular respiration: glycolysis, the Krebs cycle, and the electron transport system. Glycolysis takes place in the cytoplasm of the cell, and the Krebs cycle and electron-transport system take place in mitochondria. Organisms that have oxygen can perform aerobic cellular respiration. Organisms and cells that do not use oxygen perform anaerobic cellular respiration (fermentation) and can use only the glycolysis pathway. Aerobic cellular respiration yields much more ATP than anaerobic cellular respiration. Glycolysis and the Krebs cycle serve as a molecular interconversion system: Fats, proteins, and carbohydrates are interconverted according to the cell's needs.

Key Terms

Use the interactive flash cards on the **Concepts in Biology,** *12/e website to help you learn the meaning of these terms.*

acetyl 116
alcoholic fermentation 122
aerobic cellular respiration 113
autotrophs 112
cellular respiration 112
chemosynthesis 112

electron-transport system (ETS) 115
fermentation 122
glycolysis 114
heterotrophs 112
Krebs cycle 114
lactic acid fermentation 123

Basic Review

1. Organisms that are able to use basic energy sources, such as sunlight, to make energy-containing organic molecules from inorganic raw materials are called

   a. autotrophs.

   b. heterotrophs.

   c. aerobic.

   d. anaerobic.

2. Cellular respiration processes that do not use molecular oxygen are called

   a. heterotrophic.

   b. anaerobic.

   c. aerobic.

   d. anabolic.

3. The chemical activities that remove electrons from glucose result in the glucose being

   a. reduced.

   b. oxidized.

   c. phosphorylated.

   d. hydrolysed.

4. The positively charged hydrogen ions that are released from the glucose during cellular respiration eventually combine with _____ ion to form _____.

   a. another hydrogen, a gas

   b. a carbon, carbon dioxide

   c. an oxygen, water

   d. a pyruvic acid, lactic acid

5. The Krebs cycle and ETS are biochemical pathways performed in which eukaryotic organelle?

   a. nucleus

   b. ribosome

   c. chloroplast

   d. mitochondria

6. In a complete accounting of all the ATPs produced in aerobic cellular respiration, there are a total of _____ ATPs: _____ from the ETS, _____ from glycolysis, and _____ from the Krebs cycle.

   a. 36, 32, 2, 2

   b. 38, 34, 2, 2

   c. 36, 30, 2, 4

   d. 38, 30, 4, 4

7. Anaerobic pathways that oxidize glucose to generate ATP energy by using an organic molecule as the ultimate hydrogen acceptor are called

   a. fermentation.

   b. reduction.

   c. Krebs.

   d. electron pumps.

8. When skeletal muscle cells function anaerobically, they accumulate the compound _____, which causes muscle soreness.

   a. pyruvic acid

   b. malic acid

   c. carbon dioxide

   d. lactic acid

9. Each molecule of fat can release _____ of ATP, compared with a molecule of glucose.

   a. smaller amounts

   b. the same amount

   c. larger amounts

   d. only twice the amount

10. Some organisms excrete ammonia directly; others convert ammonia into other nitrogen-containing compounds, such as

   a. urea or uric acid.

   b. carbon dioxide.

   c. sweat.

   d. fat.

Answers

1. a 2. b 3. b 4. c 5. d 6. a 7. a 8. d 9. c 10. a

Concept Review

*Answers to the Concept Review questions can be found on the **Concepts in Biology**, 12/e website.*

6.1 Energy and Organisms

1. How do autotrophs and heterotrophs differ?
2. What is chemosynthesis?
3. How are respiration and photosynthesis related to autotrophs and heterotrophs?

6.2 Aerobic Cellular Respiration—an Overview

4. Aerobic cellular respiration occurs in three stages. Name these and briefly describe what happens in each stage.
5. Which cellular organelle is involved in the process of respiration?

6.3 The Metabolic Pathways of Aerobic Cellular Respiration

6. For glycolysis, the Krebs cycle, and the electron-transport system, list two molecules that enter and two molecules that leave each pathway.

7. How is each of the following involved in aerobic cellular respiration: NAD^+, pyruvic acid, oxygen, and ATP?

6.4 Aerobic Cellular Respiration in Prokaryotes

8. How is aerobic cellular respiration different between prokaryotic and eukaryotic organisms?

6.5 Anaerobic Cellular Respiration

9. Describe how glycolysis and the Krebs cycle can be used to obtain energy from fats and proteins.

6.6 Metabolic Processing of Molecules Other Than Carbohydrates

10. What are the differences between fat and protein metabolism biochemical pathways?
11. Describe how carbohydrates, fats, and proteins can be interconverted from one to another.

Thinking Critically

*For guidelines to answer this question, visit the **Concepts in Biology**, 12/e website.*

Picture yourself as an atom of hydrogen tied up in a molecule of fat. You are present in the stored fat of a person who is starving. Trace the biochemical pathways you would be part of as you moved through the process of aerobic cellular respiration. Be as specific as you can in describing your location and how you got there, as well as the molecules of which you are a part. Of what molecule would you be a part at the end of this process?

Biochemical Pathways— Photosynthesis

Evidence for the process of photosynthesis, the most important chemical reaction on Earth, dates back about 3 billion years. There has been plenty of time for the process to evolve into a highly efficient method of capturing and converting light energy to ATP, which is used to generate organic compounds useful to all forms of life. The light-conversion processes that nature has fashioned are estimated to be four times more efficient than the best human-made solar cells. Because it is unclear how we will meet our future energy needs, scientists are taking their lead from plants and have begun to explore building *artificial photosynthetic systems*. Plants act like little solar cells that convert light into usable energy. This energy is then used to manufacture organic molecules. In plants, oxygen is the by-product of this process, but researchers believe that, in the near future, they will be able

to change this reaction to produce hydrogen in human-made solar cells. Their goal is to create an artificial system, based on the chemistry of photosynthesis, to produce hydrogen or other fuels. Hydrogen burns cleanly and produces only water as a waste, so it makes an excellent alternative to fossil fuels, such as gasoline and coal. In addition, artificial photosynthesis would use carbon dioxide, helping remove excess carbon dioxide from the environment and control global warming.

Scientists believe this is possible based on their detailed understanding of photosynthesis. There now is excellent information about the many chemical steps involved in photosynthesis and where they occur in the cell. This chapter reveals the scientific knowledge that serves as a background for such forward-thinking scientists.

CHAPTER OUTLINE

7.1 Photosynthesis and Life

Although there are hundreds of different chemical reactions taking place within organisms, this chapter will focus on the reactions involved in the processes of photosynthesis. Recall from chapter 4 that, in photosynthesis, organisms such as green plants, algae, and certain bacteria trap radiant energy from sunlight and convert it into the energy of chemical bonds in large molecules, such as carbohydrates. Organisms that are able to make energy-containing organic molecules from inorganic raw materials are called autotrophs. Those that use light as their energy source are more specifically called *photosynthetic* autotrophs or photoautotrophs.

Among prokaryotes, there are many bacteria capable of carrying out photosynthesis. For example, the cyanobacteria are all capable of manufacturing organic compounds using light energy.

Among the eukaryotes, a few protozoa and all algae and green plants are capable of photosynthesis (figure 7.1). Photosynthesis captures energy for use by the organisms that carry out photosynthesis and provides energy to organisms that eat photosynthetic organisms. An estimated 99.9% of life on Earth relies on photosynthesis for its energy needs. Photosynthesis is also the major supplier of organic compounds, such as carbohydrates. It has been estimated that over 100 billion metric tons of sugar are produced annually by photosynthesis. Photosynthesis also converts about 1,000 billion metric tons of carbon dioxide into organic matter each year, yielding about 700 billion metric tons of oxygen. It is for these reasons that a basic understanding of this biochemical pathway is important.

7.2 An Overview of Photosynthesis

Ultimately, the energy to power all organisms comes from the sun. An important molecule in the process of harvesting sunlight is **chlorophyll,** a green pigment that absorbs light energy. Through photosynthesis, light energy is transformed to chemical-bond energy in the form of ATP. ATP is then used to produce complex organic molecules, such as glucose. It is from these organic molecules that organisms obtain energy through the process of cellular respiration. Recall from chapter 4 that, in algae and the leaves of green plants, photosynthesis occurs in cells that contain organelles called chloroplasts. Chloroplasts have two distinct regions within them: the grana and the stroma. **Grana** consist of stacks of membranous sacs containing chlorophyll, and the **stroma** are the spaces between membranes (figure 7.2).

The following equation summarizes the chemical reactions photosynthetic organisms use to make ATP and organic molecules:

light energy + carbon dioxide + water → glucose + oxygen

$$\text{light energy} + 6\,CO_2 + 6\,H_2O \rightarrow C_6H_{12}O_6 + 6\,O_2$$

There are three distinct events in the photosynthetic pathway:

1. **Light-capturing events.** In eukaryotic cells, photosynthesis takes place within chloroplasts. Each chloroplast is surrounded by membranes and contains chlorophyll, along with other pigments. Chlorophyll and the other pigments absorb specific wavelengths of light.

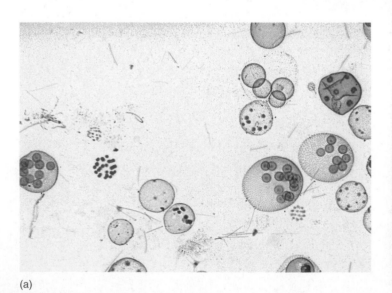

(a)

(b)

FIGURE 7.1 Photosynthetic Autotrophs
(a) Algae *(b)* Plants

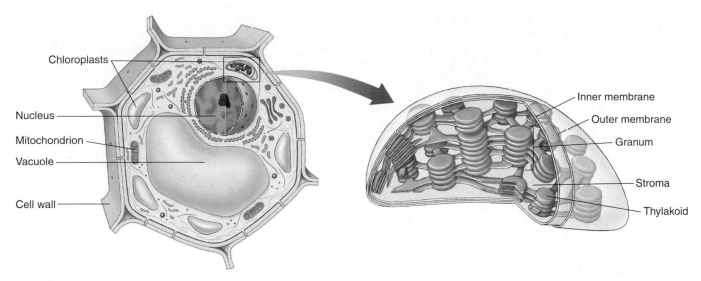

FIGURE 7.2   Photosynthesis and the Structure of a Leaf

The leaves of plants are composed of layers of cells that contain chloroplasts and mitochondria. It is the chloroplasts that contain chlorophyll and that are the site of photosynthesis. The chlorophyll molecules are actually located within chloroplasts in stacks of membranous sacs called *grana*. The fluid-filled space in which the grana are located is called the *stroma* of the chloroplast.

When specific amounts of light are absorbed by the photosynthetic pigments, electrons become "excited." With this added energy, these excited electrons can enter into the chemical reactions responsible for the production of ATP. These reactions take place within the grana of the chloroplast.

2. **Light-dependent reactions.** Light-dependent reactions use the excited electrons produced by the light-capturing events. Light-dependent reactions are also known as *light reactions*. During these reactions, excited electrons from the light-capturing events are used to produce ATP. As a by product, hydrogen and oxygen are also produced. The oxygen from the water is released to the environment as O_2 molecules. The hydrogens are transferred to the electron carrier coenzyme $NADP^+$ to produce NADPH. ($NADP^+$ is similar to NAD^+, which was discussed in chapter 5) These reactions also take place in the grana of the chloroplast. However, the NADPH and ATP leave the grana and enter the stroma, where the light-independent reactions take place.

3. **Light-independent reactions.** These reactions are also known as *dark reactions*, because light is not needed for them to occur. During these reactions, ATP and NADPH from the light-dependent reactions are used to attach CO_2 to a 5-carbon molecule, already present in the cell, to manufacture new, larger organic molecules. Ultimately, glucose ($C_6H_{12}O_6$) is produced. These light-independent reactions take place in the stroma in either the light or dark, as long as ATP and NADPH are available from the light-dependent stage. When the ATP and NADPH give up their energy and hydrogens, they turn back into ADP and $NADP^+$. The ADP and the

$NADP^+$ are recycled back to the light-dependent reactions to be used over again.

The process of photosynthesis can be summarized as follows. During the light-capturing events, *light energy* is captured by chlorophyll and other pigments, resulting in excited electrons. The energy of these excited electrons is used during the light-dependent reactions to disassociate *water* molecules into hydrogen and oxygen, and the *oxygen* is released. Also during the light-dependent reactions, ATP is produced and $NADP^+$ picks up hydrogen released from water to form NADPH. During the light-independent reactions, ATP and NADPH are used to help combine *carbon dioxide* with a 5-carbon molecule, so that ultimately organic molecules, such as *glucose*, are produced (figure 7.3).

7.3   The Metabolic Pathway of Photosynthesis

This discussion is divided into two levels: a fundamental description and a detailed description. It is a good idea to begin with the simplest description and add layers of understanding as you go to additional levels. Ask your instructor which level is required for your course of study.

Fundamental Description

Light-Capturing Events

Light energy is used to drive photosynthesis during the light-capturing events. Visible light is a combination of many different wavelengths of light, seen as different colors. Some of these colors appear when white light is separated into its

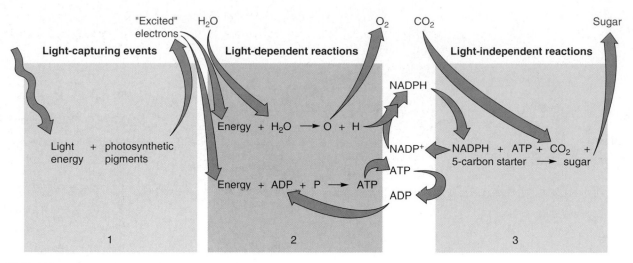

Light-capturing events

Light + photosynthetic
energy   pigments

1

Light-dependent reactions

"Excited" electrons H_2O

Energy + H_2O → O + H

NADPH

NADP$^+$

Energy + ADP + P → ATP

ATP

ADP

2

Light-independent reactions

O_2 CO_2 Sugar

NADPH

NADPH + ATP + CO_2 +
5-carbon starter → sugar

3

FIGURE 7.3 Photosynthesis Overview

Photosynthesis is a complex biochemical pathway in plants, algae, and certain bacteria. Sunlight, along with CO_2 and H_2O, is used to make organic molecules, such as sugar. This illustrates the three parts of the process: (1) the light-capturing events, (2) the light-dependent reactions, and (3) the light-independent reactions. Notice that the end products of the light-dependent reactions, NADPH and ATP, are necessary to run the light-independent reactions, whereas the water and carbon dioxide are supplied from the environment.

colors to form a rainbow. The colors of the electromagnetic spectrum that provide the energy for photosynthesis are correlated with different kinds of light-energy-absorbing pigments. The green chlorophylls are the most familiar and abundant. There are several types of this pigment. The two most common types are chlorophyll *a* and chlorophyll *b*. Both absorb strongly in the red and blue portions of the electromagnetic spectrum, although in slightly different portions of the spectrum (figure 7.4).

Chlorophylls reflect green light. That is why we see chlorophyll-containing plants as predominantly green. Other pigments common in plants are called **accessory pigments.** These include the *carotenoids* (yellow, red, and orange). They absorb mostly blue and blue-green light while reflecting the oranges and yellows. The presence of these pigments is generally masked by the presence of chlorophyll, but in the fall, when chlorophyll disintegrates, the reds, oranges, and yellows show

FIGURE 7.4 The Electromagnetic Spectrum, Visible Light, and Chlorophyll

Light is a form of electromagnetic energy that can be thought of as occurring in waves. The shorter the wavelength, the more energy it contains. Humans are capable of seeing only waves that are between about 400 and 740 nanometers (nm) long. Chlorophyll *a* (the solid graph line) and chlorophyll *b* (the dotted graph line) absorb different wavelengths of light energy.

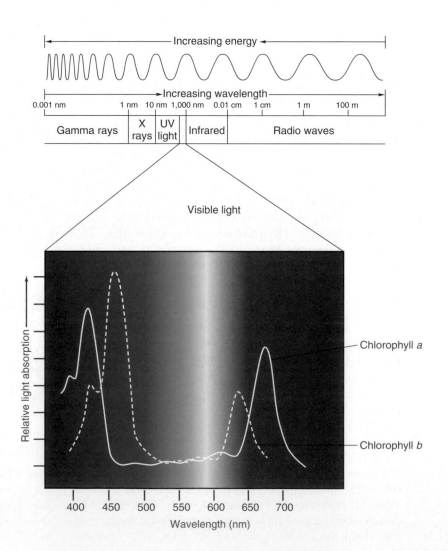

Carotenoids in Tomato

through. Accessory pigments are also responsible for the brilliant colors of vegetables, such as carrots, tomatoes, eggplant, and peppers. Photosynthetic bacteria and various species of algae have other kinds of accessory pigments not found in plants. Having a combination of different pigments, each of which absorbs a portion of the light spectrum hitting it, allows the organism to capture much of the visible light that falls on it.

Any cell with chloroplasts can carry on photosynthesis. However, in most plants, leaves are specialized for photosynthesis and contain cells that have high numbers of chloroplasts (figure 7.5). Chloroplasts are membranous, saclike organelles containing many thin, flattened sacs.

These sacs, called **thylakoids,** contain chlorophylls, accessory pigments, electron-transport molecules, and enzymes. They are stacked in groups called *grana* (singular, *granum*). Recall that the fluid-filled spaces between the grana

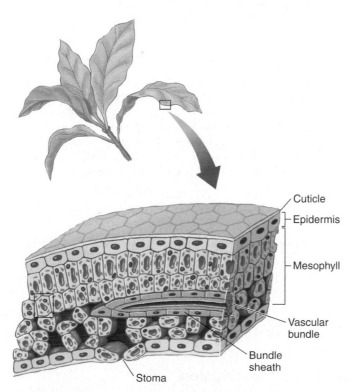

Cuticle

Epidermis

Mesophyll

Vascular bundle

Bundle sheath

Stoma

FIGURE 7.5 **Photosynthesis and the Structure of a Plant Leaf**
Plant leaves are composed of layers of cells that contain chloroplast, which contain chlorophyll.

are called the stroma of the chloroplast. The structure of the chloroplast is directly related to both the light capturing and the energy-conversion steps of photosynthesis. In the light-capturing events, the pigments (e.g., chlorophyll), which are embedded in the membranes of the thylakoids, capture light energy and some of the electrons of pigments become excited. The chlorophylls and other pigments involved in trapping sunlight energy and storing it are arranged into clusters called **photosystems.** By clustering the pigments, photosystems serve as energy-gathering, or energy-concentrating, mechanisms that allow light to be collected more efficiently to excite the electrons to higher energy levels. The light-capturing events result in

photons of light energy → excited electrons from chlorophyll

Light-Dependent Reactions

The light-dependent reactions of photosynthesis take place in the thylakoid membranes inside the chloroplast. The excited electrons from the light-capturing events are passed to protein molecules in the thylakoid membrane. The electrons are passed through a series of electron-transport steps, which result in protons being pumped into the cavity of the thylakoid. When the protons pass back out through the membrane to the outside of the thylakoid, ATP is produced. This is very similar to the reactions that happen in the electron-transport system of aerobic cellular respiration. In addition, the chlorophyll that just lost its electrons to the chloroplast's electron transport system regains electrons from water molecules. This results in the production of hydrogen ions, electrons, and oxygen gas. The next light-capturing event will excite this new electron and send it along the electron transport system. As electrons finish moving through the electron transport system, the coenzyme $NADP^+$ picks up the electrons and is reduced to NADPH. The hydrogen ions attach because, when $NADP^+$ accepts electrons, it becomes negatively charged ($NADP^-$). The positively charged H^+ are attracted to the negatively charged $NADP^-$. The oxygen remaining from the splitting of water molecules is released into the atmosphere, or it can be used by the cell in aerobic cellular respiration, which takes place in the mitochondria of plant cells. The ATP and NADPH molecules move from the grana, where the light-dependent reactions take place, to the stroma, where the light-independent reactions take place.

A generalized reaction that summarizes the light-dependent reactions follows:

excited
electrons + H_2O + ADP + $NADP^+$ → ATP + NADPH + O_2

Light-Independent Reactions

The ATP and NADPH provide energy, electrons and hydrogens needed to build large, organic molecules. The light-independent reactions are a series of oxidation-reduction reactions, which combine hydrogen from water (carried by NADPH) with carbon dioxide from the atmosphere to form simple organic molecules, such as sugar. As CO_2 diffuses into the chloroplasts, the

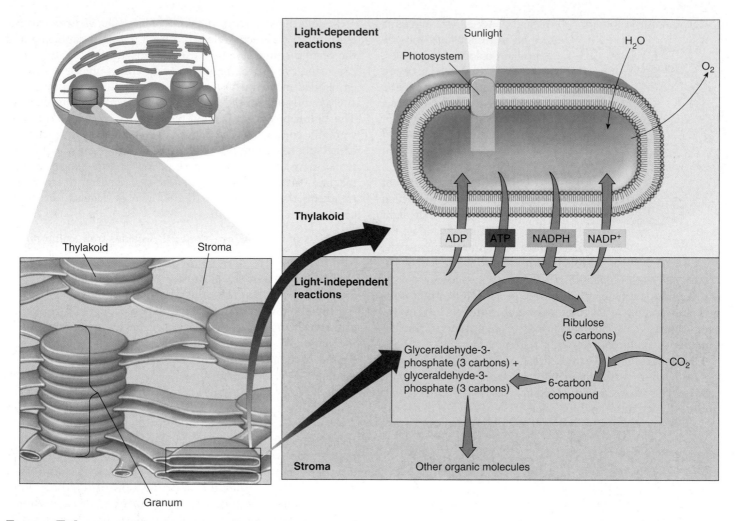

FIGURE 7.6 Photosynthesis: Fundamental Description

The process of photosynthesis involves light capturing events by chlorophyll and other pigments. The excited electrons are used in the light-dependent reactions to split water, releasing hydrogens and oxygen. The hydrogens are picked up by NAD^+ to form NADPH and the oxygen is released. Excited electrons are also used to produce ATP. The ATP and NADPH leave the thylakoid and enter the stroma of the chloroplast, where they are used in the light-independent reactions to incorporate carbon dioxide into organic molecules. During the light-independent reactions, carbon dioxide is added to a 5-carbon ribulose molecule to form a 6-carbon compound, which splits into glyceraldehyde-3-phosphate. Some of the glyceraldehyde-3-phosphate is used to regenerate ribulose and some is used to make other organic molecules. The ADP and $NADP^+$ released from the light-independent reactions stage return to the thylakoid to be used in the synthesis of ATP and NADPH again. Therefore, each stage is dependent on the other.

enzyme **ribulose bisphosphate carboxylase (RuBisCo)** speeds the combining of the CO_2 with an already present 5-carbon sugar, **ribulose.** NADPH then donates its hydrogens and electrons to complete the reduction of the molecule. The resulting 6-carbon molecule is immediately split into two 3-carbon molecules of **glyceraldehyde-3-phosphate.** Some of the glyceraldehyde-3-phosphate molecules are converted through a series of reactions into ribulose. Thus, these reactions constitute a cycle, in which carbon dioxide and hydrogens are added and glyceraldehyde-3-phosphate and the original 5-carbon ribulose are produced. The plant can use surplus glyceraldehyde-3-phosphate for the synthesis of glucose. The plant can also use glyceraldehyde-3-phosphate to construct a wide variety of other organic molecules (e.g., proteins, nucleic

acids), provided there are a few additional raw materials, such as minerals and nitrogen-containing molecules (figure 7.6).

The activities of the light-independent reactions can be summarized as follows:.

$$ATP + NADPH + ribulose + CO_2 \longrightarrow$$
$$ADP + NADP^+ + \text{complex organic molecule} + ribulose$$

Detailed Description

Light-Capturing Events

The energy of light comes in discrete packages, called *photons*. Photons of light having different wavelengths have different amounts of energy. A photon of red light has a different amount of energy than a photon of blue light. Pigments of different kinds are able to absorb photons of certain wavelengths of light. Chlorophyll absorbs red and blue light best and reflects green light. When a chlorophyll molecule is struck by and absorbs a photon of the correct wavelength, its electrons become excited to a higher energy level. This electron is replaced when chlorophyll takes an electron from a water molecule. The excited electron goes on to form ATP. The reactions that result in the production of ATP and the splitting of water take place in the thylakoids of chloroplasts. There are many different molecules involved, and most are embedded in the membrane of the thylakoid. The various molecules involved in these reactions are referred to as *photosystems*. A photosystem is composed of (1) an *antenna complex,* (2) a *reaction center,* and (3) other enzymes necessary to store the light energy as ATP and NADPH.

The antenna complex is a network of hundreds of chlorophyll and accessory pigment molecules, whose role is to capture photons of light energy and transfer the energy to a specialized portion of the photosystem known as the reaction center. When light shines on the antenna complex and strikes a chlorophyll molecule, an electron becomes excited. The energy of the excited electron is passed from one pigment to another through the antenna complex network. This series of excitations continues until the combined energies from sev-

eral excitations are transferred to the reaction center, which consists of a complex of chlorophyll *a* and protein molecules. An electron is excited and passed to a primary electron acceptor molecule, oxidizing the chlorophyll and reducing the acceptor. Ultimately, the oxidized chlorophyll then has its electron replaced with another electron from a different electron donor. Exactly where this replacement electron comes from is the basis on which two different photosystems have been identified—photosystem I and photosystem II.

In summary, the light-capturing reactions take place in the thylakoids of the chloroplast:

1. Chlorophyll and other pigments of the antenna complex capture light energy and produce excited electrons.
2. The energy is transferred to the reaction center.
3. Excited electrons from the reaction center are transferred to a primary electron acceptor molecule.

Light-Dependent Reactions

Both photosystems I and II have antenna complexes and reaction centers and provide excited electrons to primary electron acceptors. However, each has slightly different enzymes and other proteins associated with it, so each does a slightly different job. In actuality, photosystem II occurs first and feeds its excited electrons to photosystem I (figure 7.7). One special

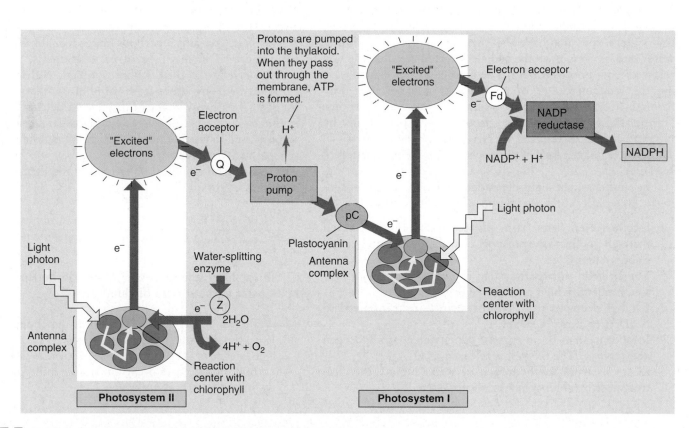

FIGURE 7.7 **Photosystems II and I and How They Interact: Detailed Description**
Although light energy strikes and is absorbed by both photosystem II and I, what happens and how they interconnect are not the same. Notice that the electrons released from photosystem II end up in the chlorophyll molecules of photosystem I. The electrons that replace those "excited" out of the reaction center in photosystem II come from water.

feature of photosystem II is that there is an enzyme in the thylakoid membrane responsible for splitting water molecules ($H_2O \rightarrow 2\,H + O$). The oxygen is released as O_2 and the electrons of the hydrogens are used to replace the electrons that had been lost by the chlorophyll. The remaining hydrogen ions (protons) are released to participate in other reactions. Thus, in a sense, the light energy trapped by the antenna complex is used to split water into H and O. The excited electrons from photosystem II are sent through a series of electron-transport reactions, in which they give up some of their energy. This is similar to the electron-transport system of aerobic cellular respiration. After passing through the electron-transport system, the electrons are accepted by chlorophyll molecules in photosystem I. While the electron-transport activity is happening, protons are pumped from the stroma into the space inside the thylakoid. Eventually, these protons move back across the membrane. When they do, ATPase is used to produce ATP (ADP is phosphorylated to produce ATP). Thus, a second result of this process is that the energy of sunlight has been used to produce ATP.

The connection between photosystem II and photosystem I involves the transfer of electrons from photosystem II to photosystem I. These electrons are important because photons (from sunlight) are exciting electrons in the reaction center of photosystem I and the electrons from photosystem II replace those lost from photosystem I.

In photosystem I, light is trapped and the energy is absorbed in the same manner as in photosystem II. However, this system does not have the enzyme involved in splitting water into oxygen, protons, and electrons; therefore, no O_2 is released from photosystem I. The high-energy electrons leaving the reaction center of photosystem I make their way through a different series of oxidation-reduction reactions. During these reactions, the electrons are picked up by $NADP^+$, which is reduced to NADPH (review figure 7.7). Thus, the primary result of photosystem I is the production of NADPH.

In summary, the light-dependent reactions of photosynthesis take place in the thylakoids of the chloroplast:

1. Excited electrons from photosystem II are passed through an electron-transport chain and ultimately enter photosystem I.
2. The electron-transport system is used to establish a proton gradient, which produces ATP.
3. Excited electrons from photosystem I are transferred to $NADP^+$ to form NADPH.
4. In photosystem II, an enzyme splits water into hydrogen and oxygen. The oxygen is released as O_2.
5. Electrons from the hydrogen of water replace the electrons lost by chlorophyll in photosystem II.

Light-Independent Reactions

The light-independent reactions take place within the stroma of the chloroplast. The materials needed for the light-independent reactions are ATP, NADPH, CO_2, and a 5-carbon starter molecule called *ribulose* (gets the reaction going). The first two ingredients (ATP and NADPH) are made available from the light-dependent reactions, photosystems II and I. The carbon dioxide molecules come from the atmosphere, and the ribulose starter molecule is already present in the stroma of the chloroplast from previous reactions.

Carbon dioxide is said to undergo *carbon fixation* through the **Calvin cycle** (named after its discoverer, Melvin Calvin). In the Calvin cycle, ATP and NADPH from the light-dependent reactions are used, along with carbon dioxide, to synthesize larger, organic molecules. As with most metabolic pathways, the synthesis of organic molecules during the light-independent reactions requires the activity of several enzymes to facilitate the many steps in the process. The fixation of carbon begins with carbon dioxide combining with the 5-carbon molecule ribulose to form an unstable 6-carbon molecule. This reaction is carried out by the enzyme ribulose bisphosphate carboxylase (RuBisCo), reportedly the most abundant enzyme on the planet. The newly formed 6-carbon molecule immediately breaks down into two 3-carbon molecules, each of which then undergoes a series of reactions involving a transfer of energy from ATP and a transfer of hydrogen from NADPH. The result of this series of reactions is two glyceraldehyde-3-phosphate molecules. Because glyceraldehyde-3-phosphate contains 3 carbons and is formed as the first stable compound in this type of photosynthesis, this is sometimes referred to as the C3 photosynthetic pathway. Outlook 7.1 describes some other forms of photosynthesis that do not use the C3 pathway. Some of the glyceraldehyde-3-phosphate is used to synthesize glucose and other organic molecules, and some is used to regenerate the 5-carbon ribulose molecule, so this pathway is a cycle (figure 7.8).

The general chemical equation for the light-independent reactions follows:

$$CO_2 + ATP + NADH + \text{5-carbon starter (ribulose)}$$
$$\longrightarrow \text{glyceraldehyde-3-phosphate} + NADP^+\ ADP + P$$

In summary, the reactions of the light-independent event takes place in the stroma of chloroplasts:

1. ATP and NADPH from the light-dependent reactions leave the grana and enter the stroma.
2. The energy of ATP is used in the Calvin cycle to combine carbon dioxide to a 5-carbon starter molecule (ribulose) to form a 6-carbon molecule.

OUTLOOKS 7.1

The Evolution of Photosynthesis

It is amazing that the processes of photosynthesis in prokaryotes and eukaryotes are so similar. The evolution of photosynthesis goes back over 3 billion years, when all life on Earth was prokaryotic and occurred in organisms that were aquatic. (There were no terrestrial organisms at the time.) Today, some bacteria perform a kind of photosynthesis that does not result in the release of oxygen. In general, these bacteria produce ATP but do not break down water to produce oxygen. Perhaps these are the descendents of the first organisms to carry out a photosynthetic process, and oxygen-releasing photosynthesis developed from these earlier forms of photosynthesis.

Evidence from the fossil record shows that, beginning approximately 2 billion years ago, oxygen was present in the atmosphere. Eukaryotic organisms had not yet developed, so the organisms responsible for producing this oxygen would have been prokaryotic. Today, many kinds of cyanobacteria perform photosynthesis in essentially the same way as plants, although they use a somewhat different kind of chlorophyll. As a matter of fact, it is assumed that the chloroplasts of eukaryotes are evolved from photosynthetic bacteria. Initially, the first eukaryotes to perform photosynthesis would have been various kinds of simple members of the kingdom Protista (various kinds of algae). Today, certain kinds of algae (red algae, brown algae, green algae) have specific kinds of chlorophylls and other accessory pigments different from the others. Because the group known as the green algae has essentially the same chlorophylls as plants, it is assumed that plants are derived from this aquatic group. endosymbiosis, p. 412

The evolution of photosynthesis did not stop once plants came on the scene, however. Most plants perform photosynthesis in the manner described in this chapter. Light energy is used to generate ATP and NADPH, which are used in the Calvin cycle to incorporate carbon dioxide into glyceraldehyde-3-phosphate. Because the primary product of this form of photosynthesis is a

Algae

Corn

Jade Plant

3-carbon molecule of glyceraldehyde-3-phosphate, it is often called C3 photosynthesis. Among plants, there are two interesting variations of photosynthesis, which use the same basic process but add interesting twists.

C4 photosynthesis is common in plants like grasses, such as corn (maize), crab grass, and sugar cane that are typically subjected to high light levels. In these plants, carbon dioxide does not directly enter the Calvin cycle. Instead, the fixation of carbon is carried out in two steps, and two kinds of cells participate. It appears that this adaptation allows plants to trap carbon dioxide more efficiently from the atmosphere under high light conditions. Specialized cells in the leaf capture carbon dioxide and convert a 3-carbon compound to a 4-carbon compound. This 4-carbon compound then releases the carbon dioxide to other cells, which use it in the normal Calvin cycle typical of the light-independent reactions. Because a 4-carbon molecule is formed to "store" carbon, this process is known as C4 photosynthesis.

Another variation of photosynthesis is known as Crassulacean acid metabolism (CAM), because this mechanism was first discovered in members of the Crassulaceae family of plants. (A common example, *Crassula*, is known as the jade plant.) CAM photosynthesis is a modification of the basic process of photosynthesis that allows photosynthesis to occur in arid environments while reducing the potential for water loss. In order for plants to take up carbon dioxide, small holes in the leaves (stomates) must be open to allow carbon dioxide to enter. However, relative humidity is low during the day and plants would tend to lose water if their stomates were open. CAM photosynthesis works as follows: During the night, the stomates open and carbon dioxide can enter the leaf. The chloroplasts trap the carbon dioxide by binding it to an organic molecule, similar to what happens in C4 plants. The next morning, when it is light (and drier), the stomates close. During the day, the chloroplasts can capture light and do the light-dependent reactions. They then use the carbon stored the previous night to do the light-independent reactions.

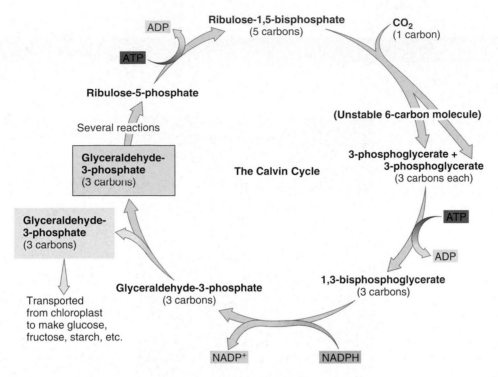

FIGURE 7.8 The Calvin Cycle: Detailed Description

During the Calvin cycle, ATP and NADPH from the light-dependent reactions are used to attach CO_2 to the 5-carbon ribulose molecule. The 6-carbon molecule formed immediately breaks down into two 3-carbon molecules. Some of the glyceraldehyde-3-phosphate formed is used to produce glucose and other, more complex organic molecules. In order to accumulate enough carbon to make a new glucose molecule, the cycle must turn six times. The remaining glyceraldehyde-3-phosphate is used to regenerate the 5-carbon ribulose to start the process again.

3. The 6-carbon molecule immediately divides into two 3-carbon molecules of glyceraldehyde-3-phosphate.
4. Hydrogens from NADPH are transferred to molecules in the Calvin cycle.
5. The 5-carbon ribulose is regenerated.
6. ADP and NADP+ are returned to the light-dependent reactions.

Glyceraldehyde-3-Phosphate: The Product of Photosynthesis

The 3-carbon glyceraldehyde-3-phosphate is the actual product of the process of photosynthesis. However, many textbooks show the generalized equation for photosynthesis as

$$CO_2 + H_2O + light \rightarrow C_6H_{12}O_6 + O_2$$

making it appear as if a 6-carbon sugar (hexose) were the end product. The reason a hexose ($C_6H_{12}O_6$) is usually listed as the end product is simply because, in the past, the simple sugars were easier to detect than was glyceraldehyde-3-phosphate.

Several things can happen to glyceraldehyde-3-phosphate. If a plant goes through photosynthesis and produces

12 glyceraldehyde-3-phosphates, 10 of the 12 are rearranged by a series of complex chemical reactions to regenerate the 5-carbon ribulose needed to operate the light-independent reactions stage. The other two glyceraldehyde-3-phosphates can be considered profit from the process. The excess glyceraldehyde-3-phosphate molecules are frequently changed into a hexose, so the scientists who first examined photosynthesis chemically saw additional sugars as the product and did not realize that glyceraldehyde-3-phosphate is the initial product.

Cells can do a number of things with glyceraldehyde-3-phosphate, in addition to manufacturing hexose (figure 7.9). Many other organic molecules can be constructed using glyceraldehyde-3-phosphate as the basic construction unit. Glyceraldehyde-3-phosphate can be converted to glucose molecules, which can be combined to form complex carbohydrates, such as starch for energy storage or cellulose for cell wall construction. In addition, other simple sugars can be used as building blocks for ATP, RNA, DNA, and other carbohydrate-containing materials.

The cell can convert the glyceraldehyde-3-phosphate into lipids, such as oils for storage, phospholipids for cell membranes, or steroids for cell membranes. The glyceraldehyde-3-phosphate can serve as the carbon skeleton for the construction of the amino acids needed to form proteins.

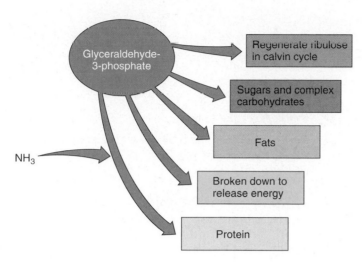

FIGURE 7.9 **Uses for Glyceraldehyde-3-Phosphate**
The glyceraldehyde-3-phosphate that is produced as the end
product of photosynthesis has a variety of uses. The plant cell can
make simple sugars, complex carbohydrates, or even the original
5-carbon starter from it. The glyceraldehyde-3-phosphate can also
serve as an ingredient of lipids and amino acids (proteins). In
addition, it provides a major source of metabolic energy when it is
sent through the respiratory pathway.

Almost any molecule that a green plant can manufacture
begins with this glyceraldehyde-3-phosphate molecule.
Finally, glyceraldehyde-3-phosphate can be broken down
during cellular respiration. Cellular respiration releases the
chemical-bond energy from glyceraldehyde-3-phosphate and
other organic molecules and converts it into the energy of
ATP. This conversion of chemical-bond energy enables the
plant cell and the cells of all organisms to do things that
require energy, such as grow and move materials.

7.4 Other Aspects of Plant Metabolism

Photosynthetic organisms are able to manufacture organic
molecules from inorganic molecules. Once they have the basic
carbon skeleton, they can manufacture a variety of other complex molecules for their own needs—fats, proteins, and complex carbohydrates are some of the more common. However,
plants produce a wide variety of other molecules for specific
purposes. Among the molecules they produce are compounds
that are toxic to animals that use plants as food. Many of
these compounds have been discovered to be useful as medicines. Digitalis from the foxglove plant causes the hearts of
animals that eat the plant to malfunction (figure 7.10).
However, it can be used as a medicine in humans who have
certain heart ailments. Molecules that paralyze animals have
been used in medicine to treat specific ailments and relax
muscles, so that surgery is easier to perform. Still others have
been used as natural insecticides.

FIGURE 7.10 **Foxglove**
Foxglove produces the compound cardenolide digitoxin, a valuable
medicine in the treatment of heart disease. The drug containing this
compound is known as digitalis.

Vitamins are another important group of organic molecules derived from plants. Vitamins are organic molecules
that we cannot manufacture but must have in small amounts
to maintain good health. The vitamins we get from plants are
manufactured by them for their own purposes. By definition,
they are not vitamins to the plant, because the plant makes
them for its own use. However, because we cannot make
them, we rely on plants to synthesize these important molecules for us, and we consume them when we eat foods containing them.

7.5 Interrelationships Between Autotrophs and Heterotrophs

The differences between autotrophs and heterotrophs were
described in chapter 6. Autotrophs are able to capture energy
to manufacture new organic molecules from inorganic molecules. Heterotrophs must have organic molecules as starting
points. *However, it is important to recognize that all organisms must do some form of respiration.* Plants and other
autotrophs obtain energy from food molecules, in the same
manner as animals and other heterotrophs—by processing
organic molecules through the respiratory pathways. This

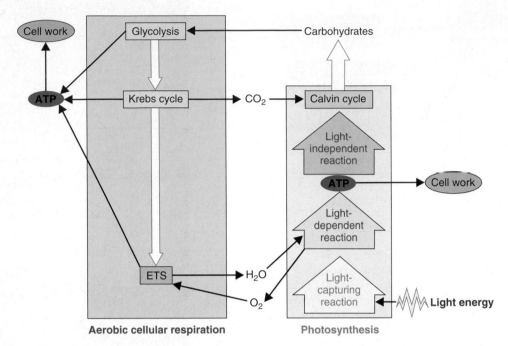

FIGURE 7.11 The Interdependence of Photosynthesis and Aerobic Cellular Respiration
Although both autotrophs and heterotrophs carry out cellular respiration, the photosynthetic process that is unique to photosynthetic autotrophs provides essential nutrients for both processes. Photosynthesis captures light energy, which is ultimately transferred to heterotrophs in the form of carbohydrates and other organic compounds. Photosynthesis also generates O_2, which is used in aerobic cellular respiration. The ATP generated by cellular respiration in both heterotrophs, (e.g., animals) and autotrophs (e.g., plants) is used to power their many metabolic processes. In return, cellular respiration supplies two of the most important basic ingredients of photosynthesis, CO_2 and H_2O.

means that plants, like animals, require oxygen for the ETS portion of aerobic cellular respiration.

Many people believe that plants only give off oxygen and never require it. Actually, plants do give off oxygen in the light-dependent reactions of photosynthesis, but in aerobic cellular respiration they use oxygen, as does any other organism. During their life spans, green plants give off more oxygen to the atmosphere than they take in for use in respiration. The surplus oxygen given off is the source of oxygen for aerobic cellular respiration in both plants and animals. Animals

are dependent on plants not only for oxygen but ultimately for the organic molecules necessary to construct their bodies and maintain their metabolism (figure 7.11).

Thus, animals supply the raw materials—CO_2, H_2O, and nitrogen—needed by plants, and plants supply the raw materials—sugar, oxygen, amino acids, fats, and vitamins—needed by animals. This constant cycling is essential to life on Earth. As long as the sun shines and plants and animals remain in balance, the food cycles of all living organisms will continue to work properly.

Summary

Sunlight supplies the essential initial energy for making the large organic molecules necessary to maintain the forms of life we know. Photosynthesis is the process by which plants, and some bacteria and protists, use the energy from sunlight to produce organic compounds. Cellular respiration converts these compounds into ATP, the fuel used by all living things. In the light-capturing events of photosynthesis, plants use chemicals, such as chlorophyll, to trap the energy of sunlight using photosystems. During the light-dependent reactions, they manufacture a source of chemical energy, ATP, and a source of hydrogen, NADPH. Atmospheric oxygen is released in this stage. In the light-independent reactions of photosynthesis, the ATP energy is used in a series of reactions (the Calvin cycle) to join the hydrogen from the NADPH to a molecule of carbon dioxide and form a simple carbohydrate, glyceraldehyde-3-phosphate. In subsequent reactions, plants use the glyceraldehyde-3-phosphate as a source of energy and raw materials to make complex carbohydrates, fats, and other organic molecules.

Key Terms

*Use the interactive flash cards on the **Concepts in Biology**, 12/e website to help you learn the meaning of these terms.*

accessory pigments 132
Calvin cycle 136
chlorophyll 130
glyceraldehyde-3-phosphate 134
grana 130
light-capturing events 130
light-dependent reactions 131

light-independent reactions 131
photosystems 133
ribulose 134
ribulose bisphosphate carboxylase (RuBisCo) 134
stroma 130
thylakoids 133

Basic Review

1. Which of the following is *not* able to carry out photosynthesis?
   a. algae
   b. cyanobacteria
   c. frogs
   d. broccoli

2. A _____ consists of stacs of membranous sacs containing chlorophyll.
   a. granum
   b. stroma
   c. mitochondrion
   d. cell wall

3. During the _____ reactions, ATP and NADPH are used to help combine carbon dioxide with a 5-carbon molecule, so that ultimately organic molecules, such as glucose, are produced.
   a. light-independent
   b. light-dependent
   c. Calvin cycle
   d. Krebs cycle

4. Pigments other than the green chlorophylls that are commonly found in plants are collectively known as _____. These include the carotenoids.
   a. chlorophylls
   b. hemoglobins
   c. accessory pigments
   d. thylakoids

5. This enzyme speeds the combining of CO_2 with an already present 5-carbon carbohydrate.
   a. DNAase
   b. ribulose

   c. ribulose bisphosphate carboxylase (RuBisCo)
   d. phosphorylase

6. Carbon dioxide undergoes carbon fixation, which occurs in the
   a. Calvin cycle.
   b. Krebs cycle.
   c. light-dependent reactions.
   d. photosystem I.

7. The chlorophylls and other pigments involved in trapping sunlight energy and storing it are arranged into clusters called
   a. chloroplasts.
   b. photosystems.
   c. cristae.
   d. thylakoids.

8. Light energy comes in discrete packages called
   a. quanta.
   b. lumina.
   c. photons.
   d. brilliance units.

9. The electrons released from photosystem _____ end up in the chlorophyll molecules of photosystem _____.
   a. I, II
   b. A, B
   c. B, A
   d. II, I

10. _____ are sacs containing chlorophylls, accessory pigments, electron-transport molecules, and enzymes.
    a. Thylakoids
    b. Mitochondria
    c. Photosystems
    d. Chloroplasts

Answer

1. c   2. a   3. a   4. c   5. c   6. a   7. b   8. c   9. d
10. d

Concept Review

*Answers to the Concept Review questions can be found on the **Concepts in Biology**, 12/e website.*

7.1   Photosynthesis and Life

7.2   An Overview of Photosynthesis

1. Photosynthesis is a biochemical pathway that involves three kinds of activities. Name these and explain how they are related to each other.

2. Which cellular organelle is involved in the process of photosynthesis?

7.3 The Metabolic Pathways of Photosynthesis

3. How do photosystem I and photosystem II differ in the kinds of reactions that take place?
4. What does an antenna complex do? How is it related to the reaction center?

7.4 Other Aspects of Plant Metabolism

5. Is vitamin C a vitamin for an orange tree?

7.5 Interrelationships Between Autotrophs and Heterotrophs

6. Even though animals do not photosynthesize, they rely on the sun for their energy. Why is this so?
7. What is an autotroph? Give an example.
8. Photosynthetic organisms are responsible for producing what kinds of materials?

Draw your own simple diagram that illustrates how photosynthesis and respiration are interrelated.

Thinking Critically

*For guidelines to answer this question, visit the **Concepts in Biology**, 12/e website.*

Both plants and animals carry on metabolism. From a metabolic point of view, which of the two is more complex? Include in your answer the following topics:

1. Cell structure
2. Biochemical pathways
3 Enzymes
4. Organic molecules
5. Photosynthetic autotrophy and heterotrophy

DNA and RNA
The Molecular Basis of Heredity

Type I diabetes is due to the inability to make a protein called insulin. In humans, the cells in the pancreas normally produce enough insulin to help regulate the rapidly changing levels of sugar that result from eating. A person who lives with Type I diabetes must take insulin injections with their meals as source of insulin. Unregulated blood sugar levels can cause death.

Since the 1920's, insulin isolated from pigs and cows was the common method of treating diabetes. Because of their similarities to human insulin, pig and cow insulins were effective means of regulating blood sugar levels. Unfortunately diabetic patients could develop allergic reactions to pig and cow insulins. As a result the treatment was not effective.

In the late 1990's synthetically produced human insulin became available to patients through medical prescriptions. Our understanding of the DNA molecule advanced greatly in the 1950's. Since then, scientists have developed the technology to produce human insulin in bacteria. This source of insulin is identical to human insulin and has the same regulatory effects. Bacteria are able to use human DNA to make human insulin because the code that allows cells to make proteins from information in DNA is the same in all cells. This chapter will discuss this code and how proteins, like insulin, are synthesized by cells.

CHAPTER OUTLINE

8.1 DNA and the Importance of Proteins

Chemicals are important in establishing a cell's structure and in maintaining its day-to-day activities. Four common types of macromolecules are present in cells—lipids, carbohydrates, proteins, and nucleic acids. Each of these groups of molecules has an important role in cells. Cell membranes function well because of the structure and chemical properties of phospholipids. Carbohydrates provide the cell with energy and form some cell structures. Proteins provide structure and help the cell accomplish chemical reactions to produce needed chemicals. Nucleic acids contain the information to make proteins.

macromolecules, p. 48

Proteins play a critical role in how cells successfully meet the challenges of living. Microtubules, intermediate filaments, and microfilaments maintain cell shape and aid in movement. Other types of proteins—enzymes—carry out important chemical reactions. Enzymes are so important to a cell that the cell will not live long if it cannot reliably create the proteins it needs for survival.

Most of the characteristics of multicellular organisms are the direct result of proteins in action. The cell's ability to make a particular protein comes from the genetic information stored in the cell's **deoxyribonucleic acid,** or DNA, is a nucleic acid that contains the blueprint for making the proteins the cell needs. DNA contains genes, which are specific messages that tell the cell to link particular amino acids in a specific order; that is, they determine a protein's primary structure.

8.2 DNA Structure and Function

DNA is able to accomplish two very important things for an organism. First, it is the chemical used to *pass genetic information* on to the next generation of organisms. Second, DNA determines an organism's characteristics by *controlling the synthesis of proteins*. The key to understanding how DNA accomplishes these tasks is in its chemical structure.

DNA Structure

DNA is one member of a group of molecules called *nucleic acids*. **Nucleic acids** are large polymers made of many repeating units called nucleotides. Each nucleotide is composed of a sugar molecule, a phosphate group, and a molecule called a *nitrogenous base*. DNA nucleotides contain a *deoxyribose* sugar and come in four different types. Each type of nucleotide has a different nitrogenous base: **adenine** (A), **guanine** (G), **cytosine** (C), and **thymine** (T) (figure 8.1). The nucleotides can combine into a long linear DNA molecule that can pair with another linear DNA molecule. The two paired strands of DNA form a helix, with the sugars and phosphates on the outside and the nitrogenous bases in the

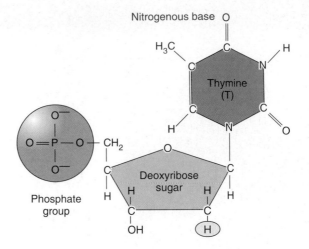

(a) DNA nucleotide

(b) The four nitrogenous bases that occur in DNA

FIGURE 8.1 DNA Structure

The nucleotide is the basic structural unit of all nucleic acids. The nucleotide consists of a sugar, a nitrogenous base, and a phosphate group. Part *(a)* shows a thymine DNA nucleotide. Notice how DNA nucleotides consist of three parts—a sugar, a nitrogenous base, and a phosphate group. *(b)* In DNA, the nitrogenous bases can be adenine, guanine, cytosine, and thymine.

inside of the helix. The nucleotides help stabilize the helical structure by forming weak chemical interactions, called hydrogen bonds. These hydrogen bonds occur only between certain combinations of nitrogenous bases. Adenine pairs with thymine and guanine pairs with cytosine (figure 8.2).

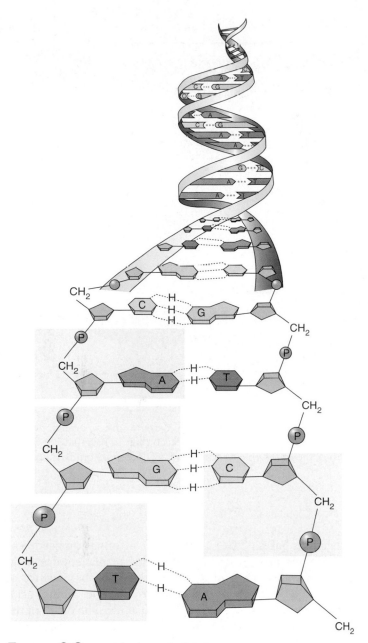

FIGURE 8.2 Double-Stranded DNA

Polymerized deoxyribonucleic acid (DNA) is a helical molecule. While the units of each strand are held together by covalent bonds, the two parallel strands are interlinked by hydrogen bonds between the paired nitrogenous bases. Four nucleotides are highlighted in the bottom of the structure.

Base Pairing in DNA Replication

When a cell grows and divides, two new cells result. Both cells need DNA to survive, so the DNA of the parent cell is copied. One copy is provided to each daughter cell. **DNA replication** is the process by which a cell makes another copy of its DNA. The process of DNA replication relies on DNA base pairing rules and many enzymes. The enzymes that repli-

cate DNA are coded for by the DNA they replicate. The general process is the same in most cells. In eukaryotic cells, this process happens simultaneously at many places along the length of the DNA.

1. DNA replication begins as enzymes, called *helicases*, bind to the DNA and separate the two strands of DNA. This forms a replication bubble (figure 8.3a and 8.3b).
2. As helicases separate the two DNA strands, another enzyme, *DNA polymerase* incorporates DNA nucleotides into the new DNA strand. Nucleotides enter each position according to base-pairing rules—adenine pairs with thymine, guanine pairs with cytosine (figure 8.3c and d).
3. In eukaryotic cells, replication process starts at the same time in several places along the DNA molecule. As the points of DNA replication meet each other, they combine and a new strand of DNA is formed (figure 8.3e). The result is two identical, double-stranded DNA molecules.

The new strands of DNA form on each of the old DNA strands. In this way, the exposed nitrogenous bases of the original DNA serve as the pattern on which the new DNA is formed. The completion of DNA replication yields two double helices, which have identical nucleotide sequences, because the DNA replication process is highly accurate. It has been estimated that there is only one error made for every 2×10^9 nucleotides. Because this error rate is so small, DNA replication is considered to be essentially error-free. A portion of the enzyme that carries out DNA replication edits the newly created DNA molecule for the correct base pairing. When an incorrect match is detected, DNA polymerase removes the incorrect nucleotide and replaces it.

Newly made DNA molecules are eventually passed on to the daughter cells. A cell does not really die when it reproduces itself; it merely starts over again. This is called the *life cycle* of a cell. When a cell divides, it redistributes its genetic information to the next generation of cells.

The Repair of Genetic Information

Errors and damage do occasionaly occur to the DNA helix. The pairing arrangement of the nitrogenous bases allows for the correction of damage to one strand of DNA by reading the remaining undamaged strand. If damage occurs to a strand that reads *AGC*, the correct information is still found in the other strand that reads *TCG*. By using enzymes to read the healthy strand, the cell can rebuild the *AGC* strand with the pairing rule that A pairs with T and G pairs with C (figure 8.4).

The DNA Code

DNA is important because it serves as a reliable way of storing information. The order of the nitrogenous bases (A, G, C, and T) in DNA is the genetic information that codes for proteins. This is similar to how letters present information in sentences. For the cell, the letters of its alphabet consist only of the

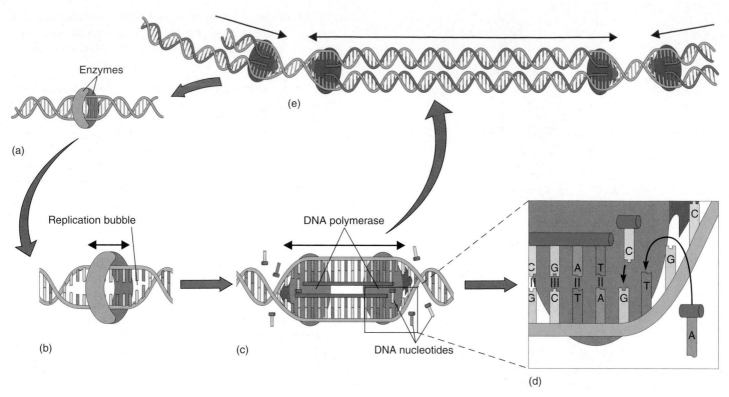

Enzymes

(a)

(e)

Replication bubble

(b)

DNA polymerase

(c)

DNA nucleotides

(d)

FIGURE 8.3 DNA Replication

(a) Enzymes bind to the DNA molecule. *(b)* The enzymes separate the two strands of DNA. *(c, d)* As the DNA strands are separated, new DNA nucleotides are added to the new strands by DNA polymerase. The new DNA strands are synthesized according to base-pairing rules for nucleic acids. *(e)* By working in two directions at once along the DNA strand, the cell is able to replicate the DNA more quickly. Each new daughter cell receives one of these copies.

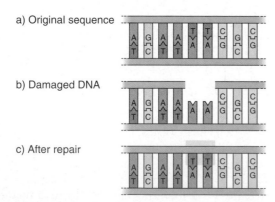

a) Original sequence

b) Damaged DNA

c) After repair

FIGURE 8.4 DNA Repair

(a) An Undamaged DNA consists of two continuous strands held together at the nitrogenous bases (A, T, G, and C). The sequence of the bases contains genetic information. *(b)* Damaged DNA has part of one strand missing. The thymine residues have been damaged and removed. *(c)* When one strand is damaged, it is possible to rebuild this strand by using the nucleotide sequence on the other side. In DNA, adenine always pairs with thymine.

nitrogenous bases A, G, C, and T. The information needed to code for one protein can be thousands of nucleotides long. The nucleotides are read in sets of three. Each sequence of three is a codeword that codes for a single amino acid. Proteins are made of a string of amino acids. The order of the amino acids corresponds to the order of the codewords (How Science Works 8.1).

8.3 RNA Structure and Function

Recall from chapter 3 that ribonucleic acid (RNA) is another type of nucleic acid important in protein production. RNA contains a *ribose* sugar. Ribose and deoxyribose differ by the chemical group that is present on one of the carbons. Ribose has an —OH group but deoxyribose has an —H group on the sugar ring. RNA contains the nitrogenous bases **uracil** (U), guanine (G), cytosine (C), and adenine (A) (figure 8.5). Note that the sets of nitrogenous bases in DNA and RNA are slightly different. RNA has uracil, whereas DNA has

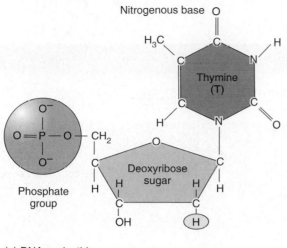

(a) DNA nucleotide

(b) RNA nucleotide

FIGURE 8.5 **RNA Nucleotide**
DNA and RNA differ from each other chemically and in the nitrogenous bases that are present in each molecule. *(a)* The deoxyribose sugar of DNA has a —H on the circled carbon. DNA also contains the nitrogenous base thymine. *(b)* The ribose sugar of RNA has an —OH on the circled carbon. RNA contains the nitrogenous base uracil instead of thymine.

thymine. Like DNA, the information in RNA is found in the order of the RNA molecule's nitrogenous bases.

Cells use DNA and RNA differently. DNA is found in the cell's nucleus; it is the original source for information to make proteins. RNA is made in the nucleus and is then taken into the cytoplasm of the cell. Once RNA is in the cytoplasm, it can help in the process of protein assembly.

The protein-coding information in RNA comes directly from DNA. RNA is made by enzymes that read the protein-coding information in DNA. Like DNA replication, RNA synthesis also follows base-pairing rules, Guanine and cytosine pair. RNA contains uracil, not thymine, so adenine in DNA pairs with uracil in RNA. The thymine in DNA still pairs with adenine in RNA.

RNA differs from DNA in some other important ways. RNA is single-stranded. When RNA is synthesized from DNA, it exists only as a single strand, not a double-stranded molecule, as is DNA. Different types of RNA molecules are classified by the way in which it used. Although all RNA is chemically the same, different types of RNA molecules have different functions. These various types of RNA all participate in making protein, but their roles are different.

8.4 Protein Synthesis

DNA and RNA are both important parts of making protein, and their molecular structures are important in understanding how this process takes place in cells. The order of the nitrogenous bases in both DNA and RNA coordinates how proteins are made.

In the cell, the DNA nucleotides are used in a cellular alphabet, which consists of only four letters. The letters of this alphabet are arranged in sets of three (*ATC, GGA, TCA, CCC,* etc.) to form code words in the DNA language. Each three-letter code word codes for a single amino acid. It is the sequence of these code words in DNA that dictates which amino acids are used and the order in which they appear in the synthesized protein. When RNA is created to aid in protein synthesis, the RNA carries these code words to the cytoplasm for use in assembling a protein. The first step in making a protein is to make the RNA for that protein.

Step One: Trancription

Transcription is the process of using DNA as a template to synthesize RNA. The enzyme **RNA polymerase** reads the sequence of DNA nucleotides and follows the base-pairing rules between DNA and RNA to build the new RNA molecule. Transcription begins in the nucleus, when enzymes separates the two strands of the double-stranded DNA. Separating the two strands of DNA exposes their nitrogenous bases, so that they can be read and paired with the RNA nucleotides. Only one of the two strands of DNA is read to create RNA for each gene. The strand of DNA that serves as a template for the synthesis of RNA is the **coding strand**. The strand of DNA that is not read directly by the enzymes is the **non-coding strand** (figure 8.6). enzymes, p. 95

The DNA molecule is very long and can code for many proteins along its length. The RNA polymerase molecule scans for the portions of DNA sequence that code for proteins. RNA polymerase does this by moving along the

HOW SCIENCE WORKS 8.1

Of Men (and Women!), Microbes, and Molecules

As recently as the 1920s, scientists did not understand the molecular basis of heredity. They understood genetics in terms of the odds that a given trait would be passed on to an individual in the next generation. This "probability" model of genetics left some questions unanswered:

- What is the nature of genetic information?
- How does the cell use genetic information?

Genetic Material Is Molecular

As is often the case in science, accidental discovery played a large role in answering questions about the nature and use of genetic information. In 1928, a medical doctor, Frederick Griffith, was studying two bacterial strains that caused pneumonia. One of the strains was extremely virulent (disease causing) and therefore killed mice very quickly. The other strain was not virulent. Griffith observed something unexpected when *dead* cells of the virulent strain were mixed with *living* cells of the nonvirulent strain: The nonvirulent strain took on the virulent characteristics of the dead strain. Genetic material had been transferred from the dead, virulent strain to the living, nonvirulent strain. This observation was the first significant step in understanding the molecular basis of genetics, because it provided scientists with a situation wherein the scientific method could be applied to ask questions and take measurements about the molecular basis of genetics. Until this point, scientists had lacked a method to provide supporting data.

This spurred the scientific community for the next 14 years to search for the identity of the "genetic molecule." A common hypothesis was that the genetic molecule would be one of the macromolecules—carbohydrates, lipids, proteins, or nucleic acids. During that period, many advances were made in

DNA Double Helix

how researchers studied cells. Many of the top minds in the field had formulated the hypothesis that the genetic molecule was protein. They had very good support for this hypothesis, too. Their argument boiled down to two ideas. The first idea is that proteins are found everywhere in the cell. It follows that, if proteins were the genetic information, they would be found wherever that information was used. The second idea is that proteins are structurally and chemically complex. They are made up of 20 different monomers (amino acids) that come in a wide variety of sizes and shapes. This complexity might account for all the genetic variety observed in nature. On the other hand, very few scientists seriously considered the notion

---

grooves of the DNA helix to find the nitrogenous bases that act as markers, or signs, that a gene is nearby. **Promoter sequences** are specific sequences of DNA nucleotides that RNA polymerase uses to find a protein-coding region of DNA and to identify which of the two DNA strands is the coding strand. Without these promoter sequences, RNA polymerase will not transcribe the gene. **Termination sequences** are DNA nucleotide sequences that indicate when RNA polymerase should finish making an RNA molecule (figure 8.7).

Transcription produces three types of RNA: *messenger RNA (mRNA)*, *transfer RNA (tRNA)*, and *ribosomal RNA (rRNA)*. Each type of RNA has a particular function in the process of protein synthesis. **Messenger RNA (mRNA)** carries the blueprint for making the necessary protein. **Transfer RNA (tRNA)** and **ribosomal RNA (rRNA)** are used to read the mRNA and bring in the necessary amino acids.

Step Two: Translation

Translation is the process of using the information in RNA to direct protein synthesis. The mRNA is read in linear sets of three nucleotides called *codons*. A **codon** is a set of three nucleotides that determines the position of a particular amino acid in the structure of a protein during translation. Each codon corresponds to a three nucleotide code word in the coding strand of the DNA. Table 8.1 shows each of the nucleotide combinations and their corresponding amino acid. For example the codon UUU corresponds to only the amino acid phenylalanine (Phe). There are 64 possible codons and only 20 commonly used amino acids. This means that some amino acids are coded for by more than 1 codon. For example, the codons UCU, UCC, UCA, and UCG all code for serine.

Recall from chapter 4 that a ribosome is a nonmembranous organelle that synthesizes proteins. A ribosome is made

HOW SCIENCE WORKS 8.1 *(continued)*

that DNA was the heritable material. After all, it was found only in the nucleus and consisted of only four monomers (nucleotides). How could this molecule account for the genetic complexity of life?

Genetic Material Is DNA

In 1944, Oswald Avery and his colleagues provided the first evidence that DNA is the genetic molecule. They performed an experiment similar to Griffith's. Avery's innovation was to use purified samples of protein, DNA, lipids, and carbohydrates from the virulent bacteria strain to transfer the virulent characteristics to the nonvirulent bacterial strain. His data indicated that DNA contains genetic information. The scientific community was highly skeptical of these results for two reasons: (1) Scientists had expected the genetic molecule to be protein, so they hadn't expected this result. More important, (2) Avery didn't know how to explain how DNA functions as the genetic molecule. Because of this mind set, Avery's data were largely disregarded on the rationale that his samples were impure. However, this objection was a factor he controlled for in his scientific design, and he reported over 99% purity in the tested DNA samples. It took 8 additional years and a different type of experiment to establish DNA as the genetic molecule.

In 1952, Alfred Hershey and Martha Chase carried out the experiment that settled the issue. Their experiment used a relatively simple genetic system—a bacterial *phage*. A phage is a type of virus that uses a bacterial cell as its host. The phage used in this experiment contained DNA. Hershey and Chase hypothesized that it was necessary for the phage's genetic information to enter the bacterial cell to create new phage. By radioactively labeling the DNA and the protein of the phage, Hershey and Chase were able to show that the DNA entered the bacterial cell, although very little protein did. They reasoned that DNA must be the genetic information.

The Structure and Function of DNA

Researchers then turned toward the issue of determining how DNA works as the heritable material. Scientists expected that the genetic molecule would have to do a number of things, such as store information, use the genetic information throughout the cell, be able to mutate, and be able to replicate itself. Their hypothesis was that the answer was hidden in the structure of the DNA molecule itself.

The investigation of how DNA functioned as the cell's genetic information took a wide variety of strategies. Some scientists looked at DNA from different organisms. They found that, in nearly every organism, the guanine (G) and cytosine (C) nucleotides were present in equal amounts. The same held true for adenine (A) and thymine (T). Later, this provided the basis for establishing the nucleic acid base-pairing rules.

Rosalind Franklin used X-ray crystallography to determine DNA's width, its helical shape, and the repeating patterns that occur along the length of the DNA molecule. Finally, two young scientists, James Watson and Francis Crick, put it all together. They simply listened to and read the information that was being discussed in the scientific community. Their key role was in the assimilation of all the data. They recognized the importance of the X-ray crystallography data in conjunction with the organic structures of the nucleotides and the data that established the base-pairing rules. Together, they created a model for the structure of DNA that accounts for all the things that a genetic molecule must do. They published an article describing this model in 1952. Ten years later, they were awarded the Nobel Prize for their work.

---

of proteins and a type of RNA called *ribosomal RNA (rRNA)*. Ribosomes usually exist in the cell as two pieces, or subunits. There is a large subunit and a small subunit. During translation, the two subunits combine and hold the mRNA between them. With the mRNA firmly sandwiched into the ribosome, the mRNA's codons are read and protein synthesis begins.

The cell has many ribosomes available for protein synthesis. Any of the ribosomes can read any of the mRNAs that come from the cell's nucleus after transcription. Some ribosomes are free in the cytoplasm, whereas others are attached to the cell's endoplasmic reticulum (ER). Proteins destined to be part of the cell membrane or packaged for export from the cell are synthesized on ribosomes attached to the endoplasmic reticulum. Proteins that are to perform their function in the cytoplasm are synthesized on unattached ribosomes. The process of translation can be broken down into three basic steps: (1) initiation, (2) elongation, and (3) termination.

Initiation

Protein synthesis begins with the small ribosomal subunit binding to a specific signal sequence on the mRNA. At this point, the first amino acid for the protein enters. Amino acids are taken to the mRNA-ribosome complex by *transfer RNA*. Transfer RNA (tRNA) is responsible for matching the correct amino acid to the codons found in the mRNA nucleotide sequence.

tRNAs are able to match amino acids to the mRNA codons because of base pairing. The portion of the tRNA that interacts with mRNA is called the **anticodon.** The anticodon of tRNA is a short sequence of nucleotides that base-pairs with the nucleotides in the mRNA molecule. The other end of the tRNA carries an amino acid. The correct match between tRNAs and amino acids is made by an enzyme in the cell.

The start codon, AUG, is the first codon that is read in the mRNA to make any protein. Since the tRNA that binds

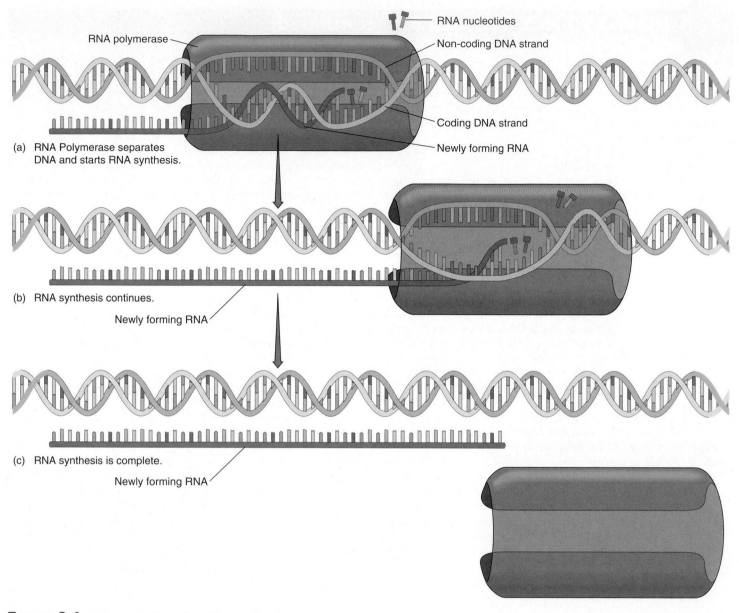

FIGURE 8.6 **A Transcription of an RNA Molecule**

This summary illustrates the basic events that occur during transcription. *(a)* An enzyme, RNA polymerase, attaches to the DNA at a point that allows it to separate the complementary strands. *(b)* As RNA polymerase moves down the DNA strand, new complementary RNA nucleotides are base-paired to one of the exposed DNA strands. The base-paired RNA nucleotides are linked together by RNA polymerase to form a new RNA molecule that is complementary to the nucleotide sequence of the DNA. *(c)* The newly formed (transcribed) RNA is then separated from the DNA molecule and used by the cell.

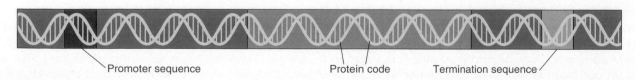

FIGURE 8.7 **Gene Structure**

There are four general regions of a gene. *(a)* The promoter sequence attracts the RNA polymerase to the site of the gene and then directs the RNA polymerase to proceed along the DNA strand in the correct direction to find the gene. *(b)* The protein-coding region contains the information needed to create a protein correctly. *(c)* The termination sequence signals the RNA polymerase to end mRNA transcription, so that the RNA can leave the nucleus to aid in translation.

TABLE 8.1

Amino Acid–Nucleic Acid Dictionary and the 20 Common Amino Acids and Their Abbreviations

These are the 20 common amino acids used in the protein synthesis operation of a cell. Each has a known chemical structure and is coded for by specific mRNA codons.

First letter	Second letter U	Second letter C	Second letter A	Second letter G	Third letter
U	UUU UUC } Phe — Phenylalanine UUA UUG } Leu Leucine	UCU UCC UCA UCG } Ser — Serine	UAU UAC } Tyr — Tyrosine UAA UAG Stop	UGU UGC } Cys — Cysteine UGA Stop UGG } Trp — Tryptophan	U C A G
C	CUU CUC CUA CUG } Leu — Leucine	CCU CCC CCA CCG } Pro- Proline	CAU CAC } His — Histidine CAA CAG } Gln — Glutamine	CGU CGC CGA CGG } Arg — Arginine	U C A G
A	AUU AUC AUA } Ile — Isoleucine AUG Start / Met — Methionine	ACU ACC ACA ACG } Thr — Threonine	AAU AAC } Asn — Asparagine AAA AAG } Lys — Lysine	AGU AGC } Ser — Serine AGA AGG } Arg — Arginine	U C A G
G	GUU GUC GUA GUG } Val — Valine	GCU GCC GCA GCG } Ala — Alanine	GAU GAC } Asp — Aspartic Acid GAA GAG } Glu — Glutamic Acid	GGU GGC GGA GGG } Gly — Glycine	U C A G

to the AUG codon carries the amino acid methionine, the first amino acid of every protein is methionine (figure 8.8). After the methionine-tRNA molecule is lined up over the start codon, the large subunit of the ribosome joins the small subunit. When the two subunits are together, with the mRNA in the middle, the ribosome is fully formed. The process of forming the rest of the protein is ready to begin.

Elongation

Once protein synthesis is started, the ribosome coordinates three events, which result in adding an amino acid to the growing protein. In this way, a ribosome is like an assembly line that organizes the steps of a complicated assembly process. A new tRNA arrives at the ribosome with its amino acid. The ribosome adds the new amino acid to the growing protein (figure 8.9).

Termination

The ribosome will continue to add one amino acid after another to the growing protein unless it encounters a stop signal (figure 8.10). The stop signal, in the mRNA, takes the form of a codon. The codons UAA, UAG, and UGA are all stop signals. When any of these three codons appear during the elongation process, a release factor enters the ribosome. The release factor causes the ribosome to release the protein, when the ribosome tries to add the release factor to the growing protein. When the protein releases, the ribosomal subunits separate and release the mRNA. The mRNA can be used by another ribosome to make another copy of the protein or can be broken down by the cell to prevent any more of the protein from being made.

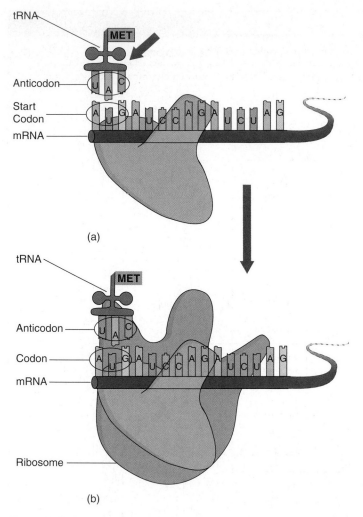

(a)

(b)

FIGURE 8.8 Initiation

(a) An mRNA molecule is positioned in the ribosome so that two codons are in position for transcription. The first of these two codons (AUG) is the initiation codon. The start tRNA aligns with the start codon. *(b)* The large subunit of the ribosome joins the small subunit. The ribosome is now assembled and able to translate the mRNA.

The Nearly Universal Genetic Code

The code for making protein from DNA is used by nearly all cells. Bacteria, protists, plants, fungi and animals all use DNA to store their genetic information. They all transcribe the information in DNA to RNA. They all translate the RNA using ribosomes and the same codon table. In eukaryotic cells, transcription always occurs in the nucleus, and translation always occurs in the cytoplasm (figure 8.11).

The similarity of protein synthesis in all cells strongly argues for a common origin of all life forms. It also creates very exciting opportunities for biotechnology. It is now possible to synthesize human proteins, such as insulin, in bacte-

ria, because bacteria and humans use the same code to make proteins. The production of insulin in this way can help create a cheap source of medication for many of those who suffer from diabetes. genetically modified organisms, p. 227

However, not all genetic information flows from DNA to RNA to proteins. Some viruses use RNA to store their genetic information. These viruses are called retroviruses. An example of a retrovirus is the *human immunodeficiency virus, HIV.* Retroviruses use their RNA to make DNA. This DNA is then used to transcribe more RNA. This RNA is then used to make proteins Outlook 8.1.

8.5 The Control of Protein Synthesis

Cells have many protein-coding sequences. **Gene expression** occurs when a cell transcribes and translates a gene. Cells do not make all their proteins at once, because it would be a great waste of resources. Cells control which genes are used to make proteins. In fact, the differences between the types of cells in the human body are due to the differences in the proteins produced. Cells use many ways to control gene expression in response to environmental conditions. Some methods help increase or decrease the amount of enzyme. Other methods help change the quality of the protein.

Controlling Protein Quantity

An enzyme's activity can be regulated by controlling how much of that enzyme is made. The cell regulates the amount of protein that is made by changing how much mRNA is available for translation. The cell can use several strategies to control how much mRNA is translated:

1. The genetic material of humans and other eukaryotic organisms consists of strands of coiled, double-stranded DNA, which has histone proteins attached along its length. The histone proteins and DNA are not arranged randomly but, rather, come together in a highly organized pattern. The double-stranded DNA spirals around repeating clusters of eight histone spheres. Histone clusters with their encircling DNA are called **nucleosomes** (figure 8.12a). These coiled DNA strands with attached proteins become visible during cell division and are called **nucleoproteins** or **chromatin fibers.** Condensed like this, a chromatin fiber is referred to as a **chromosome** (figure 8.12b Outlook 8.2). The degree to which the chromatin is coiled provides a method for long-term control of protein expression. In tightly coiled chromatin, the promoter sequence of the gene is tightly bound, so that RNA polymerase cannot attach and initiate transcription. Loosely packaged chromatin exposes the promoter sequence, so that transcription can occur.

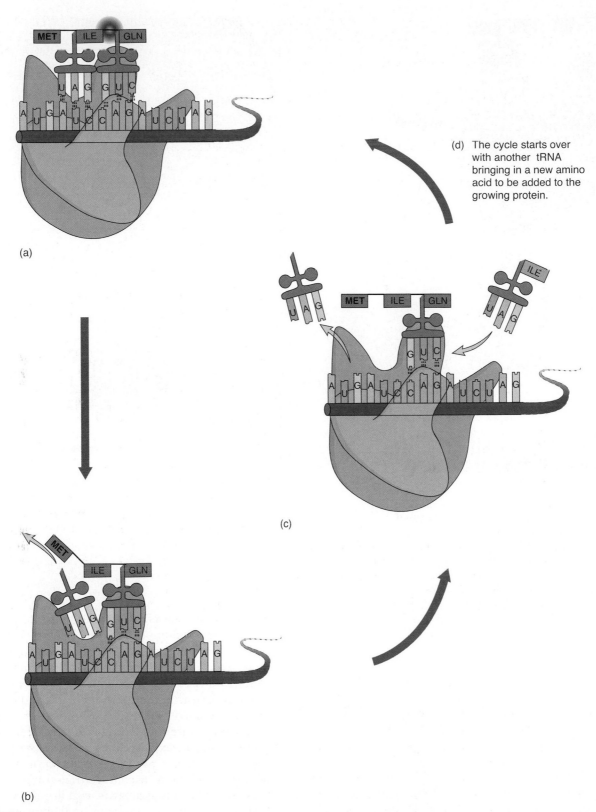

(d) The cycle starts over with another tRNA bringing in a new amino acid to be added to the growing protein.

(a)

(c)

(b)

FIGURE 8.9 Elongation

(a) The two tRNAs align the amino acids (ILE and GLN) so that they can be chemically attached to one another by a peptide bond.
(b) Once the bond is formed, the first tRNA detaches from its position on the mRNA. *(c)* The ribosome moves down one codon on the mRNA. Another tRNA now aligns so that the next amino acid (ILE) can be added to the growing protein. *(d)* The process continues with a new tRNA, a new amino acid, and the formation of a new peptide bond.

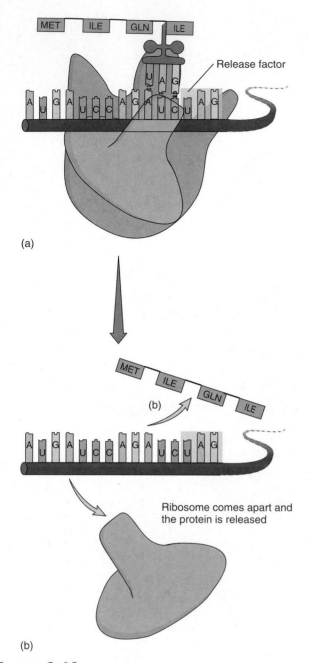

(a)

(b)

Ribosome comes apart and
the protein is released

(b)

FIGURE 8.10 Termination

(a) A release factor will move into position over a termination codon—here, UAG. *(b)* The ribosome releases the completed amino acid chain. The ribosome disassembles and the mRNA can be used by another ribosome to synthesize another protein.

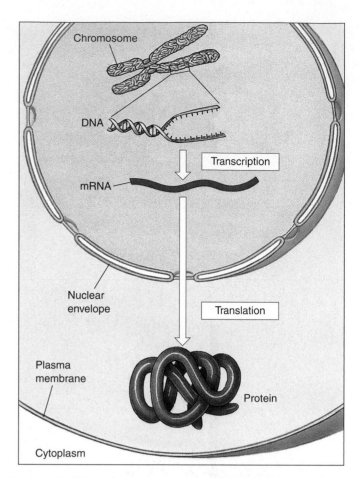

FIGURE 8.11 **Summary of Protein Synthesis**

The genetic information in DNA is rewritten in the nucleus as RNA in the nucleus during transcription. The mRNA moves from the nucleus to the cytoplasm, where the genetic information is read during translation by the ribosome.

2. **Enhancer** and **silencer sequences** are DNA sequences that regulate gene expression by (acting as binding sites for) proteins that also affect the ability of RNA polymerase to transcribe a specific protein. Enhancer sequences increase protein synthesis by helping increase transcription. Silencer sequences decrease transcription. These DNA sequences are unique, because they do not need to be close to the promoter to function and they are not transcribed.

3. **Transcription factors** are proteins that control how available a promoter sequence is for transcription. Transcription factors bind to DNA around a gene's promoter sequence and influence RNA polymerase's ability to start transcription (figure 8.13). There are many transcription factors in the cell. Eukaryotic transcription is so tightly regulated that transcription factors always guide RNA polymerase to the promoter sequence. A particular gene will not be expressed if its specific set of transcription factors is not available. Prokaryotic cells also use proteins to block or encourage transcription, but not to the extent that this strategy is used in eukaryotic cells.

4. Cells regulate gene expression by limiting the length of time that mRNA is available for translation. Enzymes in the cell break down the mRNA, so that it can no longer be used to synthesize protein. The time that a given mRNA molecule lasts in a cell is dependant on the nucleotide sequences in the mRNA itself. These sequences are in areas of the mRNA that do not code for protein.

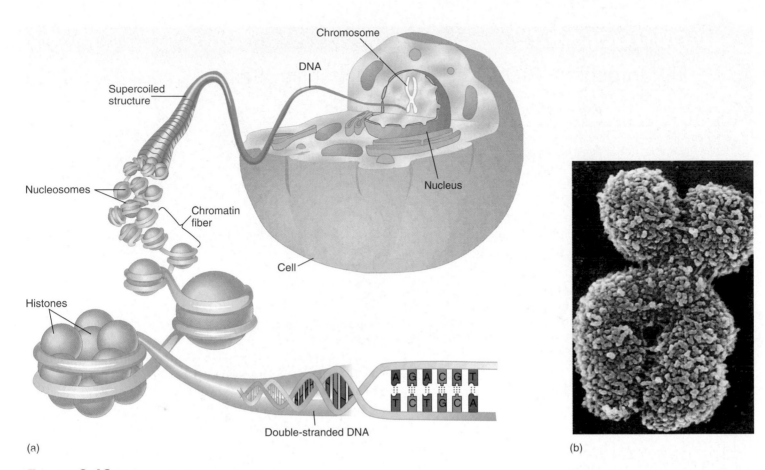

(a)

(b)

FIGURE 8.12 Eukaryotic Genome Packaging

(a) Eukaryotic cells contain DNA in their nuclei that takes the form of a double-stranded helix. The two strands fit together and are bonded by weak hydrogen bonds formed between the complementary, protruding nitrogenous bases according to the base-pairing rule. To form a chromosome, the DNA molecule is wrapped around a group of several histone proteins. Together, the histones and the DNA form a structure called the nucleosome. The nucleosomes are packaged to form a chromosome. *(b)* During certain stages in the reproduction of a eukaryotic cell, the nucleoprotein coils and "supercoils," forming tightly bound masses. When stained, these are easily seen through the microscope. In their supercoiled form, they are called *chromosomes*, meaning "colored bodies."

One Gene, Many Proteins

One of the most significant differences between prokaryotic and eukaryotic cells is that eukaryotic cells can make more than one type of protein from a single protein-coding region. Eukaryotic cells are able to do this because the protein-coding regions of eukaryotic genes are organized differently than the genes found in prokaryotic (bacterial) cells. The fundamental difference is that the protein-coding regions in prokaryotes are continuous, whereas eukaryotic protein-coding regions are not. Scattered throughout the protein-coding sequence of eukaryotes are many sequences, called **introns**, that do not code for protein. The remaining sequences, which are used to code for protein, are called **exons.** After the protein-coding region of a eukaryotic gene is transcribed into mRNA, the introns in the mRNA are cut out and the remaining exons are

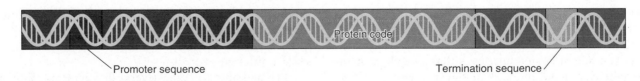

Promoter sequence   Termination sequence

FIGURE 8.13 Transcription Factors

Transcription factors bind to the DNA (shown in red) around the promoter of a gene to aid RNA polymerase in starting transcription. The other regions of a gene that are shown are the protein-coding region and the termination sequence.

OUTLOOKS 8.1

HIV Infection, AIDS, and Reverse Transcriptase

Acquired Immuno Deficiency Syndrome is caused by a retrovirus called human immunodeficiency virus (HIV). HIV is a spherical virus with an outer membrane, an inside protein coat, and an RNA core. Its genetic material is RNA, not DNA. Genes are carried from one generation to the next as RNA molecules. This is not the case in humans and most other organisms in which DNA is the genetic material.

Human Sequence

$$\text{DNA} \quad \rightarrow \quad \text{RNA} \quad \rightarrow \quad \text{protein}$$
$$\text{transcriptase}$$

However, once having entered a suitable, susceptible host cell, HIV must produce a DNA copy of its RNA in order to integrate into the host cell's chromosome. Only then can it become an active, disease-causing parasite. The use of RNA to produce DNA is contrary, reverse, or retro to the usual RNA-forming process controlled by the enzyme transcriptase. Humans do not have the genetic capability to manufacture the enzyme necessary to produce DNA from RNA—reverse transcriptase.

Retrovirus Sequence

$$\text{RNA} \quad \rightarrow \quad \text{DNA}$$
$$\text{reverse}$$
$$\text{transcriptase}$$

When HIV carries out gene replication and protein synthesis, the process is diagrammed

viral RNA → | viral DNA in human cell | → | viral RNA from human cell | → protein | new virus
reverse RNA transcriptase from virus | RNA polymerase from host cell | other enzymes from host cell

This reproductive cycle has two important implications. First, the presence of reverse transcriptase in a human can be looked upon as an indication of retroviral infection because reverse transcriptase is not manufactured by human cells. However, because HIV is only one of several types of retroviruses, the presence of the enzyme in an individual does not necessarily indicate an HIV infection. It only indicates a type of

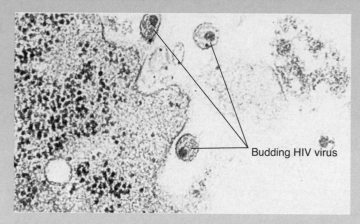

Budding HIV Virus

This electron micrograph shows HIV viruses leaving the cell. These viral particles can now infect another cell and continue the viral replication cycle unless medications prevent this from happening.

retroviral infection. Second, interference with reverse transcriptase will frustrate the virus's attempt to integrate into the host's DNA.

Drug treatments for HIV take advantage of vulnerable points in the retrovirus's life cycle. The drugs fall into three basic categories: (1) blockers, (2) reverse transcriptase inhibitors, and (3) protease inhibitors. Drugs classified as blockers prevent the virus from entering the cell. Reverse transcriptase inhibitors interfere with the crucial step of converting the virus's RNA genetic information from RNA to DNA. Protease inhibitors interfere with the modification of viral proteins after translation. This prevents a virus from completely forming and infecting another cell. A treatment regime can include a drug from each of these categories. This combination of drugs is called a *cocktail* and is much more effective against HIV than is any one of the drugs alone. However, HIV mutates very quickly, because its genetic information is stored as a single-stranded molecule of RNA. This high mutation rate quickly produces strains of the HIV virus that are resistant to these drugs. evolution, p. 260

---

spliced together, end to end, to create a shorter version of the mRNA. It is this shorter version that is used during translation to produce a protein (figure 8.14).

One advantage of having introns is that a single protein-coding region can make more than one protein. Scientists originally estimated that humans had 80,000 to 100,000 genes. This was based on techniques that allowed them to estimate the number of different proteins found in humans.

When the human genome was characterized, scientists were surprised to find that humans have only about 20,000 genes. This suggests that many of our genes are capable of making several different proteins.

It is possible to make several different proteins from the same protein-coding region by using different combinations of exons. **Alternative splicing** is the process of selecting which exons will be retained during the normal process of splicing.

OUTLOOKS 8.2

Telomeres

Each end of a chromosome contains a sequence of nucleotides called a **telomere.** In humans, these chromosome "caps" contain the following nucleotide base-pair sequence:

TTAGGG

AATCCC

This sequence is repeated many times.

Telomeres are very important segments of the chromosome. They are required for chromosome replication; they protect the chromosome from being destroyed by dangerous DNAase enzymes (enzymes that destroy DNA); and they keep chromosomes from bonding to one another end to end. Evidence shows that the loss of telomeres is associated with cell "aging," whereas not removing them has been linked to cancer. Every time a cell reproduces itself, it loses some of its telomeres. However, in cells that have the enzyme telomerase, new telomeres are added to the ends of the chromosome each time the cells divide. Therefore, cells that have telomerase do not age as other cells do, and cancer cells are immortal because of this enzyme. Telomerase enables chromosomes to maintain, if not increase, the length of telomeres from one cell generation to the next.

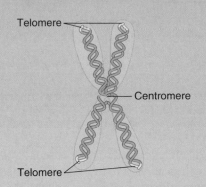

Telomees

The yellow regions on this drawing of a chromosome indicate where the telomeres are. Each time DNA is replicated, the telomeres get shorter. The shortening is associated with the aging of cell lines. Cells that have the enzyme telomerase are able to add to the length of the telomeres, and the cell lines are immortal.

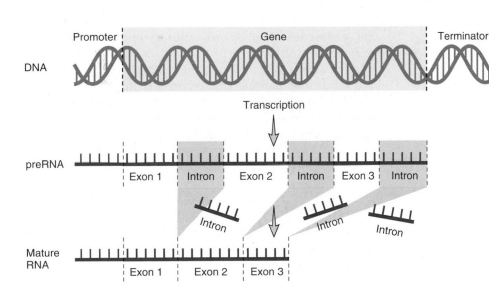

FIGURE 8.14 Transcription of mRNA in Eukaryotic Cells

This is a summary of the events that occur in the nucleus during the manufacture of mRNA in a eukaryotic cell. Notice that the original nucleotide sequence is first transcribed into an RNA molecule, which is later "clipped" and then rebonded to form a shorter version of the original. It is during this time that the introns are removed.

Alternative splicing can be a very important part of gene regulation. One protein-coding region in fruit flies, *sex-lethal*, can be spliced into two different forms. One form creates a full-sized, functional protein. The other form creates a very small protein with no function. For the fruit fly, the difference between the two alternatively spliced forms of *sex-lethal* is the difference between becoming male or becoming female fruit fly (figure 8.15).

8.6 Mutations and Protein Synthesis

A **mutation** is any change in the DNA sequence of an organism. Mutations can occur for many reasons, including errors during DNA replication. Mutations can also be caused by

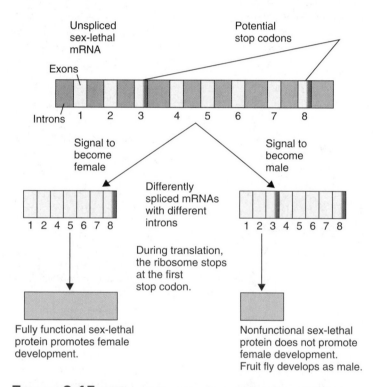

FIGURE 8.15 Different Proteins from Alternative Splicing
Male and female fruit flies produce the same unspliced mRNA from the *sex-lethal* gene. A cellular signal determines if the fruit fly will develop as a female or a male. The manner in which the sex-lethal mRNA is spliced depends on the signal that is received. Females remove the third exon from the sex-lethal mRNA, whereas males leave the third exon in. This is alternative splicing. The female-specific mRNA can be translated by ribosomes to make a fully functional sex-lethal protein. This protein promotes female body development. The male-specific mRNA contains a stop codon in the third exon. This causes the ribosome to stop synthesis of the sex-lethal protein early. The resulting protein is small and has no function. With no sex-lethal protein activity, the fruit fly develops as a male.

external factors, such as radiation, carcinogens, drugs, or even some viruses. It is important to understand that not all mutations cause a change in an organism. If a mutation occurs away from the protein-coding sequence and the DNA sequences that regulate its expression, it is unlikely that the change will be harmful to the organism. On occasion, the changes that occur because of mutations can be helpful.

Point Mutations

A **point mutation** is a change in a single nucleotide of the DNA sequence. A **silent mutation** is a change of a single nucleotide that does not cause a change in the amino acids used to build a protein. An example of a silent mutation is a change in DNA that results in the following change in an mRNA codon: UUU to UUC. The mutation from U to C still results in the amino acid phenylalanine being used to construct the protein. Another example is shown in figure 8.16.

Another type of point mutation, a **nonsense mutation**, causes a ribosome to stop protein syntheses by introducing a stop codon too early. A nonsense mutation would be caused if a codon were changed from CAA (glutamine) to UAA (stop). This type of mutation prevents a functional protein from being made, because it is terminated too soon.

Other types of point mutations can change how a protein functions. A **missense mutation** causes the wrong amino acid to be used in making a protein. A sequence change that resulted in the codon change from UUU to GUU would use valine instead of phenylalanine. Proteins rely on their shape for their function. Their active sites have very specific chemical properties. The shapes and chemical properties of enzymes are determined by the correct sequence of various types of amino acids. Substituting one amino acid for another can create an abnormally functioning protein.

The condition known as sickle-cell anemia provides a good example of the proteins that can be caused by a simple missense mutation. Hemoglobin is a protein in red blood cells that is responsible for carrying oxygen to the body's cells. Normal hemoglobin molecules are composed of four separate different proteins. The proteins are arranged with respect to each other so that they are able to hold an iron atom. The iron atom is the portion of hemoglobin that binds the oxygen.

In normal individuals, the amino acid sequence of the hemoglobin protein begins like this:

Val-His-Leu-Thr-Pro-Glu-Glu-Lys . . .

In some individuals, a single nucleotide of the hemoglobin gene has been changed. The result of this change is a hemoglobin protein with an amino acid sequence of

Val-His-Leu-Thr-Pro-Val-Glu-Lys . . .

Glutamic acid (Glu) is coded by two codons, GAA and GAG. Valine is also coded by two codons, GUA and GUG. The change that causes the switch from glutamic acid to valine is a missense mutation. With this small change, the parts of the hemoglobin protein do not assemble correctly under low oxy-

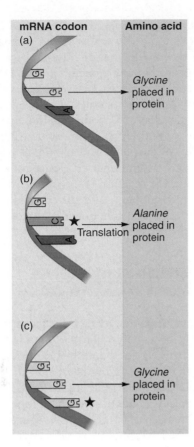

mRNA codon   Amino acid

(a)

Glycine placed in protein

(b)

Alanine placed in protein
← Translation

(c)

Glycine placed in protein

FIGURE 8.16 Silent DNA Mutations

A nucleotide substitution changes the protein only if the changed codon results in a different amino acid being substituted into a protein chain. (a) In the example, the original codon, GGA, calls for the amino acid glycine. (b) An observable mutation occurs when the nucleotide in the second position of the codon is changed. It now reads GCA. GCA calls for the amino acid alanine. (c) A *silent* mutation is shown where the third position of the codon is changed. The codon GGG calls for the same amino acid as the original version (GGA). Because the proteins produced in the first and third examples will be identical in amino acid sequence, they will function the same also.

gen levels. Many hemoglobin molecules stick together and cause the red blood cells to have a sickle shape, rather than their normal round, donut shape (figure 8.17). The results can be devastating:

- The red blood cells do not flow smoothly through the capillaries, causing the red blood cells to tear and be destroyed. This results in anemia.
- Their irregular shapes cause them to clump, clogging the blood vessels. This prevents oxygen from reaching the oxygen-demanding tissues. As a result, tissues are damaged.
- A number of physical disabilities may result, including weakness, brain damage, pain and stiffness of the joints, kidney damage, rheumatism, and, in severe cases, death.

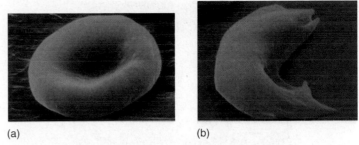

(a)   (b)

FIGURE 8.17 Normal and Sickled Red Blood Cells

(a) A normal red blood cell and *(b)* a cell having the sickle shape. This sickling is the result of a single amino acid change in the hemoglobin molecule.

Insertions and Deletions

Several other kinds of mutations involve more than a change in a single nucleotide. Insertions and deletion are different from point mutations because they change the DNA sequence by removing and adding nucleotides. An **insertion mutation** adds one or more nucleotides to the normal DNA sequence. A **deletion mutation** removes one or more nucleotides. The effect that insertions and deletions have on protein synthesis is called *frameshift*. Ribosomes read the mRNA three nucleotides at a time. This set of three nucleotides is called a reading frame. A **frameshift** occurs when insertions or deletions cause the ribosome to read the wrong sets of three nucleotides. Consider the example shown in figure 8.18.

Some viruses, such as HIV, can insert their genetic code into the DNA of their host organism. When this happens, the presence of the new viral sequence may interfere with the cells' ability to use genetic information in that immediate area, because the virus's genetic information becomes an

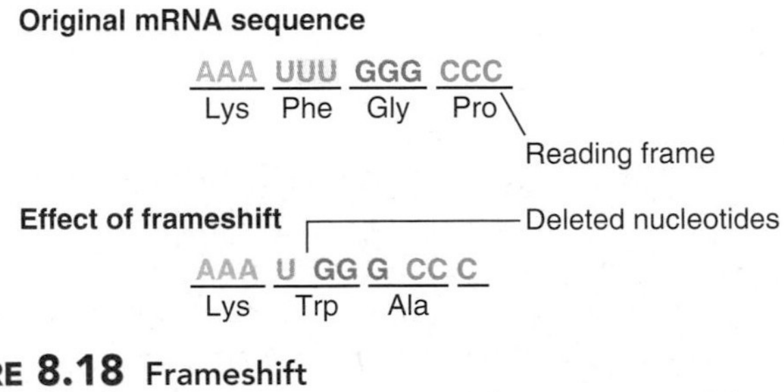

Original mRNA sequence

AAA UUU GGG CCC
Lys   Phe   Gly   Pro
                          Reading frame

Effect of frameshift ⌐――――― Deleted nucleotides
AAA U GG G CC C
Lys   Trp   Ala

FIGURE 8.18 Frameshift

A frameshift causes the ribosome to read the wrong set of three nucleotides on the mRNA. Proteins produced by this type of mutation usually bear little resemblance to the normal protein that is usually produced. In this example, the normal sequence is shown for comparison with the mutated sequence that is missing two uracil nucleotides. The underlining identifies the codons that are read by the ribosome. A normal protein is made until after the deletion is encountered.

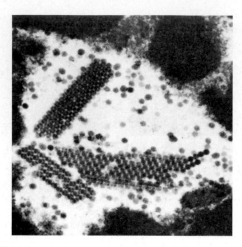

FIGURE 8.19 HPV

Genital warts are caused by the human papillomavirus (HPV). Over 70 papillomaviruses are shown in this photo, taken through an electron microscope. Several HPV strains have been associated with a higher than normal incidence of cancer. This is because HPV creates insertion mutations in the cells it infects.

insertion mutation. In the case of some retroviruses, such as the human papillomavirus (HPV), the insertion mutations can transform normal cells into cancerous ones (figure 8.19).

Chromosomal Aberrations

A **chromosomal aberration** is a major change in DNA that can be observed at the level of the chromosome. There are four types of aberrations: *inversions, translocations, duplications,* and *deletions*. An **inversion** occurs when a chromosome is broken and a piece becomes reattached to its original chromosome, but in a flipped orientation. A **translocation** occurs when one broken segment of DNA becomes integrated into a different chromosome. **Duplications** occur when a portion of a chromosome is replicated and attached to the original section in sequence. **Deletion aberrations** result when a broken piece becomes lost or is destroyed before it can be reattached. All of these aberrations are considered mutations. Because of the large segments of DNA that are involved with these types of mutations, many genes can be affected.

Many other forms of mutations affect DNA. Some damage to DNA is so extensive that the entire strand is broken, resulting in the synthesis of abnormal proteins or a total lack of protein synthesis. A number of experiments indicate that many street drugs, such as lysergic acid diethylamide (LSD), are mutagenic agents that cause DNA to break.

Mutations and Inheritance

Mutations can be harmful to the individual who first gains the mutation, but changes in the structure of DNA may also have harmful effects on the next generation if they occur in the sex cells. Sex cells transmit genetic information from one generation to the next. Mutations that occur to DNA mole-

cules can be passed on to the next generation only when the mutation is present in cells such as sperm and egg. In the next several chapters, we will look at how DNA is inherited. As you read the next chapters remember that DNA codes for proteins. Genetic differences between individuals are the result of slightly different enzymes. These enzymes help cells carry out such tasks as (1) producing the enzymes required for the digestion of nutrients; (2) manufacturing enzymes that will metabolize the nutrients and eliminate harmful wastes; (3) repairing and assembling cell parts; (4) reproducing healthy offspring; (5) reacting to favorable and unfavorable changes in the environment; and (6) coordinating and regulating all of life's essential functions. If any of these tasks are not performed properly, the cell will die.

Summary

The successful operation of a living cell depends on its ability to use accurately the genetic information found in its DNA. DNA replication results in an exact doubling of the genetic material. The process virtually guarantees that identical strands of DNA will be passed on to the next generation of cells. The production of protein molecules is under the control of the nucleic acids, the primary control molecules of the cell. The sequence of the bases in the nucleic acids, DNA and RNA, determines the sequence of amino acids in the protein, which in turn determine the protein's function. Protein synthesis involves the decoding of the DNA into specific protein molecules and the use of the intermediate molecules, mRNA and tRNA, at the ribosome. The process of protein synthesis is controlled by regulatory sequences in the nucleic acids. Errors in any of the protein coding sequences in DNA may produce observable changes in the cell's functioning and can lead to cell death.

Key Terms

Use interactive flash cards on the **Concepts in Biology,** *12/e website to help you learn the meanings of these terms.*

inversion 160

messenger RNA (mRNA) 148

missense mutation 158

mutation 158

non-coding strand 147

nonsense mutation 158

nucleoproteins (chromatin fibers) 152

nucleic acids 144

nucleosomes 152

point mutation 158

promoter sequences 148

ribosome RNA (rRNA) 148

RNA polymerase 147

silencer sequences 154

silent mutation 158

telomere 157

termination sequences 148

thymine 144

transcription 147

transcription factors 154

transfer RNA (tRNA) 148

translation 148

translocation 160

uracil 146

Basic Review

1. Genetic information is stored in what chemical?

    a. proteins

    b. lipids

    c. nucleic acids

    d. sugars

2. The difference between ribose and deoxyribose is

    a. the number of carbon atoms.

    b. an oxygen atom.

    c. one is a sugar and one is not.

    d. they are the same molecule.

3. The nitrogenous bases in DNA

    a. hold the two DNA strands together.

    b. link the nucleotides together.

    c. are part of the genetic blueprint.

    d. Both a and c are correct.

4. Transcription copies genetic information

    a. from DNA to RNA.

    b. from proteins to DNA.

    c. from DNA to proteins.

    d. from RNA to proteins.

5. RNA polymerase starts synthesizing mRNA because

    a. it finds a promoter sequence.

    b. transcription factors interact with RNA polymerase.

    c. the gene is in a region of loosely packed chromatin

    d. All of the above are true.

6. Under normal conditions, translation

    a. does not start until the stop codon.

    b. reads in sets of three nucleotides called codons.

    c. ends at the end of the mRNA.

    d. All of the above statements are true.

7. The function of tRNA is to

    a. be part of the ribosome's subunits.

    b. carry the genetic blueprint.

    c. carry an amino acid to a working ribosome.

    d. Both a and c are correct.

8. Enhancers

    a. make ribosomes more efficient at translation.

    b. prevent mutations from occurring.

    c. increase the transcription of specific genes.

    d. slow aging.

9. The process that joins exons from mRNA is called

    a. silencing.

    b. splicing.

    c. transcription.

    d. translation.

10. A deletion of a single base in the protein-coding sequence of a gene will likely create

    a. no problems.

    b. a faulty RNA polymerase.

    c. a tRNA.

    d. a frameshift.

Answers

1. c   2. b   3. d   4. a   5. d   6. b   7. c   8. c   9. b   10. d

Concept Review

Answers to Concept Review questions can be found on the Concepts in Biology, 12/e website.

8.1   DNA and the Importance of Proteins

1. What is the product of transcription? Translation?

2. What is a gene?

8.2   DNA Structure and Function

3. Why is DNA replication necessary?

4. What is DNA polymerase, and how does it function?

8.3   RNA Structure and Function

5. What are the differences between DNA and RNA?

8.4   Protein Synthesis

6. How does DNA replication differ from the manufacture of an RNA molecule?

7. If a DNA nucleotide sequence is TACAAAGCA, what is the mRNA nucleotide sequence that would base-pair with it?

8. What amino acids would occur in the protein chemically coded by the sequence of nucleotides in question 7?
9. How do tRNA, rRNA, and mRNA differ in function?
10. What are the differences among a nucleotide, a nitrogenous base, and a codon?
11. List the sequence of events that takes place when a DNA message is translated into protein.

8.5 Control of Protein Synthesis

12. Provide two examples of how a cell uses transcription to control gene expression.
13. Provide an example of why it is advantageous for a cell to control gene expression.

8.6 Mutations and Protein Synthesis

14. Both chromosomal and point mutations occur in DNA. In what ways do they differ?
15 What is a silent mutation? Provide an example.

Thinking Critically

For guidelines to answer this question, visit the **Concepts in Biology,** *12/e website.*

A friend of yours gardens for a hobby. She has noticed that she has a plant that no longer produces the same color of flower it did a few years ago. It used to produce red flowers; now, the flowers are white. Consider that petal color in plants is due to an enzyme. No color suggests no enzyme activity. Using what you know about genes, protein synthesis, and mutations, hypothesize what may have happened to cause the change in flower color. Identify several possibilities; then, identify what you would need to know to test your hypothesis.

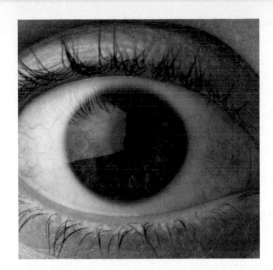

Cell Division— Replication and Reproduction

Retinoblastoma is a form of eye cancer that can affect children under 2 years of age, but early detection and therapy improvements have greatly increased the likelihood of survival. Cancer is a problem with controlling how cells divide and replace themselves. When cells divide too often, a tumor forms. Retinoblastoma is a particularly interesting form of cancer, because it has been linked to mutations in the *RB* gene on chromosome 13. This gene codes for an 816 amino acid protein that is a transcription factor—a protein that regulates the synthesis of other proteins. If no RB protein

activity is present, cancer can develop. Individuals with a mutant RB gene also have an increased chance of passing on this disposition for cancer to their children. Many other cancers are probably inherited as well.

Understanding the nature of cell division and how it is controlled can help us make better decisions regarding the early detection and treatment of cancer. Understanding how cell division influences human genetics can also help us make better choices about having children and the health concerns these children may have.

CHAPTER OUTLINE

9.1 The Importance of Cell Division

Two fundamental characteristics of life are the ability to grow and the ability to reproduce. Both of these characteristics depend on cells' ability to divide. **Cell division** is the process in which a cell becomes two new cells. Cell division serves many purposes. For single-celled organisms, it is a method of increasing their numbers. For multicellular organisms, it is a process that leads to growth, the replacement of lost cells, the healing of injuries, and the formation of reproductive cells. These reproductive cells lead to new organisms, which in turn grow by cell division. characteristics of life, p. 12

There are three types of cell division, each involving a parent cell. The first type of cell division is *binary fission*. **Binary fission** is a method of cell division used by prokaryotic cells. During binary fission, the single loop of DNA replicates, the two loops separate, and a new cell membrane forms between the two DNA molecules. This ensures that each of the daughter cells receives the same information that was possessed by the parent cell. The second type of cell division, **mitosis,** is a method of eukaryotic cell division; it also results in daughter cells that are genetically identical to the parent cell. Eukaryotic cells have several chromosomes that are replicated and divided by complex processes between two daughter cells. The third type of cell division is **meiosis,** a method of eukaryotic cell division that results in daughter cells that have half the genetic information of the parent cell.

Asexual Reproduction

For single-celled organisms, binary fission and mitosis are methods of *asexual reproduction*. **Asexual reproduction** requires only one parent and results in two organisms that are genetically identical to the parent.

In multicellular organisms, mitosis produces new cells that

- Cause growth by increasing the number of cells
- Replace lost cells
- Repair injuries

In each case, the daughter cells require the same genetic information that was present in the parent cell. With the parent cell's DNA, the daughter cells are able to participate in the appropriate metabolic activities. The daughter cells may not be able to do this if they lack some of the parent cell's genetic information.

Sexual Reproduction

Another form of reproduction requires two parents to donate genetic information toward creating a new organism. **Sexual reproduction** is the combining of genetic information from two parents. The result of sexual reproduction is a genetically unique individual.

Meiosis is the process that produces the cells needed for sexual reproduction. Meiosis is different from mitosis; in meiosis, the reproductive cells receive only part of the parent cell's genetic information. When reproductive cells (eggs and sperm) join, the amount of genetic information is returned to normal.

Understanding the purposes of cell division is an important part of understanding how cell division ensures that the daughter cells inherit the correct genetic information. The rest of this chapter takes a closer look at how cell division accomplishes the correct distribution of genetic information. The genetic information, DNA, in eukaryotic cells is in chromosomes. When we follow the activities of chromosomes we follow the distribution of DNA.

9.2 The Cell Cycle and Mitosis

The **cell cycle** consists of all the stages of growth and division for a eukaryotic cell. All cells go through the same basic life cycle, but they vary in the amount of time they spend in the various stages. The cell's life cycle is a continuous process without a beginning or an end. As cells complete one life cycle, they begin the next.

The cell cycle includes the stages in which the cell spends its time engaged in metabolism, as well as the stages of cell division. **Interphase** is the longest stage of the cell cycle. During interphase, the cell engages in metabolic activities and prepares for the next cell division. As a cell moves from interphase, it moves into the stages of cell division. Mitosis is the portion of the cell cycle in which the cell divides its genetic information. Scientists split interphase and mitosis into smaller steps in order to describe how the cell divides in more detail. Interphase contains three distinct phases of cell activity—G_1, S, and G_2. During each of these parts of interphase, the cell is engaged in specific activities needed to prepare for cell division (figure 9.1).

G₁ of Interphase

During the G_1 stage of interphase, the cell gathers nutrients and other resources from its environment. Gathering nutrients allows the cell both to grow in volume and to carry out

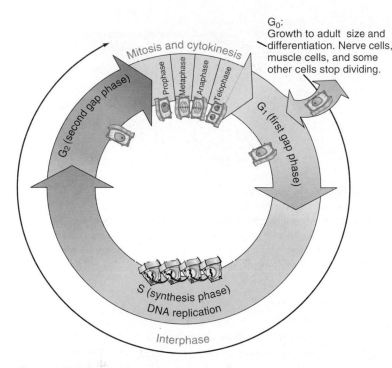

FIGURE 9.1 The Cell Cycle
Cells spend most of their time in interphase. During G_1 of interphase, the cell produces tRNA, mRNA ribosomes, and enzymes for everyday processes. During the S phase of interphase, the cell synthesizes DNA to prepare for division. During G_2 of interphase, the cell produces the proteins required for the spindles. The nucleus is replicated in mitosis and two cells are formed by cytokinesis. Once some organs, such as the brain, have completely developed, certain types of cells, such as nerve cells, enter the G_0 stage.

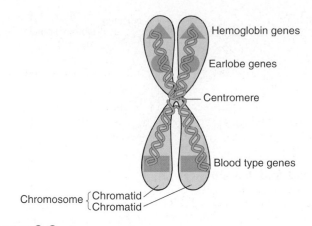

FIGURE 9.2 Chromosomes
During interphase, when chromosome replication occurs, the original double-stranded DNA unzips to form two identical double strands, which remain attached at the centromere. Each of these double strands is a chromatid. The two identical chromatids of the chromosome are sometimes termed a dyad, to reflect that there are two double-stranded DNA molecules, one in each chromatid. The DNA contains the genetic data.

its usual metabolic roles, such as producing tRNA, mRNA, ribosomes, enzymes, and other cell components. These activities allow the cell to perform its normal functions. This principle applies to single-celled eukaryotic organisms and multicellular organisms. In multicellular organisms, the normal metabolic functions may be producing proteins for muscle contraction, photosynthesis, or glandular-cell secretion.

Often, a cell stays in G_1 for an extended period. This is a normal process. For cells that remain in the G_1 stage for a long time, the stage is often renamed the G_0 stage, because the cell is not moving forward through the cell cycle. In the G_0 stage, cells may become differentiated, or specialized in their function, such as becoming nerve cells or muscle cells. The length of time cells stay in G_0 varies. Whereas some cells

entering the G_0 stage remain there more or less permanently (e.g., nerve cells), others can move back into the cell cycle and continue toward mitosis.

If a cell is going to divide, it commits to undergoing cell division during G_1 and moves to the S stage.

The S Stage of Interphase

During the S stage of interphase, DNA synthesis (replication) occurs. With two copies of the genetic information, the cell can distribute copies to the daughter cells as chromosomes.

The DNA in chromosomes is wrapped around histone proteins to form nucleosomes. The nucleosomes are coiled into chromatin. **Chromatin** is DNA wrapped around histone proteins. The individual chromatin strands are too thin and tangled to be seen. When chromatin becomes coiled, it becomes visible as a chromosome. As chromosomes become more visible at the beginning of mitosis, we may observe two threadlike parts lie side by side. Each parallel thread is called a *chromatid* (figure 9.2). A **chromatid** is one of two parallel parts of a chromosome. Each chromosome that contains one DNA molecule. After DNA synthesis, the chromosome contains two DNA molecules, one in each chromatid. **Sister chromatids** are the 2 chromatids of a chromosome that were

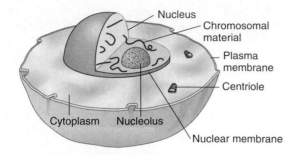

FIGURE 9.3 Interphase

Growth and the production of necessary organic compounds occur during this phase. If the cell is going to divide, DNA replication also occurs during interphase. The individual chromosomes are not visible, but a distinct nuclear membrane and nucleolus are present. (Some cells have more than one nucleolus.)

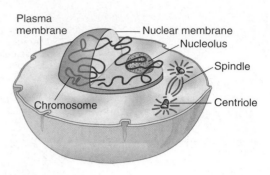

FIGURE 9.4 Early Prophase

Chromosomes begin to appear as thin, tangled threads and the nucleolus and nuclear membrane are present. The two sets of microtubules, known as the centrioles, begin to separate and move to opposite poles of the cell. A series of fibers, known as the spindle, will shortly begin to form.

produced by replication and that contain the same DNA. The centromere is the point where the sister chromosomes are attached. DNA replication, p. 145 chromosome structure, p. 152

The G₂ Stage of Interphase

The final stage of interphase is G_2. During the G_2 stage, final preparations are made for mitosis. The cell makes the cellular components it will need to divide successfully, such as the proteins it will use to move the chromosomes. At this point in the cell cycle, the nuclear membrane is intact. The chromatin has replicated, but it has not coiled and so the individual chromosomes are not yet visible (figure 9.3). The nucleolus, the site of ribosome manufacture, is also still visible during the G_2 stage.

9.3 Mitosis—Cell Replication

When eukaryotic cells divide, two events occur. 1) The replicated genetic information of a cell is equally distributed in mitosis. 2) After the contents of the nucleus is divided, the cytoplasm of the cell also divides into two new cells. This division of the cell's cytoplasm is called **cytokinesis**—cell splitting.

The individual stages of mitosis transition seamlessly from one to the next. For purposes of study and communication, scientists have divided the process of mitosis into four stages, based on recognizable events. These four phases are *prophase, metaphase, anaphase,* and *telophase.*

Prophase

As the G_2 phase of interphase ends, mitosis begins. **Prophase** is the first stage of mitosis. One of the first visible changes that identifies when the cell enters prophase is that the thin,

tangled chromatin present during interphase gradually coils and thickens, becoming visible as separate chromosomes consisting of 2 chromatids (figure 9.4). As the nucleus disassembles during prophase, the nucleolus is no longer visible.

As the cell moves toward the end of prophase, a number of other events also occur in the cell (figure 9.5). One of these events is the formation of the *spindle* and its *spindle fibers.* The **spindle** is a structure, made of microtubules, that reaches across the cell from one side to the other. The **spindle fibers** consist of microtubules and are the individual strands of the spindle. They physically interact with the chromosomes at the centromere. As prophase proceeds and the nuclear membrane gradually disassembles, the spindle fibers attach to the chromosomes. Spindle fibers must reach the chromosomes so that the spindle fibers can move chromosomes during later stages of mitosis. microtubules, p. 81

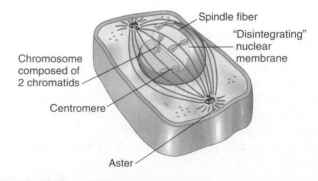

FIGURE 9.5 Late Prophase

In late prophase, the chromosomes appear as 2 chromatids (a dyad) connected at a centromere. The nucleolus and the nuclear membrane have disassembled. The centrioles have moved farther apart, the spindle is produced, and the chromosomes are attached to the spindle fibers.

One difference between plant and animal cell division can be observed in prophase. In animal cells, the spindle forms between *centrioles*. In plants, the spindle forms without centrioles. **Centrioles** make up a cellular organelle comprised of microtubules. Centrioles replicate during the G_2 stage of interphase and begin to move to opposite side of the cell during prophase. As the centrioles migrate, the spindle is formed between them and eventually stretches across the cell, so that spindle fibers encounter chromosomes when the nuclear membrane disassembles. Plant cells do not form their spindle between centrioles, but the spindle still forms during prophase.

Another significant difference between plant and animal cells is the formation of *asters* during mitosis. **Asters** are microtubules that extend from the centrioles to the plasma membrane of an animal cell. Whereas animal cells form asters, plant cells do not. Some scientists hypothesize that asters help brace the centriole against the animal cell membrane by making the membrane stiffer. This might help in later stages of mitosis, when the spindle fibers and centrioles may need firm support to help with chromosome movement. It is believed that plant cells do not need to form asters because this firm support is provided by their cell walls.

Metaphase

During **metaphase,** the second stage of mitosis, the chromosomes align at the equatorial plane. There is no nucleus present during metaphase, because the nuclear membrane has disassembled and the spindle, which started to form during prophase, is completed. The chromosomes are at their most tightly coiled, are attached to spindle fibers and move along the spindle fibers until all their centromeres align along the equatorial plane of the cell (figure 9.6). At this stage in mitosis, each chromosome still consists of 2 chromatids attached at the centromere.

To understand the arrangement of the chromosomes during metaphase, keep in mind that the cell is a three-dimensional object. A view of a cell in metaphase from the side is an equatorial view. From this perspective, the chromosomes appear as if they were in a line. If we viewed the cell from a pole, looking down on the equatorial plane, the chromosomes would appear scattered about within the cell, even though they were all in a single plane. In late metaphase, each chromosome splits at the centromeres as the sister chromatids separate.

Anaphase

Anaphase is the third stage of mitosis. The nuclear membrane is still absent and the spindle extends from pole to pole. The sister chromatids of each chromosome separate as they move along the spindle fibers toward opposite poles (figure 9.7). When this separation of chromatids occurs, the chromatids become known as separate daughter chromosomes.

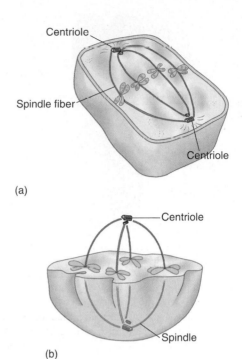

(a)

(b)

FIGURE 9.6 Metaphase

During metaphase, the chromosomes are pulled by the spindle and align at the equatorial plane. The equatorial plane is the region in the middle of the cell. Notice that each chromosome still consists of 2 chromatids. *(a)* When viewed from the edge of the plane, the chromosomes appear to be lined up. *(b)* When viewed from another angle, the chromosomes appear to be spread apart, as if on a tabletop.

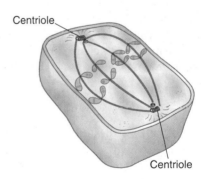

FIGURE 9.7 Anaphase

The pairs of chromatids separate after the centromeres replicate. The chromatids, now called daughter chromosomes, are separating and moving toward the poles.

The sister chromatids begin to separate because two important events occur. The first is that enzymes in the cell digest the portions of the centromeres that hold the 2 chromatids together. The second event is that the chromatids begin to separate. Chromatids move because the poles of the spindle fibers that are attached to the chromatids move apart from each other and because the proteins at the centromere of each chromosome pull the chromatids along the spindle

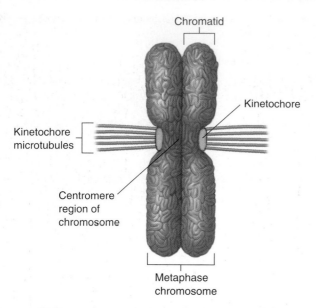

FIGURE 9.8 Kinetochore

The kinetochore on the chromosome is where the spindle fibers bind to the chromosome. During anaphase, the two chromatids separate from each other as (each) kinetochore shortens the spindle fiber (to which it is attached), pulling the chromosome toward the centrioles.

fibers toward the poles. The **kinetochore** is a protein attached to each chromatid at the centromere (figure 9.8). The kinetochore causes the shortening of the spindle fibers that are attached to it. By shortening the spindle fibers, the kinetochore pulls its chromatid toward the centriole.)

The two sets of daughter chromosomes migrating to opposite poles during anaphase have equivalent genetic information. This is true because the two chromatids of each chromosome, now called daughter chromosomes, were produced by DNA replication during the S stage of Interphase. Thus there are two equivalent sets of genetic information. Each set moves toward opposite poles.

Telophase

During **telophase,** the cell finishes mitosis. The spindle fibers disassemble. The nuclear membrane forms around the two new sets of chromosomes, and the chromosomes begin to uncoil back into chromatin, so that the genetic information found on their DNA can be read by transcriptional enzymes. The nucleolus re-forms as the cell begins to make new ribosomes for protein synthesis. The cell is preparing to reenter interphase. With the separation of genetic material into two new nuclei, mitosis is complete (figure 9.9).

Cytokinesis

At the end of telophase a cell has two nuclei. The process of mitosis has prepared the two nuclei to be passed on to the daughter cells. Next, the process of cytokinesis creates the daughter cells. Cytokinesis is the process during which the cell contents are split between the two new daughter cells.

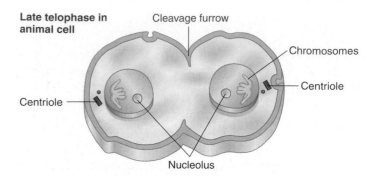

FIGURE 9.9 Telophase

During telophase, the spindle disassembles and the nucleolus forms and the nuclear membrane re-forms.

Different cell types use different strategies for achieving cytokinesis (figure 9.10). In animal cells, cytokinesis results from a *cleavage furrow*. The **cleavage furrow** is an indentation of the plasma membrane that pinches in toward the center of the cell, thus splitting the cytoplasm in two. In an animal cell, cytokinesis begins at the plasma membrane and proceeds to the center. In plant cells, a **cell plate** begins to form at the center of the cell and grows out to the cell membrane. The cell plate is made of normal plasma membrane components. It is formed by both daughter cells, so that, when complete, the two cells have separate membranes. The cell wall is then formed between the newly formed cells.

The completion of mitosis and cytokinesis marks the end of one round of cell division. Each of the newly formed daughter cells then starts the cell's cycle over by entering interphase at G_1. These cells can grow, replicate their DNA, and enter another round of mitosis and cytokinesis to continue the cell cycle or can stay metabolically active without dividing by staying in G_0.

Summary

Mitosis is much more than splitting the cytoplasm of a cell into two parts (table 9.1). Much of the process is devoted to ensuring that the genetic material is split appropriately

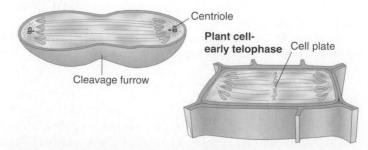

FIGURE 9.10 Cytokinesis: Animal and Plant

In animal cells, there is a pinching in of the cytoplasm, which eventually forms two daughter cells. Daughter cells in plants are formed when a cell plate separates the cell into two cells.

TABLE 9.1

Summary of the Cell Cycle

The stages of the cell cycle are shown in photographs and drawings for both animal and plant cells. The photographed animal cells are from whitefish blastulas. The photographed plant cells are from onion root tips.

Stage	Animal Cells		Plant Cells		Summary
Interphase					As the cell prepares for mitosis, the chromosomes replicate during the S phase of interphase.
Early Prophase					The replicated chromatids begin to coil into recognizable chromosomes; the nuclear membrane fragments; centrioles move to form the cell's poles; spindle fibers form; nucleolus disintegrates.
Late Prophase					
Metaphase					Chromosomes attach to spindle fibres at their centromeres and then move to the equator.
Anaphase					Chromatids, now called daughter chromosomes, separate toward the poles.
Telophase					The nuclear membranes and nucleoli re-form; spindle fibers fragment; the chromosomes unwind and change from chromosomes to chromatin.
Late Telophase					
Daughter Cells					Cytokinesis occurs and two daughter cells are formed from the dividing cells.

between the daughter cells. Sister chromatids form during DNA replication, so they represent identical genetic information. The sister chromatids are separated to each of the resulting daughter cells. By dividing the genetic information as sister chromatids, the daughter cells inherit the same genetic information that was present in the parent cell. Because the daughter cells have the same genetic information as the parent, they can replace lost cells and have access to all the same genetic information as the parent cell. With the same genetic information, the daughter cells can have the same function.

9.4 Controlling Mitosis

The cell-division process is regulated so that it does not interfere with the activities of other cells or of the whole organism. To determine if cell division is appropriate, many cells gather information about themselves and their environment. Checkpoints are times during the cell cycle when cells determine if they are prepared to move forward with cell division.

At these checkpoints, cells use proteins to evaluate their genetic health, their location in the body, and a need for more cells. Poor genetic health, the wrong location, and crowded conditions are typically interpreted as signals to wait. Good genetic health, the correct location, and uncrowded conditions are interpreted as signals to proceed with cell division.

The cell produces many proteins to gather this information and assess if cell division is appropriate. These proteins are made by one of two classes of genes. **Proto-oncogenes** code for proteins that provide signals to the cell that encourage cell division. **Tumor-supressor genes** code for proteins that provide signals that discourage cell division. A healthy cell receives signals from both groups of proteins about how appropriate it is to divide. The balance of information provided by these two groups of proteins allows for controlled cell division.

One tumor-suppressor gene is *p53*. Near the end of G_1, the protein produced by the *p53* gene identifies if the cell's DNA is damaged. If the DNA is healthy, p53 allows the cell to divide (figure 9.11a). If the p53 protein detects damaged

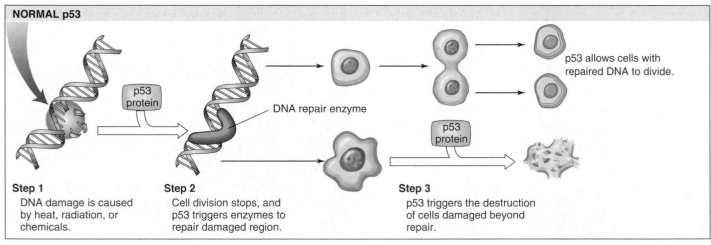

(a)

Step 1
DNA damage is caused by heat, radiation, or chemicals.

Step 2
Cell division stops, and p53 triggers enzymes to repair damaged region.

Step 3
p53 triggers the destruction of cells damaged beyond repair.

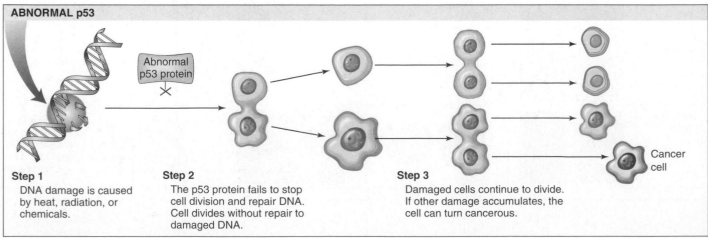

(b)

Step 1
DNA damage is caused by heat, radiation, or chemicals.

Step 2
The p53 protein fails to stop cell division and repair DNA. Cell divides without repair to damaged DNA.

Step 3
Damaged cells continue to divide. If other damage accumulates, the cell can turn cancerous.

FIGURE 9.11 The Function of p53 Protein
(a) Abnormal p53 protein stops cell division until damaged DNA is repaired. If the DNA is unrepairable, the p53 protein causes cell death.
(b) Mutated p53 allows cells with damaged DNA to divide.

DNA, it triggers other proteins to become active and repair the DNA. If the damage is too extensive for repair, the p53 protein triggers an entirely different response from the cell. The p53 protein causes the cell to self-destruct. **Apoptosis** is the process whereby a cell digests itself from the inside out. In this scenario, apoptosis prevents mutated cells from continuing to grow. Other healthy cells will undergo cell division to replace the lost cell.

Consider the implications of a mutation within the *p53* gene. If the p53 protein does not work correctly, then cells with damaged DNA may move through cell division. As these cells move through many divisions, their inability to detect damaged DNA disposes them to accumulate more mutations than do other cells. These mutations may occur in other proto-oncogenes and other tumor-supressor genes. As multiple mutations occur in the genes responsible for triggering cell division, the cell is less likely to control cell division appropriately.

9.5 | Cancer

Cancer is a disease caused by the failure to control cell division. This results in cells that divide too often and eventually interfere with normal body function. Scientists view cancer as a disease caused by mutations in the genes that regulate cell division. For example, the p53 protein is mutated in 40% of all cancers. The mutations may be inherited or may be caused by agents in the environment. For example, the tar from cigarette smoke has been directly linked to mutations in the *p53* gene. The tar in cigarette smoke is categorized as both a *mutagen* and a *carcinogen*. **Mutagens** are agents that mutate, or chemically damage, DNA. **Carcinogens** are mutagens that cause cancer.

Many agents have been associated with higher rates of cancer. The one thing they all have in common is their ability to alter DNA. Radiation and chemicals do this by chemically damaging DNA, so that the replication and transcriptional machinery can no longer read it (figure 9.12).

Radiation
X rays and gamma rays
Ultraviolet light (UV-A from tanning lamps; UV-B, the cause of sunburn)

Chemicals

Arsenic	Asbestos
Benzene	Alcohol
Dioxin	Cigarette tar
Polyvinyl chloride (PVC)	Food containing nitrates
Chemicals found in smoked meats and fish	(e.g., bacon)

Some viruses insert a copy of their genetic material into a cell's DNA. When this insertion occurs in a gene involved with regulating the cell cycle, it creates an insertion mutation, which may disrupt the cell's ability to control mitosis. Many of the viruses that are associated with higher rates of cancer are associated with a particular type of cancer (figure 9.13):

Viruses
Hepatitis B virus (HBV)
Herpes simplex virus (HSV) type II
Epstein-Barr virus
Human T-cell lymphotropic virus (HTLV-1)
Papillomavirus

Cancer
Liver cancer
Uterine cancer
Burkitt's lymphoma
Lymphomas and leukemias
Several cancers

FIGURE 9.12 Smoking
Smoking causes cancer.

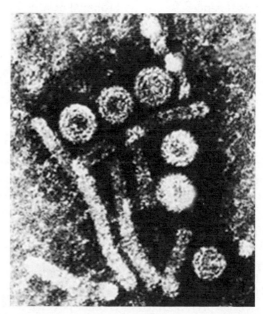

FIGURE 9.13 Cancer Causing Viruses
Cancer is both environmental and genetic. The hepatitis B virus is among the many agents that can increase the likelihood of developing cancer.

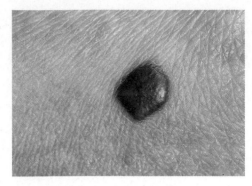

FIGURE 9.14 Skin Cancer

Malignant melanoma is a type of skin cancer. It forms as a result of a mutation in pigmented skin cells. These cells divide repeatedly, giving rise to an abnormal mass of pigmented skin cells. Only the dark area in the photograph is the cancer; the surrounding cells have the genetic information to develop into normal, healthy skin. This kind of cancer is particularly dangerous, because the cells break off and spread to other parts of the body (metastasize).

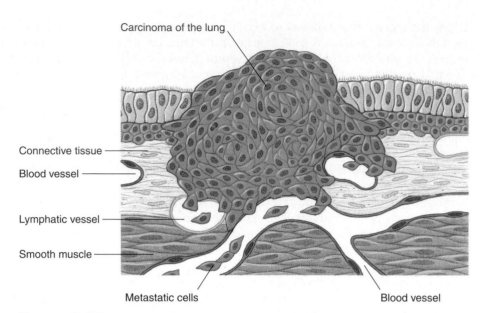

FIGURE 9.15 Metastasizing Cells

A tumor consists of cells that have lost their ability to control cell division. As these cells divide rapidly, they form a tumor and invade surrounding tissues. Cells metastasize when they reach blood vessels and are carried to other parts of the body. Once in their new locations, the cells continue to divide and form new tumors.

Because cancer is genetically based, scientists have found that a person's genetic background may leave him or her disposed to developing cancer. Predispositions to developing the following cancers have been shown to be inherited:

Leukemias   Lung cancer
Certain skin cancers   Endometrial cancer
Colorectal cancer   Stomach cancer
Retinoblastomas   Prostate cancer
Breast cancer

When uncontrolled mitotic division occurs, a group of cells forms a *tumor*. A **tumor** is a mass of cells not normally found in a certain portion of the body. A **benign tumor** is a cell mass that does not fragment and spread beyond its original area of growth. A benign tumor can become harmful, however, by growing large enough to interfere with normal body functions. Some tumors are *malignant*. **Malignant tumors** are harmful because they may spread or invade other parts of the body (figure 9.14). Cells of these tumors **metastasize,** or move from the original site and begin to grow new tumors in other regions of the body (figure 9.15).

Treatment Strategies

The Surgical Removal of Cancer

Once cancer has been detected, it is often possible to eliminate the tumor. If the cancer is confined to a few specific locations, it may be possible to remove it surgically. Many cancers of the skin or breast are dealt with in this manner. The early detection of such cancers is important, because the cancer can be removed before it has metastasized (figure 9.16). However, in some cases, surgery is impractical. Leukemia is a kind of cancer caused by the uncontrolled growth of white blood cells being formed in the bone marrow. In this situation, the cancerous cells spread throughout the body and cannot be removed surgically. Surgery is also not useful when the tumor is located

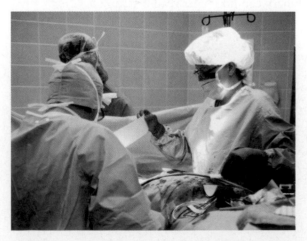

FIGURE 9.16 The Surgical Treatment of Cancer

Surgery is one option for treating cancer. Sometimes, if the cancer is too advanced or has already spread, other therapies are necessary.

where it can't be removed without destroying necessary healthy tissue. For example, removing certain brain cancers can severely damage the brain. In such cases, other treatments may be used, such as chemotherapy and radiation therapy.

Chemotherapy and Radiation Therapy

Scientists believe that chemotherapy and radiation therapy take advantage of the cell's ability to monitor cell division at the cell cycle checkpoints. By damaging DNA or preventing its replication, chemotherapy and radiation cause the target cell to stop dividing and die. Some chemotherapeutic agents disrupt parts of the cell, such as the spindle, that are critical for cell division. Most common cancers cannot be controlled with chemotherapy alone, and chemotherapy is often used in combination with radiation therapy.

Radiation therapy uses powerful X rays or gamma rays to damage the DNA of the cancer cells. At times, radiation can be used when surgery is impractical. This therapy can be applied from outside the body or by implanting radioactive "seeds" into the tumor. In both cases, a primary concern is to protect healthy tissue from the radiation's harmful effects. When radiation is applied from outside the body, a beam of radiation is focused on the cancerous cells and shields protect as much healthy tissue as possible.

Unfortunately, chemotherapy and radiation therapy can also have negative effects on normal cells. Chemotherapy may expose all the body's cells to the toxic ingredients and then weaken the body's normal defense mechanisms, because it decreases the body's ability to reproduce new white blood cells by mitosis. As a precaution against infection, cancer patients undergoing chemotherapy must be given antibiotics. The antibiotics help them defend against dangerous bacteria that might invade their bodies. Other side effects of chemotherapy include intestinal disorders and hair loss, which are caused by damage to the healthy cells in the intestinal tract and the skin that normally divide by mitosis.

Whole-body Radiation

Whole-body radiation is used to treat some leukemia patients, who have cancer of the blood-forming cells located in their bone marrow; however, not all of these cells are cancerous. A radiation therapy method prescribed for some patients involves the removal of some of their bone marrow and isolation of the noncancerous cells. The normal cells can then be grown in a laboratory. After these healthy cells have been cultured and increased in number, the patient's whole body is exposed to high doses of radiation sufficient to kill all the cancerous cells remaining in the bone marrow. Because this treatment is potentially deadly, the patient is isolated from all harmful substances and infectious microbes. They are fed sterile food, drink sterile water, and breathe sterile air while being closely monitored and treated with antibiotics. The cultured noncancerous cells are injected back into the patient. As if the cells had a memory, they migrate back to their origins in the bone marrow, establish residence, and begin regulated cell division all over again.

Because radiation damages healthy cells, it is used very cautiously. In cases of extreme exposure to radiation, people develop *radiation sickness*. The symptoms of this disease include hair loss, bloody vomiting and diarrhea, and a reduced white blood cell count. Vomiting, nausea, and diarrhea occur because the radiation kills many of the cells lining the gut and interfers with the replacement of the intestine's lining, which is constantly being lost as food travels through. Hair loss occurs because radiation prevents cell division at the hair root; these cells must divide for the hair to grow. Radiation reduces white blood cells because it prevents their continuous replacement from cells in the bone marrow and lymph nodes. When radiation strikes these rapidly dividing cells and kills them, the lining of the intestine wears away and bleeds, hair falls out, and there are very few new white blood cells to defend the body against infection.

9.6 Determination and Differentiation

The process of mitosis enables a single cell to develop into an entire body, with trillions of cells. As we have seen, the process of division is tightly regulated in the cells by genes. Genes also regulate what types of cells are produced by mitosis.

A **zygote** is the original single cell that results from the union of an egg and sperm. The zygote divides by mitosis to form genetically identical daughter cells. Mitotic cell division is repeated over and over until an entire body is formed. Although the cells in the body are genetically the same, they do not all have the same function. There are nerve cells, muscle cells, bone cells, skin cells, and many other types. Because all cells in a body have the same genetic information, the difference in the cells is not which genes they possess. The difference is in which genes they express. The proteins a cell produces defines that cell's type.

Determination is the process a cell goes through to select which genes it will eventually express on a more or less permanent basis. Determination marks the point where a cell commits to becoming a certain cell type and starts down the path of become a particular cell type. When a cell reaches the end of that path, it is said to be *differentiated*. A **differentiated** cell has become a particular cell type. Cells follow a determination pathway that leads them to differentiate into a specific cell type by selectively expressing certain genes.

Skin cells are a good example of determination and differentiation. Some skin cells produce hair; others do not. All the body's cells have the gene to produce hair, but not all cells do. When a cell starts to undergo the process of becoming a hair-producing cell, it is determined. Once the cell has become a hair-producing cell, it is differentiated. This differentiated cell is called a hair follicle cell (figure 9.17).

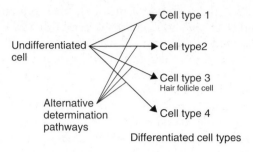

FIGURE 9.17 Determination and Differentiation
A cell starts as undetermined and undifferentiated. Specific genes are expressed to provide a cell its unique identity. Here, an undetermined cell goes through the process of determination to express the genes needed to be a hair follicle. When the process is complete and the hair follicle genes are expressed, the cell is differentiated.

9.7 Cell Division and Sexual Reproduction

For purposes of growth and repair, the goal is to create a cell that is genetically identical to the parent cell. Mitosis accomplishes this goal. However, the production of eggs and sperm requires a different approach. The cells of sexually reproducing organisms have two sets of chromosomes and thus have two sets of genetic information. One set is received from the mother's egg, the other from the father's sperm.

It is necessary for organisms that reproduce sexually to form gametes having only one set of chromosomes. If gametes contained two sets of chromosomes, the zygote resulting from their union would have four sets of chromosomes with twice the total genetic information of the parents. With each new generation, the number of chromosomes would continue to increase. Thus, eggs and sperm must be formed by a method that reduces the amount of genetic information by half.

Scientists have terms to distinguish when a cell has either one or two copies of genetic information.

Haploid cells carry only one copy of genetic information. **Diploid** cells carry two copies. Meiosis is the cell division process that generates haploid reproductive cells from diploid cells. In many sexually reproducing organisms, such as humans, meiosis takes place in the cells of organs that are devoted to reproduction—the **gonads.** The gonads in females are known as **ovaries;** in males, **testes.** Ovarian and testicular cells that divide by meiosis become reproductive cells called *gametes.* **Gamete** is a general term for reproductive cells like eggs and sperm. These gametes are also referred to as *germ cells.* Algae and plants also possess organs for sexual reproduction. Some of these are very simple. In algae such as *Spirogyra,* individual cells become specialized for gamete production. In plants, the structures are very complex. In flowering plants, the **pistil** produces eggs, or ova, and the **anther** produces pollen, which contains sperm (figure 9.18).

In sexually reproducing organisms, the life cycle involves both mitosis and meiosis. In figure 9.19, the haploid number of chromosomes is noted as n. The zygote and all the resulting cells that give rise to the adult fruit fly are diploid. The diploid number of chromosomes is noted as $2n$—mathematically, $n + n = 2n$. The egg cell and the sperm cell are produced by meiosis in female and male adult fruit flies. Notice that the male and female gamete each contain 4 chromosomes. These 4 chromosomes represent one complete copy of all the genetic information that is necessary for a fruit fly.

Fertilization is the joining of the genetic material from two haploid cells. On fertilization, each gamete contributes one copy of genetic information (one set of chromosomes) toward forming a new organism. Recall that the zygote is the diploid cell that results from the egg and sperm combining their genetic information. The zygote contains both copies of genetic information on 8 chromosomes (4 from the egg and 4 from the sperm). The zygote divides by *mitosis* and the cells grow to become an adult fruit fly, which will then produce either eggs or sperm by meiosis in its gonads. The characteristics of the fruit fly will depend on the combination of genetic information it inherits from both parents on its 8 chromosomes.

Diploid cells have two sets of chromosomes—one set from each parent. Because chromosomes contain DNA, each chromosome has many genes along its length. Each chromosome in a diploid cell can be paired to another chromosome on the basis of the genes on those chromosomes. **Homologous chromosomes** have the same order of genes along their DNA. Because of the similarity of genetic information in homologous chromosomes, homologous chromosomes are the same size and their centromeres are found in the same locations. Each parent contributes one member of each of the pairs of the homologous chromosomes. **Nonhomologous chromosomes** have different genes on their DNA (figure 9.20). The fruit fly has four pairs of homologous chromosomes—or 8 total chromosomes. Different species of organisms vary in the number of chromosomes they contain. Table 9.2 lists some organisms and their haploid and diploid chromosome numbers.

Consider the different purposes of mitosis and meiosis: Mitosis results in cells that have the same number of chromosomes as the parent cell, whereas meiosis results in cells that have half the chromosomes as the parent cell. An important question to ask is, "how are the processes of mitosis and meiosis different, so that gametes receive only half of the parent cell's chromosomes?"

9.8 Meiosis—Gamete Production

Consider a cell that has only 4 chromosomes (figure 9.21). The two from the father are shown in blue and the two from the mother are in green. Notice in figure 9.21 that there are two pairs of homologous chromosomes. Each pair consists of a green chromosome and a blue chromosome. One pair is long. The other pair is short.

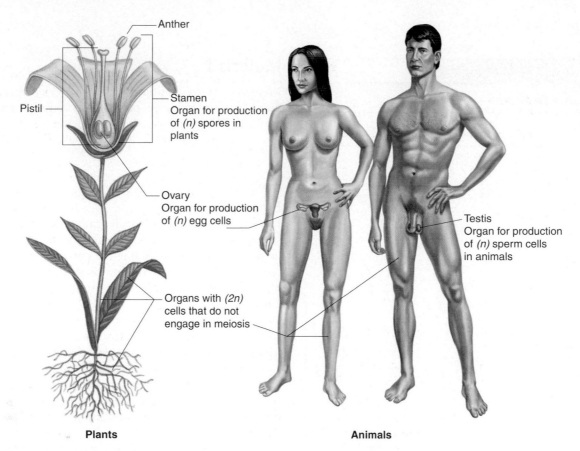

Plants   **Animals**

FIGURE 9.18 Haploid and Diploid Cells

Both plants and animals produce cells with a haploid number of chromosomes. The male anther in plants and the testes in animals produce haploid male cells, sperm. In both plants and animals, the ovaries produce haploid female cells, eggs.

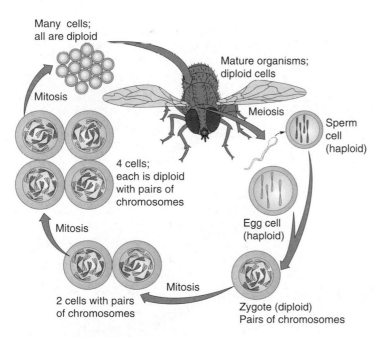

FIGURE 9.19 The Life Cycle of a Fruit Fly

The cells of this adult fruit fly have 8 chromosomes in their nuclei. In preparation for sexual reproduction, the number of chromosomes must be reduced by half, so that the gametes will have 4 chromosomes. After fertilization, this will result with an individual with the original number of 8 chromosomes. The offspring will grow and produce new cells by mitosis.

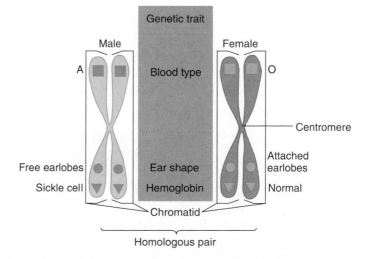

FIGURE 9.20 A Pair of Homologous Chromosomes

A pair of chromosomes are said to be homologous if they have genes for the same traits. Notice that the genes may not be identical, but the genes code for the same type of information. Homologous chromosomes are of the same length, have the same types of genes in the same sequence, and have their centromeres in the same location—one came from the male parent and the other from the female parent.

TABLE 9.2

Chromosome Numbers

Organism	Diploid Number	Haploid Number
Jumper ant	2	1
Tapeworm	4	2
Mosquito	6	3
Housefly	12	6
Onion	16	8
Rice	24	12
Tomato	24	12
Cat	38	19
Human	**46**	**23**
Chimpanzee	48	24
Potato	48	24
Horse	64	32
Dog	78	39
Stalked adder's tongue fern	1,260	630

Meiosis involves two cell divisions and produces four cells. **Meiosis I** consists of the processes that occur during the first division, and **meiosis II** consists of the processes that occur during the second division. Before meiosis occurs, the cell is in interphase of the cell cycle. As with mitosis, the interphase that precedes meiosis also includes DNA replication. Before DNA replication, chromosomes have only one chromatid. After DNA replication, chromosomes have two chromatids.

Meiosis I

Meiosis I is a **reduction division,** in which the chromosome number in the two cells produced is reduced from diploid to haploid. The sequence of events in meiosis I is divided into four phases: prophase I, metaphase I, anaphase I, and telophase I.

Prophase I

During prophase I, **synapsis** causes homologous chromosomes to move toward one another, so that the chromosomes lie next to each other. While the chromosomes are synapsed, *crossing-over* occurs. **Crossing-over** is the exchange of equivalent sections of DNA on homologous chromosomes. Crossing-over is shown in figure 9.22 as bits of blue on the green chromosome and bits of green on the blue chromosome. The crossing-over process is carefully regulated to make sure that the DNA sections that are exchanged contain equivalent information. This means that usually no information is lost or gained by either chromosome; genetic information is simply exchanged. Because the two members of each homologous pair of chromosomes came from different parents (one from the mother and one from father), there are minor differences in the DNA present on the two chromosomes. Crossing-over happens many times along the length of the homologous chromosomes.

Crossing-over is very important, because it allows a more thorough mixing of genes from one generation to the next. Without crossing-over, each of the chromosomes an organism inherits in the mother's egg would be passed on exactly as it was to the organism's offspring.

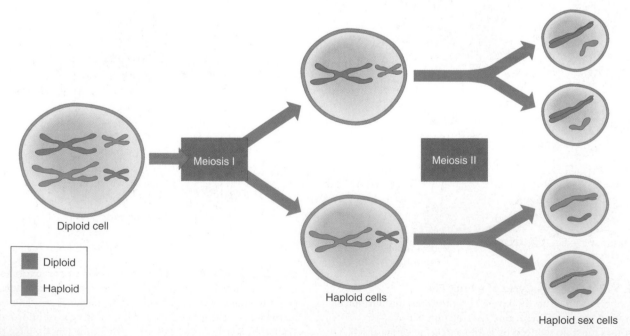

FIGURE 9.21 Meiosis

The cell division process of meiosis occurs in organisms that reproduce sexually. Meiosis occurs in two stages. The first stage, meiosis I, results in the formation of two cells. After each of these cells divides during meiosis II, four gametes are produced.

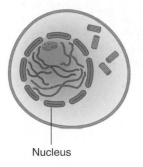

Nucleus

FIGURE 9.22 Prophase I

During prophase I, several visible changes occur as the cell prepares for division. The nuclear membrane is being broken down and the spindle begins to form. As the nuclear membrane disintegrates, the chromosomes can be moved throughout the cell. As the cell advances through prophase, the chromosomes also become more condensed and are paired as homologous pairs.

Metaphase I

In metaphase I, the centromere of each chromosome attaches to the spindle. The synapsed pair of homologous chromosomes moves into position on the cell's equatorial plane as a single unit. The orientation of the members of each pair of chromosomes is random with regard to the cell's poles. Figure 9.23 shows only one possible arrangement. An equally likely arrangement of chromosomes during this stage would be to flip the positions of two identically sized chromosomes. In the figure, this would place all of the green chromosomes on one side of the cell. The number of possible arrangements increases with the number of chromosomes present in the cell. The arrangement is determined by chance.

Anaphase I

Anaphase I is the stage during which homologous chromosomes separate (figure 9.24). During this stage, the chromosome number is *reduced from diploid to haploid*. The two members of each pair of homologous chromosomes move away from each other toward opposite poles. The direction each takes is determined by how each pair was originally oriented on the spindle.

This arrangement causes the key difference between mitosis and meiosis. In anaphase of mitosis, chromatids separate from each other. In anaphase I of meiosis, homologous chromosomes separate from each other. Each chromosome is independently attached to a spindle fiber at its centromere. Unlike the anaphase stage of mitosis, in anaphase I of meiosis the centromeres that hold the chromatids together do not divide. The chromosomes are still in their replicated form, consisting of 2 chromatids.

Because the homologous chromosomes and the genes they carry are being separated from one another, this process is called **segregation.** The way in which a single pair of homologous chromosomes segregates does not influence how other pairs of homologous chromosomes segregate. That is,

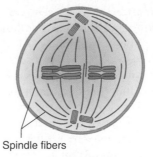

Spindle fibers

FIGURE 9.23 Metaphase I
Notice that the homologous chromosome pairs are arranged on the equatorial plane in the synapsed condition. This cell shows one way the chromosomes could be lined up; however, a second arrangement is possible.

FIGURE 9.24 Anaphase I
During this phase, one member of each homologous pair is segregated from the other member of the pair. Notice that the chromatids of the chromosomes do not separate.

Nuclear envelope

FIGURE 9.25 Telophase I
Cytokinesis occurs during telophase I. During cytokinesis, two cells are formed. Each cell is haploid, containing one set of chromosomes.

each pair segregates independently of other pairs. This is known as **independent assortment** of chromosomes. Both segregation and independent assortment are key components in understanding how to solve genetics problems. Mendel's laws, pp. 198, 201

Telophase I

Telophase I consists of changes that return the cell to an interphase-like condition (figure 9.25). The chromosomes uncoil and become long, thin threads; the nuclear membrane reforms around them; and nucleoli reappear. Following this activity, cytokinesis divides the cytoplasm into two separate cells.

Because of meiosis I, the total number of chromosomes is divided equally, so that each daughter cell has one member of each homologous chromosome pair. This means that each cell receives one-half the genetic information of the parent cell, but it has 1 chromosome of each kind and thus has one full set of chromosomes. Each chromosome is still composed of 2 chromatids joined at the centromere. The chromosome number for the cells is reduced from diploid ($2n$) to haploid (n). In the cell we have been using as our example, the number of chromosomes is reduced from 4 to 2. The four pairs of chromosomes have been distributed to the two daughter cells.

Depending on the type of cell, there may be a time following telophase I when the cell engages in normal metabolic activity corresponding to an interphase stage. Figure 9.26 summarizes the events in meiosis I.

Meiosis II

Meiosis II includes four phases: prophase II, metaphase II, anaphase II, and telophase II. The two daughter cells formed during meiosis I both continue through meiosis II, so that four cells result from the two divisions. During the time between telophase I and the beginning of meiosis II, no DNA replication occurs. The genetic information in cells starting meiosis II is the same as that in cells ending meiosis I. *The events in the division sequence of meiosis II are the same as those that occur in mitosis.*

Prophase II

Prophase II is similar to prophase in mitosis; the nuclear membrane is disassembled and the spindle apparatus begins to form. However, it differs from prophase I in that the cells are haploid, not diploid (figure 9.27).

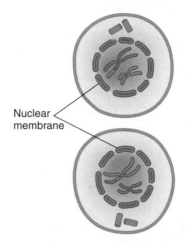

Nuclear membrane

FIGURE 9.27 Prophase II
The two daughter cells are preparing for the second division of meiosis.

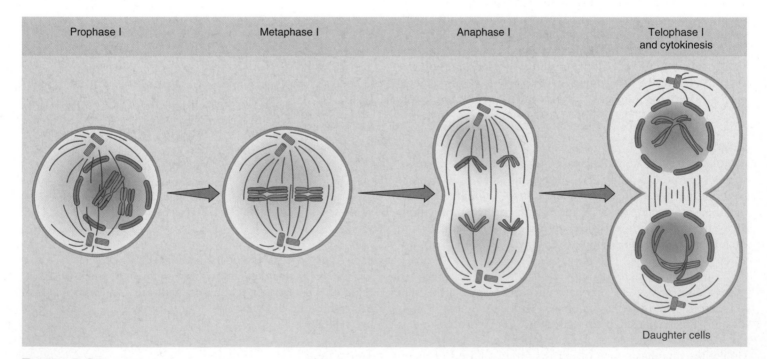

| Prophase I | Metaphase I | Anaphase I | Telophase I and cytokinesis |

Daughter cells

FIGURE 9.26 Meiosis I
The stages in meiosis I result in reduction division. This reduces the number of chromosomes in the parental cell from the diploid number to the haploid number in each of the two daughter cells.

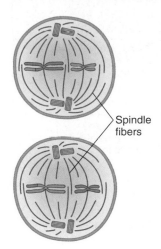

Spindle fibers

FIGURE 9.28 Metaphase II
During metaphase II, each chromosome lines up on the equatorial plane. Each chromosome is composed of 2 chromatids (a replicated chromosome) joined at a centromere.

FIGURE 9.29 Anaphase II
Anaphase II is very similar to anaphase of mitosis. The centromere of each chromosome divides and 1 chromatid separates from the other. As soon as this happens, they are no longer referred to as chromatids; each strand of nucleoprotein is now called a daughter chromosome.

Metaphase II

Metaphase II is typical of any metaphase stage, because the chromosomes attach by their centromeres to the spindle at the equatorial plane of the cell. Because pairs of homologous chromosomes are no longer together in the same cell, each chromosome moves as a separate unit (figure 9.28).

Anaphase II

Anaphase II of meiosis differs from anaphase I of meiosis in that, during anaphase II, the centromere of each chromosome divides, and the chromatids, now called *daughter chromosomes,* move to opposite poles. This is similar to mitosis (figure 9.29). There are no paired homologous chromosomes in this stage; therefore, segregation and independent assortment cannot occur as in meiosis I.

Telophase II

During telophase II, the cell returns to a nondividing condition. New nuclear membranes form, chromosomes uncoil, the spindles disappear and cytokinesis occurs (figure 9.30).

Telophase II is followed by the maturation of the four cells into gametes—either sperm or eggs. In many organisms, including humans, egg cells are produced in such a manner that three of the four cells resulting from meiosis in a female disintegrate. However, because the one that survives is randomly chosen, the likelihood of obtaining any particular combination of genes is not affected.

The events of meiosis II are summarized in figure 9.31 and table 9.3. The differences between mitosis and meiosis have been identified throughout this chapter. A comparison of these two processes appears in table 9.4.

FIGURE 9.30 Telophase II
During the telophase II stage, the nuclear membranes form, chromosomes uncoil. Cytokinesis occurs.

TABLE 9.3

Stages of Meiosis

Interphase		As the diploid (2n) cell moves from G$_0$ into meiosis, the chromosomes replicate during the S phase of interphase.
Prophase I		The replicated chromatin begins to coil into recognizable chromosomes and the homologous chromosomes synapse; chromatids may cross-over; the nuclear membrane and nucleoli fragment; centrioles move to form the cell's poles; spindle fibers are formed.
Metaphase I		Synapsed homologous chromosomes attach to the spindle fibers at their centromeres.
Anaphase I		The two members of homologous pairs of chromosomes separate from each other as they move toward the poles of the cell.
Telophase I		The two newly forming daughter cells are now haploid (n) because each contains only one of each pair of homologous chromosomes; the nuclear membranes and nucleoli re-form; spindle fibers fragment; the chromosomes unwind and change from chromosomes (composed of 2 chromatids) to chromatin.
Prophase II		Each of the two haploid (n) daughter cells from meiosis I undergoes chromatin coiling to form chromosomes, each of which is composed of 2 chromatids; the nuclear membrane fragments; centrioles move to form the cell's poles; spindle fibers form.
Metaphase II		Chromosomes attach to the spindle fibers at the centromeres and move to the equator of the cell.
Anaphase II		Centromeres replicate allowing the 2 chromatids of a chromosome to separate toward the poles.
Telophase II		Four haploid (n) cells are formed from the division of the two meiosis I cells; the nuclear membranes and nucleoli re-form; spindle fibers fragment; the chromosomes unwind and change from chromosomes to chromatin; these cells become the sex cells (egg or sperm).

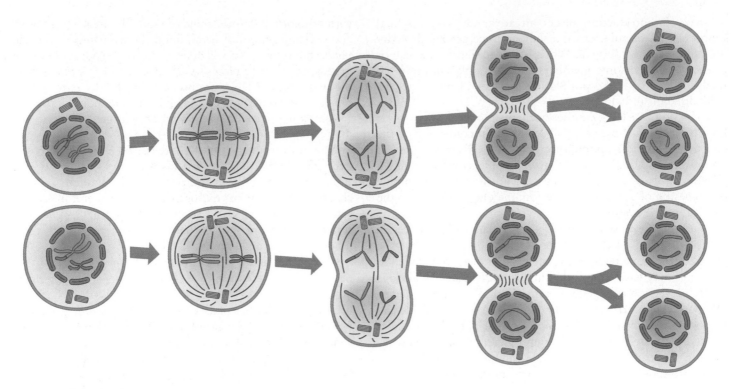

FIGURE 9.31 Meiosis II
During meiosis II, the centromere of each chromosome replicates and each chromosome divides into separate chromatids. Four haploid cells are produced, each with 1 chromatid of each kind.

TABLE 9.4	
Comparison of Mitosis and Meiosis	
Mitosis	**Meiosis**
1. One division completes the process.	1. Two divisions are required to complete the process.
2. Chromosomes do not synapse.	2. Homologous chromosomes synapse in prophase I.
3. Homologous chromosomes do not cross-over.	3. Homologous chromosomes cross-over in prophase I.
4. Centromeres divide in anaphase.	4. Centromeres divide in anaphase II but not in anaphase I.
5. Daughter cells have the same number of chromosomes as the parent cell ($2n \rightarrow 2n$ or $n \rightarrow n$).	5. Daughter cells have half the number of chromosomes as the parent cell ($2n \rightarrow n$).
6. Daughter cells have the same genetic information as the parent cell.	6. Daughter cells are genetically different from the parent cell.
7. Mitosis results in growth, the replacement of worn-out cells, and the repair of damage.	7. Meiosis results in sex cells.

9.9 Genetic Diversity—the Advantage of Sex

Cell division allows organisms to reproduce either asexually or sexually. There are advantages and disadvantages to both. Asexual reproduction always produces organisms that are genetically identical to the parent. A single organism, separated from others of its kind, can still reproduce if it can reproduce asexually. Organisms that can reproduce only sexually are at a disadvantage, because they require two different organisms to reproduce. Also, sexually reproducing populations tend to grow at a much slower rate than do asexually reproducing populations. However, asexually reproducing populations could be wiped out by a single disease or a change in living conditions, because the members of the population are genetically similar.

Sexual reproduction offers an advantage over asexual reproduction. Populations that have a large genetic diversity are more likely to survive. When living conditions change or a disease occurs, some members of the population are more likely to survive if the population consists of many, genetically different individuals.

The point of learning the mechanism of meiosis is to see how the events of meiosis and fertilization contribute to the genetic variation of a population. Haploid cells from different organisms combine to form new, unique combinations of genetic information. Each new organism, with its unique combination of genetic information, may be important to the survival of the population.

Genetic diversity in a population is due to differences in genes. Although all the members of the population have the genes for the same basic traits, the exact information coded in the genes may vary from individual to individual. An **allele** is a specific version of a gene. Examples of alleles are: A versus O blood types, dark versus light skin, normal versus sickle-cell hemoglobin, and attached versus free earlobes. Five factors create genetic diversity in offspring by creating either new alleles or new combinations of alleles: mutation, crossing-over, segregation, independent assortment, and fertilization.

Mutation

Several types of mutations were discussed in chapter 8: point mutations and chromosomal mutations. In point mutations, a change in a DNA nucleotide results in the production of a different protein. In chromosomal mutations, genes are rearranged. Both types of mutations can create new proteins. Both types of mutations increase genetic diversity by creating new alleles. DNA mutation, p. 158

Crossing-Over

The second source of variation is crossing-over. **Crossing-over** is the exchange of equivalent portions of DNA between homologous chromosomes, which occurs during prophase I while homologous chromosomes are synapsed. Remember that a chromosome is a double strand of DNA. To break chromosomes and exchange pieces of them, bonds between sugars and phosphates are broken. This is done at the same spot on both chromatids, and the two pieces switch places. After switching places, the two pieces of DNA are bonded together by re-forming the bonds between the sugar and the phosphate molecules.

Crossing-over allows new combinations of genetic information to occur. An organism receives one set of genetic information from each of its parents. As a result of crossing-over during meiosis, the organism creates unique combinations of genetic information. Each gamete contains chromosomes that have crossed-over and therefore contains some of the father's and some of the mother's genes. As a result, traits from the mother *and* from the father can be inherited on a single piece of DNA.

Examine figure 9.32 carefully to note precisely what occurs during crossing-over. This figure shows a pair of homologous chromosomes close to each other. Each gene occupies a specific place on the chromosome, its **locus**. Homologous chromosomes contain an identical order of genes, and chromosomes may contain thousands of genes.

Notice in figure 9.33 that, without crossing-over, only two kinds of genetically different gametes result. Two of the four gametes have one type of chromosome, whereas the other two have the other type of chromosome. With crossing-over, four genetically different gametes are formed. With just one cross-over, the number of genetically different gametes is doubled.

In fact, crossing-over can occur at a number of points on a chromosome; that is, there can be more than one cross-over per chromosome pair. Because crossing-over can occur at almost any point along the length of the chromosome, great variation is possible. (figure 9.34).

The closer two genes are to each other on a chromosome (i.e., the more closely they are *linked*), the more likely they will stay together, because the chance of crossing-over occurring between them is lower than if they were far apart. Thus, there is a high probability that they will be inherited together. The farther apart two genes are, the more likely it is that they will be separated during crossing-over. This fact enables biologists to map the order of gene loci on chromosomes.

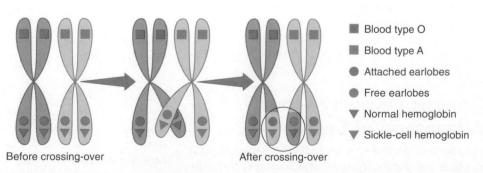

Before crossing-over   After crossing-over

■ Blood type O
■ Blood type A
● Attached earlobes
● Free earlobes
▼ Normal hemoglobin
▼ Sickle-cell hemoglobin

FIGURE 9.32 Synapsis and Crossing-Over
While pairs of homologous chromosomes are in synapsis, one part of 1 chromatid can break off and be exchanged for an equivalent part of its homologous chromatid.

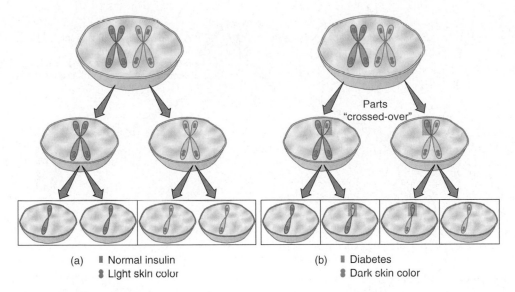

(a) ▌ Normal insulin
   ❚ Light skin color

(b) ▌ Diabetes
   ❚ Dark skin color

FIGURE 9.33 **Variations Resulting from Crossing-Over**
Cells with identical genetic information are boxed together. *(a)* These cells resulted from meiosis without crossing-over. Only two unique types of cells were produced. *(b)* These cells had one cross-over. From one cross-over, the number of genetically unique gametes doubled from two to four.

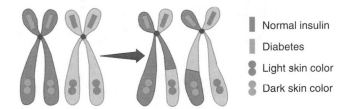

▌ Normal insulin
▌ Diabetes
❚ Light skin color
❚ Dark skin color

FIGURE 9.34 **Multiple Cross-Overs**
Crossing-over can occur several times between the chromatids of one pair of homologous chromosomes.

Segregation

Recall that segregation is the process during which the alleles on homologous chromosomes separate during meiosis I. Review figure 9.33a; in the insulin example, the different alleles of the insulin gene separate and go to different cells. Half of this individual's gametes would carry genetic information for functional insulin. The other half of the individual's gametes would carry genetic information for nonfunctional insulin. Consider if this individual's mate had the same genetic makeup. If the mate also had one normal gene for insulin production and one abnormal gene for diabetes, that person also would produce two kinds of gametes. Because of segregation, this couple could produce children that were genetically different from themselves. If both parents contributed a gamete that carried diabetes, their child would be diabetic. Other combinations of gametes would result in children without diabetes. Segregation increases genetic diversity by allowing parents to produce children that are genetically different from their parents and from their siblings. Mendel's Law of Segregation, p. 198

Independent Assortment

So far in discussing genetic diversity, we have dealt with only one pair of chromosomes. Now let's consider how variation increases when we add a second pair of chromosomes. Recall that **independent assortment** is the segregation of homologous chromosomes independent of how other homologous pairs segregate.

Figure 9.35 shows chromosomes with traits for the garden pea plant. The chromosomes carrying genes for flower color always separate from each other. The second pair of chromosomes with the information for seed texture also separates. Because the pole to which a chromosome moves is determined randomly, half the time the chromosomes divide so that the trait for purple flowers and the trait for round-smooth seeds move in one direction, whereas the trait for white flowers and the trait for wrinkled seeds move in the opposite direction. An equally likely alternative is that, the trait for purple flowers and the trait for wrinkled seeds go together, whereas the trait for white flowers and the trait for round-smooth seeds go to the other pole. With two pairs of homologous chromosomes there are four possible kinds of cells produced by meiosis.

With three pairs of homologous chromosomes, there are eight possible kinds of cells. The number of possible chromosomal combinations of gametes is calculated by using the expression 2^n, where n equals the number of pairs of chromosomes. With three pairs of chromosomes, n equals 3, so $2^n = 2^3 = 2 \times 2 \times 2 = 8$. With 23 pairs of chromosomes, as in human cells, $2^n = 2^{23} = 8,388,608$. More than 8 million genetically different kinds of sperm cells or egg cells are possible from a single human parent. This number doesn't consider the additional possible sources of variation, such as mutation and crossing-over. Thus, when genetic variation due

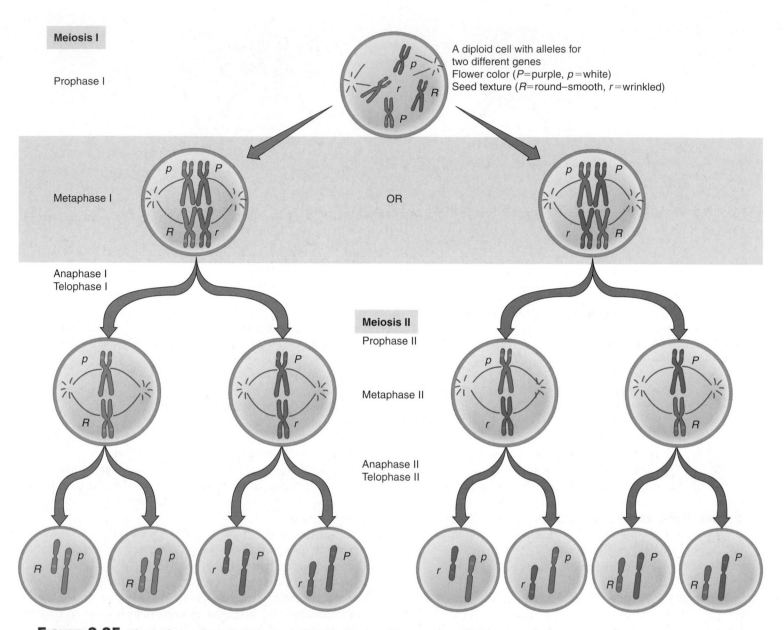

FIGURE 9.35 The Independent Assortment of Homologous Chromosome Pairs

The orientation of one pair of chromosomes on the equatorial plane does not affect the orientation of a second pair of chromosomes. Note in the figure the different ways that these two pairs of chromosomes can line up. How one homologous pair lines up does not influence how another positions itself. Just because the longer pair in the left portion of the figure is lined up with the green chromosome on the left and the blue chromosome on the right does not mean that the shorter pair has to line up with with the green chromosome on the left and the blue chromosome on the right. In this figure, *P* = purple flowers, *p* = white flowers; *R* = round-smooth seeds, *r* = wrinkled seeds.

to mutation and crossing-over is added, the number of different gametes that are possible is huge. Mendel's Law of Independent Assortment, p. 201

Fertilization

Because of the large number of genetically different gametes resulting from independent assortment, segregation, mutation, and crossing-over, an incredibly large number of types of offspring can result. Because humans can produce millions of genetically different gametes, the number of kinds of offspring possible is infinite for all practical purposes, and each offspring is unique, with the exception of identical twins.

9.10   Nondisjunction and Chromosomal Abnormalities

In the normal process of meiosis, the number of chromosomes in diploid cells is reduced to haploid. This involves segregating homologous chromosomes into separate cells during the first meiotic division. Occasionally, a pair of homologous chromosomes does not segregate properly and both chromosomes of a pair end up in the same gamete. **Nondisjunction** occurs when homologous chromosomes do not separate during cell division. As you can see in figure 9.36, two cells are missing a chromosome and the genes that were carried on it. This condition usually results in the death of the cells. The other cells have an extra copy of a chromosome. This extra genetic information may also lead to the death of the cell. Some of these abnormal cells, however, do live and develop into sperm or eggs.

If one of these abnormal sperm or eggs unites with a normal gamete, the offspring will have an abnormal number of chromosomes. In **monosomy,** a cell has just one of the pair of homologous chromosomes. The **trisomy,** a chromosome is present in three copies. All the cells that develop by mitosis from such zygotes will be either trisomic or monosomic.

It is possible to examine cells and count chromosomes. Among the easiest cells to view are white blood cells. They are dropped onto a microscope slide, so that the cells are broken open and the chromosomes are separated. Photographs are taken of chromosomes from cells in the metaphase stage of mitosis. The chromosomes in the pictures can then be cut and arranged for comparison with known samples (figure 9.37). This picture of an individual's chromosomal makeup is referred to as a *karyotype.*

One example of the effects of nondisjunction is the condition known as **Down syndrome.** If a gamete with 2 number 21 chromosomes has been fertilized by a gamete containing the typical one copy of chromosome number 21, the resulting

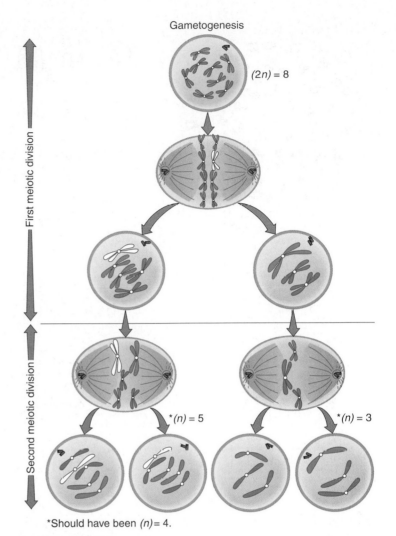

Gametogenesis

First meiotic division

Second meiotic division

$(2n) = 8$

*$(n) = 5$

*$(n) = 3$

*Should have been $(n) = 4$.

FIGURE 9.36 Nondisjunction During Gametogenesis
When a pair of homologous chromosomes fails to separate properly during meiosis I, gametogenesis results in gametes that have an abnormal number of chromosomes. Notice that two of the highlighted cells have an additional chromosome, whereas the other two are deficient by the same chromosome.

zygote has 47 chromosomes (figure 9.38). The child who developed from this fertilization has 47 chromosomes in every cell of his or her body as a result of mitosis and thus has the symptoms characteristic of Down syndrome. These include thickened eyelids, a large tongue, flattened facial features, short stature and fingers, some mental impairment, and faulty speech (figure 9.38). Premature aging is probably the most significant impact of this genetic disease. On the other hand, a child born with only 1 chromosome 21 rarely survives.

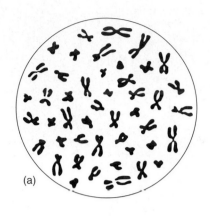

(a)

(b)

(c)

(d)

FIGURE 9.37 Human Male and Female Chromosomes

The randomly arranged chromosomes shown in the circle simulate metaphase cells spattered onto a microscope slide *(a)*. Those in parts *(b)* and *(c)* have been arranged into homologous pairs. Part *(b)* shows a male karyotype, with an X and a Y chromosome, and *(c)* shows a female karyotype, with two X chromosomes. *(d)* Notice that each pair of chromosomes is numbered and that the person from whom these chromosomes were taken has an extra chromosome number 21. The person with this trisomic condition might display a variety of physical characteristics, including slightly thickened eyelids, flattened facial features, a large tongue, and short stature and fingers. Most individuals also display some mental retardation.

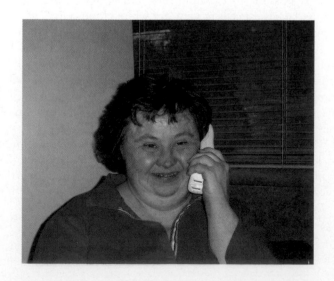

FIGURE 9.38 Down Syndrome

Every cell in the body of a person with Down Syndrome has 1 extra chromosome. With special care, planning, and training, people with this syndrome can lead happy, productive lives.

In the past, it was thought that the mother's age at childbirth played an important role in the occurrence of trisomies, such as Down syndrome. In women, gametogenesis begins early in life, but cells destined to become eggs are put on hold during meiosis I. Beginning at puberty and ending at menopause, one of these cells completes meiosis I monthly. This means that cells released for fertilization later in life are older than those released earlier in life. Therefore, it was believed that the chances for abnormalities, such as nondisjunction, increase as the mother ages. However, the evidence no longer supports this age-egg link. Currently, the increase in the frequency of trisomies with age has been correlated with a decrease in the activity of a woman's immune system. As she ages, her immune system is less likely to recognize the difference between an abnormal and a normal embryo. This means that she is more likely to carry an abnormal fetus to full term.

Figure 9.39 illustrates the frequency of the occurrence of Down syndrome births at various ages in women. Notice that the frequency increases very rapidly after age 37. For this reason, many physicians encourage couples to have their children in their early to mid-twenties, not in their late thirties or early forties. Physicians normally encourage older women who are pregnant to have the cells of their fetus checked to see if they have the normal chromosome number. It is important to know that the male parent can also contribute the extra chromosome 21.

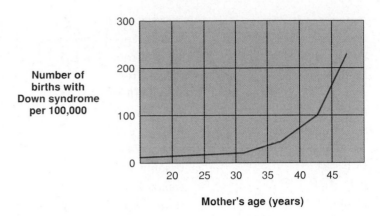

FIGURE 9.39 **Down Syndrome as a Function of a Mother's Age**
Notice that, as the age of the woman increases, the frequency of births of children with Down Syndrome increases only slightly until the age of approximately 37. From that point on, the rate increases drastically. This increase occurs because older women experience fewer miscarriages of abnormal embryos.

Summary

Cell division is necessary for growth, repair, and reproduction. Mitosis and meiosis are two important forms of cell division. Cells go through a cell cycle, which that includes a period for metabolism, DNA replication, and cell division (mitosis and cytokinesis). Interphase is the period of growth and preparation for division. Mitosis is divided into four stages: prophase, metaphase, anaphase, and telophase. During mitosis, two daughter nuclei are formed from one parent nucleus. These nuclei have identical sets of chromosomes and genes that are exact copies of those of the parent. The regulation of mitosis is important if organisms are to remain healthy. Regular divisions are necessary to replace lost cells and to allow for growth. However, uncontrolled cell division may result in cancer and disruption of the total organism's well-being.

Meiosis is a specialized process of cell division, resulting in the production of four cells, each of which has the haploid number of chromosomes. The total process involves two sequential divisions, during which one diploid cell reduces to four haploid cells. Various processes of meiosis, such as mutation, crossing-over, segregation, independent assortment, and fertilization, ensure that all sex cells are unique. Therefore, when any two cells unite to form a zygote, the zygote will also be one of a kind.

Key Terms

*Use interactive flash cards on the **Concepts in Biology**, 12/e website to help you learn the meaning of these terms.*

allele 182
anaphase 167
anther 174
apoptosis 171
asexual reproduction 164
asters 167
benign tumor 172
binary fission 164
carcinogens 171
cell cycle 164
cell division 164
cell plate 168
centrioles 167
chromatid 165
chromatin 165
cleavage furrow 168
crossing-over 176
cytokinesis 166

determination 173
differentiated 173
diploid 174
Down syndrome 185
fertilization 174
gametes 174
gonads 174
haploid 174
homologous chromosomes 174
independent assortment 177
interphase 164
kinetochore 168
locus 182
malignant tumors 172
meiosis 164
meiosis I 176
meiosis II 176

Basic Review

1. What is the key difference between mitosis and meiosis?

    a. Mitosis involves two rounds of cell division, whereas meiosis involves one round of cell divsion.

    b. DNA is not split between cells in meiosis, but this does occur during mitosis.

    c. Mitosis produces cells genetically identical to the parent, whereas meiosis produces cells with half the genetic information as the parent.

    d. None of the above are correct.

2. Which of the following is true of interphase?

    a. The chromosomes line up on the equatorial plane.

    b. The chromosomes double their chromatids.

    c. The DNA in the cell halves.

    d. All of the above are true.

3. Chromosomes are most likely to appear to be lining up near the middle of the cell during which phase of mitosis?

    a. interphase

    b. prophase

    c. metaphase

    d. telophase

4. Which of the following types of information do cells use to determine if they will divide?

    a. genetic health

    b. their current location

    c. the need for more cells

    d. All of the above are correct

5. p53 mutations lead to cancer because

    a. DNA damage is not repaired.

    b. mutated cells are allowed to grow.

    c. multiple mutations in the cell's regulatory proteins occur.

    d. All of the above are correct.

6. Haploid cells

    a. carry two copies of the genetic information.

    b. carry one copy of the genetic information.

    c. carry partial copies of the genetic information.

    d. are mutant.

7. Reduction division occurs

    a. in meiosis II.

    b. in meiosis I.

    c. in mitosis.

    d. after fertilization.

8. Genetic diversity in the gametes an individual is generated through:

    a. mitosis.

    b. independent assortment.

    c. segregation.

    d. both b and c.

9. Trisomy means

    a. that three copies of a chromosome are present.

    b. Down syndrome.

    c. that only three cells are present.

    d. none of the above.

10. A nondisjunction event occurs when

    a. homologous chromosomes did not separate correctly.

    b. non-homologous chromosomes did not separate correctly.

    c. daughter cells did not undergo cytokinesis correctly.

    d. none of the above.

Answers
1. c 2. b 3. c 4. d 5. d 6. b 7. a 8. d 9. a 10. a

Concept Review

Answers to the Concept Review questions can be found on the **Concepts in Biology,** *12/e website.*

9.1   The Importance of Cell Division

1. What are the three types of cell division?
2. What is the purpose of mitosis?

9.2   The Cell Cycle

3. What is the cell cycle?
4. What types of activities occur during interphase?

9.3   Mitosis—Cell Replication

5. Name the four stages of mitosis and describe what occurs in each stage.
6. During which stage of a cell's cycle does DNA replication occur?
7. At what phase of mitosis does a chromosome become visible?
8. List five differences between an interphase cell and a cell in mitosis.
9. Define the term *cytokinesis.*
10. What are the differences between plant and animal mitosis?
11. What is the difference between cytokinesis in plants and animals?

9.4   Controlling Cell Division

12. What are checkpoints?
13. What role does p53 have in controlling cell division?

9.5   Cancer

14. Why can radiation be used to control cancer?
15. List three factors associated with the development of cancer.

9.6   Determination and Differentiation

16. What is the difference between determination and differentiation?

9.7   Cell Division and Sexual Reproduction

17. How do haploid cells differ from diploid cells?
18. Why is meiosis necessary in organisms that reproduce sexually?
19. Define the terms *zygote, fertilization,* and *homologous chromosomes.*

20. Diagram fertilization as it would occur between a sperm and an egg with the haploid number of 3.

9.8   Meiosis—Gamete Production

21. Diagram the metaphase I stage of a cell with the diploid number of 8.
22. What is unique about prophase I?
23. During which phase do daughter chromosomes form?
24. Why is it impossible for synapsis to occur during meiosis II?
25. Can a haploid cell undergo meiosis? Why or why not?
26. List three differences between mitosis and meiosis.

9.9   Genetic Diversity—the Advantage of Sex

27. How much variation as a result of independent assortment can occur in cells with the following diploid numbers: 2, 4, 6, 8, and 22?
28. What are the major sources of variation in the process of meiosis?

9.10   Nondisjunction and Chromosomal Abnormalities

29. Define the term *nondisjunction.*
30. What is the difference between monosomy and trisomy?

Thinking Critically

For guidelines to answer this question, visit the **Concepts in Biology,** *12/e website.*

One experimental cancer therapy uses laboratory-generated antibodies to an individual's own, unique cancer cells. Radioisotopes, such as alpha-emitting radium 223, are placed in "cages" and attached to the antibodies. When these immunotherapy medications are given to a patient, the short-lived killer isotopes attach to only the cancer cells. They release small amounts of radiation for short distances; therefore, they cause little harm to healthy cells and tissues before their destructive powers are dissipated. Review the material on cell membranes, antibodies, cancer, and radiation therapy and explain the details of this treatment to a friend. (You might explore the Internet for further information.)

CHAPTER 10

Patterns of Inheritance

If you have had too many automobile accidents, your insurance company may raise your policy premium. This is because your driving history indicates that you will probably have more accidents in the future, which will cost the company more money, and it wants to be able to cover its costs. In addition, raising your rates might motivate you to become a safer driver. What if insurance companies thought that you were more likely to suffer a heart attack because others in your family had experienced heart attacks? What if you have a medical history that suggests that certain kinds of cancer run in your family? If these conditions exist, should the insurance companies be able to raise the cost of your premium?

Perhaps you know the parents of a child who was born with a disorder such as cystic fibrosis or sickle-cell anemia. Such couples often receive genetic counseling. Counseling can provide useful information. For example, there may be a 25% chance that their next child will also have this disorder. This type of information can help couples make informed decisions about whether to have more children.

How can an insurance company determine that you are at higher risk for disease? How do genetic counselors arrive at the numbers they provide their clients? Genetics uses information about how genes are organized on chromosomes and how chromosomes behave during meiosis. This information enables scientists to make predictions about how genes may be passed on to future generations. Insurance companies, genetic counselors, and you can use these predictions to make more educated decisions.

CHAPTER OUTLINE

10.1 Meiosis, Genes, and Alleles

Genetics is the branch of science that studies how the characteristics of living organisms are inherited. Classical genetics uses an understanding of meiosis to make predictions about the kinds of genes that will be inherited by the *offspring* (immediate descendants) of a sexually reproducing pair of organisms.

Various Ways to Study Genes

The previous chapters of this text used the term *gene*. In chapter 8, a gene was described as a piece of DNA with the necessary information to code for a protein. In chapter 9, on cell division, genes were described as entities that were found on chromosomes. Both of these views are correct, because the DNA with the necessary information to make a protein is packaged into a chromosome. When a cell divides, the DNA is passed on to the daughter cells in chromosomes. gene, pp. 147, 181

This chapter introduces another way to think about a gene. A gene is related to a characteristic of an organism, such as a color, a shape, or even the ability to break down a chemical. The characteristics are the result of proteins at work in the cell.

What Is an Allele?

Recall from chapter 9 that an allele is a specific version of a gene. Consider a characteristic such as earlobe shape. Some earlobes are free and some are attached (figure 10.1). These types of earlobes are two versions, or alleles, of the "earlobe-shape" gene. The two different alleles of this gene produce different versions of the same type of protein. The effect of these different proteins results in different earlobe shapes. Thus, there is an allele for free earlobes and a different allele for attached earlobes.

Genomes and Meiosis

A **genome** is a set of all the genes necessary to code for all of an organism's characteristics. In sexually reproducing organisms, a genome is diploid ($2n$) or when it has two copies of each gene. When two copies of a gene are present, the two copies need not be identical. The copies may be the same alleles, or they may be different alleles of the same gene.

The genome of a **haploid** (n) cell has only one copy of each gene. Sex cells, such as eggs and sperm, are haploid. Because sperm and eggs are haploid, they have only one version, or allele, of a gene. If the parent has two different alleles of a gene, the parent's sperm or egg can have either version of the alleles, but not both, because the sex cell is haploid. When

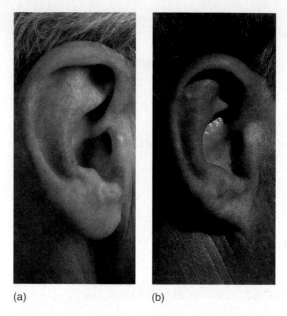

(a)   (b)

FIGURE 10.1 Genes Control Structural Features
Whether your earlobe is *(a)* free or *(b)* attached depends on the genes you have inherited. As genes express themselves, their actions affect the development of various tissues and organs. Some people's earlobes do not separate from the sides of their heads in the same manner as do those of others. How genes control this complex growth pattern and why certain genes function differently than others are yet to be clarified.

a haploid sperm (n) from a male and a haploid egg (n) from a female combine, they form a diploid ($2n$) cell, called a *zygote*. The alleles in the sperm and the alleles in the egg combine to form a new genome that is different from either of the parents. This means that each new zygote is a unique combination of genetic information. ploidy, p. 174

Meiosis is a cell's process of making haploid cells, such as eggs or sperm. Understanding the process of meiosis is extremely important to making genetic predictions. If you don't understand the cellular process of meiosis, your predictions will be less accurate. Figure 10.2 shows a pair of *homologous chromosomes* that have undergone DNA replication. After DNA replication, each homologous chromosome has two, exact copies of each allele, one on each chromatid.

When the cell containing these chromosomes undergoes meiosis I, the two homologous chromosomes go to different cells. This reduces the cell from diploid to haploid. In meiosis II, the chromatids of each chromosome are separated into different daughter cells. The cells resulting from meiosis II mature to become sperm or eggs. The probability that an allele will be passed to reproductive cells, such as sperm or eggs, is used in making predictions in genetics crosses.

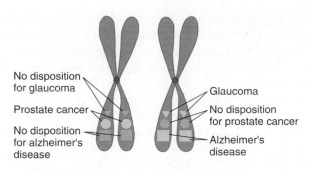

FIGURE 10.2 Homologous Chromosomes—
Human Chromosome 1

Homologous chromosomes contain genes for the same
characteristics at the same place. Different versions, or alleles, of
the genes may be present on different chromosomes. This set of
homologous chromosomes represents chromosome 1 in humans.
Chromosome 1 is known to contain genes that play a role in
glaucoma, prostate cancer, and Alzheimer's disease. Here, different
genes are shown by different shapes. Different alleles for each gene
are shown by different coloration of the shape. The three genes
shown here may be present in their normal form or in their altered,
mutant form. Different copies of the genes are sorted to different
gametes during meiosis.

10.2   The Fundamentals of Genetics

Three questions represent the biological principles behind
most of the genetics presented in this chapter:

1. What alleles do the parents have?
2. What alleles are present in the gametes that the parents
   produce?
3. What is the likelihood that any two sex cells, each with
   a particular set of alleles, will fertilize each other?

To solve genetics problems and understand biological inheritance, it is necessary to understand how to answer each of
these questions and to understand how the answer to one of
these questions can affect the others.

Phenotype and Genotype

Because diploid organisms have two copies of a gene, it is
possible for an individual to have 2 different alleles of that
gene. The **genotype** of an organism is the combination of alleles that are present in the organism's cells. The **phenotype** of
an organism is how it appears outwardly and is a result of the
organism's genotype.

Reconsider the example of earlobe type to explore the
ideas of phenotype and genotype. Earlobes can be attached or
free (review figure 10.1). If a person's earlobes are attached,

the person's *phenotype* is "attached earlobes." Likewise, if a
person's earlobes are free, his or her *phenotype* is "free earlobes." Each person has 2 alleles for earlobe type. However,
the 2 alleles do not need to be identical.

To make understanding *genotype* easier, we can use a
shorthand notation. The capital letter E can be used to represent the allele that codes for free earlobe development. A lowercase e can be used to refer to the allele that codes for
attached earlobe development. This type of coding system is
commonly used in genetics. Because each person has 2 alleles,
one person can have one of these combinations of alleles:

- (EE)—2 alleles for free earlobes
- (ee)—2 alleles for attached earlobes
- (Ee)—1 allele for free earlobes and 1 allele for
  attached earlobes

The 2 alleles will interact with each other when they are in the
same cell. Consider what happens in a cell when the allele
combination is EE, ee, or even Ee. When the cell has EE, it is
only capable of producing proteins associated with free earlobes. The organism will have free earlobes. When both alleles code for attached earlobe development (ee) that the
person will develop attached earlobes.

What does the organism look like if it has 1 allele that
codes for free earlobes and 1 allele that codes for attached
earlobes—(Ee)? In this case, the organism develops free earlobes. The E allele produces proteins for free earlobes that
"outperforms" the e allele. Therefore, E is able to dominate
the appearance of the organism. A **dominant allele** masks
the **recessive allele** in the phenotype of an organism. In the
previous example, the free earlobes allele (E) are dominant
and the attached earlobes allele (e) are recessive, because in
an (Ee) individual the phenotype that develops is free earlobes. Geneticists use the capital letter to denote that an
allele is dominant. The lowercase letter denotes the recessive
allele.

Take a closer look at the genotypes for free and attached
earlobes. Notice that organisms with attached earlobes
always have 2 e alleles (ee), whereas organisms with free earlobes might have 2 E alleles—(EE)—or both an E and an e
allele—(Ee). A dominant allele may hide a recessive allele.

The term *recessive* has nothing to do with the significance
or value of the allele—it simply describes how it is expressed
when inherited with a dominant allele. The term *recessive*
also has nothing to do with how frequently the allele is passed
on to *offspring*. **Offspring** are the descendants of a set of parents. Insert in individuals with 2 different alleles, each allele
has an equal chance of being passed on. The key on p. 194
organizes the information about how earlobe shape is inherited. This format will also be used later in this chapter to summarize information about other genes.

Gene Key
Gene: earlobe shape

Allele Symbols	Possible Genotypes	Phenotype
E = free	*EE*	Free earlobes
	Ee	Free earlobes
e = attached	*ee*	Attached earlobes

Summary: Geneticists describe an organism by its genotype and its phenotype. One rule that describes how the genotype of an organism influences its phenotype involves the principle of dominant and recessive interaction.

Application: Use the dominant and recessive principle to infer information that is not provided. Example: If a person has attached earlobes, you can infer that his or her genotype is *ee*. If a person has free earlobes, you can infer that he or she has at least 1 *E* allele. The second allele is uncertain without additional information.

Predicting Gametes from Meiosis

To predict the types of offspring that parents may produce, it is important to predict the kinds of alleles that may be in the sex cells produced by each parent. Remember that during meiosis the 2 alleles will end up in different sex cells. If an organism contains two copies of the same allele, such as in *EE* or *ee,* it can produce sex cells with only one type of allele. *EE* individuals can produce sex cells with only the *E* allele, whereas *ee* individuals can produce sex cells with only the *e* allele. The *Ee* individual can produce two different types of sex cells. Half of the sex cells carry the *E* allele. The other half carry the *e* allele. If an organism has 2 identical alleles for a characteristic and can produce sex cells with only one type of allele, the genotype of the organism is **homozygous** (*homo* = same or like). If an organism has 2 different alleles for a characteristic and can produce two kinds of sex cells with different alleles, the organism is **heterozygous** (*hetero* = different).

Gene Key
Gene: earlobe type

Allele Symbols	Possible Genotype
E = free	*EE*-homozygous
	Ee-heterozygous
e = attached	*ee*-homozygous

Phenotype	Possible Sex Cells
Free earlobes	All sex cells have *E.*
Free earlobes	Half of sexcells have *E* and half have *e.*
Attached earlobes	All sex cells have *e.*

Notice that the 2 alleles separate into separate sex cells. This is true whether the cell is homozygous or heterozygous. The

Law of Segregation states that in a diploid organism the alleles exist as two separate pieces of genetic information and that these two different pieces of genetic information, are on different chromosomes and are separated into different cells during meiosis.

Summary: When sex cells form, they receive only 1 allele for each characteristic. Homozygous organisms can produce only one kind of sex cell. In heterozygous organisms, meiosis produces two genetically different sex cells. The 2 different alleles are represented equally in the sex cells that are produced. Half the cells contain 1 of the alleles and half the cells contain the other.

Application: Make two predictions using the Law of Segregation. The first prediction describes the genetic information a sex cell can carry. The second prediction describes the expected ratios of these sex cells. If the organism is homozygous, then all sex cells will be the same. If the organism is heterozygous, then half the sex cells will carry 1 allele. The second allele will be in the other half of the sex cells.

Fertilization

Recall from chapter 9 that **fertilization** is the process of two haploid (*n*) sex cells joining to form a zygote (*2n*). The union of a sperm and an egg is fertilization. The zygote divides by mitosis to produce additional diploid cells as the new organism grows. The diploid genotype of all the cells of that organism is determined by the alleles carried by the sex cells that joined to form the zygote. sexual reproduction, p. 164

A **genetic cross** is a planned mating between two organisms. Although the cross is planned, the exact sperm and egg that join when fertilization occurs are not entirely predictable, because the process of fertilization is random. Any one of the many different sperm produced by meiosis may fertilize a given egg. However, generalizations can be made about possible results from two parents. These generalizations can be seen by drawing a diagram called a *Punnett square.* A **Punnett square** shows the possible offspring of a particular genetic cross.

Genetic crosses can be designed to investigate one or more characteristics. A **single-factor cross** is designed to look at how one genetically determined characteristic is inherited. A unique single factor cross is a *monohybrid cross.* A **monohybrid cross** is a cross between two organisms that are both heterozygous for the one observed gene. A **double-factor cross** is a genetic study in which two different genetically determined characteristics are followed from the parental generation to the offspring at the same time.

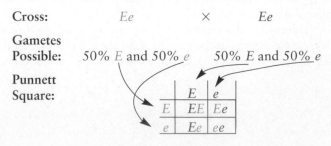

Again, consider the example of earlobe shape. The cross shown is between two heterozygous—(*Ee*)—individuals. The individuals in this cross can each produce two types of sex cells, *E* and *e*. The colors used in this monohybrid cross and Punnett square allow us to trace what happens to the sex cells from each parent. The top row lists the sex cells that can be produced by one parent, and the left-most column of the Punnett square lists the sex cells that can be produced by the other parent. The letter combinations within the four boxes represent the possible genotypes of the offspring. Each combination of letters is simply the combination of the alleles listed at the top of each column and the left of each row. There are three genotypes, *EE*, *Ee*, and *ee*. Additionally, by counting how many times each genotype is shown in the Punnett square, we can predict how frequently we expect to observe each genotype in the offspring of these parents. Here, we expect to see *Ee* twice for every time we see *EE* or *ee*. Remember that a Punnet square only generalizes. If many (*at least 30 or more*) offspring are produced from the cross, we might expect to see nearly 1/4 *EE*, 2/4 *Ee*, and 1/4 *ee*. Geneticists may abbreviate these ratios as 1:2:1. These ratios can be written in a different manner but still mean the same thing:

1/4 EE, 1/2 Ee, 1/4 ee

Summary: The outcome of a genetic cross cannot be exactly determined. The outcome can only be described by general trends.

Application: The Punnett square can be used to predict the types and ratios of offspring.

10.3 Probability vs. Possibility

Probability is the mathematical chance that an event will happen, and it is expressed as a percentage or a fraction, like the values that we identified using the Punnett square in the previous example. Probability is not the same as *possibility*. Consider the common phrase "almost anything is possible" when reading the following sentence: "It is possible for me to win the lottery, but how probable is it?" Although it is possible to win the lottery, it is extremely unlikely. When we talk about something being probable, we actually talk mathematically— in ratios and percentages—such as, "The probability of my winning the lottery is 1 in 250,000."

It is possible to toss a coin and have it come up heads, but the probability of getting a head is a more precise statement than just saying it is possible to do so. The probability of getting a head is 1 out of 2 (1/2, or 0.5, or 50%), because there are two sides to the coin, only one of which is a head. Probability can be expressed as a fraction:

$$\text{probability} = \frac{\text{the number of events that can produce a given outcome}}{\text{the total number of possible outcomes}}$$

What is the probability of cutting a deck of cards and getting the ace of hearts? The number of times the ace of hearts can occur is 1. The total number of different cards in the deck is 52. Therefore, the probability of cutting to an ace of hearts is 1/52. What is the probability of cutting to any ace? The total number of aces in the deck is 4, and the total number of cards is 52. Therefore, the probability of cutting an ace is 4/52, or 1/13.

It is also possible to determine the probability of two independent events occurring together. *The probability of two or more events occurring simultaneously is the product of their individual probabilities* (How Science Works 10.1). When two six-sided dice are thrown, it is possible that both will be 4s. What is the probability that both will be 4s? The probability of one die being a 4 is 1 out of the six sides of the die, or 1/6. The probability of the other die being a 4 is also 1/6. Therefore, the probability of throwing two 4s is

$$1/6 \times 1/6 = 1/36$$

The concepts of probability and possibility are frequently used in solving genetics problems. Consider describing the genetic contents of an individual's sex cells. Assume that the individual's genotype is *AA*. It is only *possible* for this individual to produce sex cells that carry the *A* allele. The *probability* of this occurring is 100%. Now consider an individual with the *Aa* genotype. It is *possible* for the *Aa* individual to produce *A* or *a* sex cells. The *probability* of a sex cell having *A* is 50%; only one of the two possibilities is *A*. Likewise, the probability of a sex cell having *a* is 50%; only one of the two possibilities is *a*.

In a genetics problem, the frequency with which alleles are present in gametes determines the likelihood that a couple will have children with a particular characteristic. Consider the possible fertilization events that *could* occur between an individual with the genotype *AA* and an individual with the genotype *aa* (the genetic cross *AA* × *aa*). To do this, use a Punnett square. First, predict the possible sex cells produced by each individual.

Genotype	Possible Sex Cells
AA	*A*
aa	*a*

Then, set up a Punnett Square that shows the possible fertilization events between the sex cells shown.

	a
A	*Aa*

The only possible offspring is *Aa*. The probability of obtaining an offspring with this genotype is 100%.

HOW SCIENCE WORKS 10.1

Cystic Fibrosis—What Is the Probability?

Cystic fibrosis is among the most common lethal genetic disorders that affect Caucasians in the United States. Cystic fibrosis affects nearly 30,000 people in North America. One in every 20 persons has a defective recessive allele that causes cystic fibrosis, but most of these individuals display no cystic fibrosis symptoms, because the recessive allele is masked by a normal dominant allele. Only those with two copies of the defective recessive gene develop symptoms. About 1,000 new cystic fibrosis cases are identified in the United States each year. The gene for cystic fibrosis occurs on chromosome 7; it is responsible for the manufacture of **c**ystic **f**ibrosis **t**ransmembrane **r**egulator (CFTR) protein. The CFTR protein controls the movement of chloride ions across the cell membrane.

There are many possible types of mutations in the CFTR gene. The most common mutation results in a CFTR protein with a deletion of a single amino acid. As a result, CFTR protein is unable to control the movement of chloride ions across the cell membrane. The major result is mucus filling the bronchioles, resulting in blocked breathing and frequent respiratory infections. It is also responsible for other symptoms:

1. A malfunction of sweat glands in the skin and the secretion of excess chloride ions
2. Bile duct clogging, which interferes with digestion and liver function
3. Mucus clogging the pancreas ducts, preventing the flow of digestive enzymes into the intestinal tract
4. Bowel obstructions caused by thickened stools
5. Sterility in males due to the absence of vas deferens development and, on occasion, female sterility due to dense mucus blocking sperm from reaching eggs

One in 20 people have a recessive allele for cystic fibrosis. In this group of graduates, Four or five individuals, on average, have the allele.

Consider the facts about the frequency of the cystic fibrosis gene in the population. What is the probability that any set of parents will have a child with cystic fibrosis? What is the probability that a person who carries the cystic fibrosis allele will marry someone who also has the allele?

10.4 The First Geneticist: Gregor Mendel

The inheritance patterns discussed in the section "Probability vs. Possibility" were initially described by Gregor Mendel—a member of the religious order of Augustinian monks. Mendel's (1822–1884) work was not generally accepted until 1900, however, when three men, working independently, rediscovered some of the ideas that Mendel had formulated more than 30 years earlier. Because of his early work, the study of the pattern of inheritance that follows the laws formulated by Gregor Mendel is often called **Mendelian genetics** (figure 10.3).

Heredity problems are concerned with determining which alleles are passed from parents to offspring and how likely it is that various types of offspring will be produced. Mendel performed experiments concerning the inheritance of certain characteristics in garden pea plants (*Pisum sativum*).

From his work, Mendel developed the ideas of a genetic characteristic being dominant or recessive and categorized a number of garden pea alleles by using rules of probability. Some of the phenotypes he used in his experiments are shown in table 10.1.

What made Mendel's work unique was that he initially studied only one trait at a time. In addition, he grouped the offspring by phenotype and counted them. Previous investigators had tried to follow numerous traits at the same time. This made it very difficult to follow characteristics, and they did not determine the frequency of phenotypic groups. Therefore, they were unable to see any patterns in their data.

Mendel also used clear-cut phenotypes—such as purple or white flower color, yellow or green seed pods, and tall or dwarf pea plants. He was very lucky to have chosen pea plants in his study, because they naturally self-pollinate. This means that pea plants produce both pollen and eggs and that the eggs can be fertilized by haploid nuclei from their own pollen. When self-pollination occurs in pea plants over many

(a) (b)

FIGURE 10.3 Gregor Mendel and His Pea Plant Garden

(a) Gregor Mendel was an Augustinian monk who used statistics to describe the inheritance patterns he observed in pea plants. *(b)* He carried out his investigations in this small garden of his monastery.

TABLE 10.1		
Dominant and Recessive Traits in Pea Plants		
Gene	**Dominant Allele Phenotype**	**Recessive Allele Phenotype**
Plant height	Tall	Dwarf
Pod shape	Full	Constricted
Pod color	Green	Yellow
Seed surface texture	Round	Wrinkled
Seed color	Yellow	Green
Flower color	Purple	White

generations, it is easier to develop a population of plants that is homozygous for a number of characteristics. Such a population is known as a *pure line*.

The following gene key organizes some of Mendel's findings. Remember that Mendel didn't know about DNA or even chromosomes! Mendel developed this way of thinking about genetics to explain the data he collected. He did this from the perspective of a mathematician—not a biologist.

Gene Key
Gene: Flower color

Allele Symbols	Possible Genotypes
C = Purple	*CC* = homozygous
	Cc = heterozygous
c = white	*cc* = homozygous

Phenotype	Possible Sex Cells
Purple	All sex cells have *C* (*pure line*).
Purple	Half of sex cells have *C* and half have *c*.
White	All sex cells have *c* (*pure line*)

In one experiment, Mendel took a pure line of pea plants having purple flower color, removed the male parts (anthers), and discarded them, so that the plants could not self-pollinate. He then took anthers from a pure-breeding white-flowered plant and pollinated the antherless purple flower. These plants are called the parent generation, or P_0. When the pollinated flowers produced seeds, Mendel collected, labeled, and planted them. When these seeds germinated and grew, they eventually produced flowers. The offspring of the P_0 generation are called the F_1 generation. The *F* stands for *filial*, which is Latin for relating to a son or daughter. F_1 is read as the "F-One" generation or the "first filial" generation.

Mendel's First Cross

Observed	Genetic Notation
P_0 pure breeding purple × pure breeding white	*CC* × *cc*
	$\begin{array}{c|c} & c \\ \hline C & Cc \end{array}$
F_1 100% produced purple flowers	100% *Cc*

All the F_1 plants resulting from this cross had purple flowers. One of the popular ideas of Mendel's day would have predicted that the purple and white colors would have blended, resulting in flowers that were lighter than the purple parent flowers.

Another hypothesis would have predicted that the offspring would have had a mixture of white and purple flowers. Neither of these two hypotheses was supported by Mendel's results. He observed only purple flowers from this cross.

Mendel then crossed the F_1 pea plants (all of which had purple flowers) with each other to see what the next generation would be like. Had the white-flowered characteristic been lost completely? The seeds from this mating were collected and grown. When these plants flowered, three-fourths of them produced purple flowers and one-fourth produced white flowers. This generation is called the F_2 generation.

Mendel's Second Cross

Observed	Genetic Notation
F_1 Offspring from P_0 (purple) × offspring from P_0 (purple)	$Cc \times Cc$

	C	c
C	CC	Cc
c	Cc	cc

F_2 75% produced purple flowers	25% CC + <u>50% Cc</u> 75% produced purple flowers
25% produced white flowers	25% cc produced white flowers

His experiments used similar strategies to investigate other traits. Pure-breeding tall plants were crossed with pure-breeding dwarf plants. Pure-breeding plants with yellow pods were crossed with pure-breeding plants with green pods. Mendel recognized the same pattern for each characteristic in the F_1 generation: All the offspring showed the characteristics of one parent and not the other with no blending.

After analyzing these data, Mendel identified several genetic principles:

1. Organisms have two pieces of genetic information. It is now recognized that these can be different alleles for each characteristic.
2. Because organisms have two pieces of genetic information for each characteristic, the alleles can be different. Mendel's **Law of Dominance** states that some alleles interact with each other in a dominant and recessive manner whereby the dominant allele masks the recessive allele.
3. Gametes fertilize randomly.
4. Mendel's **Law of Segregation** states that, when a diploid organism forms gametes, the two alleles for a characteristic separate from one another. In doing this, they move to different gametes and retain their individuality.

The application of Mendel's Law of Segregation may not be as apparent as the application of the Law of Dominance. The movements of chromosomes during meiosis separate the four copies (one on each chromatid) of each allele into four different sex cells. This causes only 1 allele of each gene to be present in each sex cell. We first observed this law in this chapter when we discussed sex cells and how alleles separate in both homozygous and heterozygous organisms. Finally, this law is the basis for the Punnett square, in which the possible alleles from each parent are placed in a separate row or column.

10.5 Solving Genetics Problems

Predictions about the inheritance patterns of many traits use the principles presented in this chapter. Many students become confused when they try to solve genetics problems because they are not sure where to begin or when it is appropriate to apply a principle. As a result, they move directly to drawing a Punnett square and begin filling it up with letters incorrectly. Developing an organized and consistent strategy will help solve such problems.

Single-Factor Crosses

Problem Type: Single-Factor Cross

CROSS 1: Genetics problems can range greatly in complexity and the type of information that is provided. Let's start with a genetics problem that considers a trait determined by only one gene.

The pod color of some pea plants is inherited so that green pods are dominant to yellow pods. A pea plant that is heterozygous for green pods is crossed to a pea plant that produces yellow pods. What proportion of the offspring will have green pods?

Gene Key
Gene or Condition: Pod Color

Allele Symbols	Possible Genotypes	Phenotype
G = green	GG	Green
	Gg	Green
g = yellow	gg	Yellow

The question describes a gene affecting pea pod color. The two different phenotypes are green and yellow. The question also states that "green pods are dominant to yellow pods." Because of this statement, use a capital letter to represent the *green* allele and a lowercase letter to represent the *yellow* allele. We are using the letter G, but any other letter would work. The only requirement is to use the same letter for both alleles. The gene key table shows the type of information needed to describe how the alleles for a characteristic work together. A gene key is a reference that will help solve the problem.

TABLE 10.2

Steps in Solving a Genetics Problem

Solution Pathway

Skill Use	Steps in Information Flow	The Problem
Gene key	Parental phenotypes	By reading the problem, determine that one parent is green and the other yellow. Green × Yellow
Gene key	Parental genotypes	Organisms are diploid, so 2 alleles are needed for each parent. The green parent can be either *GG* or *Gg*. However, the problem states that this parent is heterozygous. The allele combination of *Gg* is the only green heterozygous combination. The gene key shows that the only genotype that produces yellow is *gg*. Gg × gg
Segregation	Possible Sex cells	Because of the Law of Segregation, the alleles separate from each other when sex cells are formed in the parents. The *Gg* parent can produce two types of sex cells, *G* and *g*. The *gg* parent can produce only *g* sex cells. G g g
Punnett square	Offspring genotype	Set up a Punnett square to show the possible fertilization events. This square will have one column and two rows, because one parent produces only one type of gamete and the other parent produces two types. |	g
---	---
G	Gg
g	gg	 50% Gg 50% gg
Gene key	Offspring phenotype	Now, use the information in the gene key to determine the phenotypes of the offspring genotypes. *Gg* will appear as green. Yellow can be produced only by *gg*. 50% produce green pea pods (*Gg*). 50% produce yellow pea pods (*gg*).

Now, organize the actual genetic cross. Table 10.2 shows each step in a simple genetics problem and skills that may be necessary to move from one step to the next. This problem starts at the top row and works toward the bottom of the table. There are several steps in this process. The steps are represented by each row in the table. The rows are titled. "Parental Phenotype," "Parental Genotype," "Possible Sex Cells," "Offspring Genotype," and "Offspring Phenotype." The "Skill Use" column gives information about which skills to use as you move from one row to the next.

First, determine what information is provided in the question about the organisms that are involved in this cross. Identify the following pieces of information in the question.

- A pea plant with green pods is crossed to a pea plant with yellow pods.
- The green pea plant is heterozygous.
- The yellow pea plant is homozygous.

Solve the problem by using the table as a guide. These steps describe the process:

1. The first statement about the organisms being crossed is shown in the "Parental Phenotypes" row as Green × Yellow. Using this information, the gene key, and the remaining two statements, we are able to determine the  parental genotypes. The reasoning described in this process is in the "Parental Genotypes" row of table 10.2.
2. The next step is to use the parental genotypes to determine the parents' possible sex cells. It is necessary to apply Mendel's Law of Segregation to do this correctly. This process is described in the "Possible Sex Cells" row.
3. Now that the parents' gametes are identified, use a Punnett Square to predict the genotypes of the offspring. Create a Punnett square so that there is one row or column for each gamete. This process is shown in the "Offspring Genotype" row.
4. Return to the gene key to determine the offspring phenotypes from the genotypes just produced with the Punnett square.
5. Finally, remember to look at the question that was asked. In this example, the question is what proportion of the offspring will produce green pea pods. The answer is 50%.

In this example, all the information from the problem fits into the gene key and the first rows of table 10.2. Not all problems are like this. Sometimes, a problem provides information about the offspring and requests information about the parents. Table 10.2 will help you do this as well. The principles that applied as you worked down the table still apply as you go the other direction.

Problem Type: Single-Factor Cross

CROSS 2: When both parents are heterozygous and the alleles are completely dominant and recessive to each other, the predicted offspring ratio is always 3:1, or 75% of the dominant phenotype to 25% of the recessive phenotype. If the genotypes of parents are not known and the offspring have a 3:1 ratio, then geneticists frequently infer that the parents are both heterozygous for the trait being considered.

To illustrate this 3:1 pattern of inheritance, consider the disorder phenylketonuria (PKU). People with phenylketonuria are unable to convert the amino acid phenylalanine into the amino acid tyrosine. The buildup of phenylalanine in the body prevents the normal development of the nervous system. Such individuals may become mentally retarded if their disease is not controlled.

Figure 10.4 shows the metabolic pathway in which the amino acid phenylalanine is converted to the amino acid tyrosine by the enzyme phenylalanine hydroxylase. Tyrosine is then used as a substrate by other enzymes. In the abnormal pathway, the substrate phenylalanine builds up, because the enzyme phenylalanine hydroxylase does not function correctly in people with PKU. As phenylalanine levels rise, it is converted to phenylpyruvic acid, which kills nerve cells.

The normal condition is to convert phenylalanine to tyrosine. It is dominant over the condition for PKU. If both parents are heterozygous for PKU, what is the probability that they will have a child who is normal? A child with PKU?

As in the previous example, use the gene key to summarize this problem:

- There are 2 alleles. One is responsible for the normal condition and the other is for PKU.
- The normal condition is dominant over PKU. From this statement, infer that the normal condition can be either *PP* or *Pp*.

Gene Key
Gene or Condition: Phenylketonuria

Allele Symbols	Possible Genotypes	Phenotype
P = normal	*PP*	Normal
	Pp	Normal
p = phenylketonuria	*pp*	Phenylketonuria

1. The problem states that "both parents are heterozygous for PKU." This describes the parental genotypes and determines the genotypes for both parents to be *Pp*. Enter this information on the "Parental Genotypes" row. Although it is not necessary to solve the problem, you can use the gene key to determine the parental phenotypes. *Pp* individuals are normal. *Note: This genetic problem*

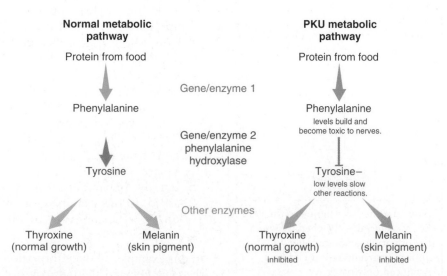

FIGURE 10.4 Phenylketonuria (PKU)

PKU is an autosomal recessive disorder located on chromosome 12. The diagram on the left shows how the normal pathway works. The diagram on the right shows an abnormal pathway. If the enzyme phenylalanine hydroxylase is not produced because of a mutated gene, the amino acid phenylalanine cannot be converted to tyrosine and is converted into phenylpyruvic acid, which accumulates in body fluids. The buildup of phenylpyruvic acid causes the death of nerve cells and ultimately results in mental retardation. Because phenylalanine is not converted to tyrosine, subsequent reactions in the pathway are also affected.

does not start with the first row of table 10.3. A genetics problem can start at any point in the table and require that you determine things about either the parents or the offspring.

2. Determine the possible sex cells. *Pp* individuals will have gametes that are *P* and *p*.
3. Determine the offspring genotypes using a Punnett square. Create your Punnett square so that there is one row or column for each gamete. Note that, in this situation, three different genotypes are produced—*PP, Pp,* and *pp*.
4. Determine the offspring phenotypes by using the gene key and combining genotypes with similar phenotypes.
5. Finally, answer the question that was asked from the problem. In this case, two questions were asked. The probability of having a normal child is 75%. The probability of having a child with PKU is 25%.

Double-Factor Crosses

Up to this point, we have worked only with single-factor crosses. Now, we will consider how to handle genetics problems with two genes—double-factor crosses. In solving double-factor crosses, it is important to consider the principle Mendel identified as the *Law of Independent Assortment*. The **Law of Independent Assortment** states that alleles of one characteristic separate independently of the alleles of another. This law is applied only when working with two genes for different characteristics that are on different chromosomes. This is an important distinction, because genes that are positioned near each other on a chromosome tend to stay together during meiosis and therefore tend to be inherited together. If genes are inherited together, they are not assorting in a random manner. Their assortment is not independent of each other.

In genetics problems, the process of predicting the sex cells that can be produced in double-factor crosses is affected by independent assortment. The following example illustrates how independent assortment works. Recall that, if an individual has the genotype *Aa*, we predict that 50% of his or her reproductive cells have the *A* allele and 50% have the *a* allele. This is an application of the Law of Segregation. If an individual has the genotype *Bb*, we can make a similar prediction with regard to the *B* and *b* alleles. What happens when we want to look simultaneously at both sets of alleles that are on two different chromosomes? What are the possible sex cells that could be produced for an individual that is *AaBb*?

In order to answer this question, we have to apply both the Law of Segregation *and* the Law of Independent Assortment. For now, we will assume that the two genes are on different chromosomes. As mentioned earlier, the law of segregation predicts that 50% of the gametes will have *A* and 50% will have *a*. Likewise 50% of the gametes will have *B* and 50% *b*. The Law of Independent Assortment says that, if a gamete receives an *A* allele, it has an equal chance of also receiving a *B* allele or a *b* allele. Thus, the sex cells that are predicted are *AB, Ab, aB,* and *ab*. Notice that

TABLE 10.3

Solution Pathway

Skill Use	Steps in Information Flow	The Problem
Gene key	Parental phenotypes	**Father × Mother** Normal × Normal
Gene key	Parental genotypes	*Pp* × *Pp*
Segregation	Possible sex cells	P P p p
Punnett square	Offspring genotype	
Gene key	Offspring phenotype	Normal Phenylketonuria 25% PP 25% Total 50% Pp 75% Total

- All the sex cells have 1 and only 1 of the 2 alleles for each characteristic.
- Each allele is found in 50% of the sex cells.
- The alleles for one characteristic are inherited independently from the other.

Note that sex cells with allele combinations such as *AA* or *AaB* are incorrect. Both of these incorrect examples have more than 1 allele for each characteristic. The first example (*AA*) should have only a single *A* and is missing a copy of the *B* gene altogether. The second example (*AaB*) should have either the *A* or the *a* allele, but not both.

Problem Type: Double-Factor Cross

In humans, the allele for free earlobes is dominant over the allele for attached earlobes. The allele for dark hair dominates the allele for light hair. If both parents are heterozygous for earlobe shape and hair color, what types of offspring can they produce, and what is the probability for each type?

Just as in a single factor cross, start by creating a gene key. You are working with two characteristics this time, so create a key for both. Remember that not all the information in the gene key is stated directly in the problem. From the problem, you should be able to identify that

- There are two genes—earlobe type and hair color.
- The free earlobe allele is dominant to the attached earlobe allele.
- The dark hair allele is dominant to the light hair allele.

From this information, you should be able to infer that

- Because the free earlobe allele is dominant, it can have two genotypes—*EE* and *Ee*.
- Because dark hair is dominant, it can have two genotypes—*HH* and *Hh*.

Gene Key
Gene: Earlobe Type

Allele Symbols	Possible Genotypes	Phenotype
E = free	*EE*	Free earlobes
	Ee	Free earlobes
e = attached	*ee*	Attached earlobes

Gene: Hair Color

Allele Symbols	Possible Genotypes	Phenotype
H = dark hair	*HH*	Dark hair
	Hh	Dark hair
h = light hair	*hh*	Light hair

1. After you have the gene key complete, move on to the cross setup in table 10.4. The problem states that "both parents are heterozygous for earlobe shape and hair color." This is a description of the parents' genotypes. It means that both parents are *EeHh*. Place this information on the "Parental Genotypes" row. Although it is not necessary, you can use the gene key to determine what the parent's phenotypes are for earlobe type and hair color.
2. Determine the possible sex cells. *EeHh* individuals will have gametes that are *EH, Eh, eH,* and *eh*. This answer uses the Law of Segregation and the Law of Independent Assortment.
3. Determine the offspring genotypes using a Punnett square. Create your Punnett square so that there is one row or column for each gamete. Your Punnett square will create a 4 × 4 grid. Fill in the genotypes as shown.
4. Determine the offspring phenotypes by using the gene key and combining genotypes with similar phenotypes. In

this problem, there will be four different groupings: (a) free earlobes and dark hair, (b) free earlobes and light hair, (c) attached earlobes and dark hair, and (d) attached earlobes and light hair.

5. Answer the question that was asked from the problem. The ratios are 9:3:3:1.

In cases where the alleles for each gene are completely dominant and recessive to each other and both parents are heterozygous for both characteristics, the predicted offspring ratio is 9:3:3:1. When scientists observe a 9:3:3:1 ratio, they suspect that both parents are heterozygous for both characteristics being considered.

<table><tr><td>**10.6**</td><td></td></tr></table>

10.6 Modified Mendelian Patterns

Mendel's principles are most clearly observed under very select conditions, in which alleles have consistent dominant/recessive interactions. So far, we have considered only a few, straightforward cases. Most, however, may not fit these fundamental patterns. This section discusses several common inheritance patterns.

Codominance

In some inheritance situations, alleles lack total dominant and recessive relationships and are both observed phenotypically to some degree. This inheritance pattern is called *codominance*. In **codominance**, the phenotype of both alleles is expressed in the heterozygous condition. Consequently, a person with the heterozygous genotype can have a phenotype very different from either of his or her homozygous parents. In problems involving codominant alleles, all capital symbols are used, and superscripts are added to represent the different alleles. The capital letters call attention to the fact that each allele can be detected phenotypically to some degree, even when in the presence of its alternative allele. For example, the coat colors (*C*) of shorthorn cattle are phenotypically red ($C^R C^R$), roan ($C^R C^W$), and white ($C^W C^W$). The roan coat is composed of individual hairs, which are either red or white. Together, they create the intermediate effect of roan. Roan coat color can be seen in several other species, including horses (figure 10.5).

Another example of codominance occurs in certain horses. A pair of codominant alleles (D^R and D^W) is known to be involved in the inheritance of these coat colors. Genotypes homozygous ($D^R D^R$) for the D^R allele are chestnut-colored (reddish); heterozygous genotypes ($D^R D^W$) are palomino-colored (golden color with lighter mane and tail). Genotypes homozygous ($D^W D^W$) for the D^W allele are almost white and called cremello.

TABLE 10.4

Solution Pathway

Skill Use	Steps in Information Flow	The Problem
Gene key	Parental phenotypes	**Father** $\times$ **Mother** Free earlobes $\times$ Free earlobes Dark hair Dark hair
Gene key	Parental genotypes	The problem states that both parents are heterozygous for both characteristics. *EeHh* $\times$ *EeHh*
Segregation independent assortment	Possible sex cells	Notice that the Law of Independent Assortment has been added as a skill that should be used for a double-factor cross. Both parents have the same genotypes, so they each produce the same types of gametes. *EH EH* *Eh Eh* *eH eH* *eh eh*

Set up a Punnett square to show the possible fertilization events.

	EH	*Eh*	*eH*	*eh*
EH	*EEHH*	*EEHh*	*EeHH*	*EeHh*
Eh	*EEHh*	*EEhh*	*EeHh*	*Eehh*
eH	*EeHH*	*EeHh*	*eeHH*	*eeHh*
eh	*EeHh*	*Eehh*	*eeHh*	*eehh*

(Skill Use: Punnett square — Step: Offspring genotype)

Count up the different genotypes and then combine them by similar phenotype using the information in the Gene Key.

(Skill Use: Gene key — Step: Offspring phenotype)

Free Earlobes and Dark Hair	Free Earlobes and Light Hair	Attached Earlobes and Dark Hair	Attached Earlobes and Light Hair
1/16—*EEHH*			
2/16—*EEHh*	1/16—*EEhh*	1/16—*eeHH*	
2/16—*EeHH*	2/16—*Eehh*	2/16—*eeHh*	1/16—*eehh*
4/16—*EeHh*			

Sums:

9/16	3/16	3/16	1/16

Incomplete Dominance

In **incomplete dominance**, the phenotype of a heterozygote is intermediate between the two homozygotes on a phenotypic gradient; that is, the phenotypes appear to be "blended" in heterozygotes. A classic example of incomplete dominance in plants is the color of the petals of snapdragons. There are 2 alleles for the color of these flowers. Because neither allele is recessive, we cannot use the traditional capital and lower-case letters as symbols for these alleles. Instead, the allele for white petals is the symbol F^W, and the one for red petals is F^R (figure 10.6).

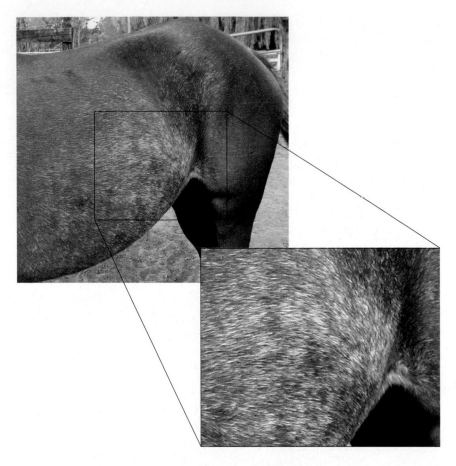

FIGURE 10.5 Codominance

The color of this breed of horse, an Arab, also displays the color called roan. Notice that there are places on the body where both white and red hairs are displayed.

FIGURE 10.6 Incomplete Dominance

The colors of these snapdragons are determined by two alleles for petal color, F^W and F^R. There are three phenotypes because of the way in which the alleles interact with one another. In the heterozygous condition, neither of the alleles dominates the other.

There are three possible combinations of these 2 alleles:

Genotype	Phenotype
$F^W F^W$	White flower
$F^R F^R$	Red flower
$F^R F^W$	Pink flower

Notice that there are only 2 different alleles, red and white, but there are three phenotypes—red, white, and pink. Both the red-flower allele and the white-flower allele partially express themselves when both are present, and this results in pink. The gene products of the 2 alleles interact to produce a blended result.

Problem Type: Incomplete Dominance

If a pink snapdragon is crossed with a white snapdragon, what phenotypes can result, and what is the probability of each phenotype? Notice that the same principles used in earlier genetics problems still apply. Only the interpretation of the gene key is altered. (Table 10.5)

TABLE 10.5

Solution Pathway

Skill Use	Steps in Information Flow	The Problem
Gene key	Parental phenotypes	Pink × White
Gene key	Parental genotypes	$F^R F^W \times F^W F^W$
Segregation	Possible sex cells	$F^R \quad F^W$ F^W
Punnett square	Offspring genotype	
Gene key	Offspring phenotype	50% pink 50% white

Offspring genotype Punnett square:

	F^W
F^R	$F^R F^W$
F^W	$F^W F^W$

Gene Key
Gene: Flower Color

Allele Symbols	Possible Genotypes	Phenotype
F^W = White flowers	$F^W F^W$	White
F^R = Red flowers	$F^R F^R$	Red
	$F^W F^R$	Pink

This cross results in two different phenotypes—pink and white. No red flowers can result, because this would require that both parents be able to contribute at least 1 red allele. The white flowers are homozygous for white, and the pink flowers are heterozygous.

Multiple Alleles

So far, we have discussed only traits that are determined by only 2 alleles: for example, *A, a*. However, there can be more than 2 different alleles for a single trait. The term **multiple alleles** refers to situations in which there are more than 2 possible alleles that control a particular trait. However, one per-son still can have only a maximum of 2 of the alleles for the characteristic. A good example of a characteristic that is determined by multiple alleles is the ABO blood type. There are 3 alleles for blood type:

Alleles*

I^A = blood has type A antigens on red blood cell surface
I^B = blood has type B antigens on red blood cell surface
i = blood type O has neither type A nor type B antigens on red blood cell surface

In the ABO system, A and B show *codominance* when they are together in an individual, but both alleles are dominant over the O allele. These 3 alleles can be combined as pairs in six ways, resulting in four phenotypes:

Genotype	*Phenotype*
$I^A I^A$	Blood type A
$I^A i$	Blood type A
$I^B I^B$	Blood type B
$I^B i$	Blood type B
$I^A I^B$	Blood type AB
ii	Blood type O

Multiple-allele problems are worked as single-factor problems.

Problem Type: Multiple Alleles

One aspect of blood type is determined by 3 alleles—A, B, and O. Allele A and allele B are codominant. Allele A and allele B are both dominant to allele O. A male heterozygous with blood type A and a female heterozygous with blood type B have a child. What are the possible phenotypes of their offspring?

Gene Key
Gene: Blood Type

Allele Symbols	Possible Genotypes	Phenotype
i = Type O	ii	Type O
I^A = Type A	$I^A I^A$	Type A
	$I^A i$	Type A
I^B = Type B	$I^B I^B$	Type B
	$I^B i$	Type B
	$I^A I^B$	Type AB

The solution for this problem is shown in Table 10.6.

---

*The symbols, I and i stand for the technical term referring to the antigenic carbohydrates attached to red blood cells, the immunogens. These alleles are located on human chromosome 9. The ABO blood system is not the only one used to type blood. Others include the Rh, MNS, and Xg systems.

TABLE 10.6

Solution Pathway

Skill Use	Steps in Information Flow	The Problem
Gene key	Parental phenotypes	Type A × Type B
Gene key	Parental genotypes	$I^A i$ × $I^B i$
Segregation	Possible sex cells	I^A I^B i i
Punnett square	Offspring genotype			I^B	i
I^A	$I^A I^B$	$I^A i$
i	$I^B i$	$i i$	
Gene key	Offspring phenotype	25% Type AB ($I^A I^B$) 25% Type A ($I^A i$) 25% Type B ($I^B i$) 25% Type O (ii)

Polygenic Inheritance

Thus far, we have considered phenotypic characteristics that are determined by single genes. However, some characteristics are determined by the interaction of several genes. This is called *polygenic inheritance*. In **polygenic inheritance,** a number of different pairs of alleles combine their efforts to determine a characteristic. Skin color in humans is a good example of this inheritance pattern. According to some experts, genes for skin color are located at a minimum of three loci. At each of these loci, the allele for dark skin is dominant over the allele for light skin. Therefore, a wide variety of skin colors is possible, depending on how many dark-skin alleles are present (figure 10.7). The number of total dark-skin alleles (capital *D* in figure 10.7) from all three genes determines skin color.

Polygenic inheritance is common with characteristics that show great variety within the population. Some obvious polygenic traits in humans are height, skin color, eye color, and intelligence. The many levels of height, skin color, eye color, and intelligence makes it difficult to separate individuals into meaningful categories. There is an entire range of expression for polygenic characteristics. For example, height in humans ranges from tall to short, with many intermediate heights. Eye color varies in some populations from deep brown to the lightest blue. Although it is still unclear how many genes are involved in determining these characteristics, at least two or three different genes have been identified. (Outlooks 10.1). Polygenic traits are different from a characteristic such as blood type, in which there are a limited number of well-defined phenotypes (A, B, O, AB).

Locus 1	$d^1 d^1$	$d^1 D^1$	$d^1 D^1$	$D^1 D^1$	$D^1 d^1$	$D^1 d^1$	$D^1 D^1$
Locus 2	$d^2 d^2$	$d^2 d^2$	$d^2 D^2$	$D^2 d^2$	$D^2 d^2$	$D^2 D^2$	$D^2 D^2$
Locus 3	$d^3 d^3$	$d^3 d^3$	$d^3 d^3$	$d^3 d^3$	$D^3 d^3$	$D^3 D^3$	$D^3 D^3$
Total number of dark-skin genes	0	1	2	3	4	5	6

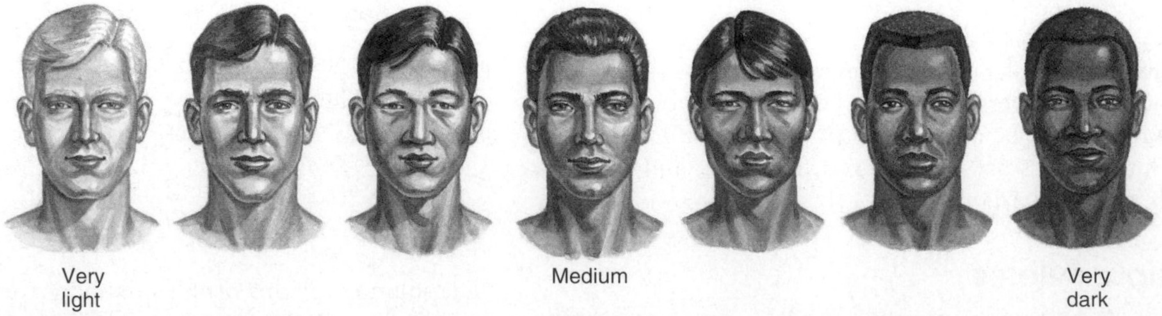

FIGURE 10.7 Polygenic Inheritance
Skin color in humans is an example of polygenic inheritance. The darkness of the skin is determined by the number of dark-skin alleles a person inherits from his or her parents. These alleles are additive and are found in several different genes.

OUTLOOKS 10.1

The Inheritance of Eye Color

It is commonly thought that eye color is inherited in a simple dominant/recessive manner, in which brown eyes are considered dominant over blue eyes. However, the real pattern of inheritance is more complicated than this. Eye color is determined by the amount of a brown pigment, melanin, present in the iris of the eye. If there is a large quantity of melanin on the anterior surface of the iris, the eyes are dark. Black eyes have a greater quantity of melanin than do brown eyes.

If melanin is absent from the front surface of the iris, the eyes appear blue, not because of a blue pigment but because blue wavelengths of light are reflected from the iris. The iris appears blue for the same reason that deep bodies of water tend to appear blue. There is no blue pigment in the water, but blue wavelengths of light are returned to the eye from the water. Just as black and brown eyes are determined by the amount of pigment present, colors such as green, gray, and hazel are produced by the various amounts of melanin in the iris. If a very small amount of brown melanin is present in the iris, the eye tends to appear green, whereas relatively large amounts of melanin produce hazel eyes.

Several genes are probably involved in determining the quantity and placement of melanin. These genes interact in such a way that a wide range of eye color is possible. Eye color is probably determined by polygenic inheritance, just as skin color and height are. Some newborn babies have blue eyes that later become brown. This is because their irises have not yet begun to produce melanin.

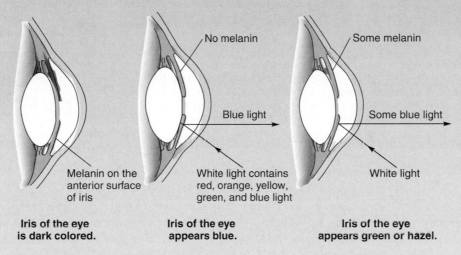

Iris of the eye is dark colored.

Iris of the eye appears blue.

Iris of the eye appears green or hazel.

Blue eyes are due to a lack of pigment, not the presence of blue pigment. In blue eyes, blue light is reflected while other colors are absorbed. Green eyes absorb some blue light.

Pleiotropy

Even though a single gene may produce only one type of protein, it often has a variety of effects on the phenotype of a person. The term **pleiotropy** (*pleio* = changeable) describes the multiple effects a gene has on a phenotype. A good example of pleiotropy—PKU—has already been discussed. In addition to the mental retardation phenotype, several other phenotypes are associated with PKU. Whereas mental retardation is caused by the buildup of phenylpyruvic acid, other phenotypes are caused by a lack of tyrosine, the next product in the pathway. Tyrosine is used by the human body to create two other important molecules—growth hormone and melanin. Growth hormone is needed for normal growth, and melanin is a skin pigment. Individuals with PKU have low levels of tyrosine because of the faulty enzyme; this results in abnormal growth and unusually pale skin, in addition to the presence of phenylpyruvic acid that can cause mental retardation.

Another example of pleiotropy is *Marfan syndrome*. This syndrome is a disorder of the body's connective tissue, but it can also have effects in many other organs. (Consider the phenotypic characteristics of the individual shown in figure 10.8. Some feel that the former U.S. president Abraham Lincoln also

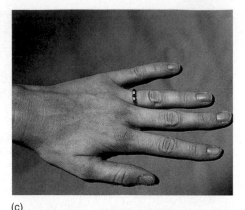

(c)

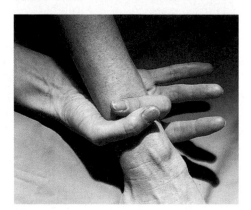

(a)　　　　　　　　　　　(b)

(d)

FIGURE 10.8 Marfan Syndrome

It is estimated that about 40,000 (1 out of 10,000) people in the United States have this autosomal dominant abnormality. Notice the common lanky appearance to the body and face of *(a)* this person with Marfan syndrome and *(b)* former U.S. president Abraham Lincoln. Photos *(c)* and *(d)* illustrate their unusually long fingers.

had Marfan syndrome. Do you see similarities?) The symptoms of Marfan syndrome generally include the following:

Skeletal
- Long arms and legs, disproportionate in length to the body
- Abnormally long fingers
- Skinniness
- Curvature of the spine
- Abnormally shaped chest, chest caves in or protrudes outward

Eye problems
- Nearsightedness

Heart and Aortic Problems
- Weak or defective heart valves
- Weak blood vessels that rupture
- Inflammation of the heart

Lung and Breathing Problems
- Collapsed lungs
- Long pauses in breathing during sleep (sleep apnea)

Both PKU and Marfan syndrome are examples of alleles that have many different effects in an organism. Cystic fibro-

sis also shows pleiotropy. Review How Science Works 10.1—what information there supports this statement?

10.7 Linkage

Although Mendel's insight into the nature of inheritance was extremely important, there were many aspects of inheritance that Mendel did not explain. **Linkage** is a situation in which the genes for different characteristics are inherited together more frequently than would be predicted by probability. Linkage can be explained by examining chromosomes.

Linkage Groups

Each chromosome has many genes located along its length. Mendel's inheritance patterns don't really describe the inheritance patterns of individual genes; they describe the inheritance patterns of chromosomes. Homologous chromosomes separate from each other (segregation). Non-homologous chromosomes separate from each other independently (independent assortment.) Because each chromosome has many genes on it, these genes tend to be inherited as a group. A **linkage group** is a set of genes located on the same chromosome. This means that they tend to be inherited together. The process of crossing-over, which occurs during prophase I of meiosis I, may split up these linkage groups. Crossing-over happens between homologous chromosomes donated by the mother and the father and results in a mixing of the allele combinations in gametes. This means that the child can have gene combinations not found in either parent alone. The closer two genes are to each other on a chromosome, the less likely crossing-over will occur between them and the more probable it is that they will be inherited together.

Autosomal Linkage

People and many other organisms have two types of chromosomes. **Autosomes** are chromosomes that are not directly involved in sex determination; they have the same kinds of genes on both members of the homologous pair of chromosomes. Of the 23 pairs of human chromosomes, 22 are autosomes. **Sex chromosomes** control the sex of an organism. An example of autosomal linkage is found in figure 10.9. The three genes listed in this figure are on the same chromosome. If the genes sit closely enough to each other, they are likely to be inherited together.

Sex Linkage

Recall that there are several kinds of chromosomes, that each chromosome carries genes unique to it, and that these genes are found at specific places on the chromosome. Furthermore, diploid organisms have homologous pairs of chromosomes. Genes determine sexual characteristics in the same manner as other types of characteristics. In many organisms, special sex chromosomes carry sex-determining genes. Sex chromosomes are different between males and females of the same species. Autosomes carry the same genes in both sexes of a species. In humans, all other mammals, and some other organisms (e.g., fruit flies), the sex of an individual is determined by the presence of a certain chromosome combination. In mammals, the genes that determine maleness are located on a small chro-

mosome known as the Y *chromosome*. The Y chromosome behaves as if it and another larger chromosome, known as the X *chromosome*, were homologous chromosomes. Males have 1 X and 1 Y chromosome. Females have 2 X chromosomes.

The sex of some animals is determined in a completely different way. In bees, for example, the females are diploid and the males are haploid. Other plants and animals have still other chromosomal mechanisms for determining their sex (Outlooks 10.2). Chromosomal determination of sex, p. 597

Sex linkage occurs when genes are located on the chromosomes that determine the sex of an individual. The Y chromosome is much shorter than the X chromosome and has fewer genes for traits than found on the X chromosome (figure 10.10). Therefore, the X chromosome has many genes for which there is no matching gene on the Y chromosome. Some genes appear on both the X chromosome and Y chromosome. Other genes, however, are found only on the X chromosome or only on the Y chromosome. Females have two copies of the genes that are found only on the X chromosomes. Because males have both a Y chromosomes with few genes on it and the X chromosome, many of the recessive characteristics present on the X chromosome appear more frequently in males than in females, who have 2 X chromosomes. Unusual sex-linked inheritance patterns occur because certain genes are found on only 1 of the 2 sex chromosomes. This is different from genes found on autosomes, because these genes are found on both of the homologous chromosomes. Genes found only on the X chromosome are said to be **X-linked genes.** Genes found only on the Y chromosome are said to be **Y-linked genes.**

Female phenotypes can be affected by the dominant and recessive allele interactions that Mendel identified. Males present a different case. Males only have one copy of the genes that are found on the X chromosome, because they have only 1 X chromosome. This 1 allele determines the male's phenotype. Some X-linked genes can result in abnormal traits, such as *color deficiency, hemophilia, brown teeth,* and at least two forms of *muscular dystrophy* (Becker's and Duchene's).

Use the following problem as an example of how to work with X-linked genes. Notice that the same basic format is followed as in previous genetics problems. The major difference is that chromosomes are represented in this problem. Here, an *X* represents the X chromosome and a *Y* represents the Y chromosome. Genes that are linked to the X chromosome are shown as superscripts. The *X* and its superscript should be treated as a single allele. You have used superscripts before in a genetics problem to look at incomplete dominance and codominance.

Problem Type: X-Linked

In humans, the allele for normal color vision is dominant and the allele for color deficiency is recessive. Both alleles are X-linked. People who cannot detect the difference between certain colors, such as between red and green are described as having "color-defective vision." A male who has normal vision mates with a female who is heterozygous for normal color vision.

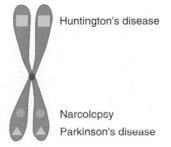

Huntington's disease

Narcolepsy

Parkinson's disease

FIGURE 10.9 Chromosome
These are just three genes that are found on human chromosome number 4. Because all these genes are found on one chromosome or strand of DNA, they are considered to be members of one linkage group. Each chromosome represents a group of linked genes.

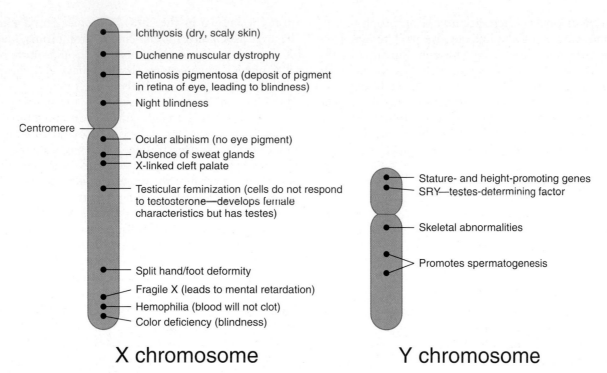

X chromosome

Y chromosome

FIGURE 10.10 Sex Chromosomes
The human X chromosome contains over 1,400 genes and over 150 million base pairs, of which approximately 95% have been determined. The human Y chromosome contains about 200 genes and about 50 million base pairs, of which approximately 50% have been determined. A number of the genes linked on these chromosomes are listed.

What type of children can they have in terms of these traits, and what is the probability for each type?

Gene Key
Gene: Color Vision

Allele Symbols

X^B = normal color vision
X^b = color-deficient vision
Y = no gene for color vision

Possible Genotypes

Females (X^BX^B or X^BX^b)
Males (X^BY)
Females (X^bX^b)
Males (X^bY)

Phenotype

Female with normal color vision
Male with normal color vision
Female with color-defective vision
Male with color-defective vision

Note that, in solving sex-linked problems, the general process is the same as that for other genetics problems. (table 10.7) The only significant difference is that the alleles

TABLE 10.7

Solution Pathway

Skill Use	Steps in Information Flow	The Problem
Gene key	Parental phenotypes	Father: normal vision × Mother heterozygote for color vision
Gene key	Parental genotypes	$X^BY \times X^BX^b$
Segregation	Possible sex cells	X^B X^B Y X^b
Punnett square	Offspring genotype	<table><tr><td></td><td>X^B</td><td>X^b</td></tr><tr><td>X^B</td><td>X^BX^B</td><td>X^BX^b</td></tr><tr><td>Y</td><td>X^BY</td><td>X^bY</td></tr></table>
Gene key	Offspring phenotype	50% normal females (½ of these are carriers) 25% normal males 25% color-deficient males

OUTLOOKS 10.2

The Birds and the Bees . . . and the Alligators

The determination of an individual's sex depends on the kind of organism it is. For example, in humans, the physical features that result in maleness are triggered by a gene on the Y chromosome. The lack of a Y chromosome results in a female individual. In other organisms, sex is determined by other combinations of chromosomes or environmental factors.

Organism	Sex Determination
Mammals	Sex is chromosomally determined: XY individuals are male.
Birds	Sex is chromosomally determined: XY individuals are female. Rather than XY the letters WZ are used in birds.
Bees	Males (drones) are haploid and females (workers or queens) are diploid.
Certain species of alligators, turtles, and lizards	Egg incubation temperatures cause hormonal changes in the developing embryo; higher incubation temperatures cause the developing brain to shift sex in favor of the individual becoming a female.
Boat shell snails	Males can become females but will remain male if they mate and remain in one spot.
Shrimp, orchids, and some tropical fish	Males convert to females; on occasion, females convert to males, probably to maximize breeding.
African reed frog	Females convert to males, probably to maximize breeding.

are listed as superscripts to the chromosomes, so that the gender and phenotypes can be determined in the last few steps of the problem.

10.8 Other Influences on Phenotype

You might assume that the dominant allele is always expressed in a heterozygous individual; however, it is not so that simple. As in other areas of biology, there are exceptions. For example, the allele for six fingers (*polydactylism*) is dominant over the allele for five fingers in humans. Some people who have received the allele for six fingers have a fairly complete sixth finger; in others, it may appear as a little stub. In some cases, this dominant characteristic is not expressed or

perhaps shows on only one hand. Thus, there may be variation in the degree to which an allele expresses itself *in an individual*. Geneticists refer to this as *variable expressivity*.

Both internal and external environmental factors can influence the expression of genes. A characteristic whose expression is influenced by internal gene-regulating mechanisms is that of male-pattern baldness (figure 10.11). In males with a genetic disposition to balding, the enzyme 5-alpha-reductase is produced in high levels. 5-alpha-reductase uses testosterone in males to produces dihydrotestosterone (DHT). DHT slows down blood supply to the hair follicle and causes baldness. In nonbalding males, 5-alpha-reductase is produced at lower levels, DHT is not produced, and baldness does not occur. The internal environment in females has lower levels of testosterone, so DHT is not produced at high levels even if the 5-alpha-reductase is expressed. Differences

FIGURE 10.11 Baldness and the Expression of Genes
It is a common misconception that males have genes for baldness and females do not. Male-pattern baldness is a sex-influenced trait, in which both males and females possess alleles coding for baldness. These genes are turned on by high levels of the hormone testosterone. This is an example of an internal gene-regulating mechanism.

in the internal environment of males and females alter the phenotype. An example of external environmental factors that affect gene expression is sunlight. Genes for freckles do not show themselves fully unless a person's skin is exposed to sunlight (figure 10.12).

Diet is an external environmental factor that can influence the phenotype of an individual. *Diabetes mellitus*, a metabolic disorder in which glucose is not properly metabolized and is passed out of the body in the urine, has a genetic basis. Some people who have a family history of diabetes are thought to have inherited the trait for this disease. Evidence indicates that they can delay the onset of the disease by reducing the amount of sugar in their diet. This change in the external environment influences gene expression in much the same way that sunlight affects the expression of freckles in humans. Similarly, diet is known to affect how the genes for intelligence, pigment production, and body height are expressed. Children who are deprived of protein during their growing years are likely to have reduced intelligence, lighter skin, and shorter overall height than children with adequate protein in their diet.

Summary

Genes are units of heredity composed of specific lengths of DNA that determine the characteristics an organism displays. Specific genes are at specific loci on specific chromosomes.

FIGURE 10.12 The Environment and Gene Expression
The expression of many genes is influenced by the environment. The allele for dark hair in the cat is sensitive to temperature and expresses itself only in the parts of the body that stay cool. The allele for freckles expresses itself more fully when a person is exposed to sunlight.

Mendel described the general patterns of inheritance in his Law of Dominance, his Law of Segregation, and his Law of Independent Assortment. Punnett squares help us predict graphically the results of a genetic cross. The phenotype displayed by an organism is the result of the environment's effect on the genes' ability to express themselves. Diploid organisms have two genes for each characteristic. The alternative forms of genes for a characteristic are called alleles. There may be many different alleles for a particular characteristic. Organisms with 2 identical alleles for a characteristic are homozygous; those with different alleles are heterozygous. Some alleles are dominant over other alleles, which are recessive. Sometimes, 2 alleles do not show dominance and recessiveness but, rather, both express themselves. Codominance and lack of dominance are examples. Often, a gene has more than one recognizable effect on the phenotype of the organism. This situation is called pleiotropy. Some characteristics are polygenic and are determined by several pairs of alleles acting together to determine one recognizable characteristic. In humans and some other animals, males have an X chromosome with a normal number of genes and a Y chromosome with fewer genes. Although the X and Y chromosomes are not identical, they behave as a pair of homologous chromosomes. Because the Y chromosome is shorter than the X chromosome and has fewer genes, many of the recessive characteristics present on the X chromosome appear more frequently in males than in females, who have 2 X chromosomes. The degree of expression of many genetically determined characteristics is modified by the internal or external environment of the organism.

Key Terms

Use interactive flash cards, on the **Concepts in Biology,** *12/e website to help you learn the meaning of these terms.*

autosomes 209	linkage 208
codominance 202	linkage group 208
dominant allele 193	Mendelian genetics 196
double-factor cross 194	monohybrid cross 194
fertilization 194	multiple alleles 205
genetic cross 194	offspring 193
genetics 192	phenotype 193
genome 192	pleiotropy 207
genotype 193	polygenic inheritance 206
haploid 192	probability 195
heterozygous 194	Punnett square 194
homozygous 194	recessive allele 193
incomplete dominance 203	sex chromosomes 209
Law of Dominance 198	sex linkage 209
Law of Independent Assortment 201	single-factor cross 194
	X-linked genes 209
Law of Segregation 194	Y-linked genes 209

Basic Review

1. Homologous chromosomes
   a. have the same genes in the same places.
   b. are identical.
   c. have the same alleles.
   d. All of the above are correct.

2. Phenotype is the combination of alleles that an organism has, whereas genotype is its appearance. (T/F)

3. A homozygous organism
   a. has the same alleles at a locus.
   b. has the same alleles at a gene.
   c. produces gametes that contain the same allele.
   d. All of the above are correct.

4. Segregation happens during meiosis. (T/F)

5. Independent assortment is concerned with the segregation of alleles of the same gene. (T/F)

6. Genes that are found only on the X chromosome in humans most consistently illustrate
   a. pleiotropy.
   b. the concept of diploid organisms.
   c. sex-linkage.
   d. All of the above are correct.

7. Double-factor crosses
   a. follow 2 alleles for 1 gene.
   b. follow the alleles for 2 genes.
   c. look at up to 4 alleles for 1 gene.
   d. None of the above are correct.

8. Mendelian principles apply when genes are found close to each other on the same chromosome. (T/F)

9. _____ occur when there are more than 2 alleles for a given gene.

10. Dominant alleles mask _____ alleles in heterozygous organisms.

Answers
1. a 2. F 3. d 4. T 5. F 6. c 7. b 8. F
9. Multiple alleles   10. recessive

Concept Review

Answers to the Concept Review questions can be found on the **Concepts in Biology,** *12/e website.*

10.1 Meiosis, Genes, and Alleles

1. What is meant by the symbols *n* and *2n*?

10.2 The Fundamentals of Genetics

2. Distinguish between phenotype and genotype.
3. What types of symbols are typically used to express genotypes?
4. How many kinds of gametes are possible with each of the following genotypes?
   a. *Aa*
   b. *AaBB*
   c. *AaBb*
   d. *AaBbCc*

10.3 Probability vs. Possibility

5. What is the difference between probability and possibility?
6. In what mathematical forms might probability be expressed?

10.4 The First Geneticist: Gregor Mendel

7. In your own words, describe Mendel's Law of Segregation.

10.5 Solving Genetics Problems

8. What does it mean when geneticists use the term *independent assortment*?
9. What is a Punnett square?
10. What is the probability of each of the following sets of parents producing the given genotypes in their offspring?

Parents			Offspring
a. *AA*	×	*aa*	*Aa*
b. *Aa*	×	*Aa*	*Aa*
c. *Aa*	×	*Aa*	*aa*
d. *AaBb*	×	*AaBB*	*AABB*
e. *AaBb*	×	*AaBB*	*AaBb*
f. *AaBb*	×	*AaBb*	*AABB*

11. What possible combinations of parental genotypes could produce an offspring with the genotype *Aa*?
12. In certain pea plants, the allele *T* for tallness is dominant over *t* for shortness.
    a. If a homozygous tall and homozygous short plant are crossed, what will be the phenotype and genotype of the offspring?
    b. If both individuals are heterozygous, what will be the phenotypic and genotypic ratios of the offspring?

10.6 Modified Mendelian Patterns

13. What is the difference between the terms *dominant* and *codominant*?

14. What is the probability of a child having type AB blood if one of the parents is heterozygous for A blood and the other is heterozygous for B? What other genotypes are possible in this child?

10.7 Linkage

15. What is a linkage group?
16. Provide examples of genes that are linked.

10.8 Other Influences on Phenotype

17. What type of factor can cause a dominant allele to not be expressed?

Thinking Critically

*For guidelines to answer this question, visit the **Concepts in Biology**, 12/e website.*

The breeding of dogs, horses, cats, and many other domesticated animals is done with purposes in mind—that is, producing offspring that have specific body types, colors, behaviors, and athletic abilities. Cows are bred to produce more meat or milk. Many grain crops are bred to produce more grain per plant. Similarly, some people have the muscle development to be great baseball players, whereas others cannot hit the ball. Some have great mathematical skills, whereas others have a tough time adding 2 + 2. How do you think you have been genetically programmed? What are your strengths? As a parent or child, what frustrations have you experienced in teaching or learning? What are the difficulties in determining which of your traits are genetic and which are not?

11

Applications of Biotechnology

In 1988, a baker in England was the first person in the world to be convicted of a crime on the basis of DNA evidence. Colin Pitchfork's crime was the rape and murder of two girls. The first murder occurred in 1983. The initial evidence in this case consisted of the culprit's body fluids, which contained his proteins and DNA. On the basis of using proteins, the police were able to create a molecular description of the culprit. The problem was that this description matched 10% of the males in the local population. On this basis, they were unable to identify just one person. In 1986, there was another murder, which closely matched the details of the 1983 killing. Another male, Richard Buckland, was the prime suspect for the second murder. In fact, while being questioned, Buckland admitted to the most recent killing but had no knowledge of the first killing. The clues still did not point consistently to a single person. The scientists at a nearby university had been working on a new forensic technique—DNA fingerprinting.

To track down the killer, police asked local men to donate blood or saliva samples. Between 4,000 and 5,000 local men participated in the dragnet. None of the volunteers matched the culprit's DNA. Interestingly, Buckland's DNA did not match the culprit's DNA, either. He was later released, because his confession was false. It wasn't until after someone reported that Colin Pitchfork had asked a friend to donate a sample for him and offered to pay several others to do the same that police arrested Pitchfork. Pitchfork's DNA matched that of the killer's.

This is a good example of how biotechnology helps the search for truth within the justice system. The additional evidence from DNA was able to provide key information to identify the culprit.

CHAPTER OUTLINE

11.1 Why Biotechnology Works

The discovery of DNA's structure in 1953 opened the door to a new era of scientific investigation. **Biotechnology** is a collection of techniques that provide the ability to manipulate the genetic information of an organism *directly*. As a result, scientists can accomplish tasks that were unthinkable just 50 years ago. The field of biotechnology has enabled scientists to produce drugs more cheaply than before; to correct genetic mutations; to create cells that are able to break down toxins and pollutants in the environment; and to develop more productive livestock and crops. Biotechnology promises more advances in the near future.

The key to understanding biotechnology is understanding the significant role that DNA plays in determining the genetic characteristics of an organism. In the cell's nucleus, chromosomes are made of DNA and histone proteins. The genetic information for the cell is the sequence of nucleotides in the DNA molecule. Genes are regions of the DNA's nucleotide sequence that contain the information to direct the synthesis of specific proteins. In turn, these proteins produce the characteristics of the cell and organism when the gene is expressed by transcription and translation.

$$DNA \quad \rightarrow \quad proteins \quad \rightarrow \quad phenotype\ of$$
$$in\ nucleus \qquad\quad in\ cells \qquad\qquad organism$$

The nearly universal connection among DNA, protein expression, and the organism's phenotype is central to biotechnology. If an organism has a unique set of phenotypes, it has a unique set of DNA sequences. The more closely related organisms are, the more similar are their DNA sequences.

11.2 Comparing DNA

Comparisons of DNA can be accomplished in two ways. A frequently used technique is to compare patterns that are created when enzymes are used to cut DNA; genetically different organisms have different patterns. This process is called *DNA fingerprinting*. Scientists look for indications of DNA sequence differences by examining the lengths of DNA fragments that are generated when DNA is cut into pieces by enzymes. This approach scans stretches of DNA quickly, but not as closely as the second method in which researchers look at the DNA nucleotide sequences directly. This process is called *DNA sequencing*, and traditionally has been a time consuming process. Because both approaches have advantages and disadvantages, scientists choose between them, depending on their needs.

DNA Fingerprinting

DNA fingerprinting is a technique that uniquely identifies individuals on the basis of DNA fragment lengths. Because no two people have the same nucleotide sequences, they do not produce the same lengths of DNA fragments when their DNA is cut with enzymes. Even looking at the many pieces of DNA

that are produced in this manner is too complex. Therefore, scientists don't look at all the possible fragments but, rather, focus on differences found in pieces of DNA that form repeating patterns in the DNA. By focusing on these regions with repeating nucleotide sequences, it is possible to determine whether samples from two individuals have the same number of repeating segments.

DNA Fingerprinting Techniques

In the scenario presented at the beginning of this chapter, a crime was committed and the scientists had evidence in the form of body fluids from the criminal. These body fluids contained cells with the criminal's DNA. The DNA in these cells was used as a template to produce enough additional DNA for analysis. The **polymerase chain reaction** (**PCR**) is a technique used to generate large quantities of DNA from minute amounts. Using PCR and the culprit's DNA, scientists were able to replicate regions of human DNA that are known to vary from individual to individual (How Science Works 11.1). Scientists target areas with *variable number tandem repeats*. **Variable number tandem repeats** (**VNTRs**) are sequences of DNA that are variably repeated from one individual to another. For example, in a given region of DNA, one person may have 4 repeats, whereas another may have 20 repeats (figure 11.1).

To detect the varying number of VNTRs, the replicated DNA sample is cut into smaller pieces with **restriction enzymes** (How Science Works 11.2). **Restriction sites** are DNA nucleotide sequences that attract restriction enzymes. When the restriction enzymes bind to a restriction site, the enzyme cuts the DNA molecule into two molecules. **Restriction fragments** are the smaller DNA fragments that are generated after the restriction enzyme has cut the selected

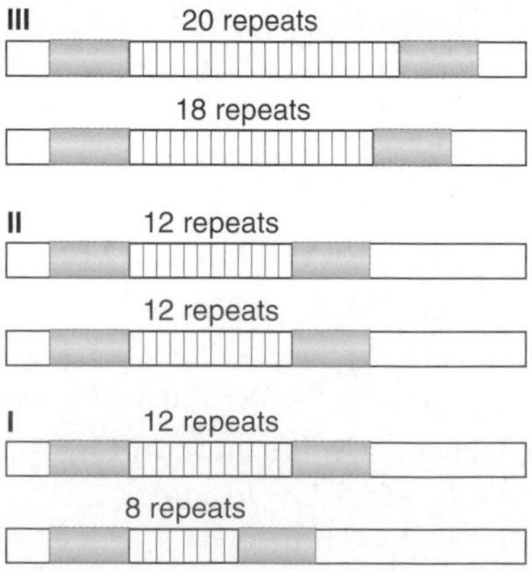

FIGURE 11.1 Variable Number Tandem Repeats
Variable number tandem repeats (VNTRs) are short sequences of DNA that are repeated often. The repeated sequences are found end-to-end. This illustration shows the VNTRs for three individuals. The individual in I has 8 repeats on 1 chromosome and 12 on the homologous chromosome. They are heterozygous. The individual in II is homozygous for 12 repeats. The individual in III is heterozygous for a different number of repeats—18 and 20.

DNA into smaller pieces. In DNA fingerprinting, scientists look for different lengths of restriction fragments as an indicator of differences in VNTRs.

Electrophoresis is a technique that separates DNA fragments on the basis of size (How Science Works 11.3). The shorter DNA molecules migrate more quickly than the long molecules. As differently sized molecules are separated, a banding pattern is generated. Each band is a differently sized restriction fragment. Each person's unique DNA banding pattern is called a DNA fingerprint (figure 11.2). The process of DNA fingerprinting includes the following basic stages:

1. DNA is obtained from a source, which may be small as one cell.

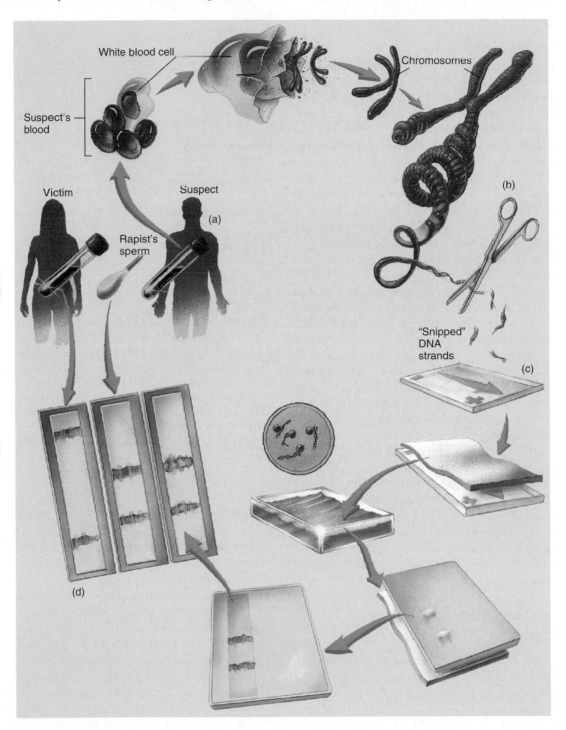

FIGURE 11.2 DNA Fingerprints

Because every person's DNA is unique *(a)*, when samples of an individual's DNA are subjected to restriction enzymes, the cuts occur in different places and DNA molecules of different sizes result. *(b)* Restriction enzymes can cut DNA at places where specific sequences of nucleotides occur. When the DNA fragments are separated by electrophoresis, *(c)* the smaller fragments migrate more quickly than the larger fragments. This produces a pattern, called a DNA fingerprint, that is unique and identifies the person who provided the DNA. *(d)* The victim's DNA is on the left. The rapist's DNA is in the middle. The suspect's DNA is on the right. The match in banding patterns between the suspect and the rapist indicates that they are the same.

2. PCR is used to make many copies of portions of the DNA, called VNTRs.

3. Restriction enzymes are used to cut the VNTR DNA into pieces.

4. The pieces are separated by electrophoresis.

5. Comparisons between patterns can be made.

DNA Fingerprinting Applications

With DNA fingerprinting, the more similar the banding patterns are from two different samples, the more likely the two samples are from the same person. The less similar the patterns, the less likely the two samples are from the same person. In criminal cases, DNA samples from the crime site can be compared with those taken from suspects. If 100% of the banding pattern matches, it is highly probable that the suspect was at the scene of the crime and is the guilty party. The same procedure can be used to confirm a person's identity, as in cases of amnesia, murder, or accidental death.

DNA fingerprinting can be used in legal (paternity) cases that determine the biological father of a child. A child's DNA is a unique combination of both the mother's DNA and the father's DNA. The child's DNA fingerprint is unique, but all the bands in the child's DNA fingerprint should be found in either the mother's or the father's fingerprint. To determine paternity, the child's DNA, the mother's DNA, and DNA from the man who is alleged to be the father are collected. The DNA from all three is subjected to PCR, restriction enzymes, and electrophoresis. During analysis of the banding patterns, scientists account for the child's banding pattern by looking at the mother and the presumed father. Bands that are common to both the biological mother and the child are identified and eliminated from further consideration. If all the remaining bands can be matched to the presumed father, it is extremely likely that he is the father (figure 11.3). If there are bands that do not match the presumed father's, then there are one of two conclusions: (1) The presumed father is not the child's biological father, or (2) the child has a new mutation that accounts for the unique band. (The latter possibility can usually be ruled out by considering multiple regions of DNA, because it is extremely unlikely that the child will have multiple new mutations.)

Gene Sequencing and The Human Genome Project

The Human Genome Project (HGP) was a 13-year effort to determine the normal or healthy human DNA sequence. Work began in 1990. It was first proposed in 1986 by the U.S. Department of Energy (DOE) and was cosponsored soon after by the National Institutes of Health (NIH). These agencies were the main research agencies within the U.S. government responsible for developing and planning the project. Estimates are that the United States spent over 3 billion publicly funded dollars from the Department of Energy and National Institutes of Health toward the Human Genome Project.

Many countries contributed both funds and labor resources to the Human Genome Project. At least 17 coun-

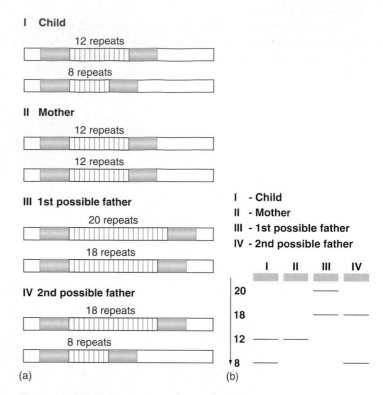

FIGURE 11.3 Paternity Determination
(a) This illustration shows the VNTRs for four different individuals—a child, the mother, one possible father, and a second possible father. *(b)* Using PCR, electrophoresis, and DNA fingerprinting analysis, it is possible to identify the child's father. The mother possesses the "12" band and has passed that to her child. The child's other band must have come from the father. Of the two men under consideration, only man IV has the "8" band. Man IV is the father. If man IV is the father, why doesn't the child have an "18" band?

tries other than the United States participated, including Australia, Brazil, Canada, China, Denmark, the European Union, France, Germany, Israel, Italy, Japan, Korea, Mexico, the Netherlands, Russia, Sweden, and the United Kingdom. The Human Genome Project is one of the most ambitious projects ever undertaken in the biological sciences.

The data that these countries produced are stored in powerful computers, so that the information can be shared. To get an idea of the size of this project, consider that a human Y chromosome (one of the smallest of the human chromosomes) is composed of nearly 60 million paired nucleotides. The larger X chromosome may be composed of 150 million paired nucleotides. The entire human genome consists of 3.12 billion paired nucleotides. That is roughly the same number as all the letter characters found in about 2,000 copies of this textbook.

Human Genome Project Techniques

Two kinds of work progressed simultaneously to determine the sequence of the human genome. First, *physical maps* were constructed by determining the location of specific "markers" (known sequences of bases) and the proximity of these markers to genes (figure 11.4). This physical map was used to organize the vast amount of data produced by the second technique, which was for the labs to determine the exact order of nitrogenous bases of the DNA for each chromosome. Techniques exist for determining base sequences (How Science Works 11.4). The challenge is storing and organizing the information from these experiments, so that the data can be used and searched.

A slightly different approach was adopted by Celera Genomics, a private U.S. corporation. Although Celera Genomics started behind the labs funded by the Department of Energy and National Institute of Health it was able to catch up and completed its sequencing at almost the same time as the government-sponsored programs by developing new techniques. Celera jumped directly to determining the DNA sequence of small pieces of DNA without the physical map. It then used computers to compare and contrast the short sequences, so that it could put them together and

assemble the longer sequence. The benefit of having these two organizations as competitors was that, when they finished their research, they could compare and contrast results. Amazingly, the discrepancies between their findings were declared insignificant.

Human Genome Project Applications

Recently, the first efforts to determine the nucleotide sequences of the human genome were completed. Many scientists feel that advances in medical treatments will occur more quickly because this information is available. The first draft of the human genome was completed early in 2003, when the complete nucleotide sequence of all 23 pairs of human chromosomes was determined. By sequencing the human genome, it is as if we have now identified all the words in the human gene dictionary. Continued analysis will provide the definitions for these words—what these words tell the cell to do.

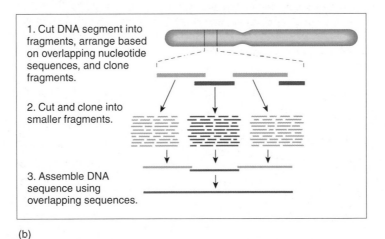

(a)

(b)

FIGURE 11.4 Genes Known to Be on Human Chromosome 21

(a) The genetic map shows the approximate positions of disease-associated genes known to be on human chromosome 21. *(b)* To determine the nucleotide sequence of this chromosome, scientists first created a physical map, which consisted of many small pieces. The DNA sequence of these pieces was determined. After this, the DNA sequence of these many small pieces are "knit" together to generate the nucleotide sequence for the entire chromosome. The nucleotide sequence of each of the genes in *a* can be identified within the assembled sequence.

HOW SCIENCE WORKS 11.1

Polymerase Chain Reaction

Polymerase chain reaction (PCR) is a laboratory procedure for copying selected segments of DNA from larger DNA molecules. With PCR, a single cell can provide enough DNA for analysis and identification. Targeting specific DNA sequences for replication enables biochemists to manipulate DNA more easily. This is like increasing the one "needle in the haystack" to such large numbers that it is easy to find, recognize, and work with.

Scientists start with a sample of DNA that contains the desired DNA region. The types of samples that can be used include semen, hair, blood, bacteria, protozoa, viruses, mummified tissue, and frozen cells. PCR is an artificial form of the cellular DNA replication process and requires similar components. The DNA from the sample specimen serves as the template for replication. Free DNA nucleotides are used to assemble new strands of DNA.

DNA primers are short stretches of single-stranded DNA, which are used to direct the DNA polymerase to replicate only certain regions of the DNA. These primer molecules are specifically designed to flank the ends of the target sequence and point the DNA polymerase to the region between the primers. This is accomplished by heating the target DNA, so that the two strands of DNA fall away from each other. This process is called *denaturation*. Once the nitrogenous bases on the target sequence are exposed and the reaction cools, the primers are able to attach to the template molecule. The primers *anneal* to the template.

Purified DNA polymerase is the enzyme that drives the DNA replication process. The presence of the primer, attached to the DNA template and added nucleotides, serves as the substrate for the DNA polymerase. Once added, the polymerase extends the DNA molecule from the primer down the length of the DNA. Extension continues until the polymerase runs out of template DNA. The enzyme bonds the new DNA nucleotides to the strand, replicating the molecule as it goes. It stops when it reaches the other end, having produced a new copy of the target sequence.

The elegance of PCR is that it allows the exponential replication of DNA. Exponential, or logarithmic, growth is a doubling in number with each round. With just one copy of template DNA, there will be a total of two copies at the end of one replication cycle. During the second round, both copies are used as a template. At the end of the second round, there is a total of 4 copies. The number of copies of the target DNA increases very quickly. With each round of replication, the number doubles—8, 16, 32, 64. Each round of replication takes only minutes. Thirty rounds of replication in PCR can be performed within 2.5 hours. Starting with just one copy of DNA and 30 rounds of replication, it is possible to produce over half a billion copies of the desired DNA segment.

Because this technique can create useful amounts of DNA from very limited amounts, it is a very sensitive test for the presence of specific DNA sequences. Frequently, the presence of a DNA sequence indicates the presence of an infectious agent or a disease-causing condition. Using PCR, scientists have been able to:

1. More accurately diagnose such diseases as sickle-cell anemia, cancer, Lyme disease, AIDS, and Legionnaires' disease.
2. Perform highly accurate tissue typing for matching organ-transplant donors and recipients.
3. Help resolve criminal cases of rape, murder, assault, and robbery by matching suspect DNA to that found at the crime scene.
4. Detect specific bacteria in environmental samples.
5. Monitor the spread of genetically engineered microorganisms in the environment.
6. Check water quality by detecting bacterial contamination from feces.
7. Identify viruses in water samples.
8. Identify disease-causing protozoa in water.
9. Determine specific metabolic pathways and activities occurring in microorganisms.
10. Determine subspecies, distribution patterns, kinships, migration patterns, evolutionary relationships, and rates of evolution of long extinct species.
11. Accurately settle paternity suits.
12. Confirm identity in amnesia cases.
13. Identify a person as a relative for immigration purposes.
14. Provide the basis for making human antibodies in specific bacteria.
15. Possibly provide the basis for replicating genes that could be transplanted into individuals suffering from genetic diseases.
16. Identify nucleotide sequences peculiar to the human genome.

HOW SCIENCE WORKS 11.1 *(continued)*

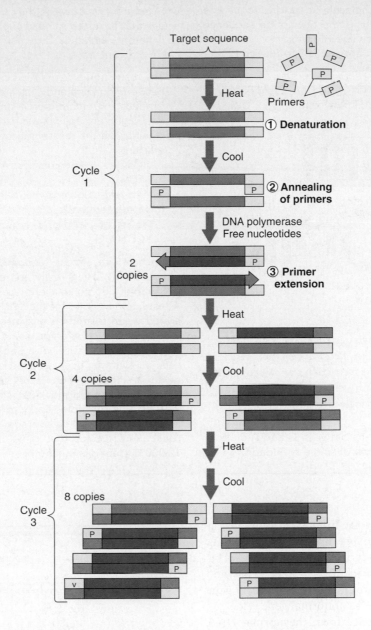

PCR Replication

During cycle one of PCR, the template DNA is denatured, so that the two strands of DNA separate. This allows the primers to attach (anneal) to the template DNA. DNA polymerase and DNA nucleotides, which are present for the reaction, create DNA by extending from the primers. During cycle two, the same process occurs again, but the previous round of replication has made more template available for further replication. Each subsequent cycle essentially doubles the amount of DNA.

Restriction Enzymes

Restriction enzymes are enzymes found in bacteria cells that cut in the middle of the DNA molecule at specific sequences of nucleotides. A restriction site is the nucleotide sequence that defines where the restriction enzyme cuts.

A commonly used restriction enzyme is Eco RI. Eco RI recognizes the sequence

```
...GAATTC...              ...G           AATTC...
...CTTAAG...   and cuts   ...CTTAA  and        G...
```

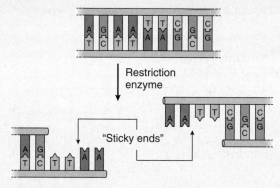

Restriction
enzyme

"Sticky ends"

Eco RI Restriction Site

Double-stranded DNA can be cut by restriction enzymes that recognize specific sequences of DNA. Eco RI, a restriction enzyme, recognizes GAATTC as its restriction site. When Eco RI cuts DNA, it generates sticky ends on the new DNA fragments. Sticky ends are small stretches of single-stranded DNA.

When Eco RI cuts the DNA, it cuts between G and A on both strands. This leaves a small stretch of single-stranded DNA on

each new molecule. These ends can be used to reassemble DNA sequences, because they can bind with other DNA molecules that have the same sequence.

Different restriction enzymes recognize different sites and cut in different ways:

- The restriction enzyme Alu I cuts the following sequence between guanine and cytosine. Notice that Alu I does not produce a stretch of single-stranded DNA when it cuts. Restriction enzymes that cut in this manner can be used for cloning, but the reassembly of DNA sequences is less efficient.

```
. . . AGCT . . .  and cuts  . . . AG         CT . . .
. . . TCGA . . .            . . . TC   and   GA . . .
```

- A third restriction enzyme, Bam HI, recognizes the following sequence:

```
. . . GGATCC  and cuts  . . . G          GATCC . . .
. . . CCTAGG            . . . CCTAG  and        G . . .
```

Restriction sites occur randomly along the DNA molecule. Consequently, using restriction enzymes breaks up the large DNA molecule into smaller pieces of DNA with different sizes. A restriction fragment is the smaller piece of DNA that is generated when restriction enzymes are used. Scientists predict that Alu I cuts, *on average*, every 256 nucleotides, because its restriction site consists of a sequence of four bases. If Alu I cuts, on average, every 256 nucleotides, Alu I restriction fragments are, on average, 256 nucleotides long; however, the fragments can be bigger or smaller.

The same principles apply to each restriction enzyme. Eco RI and Bam HI cut, on average, every 4,096 nucleotides but result in molecules that generally range between 250 and 12,000 nucleotides in length.

The information provided by the human genome project will be extremely useful in diagnosing diseases and providing genetic counseling to those considering having children. This kind of information would also identify human genes and proteins as drug targets and create possibilities for new gene therapies. Once it is known where an abnormal gene is located and how it differs in base sequence from the normal DNA sequence, steps could be taken to correct the abnormality. Further defining the human genome project will also result in the discovery of new families of proteins and will help explain basic physiological and cell biological processes common to many organisms. All this information will increase the breadth and depth of the understanding of basic biology.

It was originally estimated that there were between 100,000 and 140,000 genes in the human genome, because scientists were able to detect so many different proteins. DNA sequencing data indicate that there are only about 20,000 protein-coding genes—only about twice as many as in a worm or a fly. Our genes are able to generate several different proteins per gene because of alternative splicing (figure 11.5). Alternative splicing occurs much more frequently than previ-

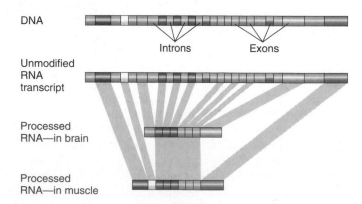

FIGURE 11.5 Different Proteins—One Gene

This illustration shows a stretch of DNA that contains a gene. Protein-coding regions of this gene are shown in different colors. Introns that do not code for protein are shown in a single color—rust. Alternative splicing allows different tissue to use the same gene but make slightly different proteins. The gray bands show how some exons are used to form both proteins, whereas other exons are used on only one protein. When a dideoxyribonucleotide binds to a DNA chain it prevents further extension of the DNA chain.

HOW SCIENCE WORKS 11.3

Electrophoresis

Electrophoresis is a technique used to separate molecules, such as nucleic acids, proteins, or carbohydrates. Electrophoresis separates nucleic acids on the basis of size. DNA is too long for scientists to work with when taken directly from the cell. To make the DNA more manageable, scientists cut the DNA into smaller pieces. Restriction enzymes are frequently used to cut large DNA molecules into smaller pieces (review How Science Works 11.2). After the DNA is broken into smaller pieces, electrophoresis is used to separate differently sized DNA fragments.

Electrophoresis uses an electric current to move DNA through a gel matrix. DNA has a negative charge because of the phosphates that link the nucleotides. In an electrical field, DNA migrates toward the positive pole. The speed at which DNA moves through the gel depends on the length of the DNA molecule. Longer DNA molecules move more slowly through the gel matrix than do shorter DNA molecules.

When scientists work with small areas of DNA, electrophoresis allows them to isolate specific stretches of DNA for other applications.

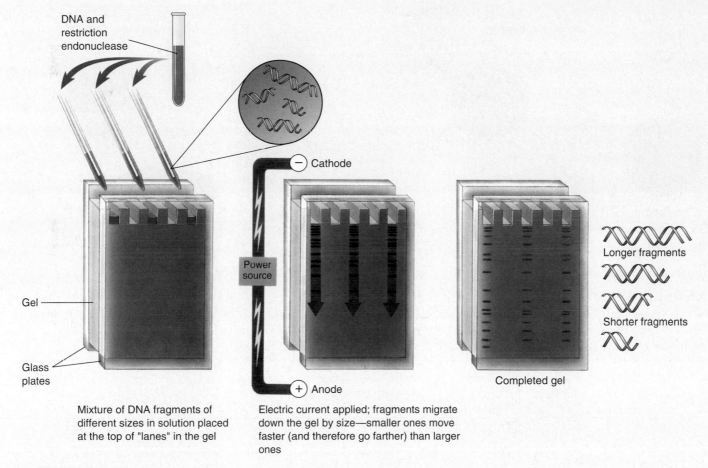

Gel Electrophoresis

Gel electrophoresis separates DNA molecules on the basis of size. DNA is cut by restriction enzymes and then separated by using a gel and an electrical field. Shorter fragments move more quickly through the gel than do the larger fragments.

HOW SCIENCE WORKS 11.4

DNA Sequencing

DNA sequencing uses electrophoresis to separate DNA fragments of different lengths. A DNA synthesis reaction is set up that includes DNA from the region being investigated. The reaction also includes (1) DNA polymerase, (2) a specific DNA primer, (3) all DNA nucleotides (G, A, T, and C), and (4) a small amount of 4 kinds of chemically altered DNA nucleotides. DNA polymerase is the enzyme that synthesizes DNA in cells by using DNA nucleotides as a substrate. The DNA primer gives the DNA polymerase a single place to start the DNA synthesis reaction. All of these components work together to allow DNA synthesis in a manner very similar to cellular DNA replication.

The DNA sequencing process also adds nucleotides that have been chemically altered in two ways. 1) The altered nucleotides are called dideoxyribonucleosides because they contain a dideoxyribose sugar rather than the normal deoxyribose. Dideoxyribose has one less oxygen in its structure than deoxyribose. 2) The four kinds of dideoxyribonucleotides (A, T, G, C) are each labeled with a different flourescent dye so that each of the four nucleotides is colored differently. During DNA sequencing, the polymerase randomly incorporates a normal DNA nucleotide or a dideoxyribonucleotide. When the dideoxyribonucleoside is used, two things happen: (1) No more nucleotides can be added to the DNA strand, and (2) the DNA strand is now tagged with the fluorescent label of the dideoxynucleotide that was just incorporated.

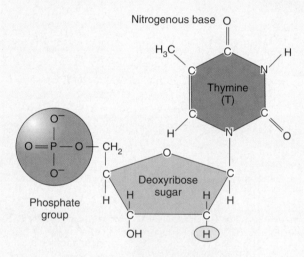

(a) Deoxyribonucleic acid

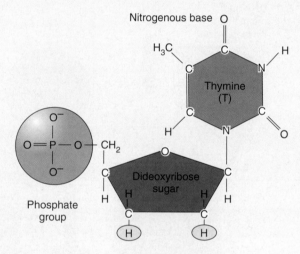

(b) Dideoxyribonucleic acid

Dideoxyribonucleic Acids
Dideoxyribonucleotides are used in sequencing reactions because they lack an oxygen that is necessary to continue DNA synthesis. DNA polymerase uses either form of nucleotide in a sequencing reaction. (*a*) If a deoxyribonucleotide is used, DNA synthesis continues. (*b*) If a dideoxyribonucleotide is used, DNA synthesis stops for that molecule.

ously expected. Knowing this information provides insights into the evolution of humans and will make future efforts to work with the genome through bioengineering much easier.

There is a concern that, as our genetic makeup becomes easier to determine, some people may attempt to use this information for profit or political power. Consider that some health insurance companies refuse to insure people with "preexisting conditions" or those at "genetic risk" for certain abnormalities. Refusing to provide coverage would save these companies the expense of future medical bills incurred by "less than perfect" people. While this might be good for insurance companies, it raises major social questions about fair and equal treatment and discrimination.

Another fear is that attempts may be made to "breed out" certain genes and people from the human population to create a "perfect race." Intentions such as these superficially

HOW SCIENCE WORKS 11.4 *(continued)*

As a group, the DNA molecules that are created by this technique have the following properties:

1. They all start at the same point, because they all started with the same primer.
2. Many DNA molecules have been stopped at each nucleotide in the sequence of the sample DNA.

3. DNA molecules of the same size (number of nucleotides) are labeled with the same color of fluorescent dye.

Electrophoresis separates this collection of molecules by size, because the shortest DNA molecules move fastest. The DNA sequence is determined by reading the color sequence from the shortest DNA molecules to the longest DNA molecules. The pattern of colors match the order of the nucleotides in the DNA.

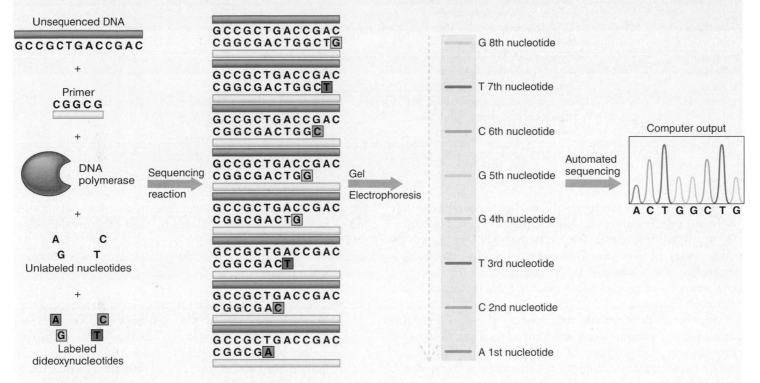

DNA Sequencing

DNA sequencing involves the use of the template DNA that will be sequenced, a primer that provides a starting point for the DNA polymerase to begin DNA synthesis and DNA nucleotides. A mixture of the normal deoxyribonucleotides are labeled with a fluorescent marker. Each type of nucleotide has a different color of marker. The fluorescent marker is used to determine which type of nucleotide was added to the end of the DNA strand. After synthesis, the DNA strands are separated by gel electrophoresis on the basis of size. The color pattern that is generated by the sequencing gel is the order of the nucleotides. Automated sequencing is done by using a laser beam to read the colored bands. A printout is provided as peaks of color to show the order of the nucleotides.

---

appear to have good intentions, but historically they have been used by many groups to justify discrimination against groups of individuals or even to commit genocide.

Other Genomes

While some scientists continue sequencing the human genome, others are sequencing the genomes of other organisms. Representatives of each major grouping of organisms have been investigated, and the DNA sequence data have been made available to the general public through a central-

ized government web site. This centralized database has made the exchange and analysis of scientific information easier than ever (table 11.1). The information gained from these studies might lead to new treatments for disease and a better understanding of evolutionary relationships.

Patterns in Protein-Coding Sequences

As scientists sequenced the human genome and compared it with other genomes, certain types of genetic repetitions became apparent.

TABLE 11.1

Completed and Current Genome Projects

The Human Genome Project has sparked major interest in nonhuman genomes. The investigation of some genomes has been very organized. Other investigations have been less directed, whereby only sequences of certain regions of interest have been reported. Regardless, information on many genomes is available at the National Center for Biotechnology Information web site.

Taxonomic Group	Genome Examples	Number of Different Genomes Represented
Viruses and retroviruses and bacteriphages	Herpes virus, human papillomavirus, HIV	Over 1,560
Eubacteria	Anthrax species, *Clamydia* species, *Escherichia coli*, *Pseudomonas* species, *Salmonella* species	Over 200
Archaea	*Halobacterium* species, *Methanococcus* species, *Pyrococcus* species, *Thermococcus* species	Over 21
Protists	*Cryptosporidium* species, *Entamoeba histolytica*, *Plasmodium* species	Over 45
Fungi	Yeast, *Aspergillus*, *Candida*	Over 70
Plants	Thale cress *Arabidopsis thaliana*, tomato, lotus, rice	Over 20
Animals	Bee, cat, chicken, chimp, cow, dog, frog, fruit fly, mosquito, nematode, pig, rat, sea urchin, sheep, zebra fish	Over 100
Cellular organelles	Mitochondria, chloroplasts	

Tandem clusters are grouped copies of the same gene that are found on the same chromosome (figure 11.6). For example, the DNA that codes for ribosomal RNA is present in many copies in the human genome. From an evolutionary perspective, the advantage to the cell is the ability to create large amounts of gene product quickly from the genes found in tandem clusters.

Segmental duplications are groups of genes that are copied from 1 chromosome and moved as a set to another chromosome. These types of gene duplications allow for genetic backups of information. If either copy is mutated, the remaining copy can still provide the necessary gene product sufficient for the organism to live. Because the function of the mutated copy of the gene is being carried out by the normal gene, the mutated copy may take on new function if it accumulates additional mutations. This can allow evolution to occur more quickly (figure 11.6b).

Multigene families are groups of different genes that are closely related. When members of multigene families are closely inspected, it is clear that certain regions of the genes carry similar nucleotide sequences. Hemoglobin is a member of the globin gene family. There are several different hemoglobin genes in the human genome. Evolutionary patterns can be tracked at the molecular level by examining gene families across species. The portions of genes that show very little change across many species represent portions of the protein that are important for function. Scientists reason that regions that are important for function will be intolerant of change and stay unaltered over time. Again, using hemoglobin as an example, it is possible to compare the hemoglobin genes of different organisms to identify specific changes in

the gene. Such comparisons can lead to a better understanding of how organisms are related to each other evolutionarily (figure 11.6c).

The following are a few more interesting facts obtained by comparing genomes:

- Eukaryotic genomes are more complex than prokaryotic genomes. Eukaryotic genomes are, on average, nearly twice the size of prokaryotic genomes. Eukaryotic genomes devote more DNA to regulating gene expression. Only 1% of human DNA actually codes for protein.

- The number of genes in a genome is not a reflection of the size of an organism. Humans possess roughly 21,000 genes. Roundworms have about 26,000 genes, and rice plants possess 32,000 to 55,000 genes.

- Eukaryotes create multiple proteins from their genes because of alternative splicing. Prokaryotes do not. Nearly 25% of human DNA consists of intron sequences, which are removed during splicing. On average, each human gene makes between 4.5 and 5 different proteins because of alternative splicing.

- There are numerous, virtually identical genes found in very distantly related organisms—for example, mice, humans, and yeasts.

- Hundreds of genes found in humans and other eukaryotic organisms appear to have resulted from the transfer of genes from bacteria to eukaryotes at some point in eukaryotic evolution.

- Chimpanzees have 98–99% of the same DNA sequence as humans. All the human "races" are

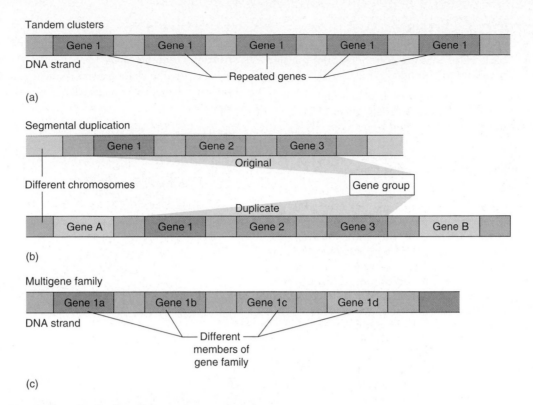

FIGURE 11.6 Patterns in Protein Coding Sequences
(a) Tandem clusters are identical or nearly identical repeats of one gene. *(b)* Segmental duplications are duplications of sets of genes. These may occur on the same chromosome or different chromosomes. *(c)* Multigene families are repeats of similar genes. The genes are similar because regions are conserved from gene to gene, but many regions have changed significantly.

about 99.9% identical at the DNA level. In fact, there is virtually no scientific reason for the concept of "race," because there is much greater variation within a so-called race than there is among the so-called races.

- Genes are unequally distributed between chromosomes and unequally distributed along the length of a chromosome.

Patterns in Non-Coding Sequence
Protein-coding DNA is not the only reason for examining DNA sequences. The regions of DNA that do not code for protein are more important than once thought. Many non-coding sequences are involved with the regulation of gene expression. Eukaryotic genomes have transposable elements, which are able to leave one position on the chromosome and move to an entirely different location. Even among eukaryotes, humans have an unusually high proportion of this type of DNA. Transposable elements account for 45% of human DNA, and this feature appears to be a significant difference from other primates.

New Fields of Knowledge
The ability to make comparisons of the DNA of organisms has led to the development of three new fields in biology—*genomics, transcriptomics,* and *proteomics.* **Genomics** is the

comparison of the genomes of different organisms to identify similarities and differences. Species relatedness and gene similarities can be determined from these studies. When the DNA sequence of a gene is known, **transcriptomics** looks at when, where, and how much mRNA is expressed from a gene. Finally, **proteomics** examines the proteins that are predicted from the DNA sequence. From these types of studies, scientists are able to identify gene families that can be used to determine how humans have evolved at a molecular level. They can also examine how genes are used in an organism throughout its body and over its life span. They can also better understand how a protein works by identifying common themes from one protein to the next.

11.3 The Genetic Modification of Organisms

For thousands of years, civilizations have attempted to improve the quality of their livestock and crops. Cows that produce more milk or more tender meat were valued over those that produced little milk or had tough meat. Initial attempts to develop improved agricultural stocks were limited to selective breeding programs, in which only the organisms with the desired characteristics were allowed to breed. With a

Cloning Genes

There are several basic steps that are common in the transfer of DNA from one organism to another:

1. The source DNA is cut into a usable size by using restriction enzymes (review How Science Works 11.2).
2. The cut DNA fragments are attached to a carrier DNA molecule.
3. The carrier DNA molecule, with its attached source DNA, is moved into an appropriate cell for the carrier DNA. In the cell, the new DNA is replicated or expressed.

Because it is isolated directly from a large number of cells, the source DNA usually consists of many copies of an organism's genome. After the source DNA is cut with restriction enzymes, it is a collection of many small DNA fragments. Isolating the small portion of DNA that contains the gene of interest can be difficult. The gene of interest is found on only a few of these fragments. To identify those fragments, scientists must search the entire collection. The search involves several steps.

The first step is to attach every fragment of source DNA to a carrier DNA molecule. A **vector** is the term scientists use to describe a carrier DNA molecule. Vectors usually contain special DNA sequences that facilitate attachment to the fragments of source DNA. Vectors also contain sequences that promote DNA replication and gene expression.

A *plasmid* is one example of a vector that is used to carry DNA into bacterial cells. A **plasmid** is a circular piece of DNA that is found free in the cytoplasm of some bacteria. Therefore, the plasmid must be cut with a restriction enzyme, so that the plasmid DNA will have sticky ends, which can attach to the source DNA. The enzyme *ligase* creates the covalent bonds between the plasmid DNA and the source DNA, so that a new plasmid ring is formed with the source DNA inserted into the ring. The plasmid and its inserted source DNA is recombinant DNA. Because there are many different source DNA fragments, this process results in many different plasmids, each with a different piece of source DNA. All of these recombinant DNA plasmids constitute a **DNA library** for the entire source genome.

Cutting Genomic DNA

The first step in cloning a specific gene is to cut the source DNA into smaller, manageable pieces with restriction enzymes.

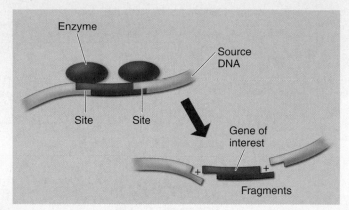

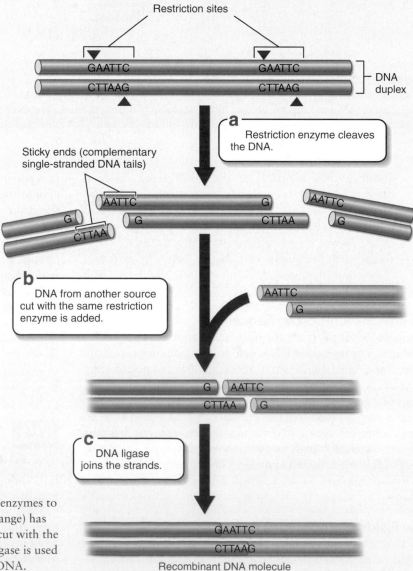

Creating Recombinant DNA

The source DNA is cut with restriction enzymes to create sticky ends. The vector DNA (orange) has compatible sticky ends, because it was cut with the same restriction enzyme. The enzyme ligase is used to bond the source DNA to the vector DNA.

The second step in the process is to mix the DNA library with bacterial cells that will take up the DNA molecules. Transformation occurs when a cell gains genetic information from its environment. Each transformed bacterial cell carries a different portion of the source DNA from the DNA library. These cells can be grown and isolated from one another.

The third step is to screen the DNA library contained within the many different transformed bacterial cells to find those that contain the DNA fragment of interest. Once the bacterial cells with the desired recombinant DNA are identified, the selected cells can be reproduced and, in the process, the desired DNA is cloned.

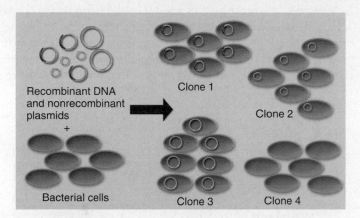

Transformation

Bacterial cells pick up the recombinant DNA and are transformed. Different cells pick up plasmids with different genomic DNA inserts.

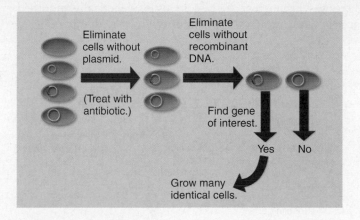

Screening the DNA Library

A number of techniques are used to eliminate cells that do not carry plasmids with attached source DNA. Once these cells are eliminated from consideration, the remainder are screened to find the genomic DNA fragments of interest.

greater understanding of genetics and mutations, scientists irradiated cells to produce mutations, which were then screened to determine if they were desirable.

Although this approach was a very informative way to learn about the genetics of an organism, it lacked the ability to create a specific desired change. Creating mutations is a very haphazard process. However, today the results are achieved in a much more directed manner using biotechnology's ability to transfer DNA from one organism to another. **Transformation** takes place when a cell gains new genetic information from its environment. Once new DNA sequences are transferred into a host cell, the cell is genetically altered and begins to read the new DNA and produce new cell products, such as enzymes. The resulting new form of DNA is called **recombinant DNA.**

A **clone** is an exact copy of biological entities, such as genes, organisms, or cells. The term refers to the outcome, not the way the results are achieved. Many whole organisms "clone" themselves simply by how they reproduce; bacteria divide by cell division and produce two genetically identical cells. Strawberry plants clone themselves by sending out runners and establishing new plants. Many varieties of fruit trees and other plants are cloned by making cuttings of the plant and rooting the cuttings.

With the development of advanced biotechnology techniques, it is now possible to clone specific genes from an organism.

Genetically Modified Organisms

Genetically modified (GM) organisms contain recombinant DNA. Viruses, bacteria, fungi, plants, and animals are examples of organisms that have been engineered so that they contain genes from at least one unrelated organism.

As this highly sophisticated procedure has been refined, it has become possible to splice genes quickly and accurately from a variety of species into host bacteria or other host cells by a process called *gene cloning* (How Science Works 11.5). Genetically modified organisms are capable of expressing the protein-coding regions found on recombinant DNA. This makes possible the synthesis of large quantities of proteins. For example, recombinant DNA procedures are responsible for the production of:

- Human insulin, used in the control of diabetes (figure 11.7)
- Nutritionally enriched "golden rice," capable of supplying poor people in less developed nations with beta carotene, which is missing from normal rice

- Interferon, used as an antiviral agent
- Human growth hormone, used to stimulate growth in children lacking this hormone
- Somatostatin, a brain hormone implicated in growth

Many more products have been manufactured using these methods. Genetically modified cells are not only used as factories to produce chemicals but also for their ability to break down many toxic chemicals. **Bioremediation** is the use of living organisms to remove toxic agents from the environment. There has been great success in using genetically modified bacteria to clean up oil spills and toxic waste dumps.

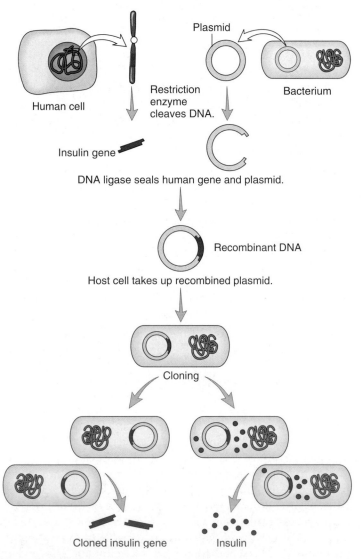

FIGURE 11.7 **Human Insulin from Bacteria**
The gene-cloning process is used to place a copy of the human insulin gene into a bacterial cell. As the bacterial cell reproduces, the human DNA it contains is replicated along with the bacterial DNA. The insulin gene is expressed along with the bacterial genes and the colony of bacteria produces insulin. This bacteria-produced human insulin is both more effective and cheaper than previous therapies, which involved obtaining insulin from the pancreas of slaughtered animals.

Genetically Modified Foods

Although some chemicals have been produced in small amounts from genetically engineered microorganisms, crops such as turnips, rice, soybeans, potatoes, cotton, corn, and tobacco can generate tens or hundreds of kilograms of specialty chemicals per year. Such crops have the potential of supplying the essential amino acids, fatty acids, and other nutrients now lacking in the diets of people in underdeveloped and developing nations. Researchers have also shown, for example, that turnips can produce interferon (an antiviral agent), tobacco can create antibodies to fight human disease, oilseed rape plants can serve as a source of human brain hormones, and potatoes can synthesize human serum albumin that is indistinguishable from the genuine human blood protein (figure 11.8).

FIGURE 11.8 **Applications of Genetically Modified Organisms**
(a) All four of these petunia plants have been exposed to a weed killer. The two plants in the top of the image were genetically modified to be resistant to the herbicide. The two plants in the bottom of the image were not changed genetically. Because the genetically modified plants are resistant to the herbicide, they can be sprayed with herbicide and live, whereas the weeds growing in the vicinity are killed. This results in less time spent weeding the plants. *(b)* Normal rice does not produce significant amounts of beta-carotene. Beta-carotene is a yellow-orange compound needed in the diet to produce vitamin A. *(c)* Genetically modified "golden rice" can provide beta-carotene to populations that have no other sources of this nutrient.

Many GM crops also have increased nutritional value yet can be cultivated using traditional methods. There are many concerns regarding the development, growth, and use of GM foods. Although genetically modified foods are made of the same building blocks as any other type of food, the public is generally wary. Countries have refused entire shipments of GM foods that were targeted for hunger relief. We may, eventually come to a point where we can no longer choose to avoid GM foods however. As the world human population continues to grow, GM foods may be an important part of meeting the human population's need for food. The following are some of the questions being raised about genetically modified food:

- Is tampering with the genetic information of an organism ethical?
- What safety precautions should be exercised to avoid damaging the ecosystems in which GM crops are grown?
- What type of approval should these products require before they are sold to the public?
- Is it necessary to label these foods as genetically modified?

Gene Therapy

The field of biotechnology allows scientists and medical doctors to work together and potentially cure genetic disorders. Unlike contagious diseases, genetic diseases cannot be transmitted, because they are caused by a genetic predisposition for a particular disorder—not separate, disease-causing organisms, such as bacteria and viruses. **Gene therapy** involves inserting genes, deleting genes, and manipulating the action of genes in order to cure or lessen the effect of genetic diseases. These therapies are very new and experimental. While these lines of investigation create hope, many problems must be addressed before gene therapy becomes a reliable treatment for many disorders.

The strategy for treating someone with gene therapy varies, depending on the disorder. When designing a gene therapy treatment, scientists have to ask exactly what the problem is. Is the mutant gene not working at all? Is it working normally but there is too little activity? Is there too much protein being made? Or is the gene acting in a unique, new manner? If there is no gene activity or too little gene activity, the scientists need to introduce a more active version of the gene. If there is too much activity or if the gene is engaging in a new activity, this excess activity must first be stopped and then the normal activity restored.

To stop a mutant gene from working, scientists must change it. This typically involves inserting a mutation into the protein-coding region of the gene or the region that is necessary to activate the gene. Scientists have used some types of viruses to do this in organisms other than humans. The difficulty in

this technique is to mutate only that one gene without disturbing the other genes and creating more mutations in other genes. Developing reliable methods to accomplish this is a major focus of gene therapy. Once the mutant gene is silenced, the scientists begin the work of introducing a "good" copy of the gene. Again, there are many difficulties in this process:

- Scientists must find a way of returning the corrected DNA to the cell.
- The corrected DNA must be made a part of the cell's DNA, so that it is passed on with each cell division, it doesn't interfere with other genes, and it can be transcribed by the cell as needed (figure 11.9).
- Cells containing the corrected DNA must be reintroduced to the patient.

Many of these techniques are experimental, and the medical community is still evaluating their usefulness in treating many disorders, as well as the risks these techniques pose to the patient.

The Cloning of Organisms

Cloning does not always refer to exchanging just a gene. Another type of cloning is the cloning of an entire organism. In this case, the goal is to create a new organism that is genetically identical to the previous organism. Whereas cloning of multicellular organisms, such as Protists, plants, fungi and many kinds of invertebrate animals, often occur naturally during asexual reproduction and is duplicated easily in laboratories. The technique used to accomplish cloning in most vertebrates is called *somatic cell nuclear transfer*. **Somatic cell nuclear transfer** removes a nucleus from a cell of the organism that will be cloned. After chemical treatment, that nucleus is placed into an egg cell that has had it original nucleus removed. The egg cell will use the new nucleus as genetic information. In successful cloning experiments with mammals, an electrical shock is used to stimulate the egg to begin to divide as if it were a normal embryo. After transferring the fertilized egg into a uterus, the embryo grows normally. The resulting organism is genetically identical to the organism that donated the nucleus.

In 1996, a team of scientists from Scotland successfully carried out somatic cell nuclear transfer for the first time in sheep. The nucleus was taken from the mammary cell of an adult sheep. The embryo was transplanted into a female sheep's uterus, where it developed normally and was born (figure 11.10). This cloned offspring was named Dolly. This technique has been applied to many other animals, such as monkeys, goats, pigs, cows, mice, mules, and horses, and has been used successfully on humans. However, for ethical reasons, the human embryo was purposely created with a mutation that prevented the embryo from developing fully. The success rate of cloning animals is still very low for any

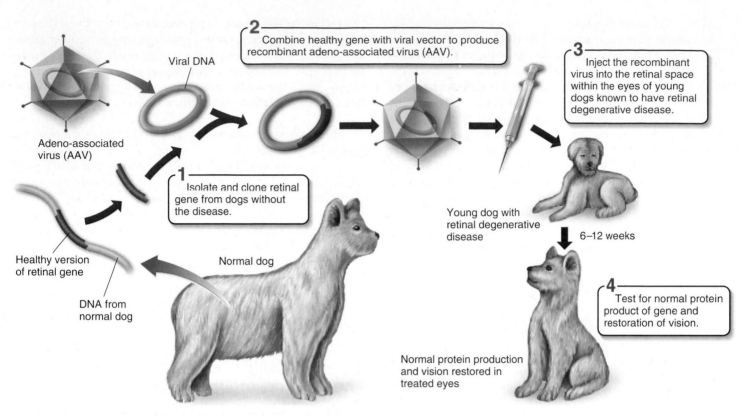

FIGURE 11.9 Gene Therapy

One method of introducing the correct genetic information to a cell is to use a virus as a vector. Here, a dog is treated for a degenerative disorder of the retina. The normal gene is spliced into the viral genome. The virus is then used to infect the defective retinal cells. When the virus infects the retinal cells, it carries the functional gene into the cell.

animal, however; only 3–5% of the transplanted eggs develop into adults (figure 11.11).

A cloning experiment has great scientific importance, because it represents an advance in scientists understanding of the processes of *determination* and *differentiation*. Recall that determination is the process a cell goes through to select which genes it will express. A differentiated cell has become a particular cell type because of the proteins that it expresses. Differentiation is more or less a permanent condition. The techniques that produced Dolly and other cloned animals use a differentiated cell and reverse the determination process, so that this cell is able to express all the genes necessary to create an entirely new organism. Until this point, scientists were not sure that this was possible.

11.4 Stem Cells

Scientists are using somatic cell nuclear transfer techniques to develop populations of special cells called *stem cells*. **Stem cells** are cells that have not yet completed determination or differentiation, so they have the potential to develop into many different cell types.

Scientists study stem cells to understand the processes of determination and differentiation. The ability to control determination and differentiation may allow the manipulation of an organism's cells or the insertion of cells into an organism to allow the regrowth of damaged tissues and organs in humans. This could aid in the cure or treatment of many medical problems, such as the repair of damaged heart tissue from a heart attack or damaged nerve tissue from spinal or head injuries. Some kinds of degenerative diseases occur because specific kinds of cells die or cease to function properly. Parkinson's disease results from malfunctioning brain cells, and many forms of diabetes are caused by malfunctioning cells in the pancreas. If stem cells could be used to replace these malfunctioning cells, normal function could be restored and the diseases cured.

Embryonic and Adult Stem Cells

Because embryonic stem cells have not undergone determination and diffentiation and have the ability to become *any* tissue in the body, they are of great interest to scientists. As an embryo develops, its stem cells go through the process of determination and differentiation to create all the necessary tissues. To study embryonic stem cells, scientists must remove them

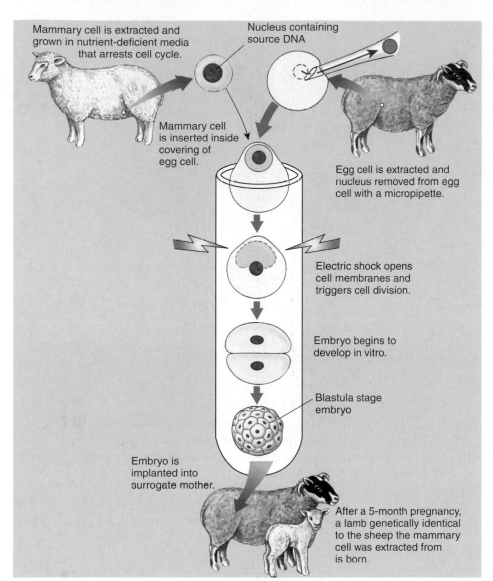

Mammary cell is extracted and grown in nutrient-deficient media that arrests cell cycle.

Nucleus containing source DNA

Mammary cell is inserted inside covering of egg cell.

Egg cell is extracted and nucleus removed from egg cell with a micropipette.

Electric shock opens cell membranes and triggers cell division.

Embryo begins to develop in vitro.

Blastula stage embryo

Embryo is implanted into surrogate mother.

After a 5-month pregnancy, a lamb genetically identical to the sheep the mammary cell was extracted from is born.

FIGURE 11.10 Cloning an Organism
The nucleus from the donor sheep is combined with an egg from another sheep. The egg's nucleus has been removed. The egg, with its new nucleus, is stimulated to grow by an electrical shock. After several cell divisions, the embryo is artificially implanted in the uterus of a sheep, which will carry the developing embryo to term.

(a)

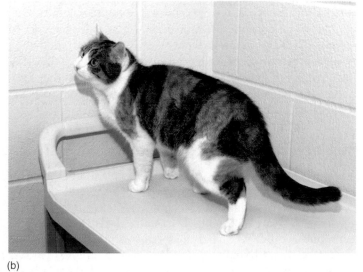

(b)

FIGURE 11.11 Success Rate in Cloning Cats
Out of 87 implanted cloned embryos, CC (Copy Cat) is the only one to survive—comparable to the success rate in sheep, mice, cows, goats, and pigs. (a) Notice that CC is completely unlike her tabby surrogate mother. (b) "Rainbow" is her genetic donor, and both are female calico domestic shorthair cats.

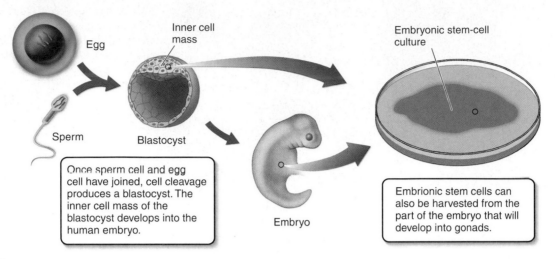

Egg

Sperm

Inner cell mass

Blastocyst

Once sperm cell and egg cell have joined, cell cleavage produces a blastocyst. The inner cell mass of the blastocyst develops into the human embryo.

Embryo

Embryonic stem-cell culture

Embrionic stem cells can also be harvested from the part of the embryo that will develop into gonads.

FIGURE 11.12 The Culturing of Embryonic Stem Cells
After fertilization of an egg with sperm, the cell begins to divide and form a mass of cells. Each of these cells has the potential to become any cell in the embryo. Embryonic stem cells may be harvested at this point or at other points in the determination process.

from embryos, destroying the embryos (figure 11.12). Because of ethical concerns regarding this practice, it is currently illegal to harvest embryonic stem cells in the United States.

As a result, scientists have explored other methods of obtaining stem cells that avoid harvesting them from embryos. Embryonic stem cells reach an intermediary level of determination at which they are committed to becoming a particular *tissue* type, but not necessarily a particular *cell* type. An example of this intermediate determination occurs when stems cells become determined to be any one of several types of nerve cells but have not yet committed to becoming any one nerve cell. Scientists call these partially determined stem cells "tissue-specific." These types of stem cells can be found in adults. One example is hematopoietic stem cells. These cells are able to become the many different types of cells found in blood—red blood cells, white blood cells, and platelets (figure 11.13). The disadvantage of using these types of stem cells is that they have already become partially determined and do not have the potential to become every cell type.

Personalized Stem Cell Lines

Scientists hope that eventually it will possible to use stem cells with the somatic cell nuclear transfer technique used for cloning a sheep. Using these techniques together may allow scientists to create a patient's personalized stem cell line. This technique would involve transferring a nucleus from the patient's cell to a human egg that has had its original nucleus removed. The human egg would be allowed to grow and develop to produce embryonic stem cells. If the process of determination and differentiation can be controlled, new tissues or even organs could be developed.

Under normal circumstances, organ transplant patients must always worry about rejecting their transplant and take

strong immunosuppressant drugs to avoid organ rejection. Tissue and organs grown from customized stem cells would have the benefit of being immunologically compatible with the patient; thus, organ rejection would not be a concern (figure 11.14).

Unfortunately, the days of customized stem cells and organ culture are still in the future. Somatic cell nuclear transfer is a very inefficient process. The use of human eggs also introduces many ethical concerns.

11.5 Biotechnology Ethics

Scientific advances frequently present society with ethical questions that must be resolved. How will new technology be used safely? Who will benefit? Should the technology be used to make a profit? Biotechnology is no different.

Many feel that biotechnology is dangerous. There are concerns about contaminating the environment with organisms that are modified genetically in the lab. What would be the impact of such contamination? Biotechnology also allows scientists to examine molecularly the genetic characteristics of an individual. How will this ability to characterize individuals be used? How will it be misused? Others feel that biotechnology is akin to playing God.

What Are the Consequences?

One way to explore the ethics of biotechnology is to weigh its pros against its cons. This method of thinking considers all the consequences and implications of biotechnology. Which outweighs the other? The benefits of nearly everything discussed in this chapter include a greater potential for better

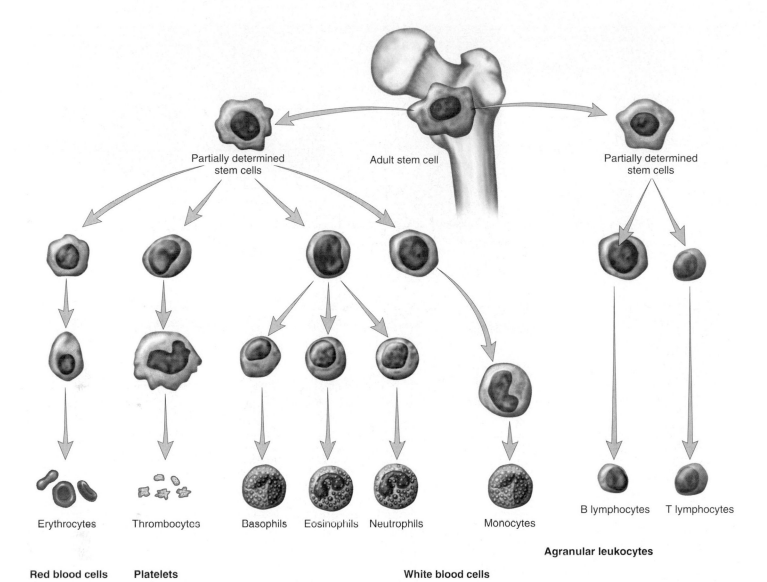

Erythrocytes   Thrombocytes   Basophils   Eosinophils   Neutrophils   Monocytes   B lymphocytes   T lymphocytes

Agranular leukocytes

Red blood cells   Platelets   **White blood cells**

FIGURE 11.13 The Differentiation of Blood Cells

One type of adult stem cell gives rise to various forms of blood cells. Here, the stem cell is found in the bone, where it divides. Some of the cells change their gene expression and become a specific cell type. The differentiated blood cells are shown across the bottom of the image.

medical treatment, cures for disease, and a better understanding of the world around us. What price must we pay for these advances?

- The development of these technologies may mean that our personal genetic information becomes public record. How might this information be misused? Insurance companies may deny coverage or charge exorbitant premiums for individuals with genetic diseases.
- Cloning technology allows the creation of genetically modified foods that increase production and are more nutritious. Is this ethical if the genetically modified organism suffers because of disorders and pain caused

by the change? Are you willing to risk the potential problems of a genetically modified organism becoming part of the ecosystem? How might the introduction of genetically modified species alter ecosystems and their delicate balance?

Is Biotechnology Inherently Wrong?

Another way to explore the ethics of biotechnology is to ask if it violates principles that are valued by society. What aspects of biotechnology threaten the principles of the Bill of Rights? Basic human rights? Religious beliefs? Quality of life issues? Animal rights issues? Which of these sets of principles should be used to help us decide if biotechnology is ethical?

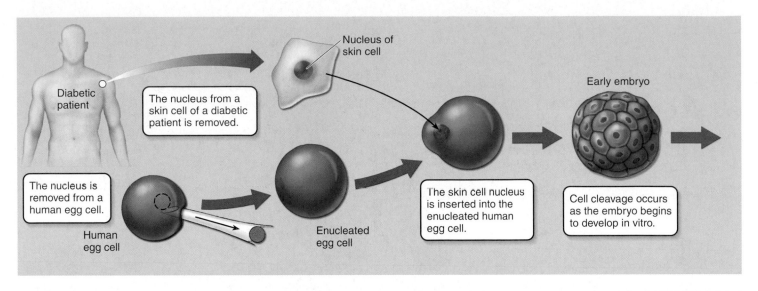

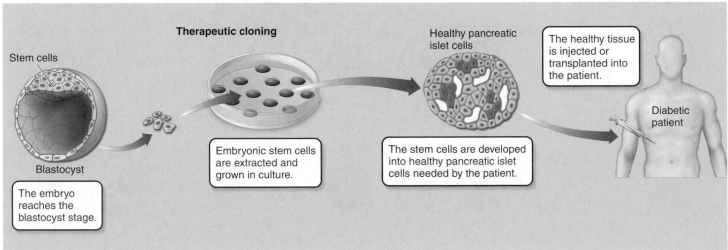

FIGURE 11.14 Customized Stem Cell Lines

One potential use of biotechnology is the production of customized stem cell lines. In this application a somatic cell from a patient would be inserted into a human egg from which the nucleus has been removed. The egg would divide and generate stem cells. These cells could then be cultured and used for therapy. In this example the stem cells could be used to create pancreatic cells to treat a diabetic patient.

- Is it inherently wrong to produce genetically modified foods? Should companies be allowed to grow genetically modified crops? Should the foods be labeled as genetically modified when sold? How does this impact you as a consumer or simply as a person?

- Is it inherently wrong to manipulate genes? Are humans wise enough to use biotechnology safely? Is this something that only God should control? Would you have your child genetically altered as a fetus to prevent a genetic disease? Would you have your child genetically altered as a fetus to enhance desirable characteristics, such as intelligence, or even to control gender? Do you feel that one situation is morally justified but the other is not?

- Is it inherently wrong to harvest embryonic stem cells? Stem cells may provide new avenues of treatment for many disorders. Although there are several sources of stem cells, the cells of most interest are embryonic stem cells. Harvesting these cells destroys the embryo. Even if the embryo is not yet aware of its environment and does not sense pain, is it ethical to use human embryos to advance the treatment of disease?

- Are we morally obligated to search for cures and treatments? Can we stop research if people still need treatment and cures?

Clearly, these are issues that our society will debate for some time. Many of these issues have been debated for decades and bring forward very strong feelings and very different world views. As you continue to hear more about biotechnology in your day-to-day life, consider how that form of biotechnology may affect you.

Summary

Advances in biotechnology are possible because organisms use a common genetic language to make proteins. New techniques, such as DNA fingerprinting and DNA sequencing, allow scientists to compare DNA directly. These techniques involve multiple steps, including polymerase chain reaction, the use of restriction enzymes, and electrophoresis. One large-scale analysis was the Human Genome Project. Scientists are hopeful that the information gained from the Human Genome Project will allow the better diagnosis and treatment of many medical conditions. The genomes of many other organisms have also been characterized, resulting in the new fields of biology called genomics, transcriptomics, and proteomics. The commonality of the genetic code allows DNA from one organism to be used by a different species. The techniques used to clone a gene and to clone and entire organism differ. Cloning a gene involves a number of techniques, including screening a DNA library. Cloning an organism involves somatic cell nuclear transfer. Stem cells have the potential to become multiple cell types. Many feel that the controlled growth of stem cells can be a medical treatment for many incurable medical conditions. The social concern surrounding biotechnology has created an ethical debate, which asks two fundamental questions:

- Do the benefits of biotechnology outweigh the problems?
- Are some aspects of biotechnology inherently wrong?

Key Terms

Use the interactive flash cards on the **Concepts in Biology,** *12/e website to help you learn the meaning of these terms.*

bioremediation 230
biotechnology 216
clone 229
DNA fingerprinting 216
DNA library 228
electrophoresis 217
gene therapy 231
genetically modified (GM) 229
genomics 227
multigene families 226
plasmid 228
polymerase chain reaction (PCR) 216
proteomics 227

recombinant DNA 229
restriction enzymes 216
restriction fragments 216
restriction sites 216
segmental duplications 226
somatic cell nuclear transfer 231
stem cells 232
tandem clusters 226
transcriptomics 227
transformation 229
variable number tandem repeats (VNTRs) 216
vector 228

Basic Review

1. Information in DNA can code for the same protein in any organism. (T/F)

2. DNA fingerprinting
   a. directly examines nucleotide sequence.
   b. examines segments of DNA, which vary in length between individuals.
   c. transfers DNA from one person to another.
   d. uses stem cells.

3. Restriction fragments
   a. are used in a technique that sequences DNA.
   b. create many copies of DNA from a small amount.
   c. are pieces of DNA cut by enzymes at specific sites.
   d. are pieces of protein cut by enzymes at specific sites.

4. A technique that separates DNA fragments of different lengths is
   a. electrophoresis.
   b. DNA sequencing.
   c. polymerase chain reaction.
   d. DNA fingerprinting.

5. The Human Genome Project
   a. was an international effort.
   b. determined the sequence of a healthy human genome.
   c. allows comparisons of the human genome with that of other organisms.
   d. All of the above are correct.

6. The term *cloning* can be applied to which of the following situations?
   a. creating an exact copy of a fragment of DNA
   b. creating a second organism that is genetically identical to the first
   c. using a restriction enzyme
   d. Both a and b are correct.

7. Which of the following terms best describes an organism that possesses a cloned fragment of DNA from another species?
   a. cloned
   b. genetically modified (GM)
   c. usable for bioremediation
   d. genomic

8. Stem cell research is controversial because
   a. of the source of stem cells.
   b. stem cells may cure certain diseases.
   c. stem cells are not yet differentiated.
   d. stem cells are not yet determined.

9. DNA libraries are
   a. stored in computers, so that they can be easily searched.
   b. are an index of various organisms.
   c. collections of DNA fragments that represent the genome of an organism.
   d. a person's unique electrophoresis banding pattern.

10. Restriction enzymes
    a. cut DNA randomly.
    b. cut DNA at specific sequences.
    c. can create sticky ends.
    d. Both b and c are correct.

Answers
1. T   2. b   3. c   4. a   5. d   6. d   7. b   8. a   9. c   10. d

Concept Review

Answers to the Concept Review questions can be found on the **Concepts in Biology,** *12/e website.*

11.1 Why Biotechnology Works?

1. Why can DNA in one organism be used to make the same protein in another organism?

11.2 Comparing DNA

2. What types of questions can be answered by comparing the DNA of two different organisms?
3. What techniques do scientists use to compare DNA?
4. What benefits does the Human Genome Project offer?

11.3 The Genetic Modification of Organisms

5. A scientist can clone a gene. An organism can be a clone. How is the use of the word *clone* different in each of these instances? How is the use of the word *clone* the same?
6. What are some of the advantages of creating genetically modified foods? What are some of the concerns?

11.4 Stem Cells

7. Embryonic stem cells are found in embryos, and adult stem cells are found in adults. In what other ways are they different?

8. What benefits does stem cell research offer? What are some of the concerns with research on stem cells?

11.5 Biotechnology Ethics

9. Match each of the following questions to the appropriate statement.
   Ethical Principles
   "What are the consequences?"
   "Is biotechnology inherently wrong?"
   Statements
   • The benefits of biotechnology more than compensate for its problems.
   • Regardless of the benefits of biotechnology, we should not tamper with organisms in this way.

Thinking Critically

For guidelines to answer this question, visit the **Concepts in Biology,** *12/e website.*

An 18-year-old college student reported that she had been raped by someone she identified as a "large, tanned white man." A student in her biology class fitting that description was said by eyewitnesses to have been, without a doubt, in the area at approximately the time of the crime. The suspect was apprehended and, on investigation, was found to look very much like someone who lived in the area and who had a previous record of criminal sexual assaults.

Samples of semen 238from the woman's vagina were taken during a physical exam after the rape. Cells were also taken from the suspect.

He was brought to trial but was found to be innocent of the crime based on evidence from the criminal investigations laboratory. His alibi—that he had been working alone on a research project in the biology lab—held up. Without PCR genetic fingerprinting, the suspect would surely have been wrongly convicted, based solely on circumstantial evidence provided by the victim and the eyewitnesses.

Place yourself in the position of the expert witness from the criminal laboratory who performed the PCR genetic fingerprinting tests on the two specimens. The prosecuting attorney has just asked you to explain to the jury what led you to the conclusion that the suspect could not have been responsible for this crime. Remember, you must explain this to a jury of 12 men and women who, in all likelihood, have little or no background in the biological sciences.

Diversity Within Species and Population Genetics

The concept of biological evolution is based in the fact that not all organisms are alike, that these difference are inheritable, and that environmental factors are the driving force of change. If organisms didn't change, they would be the same from generation to generation. The study of fossils has made it clear that variations occur in all organisms, including humans. Recently, the bones of a miniature human were discovered in a cave on the 354 kilometer- (220 mile)-long island of Flores, located east of Bali between Asia and Australia.

The skeleton is a female just 1 meter (3.3 feet) tall. She is estimated to have weighed about 25 kilograms (55 pounds) and was about 30 years old when she died about 18,000 years ago. Her size is smaller than that of African pygmies, who are 1.4 to 1.5 meters (4.6 to nearly 5 feet) tall. There is no doubt that she is not a child. Scientists believe she belongs to a newly discovered species of humans, *Homo floresiensis* but commonly called the "hobbit," after the characters in the *Lord of the Rings* books. This species had a brain smaller than a chimpanzee's, made stone tools, and hunted Komodo dragons and giant rats. The fossil *Homo floresiensis* is about 13–18,000 years old. So *Homo floresiensis* lived at the same time as *Homo sapiens*, modern humans were living on other nearby Indonesian islands. This totally unexpected discovery raises many questions. Did modern humans actually meet the "hobbit?" Were *Homo floresiensis* and *H. sapiens* related? How might they have been related to an even earlier human, *Homo erectus*? What environmental and evolutionary factors resulted in "hobbits" being so small?

CHAPTER OUTLINE

12.1 Genetics in Populations

Plants, animals, and other kinds of organisms don't exist only as genetic individuals. They also exist as part of larger, interbreeding groups. An understanding of two terms, *population* and *species,* is necessary, because these are interconnected. Recall from chapter 1 that a **population** is a group of organisms that are potentially capable of breeding naturally and are found in a specified area at the same time. *Species* is a more encompassing concept. A **species** consists of all the organisms potentially capable of breeding naturally among themselves and having offspring that also interbreed successfully. The concept of a species accounts for individuals from *different* populations that interbreed successfully. Most populations consist of only a portion of all the members of the species. For example, the dandelion population in a city park on the third Sunday in July is only a small portion of all dandelions on the planet. However, a population can also be all the members of a species—for example, the human population of the world in 2006 or all the current members of the endangered whooping cranes.

Population genetics is the study of the kinds of genes within a population, their relative numbers, and how these numbers change over time. This information is used as the basis for classifying organisms and studying evolutionary change. From the standpoint of genetics, a population consists of a large number of individuals, each with its own set of alleles. However, the populations may contain many more different alleles than any one member of the species. Any one organism has a specific genotype consisting of all the genetic information that organism has in its DNA. A diploid organism has a maximum of 2 different alleles for a gene, because it has inherited an allele from each parent. In a population, however, there may be many more than 2 alleles for a specific characteristic. In humans, there are 3 alleles for blood type (A, B, and O) within the population, but an individual can have only up to 2 of the alleles (figure 12.1).

Theoretically, all members of a population are able to exchange genetic material. Therefore, we can think of all the genetic information of all the individuals of the same group as a *gene pool*. A **gene pool** consists of all the alleles of all the individuals in a population. Because each organism is like a container of a particular set of these alleles, the gene pool contains much more genetic variation than does any one of the individuals. A gene pool is like a gum ball machine containing red, blue, yellow, and green balls (alleles). For a quarter and a turn of the knob, two gum balls are dispensed from the machine. Two red gum balls, a red and a blue, a yellow and a green, or any of the other possible color combina-

FIGURE 12.1 Genotypes of Individuals and Populations
Any individual can have only 2 alleles for a particular gene, but the population may contain different alleles.

tion may result from any one gum ball purchase. A person buying gum balls will receive no more than 2 of the 4 possible gum ball colors and only 1 of 10 possible color combinations. Similarly, individuals can have no more than 2 of the many alleles for a given gene contained within the gene pool and only 1 of several possible combinations of alleles. genetics, p. 193

12.2 The Biological Species Concept

A species is a population of organisms that share a gene pool and are reproductively isolated from other populations. This definition of a species is often called the **biological species concept;** it involves the understanding that organisms of different species do not interchange genetic information—that is, they don't reproduce with one another. *An individual organism is not a species but, rather, is a member of a species.* A clear understanding of the concept of species is important as we begin to consider how genetic material is passed around within populations as sexual reproduction takes place. It will also help in considering how evolution takes place.

If we examine the chromosomes of reproducing organisms, we find that they are equivalent in number and size and usually carry very similar groups of genes. In the final analysis, the biological species concept assumes that the genetic

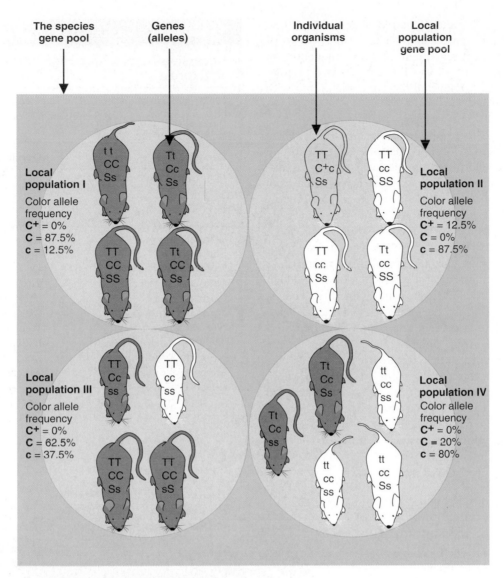

FIGURE 12.2 Genes, Populations, and Gene Pools

Each individual shown here has a specific combination of alleles that constitutes its genotype. The frequency of a specific allele varies from one local population to another. Each local population has a gene pool that is somewhat different from the others. Notice how differences in the frequencies of particular alleles in local populations affects the appearance of the individuals. *Assume that T = long tail; t = short tail; C_+ = gray color; C = brown color; c = white color; S = large size; and s = small size.*

similarity of organisms is the best way to identify a species, regardless of where or when they exist. Individuals of a species usually are not evenly distributed within a geographic region but, rather, occur in clusters as a result of factors such as geographic barriers that restrict movement or the local availability of resources. Local populations with distinct genetic combinations may differ quite a bit from one place to another. There may be differences in the kinds of alleles and the numbers of each kind of allele in different populations of the same species. Note in figure 12.2 that, within the gene pool of the species, there are 3 possible alleles for color (C^+, C, and c). However, how often these alleles appear in the population, the frequencies of these alleles, are different in the four local populations, and the difference in how often an allele occurs is reflected in the colors seen in the individuals of the population.

Gene and Allele Frequencies

In general, the term **gene frequency** is used to convey the idea that there are genetic differences between populations. The term *allele frequency* is more properly used when specifically discussing how common a particular form of a gene (allele) is compared, with other forms. **Allele frequency** is how often an allele is found in a population. For example, the frequency of the blond hair allele is high in northern Europe but low in Africa.

Allele frequency is commonly stated in terms of a percentage or a decimal fraction (e.g., 10%, or 0.1; 50%, or 0.5). It is a mathematical statement of how frequently an allele is found in a population. It is possible for two populations of the same species to have all the same alleles, but with very different frequencies.

As an example, all humans are of the same species and, therefore, constitute one, large gene pool found on Earth. There are, however, many distinct, local populations scattered around the world. These, more localized populations show many distinguishing characteristics, which have been perpetuated from generation to generation. In Africa, alleles for dark skin, tightly curled hair, and a flat nose have very high frequencies. In Europe, the allele frequencies for light skin, straight hair, and a narrow nose are the highest. People in Asia tend to have moderately colored skin, straight hair, and broad noses (figure 12.3). All three of these populations have alleles for dark skin and light skin, straight hair and curly hair, narrow noses and broad noses. The three differ, however, in the frequencies of these alleles. Once a mixture of alleles is present in a population, that mixture tends to maintain itself, unless something changes the frequencies. In other words, allele frequencies do not change without reason. With the development of transportation, more people have moved from one geographic area to another, and human allele frequencies have begun to change. Ultimately, as barriers to interracial marriage (both geographic and sociological) are leveled, the human gene pool will show fewer and fewer geographically distinct populations.

People tend to think that allele frequency has something to do with dominance and recessiveness, but this is not true. Often in a population, recessive alleles are more frequent than their dominant counterparts. Straight hair, blue eyes, and light skin are all recessive characteristics, yet they are quite common in the populations of certain European countries. See table 12.1 for other examples. What really determines the frequency of an allele in a population is the allele's value to the organisms possessing it. Dark-skin alleles are valuable to people living under the bright sun in tropical regions. These alleles are less valuable to those living in the less intense sunlight of the cooler European countries. This idea of the value of alleles and how it affects allele frequency will be dealt with more fully when the process of natural selection is discussed in chapter 13.

TABLE 12.1

Recessive Traits with a High Frequency of Expression

Many recessive characteristics are extremely common in some human populations. The corresponding dominant characteristic is also shown here.

Recessive	Dominant
Light skin color	Dark skin color
Straight hair	Curly hair
Five fingers	Six fingers
Type O blood	Type A or B blood
Normal hip joints	Dislocated hip birth defect
Blue eyes	Brown eyes
Normal eyelids	Drooping eyelids
No tumor of the retina	Tumor of the retina
Normal fingers	Short fingers
Normal thumb	Extra joint in the thumb
Normal fingers	Webbed fingers
Ability to smell	Inability to smell
Normal tooth number	Extra teeth
Presence of molars	Absence of molars

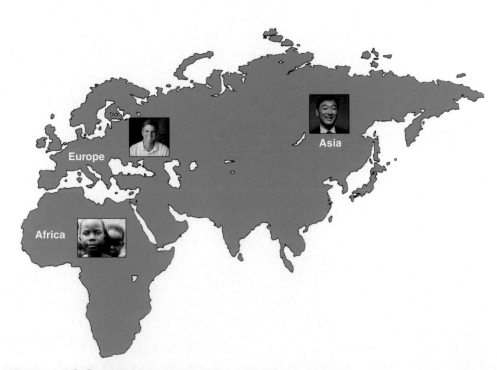

FIGURE 12.3 Allele Frequency Differences Among Humans
Different physical characteristics displayed by people from different parts of the world are an indication that allele frequencies differ as well.

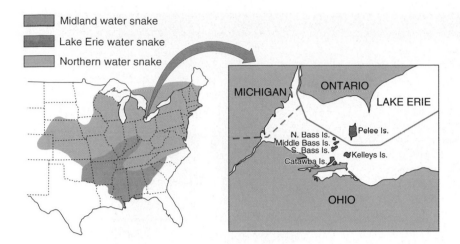

Midland water snake
Lake Erie water snake
Northern water snake

Northern Water Snake

Lake Erie Water Snake

FIGURE 12.4 Subspecies of the Water Snake *Nerodia sepidon*
The Water snake *Nerodia sepidon* is found throughout the eastern part of the United
States and extends into Canada. The northern water snake subspecies is widespread in the
north central United States and adjacent Canada and is generally brown with light
diamond-shaped patches. The Lake Erie water snake subspecies is limited to the islands in
the western section of Lake Erie. It is generally a solid color without the lighter patches.
The difference in color patterns is related to different allele frequencies for color pattern.

Subspecies, Breeds, Varieties, Strains, and Races

Within a population, genetic material is repackaged into new
individuals from one generation to the next. Often, there is
very little adding or subtracting of genetic material from a
local population of organisms, and a widely distributed
species consists of a number of more or less separate groups,
known as **subspecies** (or **breeds, varieties, strains,** or **races**).
All of these terms are used to describe various forms of
organisms that are all members of the same species.
However, certain terms are used more frequently than others, depending on one's field of interest. For example, dog
breeders use the term *breed,* horticulturalists use the term
variety, microbiologists use the term *strain,* and anthropologists use the term *race* (Outlooks 12.1). The most general
and most widely accepted term is *subspecies.* Look again at
figure 12.2. The gene pool of the species consists of all the
alleles of all individuals of the 4 separate populations. A
local population that shows differences from other local
populations is considered a subspecies. For example, water
snakes are found throughout the eastern portion of the
United States (figure 12.4). The Lake Erie water snake,
which is confined to the islands in western Lake Erie, is one
of the several, distinct populations within this species. The
northern water snakes of the mainland have light and dark
bands; the island populations do not have this banded coloration. Most island individuals have alleles for solid col-

oration; very few individuals have alleles for banded coloration. The island snakes are geographically isolated from
the main gene pool and mate only with one another. Thus,
the different color patterns shown by island snakes and
mainland snakes result from a high incidence of solid-color
alleles in the island populations and a high incidence of
banded-color alleles in the mainland populations.

12.3 How Genetic Diversity Comes About

Genetic diversity is a term used to describe genetic differences
among members of a population. *High genetic diversity* indicates many different kinds of alleles for each characteristic,
and *low genetic diversity* indicates that nearly all the individuals in the population have the same alleles. A large gene pool
with high genetic diversity is more likely to contain some
genetic combinations that will allow the organisms to adapt
to a new environment. A number of mechanisms introduce
this necessary diversity into a population.

Mutations

Mutations introduce new genetic information into a population by modifying alleles that are already present. Sometimes,
a mutation introduces a new allele into the gene pool of a

OUTLOOKS 12.1

Biology, Race, and Racism

The concept of racial difference among groups of people must be approached carefully. Three distortions can occur when people use the term *race*. First, the designation of race focuses on differences, most of which are superficial. Skin color, facial features, and hair texture are examples. Although these characteristics are easy to see, they are arbitrary, and an emphasis on them tends to obscure the fact that humans are all fundamentally the same, with minor variations in the frequency of certain alleles. A second problem with the concept of race is that it is very difficult to separate genetic from cultural differences among people. People tend to equate cultural characteristics with genetic differences. Culture is learned and, therefore, is an acquired characteristic not based on the genes a person inherits. Cultures do differ, but these differences cannot be used as a basis for claiming genetic distinctions. Third, a study of the human genome has revealed that there are usually more genetic differences within so-called racial groups than between them. Because of such distortions, the concept that humans can be divided into racial groups is no longer popular among scientists.

Cultural differences

species. At other times, a mutation may introduce an allele that was absent in a local population, although it is present in other populations of the species. All the different alleles for a trait originated as a result of mutations some time in the past and have been maintained within the gene pool of the species as they have been passed from generation to generation during reproduction. Many mutations are harmful, but very rarely one will occur that is valuable to the organism. If a mutation produces a harmful allele, the allele remain uncommon in the population. For example, the *Anopheles* mosquito is responsible for transmitting malaria in many African countries. At some point in the past, mutations occurred in the DNA of these mosquitoes that made some individuals tolerant to the insecticide Pyrethrin, even before the chemical had been used. These alleles remained very rare in these insect populations until Pyrethrin was used. Then, these alleles became very valuable to the mosquitoes that carried them. Because the mosquitoes that lacked the alleles for tolerance died when they came into contact with Pyrethrin, more of the Pyrethrin-tolerant individuals were left to reproduce the species; therefore, the Pyrethrin-tolerant alleles became much more common in these populations. Scientists have recently found up to 90% Pyrethrin resistance in *Anopheles* mosquitoes that live in several African countries. mutations, p. 158

Sexual Reproduction

Although the process of *sexual reproduction* does not create new alleles, it tends to generate new *genetic combinations* when the genetic information from two individuals mixes

during fertilization, generating a unique individual. This doesn't directly change the frequency of alleles within the gene pool, but the new member may have a unique combination of characteristics so superior to those of other members of the population that the new member will be much more successful in producing offspring. In a corn population, for example, there may be alleles for resistance to corn blight (a fungal disease) and to attack by insects. Corn plants that possess both of these characteristics will be more successful

Corn

than corn plants that have only one of these qualities. They will probably produce more offspring (corn seeds) than the others, because they will survive fungal and insect attacks. Thus, there will be a change in the allele frequency for these characteristics in future generations. meiosis, p. 174

Migration

The *migration* of individuals from one genetically distinct population to another is also an important way for alleles to be added to or subtracted from a local population. Whenever

FIGURE 12.5 Captive Breeding of the Black-Footed Ferret
In October 1985, the Wyoming Game and Fish Department, in cooperation with the U.S. Fish & Wildlife Service, started the captive breeding program for North America's most endangered mammal, the black-footed ferret. Between 1987 and 1998, the captive breeding program produced approximately 2,600 ferrets. Attention was paid to making sure that as much genetic variation was retained as possible. For example they maintain a sperm bank of particularly valuable males. As a result, the successful return of black-footed ferrets to the plains of the American West began in 1991. However biologists still fear that a lack of genetic diversity may jeopardize the populations.

an organism leaves one population and enters another, it subtracts its genetic information from the population it left and adds it to the population it joins. If it contains rare alleles, it may significantly affect the allele frequency of both populations. The extent of migration need not be great; however, as long as alleles are entering or leaving a population, the gene pool will change.

Many animal populations in zoos are in danger of dying out because of severe inbreeding (breeding with near relatives), resulting in reduced genetic diversity (figure 12.5). Often, when genetic diversity is reduced, deleterious recessive alleles in closely related mates are passed to offspring in a homozygous state, resulting in offspring that have reduced chances of survival. Most zoo managers have recognized the importance of increasing genetic diversity in their small populations of animals and have instituted programs of loaning breeding animals to distant zoos in an effort to increase genetic diversity. In effect,

they are attempting to simulate the natural migration that frequently introduces new alleles from distant populations.

The Importance of Population Size

The *size of the population* has much to do with how effective any mechanism is at generating diversity within a gene pool. The smaller the population, the less genetic diversity it can contain. Therefore, migrations, mutations, and accidental death can have great effects on the genetic makeup of a small population. For example, if a town has a population of 20 people and only 2 have brown eyes and the rest have blue eyes, what happens to those 2 brown-eyed people is more critical than if the town has 20,000 people and 2,000 have brown eyes. Although the ratio of brown eyes to blue eyes is the same in both cases, even a small change in a population of 20 could significantly change the frequency of the brown-eye allele. Often, in small populations, random events can significantly alter the gene pool when rare alleles are lost from the population. This process is called **genetic drift** because the changes are not caused by selection (figure 12.6). This idea will be discussed in greater detail in chapter 13. genetic drift, p. 272

12.4 Why Genetically Distinct Populations Exist

Many species have wide geographic distribution with reasonably distinct subspecies. There are four reasons that these subspecies developed: adaptation to local environmental conditions, the founder effect, genetic bottleneck, and barriers to movement.

Adaptation to Local Environmental Conditions

Because organisms within a population are not genetically identical, some individuals may possess genetic combinations that are valuable for survival in the local environment. As a result, some individuals find the environment less hostile than do others. The individuals with unfavorable genetic combinations leave the population more often, either by death or migration, and remove their genes from the population. Therefore, local populations that occupy sites that differ greatly from conditions at other locations would be expected to consist of individuals having gene combinations suited to local conditions. For example, White Sands National Monument in New Mexico has extensive dunes of white gypsum sand. Several of the animals that live there, such as lizards and mice, have very light coloring, which allows them to blend in with their surroundings. Other populations of the same species that do not live in such a white environment do not have the light coloring.

Many kinds of animals that live in caves lack eyes and pigment. A blind fish living in a lake is at a severe disadvantage.

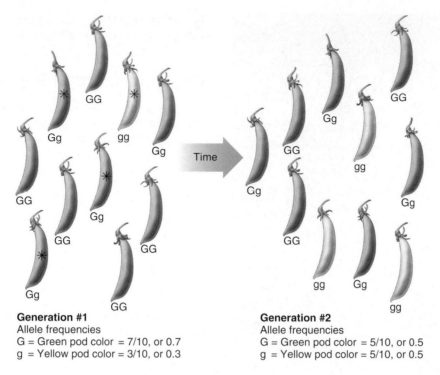

Generation #1
Allele frequencies
G = Green pod color = 7/10, or 0.7
g = Yellow pod color = 3/10, or 0.3

Generation #2
Allele frequencies
G = Green pod color = 5/10, or 0.5
g = Yellow pod color = 5/10, or 0.5

FIGURE 12.6 Genetic Drift
The gene pool of a small population may not have the same proportion of alleles as the previous generation. Notice that, in the original population of peas, the allele frequency for green pods (G) is 0.7 and for yellow pods (g) is 0.3. If among the plants chosen below, only 50% of these plants reproduce, the frequency of the allele for green pea pods will change in the next generation to 0.5 and that for yellow will be 0.5. The symbol * indicates plants that fail to reproduce. In this example, the allele frequencies changed significantly.

White Sands Earless Lizard

A blind fish living in a cave where there is no light, however, is not at the same disadvantage. Thus, these two environments might allow or encourage characteristics to be present in the two populations at different frequencies. Because it takes energy to produce eyes and pigment, individual cave fishes that do not produce these characteristics may have an advantage over the individuals that continue to spend the energy to produce these features. In addition, eyes might serve as a site for infection and those lacking eyes might have an advantage over those with eyes (figure 12.7).

FIGURE 12.7 Specialized Local Populations
Many populations of fish and other animals that live in caves where there is no light lack pigment and functional eyes. If, in the past, the genetic material that directed the production of these traits was lost or mutated, it would not have negatively affected the organism. In fact, it may be an advantage to not spend the energy to produce such unneeded traits. Hence, in many cave populations, there is a low frequency of genes for the production of eyes and pigment.

The Founder Effect

The second mechanism that creates genetically distinct populations with unique allele frequencies is the founding of a new population and, so, is called the *founder effect*. The **founder effect** is a form of genetic drift, in which a genetically distinct local population is established by a few colonizing individuals carrying with them alleles that differ in frequencies from those in their original population. The collection of alleles of a small founding population is likely to be different from that present in the larger, parent population from which it came. After all, a few individuals leaving a population would be unlikely to carry copies of all the alleles found within the original population. They may even carry an unrepresentative mixture of alleles. For example, the Lake Erie water snake that was shown in figure 12.4 may have been founded by a small number of individuals from the mainland that had a high frequency of alleles for solid coloration rather than the more typical banded pattern. (It is even possible that the island populations could have been founded by one fertilized female.) Once a small founding population establishes itself, it tends to maintain its collection of alleles, because the organisms mate only among themselves. This results in a reshuffling of alleles from generation to generation but discourages the introduction of new genetic information into the population.

Genetic Bottleneck

The third cause of local, genetically distinct populations relates to the history of the population. A **genetic bottleneck** is a form of genetic drift, in which there is a sharp reduction in population size due to a chance event that results in a reduction in genetic diversity in subsequent generations. When the size of a population is greatly reduced, some alleles will probably be lost from the population. Any subsequent increase in the size of the population by reproduction among the remaining members of the population will not replace the genetic diversity lost. For example, in the late 1800s the whooping crane population was estimated to be 1300 birds. Their numbers decreased because of hunting and industrial development and by 1941 there were only 22 wild whooping cranes left in North America. Today there are over 300 birds; however, their lack of genetic diversity puts them at risk of extinction. Thousands of species are currently undergoing genetic bottlenecks. Although some endangered species were always rare, most have experienced recent reductions in their populations and a reduction in their genetic diversity, which is a consequence of severely reduced population size.

Barriers to Movement

The fourth factor that tends to encourage the maintenance of genetically distinct populations is the presence of barriers to free movement. Animals and plants that live in lakes tend to be divided into small, separate populations by barriers of land. Whenever such barriers exist, there will very likely be differences in the allele frequencies from lake to lake, because each lake was colonized separately and the lakes' environments are not identical. Other species of organisms, such as migratory birds (e.g., robins, mallard ducks), experience few barriers; therefore, subspecies are quite rare.

12.5 Genetic Diversity in Domesticated Plants and Animals

Humans often work with small, select populations of plants and animals in order to artificially construct specific genetic combinations that are useful or desirable. This is true of plants and animals used for food. If we can produce domesticated animals and plants with genetic characteristics for rapid growth, high reproductive capacity, resistance to disease, and other desirable characteristics, we can supply ourselves with energy in the form of food. Several processes are used to develop such specialized populations of plants and animals. Most have the side effect of reducing genetic diversity.

Cloning

Recall that cloning is the process of reproducing organisms asexually, so that large numbers of genetically identical individuals are produced. These individuals are called *clones*. mitosis, p. 164

Plants are easy to work with in this manner, because we can often increase the numbers of specific organisms by asexual (without sex) reproduction. Potatoes, apple trees, strawberries, and many other plants can be reproduced by simply cutting the original plant into a number of parts and allowing these parts to sprout roots, stems, and leaves. If a single potato has certain desirable characteristics, it can be reproduced asexually. All of the individual potato plants reproduced asexually would be genetically identical and would show the same desired characteristics. Figure 12.8 shows how a clone is developed.

The cloning of most kinds of domesticated animals is much more difficult than the cloning of plants. However, in recent years, many kinds of animals have been cloned. The process involves the substitution of a nucleus from a mature animal for the nucleus of an egg. This cell is then stimulated to develop as an embryo. There is a high rate of failure, but the goals of animal cloning are the same as those of plant cloning: the production of genetically identical individuals. The first such animal produced, a sheep named Dolly, died in

(a)

(b)

(c)

(d)

(e)

FIGURE 12.8 Cloning Plants from Cuttings

All the Peperomia house plants in the last photo were produced asexually from cuttings and are identical genetically. The original plant is cut into pieces. Then, the cut ends are treated with a growth stimulant and placed in moist sand or other material. Eventually, the pieces root and become independent plants.

2003. Subsequently, monkeys, mice, and other mammals have been cloned, but the success rate is still very low.

Selective Breeding

Humans can bring together specific genetic combinations in either plants or animals by selective breeding. Because sexual reproduction tends to generate new genetic combinations rather than preserve desirable combinations, the mating of individual organisms must be controlled to obtain the desirable combination of characteristics. Selective breeding involves the careful selection of individuals with specific desirable characteristics and their controlled mating, with the goal of producing a population that has a high proportion of individuals with the desired characteristics.

Through selective breeding, some varieties of chickens have been developed that grow rapidly and are good for meat. Others have been developed to produce large numbers of eggs. Often, the development of new varieties of domesticated animals and plants involves the crossing of individuals from different populations. For this technique to be effective, the desirable characteristics in each of the two varieties should have homozygous genotypes. In small, controlled populations, it is relatively easy to produce individuals that are homozygous for a specific trait. To make two characteristics homozygous in the same individual is more difficult.

Therefore, such varieties are usually developed by crossing two different populations to collect several desirable characteristics in one organism. **Intraspecific hybrids** are organisms that are produced by the controlled breeding of separate varieties of the same species.

Occasionally, **interspecific hybrids**—hybrids between two species—are produced as a way of introducing desirable characteristics into a domesticated organism. Because plants can be reproduced by cloning, it is possible to produce an interspecific hybrid and then reproduce it by cloning. For instance,

Beefalo

the tangelo is an interspecific hybrid between a tangerine and a grapefruit. An interspecific hybrid between cattle and the American bison was used to introduce certain desirable characteristics into cattle.

Genetic Engineering

In recent years, scientific advances in understanding DNA have allowed specific pieces of genetic material to be inserted into cells. This has greatly expanded scientists' ability to modify the characteristics of domesticated plants and animals. The primary goal of genetic engineering is to manipulate particular pieces of DNA and transfer them into specific host organisms, so that they have certain valuable characteristics. These topics were dealt with in greater detail in chapter 11.

genetic engineering, p. 227

The Impact of Monoculture

Although some of the previously mentioned techniques have been used to introduce new genetic information into domesticated organisms, one of the goals of domestication is to produce organisms that have uniform characteristics. In order to achieve such uniformity, it is necessary to reduce genetic diversity. Agricultural plants have been extremely specialized through selective breeding to produce the qualities that growers want. Most agriculture in the world is based on extensive plantings of the same varieties of a species over large expanses of land. This agricultural practice is called **monoculture** (figure 12.9). It is certainly easier to manage fields in which only one kind of plant is growing, especially when herbicides, insecticides, and fertilizers are tailored to meet the needs of specific crop species. However, with monoculture comes a significant risk. Because these organisms are so similar, if a new disease comes along, most of them will be affected in the same way and the whole population may be killed or severely damaged.

Our primary food plants are derived from wild ancestors that had genetic combinations that allowed them to compete successfully with other organisms in their environment. When humans reduce genetic diversity by developing special populations with certain desirable characteristics, other valuable genetic information is lost from the gene pool. When we select specific, good characteristics, we often get harmful ones along with them. Therefore, these "special" plants and animals require constant attention. Insecticides, herbicides, cultivation, and irrigation are all used to aid the plants and animals we need. In effect, these plants are able to live only under conditions that people carefully maintain.

Because our domesticated organisms are so genetically similar, there is a great danger that an environmental change or new disease could cause great damage to our ability to produce food. In order to protect against such disasters, gene banks have been established. Gene banks consist of populations of primitive ancestors of modern domesticated plants and animals (figure 12.10). By preserving these organisms, their genetic diversity is available for introduction into our domesticated plants and animals if the need arises.

12.6 Is It a Species or Not? The Evidence

Scientists must rely on a variety of ways to identify species, because they cannot test every individual by breeding it with another to see if those individuals will have fertile offspring. Furthermore, many kinds of organisms reproduce primarily by asexual means. Because organisms that reproduce exclusively

FIGURE 12.9 Monoculture

This wheat field is an example of monoculture, a kind of agriculture in which large areas are exclusively planted with a single crop with a very specific genetic makeup. Monoculture makes it possible to use large farm machinery, but it also creates conditions that can encourage the spread of disease because the plants have reduced genetic diversity.

FIGURE 12.10 The Banking of Genes
Plant growers and breeders send genetic material, such as seeds, to the National Seed Storage Laboratory in Ft. Collins, Colorado, where it is stored at extremely cold temperatures to prevent deterioration. Gene banks will play an ever-increasing role in preserving biodiversity as the rate of extinction increases.

by asexual methods do not exchange genes with any other individuals, they do not fit the *biological species* definition very well. The standards used to identify various species include differences in morphology, behavior, metabolism, and genes.

Morphological characteristics are commonly used to differentiate species. Scientists compare the physical features of living and fossilized organisms when trying to identify a species. The idea that organisms can be classified as a species based on their structural characteristics is called the *morphological species concept*. The members of a species usually look alike, and these similarities are useful but not foolproof ways to distinguish among species. For example, the males and females of many birds look different from one another yet are the same species (figure 12.11).

Many plants have color variations or differences in leaf shape that cause them to look quite different, although they are members of the same species. Within the species, the eastern gray squirrel has black members that many people assume to be a different species because they are so different in color. A good example of the genetic diversity within a species is demonstrated by the various breeds of dogs. A Great Dane does not look very much like a Chihuahua; however, they are members of the same species (figure 12.12).

It is often difficult to distinguish among species based on morphology. For example, to most people all mosquitoes look alike, but there are many species. And, although most people think all zebras are members of the same species, there are actually three species of zebras. In some cases, experts must use detailed morphology traits, such as the vein structure in the wings of insects, to identify species.

Differences in behavior are also useful in identifying species. Some species of birds and insects are very similar structurally but can be easily identified by differences in the nature of their songs. Because it is often difficult to distin-

guish among bacteria, fungi, and other microorganisms on the basis of structure, *metabolic differences* that result in the presence or absence of specific chemicals within the organism are often used to help distinguish among species. The use of *genetic differences*—the analysis of DNA—is an even more precise way of distinguishing one species from another. Technology has allowed genes in many organisms to be sequenced and comparisons to be made. In some situations, this line of evidence has revealed remarkable differences between organisms considered to be of the same species. In other cases, it has shown extremely similar sequences in what are considered to be very different species. genetic engineering, p. 227

Because scientists use many standards—morphological, behavioral, metabolic, and genetic differences—to identify various species and none of these standards is perfect, situations frequently exist wherein individuals of two recognized species interbreed to a certain degree. For example, dogs, coyotes, and wolves have long been considered separate species. Differences in behavior and social systems tend to prevent mating among these three species. Wolves compete with coyotes and kill them when they are encountered. However, natural dog-coyote, wolf-coyote, and wolf-dog hybrids occur and the young are fertile (How Science Works 12.1). In fact, people have purposely encouraged mating between dogs and wolves for a variety of reasons. It has been demonstrated that dogs are descendants of wolves that were domesticated, so it should not be surprising that mating between wolves and dogs is easy to accomplish. Thus, because matings do occur and the offspring are fertile, should dogs and wolves be considered members of the same species? There is no simple answer.

FIGURE 12.11 Sexual Differences Within a Species
Male and female mallard ducks show strikingly different physical features.

(a)

(b)

(c)

(d)

FIGURE 12.12 Genetic Diversity in Dogs

Although these four breeds of dogs look quite different, they all have the same number of chromosomes and are capable of interbreeding. Therefore, they are considered to be members of the same species. *(a)* Great Dane, *(b)* dalmatian, *(c)* dingo, *(d)* Chihuahua. Because the extremes of these breeds rarely interbreed naturally, the question is "How long will be before they are no longer the same species?"

The species concept is an attempt to define groups of organisms that are *reproductively isolated* and, therefore, constitute a distinct unit of evolution. Some species are completely isolated from other, closely related species and do fit the definition well; some have occasional exchanges of genetic material between species and do not fit the definition as well; and some groups interbreed so much that they must be considered distinct populations of the same species. In this book, we will use the term *species*, as a population of reproductively isolated individuals, complete with the flaws and shortcomings of this definition, because it is a useful way to identify groups of organisms that have great genetic similarity and maintain a certain degree of genetic separateness from all similar organisms.

There is one other thing you need to be careful about when using the word *species*. It is both a singular and plural word, so you can talk about a single species or several species. The only way to tell how the word is being used is by assessing the context of the sentence.

12.7 Human Population Genetics

Recall from earlier in this chapter that the human gene pool consists of a number of subgroups. The particular characteristics that set one group apart from another originated many thousands of years ago, before travel was as common as it is today, and we still associate certain physical features with certain geographic areas. Although there is much more movement of people and a mixing of racial and ethnic types today, people still tend to have children with others who are of the same social, racial, and economic background and who live in the same locality.

This non-random mate selection can sometimes bring together two individuals who have alleles that are relatively rare. Information about allele frequencies within specific human subpopulations can be very important to people who wish to know the probability of having children with certain harmful genetic abnormalities. This is important if both individuals are

HOW SCIENCE WORKS 12.1

The Legal Implications of Defining a Species

The red wolf (*Canis rufus*) is an endangered species, so the U.S. Fish and Wildlife Service has instituted a captive breeding program to preserve the animal and reintroduce it to a suitable habitat in the southeastern United States, where it was common into the 1800s. Biologists have long known that red wolves hybridize with both the coyote (*Canis latrans*) and the gray wolf (*Canis lupus*), and many suspect that the red wolf is actually a hybrid between the gray wolf and coyote. Gray wolf–coyote hybrids are common in nature where one of the species is rare. Some have argued that the red wolf does not meet the definition of a species and should not be protected under the Endangered Species Act.

Museums have helped shed light on this situation by providing skulls of all three kinds of animals preserved in the early 1900s. It is known that, during the early 1900s, as the number of red wolves in the southeastern United States declined, they readily interbred with coyotes, which were very common (the gray wolf had been exterminated by the early 1900s). Some scientists believe that the skulls of the few remaining "red wolves" might not represent the true red wolf but a "red wolf" with many coyote characteristics. Studies of the structure of the skulls of red wolves, coyotes, and gray wolves show that the red wolves were recognizably different and intermediate in structure between coyotes and gray wolves. This supports the hypothesis that the red wolf is a distinct species.

DNA studies were performed using material from preserved red wolf pelts. The red wolf DNA was compared with coyote and gray wolf DNA. These studies showed that red wolves

Canis rufus

contain DNA sequences of both gray wolves and coyotes but do not appear to have distinct base sequences found only in the red wolf. This evidence supports the opinion that the red wolf is not a species but may be a population that resulted from hybridization between gray wolves and coyotes.

There is still no consensus on the status of the red wolf. Independent researchers disagree with one another and with Fish and Wildlife Service scientists, who have been responsible for developing and administering a captive breeding program and planning the reintroductions of the red wolf.

---

descended from a common ancestral tribal, ethnic, or religious group. For example, Tay-Sachs disease causes degeneration of the nervous system and the early death of children. Because it is caused by a recessive allele, both parents must pass on the allele to their child in order for the child to have the disease. By knowing the frequency of the allele in the backgrounds of both parents, the probability of their having a child with this disease can be determined.

Ashkenazi Jews have a higher frequency of the recessive allele for Tay-Sachs disease than do people of any other group of racial or social origin, and the Jewish population of New York City has a slightly higher frequency of this allele than does the worldwide population of Ashkenazi Jews (figure 12.13). Therefore, people with this background should be aware of the probability that they will have children who will develop Tay-Sachs disease, even though the allele is moving to populations other than the Ashkenazi.

Likewise, sickle-cell anemia is more common in people of specific African ancestry than in any other human subgroup (figure 12.14). Because many black slaves were brought to this country from regions where sickle-cell anemia is com-

mon, African Americans should be aware that they might carry the allele for this type of defective hemoglobin. If they carry the allele, they should consider their chances of having children with this disease. These and other cases make it very important that trained *genetic counselors* have information about allele frequencies in specific human ethnic groups, so that they can help couples with genetics questions.

12.8 Ethics and Human Population Genetics

Misunderstanding the principles of heredity and population genetics has resulted in bad public policy. Often, when there is misunderstanding there is mistrust. Even today, many prejudices against certain genetic conditions persist.

Modern genetics had its start in 1900, with the rediscovery of the fundamental laws of inheritance proposed by Mendel. For the next 40 or 50 years, this, rather simple understanding of genetics resulted in unreasonable expecta-

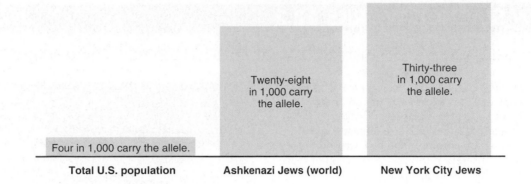

Four in 1,000 carry the allele.

Twenty-eight in 1,000 carry the allele.

Thirty-three in 1,000 carry the allele.

Total U.S. population   **Ashkenazi Jews (world)**   **New York City Jews**

FIGURE 12.13 The Frequency of the Tay-Sachs Allele

The frequency of an allele can vary from one population to another. Genetic counselors use this information to advise people of their chances of having specific alleles and of passing them on to their children.

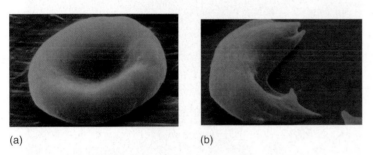

(a)   (b)

FIGURE 12.14 Normal and Sickle-Shaped Cells

Sickle-cell anemia is caused by a recessive allele, which changes one amino acid in the structure of the oxygen-carrying hemoglobin molecule within red blood cells. *(a)* Normal cells are disk-shaped. *(b)* In sickled cells, the abnormal hemoglobin molecules tend to stick to one another and distort the shape of the cells when they are deprived of oxygen.

FIGURE 12.15 A Eugenics Law

720.301 Sterilization of mental defectives; statement of policy

Sec. 1. It is hereby declared to be the policy of the state to prevent the procreation and increase in number of feebleminded and insane persons, idiots, imbeciles, moral degenerates and sexual perverts, likely to become a menace to society or wards of the state. The provisions of this act are to be liberally construed to accomplish this purpose. As amended 1962, No. 160, § 1, Eff. March 28, 1963.

This state law was enacted in 1929 and is typical of many such laws passed during the 1920s and 1930s. A basic assumption of this law is that the conditions listed are inheritable; therefore, the sterilization of affected persons would decrease the frequency of these conditions. Prior to 1962, the law also included epileptics. It was repealed in 1974.

tions on the part of both scientists and laypeople. People generally assumed that much of what a person was in terms of structure, intelligence, and behavior was inherited. This led to the passage of *eugenics laws,* whose basic purpose was to eliminate "bad genes" from the human gene pool and encourage "good genes." These laws often prevented the marriage or permitted the sterilization of people who were "known" to have "bad genes" (figure 12.15). Often, these laws were promoted as a way to save money, because sterilization would prevent the birth of future "defectives" and, therefore, would reduce the need for expensive mental institutions and prisons. Some people used these laws to legitimize racism and promote prejudice.

The writers of eugenics laws (How Science Works 12.2) overestimated the importance of genetics and underestimated the significance of environmental factors such as disease and poor nutrition. They also overlooked the fact that many genetic abnormalities are caused by recessive alleles. In most cases, the negative effects of recessive alleles can be recognized only in homozygous individuals. Removing only the homozy-

gous individuals from the gene pool would have little influence on the frequency of the "bad genes" in the population. Many recessive alleles would be masked by dominant alleles in heterozygous individuals and would continue to show up in

HOW SCIENCE WORKS 12.2

Bad Science: A Brief History of the Eugenics Movement

- **1885:** Francis Galton, cousin to Charles Darwin, proposes that human society could be improved "through better breeding." The term *eugenics* is coined; it is "the systematic elimination of undesirables to improve humanity." This would be accomplished by breeding those with "desirable" traits and preventing reproduction of those with "undesirable" traits. John Humphrey Noyes, an American sexual libertarian, molds the eugenics concept to justify polygamy: "While the good man will be limited by his conscience to what the law allows, the bad man, free from moral check, will distribute his seed beyond the legal limit."
- **1907:** Indiana is the first state to pass an involuntary sterilization law.
- **1919:** Charles B. Davenport, founder of Cold Springs Harbor Laboratory and of the Eugenics Record Office, "proves" that "pauperism" is inherited and "that being a naval officer is an inherited trait." He notes that the lack of women in the navy also "proves" that the gene is unique to males.
- **1920:** Davenport founds the American Eugenics Society. He sponsors "Fitter Families Contests," held at many state fairs around the country. The society persuades 20 state governments to authorize the sterilization of men and women in prisons and mental hospitals. The society also puts pressure on the federal government to restrict the immigration of "undesirable" races into the United States.
- **1927:** Oliver Wendel Holmes argues for the involuntary sterilization of Carrie S. Buck, an 18-year-old resident of the Virginia State Colony for Epileptics and Feeble-Minded. Buck is the first person to be selected for sterilization under the law. Buck is sterilized, even though it is later revealed that neither she nor her illegitimate daughter, Vivian, is feebleminded.
- **1931:** Involuntary sterilization measures are passed by 30 states.
- **1933–1941:** Nazi death camps, with the mass murder of Jews, Gypsies, Poles, and Russians, are established and run, resulting in the extermination of millions of people. According to the *New York Times* (August 29, 1935), "Adolf Hitler, . . . guided by the nation's anthropologists, eugenists and social philosophers, has been able to construct a comprehensive racial policy of population development and improvement. . . . It sets a pattern. . . . These ideas have met stout opposition in the Rousseauian social philosophy, . . . which bases . . . its whole social and political theory upon the patent fallacy of human equality. . . . Racial consanguinity occurs only through endogamous mating or interbreeding within racial stock . . . conditions under which racial groups of distinctly superior hereditary qualities . . . have emerged."
- **1972–1973:** Up to 4,000 sterilizations are still performed in Virginia alone, and the federal government estimates that 25,000 adults are sterilized nationwide.
- **1973:** Since March 1973, the American Eugenics Society has called itself The Society for the Study of Social Biology.

1919 Charles B. Davenport, founder of Cold Springs Harbor Laboratory and of the Eugenics Record Office, "proved" that *"pauperism"* was inherited. Also "proved that being a naval officer is an inherited trait." He noted that the lack of women in the navy also "proved" that the gene was unique to males.

- **1987:** Eugenic sterilization of institutionalized retarded persons is still permissible in 19 states, but the laws are rarely carried out. Some states enact laws that forbid the sterilization of people in state institutions.
- **Present:** Some groups and individuals still hold to the concepts of eugenics, claiming that recent evidence "proves" that traits such as alcoholism, homosexuality, and schizophrenia are genetic and, therefore, should be eliminated from the population to "improve humanity." However, the movement lacks the organization and legal basis it held in the past. Modern genetic advances, such as genetic engineering techniques and the mapping of the human genome, allow the identification of individuals with specific genetic defects. Questions about who should have access to such information and how it may be used causes renewed interest in the eugenics debate.

Genetic counseling image

future generations. In addition, we now know that most characteristics are not inherited in a simple dominant/recessive fashion; and that often many alleles at different loci cooperate in the production of a phenotypic characteristic. Furthermore, mutations occur constantly, adding new alleles to the mix. Usually, these new alleles are recessive and deleterious. Thus, essentially all individuals are carrying "bad genes."

Today, genetic diseases and the degree to which behavioral characteristics and intelligence are inherited are still important social and political issues. The emphasis, however, is on determining the specific method of inheritance and the specific biochemical pathways that result in what is currently labeled as insanity, lack of intelligence, or antisocial behavior. Although progress is slow, several genetic abnormalities have been "cured," or at least made tolerable, by medicines and control of the diet. For example, phenylketonuria (PKU) is a genetic disease caused by an abnormal biochemical pathway. If children with this condition are allowed to eat foods containing the amino acid phenylalanine, they will become mentally retarded. However, if phenylalanine is excluded from the diet, and certain other dietary adjustments are made, the children will develop normally. NutraSweet is a phenylalanine-based sweetener, so people with PKU must use caution when buying products that contain it. This abnormality can be diagnosed very easily by testing the urine of newborn infants. PKU, p. 207

Effective genetic counseling is the preferred method of dealing with genetic abnormalities. A person known to be a carrier of a "bad gene" can be told the likelihood of passing on that characteristic to the next generation before deciding whether or not to have children. In addition, *amniocentesis* (a medical procedure that samples amniotic fluid) and other tests make it possible to diagnose some genetic abnormalities early in pregnancy. If an abnormality is diagnosed, an abortion can be performed. Because abortion is unacceptable to some people, the counseling process must include a discussion of the facts about an abortion and the alternatives. Although at one

time counselors often pushed people toward specific decisions, today it is considered inappropriate for counselors to be advocates; their role is to provide information that allows individuals to make the best decisions possible for them.

Summary

All organisms with similar genetic information and the potential to reproduce are members of the same species. A species usually consists of several local groups of individuals, known as populations. Groups of interbreeding organisms are members of a gene pool. Although individuals are limited in the number of alleles they can contain, within the population there may be many different kinds of alleles for a trait. Subpopulations may have different allele frequencies from one another.

Genetically distinct populations exist because local conditions may demand certain characteristics, founding populations may have had unrepresentative allele frequencies, and barriers may prevent the free flow of genetic information from one locality to another. Distinguishable subpopulations are known as subspecies, varieties, strains, breeds, or races.

Genetic diversity is generated by mutations, which can introduce new alleles; sexual reproduction, which can generate new genetic combinations; and migration, which can subtract genetic information from, or add genetic information to, a local population. The size of the population is also important, because small populations have reduced genetic diversity.

A knowledge of population genetics is useful for plant and animal breeders and for people who specialize in genetic counseling. The genetic diversity of domesticated plants and animals has been reduced as a result of striving to produce high frequencies of valuable alleles. Clones and intraspecific hybrids are examples. Understanding allele frequencies and how they differ in various populations sheds light on why certain alleles are common in some human populations. Such understanding is also valuable in counseling members of populations with high frequencies of alleles that are relatively rare in the general population.

Key Terms

*Use the interactive flash cards on the **Concepts in Biology**, 12/e website to help you learn the meaning of these terms.*

allele frequency 241

biological species concept 240

founder effect 247

gene frequency 241

gene pool 240

genetic bottleneck 247

genetic diversity 243

genetic drift 245

interspecific hybrids 248

intraspecific hybrids 248

monoculture 249

population genetics 240

species 240

subspecies (breeds, varieties, strains, races) 243

Basic Review

1. A(n) _____ is all the individuals of the same kind of organism found within a specified geographic region and time.

2. A(n) _____ is all the alleles of all the individuals in a population.
   a. gene pool
   b. population
   c. allele pool
   d. clone

3. Which of the following does not belong?
   a. subspecies
   b. breed
   c. variety
   d. culture

4. Which of the following is a reason that genetically distinct populations exits?
   a. adaptation
   b. the founder effect
   c. cloning
   d. All of the above are correct.

5. Genetic diversity in domesticated plants and animals is affected by
   a. selective breeding.
   b. genetic engineering.
   c. cloning.
   d. All the above are correct.

6. Morphological, behavioral, metabolic, and genetic differences are all important
   a. standards used to identify species.
   b. ways of generating genetic diversity in a population.
   c. reasons for the existence of gene pools.
   d. sources of mutation.

7. A(n) _____ is a form of genetic drift in which there is a sharp reduction in population size due to a chance event that results in a reduction in genetic diversity in subsequent generations.

8. The organisms that are produced by the controlled breeding of separate varieties of the same species are often referred to as
   a. intraspecific hybrids.
   b. interspecific hybrids.
   c. mutants.
   d. clones.

9. Which of the following causes degeneration of the nervous system and the early death of children?
   a. sickle-cell anemia
   b. Tay-Sachs disease
   c. PKU
   d. eugenics

10. _____ is the term used to describe genetic differences among members of a population.

Answers
1. population   2. a   3. d   4. d   5. d   6. a
7. genetic bottleneck   8. a   9. b   10. genetic diversity

Concept Review

Answers to the Concept Review questions can be found on the Concepts in Biology, 12/e website.

12.1 Genetics in Populations

1. How do the concepts of species and genetically distinct populations differ?
2. Why does the size of a population affect the gene pool?
3. Give an example of a gene pool containing a number of separate populations.

12.2 The Biological Species Concept

4. Describe the biological species concept.
5. List three factors that change allele frequencies in a population.
6. What is meant by the terms *gene frequency* and *allele frequency*?
7. Give an example of a human characteristic that has a high frequency in Europe and a low frequency in Africa.

12.3 How Genetic Diversity Comes About

8. Why can there be greater genetic diversity within a gene pool than in an individual organism?

12.4 Why Genetically Distinct Populations Exist

9. List four processes that can lead to local, genetically distinct populations.
10. In what way are the founder effect and a genetic bottleneck similar in their effect on the genetic diversity of a local population?

12.5 Genetic Diversity in Domesticated Plants and Animals

11. How do the genetic combinations in clones and sexually reproducing populations differ?
12. How is a clone developed? What are its benefits and drawbacks?

13. How is an intraspecific hybrid formed? What are its benefits and drawbacks?
14. Why is genetic diversity in domesticated plants and animals reduced?

12.6 Is It a Species or Not? The Evidence

15. List four techniques used to distinguish one species from another.
16. Explain why the terms *reproductively isolated* and *species* are related.

12.7 Human Population Genetics

17. Give an example of a human population with a high frequency of a deleterious allele.

12.8 Ethics and Human Population Genetics

18. What forces maintain racial differences in the human gene pool?
19. What were eugenics laws? List two facts about human genetics that the advocates of eugenics failed to consider.

Thinking Critically

*For guidelines to answer this question, visit the **Concepts in Biology**, 12/e website.*

Albinism is a condition caused by a recessive allele that prevents the development of pigment in the skin and other parts of the body. Albinos need to protect their skin and eyes from sunlight. The allele has a frequency of about 0.00005. What is the likelihood that both members of a couple carry the allele? Why might two cousins or two members of a small tribe be more likely to have the allele than two nonrelatives from a larger population? If an island population has its first albino baby in history, why might it have suddenly appeared? Would it be possible to eliminate this allele from the human population? Would it be desirable to do so?

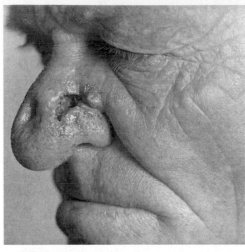

syphilis lesion

Evolution and Natural Selection

One of the more nasty sexually transmitted diseases, syphilis, has been treated since the late 1990s with the antibiotic azithromycin. Usually only four pills are needed to cure the infection. Doctors were concerned that the bacteria that cause syphilis, *Treponema pallidum*, would become resistant to the drug. In 2004, bacterial specimens were collected from patients who had been treated with azithromycin in Seattle, San Francisco, Baltimore, and Dublin, Ireland; 10% of the syphilis samples were found to be resistant to azithromycin. This shows that these populations of *Treponema pallidum* have mutations that protect them from the deadly effects of the drug. The fact that

resistant bacteria are found in cities that are geographically distant from one another demonstrates that resistant forms have become widely distributed. While taking an antibiotic, bacteria that survive because they are resistant are still capable of causing disease should they be transmitted to someone else. These resistant bacteria make up the bulk of the population infecting the new person and make their infection even harder to control with the same antibiotic. According to the Centers for Disease Control and Prevention, the incidence of syphilis decreased in the United States through the 1990s, but then climbed 19 percent from 2000 to 2003.

CHAPTER OUTLINE

13.1 The Concept of Evolution

People use the term *evolution* in many ways. We talk about the evolution of economies, fashion, and musical tastes. From a biological perspective, the word has a more specific meaning. **Evolution** is a change in the frequency of genetically determined characteristics within a population over time. Evolution can be looked at from two points of view. *Microevolution* occurs when there are minor differences in *allele frequency* between populations of the same species, as when scientists examine genetic differences between subspecies. *Macroevolution* occurs when there are major differences that have occurred over long periods that have resulted in so much genetic change that new *kinds of species* are produced (figure 13.1).

Regardless of the perspective, the way these differences are brought about are basically the same. The focus of this chapter is on the processes that result in microevolutionary change. Chapter 14 focuses on processes that lead to macroevolutionary change—that is, the development of new species. Evolution, from both perspectives, involves changes in characteristics and the genetic information that produces these characteristics over many generations.

13.2 The Development of Evolutionary Thought

For centuries, people believed that the various species of plants and animals were unchanged from the time of their creation. Although today we know this is not true, we can understand why people may have thought it was true. Because they knew nothing about DNA, meiosis, genetics, or population genetics, they did not have the tools to examine the genetic nature of species. Furthermore, the process of evolution is so slow that the results of evolution are usually not recognized during a human lifetime. It is even difficult for modern scientists to recognize this slow change in many kinds of organisms.

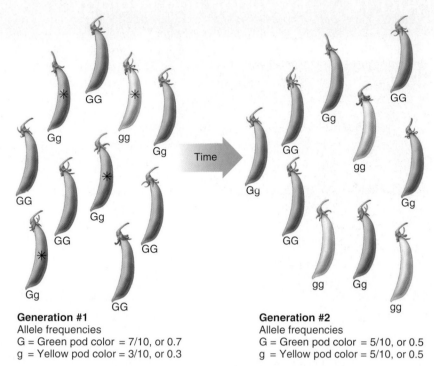

Generation #1
Allele frequencies
G = Green pod color = 7/10, or 0.7
g = Yellow pod color = 3/10, or 0.3

Generation #2
Allele frequencies
G = Green pod color = 5/10, or 0.5
g = Yellow pod color = 5/10, or 0.5

(a) **Microevolution**

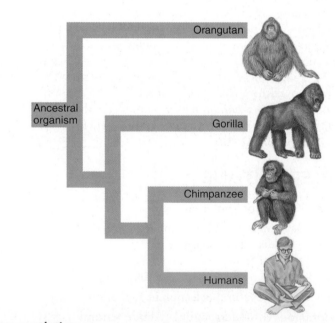

(b) **Macroevolution**

FIGURE 13.1 Microevolution and Macroevolution
(a) Microevolution occurs when gene frequencies change within the gene pool of a species. *(b)* These are relatively minor changes, compared with macroevolution changes, which result in new species from common ancestors.

Early Thinking About Evolution

In the mid-1700s, Georges-Louis Buffon, a French naturalist, wondered if animals underwent change (evolved) over time. After all, if animals didn't change, they would stay the same, and it was becoming clear from the study of fossils that changes had occurred. However, Buffon didn't come up with any suggestions on how such changes might come about. In 1809, Jean-Baptiste de Lamarck, a student of Buffon's, suggested a process by which evolution might occur. He proposed that *acquired characteristics* were transmitted to offspring. **Acquired characteristics** are traits gained during an organism's life and not determined genetically. For example, he proposed that giraffes originally had short necks but because they constantly stretched their necks to get food, their necks got slightly longer. When these giraffes reproduced, their offspring acquired their parents' longer necks. Because the offspring also stretched to eat, the third generation ended up with even longer necks. And so Lamarck's story was thought to explain why the giraffes we see today have long necks. Although we now know Lamarck's theory was wrong (because acquired characteristics are not inherited), it stimulated further thought as to how evolution might occur. From the mid-1700s to the mid-1800s, lively arguments continued about the possibility of evolutionary change. Some, like Lamarck and others, thought that change did take place; many others said that it was not even possible. It was the thinking of two English scientists that finally provided a mechanism for explaining how evolution occurs.

The Theory of Natural Selection

In 1858, Charles Darwin and Alfred Wallace suggested the theory of *natural selection* as a mechanism for evolution. The **theory of natural selection** is the idea that some individuals

Charles Darwin

Alfred Wallace

whose genetic combinations favor life in their surroundings are more likely to survive, reproduce, and pass on their genes to the next generation than are individuals who have unfavorable genetic combinations. This theory was clearly set forth in 1859 by Darwin in his book *On the Origin of Species by Means of Natural Selection, or the Preservation of Favored Races in the Struggle for Life* (How Science Works 13.1).

The theory of natural selection is based on the following assumptions about the nature of living things:

1. All organisms produce more offspring than can survive.
2. No two organisms are exactly alike.
3. Among organisms, there is a constant struggle for survival.
4. Individuals that possess favorable characteristics for their environment have a higher rate of survival and produce more offspring.
5. Favorable characteristics become more common in the species, and unfavorable characteristics are lost.

Using these assumptions, the Darwin-Wallace theory of evolution by natural selection offers a different explanation for the development of long necks in giraffes (figure 13.2):

1. In each generation, more giraffes would be born than the food supply could support.
2. In each generation, some giraffes would inherit longer necks, and some would inherit shorter necks.

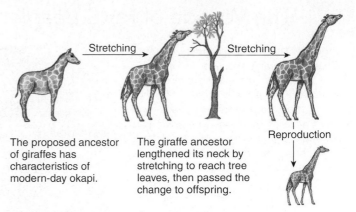

The proposed ancestor of giraffes has characteristics of modern-day okapi.

The giraffe ancestor lengthened its neck by stretching to reach tree leaves, then passed the change to offspring.

Reproduction

(a) Lamarck's theory: variation is acquired.

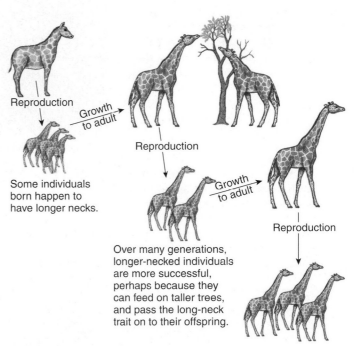

Reproduction

Growth to adult

Some individuals born happen to have longer necks.

Reproduction

Growth to adult

Reproduction

Over many generations, longer-necked individuals are more successful, perhaps because they can feed on taller trees, and pass the long-neck trait on to their offspring.

(b) Darwin's theory: variation is inherited.

FIGURE 13.2 **The Contrasting Ideas of Lamarck and the Darwin-Wallace Theory**
(a) Lamarck thought that acquired characteristics could be passed on to the next generation. Therefore, he postulated that, as giraffes stretched their necks to get food, their necks got slightly longer. This characteristic was passed on to the next generation, which would have longer necks. *(b)* The Darwin-Wallace theory states that there is variation within the population and that those with longer necks are more likely to survive and reproduce, passing on their genes for long necks to the next generation.

3. All giraffes would compete for the same food sources.
4. Giraffes with longer necks would obtain more food, have a higher survival rate, and produce more offspring.
5. As a result, succeeding generations would show an increase in the number of individuals with longer necks.

The Voyage of HMS *Beagle*, 1831–1836

Famous English naturalist Charles Darwin set forth on HMS *Beagle* in 1831, at the age of 22.

Probably the most significant event in Charles Darwin's life was his opportunity to sail on the British survey ship *Beagle*. Surveys were common at that time; they helped refine maps and chart hazards to shipping. Darwin was 22 years old and probably would not have had the opportunity, had his uncle not persuaded Darwin's father to allow him to take the voyage. Darwin was to be a gentleman naturalist and companion to the ship's captain, Robert Fitzroy. When the official naturalist left the ship and returned to England, Darwin became the official naturalist for the voyage. The appointment was not a paid position.

The voyage of the *Beagle* lasted nearly 5 years. During the trip, the ship visited South America, the Galápagos Islands, Australia, and many Pacific Islands. The *Beagle*'s entire route is shown on the accompanying map. Darwin suffered greatly from seasickness and, perhaps because of it, made extensive journeys by mule and on foot some distance inland from wherever the *Beagle* happened to be at anchor. These inland trips gave Darwin the opportunity to make many of his observations. His experience was unique for a man so young and very difficult to duplicate because of the slow methods of travel used at that time.

Although many people had seen the places that Darwin visited, never before had a student of nature collected volumes of information on them. Also, most other people who had visited these faraway places were military men or adventurers who did not recognize the significance of what they saw. Darwin's notebooks included information on plants, animals, rocks, geography, climate, and the native peoples he encountered. The natural history notes he took during the voyage served as a vast storehouse of information, which he used in his writings for the rest of his life. Because Darwin was wealthy, he did not need to work to earn a living and could devote a good deal of his time to the further study of natural history and the analysis of his notes. He was a semi-invalid during much of his later life. Many people think his ill health was caused by a tropical disease he contracted during the voyage of the *Beagle*. As a result of his experiences, he wrote several volumes detailing the events of the voyage, which were first published in 1839 in conjunction with other information related to the voyage. His volumes were revised several times and eventually were entitled *The Voyage of the Beagle*. He also wrote books on barnacles, the formation of coral reefs, how volcanos might have been involved in reef formation, and finally *On the Origin of Species*. This last book, written 23 years after his return from the voyage, changed biological thinking for all time.

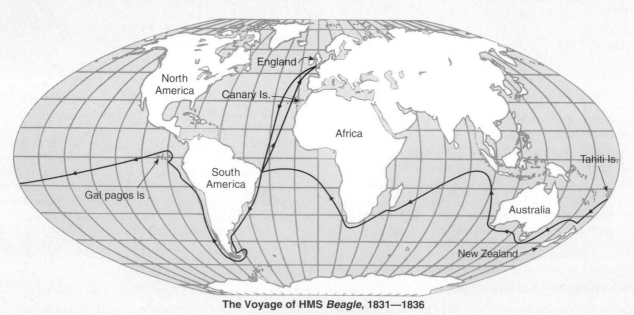

The Voyage of HMS *Beagle*, 1831—1836

0 1,000 2,000 3,000
Equatorial scale of miles

Modern Interpretations of Natural Selection

The logic of the Darwin-Wallace theory of evolution by natural selection seems simple and obvious today, but at the time Darwin and Wallace proposed their theory, the processes of meiosis and fertilization were poorly understood, and the concept of the gene was only beginning to be discussed. Nearly 50 years after Darwin and Wallace suggested their theory, the rediscovery of the work of Gregor Mendel (see chapter 10) provided an explanation for how characteristics could be transmitted from one generation to the next. Mendel's ideas of the gene explained how traits could be passed from one generation to the next. It also provided the first step in understanding mutations, gene flow, and the significance of reproductive isolation. All of these ideas are interwoven into the modern concept of evolution. If we update the five basic ideas from the thinking of Darwin and Wallace, they might look something like the following:

Garden Peas

1. An organism's ability to overreproduce results in surplus organisms.
2. Because of mutation, new, genetically determined traits enter the gene pool. Because of sexual reproduction, involving meiosis and fertilization, new genetic combinations are present in every generation. These processes are so powerful that each individual in a sexually reproducing population is genetically unique. The genetic information present is expressed in the phenotype of the organism.
3. Resources, such as food, soil nutrients, water, mates, and nest materials, are in short supply, so some individuals do without. Other environmental factors, such as disease organisms, predators, and helpful partnerships with other species, also affect survival. All the specific environmental factors that affect survival by favoring certain characteristics are called **selecting agents.**
4. Selecting agents favor individuals with the best combination of alleles—that is, those individuals are more likely to survive and reproduce, passing on more of their genes to the next generation. An organism is selected against if it has fewer offspring than other individuals that have a more favorable combination of alleles. The organism does not need to die to be selected against.
5. Therefore, alleles or allele combinations that produce characteristics favorable to survival become more common in the population and, on the average, the members of the species will be better adapted to their environment.

natural selection, p. 263

Evolution results when there are changes in allele frequency in a population. Recall that individual organisms cannot evolve—only populations can. Although evolution is a population process, the mechanisms that bring it about operate at the level of the individual.

Recall that a theory is a well-established generalization supported by many kinds of evidence. The theory of natural selection was first proposed by Charles Darwin and Alfred Wallace. Since the time it was first proposed, the theory of natural selection has been subjected to countless tests and remains the core concept for explaining how evolution occurs.

13.3   The Role of Natural Selection in Evolution

Natural selection is a primary process that brings about evolution by selecting which individuals will survive, reproduce, and pass on their genes to the next generation. These processes do not affect genes directly but do so indirectly by selecting individuals for success based on the phenotype displayed. Recall that the characteristics displayed by an organism (phenotype) are related to the genes possessed by the organism (genotype). By affecting the reproductive success of individuals, natural selection affects allele frequencies within the population. That change in allele frequency is evolution.

Three factors work together to determine how a species changes over time: *environmental factors* that affect organisms, *sexual reproduction* among the individuals in the gene pool, and the amount of *genetic diversity* within the gene

Road Kill Deer

OUTLOOKS 13.1

Common Misconceptions About the Theory of Evolution

1. **Evolution happened only in the past and is not occurring today.** In fact, there is much evidence that changes in the frequency of alleles are occurring in the populations of current species (e.g., antibiotic resistance, pesticide resistance).

2. **Evolution has a predetermined goal.** Natural selection selects the organisms that best fit the current environment. As the environment changes, so do the characteristics that have value. Random events, such as changes in sea level, major changes in climate, and collisions with asteroids, have had major influences on the subsequent natural selection and evolution.

3. **Changes in the environment cause the mutations that are needed to survive under the new environmental conditions.** Mutations are random events and are not necessarily adaptive. However, when the environment changes, mutations that were originally detrimental or neutral may have greater value. The genetic information does not change, but the environmental conditions do. In some cases, the mutation rate may increase, or there may be more frequent genetic exchanges between individuals when the environment changes, but the mutations are still random. They are not directed toward a particular goal.

4. **Individual organisms evolve.** Individuals are stuck with the genes they have inherited from their parents. Although individuals may adapt by changing their behavior or physiology, they cannot evolve; only populations can change gene frequencies.

5. **Many of the current species can be shown to be derived from other present-day species (e.g., apes gave rise to**

humans). There are few examples in which it can be demonstrated that one current species gave rise to another. Apes did not become humans, but apes and humans had a common ancestor several million years ago.

6. **Alleles that are valuable to an organism's survival become dominant.** An allele that is valuable may be either dominant or recessive. However, if it has a high value for survival, it will become *common* (more frequent). Commonness has nothing to do with dominance and recessiveness.

---

pool. In general, the reproductive success of any individual within a population is determined by how well an individual's characteristics match the demands of the environment in which it lives. **Fitness** is the success of an organism in passing on its genes to the next generation, compared with other members of its population. Just because an organism reproduces doesn't make it "fit." It can be fit only in comparison with others. Individuals whose characteristics enable them to survive and reproduce better than others in their environment have greater fitness.

Genetic diversity is important because a large gene pool with great genetic diversity is more likely to contain genetic combinations that allow some individuals to adapt to a changing environment. The characteristics of an organism are not just its structural characteristics. Behavioral, biochemical, and metabolic characteristics are also important. Scientists often use behavior, DNA differences, and other chemical differences to assess evolutionary relationships among existing organisms. However, when looking at historically extinct species, they are usually confined to using structural characteristics to guide their thinking. genetic diversity, p. 243

13.4 Common Misunderstandings About Natural Selection

There are several common misinterpretations about the process of natural selection. The first involves the phrase "survival of the fittest." Individual survival is certainly important, because those that do not survive will not reproduce. However, the more important factor is the number of descendants an organism leaves. An organism that has survived for hundreds of years but has not reproduced has not contributed any of its genes to the next generation and, so, has been selected against. Therefore, the key to being the fittest is not survival alone but, rather, survival and reproduction of the more fit organisms.

FIGURE 13.3 Tree Holes as Nesting Sites
Many kinds of birds, such as this red-bellied woodpecker, nest in holes in trees. If old and dead trees are not available, they may not be able to breed. Many people build birdhouses that provide artificial tree holes to encourage such birds to nest near their homes.

Second, the phrase "struggle for life" in the title of Darwin's book does not necessarily refer to open conflict and fighting. It is usually much more subtle than that. When a resource, such as nesting material, water, sunlight, or food, is in short supply, some individuals survive and reproduce more effectively than others. For example, many kinds of birds require holes in trees as nesting places (figure 13.3). If these are in short supply, some birds are fortunate and find a top-quality nesting site, others occupy less suitable holes, and some do not find any. There may or may not be fighting for the possession of a site. If a site is already occupied, a bird may simply fly away and look for other suitable but less valuable sites. Those that successfully occupy good nesting sites will be much more successful in raising young than will those that must occupy poor sites or those that do not find any.

Similarly, on a forest floor where there is little sunlight, some small plants may grow fast and obtain light while shading out plants that grow more slowly. The struggle for life in this instance involves a subtle difference in the rate at which the plants grow. But the plants are, indeed, engaged in a struggle, and a superior growth rate is the weapon for survival.

A third common misunderstanding involves the significance of phenotypic characteristics that are gained during the life of an organism but are not genetically determined. Although such acquired characteristics may be important to an individual's success, they are not genetically determined and cannot be passed on to future generations through sexual reproduction. Therefore, acquired characteristics are not important to the processes of natural selection. Consider an excellent tennis player's skill. Although he or she may have inherited the physical characteristics of good eyesight, tallness, and muscular coordination that are beneficial to a tennis player, the ability to play a good game of tennis is acquired through practice, not through genes. An excellent tennis player's offspring will not automatically be excellent tennis players. They may inherit some of the genetically determined physical characteristics necessary to become excellent tennis players, but the skills must be acquired through practice (figure 13.4).

Humans desire a specific set of characteristics in our domesticated animals. For example, the standard for the breed of dog known as boxers is for them to have short tails. However, the alleles for short tails are rare in this breed. Consequently, their tails are amputated—a procedure called docking. Similarly, most lambs' tails are amputated. These acquired characteristics are not passed on to the next generation. Removing the tails of these animals does not remove the genetic information for tail production from their genomes, and each generation of puppies and lambs is born with long tails.

A fourth common misconception involves understanding the relationship between the mechanism of natural selection (death, reproductive success, mate choice) and the outcomes of the selection process. Although the effects of natural selection appear at the population level, the actual selecting events take place, one at a time, at the level of the individual organism.

FIGURE 13.4 Acquired Characteristics
The ability to play an outstanding game of tennis is learned through long hours of practice. The tennis skills this person acquired by practice cannot be passed on genetically to her offspring.

13.5 What Influences Natural Selection?

Now that we have a basic understanding of how natural selection works, we can look in more detail at the factors that influence it. *Genetic diversity* within a species, the degree of *genetic expression,* and the ability of most species to *reproduce excess offspring* all exert an influence on the process of natural selection.

The Mechanisms That Affect Genetic Diversity

For natural selection to occur, there must be genetic differences among the individuals of an interbreeding population of organisms. Consider what happens in a population of genetically identical organisms. In this case, it does not matter which individuals reproduce, because the same genes will be passed on to the next generation and natural selection cannot occur. However, when genetic differences exist among individuals in a population and these differences affect fitness, natural selection can take place. Therefore, it is important to identify the processes that generate genetic diversity within a population. Genetic diversity within a population is generated by the *mutation* and *migration* of organisms and by *sexual reproduction* and *genetic recombination.* genetic diversity, p. 243

Mutation and Migration

Spontaneous mutations are changes in DNA that cannot be tied to a particular factor. Mutations may alter existing genes, resulting in the introduction of entirely new genetic information into a gene pool. It is suspected that cosmic radiation or naturally occurring mutagenic chemicals might be the cause of many of these mutations. Subjecting organisms to high levels of radiation or to certain chemicals increases the rate at which mutations occur. It is for this reason that people who work with these types of materials take special safety precautions.

Naturally occurring mutation rates are low. The odds of a gene mutating are on the order of 1 in 100,000. Most of these mutations are harmful. Rarely does a mutation occur that is actually helpful. However, in populations of millions of individuals, each of whom has thousands of genes, over thousands of generations it is quite possible that a new, beneficial piece of genetic information will come about as a result of mutation. Remember that every allele originated as a modification of a previously existing piece of DNA. For example, the allele for blue eyes may be a mutated brown-eye allele, or blond hair may have originated as a mutated brown-hair allele. In a species such as corn (*Zea mays*), there are many different alleles for seed color. Each probably originated as a mutation. Thus, mutations have been very important in introducing genetic material into species over time.

For mutations to be important in the evolution of organisms, they must be in cells that give rise to gametes (eggs or

Genetic Diversity in Corn

sperm). Mutations in other cells, such as those in the skin or liver, will affect only those cells and will not be passed on to the next generation.

Recall that migration is another way in which new genetic material can enter a population. When individuals migrate into a population from some other population, they may bring alleles that were rare or absent. Similarly, when individuals leave a population, they can remove certain alleles from the population.

Sexual Reproduction and Genetic Recombination

Sexual reproduction is important in generating new genetic combinations in individuals. Although sexual reproduction does *not* generate new genetic information, it does allow for the mixing of genes into combinations that did not occur previously. Each individual entering a population by sexual reproduction carries a unique combination of genes—half donated by the mother and half donated by the father. During meiosis, unique combinations of alleles are generated in the gametes through crossing-over between homologous chromosomes and the independent assortment of nonhomologous chromosomes. This results in millions of possible genetic combinations in the gametes of any individual. When fertilization occurs, one of the millions of possible sperm unites with one of the millions of possible eggs, resulting in a genetically unique individual. meiosis, p. 180

Genetic recombination is the mixing of genes that occurs when the genes from the male are intermingled with those from the female as a result of sexual reproduction. The new individual has a set of genes that is different from that of any other organism that ever existed. When genetic recombination occurs, a new combination of alleles may give its bearer a selective advantage, leading to greater reproductive success.

Organisms that primarily use asexual reproduction do not benefit from genetic recombination. In most cases, however, when their life history is studied closely, it is apparent that they also can reproduce sexually at certain times. Organisms that reproduce exclusively by asexual methods are not able to generate new genetic combinations but still acquire new genetic information through mutation.

As scientists have learned more about the nature of species, it has become clear that genes can be moved from one organism to another that was considered to be a different species. In some cases, it appears that whole genomes can be added when the cells of two different species combine into one cell. This process of interspecific hybridization is another method by which a species can have new genes enter its population. endosymbiosis, p. 412

The Role of Gene Expression

Even when genes are present, they do not always express themselves in the same way. *For genes to be selected for or against, they must be expressed in the phenotype of the individuals possessing them.* There are many cases of genetic characteristics being expressed to different degrees in different individuals. Often, the reason for this difference in expression is unknown.

Degrees of Expression

Penetrance is a term used to describe how often an allele expresses itself. Some alleles have 100% penetrance; others express themselves only 80% of the time. For example, there is a dominant allele that causes people to have a stiff little finger. The expression of this trait results in the tendons being attached to the bones of the finger in such a way that the finger does not flex properly. This dominant allele does not express itself in every person who contains it; occasionally, parents who do not show the characteristic in their phenotype have children that show the characteristic. **Expressivity** is a term used to describe situations in which the allele is penetrant but is not expressed equally in all individuals who have it. An example of expressivity involves a dominant allele for

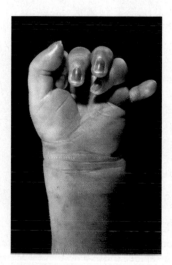

Polydactyly

six fingers or toes, a condition known as polydactyly. Some people with this allele have an extra finger on each hand; some have an extra finger on only one hand. Furthermore, some sixth fingers are well-formed with normal bones, whereas others are fleshy structures that lack bones.

Why Some Genes May Avoid Natural Selection

There are many reasons a specific allele may not feel the effects of natural selection. Some genetic characteristics can be expressed only during specific periods in the life of an organism. If an organism dies before the characteristic is expressed, it never has the opportunity to contribute to the overall fitness of the organism. Say, for example, a tree has genes for producing very attractive fruit. The attractive fruit is important because animals select the fruit for food and distribute the seeds as they travel. However, if the tree dies before it can reproduce, the characteristic may never be expressed. By contrast, genes such as those that contribute to heart disease or cancer usually have their effect late in a person's life. Because they were not expressed during the person's reproductive years, they were not selected against, because the person reproduced before the effects of the gene were apparent. Therefore, such genes are less likely to be selected against than are those that express themselves early in life.

In addition, many genes require an environmental trigger to have them expressed. If the trigger is not encountered, the gene never expresses itself. It is becoming clear that many kinds of human cancers are caused by the presence of genes that require an environmental trigger. Therefore, we try to identify triggers and prevent these negative genes from being turned on and causing disease.

When both dominant and recessive alleles are present for a characteristic, the recessive alleles must be present in a homozygous condition before they have an opportunity to express themselves. For example, the allele for albinism is recessive. There are people who carry this recessive allele but never express it, because it is masked by the dominant allele for normal pigmentation (figure 13.5).

Some genes have their expression hidden because the action of a completely unrelated gene is required before they can express themselves. The albino individual shown in figure 13.5 has alleles for dark skin and hair that will never have a chance to express themselves because of the presence of two alleles for albinism. The alleles for dark skin and hair can express themselves only if the person has the ability to produce pigment, and albinos lack that ability.

Natural Selection Works on the Total Phenotype

Just because an organism has a "good" gene does not guarantee that it will be passed on. The organism may also have "bad" genes in combination with the good, and the "good" characteristics may be overshadowed by the "bad" characteristics. All individuals produced by sexual reproduction probably have certain genetic characteristics that are extremely valuable for survival and others that are less valuable or

FIGURE 13.5 Gene Expression
Genes must be expressed to allow the environment to select for or
against them. The recessive allele *c* for albinism shows itself only in
individuals who are homozygous for the recessive characteristic.
The man in this photo is an albino who has the genotype *cc*. The
characteristic is absent in those who are homozygous dominant and
is hidden in those who are heterozygous. The dark-skinned
individuals could be either *Cc* or *CC*. However, because the albino
individual cannot produce pigment, characteristics for dark skin
and dark hair cannot be expressed.

harmful. However, natural selection operates on the total
phenotype of the organism. Therefore, it is the combination
of characteristics that is evaluated—not each characteristic
individually. For example, fruit flies may show resistance to
insecticides or lack of it, may have well-formed or shriveled
wings, and may exhibit normal vision or blindness. An individual with insecticide resistance, shriveled wings, and normal vision has two good characteristics and one negative
one, but it would not be as successful as an individual with
insecticide resistance, normal wings, and normal vision.

The Importance of Excess Reproduction

A successful organism reproduces at a rate in excess of that
necessary to merely replace the parents when they die (figure 13.6). For example, geese have a life span of about 10
years; on average, a single pair can raise a brood of about eight
young each year. If these two parent birds and all their offspring
were to survive and reproduce at this rate for a 10-year period,
there would be a total of 19,531,250 birds in the family.

However, the size of goose populations and most other
populations remains relatively constant over time. Minor
changes in number may occur but, if the environment remains
constant, a population does not experience dramatic increases
in population size. A high deathrate tends to offset the high
reproductive rate and population size remains stable. But this

is not a "static population." Although the total number of
organisms in the species may remain constant, the individuals
that make up the population change. It is this extravagant
reproduction that provides the large surplus of genetically
unique individuals that allows natural selection to take place.

If there are many genetically unique individuals within a
population, it is highly probable that some individuals will
survive to reproduce even if the environment changes somewhat, although the gene frequency of the population may be
changed to some degree. For this to occur, members of the
population must be eliminated in a non-random manner.
Even if they are not eliminated, some may have greater reproductive success than others. The individuals with the greatest
reproductive success will have more of their genetic information present in the next generation than will those that die or
do not reproduce very successfully. Those that are the most
successful at reproducing are those that are, for the most part,
better suited to the environment.

13.6 The Processes That Drive Selection

Several mechanisms allow for the selection of certain individuals for successful reproduction. If predators must pursue
swift prey organisms, the faster predators will be selected for,
and the selecting agent is the swiftness of available prey. If
predators must find prey that are slow but hard to see, the
selecting agent is the camouflage coloration of the prey, and

FIGURE 13.6 Reproductive Potential
The ability of a population to reproduce greatly exceeds the
number necessary to replace those who die. Here are some
examples of the prodigious reproductive abilities of some species.

keen eyesight is selected for. If plants are eaten by insects, the production of toxic materials in the leaves is selected for. All selecting agents influence the likelihood that certain characteristics will be passed on to subsequent generations.

Differential Survival

As stated previously, the phrase "survival of the fittest" is often associated with the theory of natural selection. Although this is recognized as an oversimplification of the concept, survival is an important factor in influencing the flow of genes to subsequent generations. If a population consists of a large number of genetically and phenotypically different individuals it is likely that some of them will possess characteristics that make their survival difficult. Therefore, they are likely to die early in life and not have an opportunity to pass on their genes to the next generation.

Charles Darwin described several species of finches on the Galápagos Islands, and scientists have often used these birds in scientific studies of evolution (figure 13.7). On one of the islands, scientists studied one of the species of seed-eating ground finches, *Geospiza fortis*. They measured the size of the animals and the size of their bills and related these characteristics to their survival. They found the following: During a drought, the birds ate the smaller, softer seeds more readily than the larger, harder seeds. As the birds consumed the more easily eaten seeds, only the larger, harder seeds remained. During the drought, mortality was extremely high. When scientists looked at ground finch mortality, they found that the larger birds with stronger, deeper bills survived better than the smaller birds with weaker, narrower bills. They also showed that the offspring of the survivors tended to show larger body and bill size as well. The lack of small, easily eaten seeds resulted in selection for larger birds with stronger bills, which could crack open larger, tougher seeds. Table 13.1 shows data on two of the parameters measured in this study.

As another example of how differential survival can lead to changed gene frequencies, consider what has happened to many insect populations as humans have subjected them to a variety of insecticides. Because there is genetic diversity within all species of insects, an insecticide that is used for the first time on a particular species kills all the exposed individuals that are genetically susceptible. However, individuals with slightly different genetic compositions and those not exposed may not be killed by the insecticide.

TABLE 13.1		
Changes in Body Structure of *Geospiza fortis*		
	Before Drought	**After Drought**
Average Body Weight	16.06 g	17.13 g
Average Bill Depth	9.21 mm	9.70 mm

Suppose that, in a population of a particular species of insect, 5% of the individuals have genes that make them resistant to a specific insecticide. The first application of the insecticide could, therefore, kill a majority of those exposed. However, tolerant individuals and those that escaped exposure would then constitute the remaining breeding population. This would mean that many more insects in the second generation would be tolerant. The second use of the insecticide on this population would not be as effective as the first. With continued use of the same insecticide, each generation would become more tolerant, because the individuals that were not tolerant were being eliminated and those that could tolerate the toxin passed on their genes for tolerance to their offspring.

Many species of insects produce a new generation each month. In organisms with a short generation time, 99% of the population could become resistant to an insecticide in just 5 years. As a result, the insecticide would no longer be useful in controlling the species. As a new selecting agent (the insecticide) is introduced into the insect's environment, natural selection results in a population that is tolerant of the insecticide.

The same kind of selection process has occurred with herbicides. Within the past 50 years, many kinds of herbicides have been developed to control weeds in agricultural fields. After several years of use, a familiar pattern develops as more and more species of weeds show resistance to the herbicide. Figure 13.8 shows several kinds of herbicides and the number of weed species that have become resistant over time. In each weed species, there has been selection for the individuals that have the genetic information that allows them to tolerate the presence of the herbicide.

Differential Reproductive Rates

Survival alone does not always ensure reproductive success. For a variety of reasons, some organisms are better able to use the available resources to produce offspring. If an individual leaves 100 offspring and another leaves only 2, the first organism has passed on more copies of its genetic information to the next generation than has the second. If we assume that all 102 offspring have similar survival rates, the first organism has been selected for, and its genes have become more common in the subsequent population.

Scientists have studied the gene frequencies for the height of clover plants. Two identical fields of clover were planted

Medium ground
finch
(*G. fortis*)

FIGURE 13.7 Darwin's Finches

Ten species of finches were described by Darwin. This is *Geospiza fortis*, a medium sized gound finch with a bill able to crush seeds which serves as its primary food.

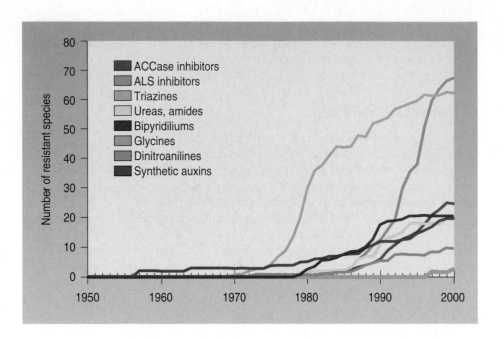

FIGURE 13.8 **Evolutionary Change**
Populations of weed plants that have been subjected repeatedly to herbicides often develop resistant populations. The individual weed plants that have been able to resist the effects of the herbicide have lived to reproduce and pass on their genes for resistance to their offspring; thus, resistant populations of weeds have developed.

and cows were allowed to graze in one of them. The cows acted as a selecting agent by eating the taller plants first. These tall plants rarely got a chance to reproduce. Only the shorter plants flowered and produced seeds. After some time, seeds were collected from both the grazed and the ungrazed fields and grown in a greenhouse under identical conditions. The average height of the plants from the ungrazed field was compared with that of the plants from the grazed field. The seeds from the ungrazed field produced some tall, some short, but mostly medium-sized plants. However, the seeds from the grazed field produced many more short plants than medium or tall ones. The cows had selectively eaten the tall plants. Because the flowers are at the tip of the plant, the tall plants were less likely to successfully reproduce, even though they were able to survive grazing by cows.

Differential Mate Choice—Sexual Selection

Sexual selection occurs within animal populations when some individuals are chosen as mates more frequently than others. Obviously, those that are frequently chosen have more opportunities to pass on more copies of their genetic information than those that are rarely chosen. The characteristics of the more frequently chosen individuals may involve general characteristics, such as body size or aggressiveness, or specific, conspicuous characteristics attractive to the opposite sex.

For example, male red-winged blackbirds establish territories in cattail marshes, where females build their nests. A male will chase out all other males, but not females. Some blackbird territories are large and others are small; some males have none. Although it is possible for any male to mate, those that have no territory are least likely to mate. Those that defend large territories may have two or more females

nesting in their territories and are very likely to mate with those females. It is unclear exactly why females choose one male's territory over another, but the fact is that some males are chosen as mates and others are not.

In other cases, it appears that females select males that display specific, conspicuous characteristics. Certain male birds, such as peacocks, have very conspicuous tail feathers (figure 13.9). Those with spectacular tails are more likely to mate and have offspring. Darwin was puzzled by such cases, because the large, conspicuous tail should have been a disadvantage to the bird. Long tails require energy to produce, make it more difficult to fly, and make it more likely that predators will capture the individual. The current theory that seeks to explain this paradox involves female choice. If the females have an innate (genetic) tendency to choose the most elaborately decorated males, genes that favor such plumage will be regularly passed on to the next generation.

Red Winged Blackbird

FIGURE 13.9 Mate Selection

In many animal species the males display very conspicuous characteristics that are attractive to females. Because the females choose the males they will mate with, those males with the most attractive characteristics will have more offspring and, in future generations, there will be a tendency to enhance the characteristic. With peacocks, those individuals with large colorful displays are more likely to mate.

13.7 Patterns of Selection

Understanding the nature of the environment is key to determining how natural selection will affect how a species will change. In general, three forms of selection have been identified: stabilizing selection, directional selection, and disruptive selection (figure 13.10).

Stabilizing Selection

Stabilizing selection occurs when individuals at the extremes of the range of a characteristic are consistently selected against. This kind of selection is very common. If the environment is stable, most of the individuals show characteristics that are consistent with the demands of the environment. For example, for many kinds of animals, there is a range of color possibilities. Suppose a population of mice has mostly brown individuals and a few white or black ones. If the white or black individuals are more conspicuous and are consistently more likely to be discovered and killed by predators, the elimination of the extreme forms will result in a continued high frequency of the brown form. Many kinds of marine animals, such as horseshoe crabs and sharks, have remained unchanged for thousands of years. The marine environment is relatively constant and probably favors stabilizing selection.

Directional Selection

Directional selection occurs when individuals at one extreme of the range of a characteristic are consistently selected for. This kind of selection often occurs when there is a consistent change in the environment in which the organism exists. For example, when a particular insecticide is introduced to control a certain species of pest insect, there is consistent selection for individuals that have alleles for resistance to the insecticide. Because of this, there is a shift in the original allele frequency, from one in which the alleles for resistance to the insecticide were rare to one in which most of the population has the alleles for resistance. Similarly, changes in climate, such as long periods of drought, can consistently select for individuals that have characteristics that allow them to survive in the dryer environment, and a change in allele frequency can result.

Disruptive selection

Disruptive selection occurs when both extremes of a range for a characteristic are selected for and the intermediate condition is selected against. This kind of selection is likely to happen when there are sharp differences in the nature of the environment where the organisms live. For example, there are many kinds of insects that feed on the leaves of trees. Many of these insects have colors that match the leaves they feed on. Suppose the species of insect ranges in color from light green

271

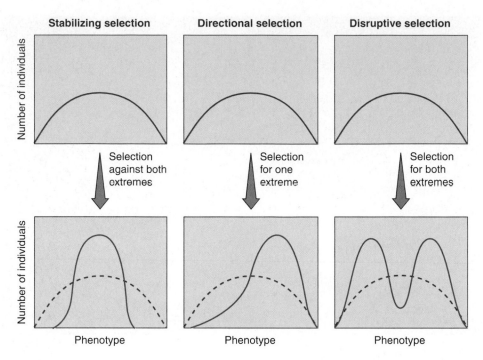

FIGURE 13.10 Patterns of Selection

In stabilizing selection, both extremes for a characteristic are selected against. Thus, gene frequencies do not change, and the original range of phenotypes is maintained. In directional selection, one extreme is selected for and the other extreme is selected against. This results in a shift in gene frequency and range of phenotype in a direction toward one extreme and away from the other. In disruptive selection, both extremes are selected for and the intermediate condition is selected against. This leads to the development of two distinct phenotypes with different gene frequencies.

to dark green, and medium green is the most common. If a particular species of insect had some individuals that fed on plants with dark green leaves, whereas other individuals fed on plants with light green leaves, medium green insects could be selected against and the two extremes selected for, depending on the kind of plant they were feeding on.

13.8 Evolution Without Selection— Genetic Drift

Recall from chapter 12 that Genetic drift is a significant change in the frequency of an allele that is not the result of natural selection. Gene-frequency differences that result from chance are more likely to occur in small populations than in large populations. Because they result from random events, such changes are not the result of natural selection or sexual selection. For example, a population of 10 organisms, of which 20% have curly hair and 80% have straight hair, is significantly changed by the death of 1 curly-haired individual. Often, the characteristics affected by genetic drift do not appear to have any adaptive value to the individuals in the population. However, in extremely small populations, vital genes may be lost.

Occasionally, a population has unusual colors, shapes, or behaviors, compared with other populations of the same

species. Such unusual occurrences are associated with populations that are small or have passed through a genetic bottleneck in the past. In large populations, any unusual shifts in gene frequency in one part of the population usually would be counteracted by reciprocal changes in other parts of the population. However, in small populations, the random distribution of genes to gametes may not reflect the percentages present in the population. For example, consider a situation in which there are 100 plants in a population and 10 have dominant alleles for patches of red color, whereas the others do not. If in those 10 plants the random formation of gametes resulted in no red alleles present in the gametes that were fertilized, the allele could be eliminated. Similarly, if all those plants with the red allele happened to be in a hollow that was subjected to low temperatures, they could be killed by a late frost and would not pass on their alleles to the next generation. Therefore, the allele would be lost, but the loss would not be the result of natural selection.

Consider the example of cougars in North America. Cougars require a wilderness setting for success. As Europeans settled the land over the past 200 years, the cougars were divided into small populations in the places where relatively undisturbed habitat still existed. The Florida panther is an isolated population of cougars found in the Everglades. The next nearest population of cougars is in Texas. Because the Florida panther is on the endangered species list, efforts have been made to ensure its continued existence in the Everglades.

However, the population is small and studies show that it has little genetic diversity. A long period of isolation and a small population created conditions that led to this reduced genetic diversity. The accidental death of a few key individuals could have resulted in the loss of valuable genes from the population. The general health of the individuals in the population is poor and reproductive success is low. In 1995, wildlife biologists began a program of introducing individuals from the Texas population into the Florida population. The purpose of the program is to reintroduce the genetic diversity lost during the long period of isolation. The program appears to be working, because there has been an increase in genetic diversity within the population (figure 13.11).

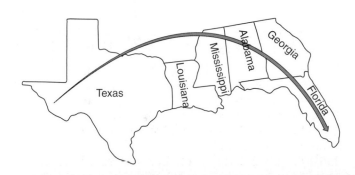

FIGURE 13.11 The Florida Panther
The Florida panther (cougar) is confined to the Everglades at the southern tip of Florida. The population is small, is isolated from other populations of cougars, and prior to 1995 had lost about 50% of its genetic diversity. In 1995, 8 female cougars from Texas were introduced into the Everglades to increase the genetic diversity of the Florida panther population. The plan seems to be working. By 2002, the population had increased to about 80 adult cougars and the population had shown about 20% genes from the Texas subpopulation. Certain obvious characteristics, such as kinked tails and cowlicks, that were common in the pre 1995 population are now less common.

13.9 Gene-Frequency Studies and the Hardy-Weinberg Concept

In the early 1900s, an English mathematician, G. H. Hardy, and a German physician, Wilhelm Weinberg, recognized that it was possible to apply a simple mathematical relationship to the study of gene frequencies. Their basic idea was that, if certain conditions existed, gene frequencies would remain constant and the distribution of genotypes could be described by the relationship $p^2 + 2pq + q^2 = 1$, where p^2 represents the frequency of the homozygous dominant genotype, $2pq$ represents the frequency of the heterozygous genotype, and q^2 represents the frequency of the homozygous recessive genotype. Constant gene frequencies over several generations would imply that evolution were *not* taking place. Changing gene frequencies would indicate that evolution were taking place.

The conditions necessary for gene frequencies to remain constant are the following:

1. Mating must be completely random.
2. Mutations must not occur.
3. The migration of individual organisms into and out of the population must not occur.
4. The population must be very large.
5. All genes must have an equal chance of being passed on to the next generation. (Natural selection is not occurring.)

The **Hardy-Weinberg concept** states that gene frequencies will remain constant if these five conditions are met. The concept is important, because it allows a simple comparison of allele frequency to indicate if genetic changes are occurring within a population. Two different populations of the same species can be compared to see if they have the same allele frequencies, or populations can be examined at different times to see if allele frequencies are changing.

Determining Genotype Frequencies

It is possible to apply the Punnett square method from chapter 10 to an entire gene pool to illustrate how the Hardy-Weinberg concept works. Consider a gene pool composed of only 2 alleles, *A* and *a*. Of the alleles in the population, 60% (0.6) are *A* and 40% (0.4) are *a*. In this hypothetical gene pool, we do not know which individuals are male or female and we do not know their genotypes. With these allele frequencies, how many of the individuals would be homozygous dominant (*AA*), homozygous recessive (*aa*), and heterozygous (*Aa*)? To find the answer, we treat these alleles and their frequencies as if they were individual alleles being distributed into sperm and eggs. The sperm produced by the males of the population will be 60% (0.6) *A* and 40% (0.4) *a*. The females will produce eggs with the same relative frequencies. We can now set up a Punnett square as shown at the top of page 274.

		Possible female gametes	
		A = 0.6	a = 0.4
Possible male gametes	A = 0.6	Genotype of offspring AA = 0.6 x 0.6 = 0.36 = 36%	Genotype of offspring Aa = 0.6 x 0.4 = 0.24 = 24%
	a = 0.4	Genotype of offspring Aa = 0.4 x 0.6 = 0.24 = 24%	Genotype of offspring aa = 0.4 x 0.4 = 0.16 = 16%

The Punnett square gives the frequency of occurrence of the three possible genotypes in this population: $AA = 36\%$, $Aa = 48\%$, and $aa = 16\%$.

If we use the relationship $p^2 + 2pq + q^2 = 1$, p^2 is the frequency of the AA genotype, $2pq$ is the frequency of the Aa genotype, and q^2 is the frequency of the aa genotype. Then, $p^2 = 0.36$ and p would be the square root of 0.36, which is 0.6—our original frequency for the A allele. Similarly, $q^2 = 0.16$ and q would be the square root of 0.16, which is 0.4. In addition, $2pq$ would equal $2 \times 0.6 \times 0.4 = 0.48$. If this population were to reproduce randomly, it would maintain an allele frequency of 60% A and 40% a alleles. It is important to understand that Hardy-Weinberg conditions rarely exist; therefore, there are usually changes in gene frequency over time or genetic differences in separate populations of the same species. If gene frequencies are changing, evolution is taking place.

Why Hardy-Weinberg Conditions Rarely Exist

Random mating does not occur for a variety of reasons. Many species are divided into smaller portions, which that are isolated from one another so that no mating with other segments occurs during the lifetime of the individuals. In human populations, these isolations may be geographic, political, or social. In addition, some individuals may be chosen as mates more frequently than others because of the characteristics they display. Therefore, the Hardy-Weinberg conditions are seldom met, because non-random mating is a factor that leads to changing gene frequencies.

Spontaneous mutations occur. Totally new kinds of alleles are introduced into a population, or 1 allele is converted into another, currently existing allele. Whenever an allele is changed, 1 allele is subtracted from the population and a different allele is added, thus changing the allele frequency in the gene pool. Mutations in disease-causing organisms may have significant impacts (Outlooks 13.2).

Immigration and emigration of individual organisms are common. When organisms move from one population to another, they carry their genes with them. Their genes are subtracted from the population they left and added to the population they enter, thus changing the gene pool of both populations. It is important to understand that migration is common for plants as well as animals. In many parts of the world, severe weather disturbances have lifted animals and plants (or their seeds) and moved them over great distances, isolating them from their original gene pool. In other instances, organisms have been distributed by floating on debris on the

Migrating Dandelion Seeds

surface of the ocean. As an example of how important immigration and emigration, consider the tiny island of Surtsey (3 km²), which emerged from the sea as a volcano near Iceland in 1963 and continued to erupt until 1967. The new island was declared a nature preserve and has been surveyed regularly to record the kinds of organisms present. The nearest possible source of new organisms is about 20 kilometers away. The first living thing observed on the island was a fly seen less than a year after the initial eruption. By 1965, the first flowering plant had been found and, by 1996, 50 species of flowering plants had been recorded on the island. In addition, several kinds of sea birds nest on the island.

Populations are not infinitely large, as assumed by the Hardy-Weinberg concept. If numbers are small, random events to a few organisms might alter gene frequencies from what was expected. Consider coin flipping as an analogy. Coins have two surfaces, so, if you flip a coin once, there is a 50:50 chance that the coin will turn up heads. If you flip two coins, you may come up with two heads, two tails, or one head and one tail. Only one of these possibilities gives the theoretical 50:50 ratio. To come closer to the statistical probability of flipping 50% heads and 50% tails, you would need to flip many coins at the same time. The more

OUTLOOKS 13.2

The Development of New Viral Diseases

Mutation is necessary if evolution is to take place. As parasites and their hosts interact, they constantly react to each other in an evolutionary fashion. Hosts develop new mechanisms to combat parasites, and parasites develop new mechanisms to overcome the hosts' defenses. One of the mechanisms viruses use is a high rate of mutation. This ability to mutate has resulted in many new, serious human diseases. In addition, many new diseases arise when viruses that cause disease in another animal are able to establish themselves in humans. Many kinds of influenza originated in pigs, ducks, or chickens and were passed to humans through close contact with infected animals or by eating infected animals. In many parts of the world, these domesticated animals live in close contact with humans (often in the same building), making conditions favorable for the transmission of animal viruses to humans.

Each year, mutations result in new varieties of influenza and colds, which pass through the human population. Occasionally, the new varieties are deadly. In 1918, a new variety of influenza virus originated in pigs in the United States and spread throughout the world. During the 1918–1919 influenza pandemic that followed, 20 to 40 million people died. In 1997 in Hong Kong, a new kind of influenza was identified that killed 6 of the 18 people infected. When public health officials discovered the virus had come from chickens, they ordered the slaughter of all the live chickens in Hong Kong, which stopped the spread of the disease.

In early 2003, an outbreak of a new viral disease, known as *severe acute respiratory syndrome (SARS)*, originated in China. SARS is a variation of a coronavirus, a class of virus commonly associated with the common cold, but it causes severe symptoms and, if untreated, can result in death. In June 2003, the SARS virus was isolated from an animal known as the masked palm civet (*Paguma larvata*). This animal is used for food in China and is a possible source of the virus that caused SARS in humans. However, other animals have also tested positive for the virus. The disease spread rapidly to several countries as people traveled by airplane from China to other parts of the world. A recognition of the seriousness of the disease and the isolation of infected persons prevented further spread, and this new disease was brought under control. However, if the virus still exists in some unknown wild animal host, it could reappear in the future.

In addition to cold and influenza, other kinds of diseases often make the leap from nonhuman to human hosts. AIDS is a disease caused by the human immunodeficiency virus (HIV), which have been traced to a similar virus in monkeys and in June 2003, public health officials announced that the monkey-pox virus was causing disease in people in Wisconsin, Illinois, and Indiana. The monkeypox virus was traced to a shipment of prairie dogs that were sold as pets. It is likely that the prairie dogs became infected with the virus through contact with an infected Gambian giant rat, which was part of a shipment of animals from Ghana in April 2003. Thus, it appears that the disease entered North America from Africa by way of the Gambian giant rat and was transferred to prairie dogs and subsequently to people.

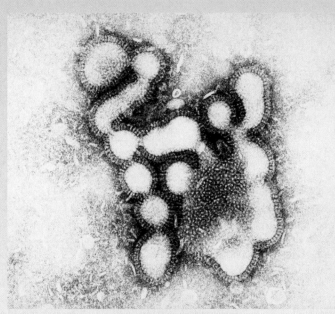

Influenza Virus at 295,000 Magnification

---

coins you flip, the more likely it is that you will end up with 50% of all the coins showing heads and the other 50% showing tails. The number of coins flipped is important. The same is true of gene frequencies. The smaller the population, the more likely it is that random events will alter the gene pool—that is, the more likely it is that genetic drift will occur.

Finally, *genes are not all equally likely to be passed to the next generation.* It is important to understand that genes differ in their value to the species. Some genes result in characteristics that are important to survival and reproductive success. Other genes reduce the likelihood of survival and reproduction. For instance, many animals have cryptic color patterns that make them difficult to see. The genes that determine the cryptic color pattern would be selected for (favored), because animals that are difficult to see are not killed and eaten as often as those that are easy to see. Recall that albinism is the inability to produce pigment, so that the individual's color is white. White animals are conspicuous, so we might expect them to be discovered more easily by predators (figure 13.12). Because not all genes have equal value, natural selection will operate and some genes will be more likely to be passed on to the next generation.

FIGURE 13.12 Albino Animals in the Wild
Predators are more likely to spot an albino than a member of the species with normal coloration. Albinism prevents the prey from blending in with its surroundings.

Using the Hardy-Weinberg Concept to Show Allele-Frequency Change

We can return to our original example of alleles A and a to determine how natural selection based on differences in survival can result in allele-frequency changes in only one generation. Again, assume that the parent generation has the following genotype frequencies: $AA = 36\%$, $Aa = 48\%$, and $aa = 16\%$, with a total population of 100,000 individuals. Suppose that 50% of the individuals having at least one A allele do not reproduce because they are more susceptible to disease. The parent population of 100,000 would have 36,000 individuals with the AA genotype, 48,000 with the Aa genotype, and 16,000 with the aa genotype. Because only 50% of those with an A allele reproduce, only 18,000 AA individuals and 24,000 Aa individuals will reproduce. All 16,000 of the aa individuals will reproduce, however. Thus, there is a total reproducing population of only 58,000 individuals out of the entire original population of 100,000. What percentage of A and a will go into the gametes produced by these 58,000 individuals?

The percentage of A-containing gametes produced by the reproducing population will be 31% from the AA parents and 20.7% from the Aa parents (table 13.2). The frequency of the A allele in the gametes is 51.7% (31% + 20.7%). The percentage of a-containing gametes is 48.3% (20.7% from the Aa parents plus 27.6% from the aa parents). The original parental allele frequencies were $A = 60\%$ and $a = 40\%$. These have changed to $A = 51.7\%$ and $a = 48.3\%$. More individuals in the population will have the aa genotype, and fewer will have the AA and Aa genotypes.

If this process continued for several generations, the allele frequency would continue to shift until the A allele became rare in the population (figure 13.13). This is natural selection in action. Differential reproduction rates have changed the frequency of the A and a alleles in this population.

TABLE 13.2

Differential Reproduction

The percentage of each genotype in the offspring differs from the percentage of each genotype in the original population as a result of differential reproduction.

Original Frequency of Genotypes	Total Number of Individuals Within a Population of 100,000 with Each Genotype	Number of Each Genotype Not Reproducing Subtracted from the Total	Total of Each Genotype in the Reproducing Population of 58,000 Following Selection	New Percentage of Each Genotype in the Reproducing Population
$AA = 36\%$	36,000	36,000 −18,000 18,000	18,000	$\dfrac{18,000}{58,000} = 31.0\%$
$Aa = 48\%$	48,000	48,000 −24,000 24,000	24,000	$\dfrac{24,000}{58,000} = 41.4\%$
$aa = 16\%$	16,000	16,000 − 0 16,000	16,000	$\dfrac{16,000}{58,000} = 27.6\%$
100%	100,000		58,000	100.0%

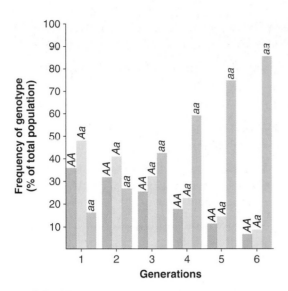

FIGURE 13.13 Changing Allele Frequency

If 50% of all individuals with the genotypes *AA* and *Aa* do not reproduce in each generation, the frequency of the *a* allele will increase, whereas the frequency of the *A* allele will decrease. Consequently, the *aa* genotype will increase in frequency, whereas that of the *AA* and *Aa* genotypes will decrease.

13.10 A Summary of the Causes of Evolutionary Change

At the beginning of this chapter, evolution was described as a change in allele frequency over time. It is clear that several mechanisms operate to bring about this change. Mutations can either change one allele into another or introduce an entirely new piece of genetic information into the population.

Immigration can introduce new genetic information if the entering organisms have genetic information that was not in the population previously. Emigration and death remove genes from the gene pool. Natural selection systematically filters some genes from the population, allowing other genes to remain and become more common. The primary mechanisms involved in natural selection are differences in deathrates, reproductive rates, and the rate at which individuals are selected as mates (figure 13.14). In addition, gene frequencies are more easily changed in small populations, because events such as death, immigration, emigration, and mutation can have a greater impact on a small population than on a large population.

Summary

At one time, people thought that all organisms were unchangeable. Lamarck suggested that change does occur and thought that acquired characteristics could be passed from generation to generation. Darwin and Wallace proposed the theory of natural selection as the mechanism that drives evolution. All populations of sexually reproducing organisms naturally exhibit genetic diversity among individuals as a result of mutation and the genetic recombination resulting from meiosis and fertilization. The genetic differences are reflected in phenotypic differences among individuals. These genetic differences are important in changing environments, because natural selection must have genetic diversity to select from. Natural selection by the environment results in better-suited organisms having greater numbers of offspring than those that are less well off genetically. Not all genes are equally expressed. Some express themselves only during specific periods in the life of

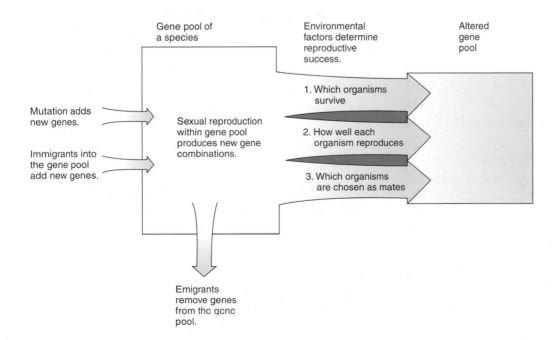

FIGURE 13.14 Processes That Influence Evolution

Several processes cause gene frequencies to change. New genetic information enters populations through immigration and mutation. Genetic information leaves populations through death and emigration. Natural selection operates within populations through death, mate selection, and rates of reproduction. Genetic drift can also result in evolutionary change but is not shown in this diagram.

an organism, and some are recessive alleles that show themselves only when in the homozygous state. Characteristics that are acquired during the life of an individual and are not determined by genes cannot be raw material for natural selection. Sexual selection occurs when specific individuals are chosen as mates to the exclusion of others. In addition to natural selection and sexual selection, genetic drift in small populations can lead to evolutionary change.

Selecting agents change the gene frequencies of the population if the conditions of the Hardy-Weinberg concept are violated. The conditions required for the Hardy-Weinberg equilibrium are random mating, no mutations, no migration, large population size, and no selective advantage for any genes. These conditions are met only rarely, however, so that, after generations of time, the genes of the more favored individuals will make up a greater proportion of the gene pool. The process of natural selection allows the maintenance of a species in its environment, even as the environment changes.

Key Terms

Use the interactive flash cards on the **Concepts in Biology,** *12/e website to help you learn the meaning of these terms.*

acquired characteristics 260
directional selection 271
disruptive selection 271
evolution 260
expressivity 267
fitness 264
genetic recombination 266

Hardy-Weinberg concept 273
penetrance 267
selecting agents 263
sexual selection 270
spontaneous mutations 266
stabilizing selection 271
theory of natural selection 261

Basic Review

1. Which of the following is not true?
   a. All organisms produce more offspring than can survive.
   b. All organisms of the same species are exactly alike.
   c. Among organisms, there is a constant struggle for survival.
   d. Individuals that possess favorable characteristics for their environment have a higher rate of survival and produce more offspring.

2. _____ characteristics are traits gained during an organism's life and not determined genetically.
   a. Genetic
   b. Acquired
   c. Sexual
   d. Dominant

3. Who proposed the theory of natural selection?
   a. Darwin and Wallace
   b. Buffon
   c. Lamarck
   d. Hardy and Weinberg

4. _____ is the success of an organism in passing on its gene to the next generation, compared with other members of its population.

5. _____ occurs within animal populations when some individuals are chosen as mates more frequently than others.
   a. Stabilizing selection
   b. Disruptive selection
   c. Sexual selection
   d. Normative selection

6. _____ is how often an allele expresses itself when present.
   a. Expressivity
   b. Dominance
   c. Fitness
   d. Penetrance

7. Specific environmental factors that favor certain characteristics are called
   a. selecting agents.
   b. hurricanes.
   c. mutations.
   d. disruptive factors.

8. _____ is a significant change in the frequency of an allele that is not the result of natural selection.

9. The conditions necessary for gene frequencies to remain constant include all the following except
   a. mating must be completely random.
   b. mutations must not occur.
   c. the migration of individual organisms into and out of the population must not occur.
   d. the population must be very small.

10. _____ occurs when there are minor differences in allele frequency between populations of the same species, as when genetic differences between subspecies are examined.

Answers
1. b 2. b 3. a 4. Fitness 5. c 6. d 7. a
8. Genetic drift 9. d 10. Microevolution

Concept Review

Answers to the Concept Review questions can be found on the Concepts in Biology, 12/e website.

13.1 The Concept of Evolution

1. Describe the biological meaning of the word *evolution*.

13.2 The Development of Evolutionary Thought

2. Why has Lamarck's theory been rejected?
3. List five assumptions about the nature of living things that support the concept of evolution by natural selection.

13.3 The Role of Natural Selection in Evolution

4. Define *natural selection*.
5. What is fitness, and how is it related to reproduction?

13.4 Common Misunderstandings About Natural Selection

6. Why are acquired characteristics of little interest to evolutionary biologists?
7. In what way are the phrases "survival of the fittest" and "struggle for existence" correct? In what ways are they misleading?

13.5 What Influences Natural Selection?

8. What factors can contribute to diversity in the gene pool?
9. Why is overreproduction necessary for evolution?
10. Why is sexual reproduction important to the process of natural selection?
11. How might a harmful allele remain in a gene pool for generations without being eliminated by natural selection?

13.6 The Processes That Drive Natural Selection

12. A gene pool has equal numbers of alleles *B* and *b*. Half of the *B* alleles mutate to *b* alleles in the original generation. What will the allele frequencies be in the next generation?
13. List three factors that can lead to changed gene frequencies from one generation to the next.

14. Give two examples of selecting agents and explain how they operate.

13.7 Patterns of Selection

15. Distinguish among stabilizing, directional, and disruptive selection.

13.8 Evolution Without Selection—Genetic Drift

16. Why is genetic drift more likely in small populations?
17. Give an example of genetic drift.

13.9 Gene-Frequency Studies and the Hardy-Weinberg Concept

18. The Hardy-Weinberg concept is only theoretical. What factors do not allow it to operate in a natural gene pool?
19. The smaller the population, the more likely it is that random changes will influence gene frequencies. Why is this true?

13.10 A Summary of the Causes of Evolutionary Change

20. Why is each of the following important for an understanding of evolution: mutation, migration, sexual reproduction, selective agents, and population size?

Thinking Critically

For guidelines to answer this question, visit the Concepts in Biology, 12/e website.

Penicillin was introduced as an antibiotic in the early 1940s. Since that time, it has been found to be effective against the bacteria that cause gonorrhea, a sexually transmitted disease. The drug acts on dividing bacterial cells by preventing the formation of a new protective cell wall. Without the wall, the bacteria can be killed by normal body defenses. Recently, a new strain of this disease-causing bacterium has been found. This bacterium produces an enzyme that metabolizes penicillin. How can gonorrhea be controlled now that this organism is resistant to penicillin? How did a resistant strain develop? Include the following in your consideration: DNA, enzymes, selecting agents, and gene-frequency changes.

14

The Formation of Species and Evolutionary Change

Extinction has been a common occurrence, affecting plants, animals, and microbes alike. Some biologists have speculated that as high as 99% of all animals ever to have existed on Earth are now extinct. However, every once in a while, an organism thought to have been extinct shows up unexpectedly. Once such animal, a mammal known as the almiqui, was thought to have been driven to extinction as a result of urban development, deforestation, and the importation of dogs to the largest Caribbean island, Cuba. The almiqui was first discovered in 1861. It has been seen only a few times since then. Only 37 of these animals have ever been captured since their discovery; three were captured in 1974 and 1975. The most recent was captured in 2001. These animals can weigh up to 1 kg. Nocturnal members of the family Solenodontidae (solenodons), they

feed on insects, millipedes, earthworms, and termites. There is only one surviving genus, *Solenodon*, and two species. Many animals that require very specific environmental conditions and foods are unable to find suitable living habitats as humans develop more and more land for agriculture and housing. The inability to adapt to new or changing conditions can result in extinction. If a population changes to adapt better to a new environment over time, the population is evolving.

How is it that an animal such as the almiqui manages to survive on an island undergoing urbanization? Is the almiqui evolving or is it on the verge of extinction? Why do so many organisms experience extinction? This chapter explores the concepts surrounding natural selection, evolution, and extinction.

CHAPTER OUTLINE

14.1 Evolutionary Patterns at the Species Level

Chapter 13 focused on the concept of microevolution—that is, minor differences in allele frequency between populations of the same species, as when genetic differences between subspecies are examined. This chapter focuses on macroevolution, the major differences that have occurred over long periods that have resulted in so much genetic change that new kinds of species are produced. Recall that a species is a population of organisms whose members have the potential to interbreed naturally and to produce fertile offspring but *do not* interbreed with other groups.

California Condor

There are two key ideas within this definition. First, a species is a population of organisms. An individual is not a species. An individual can only be a member of a group that is a species. The human species, *Homo sapiens,* consists of over 6.5 billion individuals, whereas the endangered California condor species, *Gymnogyps californianus,* consists of about 200 individuals.

Second, the definition takes into consideration the ability of individuals within the group to produce fertile offspring. Obviously, not every individual can be checked to see if it is capable of mating with any other individual that is similar to it, so we must make some judgment calls. Can most individuals within the population interbreed to produce fertile offspring? Although all humans are of the same species, some individuals are sterile and cannot reproduce. However, we don't exclude them from the human species because of this. If they were not sterile, they would have the potential to inter-

breed. Although humans normally choose mating partners from their own local ethnic groups, humans from all parts of the world are potentially capable of interbreeding. This is known to be true because of the large number of instances of reproduction involving people of different ethnic backgrounds. The same is true for many other species that have local subpopulations but have a wide geographic distribution, so how do we know if two populations really do belong to the same species?

Gene Flow

One way to find out if two populations belong to the same species is to think about **gene flow,** the movement of genes from one generation to the next as a result of reproduction or from one region to another by migration. Two or more populations that demonstrate gene flow among them constitute a single species. On the other hand, two or more populations that do not have gene flow through reproduction, when given the opportunity, are generally considered to be different species. Some examples will clarify this working definition.

Donkeys and horses are thought to be two different species, even though they can be mated and produce offspring, called mules (figure 14.1). Because mules are nearly always sterile and do not produce offspring, this is not considered to be gene flow, so donkeys and horses are considered separate species. Similarly, lions and tigers can be mated in zoos to produce offspring. However, this does not happen in nature, so gene flow does not occur naturally; thus, they are also considered two separate species.

Genetic Similarity

Another way to find out if two organisms belong to different species is to determine their degree of genetic similarity. Advances in molecular genetics have allowed scientists to

(a)

(b)

(c)

FIGURE 14.1 Hybrid Sterility

Even though they do not do so in nature, *(a)* horses and *(b)* donkeys can be mated. The offspring is called a *(c)* mule and is sterile. Because the mule is sterile, the horse and the donkey are considered to be of different species.

Akodon field mouse

examine the sequence of bases in the genes present in individuals from a variety of populations. Those that have a great deal of similarity in their nitrogenous base sequences are assumed to have resulted from populations that have exchanged genes through sexual reproduction in the recent past. If there are significant differences, the two populations have probably not exchanged genes recently and are more likely to be members of separate species. There is no universal rule that states the smallest allowable genetic difference between two species. For example, genetic similarity between two South American field mice, *Akodon dolores* and *A. molinae,* have been analyzed to determine if they are, in fact, members of the same species located in different geographic regions. Experts examining the genetic differences concluded that they are the same species but different geographic races. The interpretation of the results obtained by examining genetic differences still requires the judgment of experts. Although this technique is probably not the ultimate method to settle every dispute related to the identification of species, it is an important tool.

14.2 How New Species Originate

Speciation is the process of generating new species. When evolutionary biologists look at the evolutionary history of living things, they see that new species have arisen continuously for as long as life has been on Earth. *Fossils* are often used as evidence of the past evolution of organisms. A **fossil** is any evidence of an organism of a past geologic age, such as a preserved skeleton or body imprint. The fossil record shows that huge numbers of new species have originated and that most species have gone extinct. Two mechanisms are probably responsible for the vast majority of speciation events: geographic isolation, and polyploidy (instant speciation).

Speciation by Geographic Isolation

Geographic isolation occurs when a portion of a species becomes totally isolated from the rest of the gene pool by geographic distance. In order for geographic isolation to lead to speciation, the following steps are necessary: (1) the *population isolation* of a subpopulation of a species; (2) *genetic divergence*—that is, a change in the allele frequencies of the isolated subpopulation from the rest of the species; and (3) *reproductive isolation* of the new species from its parent species.

Geographic Isolation

There are at least three ways that populations can become geographically isolated (figure 14.2). First, the *colonization of a distant area* by one or only a few individuals can lead to the establishment of a population far from the center of their

FIGURE 14.2 **Geographically Isolated Populations**
(a) The colonization of distant areas by one or a few individuals can establish a population far from the center of their home population. *(b)* Barriers to movement can split an ancestral population into two isolated groups. *(c)* Extinction of intermediate populations can leave the remaining populations reproductively isolated from one another.

home population. If these colonies are so far from their home populations that there is no gene flow between them, they are also genetically isolated.

Second, speciation occurs if as *geographic barrier* totally isolates a subpopulation from the rest of the species. The uplifting of mountains, the rerouting of rivers, and the formation of deserts may separate one portion of a gene pool from another. For example, two kinds of squirrels are found on opposite sides of the Grand Canyon. Some people consider them to be separate species; others consider them to be different, isolated subpopulations of the same species. Even small changes can cause geographic isolation in species that have little ability to move. A fallen tree, a plowed field, or even a new freeway may effectively isolate populations within such species. Snails in two valleys separated by a high ridge have been found to be closely related but different species. The snails cannot get from one valley to the next because of the height and climatic differences presented by the ridge.

Third, the *extinction of intermediate populations* can leave the remaining populations reproductively isolated from one another for periods long enough for them to develop into separate species. For example, the *range* of an organism is the geographic area over which a species can be found. As a species expands its range, some intermediate populations may go extinct so that portions of the original population become separated from the rest. Thus, many species are made up of several, smaller populations that display characteristics significantly different from those of other local populations. Many of these differences are adaptations to local environmental conditions. If one of these smaller subpopulation in the middle becomes extinct, the distance between the more extreme isolated populations may to great for sexual reproduction to occur.

Genetic Diversity

Genetic divergence is necessary for new species to develop. Differences in environments and natural selection play very important roles in the process of forming new species. Following separation from the main portion of the gene pool by geographic isolation, the organisms within a small, local population are likely to experience different environmental conditions. If, for example, a mountain range has separated a species into two populations, one population may receive more rain or more sunlight than the other. These environmental differences act as natural selecting agents on the two gene pools and, over a long period of time, account for different genetic combinations in the two places. Furthermore, different mutations may occur in the two isolated populations, and each may generate unique combinations of genes as a result of sexual reproduction. This is particularly true if one of the populations is very small. As a result, the two populations may show differences in color, height, enzyme production, time of seed germination, or many other characteristics. Over a long period of time, the genetic differences that accumulate may result in subspecies that are significantly modified structurally, physiologically, or behaviorally. In some cases the changes may be so great that new species result.

Reproductive isolation, or *genetic isolation,* has occurred if the genetic differences between the two populations have become so great that reproduction cannot occur between members of the two populations if they are brought together. At this point, speciation has occurred. In other words, the process of speciation can begin with the geographic isolation of a portion of the species, but new species are generated only if isolated populations become separate from one another *genetically* and gene flow is not reestablished if the geographic barrier is removed.

Polyploidy: Instant Speciation

Another important mechanism known to generate new species is polyploidy. **Polyploidy** is a condition of having multiple sets of chromosomes, rather than the normal haploid or

FIGURE 14.3 Speciation Without Geographic Isolation
Mutations of some individuals could lead to a change in the time of flower production in a population of plants. Should these mutant plants not release pollen at a time when others in the population are fertile, they would not be able to interbreed with their neighbors. Given enough time, such reproductively isolated plants could lead to the evolution of a new species, even though they are not geographically isolated.

diploid number. The increase in the number of chromosomes can result from abnormal mitosis or meiosis in which the chromosomes do not separate properly. For example, if a cell had the normal diploid chromosome number of six ($2n = 6$), and the cell went through mitosis but did not divide into two cells, it would then contain 12 chromosomes. It is also possible that a new polyploid species could result from crosses between two species followed by a doubling of the chromosome number (figure 14.3). Because the number of chromosomes of the polyploidy is different from that of the parent, successful reproduction between the polyploid and the parent is usually not possible. This is because meiosis would result in gametes that had chromosome numbers different from those of the original, parent organism. In one step, the polyploid could be isolated reproductively from its original species.

A single polyploid plant does not constitute a new species. However, because most plants can reproduce asexually, they can create an entire population of organisms that have the same polyploid chromosome number. The members of this population would probably be able to undergo normal meiosis and would be capable of sexual reproduction among themselves. In effect, a new species can be created within a couple of generations. Some groups of plants, such as the grasses, may have 50% of their species produced as a result of polyploidy. Many economically important species are polyploids. Cotton, potatoes, sugarcane, broccoli, wheat, and many garden flowers are examples. Although it is rare in animals, polyploidy is found in some insects, fishes, amphibians,

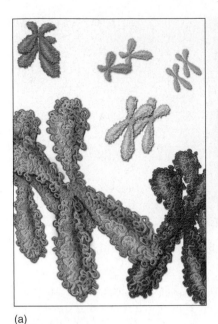
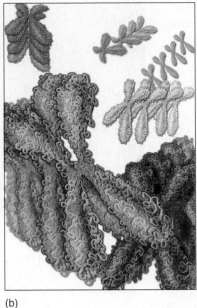

(a)   (b)   (c)

FIGURE 14.4 Polyploidy

(a) The chromosomes of cell is, $2n = 26$. *(b)* What happens when such a cell undergoes polyploidy and doubles its chromosome count—that is, $4n = 52$. *(c)* Evidence is very strong that the ancestors of today's sugarcane (*Saccharum officinarum*) plants had a haploid number of 10, $n = 10$. As a result of polyploidy, sugarcane plants have 8 times that number, $8n = 80$.

and reptiles. Certain lizards have only female individuals and lay eggs, which develop into additional females. Various species of these lizards appear to have developed by polyploidy. To date, the only mammal found to be polyploid is a rat (*Tympanoctomys barrerae*) found in Argentina and is tetraploid, $4n = 204$. species, p. 240

Other Speciation Mechanisms

Speciation can also occur without geographic isolation or polyploidy. Any process that can result in the reproductive isolation of a portion of a species can lead to the possibility of speciation. For example, within populations, some individuals may breed or flower at a somewhat different time of the year (figure 14.4). If the difference in reproductive time is genetically based, different breeding populations could be established that could eventually lead to speciation. Among animals, variations in the genetically determined behaviors related to courtship and mating could effectively separate one species into two or more separate breeding populations. In plants, genetically determined incompatibility of the pollen of one population of flowering plants with the flowers of other populations of the same species could lead to separate species. Although there are many examples of these kinds of speciation mechanisms, geographic isolation and polyploidy are considered the primary mechanisms for speciation.

14.3 The Maintenance of Reproductive Isolation Between Species

For a new species to continue to exist, it must reproduce and continue to remain genetically isolated from other, similar species. The speciation process involves the development of **reproductive,** or **genetic, isolating mechanisms.** These mechanisms prevent matings between members of two different species and, therefore, help maintain distinct species. There are several mechanisms for maintaining reproductive (genetic) isolation:

1. **Habitat preference,** or **ecological isolating mechanisms,** occur when two species do not have the opportunity to interbreed because they typically live in different ecological settings. For example, in central Mexico, two species of robin-sized birds, called *towhees,* live in the same general region. However, the collared towhee lives on the mountainsides in the pine forest, whereas the spotted towhee is found at lower elevations in oak forests. Geography presents no barriers to these birds. They are capable of flying to each other's habitats, but they do not. Therefore, they are reproductively isolated because of the habitats they prefer. Similarly, areas with wet soil have different species of plants than nearby areas with drier soils.

2. **Seasonal isolating mechanisms** (differences in the time of year at which reproduction takes place) are effective genetic isolating mechanisms. Some plants flower only in the spring, whereas other species that are closely related flower in midsummer or fall; therefore, the two species are not very likely to pollinate one another. Among insects, there are examples of similar spacing of the reproductive periods of closely related species, so that they do not overlap.

3. **Behavioral isolating mechanisms** occur when inborn behavior patterns prevent breeding between species. The mating calls of frogs and crickets are highly specific. The sound pattern produced by the males is species-specific and invites only females of the same species to engage in mating. The females have a built-in response to the particular species-specific call and mate only with those that produce the correct call. The courtship behavior of birds involves both sound and visual signals that are species-specific. For example, groups of male prairie chickens gather on meadows shortly before dawn in the early summer and begin their dances. The air sacs on both sides of the neck are inflated, so that the brightly colored skin is exposed. Their feet move up and down very rapidly and their wings are spread out and quiver slightly (figure 14.5). This combination of sight and sound is attractive to females. When the females arrive, the males compete for the opportunity to mate with them. Other, related species of birds conduct their own similar, but distinct, courtship displays. The differences among the dances are great enough that a female can recognize the dance of a male of her own species. Behavioral isolating mechanisms such as these occur among other types of animals as well. The strutting of a

FIGURE 14.5 Courtship Behavior (Behavioral Isolating Mechanism)
The dancing of a male prairie chicken attracts female prairie chickens, but not females of other species. This behavior tends to keep prairie chickens reproductively isolated from other species.

peacock, the fin display of a beta (Siamese fighting) fish, and the flashing light patterns of "lightning bugs" of various species are all examples of behaviors that help individuals identify members of their own species and prevent different species from interbreeding (figure 14.6).

4. **Mechanical,** or **morphological, isolating mechanisms** involve differences in the structure of organisms. The specific shapes of the structures involved in reproduction may prevent different species from interbreeding. Among insects, the structure of the penis and the reciprocal structures of the female fit like a lock and key; therefore, breed-

(a)

(b)

FIGURE 14.6 Animal Communication – Identifying Members of Their Own Species
Most animals use specific behaviors to communicate with others of the same species. (a) The croaking of a male toad is specific to its species and is different from that of males of other species. (b) The visual displays of beta (Siamese fighting) fish communicate to others of the same species.

ing between different species is very difficult. Similarly, the shapes of flowers may permit only certain animals to carry pollen from one flower to the next.

5. **Biochemical isolating mechanisms** occur when molecular incompatibility prevents successful mating. A vast number of biochemical activities take place around the union of egg and sperm. Molecules on the outside of the egg or sperm may trigger events that prevent their union if they are not from the same species. Among plants, biochemical interactions between the pollen and the receiving flower prevent the germination of the pollen grain and, therefore, prevent sexual reproduction between two closely related species.

6. **Hybrid inviability,** or **infertility mechanisms,** prevent the offspring of two different species from continuing to reproduce. This can occur in three ways: (a) the embryos of such a mating may not develop properly and die; (b) if offspring are produced, they die before they can reproduce; or (c) such hybrids may be sterile or have greatly reduced fertility (review figure 14.1).

14.4 Evolutionary Patterns Above the Species Level

The development of a new species is the smallest irreversible unit of evolution. Because the exact conditions present when a species came into being will never exist again, it is unlikely that it will evolve back into an earlier stage in its development. Furthermore, because species are reproductively isolated from one another, they usually do not combine with other species to make something new; they can only diverge further. When life in the past is compared with our current diversity of life, many different evolutionary patterns emerge.

Divergent Evolution

Divergent evolution is a basic evolutionary pattern in which individual speciation events cause successive branches in the evolution of a group of organisms. This basic pattern is well illustrated by the evolution of the horse, shown in figure 14.7. Each of the many branches of the evolutionary history of the

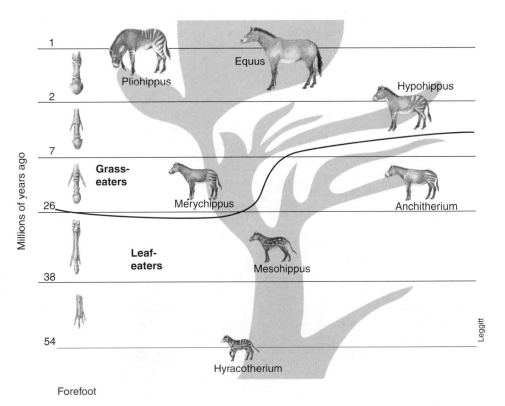

FIGURE 14.7 Divergent Evolution

In the evolution of the horse, many speciation events followed one after another. What began as a small, leaf-eating, four-toed animal of the forest evolved into a large, grass-eating, single-toed animal of the plains. There are many related species alive today (e.g., horse, donkey, zebra), but early ancestral types are extinct.

horse began with a speciation event that separated one species into two species as each separately adapted to local conditions. Changes in the environment from moist forests to drier grasslands would have set the stage for change. The modern horse, with its large size, single toe on each foot, and teeth designed for grinding grasses, is thought to be the result of accumulated genetic changes beginning from a small, dog-sized animal with four toes on its front feet, three toes on its hind feet, and teeth designed for chewing leaves and small twigs. Even though we know much about the evolution of the horse, there are still many gaps that need to be filled before we have a complete evolutionary history (Outlooks 14.1).

Extinction

Extinction is the loss of a species. It is a common pattern in the evolution of organisms. Notice in figure 14.7 that most of the species that developed during the evolution of the horse are extinct. This is typical. Most of the species that have ever existed are extinct. Estimates of extinction are around 99%; that is, 99% or more of all the species that ever existed are extinct. Given this high rate of extinction, we can picture current species of organisms as the product of much evolutionary experimentation, most of which resulted in failure. This is not the complete picture, though. Recall from chapter 13 that organisms are continually being subjected to selection pressures that lead to a high degree of adaptation to a particular set of environmental conditions. Organisms become more and more specialized. However, the environment does not remain constant; it often changes in such a way that the species that were originally present are unable to adapt to the new set of conditions. The early ancestors of the modern horse were well adapted to a moist tropical environment, but, when the climate became drier, most were no longer able to survive. Only some kinds had the genes necessary to lead to the development of modern horses.

Furthermore, many extinct species were very successful organisms for millions of years. They were not failures for their time but simply did not survive to the present. It is also important to realize that many currently existing organisms will eventually become extinct. Thus, the basic evolutionary pattern is one of divergence with a great deal of extinction. Although divergence and extinction are dominant themes in the evolution of life, adaptive radiation and convergent evolution are two other important evolutionary patterns.

Adaptive Radiation

Adaptive radiation is an evolutionary pattern characterized by a rapid increase in the number of kinds of closely related species. Adaptive radiation results in an evolutionary explosion of new species from a common ancestor. There are basically two situations thought to favor adaptive radiation. One is a condition in which an organism invades a previously unexploited environment. For example, at one time, there were no animals on the landmasses of the Earth. The amphibians were the first vertebrate animals able to spend part of their lives on land. Fossil evidence shows that a variety of amphibians evolved rapidly and exploited several kinds of lifestyles.

Another good example of adaptive radiation is found among the finches of the Galápagos Islands, located 1,000 kilometers west of Ecuador in the Pacific Ocean. These birds were first studied by Charles Darwin. Because these islands are volcanic and arose from the ocean floor, it is assumed that they have always been isolated from South America and originally lacked finches and other land-based birds. It is thought that one kind of finch arrived from South America to colonize the islands and that adaptive radiation from the common ancestor resulted in the many kinds of finches found on the islands today (figure 14.8). Although the islands are close to one another, they are quite diverse. Some are dry and treeless, some

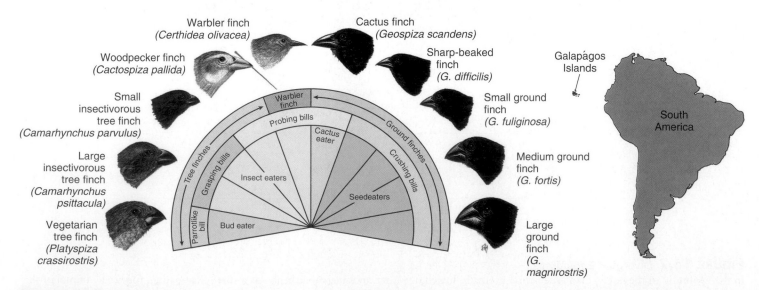

FIGURE 14.8 Adaptive Radiation
When Darwin discovered the finches of the Galápagos Islands, he thought they might all have derived from one ancestor that arrived on these relatively isolated islands. If they were the only birds to inhabit the islands, they could have evolved very rapidly into the many types shown here. The drawings show the specializations of beaks for different kinds of food.

OUTLOOKS 14.1

Evolution and Domesticated Dogs

During evolution, domesticated dogs branched off from wolves into three general types: hunters, herders, and guard dogs. As a result of selective breeding, there are now about 400 breeds of dog worldwide, and about 99% of their genes are the same. Of these hundreds of breeds, which have maintained the greatest genetic similarity to their ancient ancestors?

DNA analysis indicates that Siberian huskies, Chinese shar-pei's, and African basenjis are the most closely related to wolves. This means that the wolf had within its genome the genetic foundation for all the breeds that exist today.

This DNA analysis presented data for two Egyptian breeds, the ibizan and the pharaoh hounds. Based on the DNA study alone, it was believed that the breed had been developed in the past 200 to 300 years. However, when the breed's documented history was considered, it became evident that it was most likely not developed in the recent past. Combining both the DNA evidence and breeding history, the current explanation is that the ancient Egyptian dogs were influenced by interbreeding with more modern dogs.

have moist forests, and others have intermediate conditions. Conditions were ideal for several speciation events. Because the islands were separated from one another, the element of geographic isolation was present. Because environmental conditions on the islands were quite different, particular characteristics in the resident birds would have been favored. Furthermore, the absence of other kinds of birds meant that there were many lifestyles that had not been exploited.

In the absence of competition, some of these finches took roles normally filled by other kinds of birds elsewhere in the world. Although finches are normally seed-eating birds, some of the Galápagos finches became warblerlike insect-eaters, others became leaf-eaters, and one uses a cactus spine as a tool to probe for insects.

The second situation that can favor adaptive radiation is one in which a type of organism evolves a new set of characteristics that enables it to displace organisms that previously filled specific roles in the environment. For example, although amphibians were the first vertebrates to occupy land, they lived only near freshwater, where they would not dry out and could lay eggs, which developed in the water. They were replaced by reptiles with such characteristics as dry skin, which prevented the loss of water, and an egg that could develop on land. The adaptive radiation of reptiles was extensive. They invaded most terrestrial settings and even evolved forms that flew and lived in the sea. With the extinction of dinosaurs and many other reptiles, birds and mammals went through a similar radiation. Perhaps the development of homeothermism (the ability to maintain a constant body tem-

perature) had something to do with the success of birds and mammals. Figure 14.9 shows the sequence of radiations that occurred within the vertebrate group. The number of species of amphibians and reptiles has declined, whereas the number of species of birds and mammals has increased.

Convergent Evolution

Convergent evolution is an interesting evolutionary pattern that involves the development of similar characteristics in organisms of widely different evolutionary backgrounds. This pattern often leads people to misinterpret the evolutionary history of organisms. For example, many kinds of plants that live in desert situations have sharp, pointed structures, such as spines or thorns, and lack leaves during much of the year. Superficially, the sharp, pointed structures may resemble one another to a remarkable degree but may have a completely different evolutionary history. Some of these structures are modified twigs, others are modified leaves, and still others are modifications of the surface of the stem. The presence of sharp, pointed structures and the absence of leaves are adaptations to a desert type of environment: The thorns and spines discourage herbivores and the absence of leaves reduces water loss.

Another example is animals that survive by catching insects while flying. Bats, swallows, and dragonflies all obtain food in this manner. They have wings, good eyesight or hearing to locate flying insects, and great agility and speed in flight, but they are evolved from quite different ancestors

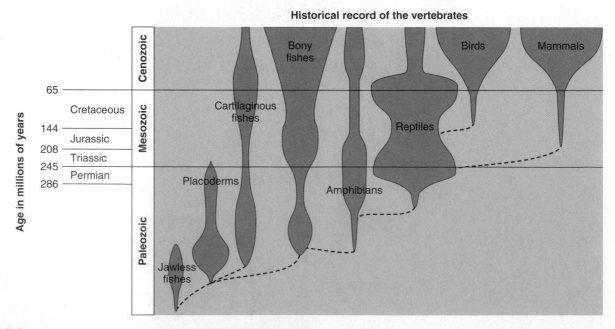

Historical record of the vertebrates

FIGURE 14.9 Adaptive Radiation in Terrestrial Vertebrates
The amphibians were the first vertebrates to live on land. They were replaced by the reptiles, which were better adapted to land. The reptiles, in turn, were replaced by the adaptive radiation of birds and mammals. (Note: The width of the colored bars indicates the number of species present.)

(figure 14.10). At first glance, they may appear very similar and perhaps closely related, but a detailed study of their wings and other structures shows that they are different kinds of animals. They have simply converged in structure, the type of food they eat, and their method of obtaining food. Likewise, whales, sharks, and tuna appear to be similar—they all have a streamlined shape, which aids in rapid movement through the water; a dorsal fin, which helps prevent rolling; fins or flippers for steering; and a large tail, which provides power for swimming. However, they are quite different kinds of animals that happen to live in the open ocean, where they pursue other animals as prey. Their structural similarities are adaptations to being fast-swimming predators.

Homologous or Analogous Structures

To help evolutionary biologists distinguish between structures that are the result of convergent evolution and those that are not, they try to determine if the structures evolved from a common ancestor. **Homologous structures** are similar structures in different species that have the same general function indicating that they been derived from a common ancestor. Thus, the wing of a bat, the front leg of a horse, and the arm of a human show the same basic pattern of bones, but they are extreme modifications of the same basic evolutionary structure (figure 14.11). On the other hand, structures that have the same function (such as the wing of a butterfly and the

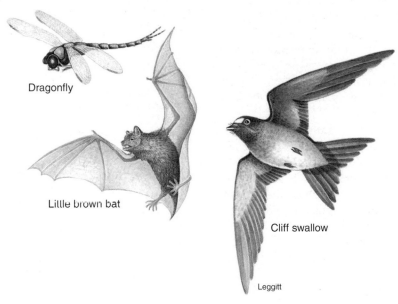

FIGURE 14.10 Convergent Evolution
All of these animals have evolved wings as a method of movement and capture insects for food as they fly. However, flight originated independently in each of them.

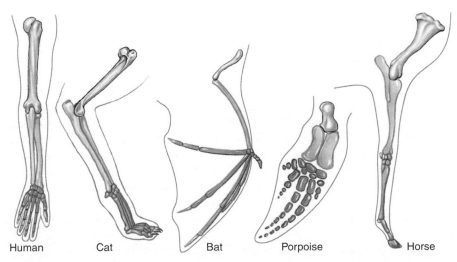

FIGURE 14.11 Maintaining Traits Through Time – Homologous Structures
Although these body parts are considerably different in structure and function, the same bones are present in the forelimbs of humans, cats, bats, porpoises, and horses.

wing of a bird) but different evolutionary backgrounds are called **analogous structures** and are the result of convergent evolution (review figure 14.10).

14.5 Rates of Evolution

Although it is commonly thought that evolutionary change takes long periods of time, rates of evolution can vary greatly. Remember that natural selection is driven by the environment. If the environment is changing rapidly, changes in organisms should be rapid. Periods of rapid environmental change also result in extensive episodes of extinction. During some periods in the history of the Earth when little environmental change was taking place, the rate of evolutionary change was probably slow. Nevertheless, when we talk about evolutionary time, we are generally thinking in thousands or millions of years. Although both of these time periods are long compared with the human life span, the difference between thousands of years and millions of years in the evolutionary time scale is still significant.

The fossil record shows many examples of gradual changes in the physical features of organisms over time. For example, the extinct humanoid fossil *Homo erectus* shows a gradual increase in the size of the skull, a reduction in the size of the jaw, and the development of a chin over about a million years. The accumulation of these changes could result in such extensive change from the original species that we would consider the current organism to be a different species from its ancestor. (Many believe that *Homo erectus* became modern humans, *Homo sapiens*.) This is such a common feature of the evolutionary record that biologists refer to this kind of evolutionary change as *gradualism* (figure 14.12a). **Gradualism** is a model for evolutionary change that evolution occurred slowly by accumulating small changes over a long period of time. Charles Darwin's view of evolution was based on gradual changes in the features of specific species he observed in his studies of geology and natural history. However, as early as the 1940s, some biologists began to challenge gradualism as the only model for evolutionary change. They pointed out that the fossils of some species were virtually unchanged over millions of years. If gradualism were the only explanation for how species evolved, then gradual changes in the fossil record of a species would always be found. Furthermore, some organisms appear suddenly in the fossil record and show rapid change from the time they first appeared. There are many modern examples of rapid evolutionary change; the development of pesticide resistance in insects and antibiotic resistance in various bacteria has occurred recently.

In 1972, two biologists, Niles Eldredge of the American Museum of Natural History and Stephen Jay Gould of Harvard University, proposed a very different idea. **Punctuated equilibrium** is their hypothesis that evolution occurs in spurts of rapid change, followed by long periods with little evolutionary change (figure 14.12b). The punctuated equilibrium con-

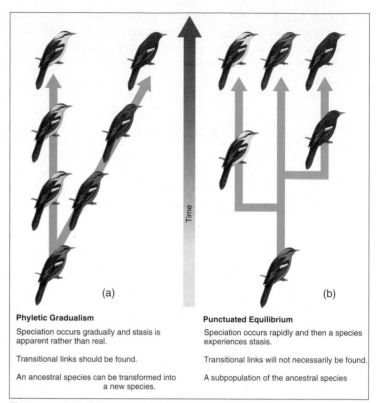

Phyletic Gradualism

Speciation occurs gradually and stasis is apparent rather than real.

Transitional links should be found.

An ancestral species can be transformed into a new species.

Punctuated Equilibrium

Speciation occurs rapidly and then a species experiences stasis.

Transitional links will not necessarily be found.

A subpopulation of the ancestral species

FIGURE 14.12 Gradualism vs. Punctuated Equilibrium
(a) Gradualism is the evolution of new species from the accumulation of a series of small changes over a long period of time. *(b)* Punctuated equilibrium is the evolution of new species from a large number of changes in a short period of time. Note that in both instances the ancestral bird has evolved into three species. However, it is possible that they were produced by different processes.

cept is a companion hypothesis to gradualism and suggests a different way of achieving evolutionary change. Punctuated equilibrium proposes that, rather than one species slowly accumulating changes to become a different descendant species, there is a rapid evolution of several closely related species from isolated populations. This would produce a number of species, that would compete with one another as the environment changed. Many of these species would become extinct and the fossil record would show change.

Another way to look at gradualism and punctuated equilibrium is to assume that both occur. It is clear from the fossil record that there were periods in the past when there was very rapid evolutionary change, compared with other times. Also, some environments, such as the ocean, have been relatively stable, whereas others, such as the terrestrial environment, have changed significantly. Many marine organisms have remained unchanged for hundreds of millions of years, but there has been great change in the kinds of terrestrial organisms in the past few million years. Thus, it is possible that both gradualism and punctuated equilibrium have operated. The important contribution of punctuated equilibrium is that there can be alternative ways of interpreting the fossil record and that the pace of evolution can be quite variable. However, both approaches take into account the importance of genetic diversity as the raw material for evolution and the mechanism of natural selection as the process of determining which gene combinations fit the environment. The gradualists point to the fossil record as proof that evolution is a slow, steady process. Those who support punctuated equilibrium point to the gaps in the fossil record as evidence that rapid change occurs.

14.6 The Tentative Nature of the Evolutionary History of Organisms

Tracing the evolutionary history of an organism back to its origins is a very difficult task, because most of its ancestors no longer exist. Scientists may be able to look at fossils of extinct organisms but must keep in mind that the fossil record is incomplete and provides only limited information about the biology of the organism represented in that record. However, new fossils are always being discovered.

There are three reasons that the fossil record is incomplete. First, the likelihood that an organism will become a fossil is low. Most organisms die and decompose, leaving no trace of their existence. (Today, road-killed opossums are not likely to become fossils, because they will be eaten by scavengers, repeatedly run over, or decompose by the roadside.) Second, in order to form a fossil, the dead organism must be covered by sediments, dehydrated, or preserved in some other way.

Several factors increase the likelihood that an organism will be found in the fossil record:

- The presence of hard body parts that resist decomposition.
- Marine organisms can be covered by sediment on the bottom.
- Fossils of more recent organisms are less likely to have been destroyed by geological forces.
- Fossils of large organisms are easier to find.
- Organisms that were extremely common are more likely to show up in the fossil record.

For example, trilobites are very common in the fossil record. They were relatively large marine organisms, with hard body parts, that were extremely abundant. However, fossils of soft-bodied, extremely ancient, wormlike organisms are rare.

Third, the discovery of fossils is often accidental. It is impossible to search through all the layers of sedimentary rock on the entire surface of the Earth. Therefore, scientists will continue to find new fossils that will extend the known information about ancient life into the foreseeable future. But there can be no question that evolution occurred in the past and continues to occur today (How Science Works 14.1).

Scientists may know a lot about the structure of the bones and teeth or the stems and leaves of an extinct ancestor from fossils but know almost nothing about its behavior, physiology, and natural history. Biologists must use a great deal of indirect evidence to piece together the series of evolutionary steps that led to a current species. Figure 14.13 is typical of evolutionary diagrams that illustrates how time and structural changes are related in the evolution of birds, mammals, and reptiles. There will always be new information that will require changes in thinking, and equally reputable scientists will disagree on the evolutionary processes or the sequence of events that led to a specific group of organisms. classification, p. 422

14.7 Human Evolution

There is intense curiosity about how our species (*Homo sapiens*) came to be, and the evolution of the human species remains an interesting and hot topic. Human beings are classified as mammals belonging to a group known as primates. Primates are thought to have come into existence approximately 66 million years ago. They include animals with enlarged, complex brains; five digits, with nails, on the hands and feet; and hands and feet adapted for grasping. Their bodies, except those of humans, are covered with hair. There are two groups of primates, the prosimians—lemurs and tarsiers—and the anthropoids—monkeys, apes, and humans (figure 14.14).

Because all of our close evolutionary relatives are extinct, it is difficult to visualize our evolutionary development, and we tend to think we are unique and not subject to the laws of nature. However, humans show genetic diversity, experience mutations, and are subject to the same evolutionary forces as other organisms.

Scientists use several kinds of evidence to try to sort out our evolutionary history. Fossils of various kinds of prehuman and ancient human ancestors have been found, but many of these are only fragments of skeletons, which, are difficult to interpret and are hard to date. Stone tools of various kinds have also been found that are associated with prehuman and early human sites. Finally, other aspects of the culture of our ancestors have been found in burial sites, including cave paintings and ceremonial objects. Various methods have been used to date these findings. Some can be dated quite accurately, whereas others are more difficult to pinpoint. When fossils are examined, anthropologists can identify differences in the structures of bones that are consistent with changes in species. Based on the amount of change they see and the ages of the fossils, scientists make judgments about the species to which the fossil belongs.

As new discoveries are made, experts' opinions will change, and our evolutionary history may become clearer as old ideas are replaced. Scientists must also review and make changes in the terminology they use to refer to our ancestors. The term *hominin* now refers to humans and their humanlike

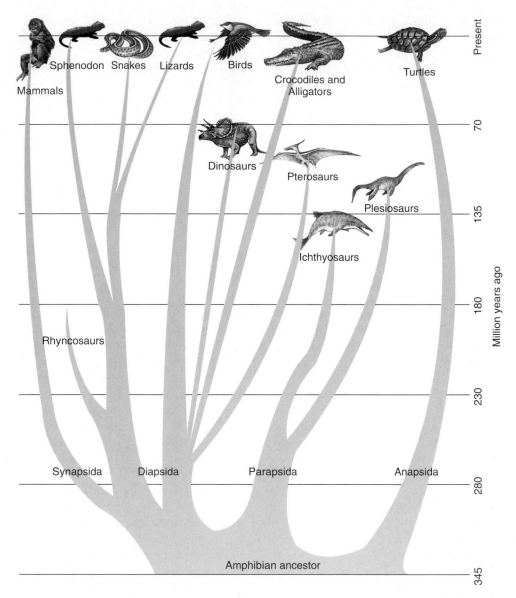

FIGURE 14.13 An Evolutionary Diagram
This diagram shows how present-day reptiles, birds, and mammals are thought to have evolved from primitive amphibian ancestors. Notice that an extremely long period of time is involved (over 300 million years) and that many of the illustrated species are extinct.

ancestors, whereas previously the term *hominid* was used. The term *hominid* now refers to the broader group that includes all humanlike organisms plus the African great apes—gorillas, orangutans, chimpanzees, and bonobos. When you read material about this topic, you will need to determine how the terms are being used. Although there is no clear picture of how humans evolved, the fossil record shows that humans are a relatively recent additions to the forms of life, and several points are well accepted:

1. There is a great deal of fossil evidence that several species of the genera *Australopithecus* and *Paranthropus* were common in Africa beginning at about 4 million years ago. These organisms walked upright and are often referred to collectively as australopiths.

2. Based on fossil evidence, it appears that the climate of Africa was becoming drier during much of the time during which the evolution of humans was occurring.

3. The earliest *Australopithecus* fossils are from about 4.2 million years ago. Earlier fossils, such as *Ardipithecus* and the recently discovered *Orrorin* and *Sahelanthropus,* may be ancestral to *Australopithecus*. *Australopithecus* and *Paranthropus* were herbivores and walked upright. Their fossils and those of earlier organisms, such as *Ardipithecus, Orrorin,* and *Sahelanthropus,* are found only in Africa.

4. The australopiths were sexually dimorphic, with the males much larger than the females, and they had relatively small brains (a cranial capacity of 530 cm^3 or less—about the size of a standard softball).

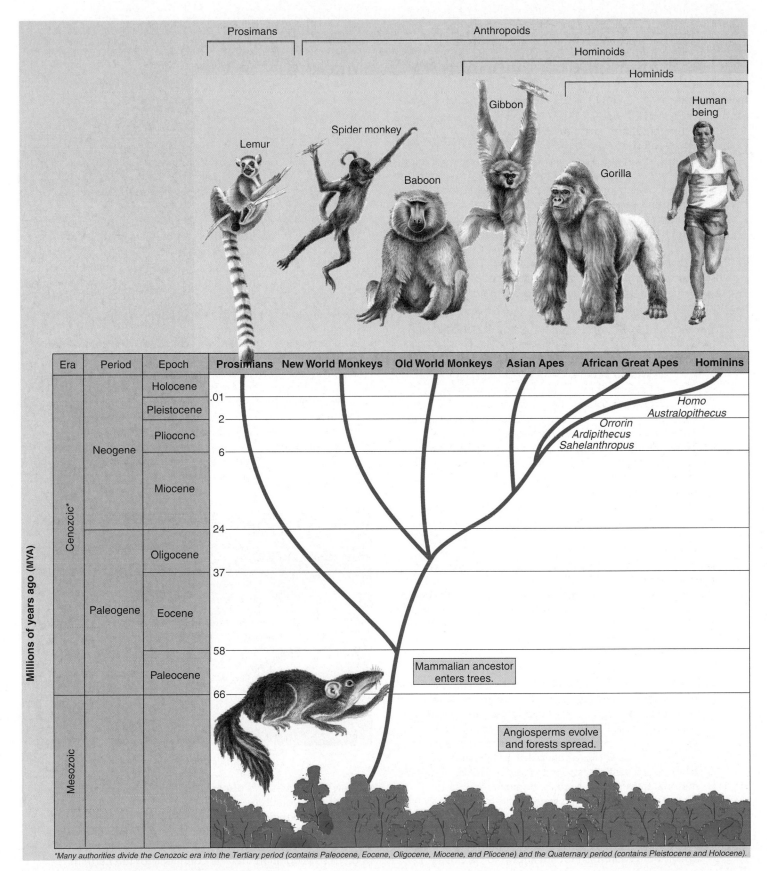

FIGURE 14.14 Where We Are on the Tree

As new evidence is uncovered, scientists reconsider how that information best fits into the hypothesis on the evolution of humans. As the pieces of the puzzle are reexamined and rearranged, we see a clearer picture of where human beings fit in the scheme of life.

HOW SCIENCE WORKS 14.1

Accumulating Evidence for Evolution

The theory of evolution has become the major unifying theory of the biological sciences. Those in medicine understand the dangers of mutations, the similarity in function of the same organ in related species, and the way in which the environment can interfere with the preprogrammed process of embryological development. Agricultural science has demonstrated the importance of selecting specific genetically determined characteristics for passage into new varieties of crop plants and animals. The concepts of mutation, selection, and evolution are so fundamental to an understanding of what happens in biology that we often forget to take note of the many kinds of observations that support the theory of evolution. The following list describes some of the more important pieces of evidence that support the idea that evolution has been a major force in shaping the nature of living things.

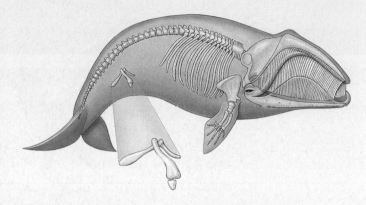

1. Species and populations are not genetically fixed. Change occurs in individuals and populations.
   a. Mutations cause slight changes in the genetic makeup of an individual organism.
   b. Different populations of the same species show adaptations suitable for their local conditions.
   c. Changes in the characteristics displayed by species can be linked to environmental changes.
   d. The selective breeding of domesticated plants and animals indicates that the shape, color, behavior, metabolism, and many other characteristics of organisms can be selected for.
   e. The extinction of poorly adapted species is common.
2. All evidence suggests that, once embarked on a particular evolutionary road, structures, behaviors, or physiological processes are not abandoned, only modified. New organisms are formed by the modification of ancestral species, not by major changes. The following list supports the concept that evolution proceeds by the modification of previously existing structures and processes, rather than by catastrophic change.
   a. All species use the same DNA code.
   b. All species use the same left-handed (levorotary) amino acid building blocks.
   c. It is difficult to eliminate a body part when it is part of a developmental process controlled by genes. A vestigial structure is an organ that is functionless and generally reduced in size. It resembles the fully functioning organs found in related organisms. Structures such as the human appendix, whale pelvic bones, and human tailbone are vestigial structures and evidence of genetic material from previous stages in evolution.
   d. The embryological development of related animals is similar, regardless of the peculiarities of adult anatomy. All vertebrate embryos have an early stage that contains gill-like structures.
   e. Species of organisms that are known to be closely related show greater similarity in their DNA than do those that are distantly related.

3. Several aspects of the fossil record support the concept of evolution.
   a. The nature of the Earth has changed significantly over time.
   b. The fossil record shows vast changes in the kinds of organisms present on Earth. New species appear and most go extinct. This is evidence that living things change in response to changes in their environment.
   c. The fossils found in old rocks do not reappear in younger rocks. Once an organism goes extinct, it does not reappear, but new organisms arise that are modifications of previous organisms.

4. New techniques and discoveries invariably support the theory of evolution.
   a. The recognition that the Earth was formed billions of years ago supports the slow development of new kinds of organisms.
   b. The recognition that the continents of the Earth have separated helps explain why organisms on Australia are so different from those elsewhere.

HOW SCIENCE WORKS 14.1 (continued)

c. The discovery of DNA and how it works helps explain mutation and demonstrates the genetic similarity of closely related species.

d. Similarities in protein structure can indicate the degree of relatedness among organisms. For example, the greater the evolutionary distance from humans, the greater is the number of amino acid differences in the vertebrate hemoglobin polypeptide.

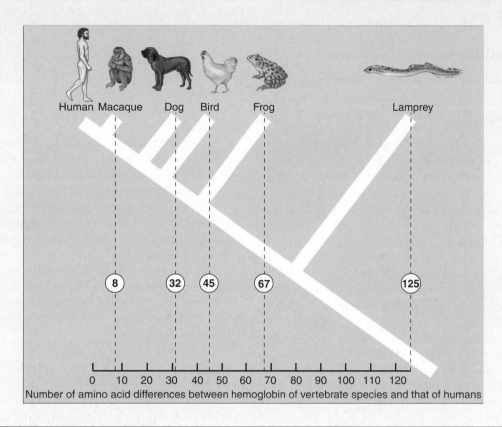

Number of amino acid differences between hemoglobin of vertebrate species and that of humans

5. Several species of the genus *Homo* became prominent in Africa beginning at about 2.2 million years ago and appear to have made a change from a primarily herbivorous diet to a carnivorous or omnivorous diet.

6. All members of the genus *Homo* have relatively large brains compared with the australopiths. The cranial capacity ranges from about 650 cm^3 for early fossils of *Homo* to about 1,450 cm^3 for modern humans. These organisms are also associated with various degrees of stone tool construction and use. Some of the australopiths may also have constructed stone tools.

7. Fossils of several later species of the genus *Homo* have larger brains and are found in Africa, Europe, and Asia, but not in Australia or the Americas. Only *Homo sapiens* is found in Australia and the Americas.

8. Since the fossils of *Homo* species found in Asia and Europe are generally younger than the early *Homo* species found in Africa, it is assumed that they moved from Africa to Europe and Asia from Africa.

9. Differences in size between males and females are less prominent in the members of the genus *Homo*, so perhaps there were fewer sexual difference in activities than with the australopiths.

When scientists put all these bits of information together, they can construct the following scenario for the likely evolution of our species. Early primates, such as modern monkeys, chimpanzees, and apes, were adapted to living in forested areas, where their grasping hands, opposable thumbs, big toes, and wide range of shoulder movement, allow them to move freely in the trees. As the climate became drier, grasslands replaced the forests. Early hominins were adapted to drier conditions. Walking upright was probably an adaptation to these conditions. Most later hominins had large brains and used tools. A recent find, however, illustrates just how tentative the understanding of human evolution really is. The discovery of *Homo floresiensis*, known as the "hobbit" because of its small size, places this organism in the same time period as *H. sapiens*. However *H. floresiensis* is extinct. Therefore, *the sole surviving member of this evolutionary process is* Homo sapiens.

The Australopiths

Various species of *Australopithecus* and *Paranthropus* were present in Africa from about 4.4 million years ago until about 1 million years ago. There are few fossils of these early humanlike organisms, however, and many fossils are fragments of organisms. This has led to much speculation and argument among experts about the specific position each fossil has in the evolutionary history of humans. However, from examinations of the fossil bones of the leg, pelvis, and foot, it is apparent that the australopiths were relatively short (males, 1.5 meters or less; females, about 1.1 meters) and stocky and walked upright, as humans do.

An upright posture had several advantages in a world that was becoming drier. It allowed for more rapid movement over long distances and the ability to see longer distances, and it reduced the amount of heat gained from the sun. In addition, upright posture freed the arms for other uses, such as carrying and manipulating objects and using tools. The various species of *Australopithecus* and *Paranthropus* shared these characteristics and, based on the structure of their skulls, jaws, and teeth, appear to have been herbivores with relatively small brains.

The Genus *Homo*

About 2.5 million years ago, the first members of the genus *Homo* appeared on the scene. There is considerable disagreement about how many species there were, but *Homo habilis* is one of the earliest. *H. habilis* had a larger brain (650 cm^3) and smaller teeth than the australopiths and made much more use of stone tools. Some experts feel that it was a direct descendant of *Australopithecus africanus*. Many feel that *H. habilis* was a scavenger that made use of group activities, tools, and higher intelligence to commandeer the kills made by other carnivores. This higher-quality diet would have supported the metabolic needs of a larger brain.

About 1.8 million years ago, *Homo ergaster* appeared. It was much larger, up to 1.6 meters tall, than *H. habilis,* which,

was about 1.3 meters tall and had a much larger brain (a cranial capacity of 850 cm^3). A little later, a similar species (*Homo erectus*) appeared in the fossil record. Some consider *H. ergaster* and *H. erectus* to be variations of the same species. Along with their larger brains and body size, *H. ergaster* and *H. erectus* distinguished from earlier species by their extensive use of stone tools. Hand axes were manufactured and used to cut the flesh of prey and crush the bones for marrow. These organisms appear to have been predators, whereas *H. habilis* was a scavenger. The use of meat as food allows animals to move about more freely, because appropriate food is available almost everywhere. By contrast, herbivores are often confined to places that have foods appropriate to their use: fruits for fruit eaters, grass for grazers, forests for browsers, and so on. In fact, fossils of *H. erectus* have been found in the Middle East and Asia, as well as Africa. Most experts feel that *H. erectus* originated in Africa and migrated through the Middle East to Asia.

At about 800,000 years ago, another hominin, classified as *Homo heidelbergensis,* appeared in the fossil record. Since fossils of this species are found in Africa, Europe, and Asia, it appears that they constitute a second wave of migration of early *Homo* from Africa to other parts of the world. Both *H. erectus* and *H. heidelbergensis* disappear from the fossil record as two new species (*Homo neanderthalensis* and *Homo sapiens*) become common.

The Neandertals were primarily found in Europe and adjoining parts of Asia but were not found in Africa. Therefore, many experts feel the Neandertals are descendants of *H. heidelbergensis*, which was common in Europe (How Science Works 14.2).

Two Points of View on the Origin of *Homo sapiens*

Homo sapiens is found throughout the world and is now the only species remaining of a long line of ancestors. Two theories seek to explain the origin of *Homo sapiens*. One theory, known as the **out-of-Africa hypothesis**, states that modern humans (*Homo sapiens*) originated in Africa, as did several other similar species. They migrated from Africa to Asia and Europe and displaced species such as *H. erectus* and *H. heidelbergensis,* which had previously migrated into these areas. The other theory, known as the **multiregional hypothesis**, states that *H. erectus* evolved into *H. sapiens*. During a period of about 1.7 million years, fossils of *Homo erectus* showed a progressive increase in the size of the cranial capacity and reduction in the size of the jaw, so it is difficult to distinguish *H. erectus* from *H. heidelbergensis* and *H. heidelbergensis* from *H. sapiens*. Proponents of this hypothesis believe that *H. heidelbergensis* is not a distinct species but,

HOW SCIENCE WORKS 14.2

Neandertals—*Homo neanderthalensis* or *Homo sapiens*??

An ongoing controversy surrounds the relationship between the Neandertals and other forms of prehistoric humans. One position is that the Neandertals were a small, separate race or subspecies of human that could have interbred with other humans and may have disappeared because their subspecies was eliminated by interbreeding with more populous, more successful groups. (Many small, remote tribes have been eliminated in the same way in recent history.) Others maintain that the Neandertals show such great difference from other early humans that they must be a different species and became extinct because they could not compete with more successful *Homo sapiens* immigrants from Africa. (The names of these ancient people typically are derived from the place where the fossils were first discovered. For example, the Neandertals were first found in the Neander Valley of Germany, and the Cro-Magnons, considered to be *Homo sapiens,* were initially found in the Cro-Magnon caves in France.)

The use of molecular genetic technology has shed some light on the relationship of the Neandertals to other kinds of humans. Examination of the mitochondrial DNA obtained from the bones of a Neandertal individual reveals that there are significant differences between the Neandertals and other kinds

of early humans. This greatly strengthens the argument that the Neandertals were a separate species, *Homo neanderthalensis.*

---

rather, an intermediate between the earlier *H. erectus* and *H. sapiens*. According to this theory, various subgroups of *H. erectus* existed throughout Africa, Asia, and Europe and interbreeding among the various groups gave rise to the various races of humans we see today.

Another continuing puzzle is the relationship of *Homo sapiens* to the Neandertals. Some people consider the Neandertals to be a subgroup of *Homo sapiens* specially adapted to life in the harsh conditions found in postglacial Europe. Others consider them to be a separate species, *Homo neanderthalensis*. The Neandertals were muscular, had a larger brain capacity than modern humans, and had many elements of culture, including burials. The cause of their disappearance from the fossil record at about 25,000 years ago remains a mystery. Perhaps a change to a warmer climate was

responsible. Perhaps contact with *Homo sapiens* resulted in their elimination either through hostile interactions or interbreeding with *H. sapiens,* resulting in their absorption into the larger *H. sapiens* population.

Large numbers of fossils of prehistoric humans have been found in all parts of the world. Many of these show evidence of a collective group memory, called *culture*. Cave paintings, carvings in wood and bone, tools of various kinds, and burials are examples. These are also evidence of a capacity to think and invent, as well as "free time" to devote to things other than gathering food and other necessities of life. We may never know how we came to be, but we will always be curious and will continue to search and speculate about our beginnings. Figure 14.15 summarizes the current knowledge of the historical record of humans and their relatives.

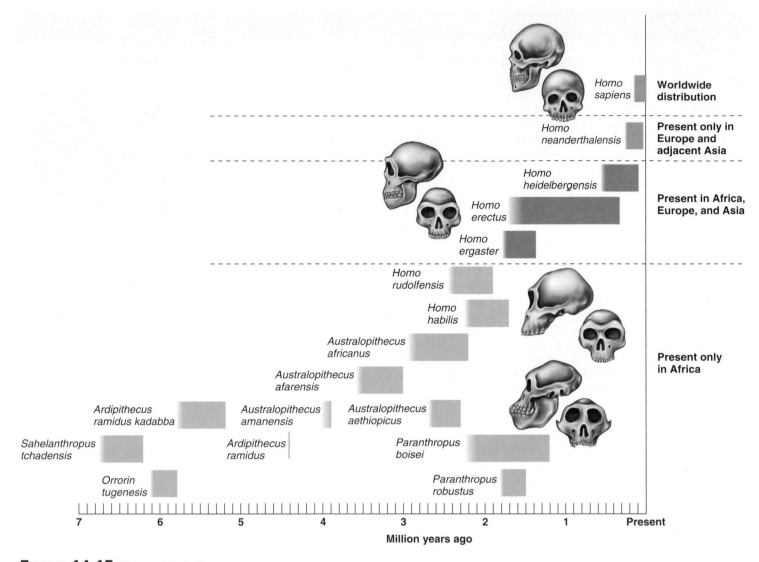

FIGURE 14.15 Human Evolution

This diagram shows the various organisms thought to be relatives of humans. The bars represent approximate times the species are thought to have existed. Notice that (1) all species are extinct today except for modern humans, (2) several species of organisms coexisted for extensive periods, (3) all the older species are found only in Africa, and (4) more recent species of *Homo* are found in Europe and Asia, as well as Africa.

Summary

Populations are usually genetically diverse. Mutations, meiosis, and sexual reproduction tend to introduce genetic diversity into a population. Organisms with wide geographic distribution often show different gene frequencies in different parts of their range. A species is a group of organisms that can interbreed to produce fertile offspring. The process of speciation usually involves the geographic separation of the species into two or more isolated populations. While they are separated, natural selection operates and each population adapts to its environment. If this generates enough change, the two populations may become so different that they can-

not interbreed. Similar organisms that have recently evolved into separate species normally have genetic (reproductive) isolating mechanisms to prevent interbreeding. Some of these are habitat preference, seasonal isolating mechanisms, and behavioral isolating mechanisms. Plants have a special way of generating new species by increasing their chromosome numbers as a result of abnormal mitosis or meiosis. Organisms that have multiple sets of chromosomes are called polyploids.

Evolution is basically a divergent process on which other patterns can be superimposed. Adaptive radiation is a very rapid divergent evolution; convergent evolution involves the development of superficial similarities among widely different organisms. The rate at which evolution has occurred probably varies. The fossil record shows periods of rapid change interspersed with periods of little change. This has caused some to look for mechanisms that could cause the sudden appearance of large numbers of new species in the fossil record, which challenge the traditional idea of slow, steady change accumulating enough differences to cause a new species to be formed. The early evolution of humans has been difficult to piece together because of the fragmentary evidence. Beginning about 4.4 million years ago, the earliest forms of *Australopithecus* and *Paranthropus* showed upright posture and other humanlike characteristics. The structure of the jaw and teeth indicates that the various kinds of australopiths were herbivores. *Homo habilis* had a larger brain and appears to have been a scavenger. Several other species of the genus *Homo* arose in Africa. These forms appear to have been carnivores. Some of these migrated to Europe and Asia. The origin of *Homo sapiens* is in dispute. It may have arisen in Africa and migrated throughout the world or evolved from earlier ancestors found throughout Africa, Asia, and Europe.

Key Terms

Use the interactive flash cards on the **Concepts in Biology,** *12/e website to help you learn the meaning of these terms.*

adaptive radiation 288

analogous structures 292

behavioral isolating
  mechanisms 286

biochemical isolating
  mechanisms 287

convergent evolution 290

divergent evolution 287

extinction 288

fossil 283

gene flow 282

geographic isolation 283

gradualism 292

habitat preference
  (ecological isolating
  mechanisms) 285

homologous structures 291

hybrid inviability (infertility
  mechanisms) 287

mechanical (morphological)
  isolating mechanisms 286

multiregional hypothesis 298

out-of-Africa hypothesis 298

polyploidy 284

punctuated equilibrium 292

reproductive (genetic)
  isolating mechanisms 285

seasonal isolating
  mechanisms 286

speciation 283

Basic Review

1. _____ is the movement of genes from one generation to the next as a result of reproduction or from one region to another by migration.

2. A(n) _____ is any remains of an organism of a past geologic age, such as a preserved skeleton or body imprint.

3. Which of the following steps are not necessary for speciation to occur?
   a. geographic isolation
   b. genetic divergence
   c. reproductive isolation
   d. hybrid viability

4. The _____ of an organism is the geographic area over which a species can be found,
   a. range
   b. region
   c. pasture
   d. geographic location

5. If individuals overcome the geographic barrier, they may not have accumulated enough _____ to prevent reproductive success.
   a. mutations
   b. genetic differences
   c. barriers
   d. sexual differences

6. _____ is a condition of having multiple sets of chromosomes, rather than the normal haploid or diploid number.

7. Differences in the time of the year at which reproduction takes place are called
   a. geographic isolating mechanisms.
   b. hybrid isolating mechanisms.
   c. seasonal isolating mechanisms.
   d. physical isolating mechanisms.

8. Many people talk about the "male species" therefore, sexual reproduction is the same as speciation. (T/F)

9. The term _____ is now used to refer to humans and their humanlike ancestors.
   a. *hominid*
   b. *anthropoid*
   c. *hominoid*
   d. *hominin*

10. The scientific name for modern human beings is
   a. *Homo habilis.*
   b. *Homo neanderthalensis.*
   c. *H. erectus.*
   d. *Homo sapiens.*

Answers
1. Gene flow 2. fossil 3. d 4. a 5. b
6. Polyploidy 7. c 8. f 9. d 10. d

Concept Review

*Answers to the Concept Review questions can be found on the **Concepts in Biology,** 12/e website.*

14.1 Evolutionary Patterns at the Species Level

1. How is the concept of gene flow related to the species concept?

14.2 How New Species Originate

2. How does speciation differ from the formation of subspecies?
3. Why aren't mules considered a species?
4. Can you always tell by looking at two organisms whether or not they belong to the same species?
5. Why is geographic isolation important in the process of speciation?
6. How does a polyploid organism differ from a haploid or diploid organism?
7. List the series of events necessary for speciation to occur.

14.3 The Maintenance of Reproductive Isolation Between Species

8. Describe three kinds of reproductive isolating mechanisms that prevent interbreeding between different species.
9. Give an example of seasonal isolating mechanisms, habitat preference, and behavioral isolating mechanisms.

14.4 Evolutionary Patterns Above the Species Level

10. Describe convergent evolution and adaptive radiation.

11. What are the two dominant evolutionary patterns?
12. How are analogous and homologous structures different?

14.5 Rates of Evolution

13. What is the difference between gradualism and punctuated equilibrium?

14.6 The Tentative Nature of the Evolutionary History of Organisms

14. Why is it difficult to determine the evolutionary history of a species?

14.7 Human Evolution

15. List three differences between australopiths and members of the genus *Homo*.
16. Compare the out-of-Africa hypothesis with the multiregional hypothesis of the origin of *Homo sapiens*.

Thinking Critically

*For guidelines to answer this question, visit the **Concepts in Biology,** 12/e website.*

Explain how all the following are related to the process of speciation: mutation, natural selection, meiosis, geographic isolation, changes in the Earth, and gene pool.

CHAPTER

15

Ecosystem Dynamics

The Flow of Energy and Matter

Because we buy our food in a supermarket, we don't think about how that food was produced. We are only vaguely aware that we are connected to natural systems that provide that food. What effect does our food consumption have on the world?

Consider a simple hamburger patty—no pickle, mustard, lettuce, tomato, catsup, or bun—just a meat patty. What went into its creation? We are aware that a hamburger patty is ground-up cow muscle and that the cow required a source of food to produce the muscle that was ground to make the patty. The food the cow ate consisted of a variety of plants. The plants produced the food the cow ate by absorbing minerals and water from the soil and using sunlight in the process of photosynthesis.

However, that is not the whole story, because we need to consider how our consumption of hamburger patties has changed our world. About 20% of the landmass of the United States (excluding Alaska and Hawaii) is used for raising crops. Much of this is used for animal feed. For example, about 20% of the farmland in the United States is used to raise corn, 60% of which is fed to animals. Thus, over 10% of the farmland is used to raise corn to feed to animals. In addition, about 40% of the landmass is used for grazing cattle. Thus, over 50% of all U.S. land area is affected by our consumption of meat products. That is a large impact.

CHAPTER OUTLINE

15.1 What Is Ecology?

People often use the terms *ecology* and *environment* as if they meant the same thing. Students, homeowners, politicians, planners, and union leaders speak of "environmental issues" and "ecological concerns." They speak of products that are "environmentally friendly" and activities that will do "ecological damage." However, scientists use the terms *ecology* and *environment* in a more restricted way.

Ecology is the branch of biology that studies the relationships between organisms and their environments. This is a very simple definition for a very complex branch of science. Throughout the next four chapters, we will explore many of the aspects of this interesting topic. Because the word *environment* is included in this definition of *ecology*, it is important to have a clear understanding of how an ecologist uses the term. Most ecologists define the word **environment** very broadly as anything that affects an organism during its lifetime. These environmental influences can be divided into two broad categories: biotic environmental factors and abiotic environmental factors.

Biotic and Abiotic Environmental Factors

Biotic factors are living things that affect an organism. You are affected by many different biotic factors. Your classmates, disease organisms, the food you eat, and the trees you seek for shade are all biotic factors. **Abiotic factors** are nonliving things that affect an organism. Common abiotic factors are wind, rain, the composition of the atmosphere, minerals in the soil, sunlight, temperature, and elevation above sea level (figure 15.1).

Characterizing the environment of an organism is a complex and challenging process. There are many things to be considered, and everything seems to be influenced or modified by other factors. For example, consider a fish in a stream; many environmental factors are important to its life. The temperature of the water is extremely important as an abiotic factor, but the temperature may be influenced by the presence of trees (biotic factor) along the stream bank that shade the stream and prevent the Sun from heating it. Obviously, the kind and number of food organisms in the stream are important biotic factors as well. The type of material that makes up the stream bottom and the amount of oxygen dissolved in the water are other important abiotic factors, both of which are related to how rapidly the water is flowing.

Similarly, a plant is influenced by many factors during its lifetime. The types and amounts of minerals in the soil, the amount of sunlight hitting the plant, and the amount of rainfall are important abiotic factors. The animals that eat the plant and the fungi that cause disease are important biotic factors. Each item on this list can be further subdivided. For instance, because plants obtain water from the soil, rainfall is studied in plant ecology. But even the study of rainfall is not simple. In some places, it rains only during one part of the

(a)   (b)

FIGURE 15.1 Biotic and Abiotic Environmental Factors
(a) Redwing blackbirds build their nests in cattail marshes. The cattails and the materials from which the nest is constructed are biotic factors important to the raising of young blackbirds.
(b) The irregular shape of the trees is the result of wind and snow, both abiotic factors. Snow driven by the prevailing winds tends to "sandblast" one side of the tree and prevent limb growth on that side.

year; in other places, it rains throughout the year. Some places experience hard, driving rains, whereas others experience long, misty showers. The kind of rainfall affects how much water soaks into the ground—heavy rains tend to run off into streams and be carried away, whereas gentle rain tends to sink into the soil.

Levels of Organization in Ecology

Ecologists study ecological relationships at several levels of organization. Some ecologists focus on what happens to individual organisms and how they interact with their surroundings. Others are interested in groups of organisms of the same species—called **populations** and how they change. Interacting populations of different species are called **communities,** and community ecologists are interested in how various kinds of organisms interact in a specific location. The highest level of ecological organization is the **ecosystem** which consists of all the interacting organisms in an area and their interactions with their abiotic surroundings. Figure 15.2 shows how these levels of organization are related to one another.

Ecology involves a great deal of detailed knowledge of the physical and living worlds; the relationships between organisms and their environments are quite complex. However, ecologists have developed several broad concepts to help them understand this complexity. These include trophic levels in food chains and how energy and matter flow through ecosystems.

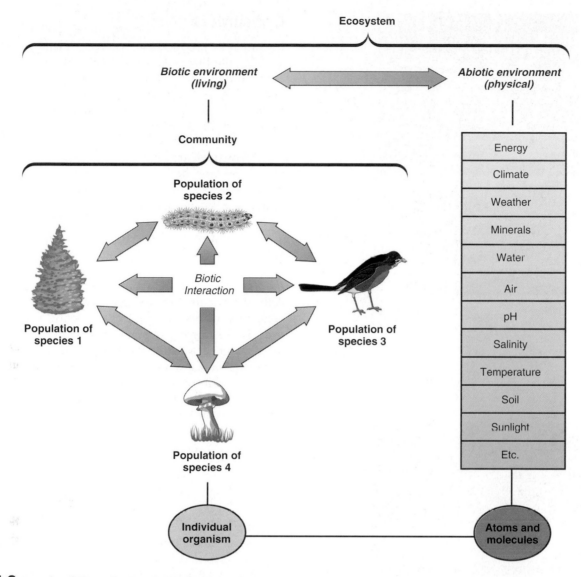

FIGURE 15.2 Levels of Organization in Ecology

Ecology is the branch of biology that studies the interactions between organisms and their environments. This study can take place at several levels, from the broad ecosystem level through community interactions to population studies and the study of individual organisms. Ecology also involves the study of the physical environment, which makes up the nonliving parts of an ecosystem.

15.2 Trophic Levels and Food Chains

One of the broad concepts of ecology is that organisms fit into categories, based on how they satisfy their energy requirements. There is a pattern in the way energy moves through an ecosystem. In general, energy flows from the Sun to plants and from plants to animals. However, there are recognizable steps in this flow of energy. Each stage in the flow of energy through an ecosystem is known as a **trophic level**. The series of organisms feeding on one another is known as a **food chain** (figure 15.3).

Producers

Producers are organisms that trap sunlight and use it to produce organic molecules through the process of photosynthesis. Photosynthesis allows these organisms to produce organic material from inorganic material. Green plants and other photosynthetic organisms, such as algae and cyanobacteria, in effect, convert sunlight energy into the energy contained within the chemical bonds of organic compounds. The creation of these chemical bonds is the result of the flow of energy from the Sun into the living matter of plants. Because producers are the first step in the flow of energy, they occupy the *first trophic level*. photosynthesis, p. 129

Quaternary carnivore	Trophic level 6
Tertiary carnivore	Trophic level 5
Secondary carnivore	Trophic level 4
Primary carnivore	Trophic level 3
Herbivore	Trophic level 2
Producer	Trophic level 1

FIGURE 15.3 Trophic Levels in a Food Chain
As one organism feeds on another organism, energy flows from one trophic level to the next. This illustration shows six trophic levels.

Consumers

Because only producers are capable of capturing energy from the Sun, all other organisms are directly or indirectly dependent on the producer trophic level to meet their energy needs. Because these dependent organisms must consume organic matter as a source of energy, they are called **consumers.** Consumers can be subdivided into several categories, based on how they obtain food.

Herbivores are animals that obtain energy by eating plants. Because herbivores obtain their energy from eating plants, they are also called **primary consumers,** and they occupy the *second trophic level.*

Carnivores are animals that eat other animals. They are also referred to as **secondary consumers.** Carnivores can be subdivided into different trophic levels, depending on what animals they eat. Animals that feed on herbivores occupy the *third trophic level* and are known as **primary carnivores.** Animals that feed on the primary carnivores are known as **secondary carnivores,** and they occupy the *fourth trophic level.* For example, a human may eat a fish that ate a frog that ate a spider that ate an insect that consumed plants for food.

Omnivores are animals that have generalized food habits and act as carnivores sometimes and herbivores other times. They are classified into different trophic levels depending on what they are eating at the moment. Most humans are omnivores.

Decomposers

Decomposers are a special category of consumers that obtain energy when they decompose the organic matter of dead organisms and the waste products of living organisms. Decomposers are usually not assigned to a trophic level, because they break down the organic matter produced by all trophic levels. Furthermore, the decomposer category includes a wide variety of organisms, such as bacteria, fungi, and other microorganisms, that feed on one another. Thus, there are food chains of decomposers (Outlooks 15.1).

Decomposers efficiently convert nonliving organic matter into simple inorganic molecules, which producers can use in the process of trapping energy. Thus, decomposers are very important components of ecosystems that cause materials to be recycled. As long as the Sun supplies the energy, elements are cycled repeatedly through ecosystems. Table 15.1 summarizes the categories of organisms within an ecosystem.

15.3 Energy Flow Through Ecosystems

The ancient Egyptians constructed elaborate tombs we call *pyramids.* The broad base of the pyramid is necessary to support the upper levels of the structure, which narrows to a point at the top. The same kind of relationship exists for the various trophic levels of ecosystems. Biologists have adopted this pyramid model as a way to think about how ecosystems

OUTLOOKS 15.1

Detritus Food Chains

Although most ecosystems receive energy directly from the Sun through the process of photosynthesis, some ecosystems obtain most of their energy from a constant supply of dead organic matter. For example, forest floors and small streams receive a rain of leaves and other bits of material that small animals use as a food source. The small pieces of organic matter, such as broken leaves, feces, and body parts, are known as *detritus*. The insects, slugs, snails, earthworms, and other small animals that use detritus as food are often called *detritivores*.

In the process of consuming leaves, detritivores break down the leaves and other organic material into smaller particles, which other organisms use for food. The smaller size also allows bacteria and fungi to colonize the dead organic matter more effectively, further decomposing the organic material and making it available to still other organisms as a food source. The bacteria and fungi, in turn, are eaten by other detritus feeders. Some biologists believe that the energy flow through detritus food chains is greatly underestimated.

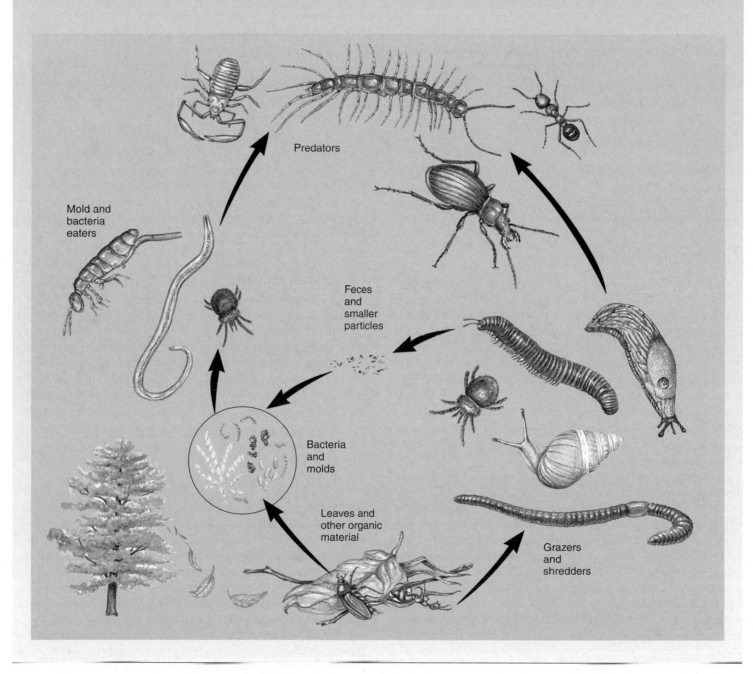

Predators

Mold and bacteria eaters

Feces and smaller particles

Bacteria and molds

Leaves and other organic material

Grazers and shredders

TABLE 15.1

Categories in an Ecosystem

Category	Description	Examples
Producers	Organisms that convert simple inorganic compounds into complex organic compounds by photosynthesis	Trees, flowers, grasses, ferns, mosses, algae, cyanobacteria
Consumers	Organisms that rely on other organisms as food, animals that eat plants or other animals	
Herbivore	Eats plants directly	Deer, goose, cricket, vegetarian human, many snails
Carnivore	Eats meat	Wolf, pike, dragonfly
Omnivore	Eats plants and meat	Rat, most humans
Scavenger	Eats food left by others	Coyote, skunk, vulture, crayfish
Parasite	Lives in or on another organism, using it for food	Tick, tapeworm, many insects
Decomposer	Returns organic compounds to inorganic compounds, is an important component in recycling	Bacteria, fungi

are organized. Most ecosystems have large quantities of producers, small quantities of herbivores, and still smaller quantities of carnivores. Because this is so common, ecologists have sought reasons to explain the relationship.

The Pyramid of Energy

Two fundamental physical laws of energy are important when looking at ecological systems from an energy point of view. The first law of thermodynamics states that energy is neither created nor destroyed. That means that we should be able to describe the amounts of energy in each trophic level and follow energy as it flows through successive trophic levels (figure 15.4). The second law of thermodynamics states that, when energy is converted from one form to another, some energy escapes as heat. This means that, as energy passes from one trophic level to the next, there is a reduction in the amount of energy in living things and an increase in the amount of heat in their surroundings. energy, p. 24

The energy within an ecosystem can be measured in several ways. One simple way is to collect all the organisms present at any trophic level and burn them. For example, all the plants in a small field (producer trophic level) can be harvested and burned. The number of calories of heat produced by burning is equivalent to the energy content of the organic material collected. Similarly, all the herbivores in the second trophic level could be collected and burned. In this way, you can get an idea of how much energy is lost as you go from the producer to the herbivore trophic level. Another way of determining the energy present is to measure the rate of photosynthesis and respiration of a group of producers. The difference between the rates of respiration and photosynthesis is the amount of energy trapped in the living material of the plants.

When we examine a wide variety of ecosystems, we find that the producer trophic level has the most energy, the herbivore trophic level has less, and the carnivore trophic level has the least. *In general, there is about a 90% loss of energy from one trophic level to the next higher level.* Actual measurements vary from one ecosystem to another. Some may lose as much as 99%, whereas other, more efficient systems may lose only 70%, but 90% is a good rule of thumb. This loss in energy content at the second and subsequent trophic levels is primarily due to the second law of thermodynamics. (Whenever energy is converted from one form to another, some energy is lost as heat.) Think of any energy-converting machine; a certain amount of energy enters the machine and a certain amount of work is done. However, it also releases a great deal of heat energy. For example, an automobile engine must have a cooling system to get rid of the heat energy produced. Similarly, electrical energy is used in an incandescent lightbulb to produce light, but the bulb also produces large amounts of heat. Although living systems are somewhat different, they follow the same energy rules.

In addition to the loss of energy as a result of the second law of thermodynamics, there is an additional loss involved in the capture and processing of food material by herbivores and carnivores. Although herbivores don't need to chase their food, they do need to travel to where food is available, then gather, chew, digest, and metabolize it (figure 15.5). All of these processes require energy. Just as the herbivore trophic level experiences a 90% loss in energy content, the higher trophic levels of primary carnivores, secondary carnivores, and tertiary carnivores also experience a reduction in the energy available to them. Figure 15.6 shows the flow of energy through an ecosystem. At each trophic level, the energy content decreases by about 90%.

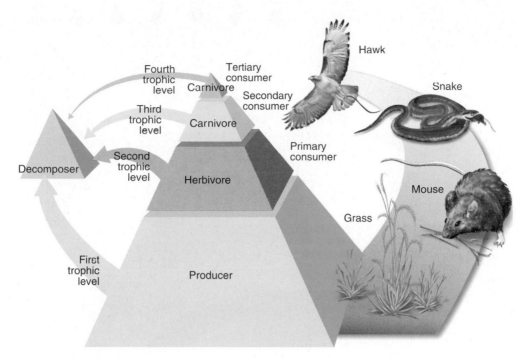

FIGURE 15.4 **Energy and Trophic Levels**
The producer trophic level has the greatest amount of energy and matter. At each successive trophic level, there is less energy and matter.

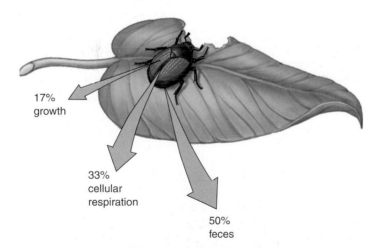

17% growth

33% cellular respiration

50% feces

FIGURE 15.5 **Energy Losses in an Herbivore**
When an insect eats a plant to obtain energy, only a small amount is actually converted to new biological tissue.

The Pyramid of Numbers

Because it is difficult to measure the amount of energy at any one trophic level of an ecosystem, scientists often use other methods to quantify trophic levels. One method is simply to count the number of organisms at each trophic level. This generally gives the same pyramid relationship, called a *pyramid of numbers* (figure 15.7). This is not a very good method to use if the organisms at the different trophic levels are of greatly differing sizes. For example, if you counted all the small insects feeding on the leaves of one large tree, you would actually get an inverted pyramid.

The Pyramid of Biomass

One way to overcome some of the problems associated with simply counting organisms is to measure the *biomass* at each trophic level. **Biomass** is the amount of living material present; it is usually determined by collecting all the organisms at one trophic level and measuring their dry weight. This eliminates the size-difference problem associated with a pyramid of numbers, because all the organisms at each trophic level are weighed. The *pyramid of biomass* also shows the typical 90% loss at each trophic level.

Although a pyramid of biomass is better than a pyramid of numbers in measuring some ecosystems, it has some short-

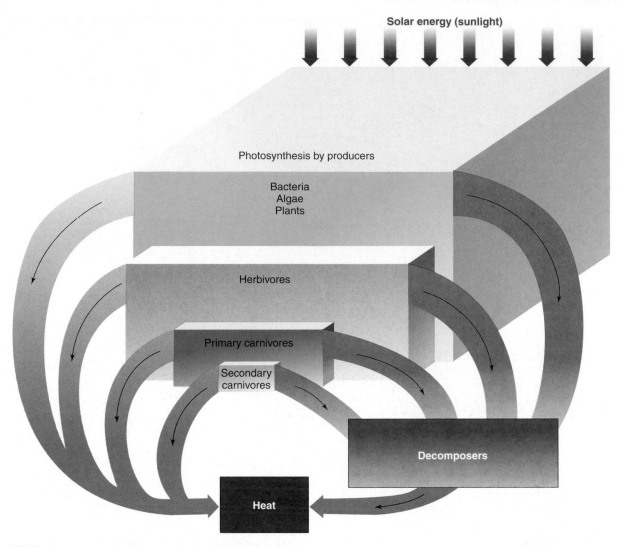

FIGURE 15.6 Energy Flow Through an Ecosystem

Energy from the Sun is captured by organisms that carry on photosynthesis. These are the producers at the first trophic level. As energy flows from one trophic level to the next, approximately 90% of it is lost. This means that the amount of energy at the producer level must be 10 times larger than the amount of energy at the herbivore level. Ultimately, all the energy used by organisms is released to the surroundings as heat.

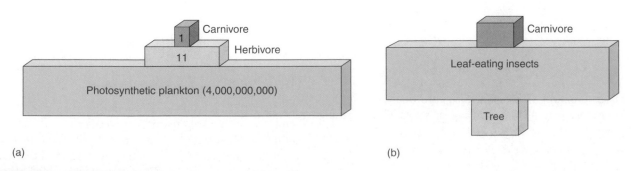

(a)   (b)

FIGURE 15.7 Pyramid of Numbers

One of the easiest ways to quantify the various trophic levels in an ecosystem is to count the number of individuals in a small portion of the ecosystem. As long as all the organisms are of similar size and live about the same length of time, this method gives a good picture of how the trophic levels are related. (*a*) The relationship among photosynthetic plankton in the ocean, the herbivores that eat them, and the carnivores that eat the herbivores is a good example. However, if the organisms at one trophic level are much larger or live much longer than those at other levels, the picture of the relationship may be distorted. (*b*) This is the relationship between forest trees and the insects that feed on them. This pyramid of numbers is inverted.

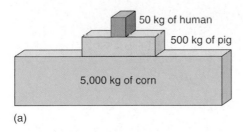

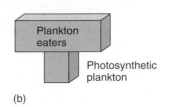

FIGURE 15.8 Pyramid of Biomass

Biomass is determined by collecting and weighing all the organisms in a small portion of an ecosystem. *(a)* This method of quantifying trophic levels eliminates the problem of different-sized organisms at different trophic levels. However, it does not always give a clear picture of the relationship among trophic levels if the organisms have widely different lengths of life. *(b)* For example, in aquatic ecosystems, many of the small producers divide several times per day. The tiny animals (zooplankton) that feed on them live much longer and tend to accumulate biomass over time. The single-celled algae produce much more living material but are eaten as fast as they are produced and, so, do not accumulate a large biomass.

comings. Some organisms tend to accumulate biomass over long periods of time, whereas others do not. Many trees live for hundreds of years; their primary consumers, insects, generally live only 1 year. Likewise, a whale is a long-lived animal, whereas its food organisms are relatively short-lived. Figure 15.8 shows two pyramids of biomass.

15.4 The Cycling of Materials in Ecosystems

Except for small amounts of matter added to the Earth from cosmic dust and meteorites, the amount of matter that makes up the Earth is essentially constant. However, sunlight energy comes to the Earth in a continuous stream, and even this is ultimately returned to space as heat energy. It is this flow of energy that drives all biological processes. Living systems use this energy to assemble organic matter and continue life through growth and reproduction. Because the amount of matter on Earth does not change, the existing atoms must be continually reused as organisms grow, reproduce, and die. In this recycling process, photosynthesis is involved in combining inorganic molecules to form the organic compounds of living things. The process of respiration by all organisms breaks down organic molecules to inorganic molecules. Decomposer organisms are particularly important in breaking down the organic remains of waste products and dead organisms. If there were no way of recycling this organic matter back into its inorganic forms, organic material would build up as the bodies of dead organisms. This is thought to have occurred millions of years ago, when the present deposits of coal, oil, and natural gas were formed. One way to get an appreciation of how various kinds of organisms interact to cycle materials is to look at a specific kind of atom and follow its progress through an ecosystem. Carbon, nitro-

gen, oxygen, hydrogen, phosphorus, and many other atoms are found in all living things and are recycled when an organism dies.

The Carbon Cycle

Carbon atoms are the central ingredients in all organic molecules. These organic molecules are formed by living things from simpler, inorganic molecules. Carbon and oxygen atoms combine to form the inorganic molecule carbon dioxide (CO_2), which is a gas found in small quantities in the atmosphere. During photosynthesis, carbon dioxide (CO_2) combines with water (H_2O) to form complex organic molecules, such as sugar ($C_6H_{12}O_6$). At the same time, oxygen molecules (O_2) are released into the atmosphere (Outlooks 15.2).
photosynthesis, p. 129; cellular respiration, p. 111

The organic matter in the bodies of plants may be used by herbivores as food. When an herbivore eats a plant, it breaks down the complex organic molecules into more simple molecules, such as simple sugars, amino acids, glycerol, and fatty acids. The herbivore can use these molecules as building blocks to construct its own body. Thus, the atoms in the herbivore's body can be traced back to the plants it ate. Similarly, when herbivores are eaten by carnivores, the same atoms are transferred from the herbivores to the carnivores. Finally, the waste products of plants and animals and the remains of dead organisms are used by decomposer organisms as sources of carbon and other atoms they need for survival. In addition, all the organisms in this cycle—plants, herbivores, carnivores, and decomposers—obtain energy (adenosine triphosphate [ATP]) from the process of respiration, in which oxygen (O_2) is used to break down organic compounds into carbon dioxide (CO_2) and water (H_2O). Thus, the carbon atoms that start out as components of carbon dioxide (CO_2) molecules pass through the bodies

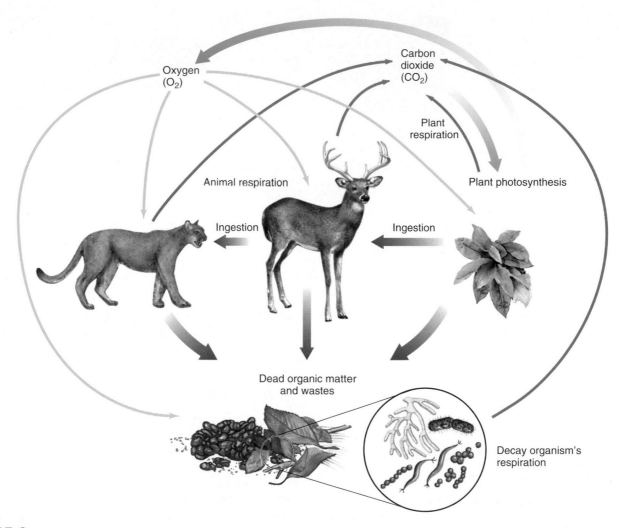

FIGURE 15.9 The Carbon Cycle

Carbon atoms are cycled through ecosystems. Carbon dioxide (green arrows) produced by respiration is the source of the carbon that plants incorporate into organic molecules when they carry on photosynthesis. These carbon-containing organic molecules (black arrows) are passed to animals when they eat plants and other animals. Organic molecules in waste products or dead organisms are consumed by decomposers. In the process, decomposers break down organic molecules into inorganic molecules. All organisms (plants, animals, and decomposers) return carbon atoms to the atmosphere as carbon dioxide when they carry on cellular respiration. Oxygen (blue arrows) is being cycled at the same time that carbon is. The oxygen is released during photosynthesis and taken up during cellular respiration.

of living organisms as parts of organic molecules and return to the atmosphere as carbon dioxide, ready to be cycled again.

The recycling of oxygen atoms is associated with the recycling of carbon atoms. The oxygen molecules (O_2) released during photosynthesis are used during respiration. Thus, photosynthesis and respiration are involved in recycling both carbon and oxygen atoms (figure 15.9).

The Hydrologic Cycle

Water molecules are the most common molecules in living things. Because all the metabolic reactions that occur in organisms take place in a watery environment, water is essen-

tial to life. During photosynthesis, the hydrogen atoms (H) from water molecules (H_2O) are added to carbon atoms to make carbohydrates and other organic molecules. At the same time, the oxygen atoms in water molecules are released as oxygen molecules (O_2). The movement of water molecules can be traced as a hydrologic cycle (figure 15.10)

Most of the forces that cause water to be cycled do not involve organisms but, rather, are the result of normal physical processes. Because of the kinetic energy possessed by water molecules, at normal Earth temperatures liquid water evaporates into the atmosphere as water vapor. This can occur wherever water is present; it evaporates from lakes, rivers, soil, and the surfaces of organisms. Because the oceans contain most of the world's water, an extremely large amount

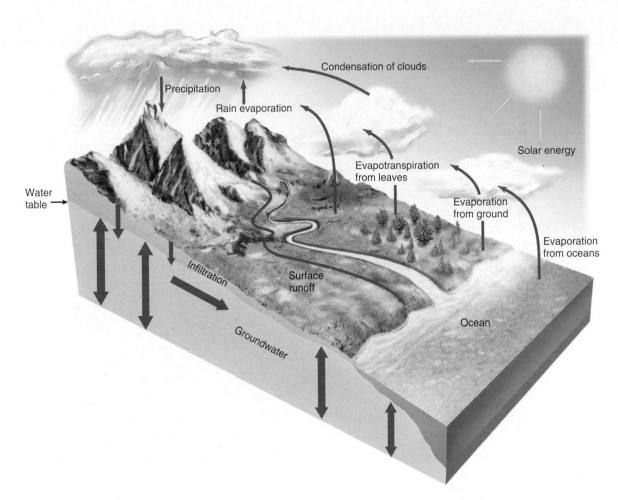

FIGURE 15.10 The Hydrologic Cycle
The cycling of water through the environment follows a simple pattern. Moisture in the atmosphere condenses into droplets, which fall to the Earth as rain or snow. Organisms use some of the water and some of it evaporates from soil and organisms, but much of it flows over the Earth as surface water or through the soil as groundwater. It eventually returns to the oceans, where it evaporates back into the atmosphere to begin the cycle again.

of water enters the atmosphere from the oceans. Water molecules also enter the atmosphere as a result of *transpiration* by plants. **Transpiration** is a process that involves the absorption of water from the soil into roots and the transport of water from the roots to leaves, where it evaporates. This movement of water carries nutrients to the leaves, and the evaporation of the water from the leaves assists in the movement of water upward in the stem.

Once the water molecules are in the atmosphere, they are moved along with other atmospheric gases by prevailing wind patterns. If warm, moist air encounters cooler temperatures, which often happens over landmasses, the water vapor condenses into droplets and falls as rain or snow. When the precipitation falls on land, some of it runs off the surface, some of it evaporates, and some penetrates into the soil. The water that runs off the surface makes its way through streams and rivers to the ocean. The water in the soil may be taken up by plants and transpired into the atmosphere, or it may become groundwater. Much of the groundwater eventually makes its

way into lakes and streams and ultimately arrives at the ocean, from which it originated.

The Nitrogen Cycle

Nitrogen (N) is another important element for living things. Nitrogen is essential in the formation of amino acids, which are needed to form proteins, and in the formation of nitrogenous bases, which are a part of ATP and the nucleic acids, DNA and RNA. Nitrogen is found as molecules of nitrogen gas (N_2) in the atmosphere. Although nitrogen gas (N_2) makes up approximately 80% of the Earth's atmosphere, it is not readily available to most organisms. Only a few kinds of bacteria are able to convert molecular nitrogen (N_2) into the nitrogen compounds that other organisms can use. Therefore, in most terrestrial ecosystems, the amount of nitrogen available limits the amount of plant biomass that can be produced. (Most aquatic ecosystems are limited by the amount of phosphorus, rather than the amount of nitrogen.) Plants use several nitrogen-containing

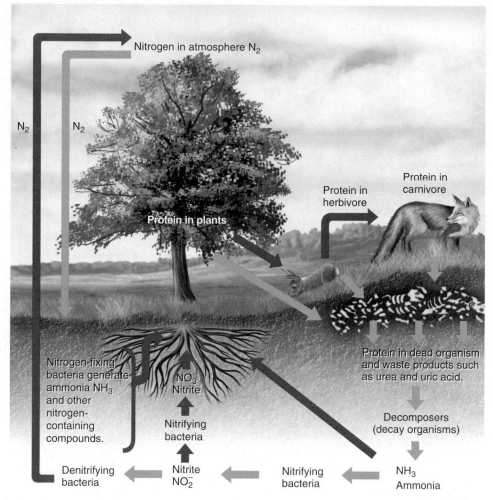

FIGURE 15.11 The Nitrogen Cycle

Nitrogen atoms are cycled through ecosystems. Atmospheric nitrogen is converted by nitrogen-fixing bacteria to nitrogen-containing compounds, which plants can use to make proteins and other compounds. Proteins are passed to other organisms when one organism is eaten by another. Dead organisms and their waste products are acted upon by decay organisms to form ammonia, which can be reused by plants and converted to other nitrogen compounds by nitrifying bacteria. Denitrifying bacteria return nitrogen as a gas to the atmosphere.

compounds to obtain the nitrogen atoms they need to make amino acids and other compounds.

Many kinds of bacteria are important in the nitrogen cycle. **Symbiotic nitrogen-fixing bacteria** live in the roots of certain kinds of plants, where they convert nitrogen gas (N_2) molecules into compounds that the plants can use to make amino acids and nucleic acids. The word *symbiotic* refers to the fact that two kinds of organisms live in close physical contact with each other. The plants that enter most commonly into this symbiotic relationship with bacteria are legumes, such as beans, clover, peas, alfalfa, and locust trees. Some other organisms, such as alder trees and even a kind of aquatic fern, can also participate in this relationship. There are also *free-living nitrogen-fixing bacteria* in the soil. **Free-living nitrogen-fixing bacteria** do not live in a close physical union with plants but release nitrogen compounds that can be taken up through the roots.

Another way in which plants get usable nitrogen compounds involves a different series of bacteria. Decomposer bacteria break down organic nitrogen-containing compounds and release ammonia (NH_3). Various kinds of **nitrifying bacteria** convert ammonia (NH_3) into nitrite-containing (NO_2^-) compounds, which in turn can be converted by other bacteria into nitrate-containing (NO_3^-) compounds. Most plants can use either ammonia (NH_3) or nitrate (NO_3^-) from the soil as building blocks for amino acids and nucleic acids.

All animals obtain nitrogen from the food they eat. The ingested proteins are broken down into their component amino acids during digestion. These amino acids can then be reassembled into new proteins characteristic of the animal. During the the animals' use of the amino acids in food, some of the nitrogen is lost in the waste products as ammonia, urea, or uric acid. As mentioned earlier, decomposer organisms act on all the dead organic matter and waste products of plants and animals to

OUTLOOKS 15.2

Carbon Dioxide and Global Warming

Humans have significantly altered the carbon cycle. As we burn fossil fuels, the amount of carbon dioxide in the atmosphere continually increases. Carbon dioxide allows light to enter the atmosphere but does not allow heat to exit. Because this is similar to what happens in a greenhouse, carbon dioxide and the other gases that have similar effects are called greenhouse gases. Therefore, many scientists are concerned that increased carbon dioxide levels will lead to a warming of the planet, causing major changes in our weather and climate and leading to the flooding of coastal cities and major changes in agricultural production. In 1988, the United Nations established the Intergovernmental Panel on Climate Change (IPCC) to assess the issue of global warming. In 2001 the panel concluded that there has been an increase in the average temperature of the Earth and that the burning of fossil fuels and humans' destruction of forests are the probable causes. There is no doubt that the amount of carbon dioxide in the atmosphere has been increasing. Despite this fact and the conclusions of the IPCC, there is still controversy about this topic, and some still doubt

that warming is occurring. At a meeting in Kyoto, Japan, in 1998, many countries agreed to reduce the amount of carbon dioxide and other greenhouse gases they release into the atmosphere. However, carbon dioxide emissions are directly related to economic activity and the energy use that fuels it. Therefore, it remains to be seen if countries will meet their goals or will succumb to economic pressures to allow the continued use of large amounts of fossil fuels. However, some countries have sought to control the change in the amount of carbon dioxide in the atmosphere by planting millions of trees and preventing the destruction of forests. The thought is that the trees carry on photosynthesis, grow, and store carbon in their bodies, leading to reduced carbon dioxide levels. At the same time, people in other parts of the world continue to destroy forests at a rapid rate. Tree planting does not offset deforestation. In addition, the trees that have been planted will ultimately die and decompose, releasing carbon dioxide back into the atmosphere, so it is not clear that this is an effective means of reducing atmospheric carbon dioxide over the long term.

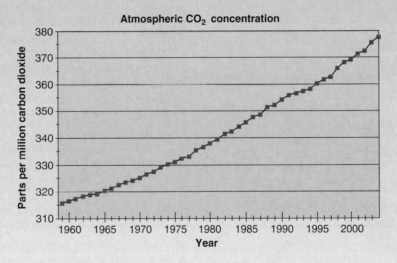

The final step in the nitrogen cycle involves the release of nitrogen gas (N_2) back into the atmosphere. **Denitrifying bacteria** can convert nitrite (NO_2^-) to nitrogen gas (N_2). Thus, in the nitrogen cycle, nitrogen from the atmosphere is passed through a series of organisms, many of which are bacteria, and ultimately returns to the atmosphere to be cycled again. However, the nitrogen cycle differs from other cycles in that a secondary cycle occurs, involving nitrifying bacteria, which recycle nitrogen compounds without returning nitrogen to the atmosphere (figure 15.11).

Because nitrogen is in short supply in most ecosystems, farmers usually find it necessary to supplement the natural

release ammonia (NH_3). Ammonia can be taken up by plants and used directly or can be acted on by nitrifying bacteria to make nitrate (NO_3^-), which also can be used by plants.

nitrogen sources in the soil to obtain maximum plant growth. This can be done in a number of ways. Alternating nitrogen-producing crops with nitrogen-demanding crops helps maintain high levels of usable nitrogen in the soil. One year, a crop such as beans or clover, which has symbiotic nitrogen-fixing bacteria in its roots, can be planted. The following year, the farmer can plant a nitrogen-demanding crop, such as corn. The use of animal wastes (manure) is another way of improving nitrogen levels. Animal wastes are broken down by decomposer bacteria and nitrifying bacteria, resulting in enhanced levels of ammonia and nitrate. Finally, farmers can use industrially produced fertilizers containing ammonia or nitrate. These compounds can be used directly by plants or converted into other useful forms by nitrifying bacteria.

The Phosphorus Cycle

Phosphorus is another kind of atom common in the structure of living things. It is present in many important biological molecules, such as DNA, and in the membrane structure of cells. In addition, animal bones and teeth contain significant quantities of phosphorus. Most of the processes involved in the phosphorus cycle are the geologic processes of erosion and deposition. The ultimate source of phosphorus atoms is rock. In nature, new phosphorus compounds are released by the erosion of rock and are dissolved in water. Plants use the dissolved phosphorus compounds to construct the molecules they need. Animals obtain phosphorus when they consume plants or other animals. When an organism dies or excretes waste products, decomposer organisms recycle the phosphorus compounds back into the soil, where they can be reused.

Phosphorus compounds that are dissolved in water are ultimately precipitated as mineral deposits. This has occurred in the geologic past and typically has involved deposits in the oceans. Geologic processes elevate these deposits and expose them to erosion, thus making phosphorus available to organisms. Animal wastes often have significant amounts of phosphorus. In places where large numbers of seabirds or bats have congregated for hundreds of years, their droppings (called *guano*) can be a significant source of phosphorus for fertilizer (figure 15.12).

In many soils, phosphorus is in short supply and must be provided to crop plants to get maximum yields. Phosphorus

is also in short supply in aquatic ecosystems. Fertilizers usually contain nitrogen, phosphorus, and potassium compounds. The numbers on a fertilizer bag indicate the percentage of each in the fertilizer. For example, a 6-24-24 fertilizer has 6% nitrogen, 24% phosphorus, and 24% potassium compounds.

In addition to carbon, nitrogen, and phosphorus; potassium and other elements are cycled within ecosystems. In an agricultural ecosystem, these elements are removed when the crop is harvested. Therefore, farmers must not only return the nitrogen, phosphorus, and potassium but also analyze for other, less prominent elements and add them to their fertilizer mixture. Aquatic ecosystems are also sensitive to nutrient levels. High levels of nitrates or phosphorus compounds often result in the rapid growth of aquatic producers. In aquaculture, such as that used to raise catfish, fertilizer is added to the body of water to stimulate the production of algae, which is the base of many aquatic food chains. Even synthetic materials that are products of modern technology are cycled through ecosystems (How Science Works 15.1).

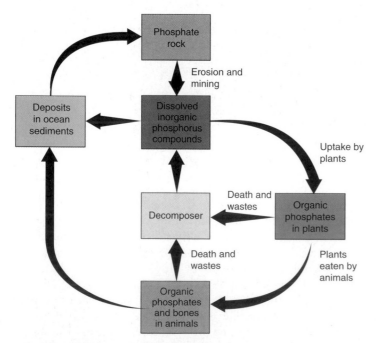

FIGURE 15.12 The Phosphorus Cycle
The primary source of phosphorus is phosphorus-containing rock. The erosion of rock and the dissolving of phosphorus compounds in water makes phosphorus available to the roots of plants. Animals obtain phosphorus in their food. Decomposers recycle phosphorus compounds back into the soil.

15.5 Human Use of Ecosystems

The extent to which humans use an ecosystem is tied to its *productivity*. **Productivity** is the rate at which an ecosystem can accumulate new organic matter. Because plants are the producers, it is their activities that are most important. Ecosystems in which the conditions are the most favorable for plant growth are the most productive. Warm, moist, sunny areas with high levels of nutrients in the soil are ideal. Some areas have low productivity because one of these essential factors is missing. Deserts have low productivity because water is scarce, arctic areas because temperature is low, and the open ocean because nutrients are in short supply. Some terrestrial ecosystems, such as forests and grasslands, have high productivity. Aquatic ecosystems, such as marshes and estuaries, are highly productive, because the waters running into them are rich in the nutrients that aquatic photosynthesizers need. Furthermore, these aquatic systems are usually shallow, so that light can penetrate through most of the water column.

The Conversion of Ecosystems to Human Use

The way humans use ecosystems has changed dramatically over the past several thousand years. Initially, humans fit into ecosystems as just another consumer. These kinds of societies are known as hunter-gatherer societies because they collect food and other needed materials directly from the plants and animals that are a natural part of the ecosystem. There are still examples of peoples who live this way (figure 15.13).

However, the development of agriculture has changed how humans interact with other organisms in ecosystems. We have altered certain ecosystems substantially to increase

HOW SCIENCE WORKS 15.1

Herring Gulls as Indicators of Contamination in the Great Lakes

Herring gulls nest on islands and other protected sites throughout the Great Lakes region. Because they feed primarily on fish, they are near the top of the aquatic food chains and tend to accumulate toxic materials from the food they eat. Eggs taken from nests can be analyzed for a variety of contaminants.

Since the early 1970s, the Canadian Wildlife Service has operated a herring gull monitoring program to assess trends in the levels of various contaminants in the eggs of herring gulls. In general, the levels of contaminants have declined as both the Canadian and U.S. governments have taken action to stop new contaminants from entering the Great Lakes. The figure

shows the trends for PCBs. PCBs are a group of organic compounds, some of which are much more toxic than others. They were formerly used as fire retardants, lubricants, insulation fluids in electrical transformers, and some printing inks. Both Canada and the United States have eliminated most uses of PCBs.

The data collected show a downward trend in the amount of PCBs present in the eggs of herring gulls. Long-term studies like this one are very important in showing slow, steady responses to changes in the environment. Without such long-term studies, scientists would be less sure of the impact of environmental clean-up activities.

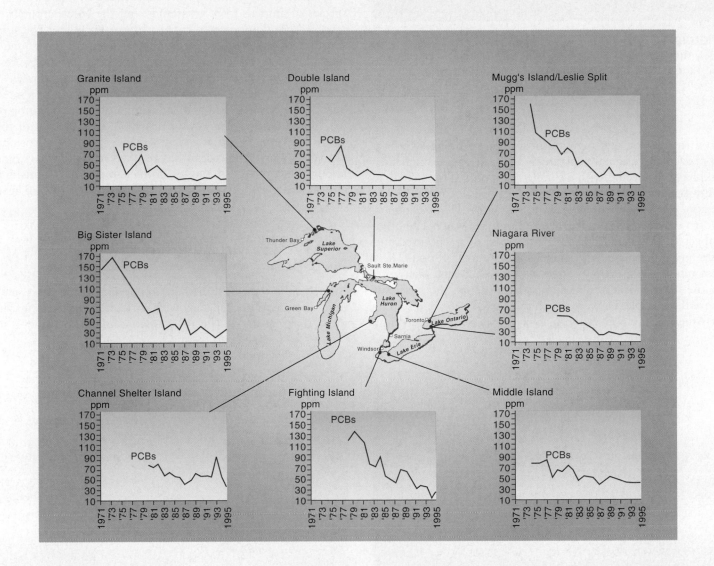

FIGURE 15.13 A Hunter-Gatherer
This Australian aboriginal hunter functions as a carnivore and uses the natural ecosystem as a source of energy and materials.

productivity for our own purposes. In so doing, we have destroyed the original ecosystem, with all of its complexity, and have replaced it with a simpler agricultural ecosystem. In addition, many of the crops we plant are not native to the region. For example, nearly all of the Great Plains region of North America has been converted to agriculture. The original ecosystem included the Native Americans, who used buffalo and other plants and animals as a source of food. There was much grass, many buffalo, and few humans. Therefore, in the Native Americans' pyramid of energy, the base was more than ample. However, with the exploitation and settling of America, the population in North America increased at a rapid rate. The top of the pyramid became larger. The food chain (prairie grass–buffalo–human) could no longer supply the food needs of the growing population. As the top of the pyramid grew, the producer base had to grow larger. Because wheat and corn yield more biomass for humans than the original prairie grasses could, the settlers' domestic grain and cattle replaced the prairie grass and buffalo. This was fine for the settlers, but devastating for the buffalo and Native Americans (figure 15.14). In similar fashion, the deciduous forests of the East were cut down and burned to provide land for crops. The crops were able to provide more food than did harvesting game and plants from the forest.

Many ecosystems, particularly the drier grasslands, cannot support the raising of crops. However, they can still be used as grazing land to raise livestock. Like the raising of crops, grazing often significantly alters the original grassland ecosystem. Some attempts have been made to harvest native species of animals from grasslands, but the species primarily involved are domesticated cattle, sheep, and goats. The substitution of the domesticated animals displaces the animals that are native to the area and alters the kinds of plants present, particularly if too many animals are allowed to graze.

Even aquatic ecosystems have been significantly altered by human activity. The Food and Agriculture Organization of the United Nations states that nearly all the fisheries of the world are being fished at capacity or overfished. Overfishing in many areas of the ocean has resulted in the loss of some important commercial species. For example, the codfishing industry along the East Coast of North America has been destroyed by overfishing.

The Energy Pyramid and Human Nutrition

Anywhere in the world the human population increases, natural ecosystems are replaced with agricultural ecosystems. In many parts of the world, the human demand for food is so

FIGURE 15.14 The Conversion of Prairie to Agricultural Production
North America's Great Plains changed from a natural prairie ecosystem to an agricultural ecosystem.

large that it can be met only if humans occupy the herbivore trophic level, rather than the carnivore trophic level. Humans are omnivores able to eat both plants and animals as food, so they have a choice. However, as the size of the human population increases, it cannot afford the 90% loss that occurs when plants are fed to animals that are in turn eaten by humans. In much of the less-developed world, the primary food is grain; therefore, the people are already at the herbivore level. It is only in the developed countries that people can afford to eat large quantities of meat. This is true from both an energy point of view and a monetary point of view. Figure 15.15 shows a pyramid of biomass having a producer base of 100 kilograms of grain. The second trophic level has only 10 kilograms of cattle because of the 90% energy loss typical when energy is transferred from one trophic level to the next (90% of the corn raised in the United States is used as cattle feed). The consumers at the third trophic level— humans, in this case—experience a similar 90% loss. Therefore, only 1 kilogram of humans can be sustained by the two-step energy transfer. There has been a 99% loss in energy: 100 kilograms of grain are necessary to sustain 1 kilogram of humans. Because much of the world's population is already feeding at the second trophic level, we cannot expect food production to increase to the extent that we could feed 10 times as many people as exist today.

It is unlikely, however, that most people are able to fulfill all their nutritional needs by eating only grains. In addition to calories, people need a certain amount of protein in their diets, and one of the best sources of protein is meat. Although protein is available from plants, the concentration is greater from animal sources. People in major parts of Africa and Asia have diets that are deficient in both calories and protein. These people have very little food, and what food they do have is mainly from plant sources. These are also the parts of the world where human population growth is the most rapid. In other words, these people are poorly nourished, and as the population increases they will probably experience greater calorie and protein deficiency. Even when people live as consumers at the second trophic level, they may still not get enough food; even if they do, it may not have the protein necessary for good health. It is important to realize that currently there is enough food in the world to feed everyone, but it is not distributed equitably for a variety of reasons. The primary reasons for starvation are political and economic. Wars and civil unrest disrupt the normal food-raising process. People leave their homes and migrate to areas unfamiliar to them. Poor people and poor countries cannot afford to buy food from the countries that have a surplus.

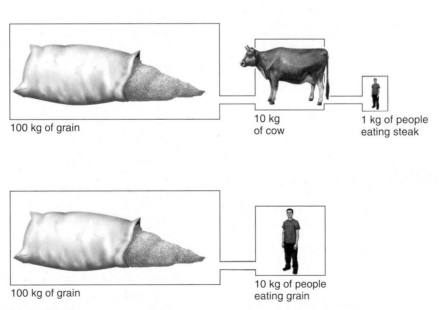

100 kg of grain

10 kg of cow

1 kg of people eating steak

100 kg of grain

10 kg of people eating grain

FIGURE 15.15 Human Pyramids of Biomass
Because approximately 90% of the energy is lost as energy passes from one trophic level to the next, more people can be supported if they eat producers directly than if they feed on herbivores. Much of the less-developed world is in this position today. Rice, corn, wheat, and other producers provide most of the food for the world's people.

Summary

Ecology is the study of how organisms interact with their environment. The environment consists of biotic and abiotic components, which are interrelated in an ecosystem. All ecosystems must have a constant input of energy from the Sun. Producer organisms are capable of trapping the Sun's energy and converting it into biomass. Consumers eat other organisms. Herbivores feed on producers and are, in turn, eaten by carnivores, which may be eaten by other carnivores. Each level in the food chain is known as a trophic level. Other kinds of organisms involved in food chains are omnivores, which eat both plant and animal food, and decomposers, which break down dead organic matter and waste products.

All ecosystems have a large producer base with successively smaller amounts of energy at the herbivore, primary carnivore, and secondary carnivore trophic levels. This is because, each time energy passes from one trophic level to the next, about 90% of the energy is lost from the ecosystem.

The amount of matter in the world does not change but, rather, is recycled. The carbon cycle involves the processes of photosynthesis and respiration in the cycling of carbon through ecosystems. Water is essential to living things, most of the cycling of water involves the physical processes of evaporation and condensation. The nitrogen cycle relies on the activities of nitrogen-fixing bacteria, nitrifying bacteria, and decomposers to cycle nitrogen through ecosystems. The phosphorus cycle involves the deposition of phosphorus-containing compounds into oceans and the geologic processes of uplift and erosion to make phosphorus available to organisms.

Humans use ecosystems to provide themselves with necessary food and raw materials. As the human population increases, most people will be living as herbivores at the second trophic level, because they cannot afford to lose 90% of the energy by first feeding it to an herbivore, which they then eat. Humans have converted most productive ecosystems to agricultural production and continue to seek more agricultural land as populations increase.

Key Terms

Use the interactive flash cards on the **Concepts in Biology, 12/e** website to help you learn the meaning of these terms.

abiotic factors 304
biomass 309
biotic factors 304
carnivores 306
community 304
consumers 306
decomposers 306
denitrifying bacteria 315
ecology 304
ecosystem 304
environment 304
food chain 305
free-living nitrogen-fixing bacteria 314
herbivores 306
nitrifying bacteria 314
omnivores 306
population 304
primary carnivores 306
primary consumers 306
producers 305
productivity 316
secondary carnivores 306
secondary consumers 306
symbiotic nitrogen-fixing bacteria 314
transpiration 313
trophic level 305

Basic Review

1. Which one of the following is an abiotic factor?
   a. a nest in a tree
   b. the water in a pond
   c. the producers in an ecosystem
   d. the fish in a pond
2. Which one of the following categories of organisms have the largest total energy and biomass?
   a. eagles, which eat fish
   b. herbivores, which eat plants
   c. organisms that carry on photosynthesis
   d. fish that eat insects
3. The carbon that plants need for photosynthesis comes from _____.
4. Symbiotic nitrogen-fixing bacteria
   a. live in association with the roots of certain plants.
   b. convert ammonia to nitrate.
   c. are found in the atmosphere.
   d. are rare.
5. The process of absorbing water from the soil and releasing it from leaves is called _____.
6. In the phosphorus cycle, phosphorus enters plants through the roots. (T/F)
7. When energy flows from one trophic level to the next, about _____ percent of the energy is lost.
8. An herbivore is at the second trophic level. (T/F)
9. Nitrogen is important in which one of the following organic molecules?
   a. sugars
   b. fats
   c. water
   d. proteins
10. Decomposers break down organic matter and release _____ and _____.

Answers
1. b 2. c 3. carbon dioxide 4. a 5. transpiration 6. T
7. 90% 8. T 9. d 10. carbon dioxide and water

Concept Review

*Answers to the Concept Review questions can be found on the **Concepts in Biology**, 12/e website.*

15.1 What Is Ecology?

1. List three abiotic factors in the environment of a fish.
2. List three biotic factors in the environment of a songbird.
3. What is the difference between the terms *ecosystem* and *environment*?

15.2 Trophic Levels and Food Chains

4. Describe two ways that decomposers differ from herbivores.
5. Name an organism that occupies each of the following trophic levels: the producer trophic level, the second trophic level, and the third trophic level.
6. What is the difference between an ecosystem and a community?

15.3 Energy Flow Through Ecosystems

7. Describe the flow of energy through an ecosystem.
8. Why is an herbivore biomass usually larger than a carnivore biomass?
9. Can energy be recycled through an ecosystem? Explain why or why not.

15.4 The Cycling of Materials in Ecosystems

10. Trace the flow of carbon atoms through a community that contains plants, herbivores, decomposers, and parasites.
11. Describe four roles that bacteria play in the nitrogen cycle.
12. Describe the flow of water through the hydrologic cycle.
13. List three ways the carbon and nitrogen cycles are similar and three ways they differ.
14. Describe the major processes that make phosphorus available to plants.

15.5 The Human Use of Ecosystems

15. Why must people who live in countries that are not able to produce surplus food eat grains?
16. Define the term *productivity*.

Thinking Critically

*For guidelines to answer this question, visit the **Concepts in Biology**, 12/e website.*

Construct a map on a piece of paper that includes the following items to show their levels of interaction. Which is the most important item? Which items are dependent on others?

People are starving.
Commercial fertilizer production requires temperatures of 900°C.
Geneticists have developed plants that grow very rapidly and require high amounts of nitrogen to germinate during the normal growing season.
Fossil fuels are stored organic matter.
The rate of the nitrogen cycle depends on the activity of bacteria.
The Sun is expected to last for several million years.
Crop rotation is becoming a thing of the past.
The clearing of forests for agriculture changes the weather in the area.

16

Community Interactions

Biological communities and human communities have many similarities. In your social community, you rely on slaughterhouse workers, physicians, farmers, garbage haulers, and many other specialists to perform services for you. In biological communities, there are also specialists with specific roles to play. Carnivores, parasites, plants, and decomposers all have special roles in communities. These two kinds of communities are interconnected. Slaughterhouse workers perform the role of carnivores and provide meat, so that we do not need to do the killing ourselves. Physicians provide knowledge that helps us avoid becoming a host for a parasitic disease. Farmers encourage the growth of plants, so that we do not need to harvest our own plants for food. Garbage haulers are similar to decomposers in that they help us get rid of our waste. This chapter discusses community interactions from a biological point of view.

CHAPTER OUTLINE

16.1 The Nature of Communities

Scientists approach the study of ecological interactions in different ways. For example, in chapter 15, we looked at ecological relationships from the point of view of ecosystems and the way energy and matter flow through them. But we can also study relationships at the community level and focus on the kinds of interactions that take place among organisms. Recall that a community consists of all the populations of different kinds of organisms that interact in a particular location.

Defining Community Boundaries

One of the first things a community ecologist must do is determine the boundaries of the community to be studied. A small pond is an example of a community with easily determined natural boundaries (figure 16.1). The water's edge naturally defines the limits of this community. We would expect to find certain animals and plants living in the pond, such as fish, frogs, snails, insects, algae, pondweeds, bacteria, and fungi. But you might ask at this point, What about the plants and animals that live at the water's edge? Are they part of the pond community? Or what about great blue herons which catch fish and frogs in the pond but build nests

atop some tall trees away from the pond? Or should we include in this community the ducks that spend the night but fly off to feed elsewhere during the day? Should the deer that comes to the pond to drink at dusk be included? What originally seemed to be a clear example of a community has become less clear-cut.

The point of this discussion is that all community boundaries are artificial. However, defining boundaries is important, because it allows us to focus on the changes that occur in a particular area, recognize patterns and trends, and make predictions.

Complexity and Stability

Each community has a particular combination of producers, consumers, and decomposers, which interact in many ways. Within the community, each species is a specialist in certain aspects of community function. One of the ways that organisms interact is by feeding on one another. Most organisms in a community participate in several food chains. When we link all of these food chains together, they become interwoven into a **food web** (figure 16.2).

One of the common features of such a complex set of interrelationships is that natural communities are relatively

FIGURE 16.1 A Pond Community
Although a pond seems an easy community to characterize, it interacts extensively with the surrounding land-based communities. Some of the organisms associated with a pond community are always present in the water (e.g., fish, pondweeds, clams); others occasionally venture from the water to the surrounding land (e.g., frogs, dragonflies, turtles, muskrats); still others are occasional or rare visitors (e.g., minks, heron, ducks).

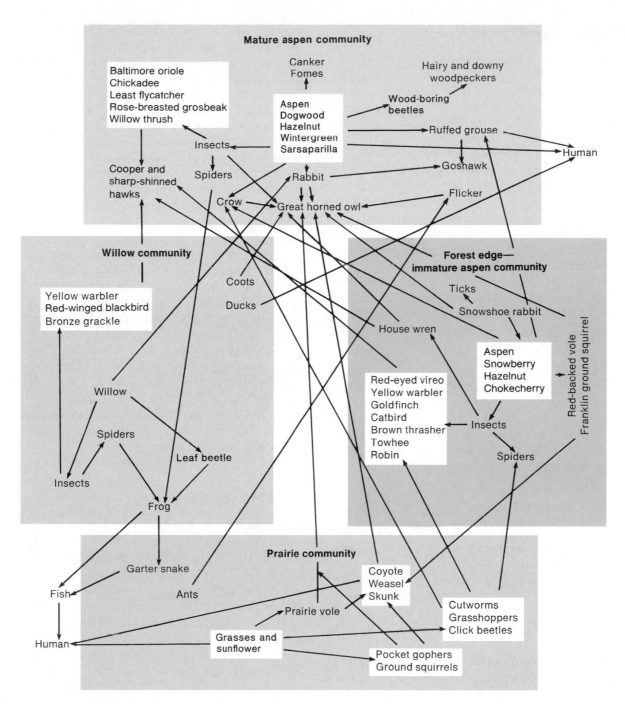

FIGURE 16.2 A Food Web

When many food chains are interlocked with one another, a food web results. The arrows indicate the direction of energy flow. Notice that some organisms are a part of several food chains—the great horned owl, in particular. Because of the interlocking nature of the food web, changing conditions may shift the way in which energy flows through this system.

stable. The fact that we are able to recognize various kinds of communities implies that there is stability. In fact, there is a relationship between the degree of complexity and the stability of a community. Complex communities with many kinds of organisms interacting in numerous ways tend to be more stable than those that have few organisms and interactions. In other words, over hundreds of years, you would see the same kinds of organisms living in a woodlot, prairie, or ponderosa pine forest. On the other hand, many human-influenced communities are very unstable. Agricultural ecosystems tend to have very few kinds of organisms and cannot maintain themselves without human involvement.

Communities Are Dynamic

Stability does not mean that there is no activity, though. For example, although the human body is relatively stable, its stability results from the constant activity of various organs. Although at the broad level communities may appear to be unchanging, there is much change occurring among the organisms present. Producers are capturing sunlight, herbivores are eating plants, carnivores are eating animals, and decomposers are recycling materials. Populations of many organisms may rise and fall as well.

However, if the numbers of a particular kind of organism in a community increase or decrease significantly, some adjustment usually occurs in the populations of other organisms within the community. For example, the populations of many kinds of small mammals (mice, voles, lemmings, and so on) vary from year to year. If there are few small mammals to eat, the populations of their predators decrease. In addition, their predators switch to different prey species, affecting other parts of the community.

16.2 Niche and Habitat

One of the ways scientists learn how communities work is by focusing on the activities and impacts of certain species. Each organism has a role to play and is involved in a complex set of interactions with other organisms.

The Niche Concept

The **niche** of an organism is its specific functional role in its community. A complete description of an organism's niche involves a detailed understanding of the impacts an organism has on its biotic and abiotic surroundings, as well as all the factors that affect the organism. A complete description of the niche of an organism is probably impossible. If we look hard enough, we discover something new we didn't understand before.

For example, the niche of an earthworm includes how an earthworm is affected by abiotic factors such as soil particle size; soil texture; and soil moisture, pH, and temperature. Its niche also includes the earthworm's biotic impacts, such as its roles as food for birds, moles, and shrews; as bait for anglers; and as a consumer of dead plant organic matter (figure 16.3). In addition, an earthworm (1) serves as a host for a variety of parasites, (2) transports minerals and nutrients from deeper soil layers to the surface, (3) incorporates organic matter into the soil, and (4) creates burrows, which allow air and water to penetrate the soil more easily. And this is only a limited sample of all the aspects of its niche.

Some organisms have rather broad niches; others, with very specialized requirements and limited roles to play, have niches that are quite narrow. The opossum (figure 16.4a) is an animal with a very broad niche. It eats a wide variety of plant and animal foods, can adjust to a wide variety of cli-

mates, is used as food by many kinds of carnivores (including humans), and produces large numbers of offspring. By contrast, the koala of Australia (figure 16.4b) has a very narrow niche. It can live only in areas of Australia with specific species of *Eucalyptus* trees, because it eats the leaves of only a few kinds of these trees. Furthermore, it cannot tolerate low temperatures and does not produce many offspring. As you might guess, the opossum is expanding its range, and the koala is endangered in much of its range.

Because an organism's niche includes a complex set of interactions with its surroundings, it is often easy to overlook important roles played by some organisms. For example, Europeans introduced cattle into Australia, where there had previously been no large, hoofed mammals. When this was done, no one thought about the effects of cow manure on the community or the significance of *dung beetles*. In parts of the world where dung beetles exist, they rapidly colonize fresh dung and cause it to be broken down. No such beetles existed in Australia; therefore, in areas of Australia where cattle were raised, a significant amount of land became covered with accumulated cow dung. This reduced the area where grass could grow and reduced productivity. The problem was eventually solved by the importation of several species of dung beetles from Africa, where large, hoofed mammals are common. The dung beetles made use of what the cattle did not digest, returning it to a form that plants could recycle more easily into plant biomass.

The Habitat Concept

The **habitat** of an organism is the kind of place or community in which it lives. Each organism has particular requirements for life and lives where the environment provides what it needs. Habitats are usually described in terms of a conspicuous or particularly significant feature. For example, the habitat of a prairie dog is usually described as a grassland and the habitat of a tuna is described as the open ocean. The habitat of a fiddler crab is sandy ocean shores and the habitat of various kinds of cacti is the desert. It is possible to describe the habitat of the bacterium *Escherichia coli* as the gut of humans and other mammals and the habitat of a fungus as a rotting log. Organisms that have very specific places in which they live simply have more restricted habitats.

16.3 Kinds of Organism Interactions

One of the important components of an organism's niche is the other living things with which it interacts. Organisms can influence one another in numerous ways. Some interactions are harmful to one or both of the organisms, whereas other interactions are beneficial. Ecologists have classified the kinds of interactions between organisms into broad categories.

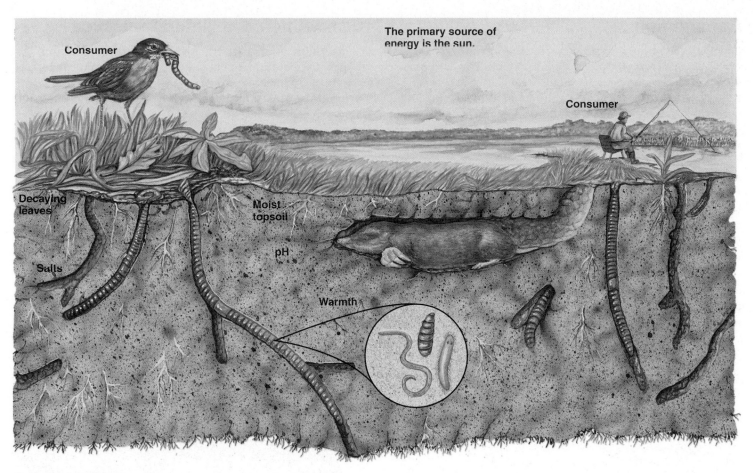

FIGURE 16.3 The Niche of an Earthworm

The niche of an earthworm involves a great many factors. It includes the fact that the earthworm is a consumer of dead organic matter, a source of food for other animals, a host to parasites, and bait for an angler. Furthermore, the earthworm loosens the soil by burrowing and "plows" the soil when it deposits materials on the surface. Additionally, the pH, texture, and moisture content of the soil have an impact on the earthworm. This is but a small part of what the niche of the earthworm includes.

(a)

(b)

FIGURE 16.4 Broad and Narrow Niches

(a) The opossum has a very broad niche. It eats a variety of foods, is able to live in a variety of habitats, and has a large reproductive capacity. It is generally extending its range in the United States. (b) The koala has a narrow niche. It feeds on the leaves of only a few species of *Eucalyptus* trees, is restricted to relatively warm, forested areas, and is generally endangered in much of its habitat in Australia.

Competition

Competition is an interaction between organisms in which both organisms are harmed to some extent. This is the most common kind of interaction among organisms. Every organism is involved in competition at most times in its life. Competition occurs whenever two organisms need a vital resource that is in short supply (figure 16.5). The vital resource may be food, shelter, nesting sites, water, mates, or space.

Intraspecific competition occurs between members of the *same* species. It can involve a snarling tug-of-war between two dogs over a scrap of food or a silent struggle between lodge pole pine seedlings for access to available light. **Interspecific competition** occurs between members of *different* species. The interaction between weeds and tomato plants in a garden is an example of interspecific competition. If the weeds are not removed, they compete with the tomatoes for available sunlight, water, and nutrients, resulting in poor growth of both the tomatos and weeds. Similarly, there is interspecific competition among the many species of carnivores (e.g., hawks, owls, coyotes, foxes) for the small mammals and birds they use for food.

Competition and Natural Selection

Competition is a powerful force for natural selection. Although competition results in harm to both organisms, there can still be winners and losers. The two organisms may not be harmed to the same extent, with the result that one gains greater access to the limited resource. Biologists have recognized that, the more similar the requirements of two species of organisms, the more intense the competition between them. This has led to the development of a general rule known as the *competitive exclusion principle.* According to the **competitive exclusion principle,** no two species of organisms can occupy the same niche at the same time. If two species of organisms do occupy the same niche, the competition will be so intense that one or more of the following processes will occur: (1) one of the two species will become extinct, (2) one will migrate to a different area where competition is less intense, or (3) the two species will evolve into slightly different niches, so that the intensity of the competition is reduced. For example, a study of the feeding habits of several kinds of warblers shows that, although they live in the same place and feed on similar organisms, their niches are

(a)

(b)

FIGURE 16.5 Competition

Whenever a needed resource is in limited supply, organisms compete for it. This competition may be between members of the same species (*intraspecific*). *(a)* These three grizzly cubs are engaging in intraspecific competition. Competition may also occur between different species (*interspecific*). *(b)* The many species of plants in the Olympic Rainforest are all competing for sunlight.

slightly different, because they feed in different places on trees (figure 16.6).

Another example, involves the many birds which catch flying insects as food. However, they do not compete directly with each other, because some feed at night, some feed high in the air, some feed only near the ground, and still others perch on branches and wait for insects to fly past. The insect-eating niche can be further subdivided by specialization in particular sizes or kinds of insects.

Predation

Predation is an interaction in which one animal captures, kills, and eats another animal. The organism that is killed is the **prey,** and the one that does the killing is the **predator.** The predator benefits from the relationship because it obtains a source of food; obviously, the prey organism is harmed. Most predators are relatively large, compared with their prey, and have specific adaptations to aid them in catching prey. Many predators, such as leopards, lions, cheetahs, hawks, squid, sharks, and salmon, use speed and strength to capture their prey. Similarly, dragonflies, bats, and swallows patrol areas where they can capture flying insects.

Although we tend to think of predators as fierce animals that chase and kill their prey, there are other styles of predation. Predators such as frogs, many kinds of lizards, and insects (e.g., praying mantis) blend in with their surroundings and strike quickly when a prey organism happens by. Wolf spiders and jumping spiders have large eyes, which help them find prey, which they pounce on and kill. The webs of other kinds of spiders serve as nets to catch flying insects. The prey are quickly paralyzed by the spider's bite and wrapped in a tangle of silk threads. Many kinds of birds, insects, and mammals simply search for slow-moving prey, such as caterpillars, grubs, aphids, slugs, snails, and similar organisms. Many kinds of marine snails are predators of other slow-moving sea creatures (figure 16.7).

Often predators are useful to humans because they control populations of organisms that do us harm. For example, snakes eat many kinds of rodents that eat stored grain and other agricultural products. Birds and bats eat insects that are agricultural pests.

It is even possible to think of a predator as having a beneficial effect on the prey species. Certainly, the *individual* organism that is killed is harmed, but the *population* can benefit. Predators can prevent large populations of prey organisms from destroying their habitat by hindering the overpopulation of prey species, or they can reduce the likelihood of epidemic disease by eating sick or diseased individuals. Furthermore, predators act as selecting agents. The individuals that fall prey to them are likely to be less well adapted than the ones that escape predation. Predators usually kill slow, unwary, poorly hidden, sick, or injured individuals. Thus, the genes that may have contributed to slowness,

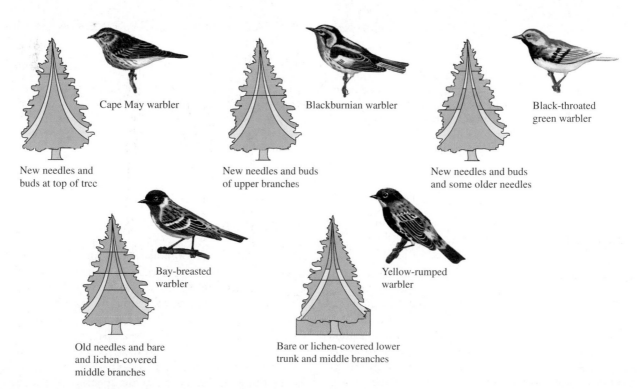

Cape May warbler

New needles and buds at top of tree

Blackburnian warbler

New needles and buds of upper branches

Black-throated green warbler

New needles and buds and some older needles

Bay-breasted warbler

Old needles and bare and lichen-covered middle branches

Yellow-rumped warbler

Bare or lichen-covered lower trunk and middle branches

FIGURE 16.6 Niche Specialization
Although all of these warbler species have similar feeding habits, they limit the intensity of competition by feeding on different parts of the tree.

(a)

(b)

(c)

FIGURE 16.7 The Predator-Prey Relationship

(a) Many predators, such as lions and cheetahs, use speed and strength to capture prey. *(b)* Other predators, such as frogs and chameleons, blend in with their surroundings, lie in wait, and ambush their prey. *(c)* Some spiders use nets to capture prey. Obviously, predators benefit from the food they obtain, to the detriment of the prey.

inattention, poor camouflage, illness, or the likelihood of being injured are removed from the gene pool.

Symbiotic Relationships

Symbiosis means "living together." Unfortunately, this word is used in several ways, none of which is very precise. However, the term symbiosis is usually used for interactions that involve a close physical relationship between two kinds of organisms. The three kinds of relationships discussed in the following sections—parasitism, commensalism, and mutualism—are often referred to as symbiotic relationships because they usually involve organisms that are physically connected to one another.

Parasitism

In **parasitism**, one organism lives in or on another living organism, from which it derives nourishment. The **parasite** derives benefit and harms the **host**, the organism it lives in or on. Parasites are smaller than their hosts. In general, they do not kill their host quickly but, rather, use it as a source of food for a long time. However, the parasite's activities may weaken the host so that it eventually dies. Parasitism is a very common kind of interrelationship. Nearly every category of living thing has some parasitic species. There are parasitic bacteria, fungi, protozoa, plants, fish, insects, worms, mites, and ticks. In fact, there are more species of parasites in the world than there are nonparasites.

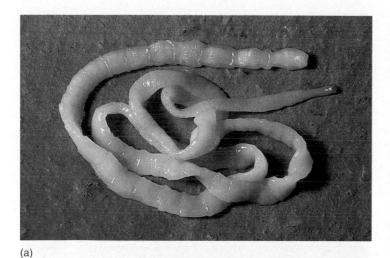

(a)

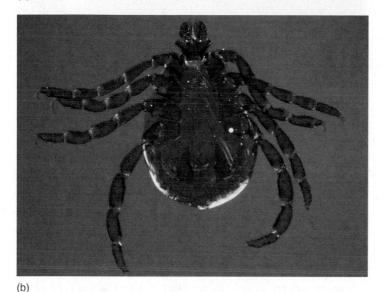

(b)

(c)

FIGURE 16.8 The Parasite-Host Relationship
Parasites benefit from the relationship because they obtain nourishment from the host. *(a)* Tapeworms are internal parasites in the guts of their host, where they absorb food from the host's gut. *(b)* The tick is an external parasite that sucks body fluids from its host. *(c)* Mistletoe is a photosynthesizing plant that also absorbs nutrients from the tissues of its host tree. The host in any of these three situations may not be killed directly by the relationship, but it is often weakened, becoming more vulnerable to predators and diseases.

There are many styles of parasitism. Parasites that live on the outside of their hosts are called **external parasites.** For example, ticks live on the outside of the bodies of animals, such as rats, turtles, and humans, where they suck blood and do harm to their hosts. **Internal parasites** live on the inside of their hosts. For example, tapeworms live in their hosts' intestines. Several kinds of plants are parasitic; mistletoe invades the tissues of the tree it is living on to derive its nourishment (figure 16.8). Some flowering plants, such as beech drops and Indian pipe, lack chlorophyll and are not able to do photosynthesis. They derive their nourishment by living on the roots of their host tree and grow above ground for a short period when they flower.

Many kinds of fungi are parasites of plants, including commercially valuable plants. Farmers spend millions of dollars each year to control fungus parasites. Many kinds of insects, worms, protozoa, bacteria, and viruses are important human parasites.

Many parasites have extremely complicated life cycles (chapter 23 discusses the life cycle of several kinds of worm parasites). In many of these life cycles, some species carry the parasite from one host to the next. Such a carrier organism is known as a **vector.** For example, the protozoan that causes malaria is carried from one human to another by certain species of mosquitos, and the bacterium *Borrelia burgdorferi*, which causes lyme disease, is carried by certain species of ticks (figure 16.9).

Special Kinds of Predation and Parasitism

Both predation and parasitism are relationships in which one member of the pair is helped and the other is harmed. But there are many kinds of common interactions in which one is harmed and the other aided that don't fit neatly into the categories of interactions dreamed up by scientists. For example,

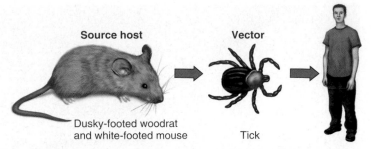

Source host   Vector

Dusky-footed woodrat
and white-footed mouse   Tick

FIGURE 16.9 Lyme Disease—Hosts, Parasites, and Vectors
Lyme disease is a bacterial disease originally identified in a small
number of people in the Old Lyme, Connecticut, area. Today, it is
found throughout the United States and Canada. The parasite
Borrelia burgdorferi, is a bacterium that can live in a variety of
mammalian hosts (e.g., humans, mice, horses, cattle, domestic cats,
and dogs). Certain ticks are vectors that suck blood from an
infected animal and carry the disease to another.

when a deer eats the leaves off a tree or a goose eats grass,
they are doing harm to the plant they are eating while deriving a benefit. In essence, these herbivores are *plant predators
or parasites* (figure 16.10). Likewise, such animals as mosquitoes, biting flies, vampire bats, and ticks take blood meals
but don't usually live permanently on the host or kill it. Are
they temporary parasites or specialized predators?

Finally, birds such as cowbirds and some species of
European cuckoos do not build nests but, rather, lay their
eggs in the nests of other species of bird, which raise these foster young rather than their own. The adult cowbird and
cuckoo often remove eggs from the host nest. In addition,
cowbird and cuckoo offspring typically push the hosts' eggs
or young out of the nest. For these reasons, typically only the
cowbird or cuckoo chick is raised by the foster parents. This
kind of relationship has been called *nest parasitism.* The surrogate parents (hosts) are harmed, and the cowbird or cuckoo
is aided by having others expend the energy needed to raise
its young (figure 16.11).

Commensalism

Commensalism is a relationship in which one organism benefits and the other is not affected. For example, sharks often
have another fish, the remora, attached to them. The remora
has a sucker on the top side of its head, which allows it to
attach to the shark for a free ride (figure 16.12a). Although the
remora benefits from the free ride and by eating leftovers from
the shark's meals, the shark does not appear to be troubled by
this uninvited guest, nor does it benefit from its presence.

Another example of commensalism is the relationship
between trees and epiphytic plants. **Epiphytes** are plants that
live on the surface of other plants but that do not derive nourishment from them (figure 16.12b). Many kinds of plants
(e.g., orchids, ferns, and mosses) use the surfaces of trees as
places to live. These kinds of organisms are particularly common in tropical rainforests. Many epiphytes derive a benefit

FIGURE 16.10 Herbivores—Plant Predators or Parasites
Herbivores have a relationship with plants that is very similar to
that of carnivores with their prey and parasites with their hosts.
The herbivores are aided and the plants they feed on are harmed.

FIGURE 16.11 Nest Parasitism
A cowbird chick in the nest is being fed by its host parents. The
cowbird benefits but the host is harmed.

(a)

(b)

FIGURE 16.12 Commensalism

In commensalism, one organism benefits and the other is not affected. *(a)* The remora shown here hitchhike a ride on the shark. They eat scraps of food left over from the shark's messy eating habits. The shark does not seem to be hindered in any way. *(b)* The epiphytic plants growing on these trees do not harm the trees but are aided by using the tree surface as a place to grow.

from the relationship because they are able to be located in the tops of the trees, where they receive more sunlight and moisture. The trees derive no benefit from the relationship, nor are they harmed; they simply serve as support surfaces for the epiphytes.

Mutualism

Mutualism is an interrelationship in which two species live in close association with one another, and both benefit. Many kinds of animals that eat plants have a mutualistic relationship with the bacteria and protozoa that live in their guts. One of the major components of plant material is the cellulose material that makes up the cell wall. Most animals are unable to digest cellulose but rely on a collection of microorganisms in their guts to perform that function. For example, mammals such as cows, goats, camels, giraffes, and sheep have specialized portions of their guts, called a rumen, in which microorganisms live. These microbes produce enzymes, known as *cellulases,* that break down the cellulose in the food the animal eats. The microorganisms benefit because the gut provides them with a moist, warm, nourishing environment in which to live. The animals benefit because the breakdown of cellulose provides nutrients the animal could not get otherwise. Termites, plant-eating lizards, and many other kinds of herbivores have similar relationships with the bacteria and protozoa living in their digestive tracts which help them digest cellulose.

Lichens and corals exhibit a more intimate kind of mutualism. Both consist of the cells of two different, intermingled organisms. Lichens consist of fungal cells and algal cells in a partnership; corals consist of the cells of the coral organism intermingled with algal cells. In both, the algae carry on photosynthesis and provide nutrients, and the fungus or coral provides a moist, fixed structure for the algae to live in.

Another kind of mutualistic relationship exists between flowering plants and insects. Bees and other insects visit flowers to obtain nectar from the blossoms (figure 16.13). Usually, the flowers are constructed so that the bees pick up pollen (sperm-containing packages) from the plant on their hairy bodies, which they transfer to the female part of the next flower they visit. Because bees normally visit many individual flowers of the same species for several minutes and ignore other species of flowers, they can serve as pollen carriers between two flowers of the same species. Plants pollinated in this manner produce less pollen than do plants that rely on the wind to transfer pollen. This saves the plant energy, because it doesn't need to produce huge quantities of pollen. It does, however, need to transfer some of its energy savings into the production of showy flowers and nectar to attract the bees. The bees benefit from both the nectar and the pollen; they use both for food.

(a)

(b)

(c)

FIGURE 16.13 Mutualism

Mutualism is an interaction between two organisms in which both benefit. *(a)* The gray lichen in this photograph consists of a mutualistic association between a fungus and an alga *(b)* Ruminant animals have a relationship with the microorganisms in their gut that helps them obtain nutrients from the plants they eat. *(c)* Insects obtain nectar from plants; the plants benefit by being pollinated. (Note the yellow pollen on the bee.)

Some plants use birds, bats, mice, beetles, flies, and other kinds of organisms to get their pollen distributed. Each kind of flower is specialized to the kind of pollinating animal. Flowers that are pollinated by bats flower at night; and many of those that are pollinated by hummingbirds have long, tubular flowers.

16.4 Types of Communities

Ecologists recognize that the world can be divided into large, regional, terrestrial communities known as *biomes*. **Biomes** are communities of organisms that are adapted to particular climatic conditions. The primary climatic factors that determine the kinds of organisms that can live in an area are the amount and pattern of *precipitation* and the *temperature* ranges typical for the region (figure 16.14). The map in figure 16.15 shows the distribution of the major biomes of the world. Each biome can be characterized by specific climatic conditions, particular kinds of organisms, and characteristic activities of the region's organisms.

Temperate Deciduous Forest

Temperate deciduous forest exists in the parts of the world that have moderate rainfall (75 to 130 centimeters per year) spread over the entire year and a relatively long summer growing season (130 to 260 days without frost). This biome, like other land-based biomes, is named for a major feature of the community, its dominant vegetation. The predominant

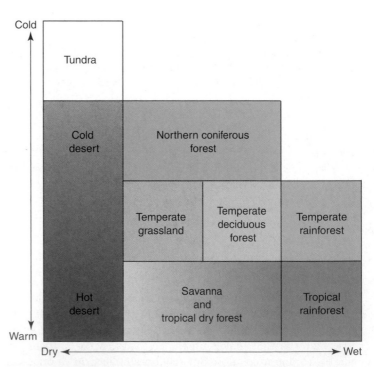

FIGURE 16.14 The Influence of Precipitation and Temperature on Vegetation

Temperature and moisture influence the kind of vegetation that can live in an area. Areas with low moisture and low temperature produce tundra vegetation. Areas with high moisture and freezing temperatures during part of the year produce deciduous or coniferous forests. Dry areas are characterized by deserts. Moderate amounts of rainfall or seasonal rainfall support grasslands or savannas. Areas with high rainfall and high temperatures support tropical rainforests.

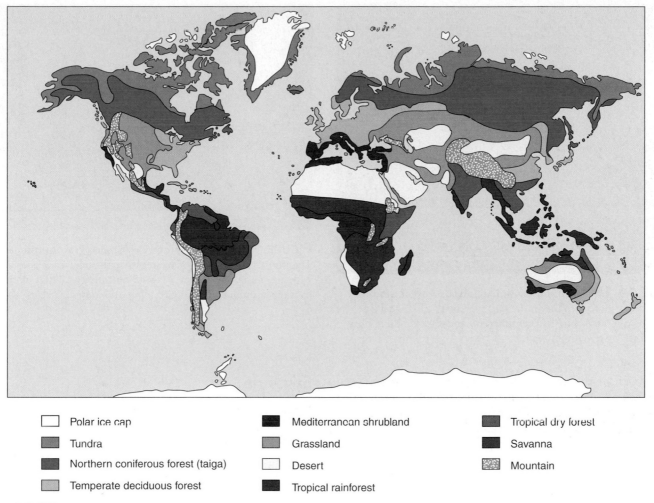

FIGURE 16.15 Biomes of the World
Major climatic differences determine the kind of vegetation that can live in a region. Associated with specialized groups of plants are particular kinds of animals. These regional communities are called biomes.

Legend:
- Polar ice cap
- Tundra
- Northern coniferous forest (taiga)
- Temperate deciduous forest
- Mediterranean shrubland
- Grassland
- Desert
- Tropical rainforest
- Tropical dry forest
- Savanna
- Mountain

plants are large trees that lose their leaves during the fall. Thus, they are called deciduous (figure 16.16). The trees typical of this biome are adapted to conditions with significant precipitation and short, mild winters. The temperate deciduous forest covers a large area from the Mississippi River to the Atlantic Coast and from Florida to southern Canada. This type of biome is also found in parts of Europe and Asia.

Because the trees are the major producers and new leaves are produced each spring, the important primary consumers in this biome are leaf-eating insects. These insects then become food for a variety of birds that raise their young in the forest during the summer and migrate to more moderate climates in the fall. Many other animals, such as squirrels, some birds, and deer, use the trees' fruits for food. Carnivores, such as foxes, hawks, and owls, eat many of the small mammals and birds typical of the region. Another feature of the temperate deciduous forest is an abundance of spring woodland wildflowers, which emerge early in the spring before the trees have leafed out.

The region where temperate deciduous forests exist is large and has somewhat different climatic conditions in various areas. Therefore, there are differences in the species of trees (and other organisms) usually found in different areas within this biome. For instance, in Maryland the tulip tree is one of the state's common large trees, whereas in Michigan it is so unusual that people plant it in lawns and parks as a decorative tree. Aspen, birch, cottonwood, oak, hickory, beech, and maple are typical trees of the temperate deciduous forest. Typical animals of this biome are many kinds of leaf-eating insects, wood-boring beetles, migratory birds, skunks, porcupines, deer, frogs, opossums, owls, and mosquitoes.

FIGURE 16.16 The Temperate Deciduous Forest Biome
This kind of biome is found in the parts of the world that have significant rainfall (75 to 130 centimeters) and short, mild winters when the trees are without leaves.

In much of this region, the natural vegetation has been removed to allow for agriculture, so the original character of the biome is gone, except where farming is not practical or the original forest has been preserved.

Temperate Grassland (Prairie)

Temperate grassland, or *prairie*, exists where the rainfall (30 to 85 centimeters per year) is not adequate to support the growth of trees; the dominant vegetation consists of various species of grasses (figure 16.17). Typical are long periods during the year when there is no rainfall. Trees are common in this biome only along streams, where they can obtain sufficient water. Interspersed among the grasses are many kinds of prairie wildflowers. The dominant animals are those that use grasses as food; large, grazing mammals (bison and pronghorn antelope); small insects (e.g., grasshoppers and ants); and rodents (e.g., mice and prairie dogs). A variety of carnivores (e.g., meadowlarks, coyotes, and snakes) feed on the herbivores. Most of the bird species are seasonal visitors to the prairie. At one time, fire was a common feature of the prairie during the dry part of the year.

In North America, temperate grasslands are found throughout much of the region between the Mississippi River and the Rocky Mountains. This kind of biome is also common in parts of Eurasia, Africa, Australia, and South America.

Today, most of the original grasslands, like the temperate deciduous forest, have been converted to agricultural uses. Grasslands that are too dry to allow for farming typically have been used as grazing land for cattle and sheep. The grazing of these domesticated animals has modified the natural vegetation, as has farming in the moister grassland regions.

FIGURE 16.17 Temperate Grassland (Prairie) Biome
This short-grass prairie of western North America is associated with an annual rainfall of 30 to 85 centimeters. This community contains a unique grouping of plant and animal species.

Savanna

Savannas are tropical biomes of central Africa, northern Australia, and parts of South America; they have distinct wet and dry seasons (figure 16.18). Although these regions may receive 100 centimeters of rainfall per year, there is an extended dry season of 3 months or more. Because of the extended period of dryness, the dominant vegetation consists of grasses. In addition, a few thorny, widely spaced, drought-resistant trees dot the landscape. Many kinds of grazing mammals are found in this biome—various species of antelope, wildebeest, and zebras in Africa; various kinds of kangaroos in Australia; and a large rodent, the capybara, in South America. Another animal common to the savanna is the termite, a colonial insect that typically builds mounds above the ground.

During the wet part of the season, the trees produce leaves, the grass grows rapidly, and most of the animals raise their young. In the African savanna, seasonal migration of the grazing animals is typical. Many of these tropical grasslands have been converted to grazing for cattle and other domesticated animals.

Desert

Deserts are very dry areas found throughout the world wherever rainfall is low and irregular. Typically, the rainfall is less than 25 centimeters per year. Some deserts are extremely hot; others can be quite cold during much of the year. The distinguishing characteristic of desert biomes is low rainfall, not high temperature. Furthermore, deserts show large daily fluctuations in air temperature. When the Sun goes down at

FIGURE 16.18 Savanna Biome

A savanna is likely to develop in tropical areas that have a rainy season and a dry season. During the dry season, fires are frequent. The fires kill tree seedlings and prevent the establishment of forests.

FIGURE 16.19 Desert Biome

The desert gets less than 25 centimeters of precipitation per year, but it contains many kinds of living things. Cacti, sagebrush, lichens, snakes, small mammals, birds, and insects inhabit the desert. All deserts are dry, and the plants and animals show adaptations that allow them to survive under these extreme conditions. In deserts where daytime temperatures are high, most animals are active only at night, when the air temperature drops significantly.

night, the land cools off very rapidly, because there is no insulating blanket of clouds to keep the heat from radiating into space.

A desert biome is characterized by scattered, thorny plants that have few or no leaves. However, often the stems are green and carry on photosynthesis (figure 16.19). Because leaves tend to lose water rapidly, the lack of leaves is an adaptation to dry conditions. Many of the plants, such as cacti, can store water in their fleshy stems. Others store water in their roots. Although this is a very harsh environment, many kinds of flowering plants, insects, reptiles, and mammals live there. The animals usually avoid the hottest part of the day by staying in burrows or other shaded, cool areas. Staying underground or in the shade also allows animals to conserve water. Many annual plants also live in this biome, but the seeds germinate and grow only after the infrequent rainstorms. When it does rain, the desert blooms.

Boreal Coniferous Forest

Boreal coniferous forest—also known as the *northern coniferous forest* or the *taiga*—is dominated by evergreen trees (conifers) that are especially adapted to withstand long, cold winters with abundant snowfall. This biome is found throughout the world in northern regions, including parts of southern Canada, extending southward along the Appalachian and Rocky Mountains of the United States; much of northern Europe, and Asia.

Typically, the growing season is less than 120 days, and rainfall ranges between 40 and 100 centimeters per year. However, because of the low average temperature, evaporation is low and the climate is humid. Most of the trees in the wetter, colder areas are spruces and firs, but some drier, warmer areas have pines. The wetter areas generally have dense stands of

small trees intermingled with many other kinds of vegetation and many small lakes and bogs (figure 16.20). In the mountains of the western United States, pine trees are widely scattered and very large, with few branches near the ground. The area has a parklike appearance, because there is very little vegetation on the forest floor. The animals characteristic of this biome include mice, snowshoe hare, lynx, bears, wolves, squirrels, moose, midges, and flies. These animals can be divided into four general categories: those that become dormant in winter (e.g., insects and bears); those that are adapted to withstand the severe winters (e.g., snowshoe hare, lynx); those that live in protected areas (e.g., mice under the snow); and those that migrate south in the fall (most birds).

Temperate Rainforest

Temperate rainforest exists in the coastal areas of northern California, Oregon, Washington, British Columbia, and southern Alaska. In these coastal areas, the prevailing winds from the west blow over the ocean and bring moisture-laden air to the coast. As the air meets the coastal mountains and is forced to rise, it cools and the moisture falls as rain or snow. Most of these areas receive 200 centimeters or more precipitation per year. This abundance of water, along with fertile soil and mild temperatures, results in a lush growth of plants.

Sitka spruce, Douglas fir, and western hemlock are typical evergreen coniferous trees in the temperate rainforest. Undisturbed (old growth) forests of this region have trees as

FIGURE 16.20 Boreal Coniferous Forest Biome

Conifers are the dominant vegetation in most of Canada, in a major part of Russia, and at high elevations in sections of western North America. The boreal coniferous forest biome is characterized by cold winters with abundant snowfall.

old as 800 years that are nearly 100 meters tall. Deciduous trees of various kinds (e.g., red alder, big leaf maple, black cottonwood) also exist in open areas where they can get enough light. All the trees are covered with mosses, ferns, and other plants that grow on the surface of the trees. The dominant color is green, because most of the surfaces have a photosynthetic organism growing on them (figure 16.21).

When a tree dies and falls to the ground, it rots in place and serves as a site for the establishment of new trees. This is such a common feature of the forest that the fallen, rotting trees are called nurse trees. Fallen trees also serve as food for a variety of insects, which are food for a variety of larger animals.

Because of the rich resource of trees, 90% of the original temperate rainforest has already been logged. Many of the remaining areas have been protected, because they are home to the endangered northern spotted owl and marbled murrelet (a seabird).

Tundra

Tundra is the most northerly biome (figure 16.22). It is characterized by extremely long, severe winters and short, cool summers. The growing season is less than 100 days and, even during the short summer, the nighttime temperatures approach 0°C. Rainfall is low (10 to 25 centimeters per year). The deeper layers of the soil remain permanently frozen, forming a layer called the *permafrost*. Because the deeper layers of the soil are frozen, when the surface thaws the water forms puddles on the surface.

Under these conditions of low temperature and short growing season, very few kinds of animals and plants can survive. No trees can live in this region. The typical plants and animals of the area are grasses, sedges, dwarf willow, some other shrubs, reindeer moss (actually a lichen), caribou, wolves, musk oxen, fox, snowy owls, mice, and many kinds of insects. Many kinds of birds are summer residents only. The tundra community is relatively simple, so any changes have drastic and long-lasting effects. The tundra is easy to injure and slow to heal; therefore, it must be treated gently; the construction of the Alaskan pipeline has left scars that might still be there 100 years from now.

Tropical Rainforest

Tropical rainforest is at the other end of the climate spectrum from the tundra. Tropical rainforests are found primarily near the equator in Central and South America, Africa, parts of southern Asia, and some Pacific islands (figure 16.23). The temperature is high (averaging about 27°C), rain falls nearly every day (typically, 200 to 1,000 centimeters per year), and there are thousands of species of plants in a small area. Balsa (a very light wood), teak (used in furniture), and ferns the size of trees are plants from the tropical rainforest. Every plant has other plants growing on it. Tree trunks are likely to be covered with orchids, many kinds of vines, and mosses. Tree frogs, bats, lizards, birds, monkeys, and an almost infinite variety of insects inhabit the rainforest. These forests are very dense, and little sunlight reaches the forest floor. When the forest is opened up (by a hurricane or the death of a large tree) and sunlight reaches the forest floor, the opened area is rapidly overgrown with vegetation.

Because plants grow so quickly in these forests, people assume the soils are fertile, and many attempts have been made to bring this land under cultivation. In reality, the soils

FIGURE 16.21 Temperate Rainforest Biome
The temperate rainforest is characterized by high levels of rainfall, which support large evergreen trees and the many mosses and ferns that grow on the surface of the trees.

FIGURE 16.22 Tundra Biome
The tundra biome is located in northern parts of North America and Eurasia. It is characterized by short, cool summers and long, extremely cold winters. A layer of soil below the surface remains permanently frozen; consequently, no large trees exist in this biome. Relatively few kinds of plants and animals can survive this harsh environment.

FIGURE 16.23 Tropical Rainforest Biome
The tropical rainforest is found in moist, warm regions of the world located near the equator. The growth of vegetation is extremely rapid. There are more kinds of plants and animals in this biome than in any other.

are poor in nutrients. The nutrients are in the organisms; as soon as an organism dies and decomposes, its nutrients are reabsorbed by other organisms. Typical North American agricultural methods, which require the clearing of large areas, cannot be used with the soil and rainfall conditions of the tropical rainforest. The constant rain falling on these fields quickly removes the soil's nutrients, so that heavy applications of fertilizer are required. Often, these soils become hardened when exposed in this way. Although most of these forests are not suitable for agriculture, large expanses of tropical rainforest are being cleared yearly because of the pressure for more farmland in the highly populated tropical countries and the desire for high-quality lumber from many of the forest trees.

The Relationship Between Elevation and Climate

The distribution of terrestrial communities is primarily related to temperature and precipitation. Air temperatures are warmest near the equator and become cooler toward the poles. Similarly, air temperature decreases as elevation increases. This means that, even at the equator, cold temper-

atures are possible on the peaks of tall mountains. Therefore, as one proceeds from sea level to the tops of mountains, it is possible to pass through a series of biomes that are similar to what one would encounter traveling from the equator to the North Pole (figure 16.24).

16.5 | Succession

Biomes consist of communities that are relatively stable over long periods of time. A relatively stable, long-lasting community is called a **climax community.** The word *climax* implies the final step in a series of events. That is just what the word means in this context, because communities can go through a series of predictable, temporary stages, which eventually result in a long-lasting, stable community. The process of changing from one type of community to another is called **succession,** and each intermediate stage leading to the climax community is known as a **successional stage** or **successional community.**

Scientists recognize two kinds of succession. **Primary succession,** occurs when a community of plants and animals

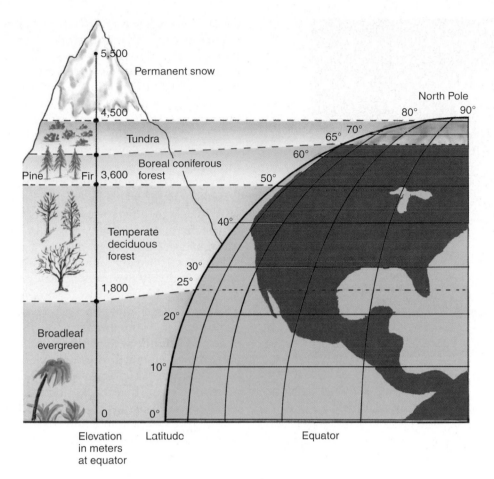

FIGURE 16.24 The Relationship Among Elevation, Latitude, and Vegetation
As one travels up a mountain, one experiences changes in climate. The higher the elevation, the cooler the climate. Even in the tropics, tall mountains can have snow on the top. Thus, it is possible to experience the same change in vegetation by traveling up a mountain as one would experience traveling from the equator to the North Pole.

develops where none existed previously. **Secondary succession,** occurs when a community of organisms is disturbed by a natural or human-related event (e.g., a hurricane, volcano, fire, forest harvest) and is returned to a previous stage in the succession.

Primary Succession

Primary succession is much more difficult to observe than secondary succession, because there are relatively few places on Earth that lack communities of organisms. The tops of mountains, newly formed volcanic rock, and rock newly exposed by erosion or glaciers can be said to lack life. However, bacteria, algae, fungi, and lichens quickly begin to grow on the bare rock surface, beginning the process of succession. The first organisms to colonize an area are often referred to as **pioneer organisms,** and the community is called a **pioneer community.**

Lichens are frequently important in pioneer communities. They are unusual organisms consisting of a combination of algae cells and fungi cells—a very hardy combination that is able to grow on the surface of bare rock (figure 16.25). Because algae cells are present, the lichen is capable of photosynthesis and can form new organic matter. Furthermore, many tiny consumer organisms can use the lichen as a food source and a sheltered place to live. The lichen's action also tends to break down the rock surface on which it grows. This fragmentation of rock by lichens is aided by the physical weathering processes of freezing and thawing, dissolution by water, and wind erosion. Lichens also trap dust particles, small rock particles, and the dead remains of lichens and other organisms that live in and on lichens. These processes of breaking down rock and trapping particles result in the formation of a thin layer of soil.

As the soil layer becomes thicker, small plants, such as mosses, may become established, increasing the rate at which energy is trapped and adding more organic matter to the soil. Eventually, the soil may be able to support larger plants that are even more efficient at trapping sunlight, and the soil-building process continues at a more rapid pace. Associated with the producers in each successional stage are a variety of small animals, fungi, and bacteria. Each change in the

FIGURE 16.25 Pioneer Organisms

Lichens growing on rock begin the process of soil formation that is necessary for the development of later successional stages.

community makes it more difficult for the previous group of organisms to maintain itself. Tall plants shade the smaller ones they replace; consequently, the smaller organisms become less common, and some disappear entirely. Only shade-tolerant species are able to grow and compete successfully in the shade of the taller plants. As this takes place, one stage succeeds the other. Figure 16.26 summarizes these changes.

Depending on the physical environment and the availability of new colonizing species, succession from this point can lead to different kinds of climax communities. If the area is dry, it might stop at a grassland stage. If it is cold and wet, a coniferous forest might be the climax community. If it is warm and wet, it may become a tropical rainforest. The rate at which succession takes place is also variable. In some warm, moist, fertile areas, the entire process might take place in less than 100 years. In harsh environments, such as mountaintops and very dry areas, it may take thousands of years.

Primary succession also occurs in the progression from an aquatic community to a terrestrial community. Lakes, ponds, and slow-moving parts of rivers accumulate organic matter. Where the water is shallow, this organic matter supports the development of rooted plants. In deeper water live only floating plants, such as water lilies, which send their roots down to the mucky bottom. In shallower water, upright, rooted plants, such as cattails and rushes, develop. As the plants contribute more organic matter to the bottom, the water level becomes shallower. Eventually, a mat of mosses, grasses, and even small trees may develop on the sur-

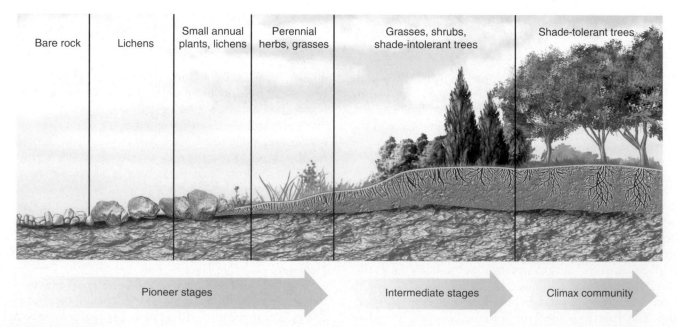

FIGURE 16.26 Primary Succession

The formation of soil is a major step in primary succession. Until soil is formed, the area is unable to support large amounts of vegetation. The vegetation modifies the harsh environment and increases the amount of organic matter that can build up in the area. As the kinds of plants change, so do the animals. The presence of plants eliminates the earlier pioneer stages of succession. If given enough time, a climax community may develop.

face along the edge of the water. If this continues for perhaps 100 to 200 years, an entire pond or lake will become filled in. More organic matter accumulates because of the large number of producers and because the depression that was originally filled with water becomes drier. This usually results in a wet grassland, which in many areas is replaced by the climax forest community typical of the area (figure 16.27).

Secondary Succession

Secondary succession occurs when a climax community is altered or destroyed by natural events or human activity. For example, when land is converted to agriculture, the original forest or grassland community is destroyed and replaced by crops. However, when agricultural land is abandoned, it returns to something like the original climax community (How Science Works 16.1). One obvious difference between primary succession and secondary succession is that, in secondary succession, there is no need to develop a soil layer. Another difference is that there is likely to be a reservoir of seeds from plants that were part of the original climax community. The seeds can survive in the soil for years in a dormant state, or they might be transported to the disturbed site from undisturbed sites nearby.

If we begin with bare soil the first year, it is likely to be invaded by a pioneer community of annual weed species. Within a year or two, perennial plants, such as grasses, become established. Because most of the weed species need bare soil for seed germination, they are replaced by the peren-

nial grasses and other plants that live in association with grasses. The more permanent grassland community is able to support more insects, small mammals, and birds than the weed community could. In regions where rainfall is low, succession is likely to stop at this grassland stage. In regions with adequate rainfall, several species of shrubs and fast-growing trees that require lots of sunlight (e.g., birch, aspen, juniper, hawthorn, sumac, pine, and spruce) become common. As the trees become larger, the grasses fail to get sufficient sunlight and die out. Eventually, shade-tolerant species of trees (e.g., beech, maple, hickory, oak, hemlock, and cedar) replace the shade-intolerant species, and a climax community results (figure 16.28).

Succession and Human Activity

Because most agricultural use of ecosystems involves replacing the natural climax community with an artificial early successional stage, it requires considerable effort on our part to prevent succession to the original climax community. This is certainly true if remnants of the original natural community are still locally available to colonize agricultural land. Small woodlots in agricultural areas of the eastern United States serve this purpose. Much of the work and expense of farming (e.g., tilling, herbicides) is necessary to prevent succession to the natural climax community. It takes a lot of energy to fight nature.

Many human-constructed lakes and farm ponds have weed problems, because they are shallow and provide ideal

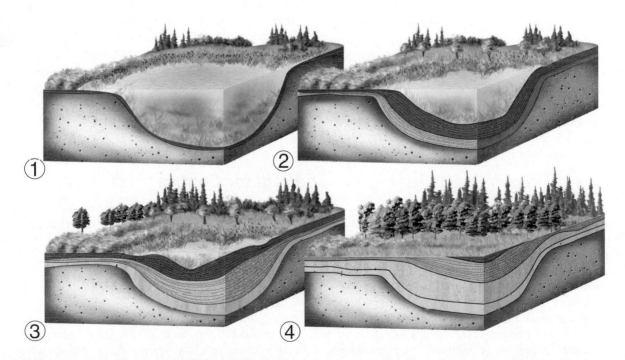

FIGURE 16.27 Succession from a Pond to a Wet Meadow
A shallow pond will slowly fill with organic matter from producers in the pond. Eventually, a floating mat will form over the pond and grasses will become established. In many areas, this will be succeeded by a climax forest.

Mature oak/hickory forest destroyed	Farmland abandoned	Annual plants	Grasses and biennial herbs	Perennial herbs and shrubs begin to replace grasses and biennials.	Pines begin to replace shrubs.	Young oak and hickory trees begin to grow.	Pines die and are replaced by mature oak and hickory trees.	Mature oak/hickory forest
		1–2 years	**3–4 years**	**4–15 years**	**5–15 years**	**10–30 years**	**50–75 years**	

FIGURE 16.28 Secondary Succession on Land
When agricultural land is abandoned, it goes through a series of changes. The general pattern is for annual weeds to become established in the first year or two following abandonment. These weeds are replaced by grasses and other perennial herbs, which are replaced by shrubs, which are replaced by trees. As the plant species change, so do the animal species.

conditions for the normal successional processes that lead to their being filled in. Often, humans do not recognize what a powerful force succession is.

16.6 The Impact of Human Actions on Communities

There is very little of the Earth's surface that has not been altered by humans activities. Often, the changes we cause have far greater impact than we might think. All organisms are interlinked in a complex network of relationships. Therefore, before changing a community, it is wise to analyze how its organisms are interrelated. This is not always easy, because there is much we still do not know about how organisms interact and how they use molecules from their environment. Several lessons can be learned from studying the effects of human activity on communities.

Introduced Species

One of the most far-reaching effects humans have had on natural communities involves the introduction of foreign species. Many of these introductions have been conscious decisions. Nearly all domesticated plants and animals in North America are introductions from elsewhere Corn, beans, sunflowers, squash, and turkeys are exceptions. Cattle, horses, pigs, goats, and many introduced grasses have significantly altered the original communities present in the Americas. Cattle have replaced the original grazers on grasslands. Pigs have become a major problem in Hawaii and many other places in the world, where they destroy the natural community by digging up roots and preventing the reproduction of native plants.

The introduction of grasses as food for cattle has resulted in the decline of many native species of grasses and other plants that were originally part of grassland communities. In Australia, the introduction of domesticated plants and animals—as well as wild animals, such as rabbits and foxes—has severely reduced the populations of many native marsupial mammals.

Accidental introductions have also significantly altered communities. Chestnut blight has essentially eliminated the American chestnut from the forests of eastern North America. Similarly, a fungal disease (Dutch elm disease) has severely reduced the number of elms. The accidental introduction of zebra mussels has significantly altered freshwater communities in eastern North America and has severely reduced the populations of native clams (Outlooks 16.1). Figure 16.29 shows two introduced species that have significantly altered communities in North America.

Predator Control

Many kinds of large predators were actively destroyed to protect livestock. Predator control was also thought to be important in the management of game species. During the formative years of wildlife management, it was thought that populations of game species could be increased if the populations of their predators were reduced. For these reasons, many states passed laws to encourage the killing of foxes, eagles, hawks, owls, coyotes, cougars, and other predators that could kill livestock or use game animals as a source of food. Often, bounties were paid to people who killed these predators (figure 16.30).

The absence of predators can lead to many kinds of problems. In many metropolitan areas, deer have become

HOW SCIENCE WORKS 16.1

The Changing Nature of the Climax Concept

When European explorers traveled across the North American continent, they saw huge expanses of land covered by recognizable communities of organisms. They found deciduous forests in the East, coniferous forests in the North, grasslands in central North America, and deserts in the Southwest. When ecologists began to explore the way in which communities develop over time, they thought of these communities as the end point, or climax, of a long journey, beginning with the formation of soil and its colonization by a variety of plants and other organisms.

As settlers removed the original forests or grasslands and converted the land to farming, they replaced the original climax communities with agricultural communities. Eventually, as poor farming practices depleted the soil, the farms were abandoned and the land was allowed to return to its original condition. This secondary succession often resulted in forests or grasslands that resembled those that had been destroyed. However, in most cases, these successional communities contained fewer species and, in some cases, were entirely different from the originals.

As ecologists recognized that there was not a fixed, predetermined community for each part of the world, they began to modify the way they looked at the concept of climax communities. The concept today is a more plastic one.

The term *climax* is still used to talk about a stable stage following a period of change, but ecologists no longer believe that land will eventually return to a "preordained" climax condition. They have also recognized in recent years that the type of climax community that develops depends on many factors other than climate. One of these is the availability of seeds to colonize new areas. Two areas with very similar climate and soil characteristics may contain different species because of the seeds available when the lands were released from agriculture. Furthermore, the only thing that differentiates a climax community from a successional one is the time scale over which change occurs. Climax communities do not change as rapidly as successional ones; however, all communities are eventually replaced, as were the swamps that produced coal deposits, the preglacial forests of Europe and North America, and the pine forests of the northeastern United States.

What should we do with the climax concept? Although it embraces a false notion that there is a specific end point to succession, it is still important to recognize that there is a predictable pattern of change during succession and that later stages in succession are more stable and longer-lasting than early stages. Whether it is called a climax community is not really important.

FIGURE 16.29 Introduced Species

Many introduced species become pests. Dandelions and mute swans are two examples.

(a)

(b)

FIGURE 16.30 Predator Control

At one time, people were paid to kill various kinds of predators. (*a*) These men receive $35 for each wolf killed. (*b*) This coyote was killed and hung on a fence because it was considered a threat to livestock.

pests. This is due to several reasons, including the fact that there are no predators, and hunting (predation by humans) either is not allowed or is impractical because of the highly urbanized nature of the area. Some municipalities have instituted programs of chemical birth control for their deer populations.

Only a few years ago, the alligator was on the endangered species list and all hunting was suspended. However, today increased numbers of alligators present a danger, particularly to pets and children. Hunting is now allowed in an effort to control the numbers of alligators, because humans are the only effective predators of large alligators.

Similarly, in Yellowstone National Park, elk, bison, and moose populations have become very large, because hunting is not allowed and predator populations have been small. In 1995, wolves were reintroduced to the park in the hope that they would help bring the elk and moose populations under control (figure 16.31). This was a controversial decision, however, because ranchers in the vicinity did not want a return of large predators that might prey on their livestock. They are also opposed to having bison, many of which carry a disease that can affect cattle, stray onto their land. The wolf population has increased significantly in Yellowstone and is having an effect on the populations of bison, elk, and moose. Regardless of the politics involved in the decision, Yellowstone is in a more natural condition today, with wolves present, than it was prior to 1995.

By contrast, the state of Alaska allows the hunting of wolves, because Alaskans believe the wolves are reducing caribou populations below optimal levels. Caribou hunting is an important source of food for Alaskan natives, and hunters who visit the state provide a significant source of income. Many groups oppose the killing of wolves in Alaska, however, they consider the policy misguided and believe it will not have a positive effect on the caribou population. They also object to the killing of wolves on ethical grounds.

Habitat Destruction

Some communities are fragile and easily destroyed by human activity, whereas others seem able to resist human interference. Communities with a wide variety of organisms that show a high level of interaction are more resistant than are those with few organisms and little interaction. In general, the more complex a community, the more likely it is to recover after being disturbed. The tundra biome is an example of a community with relatively few organisms and interactions. It is not very resistant to change and, because of its slow rate of repair, damage caused by human activity may persist for hundreds of years.

Some species are more resistant to human activity than are others. Rabbits, starlings, skunks, and many kinds of insects and plants are able to maintain high populations despite human activity. Indeed, some may even be encouraged

OUTLOOKS 16.1

Zebra Mussels: Invaders from Europe

In the mid-1980s, a clamlike organism called the zebra mussel, *Dressenia polymorpha*, was introduced into the waters of the Great Lakes. It probably arrived in the ballast water of a ship from Europe. Ballast water is pumped into empty ships to make them more stable when crossing the ocean. Immature stages of the zebra mussel were probably emptied into Lake St. Clair, near Detroit, Michigan, when a ship discharged its ballast water to take on cargo. This organism has since spread to many areas of the Great Lakes and smaller inland lakes. It has also been discovered in other parts of the United States, including the mouth of the Mississippi River. Zebra mussels attach to any hard surface and reproduce rapidly. Densities of more than 20,000 individuals per square meter have been documented in Lake Erie.

These invaders are of concern for several reasons. First, they coat the intake pipes of municipal water plants and other facilities, requiring expensive measures to clean the pipes. Second, they coat any suitable surface, preventing native organisms from using the space. Third, they introduce a new organism into the food chain. Zebra mussels filter small aquatic organisms from the water very efficiently and may remove the food organisms required by native species. Their filtering activity has significantly increased the clarity of the water in several areas in the Great Lakes. This can affect the kinds of fish pres-

ent, because greater clarity allows predator fish to find prey more easily. In addition, the clarity of the water has allowed for greater production of aquatic vegetation and algae, because light penetrates the water better and to a greater depth. There is concern that zebra mussels have significantly changed the ecological organization of the Great Lakes.

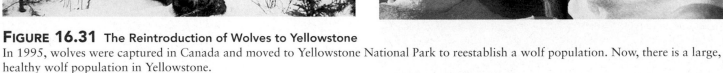

FIGURE 16.31 The Reintroduction of Wolves to Yellowstone
In 1995, wolves were captured in Canada and moved to Yellowstone National Park to reestablish a wolf population. Now, there is a large, healthy wolf population in Yellowstone.

by human activity. By contrast, whales, condors, eagles, and many plant and insect species are not able to resist human interference very well. For most of these endangered species, it is not humans' acting directly with the organisms that cause their endangerment. Very few organisms have been driven to

extinction by hunting or direct exploitation. Usually, the cause of extinction or endangerment is an indirect effect of habitat destruction as humans exploit natural communities. As humans convert land to farming, grazing, commercial forestry, development, and special wildlife management areas, natural communities are disrupted, and plants and animals with narrow niches tend to be eliminated, because they lose critical resources in their environment (Outlooks 16.2.) Figure 16.32 shows various kinds of habitat destruction, and table 16.1 lists eight endangered and threatened species and the probable causes of their difficulties.

FIGURE 16.32 Habitat Destruction
Habitat destruction by the building of cities, conversion of land to agriculture, and commercial forest practices has a major impact on the kinds of organisms that can survive in our world.

Pesticide Use

Humans have developed a variety of chemicals to control specific pests. These chemicals have a variety of names. *Herbicides* are used to kill plants. *Insecticides* are used to kill insects. *Fungicides* are used to kill fungi. Often, all these kinds of chemicals are lumped together into one large category—*pesticides*—because they are used to control various kinds of pests. Although various kinds of pesticides are valuable in controlling disease in human and domesticated animal populations and in controlling pests in agriculture, they have some negative community effects as well.

One problem associated with continual pesticide use is that pests become resistant to these chemicals. When a pesticide is used, most of the pests are killed. However, some may be able to resist its effects. When these survivors reproduce, they pass on their genes for resistance to their offspring, and the next generation is less susceptible to the pesticide. Ultimately, resistant populations develop and the pesticide is no longer useful.

Another problem associated with pesticide use is the effects they have on valuable nontarget organisms. Insecticides typically kill a wide variety of organisms other

TABLE 16.1

Endangered and Threatened Species

Species	Reason for Endangerment
Hawaiian crow (*Corvis hawaiinsis*)	Predation by introduced cat and mongoose, disease, habitat destruction
Sonora chub (*Gila ditaenia*)	Competition with introduced species in streams in Arizona and Mexico
Black-footed ferret (*Mustela nigripes*)	The poisoning of prairie dogs (their primary food)
Snail kite (*Rostrhamus sociabilis*)	Specialized eating habits (they eat only apple snails), the draining of marshes in Florida
Grizzly bear (*Ursus arctos*)	The loss of wilderness areas
California condor (*Gymnogyps californianus*)	Slow breeding, lead poisoning
Ringed sawback turtle (*Graptemys oculifera*)	The modification of habitat by the construction of a reservoir in Mississippi that reduced their primary food source
Scrub mint (*Dicerandra frutescens*)	The conversion of their habitat to citrus groves and housing in Florida

than the targeted pest species. Often, other species in the community have a role in controlling pests. Predators kill pests and parasites use them as hosts. Generally, predators and parasites reproduce more slowly than their prey or host species. Because of this, the use of a nonspecific insecticide may indirectly make controlling a pest more difficult. If such an insecticide is applied to an area, the pest is killed, but so are its predators and parasites. Because the herbivore pest reproduces faster than its predators and parasites, the pest population rebounds quickly, unchecked by natural predation and parasitism (figure 16.33).

Today, a more enlightened approach to pest control is *integrated pest management,* which uses a variety of approaches to reduce pest populations. Integrated pest management includes the use of pesticides as part of a pest control program, but it also includes strategies such as encouraging the natural enemies of pests, changing farming practices to discourage pests, changing the mix of crops grown, and accepting low levels of crop damage as an alternative to costly pesticide applications.

Biomagnification

The use of persistent chemicals has an effect on the food chain. Chemicals that do not break down are passed from one organism to the next, and organisms at higher trophic levels tend to accumulate larger amounts than the organisms they feed on. This situation is known as **biomagnification.**

The history of DDT use illustrates the problems associated with persistent organic molecules. DDT was a very effective insecticide because it was extremely toxic to insects but not very toxic to birds and mammals. It is also a very stable compound, which means that, once applied, it remained effective for a long time. However, when an aquatic area was sprayed with a small concentration of DDT, many kinds of organisms in the area can accumulate tiny quantities in their bodies. Because marshes and other wet places are good mosquito habitats, these areas were often sprayed to control these pests. Even algae and protozoa found in aquatic ecosystems accumulate persistent pesticides. They may accumulate concentrations in their cells that are 250 times more concentrated than the amount sprayed on the ecosystem. The algae and protozoa are eaten by insects, which in turn are eaten by frogs, fish, and other carnivores.

The concentration in frogs and fish may be 2,000 times the concentration sprayed. The birds that feed on the frogs and fish may accumulate concentrations that are as much as 80,000 times the original amount. Because DDT is relatively stable and is stored in the fat deposits of the organisms that take it in, what was originally a dilute concentration became more concentrated as it moved up the food chain (figure 16.34).

Before DDT use was banned in the United States in 1972 many animals at higher trophic levels died as a result of lethal concentrations of the pesticide accumulated from the food they ate. Each step in the food chain accumulated some DDT; therefore, higher trophic levels had higher concentrations. Even if they were not killed directly by DDT, many birds at higher trophic levels, such as eagles, pelicans, and osprey, suffered reduced populations. This occurred because the DDT

FIGURE 16.33 Pesticide Use
Pesticides are used to control pests that injure or compete with crops and domesticated animals. Their use has several effects on the communities of plants and animals where they are used.

interfered with the female birds' ability to produce eggshells. Thin eggshells are easily broken; thus, no live young hatched. Both the bald eagle and the brown pelican were placed on the endangered species list because their populations had dropped dramatically as a result of DDT poisoning. The ban on DDT use in the United States and Canada has resulted in an increase in the populations of both kinds of birds; the status of the bald eagle has been upgraded from endangered to threatened and that of the brown pelican has been upgraded to threatened in all states except Texas, Louisiana, Mississippi, and California.

Several other chemical compounds are also of concern today because they are biomagnified in food chains. These include polychlorinated biphenyls (PCBs), dioxins, methylmercury, and many other compounds. Because these compounds can reach high concentrations in fish that are at the top of the food chain, many states and countries publish advisories that caution people not to eat large quantities of these fish. When the production of PCBs was halted in 1979, the level of contamination in fish declined.

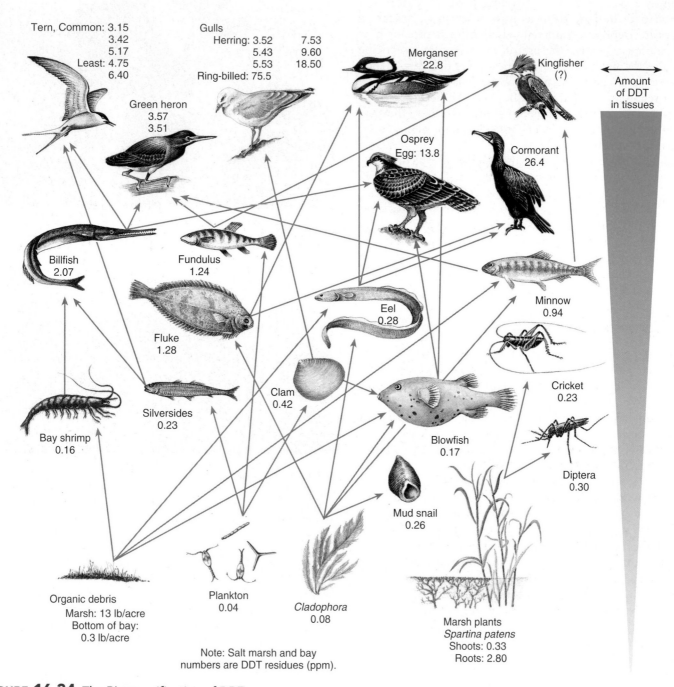

Tern, Common: 3.15
3.42
5.17
Least: 4.75
6.40

Gulls
Herring: 3.52   7.53
5.43   9.60
5.53   18.50
Ring-billed: 75.5

Merganser
22.8

Kingfisher
(?)

Amount
of DDT
in tissues

Green heron
3.57
3.51

Osprey
Egg: 13.8

Cormorant
26.4

Billfish
2.07

Fundulus
1.24

Eel
0.28

Minnow
0.94

Fluke
1.28

Cricket
0.23

Clam
0.42

Blowfish
0.17

Diptera
0.30

Silversides
0.23

Bay shrimp
0.16

Mud snail
0.26

Organic debris
Marsh: 13 lb/acre
Bottom of bay:
0.3 lb/acre

Plankton
0.04

Cladophora
0.08

Marsh plants
Spartina patens
Shoots: 0.33
Roots: 2.80

Note: Salt marsh and bay
numbers are DDT residues (ppm).

FIGURE 16.34 The Biomagnification of DDT

All the numbers shown are in parts per million (ppm). A concentration of one part per million means that, in a million equal parts of the organism, one of the parts would be DDT. Notice how the amount of DDT in the bodies of the organisms increases from producers to herbivores to carnivores. Because DDT is persistent, it builds up in the top trophic levels of the food chain.

Biodiversity "Hot Spots"

In an attempt to prioritize efforts to protect biodiversity, scientists have identified the most endangered habitats around the world. The environmental organization, Conservation International, has identified 34 biodiversity "hot spots" according to the amount of the original habitat remaining and the number of unique species present in the region. The primary cause of habitat loss is human activity. Therefore, the protection of these areas requires actions that discourage the continued modification of the habitat. Species that are found only in specific locations are said to be *endemic* to the region. Some regions have many such unique organisms and are said to have high rates of endemism. Because it is impossible to study all the organisms in an area, scientists have looked at two groups of reasonably well-known organisms, plants and vertebrate animals, and have determined rates of endemism for each of the 34 regions.

High rates of endemism are found in areas that have been isolated from other areas for a long time. In the accompanying table, you will notice that 10 such areas consist of islands or groups of islands. Other areas may be isolated because they are surrounded by other kinds of systems that have very different kinds of organisms. For example, the tropical rainforest areas of western Africa are isolated by drier savanna to the north, east, and south. Whenever ecosystems have existed in relative isolation for millions of years, there is a high likelihood that unique species will have evolved in the region.

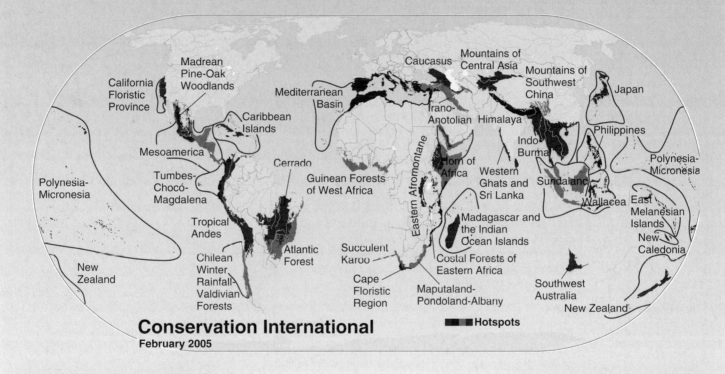

Conservation International
February 2005

Hot Spot	Characteristics	% of Original Area Remaining	Number of Endemic Plants (%)	Number of Endemic Vertebrate Animals (%)
South America				
Tropical Andes	This region is quite varied. At lower elevations (500 to 1,500 meters), tropical wet and moist forests occur. Various types of cloud forests occur at intermediate elevations (800 to 3,500 meters). At higher elevations (3,000 to 4,800 meters), grassland and scrubland ecosystems are found up to the snow line.	25%	15,000 (50%)	1,733 (40.6%)

OUTLOOKS 16.2 (continued)

Biodiversity "Hot Spots"

Hot Spot	Characteristics	% of Original Area Remaining	Number of Endemic Plants (%)	Number of Endemic Vertebrate Animals (%)
South America (continued)				
Tumbes-Chocó-Magdelena	This region is quite varied and includes: mangroves, beaches, rocky shorelines, and coastal rain forests. The only remaining coastal dry forests in South America occur in this region. The Galapagos Islands are included in this region.	24%	2,750 (25%)	364 (18.6%)
Atlantic Forest of Brazil, Paraguay, and Argentina	Coastal and inland tropical and subtropical rainforest of many types exist in this area.	8.1%	8,000 (40%)	725 (31.3%)
Brazilian Cerrado	The Cerrado is a region that receives extensive rainfall for about half the year followed by a drought. This region is the most extensive woodland/savanna region in South America. There are many different vegetation types, including tree and scrub savanna, grassland with scattered trees, and occasional patches of a dry, closed canopy forest. Many of the plants show adaptation to resist fires.	21.6%	4,400 (44%)	292 (14.5%)
Chilean Winter Rainfall-Valdivian Forests	This region has three distinct climates. A winter rainfall area can be divided into two regions. The wetter area supports Mediterranean vegetation and the drier area supports desert vegetation. The wetter west side of the Andean mountains in the southern parts of this region, support a variety of forest types.	30%	1,957 (50.3%)	107 (25.5%)
North and Central America				
Caribbean Islands	Because the climate in the Caribbean is highly variable, the vegetation differs from island to island and is quite variable on a single island. Several kinds of dry habitats are found: dry evergreen thickets, savanna, cactus scrub, and spiny shrub vegetation in a few of the driest spots dominate the dry and flat islands. In wetter areas, marsh forests, forests, and woodlands occur.	10%	6,550 (50.4%)	908 (59.5%)
California Floristic Province	The climate of the California Floristic Province consists of hot, dry summers and cool, wet winters. The habitat types are quite variable. Sagebrush, prickly pear, coastal sage scrub, chaparral, and juniper-pine woodland are found at lower elevations. At higher elevations and where there is more moisture, there are evergreen forests of cypress, Douglas fir, sequoia, and redwood. There are also dunes and salt marshes along the coast.	25%	2,124 (60.9%)	70 (10.2%)
Mesoamerica	There is great variety in the ecosystems of this region: dry forest, lowland moist forest, subtropical wet forests, rainforest, coastal swamp, and mangrove forest, with broad-leaved and coniferous forest at higher altitudes.	20%	2,941 (17.3%)	1,212 (36.6%)

OUTLOOKS 16.2 *(continued)*

Biodiversity "Hot Spots"

Hot Spot	Characteristics	% of Original Area Remaining	Number of Endemic Plants (%)	Number of Endemic Vertebrate Animals (%)
North and Central America *(continued)*				
Madrean Pine-Oak Woodlands	This region consists of the mountains of Mexico and adjacent southwestern United States. These forests form "islands" at higher elevations. Vegetation includes pine, fir, and oak forests. Some are mixed forests while others may be exclusively oak, pine, or fir.	20%	3,975 (75%)	133 (8.8%)
Europe and Asia				
Caucasus	The vegetation of the Caucasus is quite diverse, but the primary vegetation types are dry—grasslands, semidesert, desert, and arid woodlands with scattered broadleaf forests, coniferous forests, and shrublands.	27%	1,600 (25%)	54 (7.3%)
Mediterranean Basin	The climate of the Mediterranean Basin consists of cool, wet winters and hot, dry summers. Much of the region was once covered by deciduous forests, which have been modified by human activity. Today, the most widespread vegetation consists of shrubs with dry, hard leaves.	4.7%	11,700 (52%)	217 (17.5%)
Irano-Anatolian	This mountainous region has hot summers and cold winters. The climate is dry with most of the precipitation in the winter and spring. At lower elevations oak and juniper forests are typical. At higher elevations alpine and subalpine vegetation occurs.	15%	2,500 (41.7%)	54 (7.4%)
Mountains of Central Asia	The climate is arid with most of the precipitation in the winter and spring. Much of the region has desert or grassland vegetation with some forests, subalpine and alpine vegetation on some of the mountains.	20%	1,500 (27.3%)	16 (2.2%)
Africa				
Madagascar and Indian Ocean islands	Madagascar has been separate from Africa for about 160 million years. Although it is only about 370 km from Africa, Madagascar shares few species of organisms with Africa. The vegetation types are quite varied: tropical rainforest, dry deciduous forest, a spiny desert, and several high mountain ecosystems. The islands have high marine biodiversity with large areas of coral reefs.	10%	11,600 (89.2%)	1,018 (81.9%)
Coastal Forest of Eastern Africa	This is a tropical region with high temperatures and humidity which supports a mixture of vegetation types that include moist forests, dry forests, savanna woodlands, mangrove and other types of swamps.	10%	1,750 (43.8%)	113 (8.1%)
Eastern Afromontane	This mountainous area is quite diverse and includes dry woodlands, many kinds of forests, and alpine and sub-alpine vegetation. Many of the lakes have been isolated for millions of years and are home to many endemic species of fish.	10.5%	2,356 (31%)	988 (30.3%)

OUTLOOKS 16.2 (continued)

Biodiversity "Hot Spots"

Hot Spot	Characteristics	% of Original Area Remaining	Number of Endemic Plants (%)	Number of Endemic Vertebrate Animals (%)
Africa (continued)				
Guinean Forests of West Africa	This area supports a variety of forest types that include: moist forests along the coast, freshwater swamp forests, and semi-deciduous forests inland where there are prolonged dry seasons. It has the highest diversity of mammals of all the hotspots. This area is politically divided among 10 African countries. It has been highly fragmented by human activity.	15%	1,800 (20%)	422 (20.6%)
Horn of Africa	This arid region is dominated by bushland with small areas of other kinds of vegetation such as evergreen bushland, succulent shrubland, dry evergreen forest and woodland, semi-desert grassland low-growing dune and rock vegetation, and small areas of mangrove.	5%	2,750 (55%)	153 (11.5%)
Maputaland-Pondoland-Albany	This region is a warm temperate region with a mixture of unique kinds of forests, thickets, bushveld and grasslands. It has the highest diversity of trees of any hotspot. Succulents are common in parts of this region.	24.5%	1,900 (23.5%)	65 (6.0%)
Cape Floristic region	The vegetation of the Cape Floristic region is a unique broadleaved, evergreen, fire-adapted shrubland known as fynbos.	20%	6,210 (69%)	62 (10.4%)
Succulent Karoo	The Succulent Karoo consists of deserts. One part has a mild climate with winter rainfall. The other receives rain in the spring and fall. The dominant vegetation consists of shrubs, with thick, water-holding leaves. Other plants have thick, water-holding stems. Many kinds of bulb plants and annual plants bloom during the wetter periods.	29%	2,439 (38.4%)	19 (4.3%)
Asia Pacific				
Mountains of southwest China	There is a wide variety of vegetation types in the mountains and valleys of this region: broad-leaved and coniferous forests, bamboo groves, savannah, grasslands, freshwater wetlands, and alpine scrub communities.	8%	3,500 29.2%)	53 (4.7%)
Indo-Burma	A wide variety of ecosystems occur in this region: deciduous and evergreen, broad-leafed forests with patches of shrublands, woodlands, and heath forests.	5%	7,000 (51.9%)	1,048 (27.8%)
Western Ghats and Sri Lanka	The most common vegetation types are deciduous and tropical rainforests with grasslands, as well as scrub forests in lower, drier areas.	23%	3,049 (51.5%)	496 (40.2%)
Japan	Japan's mountainous terrain and numerous islands support a wide range of vegetation types. These include: alpine and subalpine vegetation, boreal mixed forests, subtropical broadleaf evergreen forests, and mangrove swamps.	20%	1,950 (34.8%)	183 (24.1%)

OUTLOOKS 16.2 (continued)

Biodiversity "Hot Spots"

Hot Spot	Characteristics	% of Original Area Remaining	Number of Endemic Plants (%)	Number of Endemic Vertebrate Animals (%)
Asia Pacific *(continued)*				
Himalaya	This region has the highest mountains in the world. The great variation in elevation leads to a great diversity of ecosystems including: grasslands, broadleaf forests, conifer forests, and alpine meadows.	25%	3,160 (31.6%)	150 (8.2%)
East Melanesian Islands	The region has a diverse range of islands. Some are the result of volcanic activity and include large mountains. Habitats include various kinds of tropical forests, swamps and grasslands. The mountainous islands often have rainforests.	30%	3,000 (37.5%)	283 (42.9%)
Philippines	Although most of the region has been modified by human activity, the original vegetation was quite varied and included lowland rainforest with patches of seasonal forest, mixed forest, savanna, and pine-dominated cloud forest. Marine biodiversity is also important, with many species of corals and related marine ecosystems.	7%	6,091 (65.8%)	591 (45.1%)
Sundaland of Sumatra, Borneo, and Java	Lowland areas are dominated by tropical rainforest. Higher elevations consist of forests with many mosses, lichens, and orchids. At the highest elevations, a shrubby forest exists, with many species of rhododendrons. Other specialized localities have special vegetation, including mangroves along the ocean, swamps on wet inland sites, and heaths on dry, rocky outcrops.	6.7%	15,000 (60%)	1,103 (39.5%)
Wallacea region of Indonesia	Wallacea consists of several islands in the eastern part of Indonesia. These islands have been isolated from Asia for about a million years. Each island in the group has unique organisms. The forests are mostly tropical rainforests, with some drier habitat types. In mountainous regions, many kinds of forests exist.	15%	1,500 (15%)	571 (41.1%)
Southwest Australia	This region is a coastal plain with winter rains and dry summers. Vegetation consists of forests, woodlands, shrublands, and heaths, but there are no grasslands.	30%	2,949 (52.9%)	81 (14.1%)
New Zealand	New Zealand is isolated from other landmasses and has a wide range of topography. Consequently, the climate varies greatly from place to place. Its isolation, varied topography, and climate have resulted in the evolution of a large number of unique plants and animals. Temperate rainforests were once the dominant vegetation, with some patches of dunes, tussock grasslands, and wetlands.	22%	1,865 (81.1%)	155 (54.4%)

OUTLOOKS 16.2 (continued)

Biodiversity "Hot Spots"

Hot Spot	Characteristics	% of Original Area Remaining	Number of Endemic Plants (%)	Number of Endemic Vertebrate Animals (%)
Asia Pacific (continued)				
New Caledonia	New Caledonia is isolated from other landmasses; it once had extensive conifer rainforests, which are now confined to a few scattered pockets in the central mountains. Today, shrubland dominates much of the island. Other vegetation types include mangroves and savannas. Marine biodiversity is also important in the coral reefs that surround the island.	27%	2,432 (74.4%)	100 (37.2%)
Polynesia and Micronesia	The islands of Polynesia and Micronesia include rocky islets, low-lying coral atolls, and volcanic islands. Because this area consists of so many types of islands, there are many vegetation types, including mangroves, coastal wetlands, tropical rainforests, cloud forests, savannas, open woodlands, and shrublands.	21.2%	3,074 (57.7%)	229 (48.6%)

Summary

Each organism in a community occupies a specific space, known as its habitat, and has a specific functional role to play, known as its niche. An organism's habitat is usually described in terms of a conspicuous element of its surroundings. The niche is difficult to describe, because it involves so many interactions with the physical environment and other living things.

Interactions between organisms fit into several categories. Predation is one organism benefiting (predator) at the expense of the organism killed and eaten (prey). Parasitism is one organism benefiting (parasite) by living in or on another organism (host) and deriving nourishment from it. Organisms that carry parasites from one host to another are called vectors. Commensal relationships exist when one organism is helped but the other is not affected. Mutualistic relationships benefit both organisms. Symbiosis is any interaction in which two organisms live together in a close physical relationship. Competition causes harm to both of the organisms involved, although one may be

harmed more than the other and may become extinct, evolve into a different niche, or be forced to migrate.

A community consists of the interacting populations of organisms in an area. The organisms are interrelated in many ways in food chains, which interlock to create food webs. Because of this interlocking, changes in one part of the community can have effects elsewhere.

Major land-based regional communities are known as biomes. The temperate deciduous forest, boreal coniferous forest, tropical rainforest, temperate grassland, desert, savanna, temperate rainforest, and tundra are biomes. Communities go through a series of predictable changes, which lead to a relatively stable collection of plants and animals. This process of change is called succession, and the resulting stable unit is called a climax community.

Organisms within a community are interrelated in sensitive ways; thus, changing one part of a community can lead to unexpected consequences. The introduction of foreign species, predator-control practices, habitat destruction, pesticide use, and biomagnification of persistent toxic chemicals all have caused unanticipated changes in communities.

Key Terms

*Use the interactive flash cards on the **Concepts in Biology,**
12/e website to help you learn the meaning of these terms.*

biomagnification 349
biomes 334
climax community 340
commensalism 332
competition 328
competitive exclusion
  principle 328
epiphytes 332
external parasites 331
food web 324
habitat 326
host 330
internal parasites 331
interspecific competition 328
intraspecific competition
  328
mutualism 333

niche 326
parasite 330
parasitism 330
pioneer community 341
pioneer organisms 341
predation 329
predator 329
prey 329
primary succession 340
secondary succession 341
succession 340
successional stage
  (successional community)
  340
symbiosis 330
vector 331

Basic Review

1. The role an organism plays in its surroundings is its
   a. niche.
   b. habitat.
   c. community.
   d. food web.
2. The kind of interrelationship between two organisms in which both are harmed is _____.
3. A desert is always be characterized by
   a. high temperature.
   b. low amounts of precipitation.
   c. few kinds of plants and animals.
   d. sand.
4. When a community is naturally changing with the addition of new species organisms and the loss of others, _____ is occurring.
5. When two organisms cooperate and both derive benefit from the relationship, it is known as commensalism. (T/F)
6. Most of the plants and animals involved in agriculture are introduced from other parts of the world. (T/F)

7. A biome that has trees adapted to long winters is known as
   a. tundra.
   b. temperate deciduous forest.
   c. northern coniferous forest.
   d. temperate rainforest.
8. If a forest is destroyed by a fire, it will eventually return to being a forest. This process is known as _____ succession.
9. A collection of organisms that interact with one another in an area is known as a
   a. community.
   b. biome.
   c. succession.
   d. biomagnification.
10. The idea that no two organisms can occupy the same niche is known as the competitive exclusion principle. (T/F)

Answers
1. a 2. competition 3. b 4. succession 5. F 6. T 7. c 8. secondary 9. a 10. T

Concept Review

*Answers to the Concept Review questions can be found on the **Concepts in Biology,** 12/e website.*

16.1 The Nature of Communities

1. Why is a food web better than a food chain as a way to describe a community?
2. As the number of insects drops in the autumn, what would you expect to happen to bird populations?

16.2 Niche and Habitat

3. List 10 items that are a part of your niche.
4. What is the difference between habitat and niche?

16.3 Kinds of Organism Interactions

5. What do parasites, commensal organisms, and mutualistic organisms have in common? How are they different?
6. Describe two situations in which competition may involve combat and two that do not involve combat.

16.4 Types of Communities

7. List a predominant abiotic factor in each of the following biomes: temperate deciduous forest, boreal coniferous forest, grassland, desert, tundra, temperate rainforest, tropical rainforest, and savanna.

8. List a dominant producer organism typical of each of the following biomes: temperate deciduous forest, boreal coniferous forest, grassland, desert, tundra, temperate rainforest, and savanna.

16.5 Succession

9. How does primary succession differ from secondary succession?
10. How does a climax community differ from a successional community?
11. Describe the steps in the secondary succession of abandoned farmland to a climax community.
12. Describe the steps in the succession of a pond to a climax community.

16.6 The Impact of Human Actions on Communities

13. Why do DDT and PCBs increase in concentration in the bodies of organisms at higher trophic levels?

14. How have past practices of predator control and habitat destruction negatively altered biological communities?
15. List three introduced species that have become pests, and explain why they became pests.

Thinking Critically

*For guidelines to answer this question, visit the **Concepts in Biology**, 12/e website.*

Farmers are managers of ecosystems. Consider a cornfield in Iowa. Describe five ways in which the cornfield ecosystem differs from the original prairie it replaced. At what trophic level does the farmer exist?

Population Ecology

Many urban areas are experiencing problems with wildlife populations. Cities such as Philadelphia and Chicago have serious problems with white-tailed deer. The deer eat shrubs and other plantings in parks and in the yards of homes. Car-deer collisions result in millions of dollars of damage yearly and many human and deer deaths.

Geese are a problem in many urban areas that have lakes, ponds, and other water features. They eat the grass around the water and leave droppings, which make the area unsightly. They often damage the grass of carefully maintained golf courses, athletic fields, and parks.

The city of Anchorage, Alaska, has a problem with moose that invade the city during the winter months. In a typical year, 150 to 200 moose are killed as a result of collisions with vehicles. The damage to vehicles is extensive when they hit an animal that weighs over 1,000 pounds. People have been severely injured as a result of being trampled by moose.

What conditions allow these populations to become so large that they become a public nuisance and cause serious economic damage? What practical steps could be taken to solve these problems? This chapter focuses on populations, their characteristics, and the conditions that lead to population growth.

CHAPTER OUTLINE

17.1 Population Characteristics

A **population** is a group of organisms of the same species located in the same place at the same time. Examples are the dandelions in a yard, the rat population in your city sewer, and the number of students in a biology class. On a larger scale, all the people of the world constitute the world human population.

The terms *species* and *population* are interrelated, because a species is a population—the largest possible population of a particular kind of organism. The term *population,* however, is often used to refer to portions of a species by specifying a space and time. For example, the size of the human population in a city changes from hour to hour during the day and varies according to the city's boundaries. Because each local population is a small portion of a species, various populations of the same species should show many kinds of differences.

Remember from chapter 12 that a species is a population of organisms capable of interbreeding. Thus, the reproduction of individuals within species or populations results in changes in the size of a population and the genes present from one generation to the next.

Gene Frequency and Gene Flow

Recall from chapter 12 that gene frequency is a measure of how often a specific gene shows up in the gametes of a pop-

ulation. Two populations of the same species often have quite different gene frequencies. For example, many populations of mosquitoes have high frequencies of insecticide-resistant genes, whereas others do not. The frequency of the genes for tallness in humans is greater in certain African tribes than in any other human population. Figure 17.1 shows that the frequency of the allele for type B blood differs significantly from one human population to another.

Recall from chapter 14 that gene flow is the movement of genes from one place to another or from one generation to another (figure 17.2). When organisms reproduce, they pass on genes from one generation to the next. Genes can also flow from one place to another as organisms migrate or are carried from one location to another. Typically, both kinds of gene flow happen together as individuals migrate to new regions and reproduce, passing on their genes to the next generation in the new area.

Age Distribution

Age distribution is the number of organisms of each age in a population (figure 17.3). Often, organisms are grouped into three general categories based on their reproductive status:

1. Prereproductive juveniles (e.g. insect larvae, plant seedlings, and babies)
2. Reproductive adults (e.g. mature insects, plants producing seeds, and humans in early adulthood)

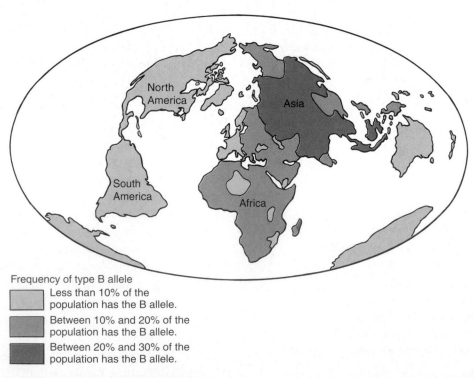

Frequency of type B allele
- Less than 10% of the population has the B allele.
- Between 10% and 20% of the population has the B allele.
- Between 20% and 30% of the population has the B allele.

FIGURE 17.1 Distribution of the Allele for Type B Blood
The allele for type B blood is not evenly distributed in the world. This map shows that the type B allele is most common in parts of Asia and has been dispersed to the Middle East and parts of Europe and Africa. There has been very little flow of the allele to the Americas.

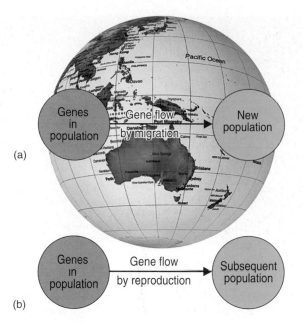

(a)

(b)

FIGURE 17.2 Gene Flow

Gene flow within a species occurs in two ways. *(a)* Genes flow from place to place when organisms migrate. *(b)* Genes flow from generation to generation as a result of reproduction.

3. Postreproductive adults no longer capable of reproduction (e.g. annual plants that have shed their seeds, salmon that have spawned, and many elderly humans)

A population is not necessarily divided into equal thirds. Some populations are made up of a majority of one age group (figure 17.4).

Many populations of organisms that live only a short time and have high reproductive rates have age distributions that change significantly in a matter of weeks or months. For example, many birds have a flurry of reproductive activity during the summer months. Therefore, samples of the population of a particular species of bird at different times during the summer would show widely different proportions of reproductive and prereproductive individuals. In early spring before they have started to nest, all the birds are reproductive adults. In late spring through midsummer, there is a large proportion of prereproductive juveniles.

Similarly, in the spring, annual plants germinate from seeds and begin to grow—all of the individuals are prereproductive juveniles. Later in the year, all the plants flower—they become reproducing adults. Finally, all the plants become postreproductive adults and die. But they have left behind

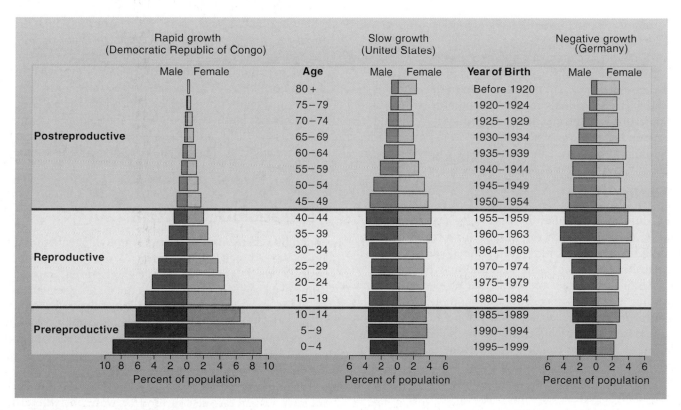

FIGURE 17.3 Age Distribution in Human Populations

The relative numbers of individuals in each of the three categories (prereproductive, reproductive, and postreproductive) are good clues to the future growth of a population. The Democratic Republic of Congo has a large number of young individuals who will become reproducing adults. Therefore, this population is likely to grow rapidly. The United States has a large proportion of reproductive individuals and a moderate number of prereproductive individuals. Therefore, this population is likely to grow slowly. Germany has a declining number of reproductive individuals and a very small number of prereproductive individuals. Therefore, its population has begun to decline.

FIGURE 17.4 Age Distribution in Selected Populations
Some populations are composed of many individuals of the same general age.

FIGURE 17.5 Sex Ratio
In some species of animals, males defend a harem of females; therefore, the sex ratio in these groups is several females per male. This male sea lion is defending a harem of several females.

seeds—prereproductive juveniles—which will produce the next generation.

Age distribution can have a major effect on how the population grows. If most of the population is prereproductive, a rapid increase in its size can be anticipated in the future as the prereproductive individuals reach sexual maturity. If most of the population is reproductive, the population should be growing rapidly. If most of the population is postreproductive, a population decline can be anticipated.

Sex Ratio

The **sex ratio** of a population is the number of males in a population compared with the number of females. In many kinds of animals, such as bird and mammal species in which strong pair-bonding occurs, the sex ratio may be nearly 1 to 1 (1:1). Among mammals and birds that do not have strong pair-bonding, sex ratios may show a larger number of females than males. This is particularly true among game species in which more males than females are shot, resulting in a higher proportion of surviving females. Because one male can fertilize several females, the population can remain large even though the females outnumber the males. In addition to these examples, many species of animals, such as bison, horses, elk, and sea lions, have mating systems in which one male maintains a harem of females. The sex ratio in these small groups is quite different from a 1:1 ratio (figure 17.5).

In many kinds of insect populations, such as bees, ants, wasps, and termites, there are many more females than males. Generally, in a colony of these organisms there is one or a few reproductive females, a large number of worker females, and very few males.

There are very few situations in which the number of males exceeds the number of females. In some human and other populations, there may be sex ratios in which the males dominate if female mortality is unusually high or if a special mechanism separates most of one sex from the other.

Many kinds of animals and plants are hermaphroditic, having both kinds of sex organs in the same body (e.g. earthworms and many flowering plants). Thus, the concept of sex ratio does not apply to them. Also, some species of animals—oysters, some fish, and others—change their sex at different times of their lives. They spend part of their lives as males and part as females.

Population Distribution

Population distribution is the way individuals within a population are arranged with respect to one another. There are basically three kinds of arrangements: even, random, and clumped. Even distributions occur under circumstances in which the organisms arrange themselves by very specific rules. In many birds that form dense breeding colonies, each nest is just out of reach of the neighbors. Random distributions are typical for many kinds of organisms with widely dispersed individuals. Clumped distributions are typical for many kinds of organisms in which they are found in large numbers near valuable resources, such as food or water, or where the offspring are associated with the parents (figure 17.6).

Population Density

Population density is the number of organisms of a species per unit area. For example, the population density of dandelions in a park can be measured as the number of dandelions

Even

Random

Clumped

FIGURE 17.6 Population Distribution

The way organisms are distributed in their habitat varies. Some are evenly distributed. Some are randomly distributed. Most show some degree of clumped distribution.

per square meter; the population density of white-tailed deer can be measured as the number of deer per square kilometer. Depending on the reproductive success of individuals in the population and the resources available, the density of populations can vary considerably. Some populations are extremely concentrated in a limited space; others are well dispersed. As reproduction occurs, the population density increases, which leads to increased competition for the necessities of life. This increases the likelihood that some individuals will migrate to other areas.

Population pressure is the concept that increased intensity of competition causes changes in the environment and

leads to the dispersal of individuals to new areas. This dispersal can relieve the pressure on the home area and lead to the establishment of new populations. Among animals, it is often the juveniles that participate in dispersal. For example, female bears generally mate every other year and abandon their nearly grown young the summer before the next set of cubs is to be born. The abandoned young bears tend to wander and disperse to new areas. Similarly, young turtles, snakes, rabbits, and many other common animals disperse during certain times of the year. That is one of the reasons so many animals are killed on the roads in the spring and fall.

If dispersal cannot relieve population pressure, there is usually an increase in the rate at which individuals die because of predation, parasitism, starvation, and accidents. For example, plants cannot relieve population pressure by dispersal. Instead, the death of weaker individuals usually results in reduced population density, which relieves population pressure. Following a fire, large numbers of lodgepole pine seeds germinate. The young seedlings form dense thickets of young trees. As the stand ages, many small trees die and the remaining trees grow larger as the population density drops (figure 17.7).

(a)

(b)

FIGURE 17.7 Changes in Population Density

(a) This population of lodgepole pine seedlings consists of a large number of individuals very close to one another. *(b)* As the trees grow, many of the weaker trees will die, the distance between individuals will increase, and the population density will be reduced.

17.2   Reproductive Capacity

Sex ratios and age distributions within a population have a direct bearing on the rate of reproduction. A population's **reproductive capacity,** or **biotic potential,** is the theoretical number of offspring that could be produced. Some species produce huge numbers of offspring, whereas others produce few per lifetime. However, the reproductive capacity is generally much larger than the number of offspring needed simply to maintain the population. For example, a female carp may produce 1 million to 3 million eggs in her lifetime; this is her reproductive capacity. However, only two or three of these offspring ever develop into sexually mature adults. Therefore, her reproductive rate is two or three offspring per lifetime and is much smaller than her reproductive capacity.

In general, there are two strategies organisms use to make certain that there will be enough offspring that live to adulthood to assure the continuation of the species. One strategy is to produce huge numbers of offspring but not provide any support for them. For example, an oyster may produce a million eggs a year, but not all of them are fertilized, and most that are fertilized die. An apple tree with thousands of flowers may produce only a few apples because the pollen that contains the sperm cells was not transferred to the female part of each flower in the process of pollination. Even after offspring are produced, mortality is usually high among the young. Most seeds that fall to the Earth do not grow, and most young animals die. Usually, however, enough survive to ensure the continuance of the species. Organisms that reproduce in this way spend large amounts of energy on the production of gametes and young, without caring for the young. Thus, the probability that any individual will reach reproductive age is small.

The second way of approaching reproduction is to produce relatively fewer individuals but provide care and protection, which ensures a higher probability that the young will become reproductive adults. Humans generally produce a single offspring per pregnancy, but nearly all of them live. In effect, energy has been channeled into the care and protection of the young, rather than into the production of incredibly large numbers of potential young. Even though fewer young are produced by animals such as birds and mammals, their reproductive capacity still greatly exceeds the number required to replace the parents when they die.

17.3   The Population Growth Curve

Because most species have a high reproductive capacity, populations tend to grow if environmental conditions permit. The change in the size of a population depends on the rate at which new organisms enter the population, compared with the rate at which they leave. **Natality** is the number of individuals added to the population by reproduction per thousand individuals in the population. **Mortality** is the number of individuals leaving a population by death per thousand individuals in the population. If a species enters a previously uninhabited area, its population will go through a typical pattern of growth. Figure 17.8 shows a **population growth curve,** which is a graph of change in population size over time. There are three recognizable portions in a population growth curve: the *lag phase,* the *exponential growth phase,* and the *stable equilibrium phase.*

The Lag Phase

The **lag phase** of a population growth curve is the period of time immediately following the establishment of a population, when the population remains small and fairly constant. During the lag phase, both natality and mortality are low.

The lag phase occurs because reproduction is not an instantaneous event. Even after animals enter an area, they must mate and produce young. This may take days or years, depending on the animal. Similarly, new plant introductions must grow to maturity, produce flowers, and set seed. Some annual plants do this in less than a year, whereas some large trees take several years of growth before they produce flowers. In organisms that take a long time to mature and produce young, such as elephants, deer, and many kinds of plants, the lag phase may be measured in years.

The Exponential Growth Phase

The **exponential growth phase** of a population growth curve is the period of time when a population is growing faster and faster. A population of mice is a good example of why populations can increase rapidly. Assuming that the population is started by a single pair—a male and female—it will take some time for them to mate and produce their first litter of offspring. This is the lag phase. However, once the first litter of young has been produced, the population is likely to increase rapidly. The first litter of young will become sexually mature and able to reproduce in a matter of weeks. If the usual litter size for a pair of mice is 4, the 4 produce 8, which in turn produce 16, and so forth. Furthermore, the original parents will probably produce an additional litter or two during this time period. Now, several pairs of mice are reproducing more than just once. With several pairs of mice reproducing, natality increases and mortality remains low; therefore, the population begins to grow at an ever-increasing rate. However, the population cannot continue to increase indefinitely. Eventually, the rate at which the population is growing will begin to level off.

The Stable Equilibrium Phase

The **stable equilibrium phase** of a population growth curve is the period of time when a population stops growing and maintains itself at a reasonably stable level. The number of organisms cannot continue to increase at a faster and faster rate, because eventually something in the environment will

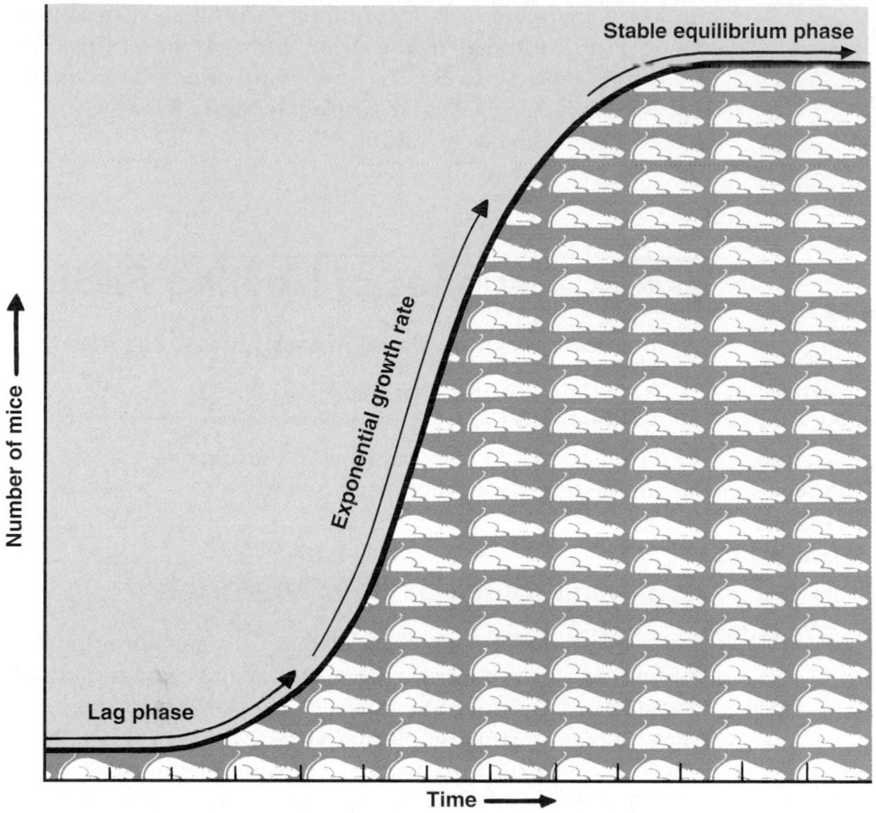

Stable equilibrium phase

Number of mice →

Exponential growth rate

Lag phase

Time →

FIGURE 17.8 A Typical Population Growth
Curve
In this mouse population, the period of time in which
there is little growth is known as the lag phase. This
is followed by a rapid increase in population as the
offspring of the originating population begin to
reproduce themselves; this is known as the
exponential growth phase. Eventually, the population
reaches a stable equilibrium phase, during which the
birthrate equals the deathrate.

become limiting and cause an increase in the number of
deaths. For animals, food, water, or nesting sites may be in
short supply, or predators or disease may kill many individuals. Plants may lack water, soil nutrients, or sunlight.
Eventually, the number of individuals entering the population
will come to equal the number of individuals leaving it by
death or migration, and the population size will become stable. Often, there is both a decrease in natality and an increase
in mortality at this point. During the stable equilibrium
phase, natality and mortality are about equal. However, the
population is still changing. Birth, death, and migration are
still going on, resulting in a changing mix of individuals.
Although there are changes in the individuals within the population, its size does not change. It is stable because there is
an equilibrium between those entering and those leaving the
population.

When ecologists look at many kinds of organisms and
how their populations change, they recognize two general
types. (1) Large organisms, which live a long time, tend to
reach a population size that can be sustained over an
extended period. These kinds of organisms follow the pattern
of population growth just described. (2) Organisms that are
small and have short life spans tend to have fluctuating populations and do not reach a stable equilibrium phase during
population growth. Their populations go through the normal
pattern of beginning with a lag phase, followed by an exponential growth phase. However, they typically reach a maximum, followed by a rapid decrease in population often called
a "crash" (figure 17.9).

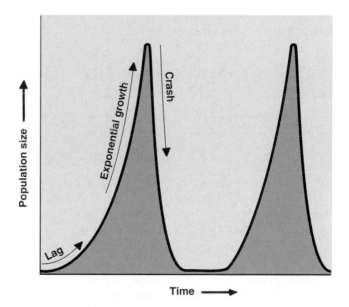

FIGURE 17.9 A Population Growth Curve for Short-Lived
Organisms
Organisms that are small and live only a short time often show this
kind of population growth curve. There is a lag phase, followed by
an exponential growth phase. However, instead of entering into a
stable equilibrium phase, the population reaches a maximum and
crashes.

17.4 Limits to Population Size

Populations cannot continue to increase indefinitely. Eventually, a factor or set of factors limits their size. The factors that prevent unlimited population growth are known as **limiting factors**. All the limiting factors that act on a population are collectively known as **environmental resistance.**

Extrinsic and Intrinsic Limiting Factors

Some factors that control populations come from outside the population and are known as **extrinsic limiting factors.** Predators, the loss of a food source, a lack of sunlight, and accidents of nature are all extrinsic factors. However, the populations of many kinds of organisms appear to be regulated by factors from within the populations themselves. Such limiting factors are called **intrinsic limiting factors.** For example, a study of rats under crowded living conditions showed that, as conditions became more crowded, abnormal social behavior became common. There was a decrease in litter size, fewer litters per year were produced, the mothers were more likely to ignore their young, and many young were killed by adults. Thus, changes in the rats' behavior resulted in lower birthrates and higher deathrates, which limit population size. In another example, the reproductive success of white-tailed deer is reduced when the deer experience a series of severe winters. When times are bad, the female deer are more likely to have single offspring than twins.

Density-Dependent and Density-Independent Limiting Factors

Density-dependent limiting factors are those that become more effective as the density of the population increases. For example, the larger a population becomes, the more likely that predators will have a chance to catch some of the individuals. A prolonged period of increasing population allows the size of the predator population to increase as well. Disease epidemics are also more common in large, dense populations, because dense populations allow for the easy spread of parasites from one individual to another. The rat example previously mentioned also illustrates a density-dependent limiting factor in operation—the amount of abnormal behavior increased as the density of the population increased. In general, whenever there is competition among the members of a population, its intensity increases as the population density increases. Large organisms that tend to live a long time and have relatively few young are most likely to be controlled by density-dependent limiting factors.

Density-independent limiting factors are population-controlling influences that are not related to the density of the population. They are usually accidental or occasional extrinsic factors in nature that happen regardless of the density of a population. A sudden rainstorm may drown many small plant seedlings and soil organisms. Many plants and animals are killed by frosts in late spring or early fall. A small pond may dry up, resulting in the death of many organisms. The organisms most likely to be controlled by density-independent limiting factors are small, short-lived organisms that can reproduce very rapidly.

17.5 Categories of Limiting Factors

Limiting factors can be placed in four broad categories:

1. Availability of raw materials
2. Availability of energy
3. Production and disposal of waste products
4. Interaction with other organisms

The Availability of Raw Materials

The availability of raw materials is an extremely important limiting factor. For example, plants require magnesium to manufacture chlorophyll, nitrogen to produce protein and water to transport materials and as a raw material for photosynthesis. If these substances are not present in the soil, the growth and reproduction of plants are inhibited. However, if fertilizer supplies these nutrients, or if irrigation supplies water, the effects of these limiting factors can be removed, and a different factor becomes limiting. For animals, the amount of water, minerals, materials for nesting, suitable burrow sites, and food may be limiting factors.

The Availability of Energy

The availability of energy is a major limiting factor, because all living things need a source of energy. The amount of light available is often a limiting factor for plants, which require light as an energy source for photosynthesis. Because all animals use other living things as sources of energy and raw materials, a major limiting factor for any animal is its food source.

The Accumulation of Waste Products

The accumulation of waste products can significantly limit the size of populations. It does not usually limit plant populations, however, because they produce relatively few wastes. However, the buildup of high levels of self-generated waste products is a problem for bacterial populations and populations of tiny aquatic organisms. As wastes build up, they become more and more toxic, and eventually reproduction stops, or the population dies out. For example, when a few bacteria are introduced into a solution containing a source of food, they go through the kind of population growth curve typical of all organisms. As expected, the number of bacteria begins to increase following a lag phase, increases rapidly

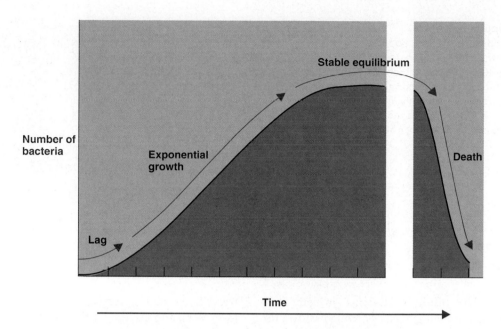

FIGURE 17.10 Bacterial Population Growth Curve

The rate of increase in the size of the population of these bacteria is typical of population growth in a favorable environment. When the environmental conditions worsen as a result of an increase in the amount of waste products, the population first levels off, then begins to decrease. This period of decreasing population size is known as the death phase.

during the exponential growth phase, and eventually reaches stability in the stable equilibrium phase. However, as waste products accumulate, the bacteria drown in their own wastes. When space for disposal is limited, and no other organisms are present that can convert the harmful wastes to less harmful products, a population decline, known as the **death phase**, follows (figure 17.10).

In small pools, such as aquariums, it is often difficult to keep organisms healthy because of the buildup of ammonia in the water from the animals' waste products. This is the primary reason that activated charcoal filters are used in aquariums. The charcoal removes many kinds of toxic compounds and prevents the buildup of waste products.

Interaction with Other Organisms

Organism interactions are also important in limiting population size. Recall from chapter 16 that organisms influence each other in many ways. Some interactions are harmful; others are beneficial.

Predation, parasitism, and competition tend to limit the growth of populations and their maximum size. Parasitism and predation usually involve interactions between two different species. Thus, they are extrinsic limiting factors. Competition between members of different species (interspecific competition) is an extrinsic limiting factor. Competition among members of the same species (intraspecific competition) is an intrinsic limiting factor and often is extremely intense.

On the other hand, many kinds of organisms perform beneficial services for others that allow the population to grow more than it would without the help. For example, decomposer organisms destroy toxic waste products, thus benefiting populations of aquatic animals. They also recycle the materials that all organisms need for growth and development. Mutualistic relationships benefit both populations involved. The absence of such beneficial organisms is a limiting factor.

Often, the population sizes of two kinds of organisms are interdependent, because each is a primary limiting factor of the other. This is most often seen in parasite-host relationships and predator-prey relationships.

A study of the population biology of the collared lemming (*Dicrostonyx groenlandicus*) in Greenland illustrates the population interactions between lemmings and four predators. Lemmings have a very high reproductive capacity, producing two or three litters per year. However, their population is held in check by four predators. Three—the snowy owl, the arctic fox, and the longtailed skua (a bird that resembles a gull)—are generalist predators, whose consumption of lemmings is directly related to the size of the lemming population. They constitute a density-dependent limiting factor for the lemming population. When lemming numbers are low, these predators seek other prey.

The fourth lemming predator is the shorttail weasel (*Mustela ermina*); it is a specialist predator on lemmings. The weasels are much more dependent on lemmings for food than are the other three predators. The weasels mate once a year, so their population increases at a slower rate than that of the lemmings. However, as the weasel population increases, it eventually become large enough that it drives down the

FIGURE 17.11 Population Cycles
In many northern regions of the world, population cycles are common. In the case of collared lemmings and shorttail weasels in Greenland, interactions between the two populations result in population cycles of about 4 years. The graph shows the population of lemmings per hectare and the number of lemming nests occupied by weasels per hectare.

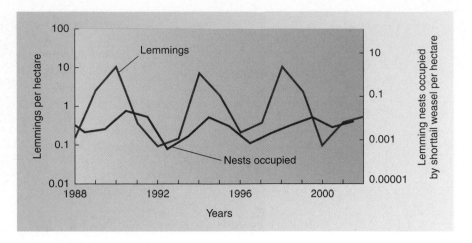

lemming population. The resulting decrease in lemmings leads to a decline in the number of weasels, which allows for the greater survival of lemmings, which ultimately leads to another cycle of increased weasel numbers (figure 17.11).

17.6 Carrying Capacity

Many populations reach a maximum size when they reach the stable equilibrium phase. This suggests that the environment sets an upper limit on population size. The **carrying capacity** is the maximum sustainable population for an area. The carrying capacity is determined by a set of limiting factors (figure 17.12).

Carrying capacity is not an inflexible number, however. Often, such environmental differences as successional changes, climatic variations, disease epidemics, forest fires, and floods can change the carrying capacity of an area for specific species. In aquatic ecosystems, one of the major factors that determine the carrying capacity is the amount of nutrients in the water. Where nutrients are abundant, the numbers of various kinds of organisms are high. Often, nutrient levels fluctuate with changes in current or runoff from the land, and plant and animal populations fluctuate as well. In addition, a change that negatively affects the carrying capacity for one species may increase the carrying capacity for another. For example, the cutting down of a mature forest followed by the growth of young trees increases the carrying capacity for deer and rabbits, which use the new growth for food, but decreases the carrying capacity for squirrels, which need mature, fruit-producing trees as a source of food and old, hollow trees for shelter. Outlooks 17.1 provides an example in which changed food habits have changed the carrying capacity of snow geese.

Wildlife management practices often encourage modifications to the environment that will increase the carrying capacity for the designated game species. The goal of wildlife managers is to have the highest sustainable population available for harvest by hunters. Typical habitat modifications include creating water holes, cutting forests to provide young growth, planting food plots, and building artificial nesting sites.

In some cases, the size of the organisms in a population also affects the carrying capacity. For example, an aquarium of a certain size can support only a limited number of fish, but the size of the fish makes a difference. If all the fish are tiny, a large number can be supported, and the carrying capacity is high; however, the same aquarium may be able to support only one, large fish. In other words, the biomass of the population makes a difference (figure 17.13). Similarly, when an area is planted with small trees, the population size is high; however, as the trees get larger, competition for nutrients and sunlight becomes more intense, and the number of trees declines as the biomass increases.

17.7 Limiting Factors to Human Population Growth

Today we hear differing opinions about the state of the world's human population. On one hand we hear that the population is growing rapidly. By contrast we hear that some countries are afraid that their populations are shrinking. Other countries are concerned about the aging of their populations, because birthrates and deathrates are low. In magazines and on television, we see that there are starving people in the world. At the same time, we hear discussions about the problem of food surpluses and obesity in many countries. Some have even said that the most important problem in the world today is the rate at which the human population is growing. Others maintain that the growing population will provide markets for goods and be an economic boon. How do we reconcile this mass of conflicting information?

It is important to realize that human populations follow the same patterns of growth and are acted on by the same kinds of limiting factors as are populations of other organisms. The growth of the human population over the past several thousand years follows a pattern that resembles the lag

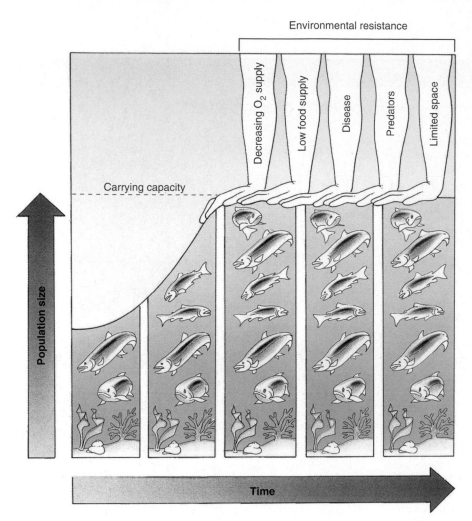

FIGURE 17.12 Carrying Capacity

A number of factors in the environment, such as food, oxygen supply, diseases, predators, and space, determine the maximum number of organisms that can be sustained in a given area—the carrying capacity of that area. The environmental factors that limit populations are collectively known as environmental resistance.

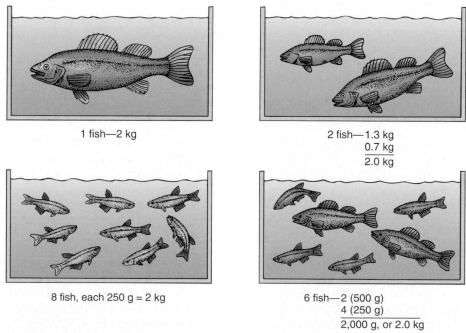

1 fish—2 kg

2 fish—1.3 kg
0.7 kg
————
2.0 kg

8 fish, each 250 g = 2 kg

6 fish—2 (500 g)
4 (250 g)
——————
2,000 g, or 2.0 kg

FIGURE 17.13 The Effect of Biomass on Carrying Capacity
Each aquarium can support a biomass of 2 kilograms of fish. The size of the population is influenced by the body size of the fish in the population.

OUTLOOKS 17.1

The Lesser Snow Goose—A Problem Population

Lesser snow geese use breeding grounds in the arctic and subarctic regions of Canada and wintering grounds on salt marshes of the Gulf of Mexico, California, and Mexico. One population of the lesser snow goose has breeding grounds around Hudson Bay and wintering grounds on the Gulf of Mexico. Because these birds migrate south from Hudson Bay through the central United States west of the Mississippi River to wintering ground on the west coast of the Gulf of Mexico, they are called the mid-continent population. Populations of the mid-continent lesser snow goose increased from about 800,000 birds in 1969 to an estimated 3.2 million in 2003, a fourfold increase. (Some experts believe the number may be closer to 6 million.) These numbers are so large that the nesting birds are destroying their breeding habitat.

What has caused this drastic population increase? A primary reason appears to be a change in feeding behavior. Originally, snow geese fed primarily on the roots of aquatic vegetation in the salt marshes of the Gulf of Mexico. Because this habitat and the food source it provided were limited, there was a natural control on the size of the population. However, many of these wetland areas have been destroyed. Simultaneously, agriculture (chiefly, rice farming) in the Gulf Coast region provided an alternative food source. The geese began to use agricultural areas for food, and they had a virtually unlimited source of food during the winter months, which led to very high winter survival. Furthermore, concerns about erosion have resulted in farmer's using increased notill and reduced tillage practice in the grain fields along their flyway. These practices leave crop residue on the fields through the fall and winter, which provides food for the fall migration, improving the survival of young snow geese. During the spring migration, the geese used the same resources to store food as body fat that they took with them to their northern breeding grounds. Other factors that may have played a role is the decline in the number of people who hunt geese and a warming of the climate, leading to greater breeding success on the cold northern breeding grounds.

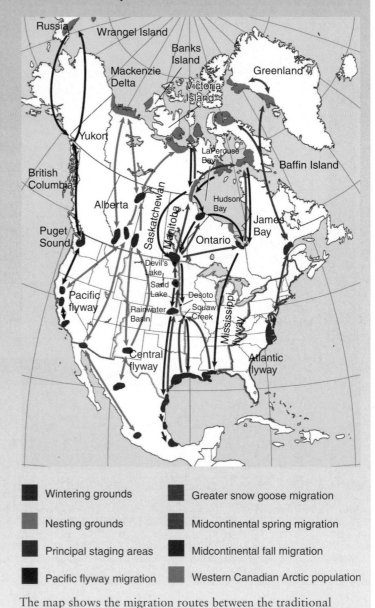

■ Wintering grounds	■ Greater snow goose migration
■ Nesting grounds	■ Midcontinental spring migration
■ Principal staging areas	■ Midcontinental fall migration
■ Pacific flyway migration	■ Western Canadian Arctic population

The map shows the migration routes between the traditional nesting and wintering areas of different snow goose populations.

and exponential growth phases of a population growth curve. It is estimated that the human population remained low and constant for thousands of years but has increased rapidly in the past few hundred years (figure 17.14). For example, it has been estimated that, before European discovery, the Native American population was about 1 million and was at or near its carrying capacity. Today, the population of North America is over 300 million people. Does this mean that humans are different from other animal species? Can the human population continue to grow forever?

The human species has an upper limit set by the carrying capacity of the environment, as does any other species. However, the human population has been able to increase astronomically because technological changes and the displacement of other species has allowed us to shift the carrying capacity upward. Much of the exponential growth phase of the human population can be attributed to improved sanitation, the control of infectious diseases, improvements in agricultural methods, and the replacement of natural ecosystems with artificial agricultural ecosystems. But even these condi-

OUTLOOKS 17.1 *(continued)*

The habitat destruction that is occurring in Canada is a result of several factors. Snow geese feed by ripping plants from the ground, and they nest in large colonies. Therefore, they can have a huge local impact on the vegetation; large areas of the coastal vegetation around Hudson Bay has been destroyed.

What can be done to control the snow goose population and protect their breeding habitat? There are two possibilities: let nature take its course or use population management tools. If nothing is done, the geese will eventually destroy so much of their breeding habitat that they will be unable to breed successfully and the population will crash. However, it might take decades for the habitat to recover. The alternative is to take action to resolve the problem. Several management activities have begun: allowing hunters to kill more geese, increasing the length of the hunting season, allowing hunts during the spring migration season in addition to the fall migration season, and allowing more geese to be harvested by Canadian native communities. In recent years, the annual harvest in the United States and Canada has increased and varied between 1 and 1.5 million geese killed. Preliminary evidence suggests that the increased harvest is starting to have an effect. However it will take several more years to determine if the population can be controlled.

This photo shows the kind of damage done by geese. The exclosures were established by Dawn R. Bazely and Robert L. Jefferies of the Hudson Bay Project.

tions have their limits. Some limiting factors will eventually cause a leveling off of our population growth curve. We cannot increase beyond our ability to get raw materials and energy, nor can we ignore the waste products we produce and the other organisms with which we interact.

Available Raw Materials

To many of us, raw materials consist simply of the amount of food available, but we should not forget that, in a technological society, iron ore, lumber, irrigation water, and silicon chips are also raw materials. However, most people of the world have much more basic needs. For the past several decades, large portions of the world's population have not had enough food (figure 17.15). Although it is biologically accurate to say that the world can currently produce enough food for everyone, there are complex political, economic, and social issues related to food production and distribution. Probably most important is the fact that the transportation of food from centers of excess to centers of need is often very difficult and expensive. Societies with excess food have been unwilling to bridge these political and economic gaps.

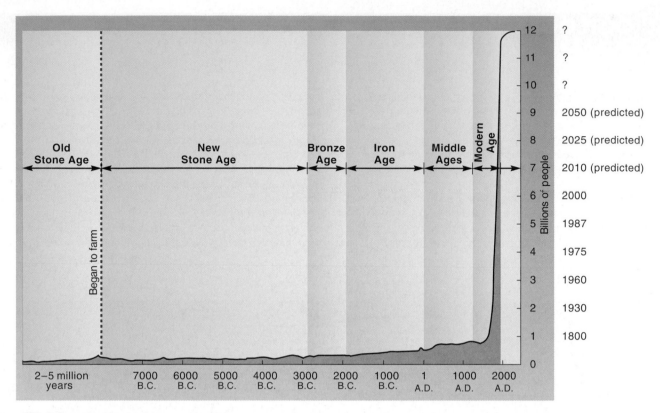

FIGURE 17.14 **Human Population Growth**
The number of humans doubled from A.D. 1800 to 1930 (from 1 billion to 2 billion), had doubled again by 1975 (4 billion), and is projected to double again (8 billion) by about 2025. How long can the human population continue to double before the Earth's ultimate carrying capacity is reached?

FIGURE 17.15 Food is a Raw Material

However, a more fundamental question is whether the world can continue to produce enough food. In 2005, the world population was growing at a rate of 1.3% per year. This amounts to about 160 new people added to the population every minute. This would result in an increase from our current population of about 6.5 billion to over 9 billion by 2050. With a continuing increase in the number of mouths to feed, it is unlikely that food production will be able to keep pace with the growth in human population (How Science Works 17.1).

A primary indicator of the status of the world food situation is the amount of grain produced for each person in the world (per capita grain production). World per capita grain production peaked in 1984.

Available Energy

The availability of energy also affects human populations. All species on Earth ultimately depend on sunlight for energy. All energy—whether power from a hydroelectric dam, fossil fuels, or solar cell—is derived from the Sun. Energy is needed for transportation, the construction and maintenance of homes, and food production. It is very difficult to develop

HOW SCIENCE WORKS 17.1

Thomas Malthus and His Essay on Population

In 1798, Thomas Robert Malthus, an Englishman, published an essay on human population, presenting an idea that was contrary to popular opinion. His basic thesis was that the human population increased in a geometric, or exponential, manner (2, 4, 8, 16, 32, 64, etc.), whereas the ability to produce food increased only in an arithmetic manner (1, 2, 3, 4, 5, 6, etc.). The ultimate outcome of these different rates would be that the population would outgrow the land's ability to produce food. He concluded that wars, famines, plagues, and natural disasters would be the means (limiting factors) of controlling the size of the human population. His predictions were hotly debated by the intellectual community of his day, and his assumptions and conclusions were attacked as erroneous and against the best interest of society. At the time he wrote the essay, the popular opinion was that human knowledge and "moral constraint" would be able to create a world that would supply all human needs in abundance. One of Malthus's basic postulates was that "commerce between the sexes" (sexual intercourse) would continue unchanged. Other philosophers of the day believed that sexual behavior would take less procreative forms and the human pop-

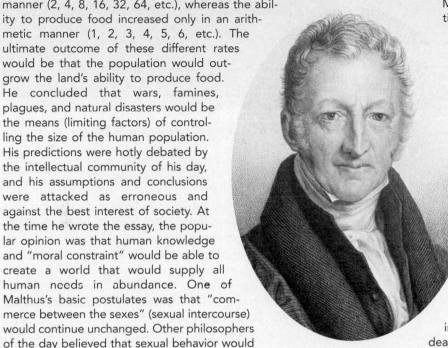

Thomas Robert Malthus

ulation would be limited. Only within the past 50 years, however, have effective conception-control mechanisms become widely accepted and used, and they are used primarily in developed countries.

Malthus did not foresee the use of contraception, major changes in agricultural production techniques, or the exporting of excess people to colonies in the Americas. These factors, as well as high deathrates, prevented the most devastating of his predictions from coming true. However, in many parts of the world, people are experiencing the forms of population control (famine, epidemic disease, wars, and natural disasters) Malthus predicated in 1798. Many people believe that his original predictions were valid—only his time scale was not correct—and that his predictions are coming true today.

Another important impact of Malthus's essay was its effect on young Charles Darwin. When Darwin read it, he saw that what was true for the human population could be applied to the whole of the plant and animal kingdoms. As over-reproduction took place, there would be increased competition for food, resulting in the death of the less fit organisms. This was an important part of his theory of natural selection.

---

unbiased, reasonably accurate estimates of global energy reserves in the form of petroleum, natural gas, and coal. Therefore, it is difficult to predict how long these reserves might last. However, the quantities are limited and the rate of use has been increasing (figure 17.16).

If the less-developed countries were to attain a standard of living equal to that of the developed nations, the global energy reserves would disappear overnight. People should realize that there is a limit to our energy resources; we are living on solar energy that was stored over millions of years, and we are using it at a rate that could deplete it in hundreds of years.

Accumulation of Wastes

One of the most talked about aspects of human activity is the problem of waste disposal. We have normal biological wastes, which can be dealt with by decomposer organisms. However, we also generate a variety of technological wastes and by-products that cannot be degraded efficiently by decomposers (figure 17.17). Most of what we call pollution

FIGURE 17.16 Humans' Energy Use

FIGURE 17.17 Human Waste Production

FIGURE 17.18 Interaction with Other Organisms

results from the waste products of technology. The biological wastes usually can be dealt with fairly efficiently by building wastewater treatment plants and other sewage facilities. Certainly, these facilities take energy to run, but they rely on decomposers to degrade unwanted organic matter to carbon dioxide and water. Earlier in this chapter, we discussed the problems aquarium organisms face when their metabolic waste products accumulate. In this situation, the organisms so "befoul their nest" that their wastes poison them. Are humans in a similar situation on a much larger scale? Are we dumping so much technological waste, much of it toxic, into the environment that we are being poisoned?

Interactions with Other Organisms

Humans interact with other organisms in as many ways as other animals do. We have parasites and, occasionally, predators. We are predators in relation to a variety of animals, both domesticated and wild. We have mutualistic relationships with many of our domesticated plants and animals, because they could not survive without our agricultural practices, and we would not survive without the food they provide. Competition is also very important. Insects and rodents compete for the food we raise, and we compete directly with many other kinds of animals for the use of ecosystems.

As humans convert more and more land to agriculture and other purposes, many other organisms are displaced (figure 17.18). Many of these displaced organisms are not able to compete successfully and must leave the area, reduce their populations, or become extinct. The American bison (buffalo), African and Asian elephants, panda, and grizzly bear are a few species that have reduced populations because they were not able to compete successfully with the human species. The passenger pigeon, Carolina parakeet, and great auk are a few that have become extinct. Our parks and natural areas have become tiny refuges for plants and animals that once occupied vast expanses

of the world. If these refuges are lost, many organisms will become extinct. What today might seem to be an insignificant organism may tomorrow be seen as a link to our very survival. We humans have been extremely successful in our efforts to convert ecosystems to our own uses at the expense of other species.

Competition with one another (intraspecific competition), however, is a different matter. Because competition is negative for both organisms, competition between humans harms humans. We are not displacing another species; we are displacing our own kind. Certainly, when resources are in short supply, there is competition. Unfortunately, it is usually the young who are least able to compete, and high infant mortality is the result. organism interactions, p. 326

17.8 Control of the Human Population—a Social Problem

Humans are different from most other organisms in a fundamental way: We are able to predict the outcome of a specific course of action. Current technology and medical knowledge are available to control the human population and improve the health and well-being of the people of the world. Why, then, does the human population continue to grow, resulting in human suffering and stressing the environment in which we live? Because we are social animals with freedom of choice, we frequently do not do what is considered best from an unemotional, unselfish, biological point of view. People make decisions based on historical, social, cultural, ethical, and personal considerations. What is best for the population as a whole may be bad for an individual.

The biggest problems associated with the control of the human population are not biological problems; rather, they require the efforts of philosophers, theologians, politicians, and sociologists. As the population increases, so do political, social, and biological problems; individual freedom diminishes, intense competition for resources intensifies, and famine and starvation becomes more common. The knowledge and technology necessary to control the human population are available, but the will is not. What will eventually limit the size of our population? Will it be lack of resources, lack of energy, accumulated waste products, competition among ourselves, or rational planning of family size?

Studies of the changes in the population growth rates of various countries indicated that a major factor in determining the size of families is the educational status of women. Regardless of other cultural differences, as girls and women become educated, they have fewer children. Several reasons have been suggested for this trend. Higher levels of education allow women to get jobs with higher pay, which makes them less dependent on males for their support. Being able to read may lead to better comprehension of how methods of birth control work. Regardless of the reasons, increasing the educational levels of women has become a major technique used by rapidly growing countries that hope to control their populations.

Summary

A population is a group of organisms of the same species in a particular place at a particular time. Populations differ from one another in gene frequency, age distribution, sex ratio, population distribution, and population density. Organisms typically have a reproductive capacity that exceeds what is necessary to replace the parent organisms when they die. This inherent capacity to overreproduce causes a rapid increase in population size when a new area is colonized. A typical population growth curve consists of a lag phase, in which the population rises very slowly, followed by an exponential growth phase in which the population increases at an accelerating rate, followed by a leveling off of the population in a stable equilibrium phase as the carrying capacity of the environment is reached. In some populations, a fourth phase, the death phase, occurs.

The carrying capacity is the maximum sustainable number of organisms an area can support. It is set by a variety of limiting factors. The availability of energy, the availability of raw materials, the accumulation of wastes, and interactions with other organisms are limiting factors. Because organisms are interrelated, population changes in one species sometimes affect the size of other populations. This is particularly true when one organism uses another as a source of food. Some limiting factors operate from outside the population and are known as extrinsic limiting factors; others are properties of the species itself and are called intrinsic limiting factors. Some limiting factors become more intense as the density of the population increases; these are known as density-dependent limiting factors. Limiting factors that are more accidental and unrelated to population density are called density-independent limiting factors.

Humans as a species have the same limits and influences that other organisms do. Our current problems of food production, energy needs, pollution, and habitat destruction are outcomes of uncontrolled population growth. However, humans can reason and predict, thus offering the possibility of population control through conscious population limitation.

Key Terms

Use the interactive flash cards on the **Concepts in Biology,** *12/e website to help you learn the meaning of these terms.*

age distribution 360	limiting factors 366
carrying capacity 368	mortality 364
death phase 367	natality 364
density-dependent limiting factors 366	population 360
	population density 362
density-independent limiting factors 366	population distribution 362
	population growth curve 364
environmental resistance 366	population pressure 363
exponential growth phase 364	reproductive capacity (biotic potential) 364
extrinsic limiting factors 366	sex ratio 362
intrinsic limiting factors 366	stable equilibrium phase 364
lag phase 364	

Basic Review

1. The number of reproducing adults in a population compared with the number of juveniles is the
   a. population density.
   b. age distribution.
   c. population distribution.
   d. gene frequency.

2. The period of time when a population is growing rapidly is known as the _____.

3. Populations grow
   a. because most species have a high reproductive capacity.
   b. when birthrates are greater than deathrates.
   c. when there are high numbers of reproductive and juvenile individuals in the population.
   d. All of the above are correct.

4. The maximum size of a population is set by limiting factors of the environment. (T/F)

5. A limiting factor that becomes more intense as the size of a population increases is known as a density-independent limiting factor. (T/F)

6. The carrying capacity

   a. for the human population has been reached.

   b. is determined by the limiting factors of the environment.

   c. is the same for all organisms.

   d. None of the above is correct.

7. When the size of a population is caused to stop growing because of competition among its members, there are _____ in action.

   a. extrinsic limiting factors

   b. density-independent limiting factors

   c. intrinsic limiting factors

   d. population distribution factors

8. The populations of all species eventually reach a stable equilibrium phase. (T/F)

9. Which one of the following populations would grow most rapidly?

   a. a population of mice in which there were twice as many males as females

   b. a population of mice that had reached its carrying capacity

   c. a population of mice in which density-dependent limiting factors were acting strongly

   d. a population that was in the lag phase

10. The human population has been increasing rapidly for the past 200 years because

    a. humans have displaced other organisms.

    b. humans have controlled many disease organisms.

    c. humans have developed improvements in agriculture.

    d. All of the above are correct.

Answers

1. b 2. exponential growth phase 3. d 4. T 5. F 6. b 7. c 8. F 9. d 10. d

5. All organisms overreproduce. What advantages does this give to the species? What disadvantages?

17.3 The Population Growth Curve

6. Draw a population growth curve. Label the lag, exponential growth, and stable equilibrium phases.

7. What causes a lag phase in a population growth curve? An exponential growth phase? A stable equilibrium phase?

17.4 Limits to Population Size

8. Differentiate between density-dependent and density-independent limiting factors. Give an example of each.

9. Differentiate between intrinsic and extrinsic limiting factors. Give an example of each.

17.5 Categories of Limiting Factors

10. List four kinds of limiting factors that help set the carrying capacity for a species.

17.6 Carrying Capacity

11. How is the carrying capacity of an environment related to the stable equilibrium phase of a population growth curve?

12. Give an example of how the carrying capacity of an environment can be related to the size of individual organisms.

17.7 Limiting Factors to Human Population Growth

13. As the human population continues to grow, what should we expect to happen to other species?

14. How does the shape of the population growth curve of humans compare with that of other kinds of animals?

17.8 Control of Human Population is a Social Problem

15. How does human population growth differ from the population growth of other kinds of organisms?

16. What forces will ultimately lead to the control of human population growth?

Concept Review

Answers to the Concept Review questions can be found on the Concepts in Biology, 12/e website.

17.1 Population Characteristics

1. Describe four ways in which two populations of the same species can differ.

2. How is population pressure related to population density?

17.2 Reproductive Capacity

3. Why do populations grow?

4. In what way do the activities of species that produce few young differ from those that produce huge numbers of offspring?

Thinking Critically

For guidelines to answer this question, visit the Concepts in Biology, 12/e website.

Review figure 17.14; note that this population growth curve has very little in common with the population growth curve shown in figure 17.8. What factors have allowed the human population to grow so rapidly? What natural limiting factors will eventually bring our population under control? What is the ultimate carrying capacity of the world? What alternatives to the natural processes of population limitation could bring human population under control? Consider the following in your answers: reproduction, death, diseases, food supply, energy, farming practices, food distribution, cultural biases, and anything else you consider relevant.

CHAPTER

18

Evolutionary and Ecological Aspects of Behavior

Many people enjoy feeding birds. Some take food to the park and feed the pigeons or ducks. Some place bird feeders in their yards. If you observe birds being fed in this manner, you can ask yourself many different kinds of questions. How do the birds find the food initially? Do they recognize certain kinds of structures as bird feeders? Do they recognize the people that give them food? How does the presence of an artificial source of food change their behavior? Why do people feed birds, not ants or mice? This chapter will provide some answers.

CHAPTER OUTLINE

18.1 # Interpreting Behavior

Behavior is how an animal acts—what it does and how it does it. When you watch a bird, a squirrel, or any other animal, its activities appear to have a purpose. Birds search for food, take flight as you approach, and build nests in which to raise young. Usually, the nests are inconspicuous or placed in difficult-to-reach spots. Likewise, squirrels collect and store nuts and acorns, "scold" when you get too close, and learn to visit sites where food is available.

Discovering the Significance of Behavior

It is not always easy to identify the significance of a behavior without careful study of the behavior pattern and its impact on other organisms (How Science Works 18.1). For example, a baby herring gull pecks at a red spot on its parent's bill. What possible value can this behavior have for either the chick or the parent? If we watch, we see that, when the chick pecks at the spot, the parent regurgitates food onto the ground, and the chick feeds (figure 18.1). This looks like a simple behavior, but there is more to it than meets the eye. Why did the chick peck to begin with? How did it "know" to peck at that spot? Why did the pecking cause the parent to regurgitate food? These questions are not easy to answer. Many people assume that the actions have the same motivation and direction as similar human behaviors, but this is not necessarily a correct assumption. For example, when a little girl points to a piece of candy and makes appropriate noises, she is indicating to her parent that she wants some candy. Is that what the herring gull chick is doing? We don't know. Although both kinds of young may get food, we don't know what the baby gull is thinking because we can't ask it.

FIGURE 18.1 Understanding Behavior
A baby herring gull causes its parent to regurgitate food onto the ground by pecking at the red spot on its parent's bill. This behavior is instinctive. The parent then picks up the food and feeds it to the baby.

Behavior is Adaptive

When we observe behavior, we must keep in mind that behavior is adaptive just like any other characteristic displayed by an animal. Behaviors are important for survival and appropriate for the environment in which an animal lives. As with animals' highly adaptive structural characteristics, behaviors are the result of a long evolutionary process. Birds that do not take flight at the approach of another animal are eaten by predators. Squirrels that do not find deposits of food are less likely to survive, and birds that build obvious nests on the ground are more likely to lose their young to predators.

Because the nervous system of animals is the product of genes, many aspects of behavior have a genetic component. In this respect, behavioral characteristics are no different from physical characteristics, such as eye structure, wing shape, or tail length. However, the evolution of behavior is much more difficult to study than structural characteristics. This is because behavior is temporary, and it is difficult to find fossils that show the development of behavior, the way fossils allow us to follow changes in structures. Fossils of footprints, nests, and specific structures such as teeth adapted for grinding give clues about the behavior of extinct animals. However, these examples represent only a few fragments of the total behavior that must have been a part of the lives of extinct animals. When scientists compare the behavior of closely related animals living today, they can see inherited behaviors that are slightly different from one another. This strongly suggests that these behaviors have evolved, just as the wings of different species of birds have evolved for different kinds of flight. In both cases (structural and behavioral) a basic pattern has been modified as the organisms adapted to their environments.

18.2 # The Problem of Anthropomorphism

Anthropomorphism is the idea that we can assign human feelings, meanings, and emotions to the behavior of animals (figure 18.2). Poets, composers, and writers have often described birdsong as the act of a "joyful songster," but is a bird singing on a warm, sunny spring day making a beautiful sound because it is happy? Students of animal behavior do not accept this idea and have demonstrated that a bird sings primarily to attract mates and to tell other birds to keep out of its territory. The barbed stinger of a honeybee remains in your skin after you are stung, and the bee tears the stinger out of its body when it flies away. The damage to its body is so great that it dies. Has the bee performed a noble deed of heroism and self-sacrifice? Was it defending its hive from you? Scientists need to know a great deal more about the behavior of bees to understand the value of such behavior to the success of the bee species. The fact that bees are social animals, like humans, makes it particularly tempting to think that they are doing things for the same reasons we are.

FIGURE 18.2 Anthropomorphism
It is tempting to think that both the girl and the dog are having the same experience. However, that would be anthropomorphic thinking. We really don't know what the dog is experiencing. We can't ask the dog what it is thinking.

If you chase a squirrel in a park it will run up a tree and chatter at you. This is often referred to as scolding; however, this is a human interpretation based on what we might do if someone disturbed us. We might yell at the other person to go away and leave us alone. However, several other possible interpretations are equally reasonable. Perhaps the purpose of the squirrel's noise making is directed at driving you away. Or perhaps its vocalizations warn other squirrels of the presence of a predator. Or perhaps it is simply a nervous reaction to being disturbed. We cannot crawl inside the brain of another animal and see what it is thinking.

In order to evaluate behavior objectively, we need to avoid anthropomorphic assumptions. A scientific study of this scolding behavior would attempt to determine what triggers the behavior, rather than jump to a conclusion that it must be for the same reasons we humans do things. Furthermore, a scientific study of this behavior would seek to determine the effect of the behavior. A scientific study of an organism's behavior looks at the behavior of an animal during its lifetime and seeks to determine the behavior's value to the animal and how that behavior may have evolved. This scientific way of thinking requires the testing of hypotheses; an anthropomorphic approach does not.

18.3 Instinctive and Learned Behavior

There are two major kinds of behaviors: instinctive and learned. **Instinctive behavior** is behavior that is inherited, automatic, and inflexible. **Learned behavior** is behavior that changes as a result of experience. Both are involved in the behavior patterns of most organisms. Learning allows an organism to generate a completely new behavior or to slightly modify existing behavior patterns. Most animals (e.g., jellyfish, clams, insects, worms, frogs, turtles) have a high proportion of instinctive behavior and very little learning. Others, such as many birds and mammals, have a great deal of learned behavior in addition to many instinctive behaviors.

The Nature of Instinctive Behavior

Instinctive behavior makes up most of the behavior in animals that have short life cycles, simple nervous systems, and little contact with their parents. These behaviors are performed correctly the first time without previous experience. Such behaviors are found in a wide range of organisms, from simple, one-celled protozoans to complex vertebrates.

Stimulus and Response

Instinctive behaviors occur as a result of a specific *response* to a particular *stimulus*. A **stimulus** is a change in the organism's internal or external environment that an animal is able to detect. A **response** is an organism's reaction to a stimulus. nervous system, p. 572

In our example of the herring gull chick, the red spot on the parent's bill serves as a stimulus for the chick. The chick responds to this spot in a genetically programmed way. The behavior is innate—it is done correctly the first time without prior experience. The chick's pecking behavior, in turn, is the stimulus for the adult bird to regurgitate food. Obviously, these behaviors have adaptive value for the survival of the young, because they leave little to chance. The young will get food automatically from the adult. Instinctive behavior has great value to a species, because it allows correct, appropriate, and necessary behavior to occur without prior experience.

An organism can respond only to stimuli it can recognize. For example, it is difficult for humans to appreciate what the world is like to a bloodhound. The bloodhound is able to identify individuals by smell, whereas we have great difficulty detecting, let alone distinguishing among, many odors. Some animals, such as dogs, deer, and mice, can see only shades of gray. Others, such as honeybees, see ultraviolet light, which is invisible to us. Some birds and other animals are able to detect the magnetic field of the Earth. And some, such as rattlesnakes, are able to detect infrared radiation (heat) from distant objects.

Instinctive Behavior Cannot be Modified

Instinctive behaviors can be very effective for the survival of individuals of a species, but only as fundamental, essential activities that do not require modification. The major drawback of instinctive behavior is that it cannot be modified when a new situation presents itself.

Over long periods of evolutionary time, genetically determined behaviors have been selected for and have been useful to most individuals of the species. However, instances of inappropriate behavior may be generated by unusual stimuli or

HOW SCIENCE WORKS 18.1

Observation and Ethology

Ethology is the study of the behavior of animals in their natural environment. Such studies do not lend themselves easily to experimentation, because it is difficult to control the conditions to fit the criteria of an experimental design. Therefore, most studies of the behavior of wild animals involve careful, extended observation to determine the significance of specific animal behaviors. Often, the meanings of behaviors can be understood only if the observer is able to identify the individuals involved. For most of us, all robins, raccoons, and opossums look alike, making it difficult for us to understand the behaviors we see between individuals.

Ethologists have used various techniques to allow for the easy identification of individuals. Some animals have variations that allow them to be identified. Wild horses differ in size and color patterns, whales have differently shaped tails, and gorillas have size, color, and facial differences. It is the observational skill of the scientist that allows them to recognize these unique features. For example, one scientist observing chipmunks was able to identify many individuals by specific characteristics of their tails, such as the shape and size of the tail, how the tail was carried, and minor differences in color. In studies of bird behavior, scientists have placed colored bands on birds' legs that can be viewed through powerful binoculars. This has allowed investigators to determine which birds in a group were dominant, how they moved through their environment, and how long they lived. The attachment of small radio transmitters to animals as small as pigeons and as large as bears and sea turtles has allowed investigators to locate and follow individuals over large areas. Although various techniques can be used to identify individual animals, they are only tools to help investigators answer questions about and decipher the meaning of the behaviors they observe. For example, in a study of leatherback turtles in Costa Rica, the ability to identify individual turtles and follow their migration has provided information on the frequency with which females nest, whether individual females prefer specific places to lay their eggs, their longevity, and population size.

unusual circumstances in which the stimulus is given. For example, many kinds of insects fly toward a source of light. Over the millions of years of insect evolution, this has been a valuable, useful behavior. It allows them easily to find their way to open space. However, the human species invented artificial lights and transparent windows, which cause the animals to engage in totally inappropriate behavior. We have all seen insects drawn to lights at night or insects inside houses, constantly flying against window panes through which sunlight is entering. This mindless, mechanical behavior seems incredibly stupid to us. Although *individual* animals die, the general behavior pattern of flying toward light is still valuable because *most* of the individuals of the species do not encounter artificial lights or get trapped inside houses, and they complete their life cycles normally.

Geese and other birds sit on nests. This behavior keeps the eggs warm and protects them from other dangers. During this incubation period, the eggs may roll out of the nest as the parents get on and off the nest. When this happens, certain species of geese roll the eggs back into the nest. If the developing young within the egg are exposed to extremes of heat or cold, they are killed; thus, the egg-rolling behavior has a significant adaptive value. If, however, the egg is taken from the goose when it is in the middle of egg-rolling behavior, the

FIGURE 18.3 Inflexible Instinctive Behavior
Geese use a specific set of head movements to roll any reasonably round object back to the nest. There are several components of this instinctive behavior, including recognition of the object and head-tucking movements. If the egg is removed during the head-tucking movements, the behavior continues as if the egg were still there. This demonstrates the inflexible nature of instinctive behavior.

goose will continue its egg rolling until it gets back to the nest, even though there is no egg to roll (figure 18.3). This is typical of the inflexible nature of instinctive behaviors. It has also been found that many other, somewhat egg-shaped structures can stimulate egg-rolling behavior. For example, beer cans and baseballs trigger egg-rolling behavior. Thus, not only are the birds unable to stop the egg rolling in midstride but also nonegg objects can generate inappropriate behavior because they have approximately an egg shape.

Instinctive Behavior Can Be Complex

Some activities are so complex that it seems impossible for an organism to be born with such abilities. For example, a spider web is not just a careless jumble of silk threads; it is so precisely made that scientists can recognize what species of spider made it by the pattern of threads in the web. However, web spinning is not a learned ability. A spider has no opportunity to learn how to spin a web, because it never observes others doing it. Furthermore, spiders do not practice several times before they get a proper, workable web. It is as if a "program" for making a particular web were in the spider's "computer" (figure 18.4).

Might these behavior patterns be the result of natural selection? Many kinds of behaviors are controlled by genes. The "computer" in our example is really the spider's DNA, and the "program" is a specific package of genes. Through the millions of years that spiders have been in existence, natural selection has modified the web-making program to refine the process. Certain genes of the program have undergone mutation, resulting in changes in behavior. Imagine various ancestral spiders, each with a slightly different program. The inherited program that gives individuals the best chance of living long enough to produce a new generation is the program selected for, and the most likely to be passed on to the next generation.

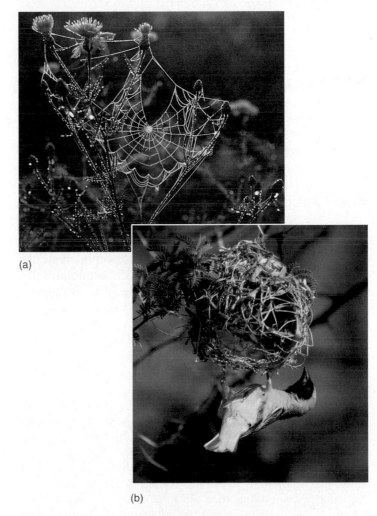

(a)

(b)

FIGURE 18.4 Complex Instinctive Behavior
The kind of web constructed by a spider is determined by instinctive behavior patterns. (a) The construction of a web is a complex behavior. (b) Instinctive behavior by this masked weaver is responsible for its construction of this complex nest.

The Nature of Learned Behavior

The alternative to preprogrammed, instinctive behavior is learned behavior. **Learning** is a change in behavior as a result of experience.

Learning is very significant in long-lived animals that care for their young. Animals that live many years are highly likely to benefit from an ability to recognize previously encountered situations and to modify their behavior accordingly. Furthermore, because the young spend time with their parents, they can imitate their parents and develop behaviors that are appropriate to local conditions. These behaviors take time to develop but have the advantage of adaptability. For learning to become a dominant feature of an animal's life, the animal must also have a memory, which requires a relatively large brain in which to store new information. This is probably why learning is a major part of the lives of only a few kinds of animals, such as the vertebrates. Nearly all human behavior is learned. Even such important behaviors as walking, communicating, feeding oneself and sexual intercourse must be learned.

FIGURE 18.5 Habituation
This moose in Anchorage, Alaska, is habituated to people.

18.4   Kinds of Learning

Scientists who study learning recognize that there are various kinds of learning: *habituation, association, exploratory learning, imprinting,* and *insight.*

Habituation

Habituation is a change in behavior in which an animal ignores an insignificant stimulus after repeated exposure to it. There are many examples of this kind of learning. Typically, wild animals flee from humans. Under many conditions, this is a valuable behavior. Wild animals that frequently encounter humans and never experience negative outcomes may learn to ignore humans. Many wild animals, such as the deer, elk, and bears in parks, have been habituated to the presence of humans and behave in a way that would be totally inappropriate in areas near the park where hunting is allowed (figure 18.5).

Similarly, loud noises usually startle humans and other animals. However, constant exposure to such sounds cause the individuals to ignoring them. As a matter of fact, the sound may become so much a part of the environment that the stopping of the sound causes a response. This kind of learning is valuable because the animal does not waste time and energy responding to a stimulus that does not have a beneficial or negative impact on the animal. Animals that respond continually to inconsequential stimuli have less time to feed and may miss other, more important stimuli.

Association

Association occurs when an animal makes a connection between a stimulus and an outcome. Associating a particular outcome with a particular stimulus is important to survival, because it allows an animal to avoid danger or take advantage of a beneficial event. If association allows the animal to get more food, avoid predators, or protect its young more effectively, this kind of learning is advantageous to the species. The association of certain shapes, colors, odors, or sounds with danger is especially valuable. There are three common kinds of association: *classical conditioning, operant (instrumental) conditioning,* and *observational learning (imitation).*

Classical Conditioning

Classical conditioning occurs when an involuntary, natural, reflexive response to a natural stimulus is transferred from the natural stimulus to a new stimulus. The response provoked by the new stimulus is called a **conditioned response.** During the period when learning is taking place, the new stimulus is given *before* or *at the same time as* the normal stimulus.

A Russian physiologist, Ivan Pavlov (1849–1936), was investigating the physiology of digestion when he discovered that dogs can transfer a natural response to a new stimulus. He was studying the production of saliva by dogs and knew that a natural stimulus, such as the presence or smell of food, would cause the dogs to start salivating. Then, he rang a bell just before presenting them with food. After a training period, the dogs began to salivate when the bell was rung even though no food was presented. The natural response (salivating) was transferred from the natural stimulus (the smell or taste of food) to a new stimulus (the sound of a bell).

Animals can also be conditioned unintentionally. Many pets anticipate their mealtimes, because their owners go through a certain set of behaviors, such as going to a cupboard or opening a can of pet food prior to putting food in a dish. It is doubtful that this kind of learning is common in wild animals, because it is hard to imagine such tightly controlled sets of stimuli in nature.

Operant (Instrumental) Conditioning

Operant (instrumental) conditioning also involves the association of a particular outcome with a specific stimulus, but it differs from classical conditioning in several ways. First, during operant conditioning, the animal learns to repeat acts that bring good results and avoid those that bring bad results. Second, the animal receives a reward or punishment *after* engaging in a particular behavior. Third, the animal's response is typically a more complicated behavior than a simple reflex. A reward that encourages a behavior is known as positive reinforcement, and a punishment that discourages a behavior is known as negative reinforcement.

The training of many kinds of animals involves this kind of conditioning. If a dog being led on a leash is given the command *heel* and is then vigorously jerked into the correct position, it eventually associates the word *heel* with assuming the correct position at the knee. This is negative reinforcement, because the animal avoids the unpleasantness of being jerked about if it assumes the correct position. Similarly, petting or giving food to a dog when it has done something correctly will positively reinforce the desired behavior. For example, placing the dog into the sitting position on the command *sit* and rewarding the dog when it performs the behavior on command is positive reinforcement.

Wild animals have many opportunities to learn through positive or negative reinforcement. As animals encounter the same stimulus repeatedly, there is an opportunity to associate the stimulus with a particular outcome. For example, many kinds of birds eat berries and other small fruits. If a distinctly colored berry has a good flavor, birds will return repeatedly to eat them. Pigeons in cities have learned to associate food with people in parks. They can even identify specific individuals who feed them regularly. Their behavior is reinforced by being fed. Many birds in urban areas have associated automobiles with food and are seen picking smashed insects from the grills and bumpers of cars. When a car drives into the area, birds immediately examine it for food.

In some of national parks, bears have associated backpacks with food. In some cases, attempts have been made to use negative reinforcement to condition the bears to avoid humans. Usually, the bears are exposed to loud or painful stimuli if they approach backpacks. Some bears that cannot be trained to avoid humans and their "equipment" are killed. Conversely, in areas where bears are hunted, they have generally been conditioned to avoid contact with humans. Similarly, if certain kinds of fruits or insects have unpleasant

FIGURE 18.6 Association Learning
Many animals learn to associate unpleasant experiences with the color or shape of offensive objects and, thus, avoid them in the future. The blue jay is eating a monarch butterfly. These butterflies contain a chemical that makes the blue jay sick. After one or two such experiences, blue jays learn not to eat monarchs.

tastes, animals learn to associate the bad tastes with the colors and shapes of the offending objects and avoid them in the future (figure 18.6). Each species of animal has a distinctive smell. If a deer or rabbit has several bad experiences with a predator that has a particular smell, it can avoid places where the smell of the predator is present.

Animals also engage in trial-and-error learning, which includes some elements of conditioning. When confronted with a problem, they try one option after another until they achieve a positive result. Once they have solved the problem, they can use the same solution repeatedly. For example, if a squirrel has a den in a hollow tree on one side of a stream and is attracted to a source of food on the other side, it may explore several routes to get across the stream. It may jump from a tree on one side of the stream to another on the opposite side. It may run across a log that spans the stream. It may wade a shallow portion of the stream. Once it has found a good pathway, it is likely to use the same pathway repeatedly. Similarly, many hummingbirds visit many different flowers during the course of a day. When they have found a series of nectar-rich flowers, they follow a particular route and visit the same flowers several times a day.

FIGURE 18.7 Observational Learning
By following and observing their mother, these cubs will learn how to fish for salmon.

Observational Learning (Imitation)

Observational learning (imitation) is a form of association; it consists of a complex set of associations formed while watching another animal being rewarded or punished after performing a particular behavior. The animal does not receive the reward or punishment itself but, rather, observes the fruits of the other's behavior. Subsequently, the observing animal may show behaviors that receive rewards and avoid behaviors that lead to punishment. It is likely that conditioning is involved in imitation, because an animal that imitates a beneficial behavior is rewarded. Observing a negative outcome to another animal is also beneficial, because it allows the observer to avoid negative consequences. Many kinds of young birds and mammals follow their parents and sample the same kinds of foods their parents eat (figure 18.7). If the foods taste good, they are positively reinforced. They may also observe warning and avoidance behaviors associated with particular predators and mimic these behaviors when the predators are present. For example, crows will mob predators, such as hawks and owls. As young crows observe older ones cawing loudly and chasing an owl, they learn the same behavior. They associate a certain kind of behavior (mobbing) with a certain kind of stimulus (owl or hawk).

Exploratory Learning

Animals are constantly moving about and sampling their environment. For animals that rely primarily on instinctive behaviors, this movement increases the likelihood that they will find food or other valuable resources but does not result in learning because they do not increase their efficiency at finding resources. Animals that have a significant amount of memory can store information about their surroundings as they wander about. In some cases, the new information may have immediate value—for example, many kinds of birds and mammals remember where certain physical features are located in their environment. They may find a food source during their wanderings and can return to it repeatedly in the future.

When you put up a bird feeder, it does not take very long before many birds are visiting the feeder on a regular basis. Birds certainly find suitable nesting sites by exploring; then, they store the location in memory. Even invertebrate animals with limited ability to store information engage in exploratory learning. For example, in spring, a queen bumblebee will fly about, examining holes in the ground. Eventually, she will find a hole in which she will lay eggs and begin to raise her first brood of young. Once she has selected a site, she must learn to recognize that spot, so that she can return to it each time she leaves to find food, or her young will die.

In other cases, the information learned is not used immediately but might be of use in the future. If an animal has an inventory of its environment, it can call on the inventory to solve problems later in life. Many kinds of animals hide food items when food is plentiful and are able to find them later when food is scarce. Even if they don't remember exactly where the food is hidden, if they always hide food in a particular kind of place, they are likely to find it later. (For example, if you needed to drive a car that you have never seen before, you would know that you need to use a key, and you would search in a particular place in the car to insert the key.) Having a general knowledge of its environment is very useful to an animal.

Many kinds of small mammals, such as mice and ground squirrels, avoid predators by scurrying under logs or other objects or into holes in the ground. Experiments with mice and owl predators show that mice that are familiar with their surroundings are more likely to escape predators than are those that are not.

Imprinting

Imprinting is a special kind of irreversible learning in which a very young animal is genetically primed to learn a specific behavior in a very short period during a specific time in its life. The time during which imprinting is possible is known as the **critical period**. This type of learning was originally recognized by Konrad Lorenz (1903–1989) in his experiments with geese and ducks. He determined that, shortly after hatching, goslings and ducklings would follow an object if the object were fairly large, moved, and made noise. Figure 18.8a shows goslings demonstrating following behavior after being imprinted on Lorenz. The critical period was short—minutes to hours immediately following hatching.

Imprinting is irreversible. Ducklings will follow only the object on which they are originally imprinted. Under normal conditions, the first large, noisy, moving object newly hatched ducklings see is their mother. Imprinting ensures that the immature birds will follow her and may learn appropriate

(a)

(b)

FIGURE 18.8 Imprinting

Imprinting is irreversible learning that occurs during a very specific part of the life of an animal. Photo *a* shows geese that were imprinted on Konrad Lorenz and exhibit the "following response" typical of this type of behavior. (*b*) Ducks exhibit the same behavior in a natural setting.

feeding, defensive tactics, and other behaviors by example. Because they are always near their mother, she can also protect them from enemies or bad weather. If animals imprint on the wrong objects, they are not likely to survive.

Since these experiments by Lorenz in the early 1930s, scientists have discovered that many young animals can be imprinted on several types of stimuli and that there are responses other than following. The way song sparrows learn their song appears to be a kind of imprinting. The young birds must hear the correct song during a specific part of their youth, or they will never be able to perform the song correctly as adults. This is true even if later in life they are surrounded by other adult song sparrows that are singing the correct song. Furthermore, the period of time when they learn the song is before they begin singing. Recognizing and performing the correct song is important, because it has a particular meaning to other song sparrows. For males, it conveys that a male song sparrow has reserved a space for himself. For females, the male's song is an announcement of the location of a male of the correct species that is a possible mate.

Mother sheep and many other kinds of mammals imprint on the odor of their offspring. They are able to identify their offspring among a group of lambs and allow only their own lambs to suck milk. Shepherds have known for centuries that they can sometimes get a mother that has lost her lambs to accept an orphan lamb if they place the skin of the mother's dead lamb over the orphan.

Many fish appear to imprint on odors in the water. Salmon are famous for their ability to return to the fresh water streams where they were hatched. Fish that are raised in artificial hatcheries can be imprinted on minute amounts of specific chemicals and be induced to return to any stream that contains the chemical.

Insight

Insight is learning in which past experiences are reorganized to solve new problems (table 18.1). When faced with a new problem, whether it is a crossword puzzle, a math problem, or any other, everyday problem, we sort through our past experiences to find those that apply. We may not even realize we are doing it, but we put these past experiences together in a new way that may solve the current problem. Because this process is internal and can be demonstrated only through a response, it is very difficult to understand exactly what goes on when insight occurs. Behavioral scientists have explored this area for many years, but the study of insight is still in its infancy.

Insight is particularly difficult to study, because it is impossible to know for sure whether a novel solution to a problem is the result of thinking it through or simply an accident. For example, a small group of Japanese macaques (monkeys) was studied on an island. They were fed by simply dumping food, such as sweet potatoes or wheat, onto the beach. Eventually, one of the macaques discovered that she could get the sand off the sweet potato by washing it in a nearby stream. She also discovered that she could sort the wheat from the sand by putting the mixture into water because the wheat would float. Are these examples of insight? We will probably never know, but it is tempting to think so. In addition, in the colony of macaques, the others soon began to display the same behavior, probably imitating the female that had first made the discovery.

TABLE 18.1

Kinds of Learning

Kind of Learning	Defining Characteristic	Ecological Significance	Example
Habituation	An animal ignores a stimulus to which it is subjected continually.	An animal does not waste time or energy by responding to unimportant stimuli.	Wild animals raised in the presence of humans lose their fear of humans.
Association	An animal learns that a particular outcome is connected with a particular stimulus.	An animal can avoid danger or anticipate beneficial events by connecting a particular outcome with a specific stimulus when that stimulus is frequently tied to a particular outcome.	Many mammalian predators associate prey with high-pitched, squealing sounds.
Classical conditioning	A new stimulus is presented in combination with a natural stimulus. The animal transfers its response from the natural stimulus to the new stimulus.	Classical conditioning probably does not happen in natural settings.	Pets can anticipate when they will be fed, because their owners' food-preparation behavior occurs shortly before food is presented.
Operant (instrumental) conditioning	An animal responds to a new stimulus. It is rewarded or punished following its response. Eventually, the animal associates the reward or punishment with the stimulus and responds appropriately to the stimulus.	Animals can avoid harmful situations that are associated with particular stimuli. They can learn to repeat behaviors that bring benefits.	Animals that associate a shape, color, smell, or sound with a predator can take refuge. Animals that associate a shape, color, smell, or sound with the presence of food will find food more readily.
Observational learning (imitation)	An animal imitates the behaviors of others.	The animal can acquire the knowledge of others by watching. This can be more rapid than learning by trial and error.	Young animals run when their mothers do and feed on the food their parents do.
Exploratory learning	An animal moves through and observes the elements of its environment.	Information stored in memory may be valuable later.	An awareness of hiding places can allow an animal to escape a predator.
Imprinting	An animal learns specific, predetermined activity at a specific time in life.	The animal quickly gains a completely developed behavior that has immediate value to its survival.	Many kinds of newborn animals follow their mothers.
Insight	An animal understands the connection between things it had no way of experiencing previously.	Information stored in an animal's memory can be used to solve new problems.	The manufacture and use of tools by humans and some other primates suggests insight is involved.

18.5 Instinct and Learning in the Same Animal

It is important to recognize that all animals have both learned and instinctive behaviors, and that a single behavior pattern can include both instinctive and learned elements. For example, biologists have raised young song sparrows in the absence of any adult birds, so that there was no song for the young birds to imitate. These isolated birds sang a series of notes similar to the normal song of the species, but not exactly correct. Birds from the same nest that were raised with their parents developed a song nearly identical to that of their parents. If bird songs were totally instinctive, there would be no difference between these two groups. It appears that all these birds inherited the ability to produce appropriate notes but the refinements of the song's melody came from experience. Therefore, the characteristic song of that species is

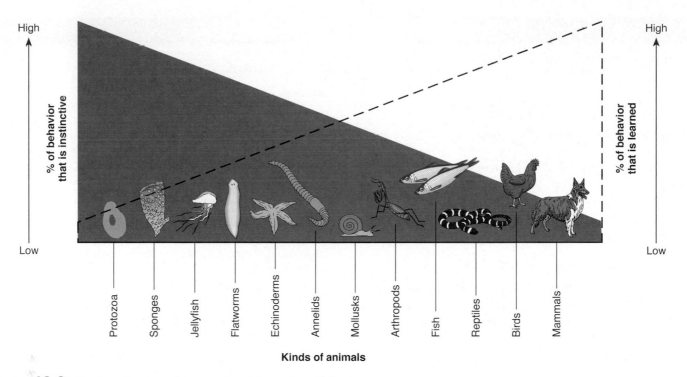

High

% of behavior that is instinctive

Low

High

% of behavior that is learned

Low

Protozoa

Sponges

Jellyfish

Flatworms

Echinoderms

Annelids

Mollusks

Arthropods

Fish

Reptiles

Birds

Mammals

Kinds of animals

FIGURE 18.9 **The Distribution of Learned and Instinctive Behaviors**
Different groups of animals show different proportions of instinctive and learned behaviors in their behavior patterns.

partly learned behavior (a change in behavior as a result of experience) and partly unlearned (instinctive). This is probably true of the behavior of many organisms; they show complex behaviors that are a combination of instinct and learning. It is important to note that many kinds of birds learn most of their songs with very few innate components. Mockingbirds are very good at imitating the songs of a wide variety of bird species in their region. They even imitate the sounds made by inanimate objects, such as the screeching of a rusty hinge.

This mixture of learned and instinctive behavior is not the same for all species. Many invertebrate animals rely on instinct for most of their behavior patterns, whereas many of the vertebrates (particularly birds and mammals) make use of a great deal of learning (figure 18.9).

When a behavior pattern has both instinctive and learned components, typically the learned components have particular value for the animal's survival. For example, most of a honeybee's behavior is instinctive; however, when it finds a new food source, it can remember the route between the hive and the food source. A bird's style of nest is instinctive, but the skill with which it builds may improve with experience. Birds' food-searching behavior is probably instinctive, but the ability to modify the behavior to exploit unusual food sources, such as bird feeders, is learned. On the other hand, honeybees cannot be taught to make products other than honey and beeswax, a robin will not build a nest in a birdhouse, and most insect-eating birds will not learn to visit bird feeders. Table 18.2 compares instinctive behaviors and learned behaviors.

18.6 Human Behavior

We tend to think of human behavior as different from that of other animals, and it is. However, we are different only in the degree to which we demonstrate different kinds of behavior.

Instinctive behavior is rare in humans. We certainly have reflexes that cause us to respond appropriately without thinking. Touching a hot object and rapidly pulling your hand away is a good example. Newborns grasp objects and hang on tightly with both their hands and feet. This kind of grasping behavior in our primitive ancestors would have allowed the child to hang onto its mother's hair as the mother and child traveled from place to place. But do we have more complicated instinctive behaviors? Newborns display several behaviors that can be considered instinctive. If you stroke the side of an infant's face, the child will turn its head toward the side touched and begin sucking movements. This is not a simple reflex behavior but, rather, requires the coordination of several sets of muscles and involves the brain. It is hard to see how this is a learned behavior, because the child does the behavior without prior experience. Therefore, it is probably instinctive. This behavior may be associated with nursing, because carrying the baby on its back would place the child's cheek against the mother's breast. Other mammals, even those whose eyes do not open for several days following birth, are able to find nipples and begin nursing shortly after birth.

Habituation is a common human experience. We readily ignore sounds that are continuous, such as the sound of air

TABLE 18.2	
Comparison of Instinct and Learning	
Instinct	**Learning**
The animal is born with the behavior.	The animal is not born with the behavior.
Instinctive behavior is genetically determined.	Learned behavior is not genetically determined, but the way in which learning occurs is at least partly hereditary.
No experience is required; the behavior is done correctly the first time.	Performance improves with experience; the behavior requires practice.
The behavior cannot be changed.	The behavior can be changed.
Memory is not important.	Memory is important.
Instinctive behavior can evolve as gene frequencies change.	Changes in learned behaviors are not the result of genetic changes. However, the ability to learn clearly has genetic components.
Instinct is typical of simple animals with short lives and little contact with their parents.	Learning is typical of more complex animals with long lives and extensive contact with parents.
Instinctive behaviors can be passed from parents to offspring only by genetic means.	Learning allows acquired behaviors to be passed from parents to offspring by cultural means.

conditioning equipment or the background music in shopping malls. Teachers recognize that it is important to change activities regularly to keep their students' attention.

Association is extremely common in humans. We associate smells with certain kinds of food, sirens with emergency vehicles, and words with their meanings. Much of the learning we do is by association. We also use positive and negative reinforcement to change behavior. We seek to reward appropriate behavior and punish inappropriate behavior (figure 18.10). We can even experience positive and negative reinforcement without actually engaging in a behavior, because we can visualize its possible consequences. Adults routinely describe consequences for children, so that they will not experience particularly harmful effects: "If you don't study for your biology exam, you'll probably fail it."

Exploratory learning is extremely common in humans. Children wander about and develop a mental picture of where things are in their environment. Exploration also involves behaviors such as picking things up, tasting things, and making sounds. Even adults explore new ideas and activities.

Imprinting in humans is more difficult to demonstrate, but there are instances in which imprinting may be taking place. Bonding between mothers and infants is thought to be an important step in the development of the mother-child relationship. Most mothers form very strong emotional attachments to their children; likewise, children are attached to their mothers, sometimes literally, as they seek to maintain physical contact with them. However, it is very difficult to show what is actually happening at this early time in the life of a child.

Language development in children may also be an example of imprinting. All children learn whatever languages are

FIGURE 18.10 Negative Reinforcement
The reprimand this recruit is receiving is an example of negative reinforcement.

spoken where they grow up. If multiple languages are spoken, they learn them all and they learn them easily. However, adults have more difficulty learning new languages, and many find it impossible to "unlearn" languages they spoke previously, so they speak new languages with an accent. This appears to meet the definition of imprinting. Learning takes place at a specific time in life (critical period), the kind of learning is preprogrammed, and what is learned cannot be unlearned. Recent research using brain-imaging technology shows that those who learn a second language as adults use two different parts of the brain for language—one part for the native language or languages they learned as children and a different part for their second language.

Insight is what our species prides itself on. We are *thinking* animals. **Thinking** is a mental process that involves memory and an ability to reorganize information. A related aspect of our thinking nature is our concept of self. We can project ourselves into theoretical situations and measure our success by thinking without needing to experience a situation. For example, we can look at the width of a small stream and estimate our ability to jump across it without needing to do the task. We are mentally able to measure our personal abilities against potential tasks and decide if we should attempt them. In the process of thinking, we come up with new solutions to problems. We invent new objects, new languages, new culture, and new challenges to solve. However, how much of what we think is really completely new, and how much is imitation? As mentioned earlier, association is a major core of our behavior, but we also are able to use past experiences, stored in our large brains, to provide clues to solving new problems.

18.7 Selected Topics in Behavioral Ecology

Of the examples so far in this chapter, some involved laboratory studies, some were field studies, and some included aspects of both. Often, these studies overlap with the field of psychology. This is particularly true for many of the laboratory studies. The science of animal behavior is a broad one, drawing on information from several fields of study, and it can be used to explore many kinds of questions. The topics that follow concentrate on the significance of behavior from ecological and evolutionary points of view. Now that we have some understanding of how organisms generate behavior, we can look at a variety of behaviors in several kinds of animals and see how they are significant for their ecological niches.

Communication

Communication is the use of signals to convey information from one animal to another. Animals communicate in many ways and for a variety of purposes. The senses of hearing, sight, taste, smell, and touch are all involved in communication. Regardless of the sensory system used, communication is the production of signals that can be received and interpreted by others of the same species. For most animals, the communication system is instinctive. However, in humans language is a learned behavior.

Sound is used by many kinds of animals for communication. Birds, frogs, alligators, crickets, wolves, and many other animals use sound to communicate information.

Sight is used to communicate in many ways. Colors, lights, body postures, and movements are all involved in communication.

The chemical senses of taste and smell are important in the communication system of many kinds of animals. **Pheromones** are chemicals produced by animals and released into the environment that trigger behavioral or developmental changes in other animals of the same species. For example, honeybees produce a pheromone that smells like bananas to humans. This pheromone triggers aggressive behavior in the bees of the hive. Many kinds of animals produce odors that communicate information about their reproductive status. Licking or touching with antennae is a way to obtain chemical information about other animals.

Touch is probably involved in the communication systems of nearly all organisms. If nothing else, the touch of one animal by another announces its presence.

Often, several of the senses are used at the same time in the process of communication. For example, when two dogs meet for the first time, they are likely to communicate by sound, sight, smell, touch, and taste. They bark or growl. They assume special body postures or bare their teeth, which are visual signals. They sniff one another. They may push one another, or one may put a leg over the other's body. Finally, if they are friendly, they are likely to lick one another.

Animals communicate for several purposes. *Advertising location* is an important function of communication. Information about location can be used to gather individuals together. The calls of frogs in the spring bring animals together at a breeding site (figure 18.11). The calls of crows

FIGURE 18.11 Communication by Sound
This tree frog is calling. The inflated vocal sac amplifies the sound he produces.

can call attention to a predator or food source. In contrast, many birds use sounds to tell others of the same species that a particular piece of the habitat is occupied and to stay away, as well as to attract potential mates.

Social structure among animals that live in groups is maintained through communication. The elaborate greeting activities of wolves are an example. There is much touching, smelling, licking, and vocalization, all of which help reinforce the status of each individual in the group.

Alarm signals alert animals to potential danger. The alarm pheromone of honeybees mentioned earlier is an example. The raised tail of a whitetailed deer is a similar signal, but deer also snort and stamp their forefeet when they are disturbed by something in their surroundings. And many birds and mammals use alarm calls to alert others to the presence of danger.

Reproduction requires communication. Information about the location and reproductive readiness of individuals is important for reproduction to be successful.

Reproductive Behavior

The reproductive behavior of many kinds of animals has been studied a great deal. Although each species has its own behaviors, some components of reproductive behavior are common to nearly all animals. In order for an animal to produce offspring that reach adulthood, several events must occur. A suitable mate must be located, mating and fertilization must take place, and the young must be provided for.

Finding Each Other

To reproduce, an animal must find individuals of the same species that are of the opposite sex. They use several techniques for this purpose. Different species of animals use different methods, but most send signals that can be interpreted by others of the same species. Depending on the species, any of the senses (sound, sight, touch, smell, or taste) may be used to identify the species, sex, and sexual receptiveness of another animal. For instance, different species of frogs produce distinct calls. Each call is a code system, which delivers a very private message meant for only one species. It is, however, meant for any member of that species near enough to hear. The call produced by male frogs, which both male and female frogs can hear, results in frogs of both sexes congregating in a limited area. Once they gather in a small pond, it is much easier to have the further communication necessary for mating to take place. Many other animals, including most birds, insects (e.g., crickets), reptiles (e.g., alligators), and some mammals, produce sounds that are important for bringing individuals together for mating.

Chemicals can also attract animals. They have the same effect as the sounds made by frogs or birds; they are just a different code system. For example, many kinds of female moths release a chemical into the air. The large, fuzzy antennae of the male moths can receive the chemical in unbelievably tiny amounts. The males fly upwind to the source of the pheromone, which is a female (figure 18.12).

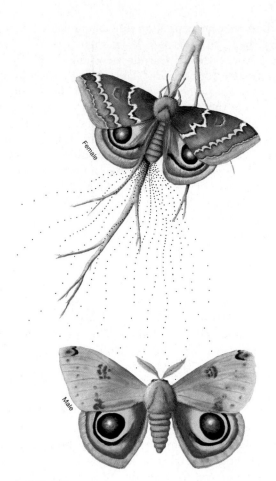

FIGURE 18.12 Chemical Communication
The female moth signals her readiness to mate and attracts males by releasing a pheromone, which attracts males from long distances downwind.

Most mammals rely on odors. Females typically produce distinct odors when they are in breeding condition. When a male happens on such an odor trail, he follows it to the female. Many reptiles also produce distinctive odors.

Visual cues are also important for many species. Brightly colored birds, insects, fish, and many other animals use specific patches of color for species identification. Conspicuous movements can also be used to attract the attention of a member of the opposite sex.

The firefly is probably the most familiar organism that uses light signals to bring males and females together. Several species may live in the same area, but each species flashes its own code. The code is based on the length of the flashes, their frequency, and their overall pattern (figure 18.13). There is also a difference between the signals given by males and those given by females. For the most part, males are attracted to and mate with females of their own species. Once male and female animals have attracted one another's attention, the second stage in successful reproduction takes place. However, in one species of firefly, the female has the remarkable ability to signal the correct answering code to species other than her

FIGURE 18.13 Firefly Communication
The pattern of light flashes, their location, and their duration all help fireflies identify members of the opposite sex who are of the appropriate species.
Courtesy of James E. Lloyd.

FIGURE 18.14 No Parental Care
The mother of these caterpillars does not provide any care for her offspring. The only provision she has made for her offspring is to lay her eggs on a suitable food plant. Most of these larvae will die.

own. After she has mated with a male of her own species, she continues to signal to passing males of other species. She is not hungry for sex, just hungry. The luckless male who responds to her "come-on" will be her dinner.

Assuring Fertilization

The second important activity in reproduction is fertilizing eggs. Many marine organisms simply release their gametes into the sea simultaneously and allow fertilization and further development to take place without any input from the parents. Sponges, jellyfishes, and many other marine animals fit into this category. Other aquatic animals congregate, so that the chances of fertilization are enhanced by the male and female being near one another as the gametes are shed. This is typical of many fish and some amphibians, such as frogs. Internal fertilization, in which the sperm are introduced into the female's reproductive tract, occurs in most terrestrial animals. Some spiders and other terrestrial animals produce packages of sperm, which the female picks up with her reproductive structures. Many of these mating behaviors require elaborate, species-specific communication prior to the mating act.

Raising Young

The third element in successful reproduction is providing the young with the resources they need to live to adulthood. Many invertebrate animals spend little energy on the care of the young, leaving them to develop on their own. Usually, the young become free-living larvae, which eat and grow rapidly. Many insects lay their eggs on the particular species of plant that the larva will feed on as it develops. Parasitic species seek out the required host in which to lay their eggs. The eggs of others may be placed in spots that provide safety until the young hatch from the egg (figure 18.14). Turtles, many fish, and some insects fit into this category. In most of these cases, however, the female lays large numbers of eggs, and most of the young die before reaching adulthood. This is an enormously expensive process: The female invests considerable energy in producing eggs but has a low success rate. Nevertheless, most species use this strategy, and it has proven to be very successful.

An alternative strategy to this "wasteful" loss of potential young is to produce fewer young but invest large amounts of energy in their care. This is not necessarily a more efficient system, because it still requires large investments by the parents in the production and care of the young. This strategy is typical of birds and mammals. Parents build nests, share in the feeding and protection of the young, and often assist the young in learning appropriate behaviors. Caring for the young requires many complex behavior patterns (figure 18.15). Most of the animals that feed and raise young are able to distinguish their own young from those of other nearby families and may

FIGURE 18.15 Parental Care

This doe provides food, in the form of milk, for her fawn. She also provides protection. Thus, a large percentage of fawns survive, compared with species in which the parents provide no care.

FIGURE 18.16 Territorial Behavior

This clown fish lives with a sea anemone. It will defend its small territory against other clown fish.

even kill the young of another family unit. Elaborate greeting ceremonies are usually performed when animals return to the nest or the den. Perhaps this has something to do with being able to identify individual young. Often, this behavior is shared among adults as well. This is true for many colonial nesting birds, such as gulls and penguins, and for many carnivorous mammals, such as wolves, dogs, and hyenas.

Care of the young also occurs in several species of fish and reptiles, such as alligators. Certain kinds of insects, such as bees, ants, and termites, have elaborate social organizations in which one or a few females (queens) produce large numbers of young, which are cared for by the queen's sterile offspring. Most of the offspring are sterile workers, but a few become new queens.

Territorial Behavior

A **territory** is the space used for food, mating, or other purposes, that an animal defends against others of the same species. **Territorial behaviors** are activities performed to secure and defend a space. These behaviors are widespread in the animal kingdom and can be seen in such diverse groups as insects, spiders, fish, reptiles, birds, and mammals (figure 18.16). On the other hand, many kinds of animals do not establish fixed territories. For example, most insects, worms, and many other invertebrates do not form territories. In addition, many kinds of migratory animals, such as tuna and wildebeest, do not establish territories or do so only during the breeding season. competition, p. 326

Territorial behavior reduces conflict

When territories are first being established, there is often much conflict among individuals. However, this eventually gives way to their use of signals that define the territory and communicate to others that the territory is occupied. For example, the male redwing blackbird has red shoulder patches, but the female does not. The male will perch on a high spot, flash his red shoulder patches, and sing to other males that venture into his territory. Most other males get the message and leave his territory; those that do not leave are attacked. He will also attack a dead and stuffed male redwing blackbird in his territory, or even a small piece of red cloth. Clearly, the spot of red is the characteristic that stimulates the male to defend his territory. Once the initial period of conflict is over, the birds tend to respect one another's boundaries. All that is required is to frequently announce that the territory is still occupied. They do so by singing from a conspicuous position in the territory. After territorial boundaries have been established, little time is required to prevent other males from venturing close. Thus, the animal may spend a great deal of time and energy securing the territory initially, but doesn't need to expend much to maintain it.

Territorial behavior involves special signals

For redwing blackbirds, singing in a conspicuous location and flashing red shoulder patches is a signal warning other male blackbirds to stay away. The use of signals replaces the need for fighting. Much of the singing behavior of other species of birds is also territorial. Many carnivorous mammals, such as foxes, weasels, cougars, coyotes, and wolves, use urine or other scents to mark the boundaries of their territories. Territorial fish use color patterns and threat postures to defend their territories. Crickets use sound and threat postures. Male bullfrogs engage in shoving matches to displace males that invade their small territories along the shoreline.

Territorial behavior allocates resources

A territory has great importance, because it reserves exclusive rights to the use of a certain space for an individual or small group of individuals, such as a pair of robins or a wolf pack. These resources may include sources of food or water, nest or den sites, or access to potential mates. Thus, establishing and maintaining a territory is a way to allocate resources among the members of a species. This is also likely to have an effect on population size, because animals that are not able to establish a territory or are forced to accept poor-quality territories will be less likely to survive and reproduce.

Territory size

Different species of animals maintain territories for different purposes. Some species maintain small territories of a few square meters, whereas others maintain territories of several square kilometers. For example, many seabirds build nests in colonies. Each pair of birds maintains an extremely small nesting territory of about 1 square meter within the colony. Each territory is just beyond the reach of the bills of the neighbors (figure 18.17). When one bird invades the nesting territory of another, the resident bird uses threat postures to drive the intruder away. If the intruder does not take the hint, it is attacked.

FIGURE 18.17 Territory Size
Colonial nesting seabirds typically have very small nest territories. Each territory is just out of pecking range of the neighbors.

Many other birds maintain territories for both nesting and gathering food resources. These territories range from a few hundred square meters, such as for robins, to several square kilometers, such as for owls. Many fish also maintain small territories.

Many large carnivores have territories that cover several square kilometers. In such cases, the signals they use must be perceived over great distances or last for a long time. The howling of wolves and the roaring of lions are signals that can be heard over long distances (figure 18.18). The use of

FIGURE 18.18 Advertising One's Territory
The howling of this wolf has several functions. One is to advertise its presence in its territory.

scent posts—places where animals urinate, defecate, or deposit other scents, transmits long-lasting signals. The "owner" of the territory does not need to be present all the time to defend its territory, but it must visit these sites regularly to renew the signal.

Dominance Hierarchy

In social animals, a kind of organization known as a *dominance hierarchy* is often observed. A **dominance hierarchy** is a relatively stable, mutually understood order of priority within the group. One individual in the group dominates all the others. A second-ranking individual dominates all but the highest-ranking individual, and so forth; the lowest-ranking individual must give way to all others within the group. Generally, the initial establishment of a dominance hierarchy involves fighting. The best fighters take the top places. However, once the hierarchy is established, it results in a more stable social unit with little conflict, because the positions of the individuals within the hierarchy are reinforced by various kinds of threats without the need for fighting. This kind of behavior is seen in barnyard chickens and is known as a *pecking order*. Figure 18.19 shows a dominance hierarchy; the lead animal has the highest ranking and the last animal has the lowest ranking.

A dominance hierarchy gives certain individuals preferential treatment when resources are scarce. The dominant individual has first choice of food, mates, shelter, water, and other resources because of its position. Animals low in the hierarchy may be malnourished or fail to mate in times of scarcity. In many social animals, such as wolves, usually only the dominant male and female reproduce. If the dominant male and female have achieved their status because they have superior characteristics, their favorable genes will probably

FIGURE 18.19 A Dominance Hierarchy
Many animals maintain order within their groups by establishing a dominance hierarchy. For example, in a group of cows or sheep walking in single file, the dominant animal is probably at the head of the line and the lowest-ranking individual is at the end.

be passed on to the next generation. Poorly adapted animals with low rank may never reproduce. Such a hierarchy frequently results in low-ranking individuals emigrating from the area. Such migrating individuals are often subject to heavy predation.

Behavioral Adaptations to Seasonal Changes

Most animals live in environments that change from time to time. There may be differences in temperature, rainfall, or availability of food. In some areas, the dry part of the year is the most stressful. In temperate areas, winter reduces many sources of food and forces organisms to adjust. Animals have several behavioral means for coping with these changes.

Metabolic changes allow many animals to avoid seasonal changes or times of environmental stress. Where drought occurs, many animals become inactive until water becomes available. Frogs, toads, and many insects remain inactive (*estivate*) underground during long periods of drought and emerge to mate when it rains. *Hibernation* in mammals is a response to cold, seasonal temperatures in which the body temperature drops and all the body processes slow down. The slowing of body processes allows an animal to survive on the food it has stored within its body as fat prior to the onset of severe environmental conditions and low food availability. Hibernation is typical of bats, marmots, and some squirrels. Animals such as insects, reptiles, and amphibians are not able to regulate their temperatures the way birds and mammals

do. Because their body temperatures drop when the environmental temperature drops, their activities are slowed during the winter and they require less food.

Migration allows many animals simply to move to areas that are less stressful. Migratory birds fly hundreds or thousands of kilometers. Many birds that nest in the north avoid winter by migrating to more southerly regions (figure 18.20). Many migratory mammals move from drier areas where food has been depleted to other moister areas where food is more available. In many cases, the migrations are triggered by instinctive responses to environmental clues, and the same migration routes are used generation after generation.

FIGURE 18.20 Seasonal Migration
Snow geese and many other birds migrate from their northern breeding grounds to milder climates during the winter. In this way, they are able to use the north for breeding and avoid the harsh winter climate.

Storing food during seasons of plenty for periods of scarcity is a common behavior pattern. These behaviors are instinctive and are seen in a variety of animals. Squirrels bury nuts, acorns, and other seeds. (They also plant trees because they never find all the seeds they bury.) Chickadees stash seeds in cracks and crevices when food is plentiful and spend many hours during the winter exploring similar places for food. Some of the food they find is food they stored. Honeybees store honey, which allows them to live through the winter when nectar is not available. This requires a rather complicated set of behaviors that coordinates the activities of thousands of bees in the hive.

Navigation and Migration

Because animals move from place to place to meet their needs, it is useful for them to be able to return to a nest, water hole, den, or favorite feeding spot. This requires some memory of their surroundings (a mental map) and a way of determining direction. Direction can be determined by such things as magnetic fields, landmarks, scent trails, or reference to the Sun or stars. If the Sun or stars are used for navigation, some sort of time sense is also needed because these bodies move in the sky. It is valuable to have information about distance as well.

Animals often use the ability to sense changes in time to prepare for seasonal changes. Away from the equator, the length of the day— the **photoperiod**—changes as the seasons change. Many birds prepare for migration and have their migration direction determined by the changing photoperiod. For example, in the fall, many birds instinctively change their behavior, store up fat, and begin to migrate from northern areas to areas closer to the equator. This seasonal migration allows them to avoid the harsh winter conditions signaled by the shortening of days. The return migration in the spring is triggered by the lengthening photoperiod. This migration requires a lot of energy, but it allows many birds to exploit temporary food resources in the north during the summer months.

In honeybees, navigation also involves communication among the various individuals that are foraging for nectar. The bees are able to communicate information about the direction and distance of the nectar source from the hive. If the source of nectar is some distance from the hive, the scout bee performs a "wagging dance" in the hive. The bee walks in a straight line for a short distance, wagging its rear end from side to side. It then circles around back to its starting position and walks the same path as before (figure 18.21). This dance is repeated many times. The direction of the straight-path portion of the dance indicates the direction of the nectar relative to the position of the Sun. For instance, if the bee walks straight upward on a vertical surface in the hive, this tells the other bees to fly directly toward the Sun. If the path is 30° to the left of vertical, the source of the nectar is 30° to the left of the Sun's position.

The duration of the entire dance and the number of waggles in its straight-path portion are positively correlated with the length of time the bee must fly to get to the nectar source. So the dance communicates the duration of flight as well as the direction. Because the recruited bees have picked up the scent of the nectar source from the dancer, they also have information about the kind of flower to visit when they arrive at the correct spot. Because the Sun is not stationary in the sky, the bee must constantly adjust its angle to the Sun. It appears that they do this with an internal clock. Bees that are prevented from going to the source of nectar or from seeing the Sun still fly in the proper direction sometime later, even though the position of the Sun is different.

Like honeybees, some daytime-migrating birds use the Sun to guide them. For nighttime migration, some birds use the stars to help them find their way. In one interesting experiment, warblers, which migrate at night, were placed in a planetarium. The pattern of stars as they appear at any season could be projected onto a large, domed ceiling. In autumn, when these birds would normally migrate southward, the stars of the autumn sky were shown on the ceiling. The birds responded with much fluttering activity at the south side of the cage, as if trying to migrate southward. Then, the experimenters projected the stars of the spring sky, even though it was autumn. The birds then attempted to fly northward, although there was less unity in their efforts to head north; the birds seemed somewhat confused. Nevertheless, the experiment showed that the birds recognized star patterns and were influenced by them.

Some birds navigate by compass direction—that is, they fly as if they had a compass in their heads. They seem to be able to sense magnetic north. Their ability to sense magnetic fields was proven at the U.S. Navy's test facility in Wisconsin. The weak magnetism radiated from this test site has changed the flight patterns of migrating birds, but it is yet to be proved that birds use the Earth's magnetism to guide their migration. Homing pigeons are famous for their ability to find their way home. They make use of a wide variety of clues, but it has been shown that one of the clues they use is magnetism. In one study, birds with tiny magnets glued to the sides of their heads were very poor navigators; others, with nonmagnetic objects attached to the sides of their heads, did not lose their ability to navigate.

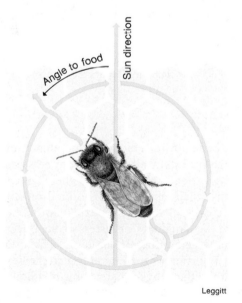

Leggitt

FIGURE 18.21 Honeybee Communication and Navigation
The direction of the straight, tail-wagging part of the dance of the honeybee indicates the direction to a source of food. The angle that this straight run makes with the vertical is the same angle the bee must fly in relation to the Sun to find the food source. The length of the straight run and the duration of each dance cycle indicate the flying time necessary to reach the food source.

Social Behavior

A **society** is a group of animals of the same species that interact with one another and in which there is a division of labor. Societies differ from simple collections of organisms by the greater specialization and division of labor in the roles displayed by the group members. The individuals performing one function cooperate with others having different special abilities. As a result of specialization and cooperation, the

FIGURE 18.22 African Wild Dog Society
African wild dogs hunt in groups and share food, which they take back to the den. Only the dominant male and female mate and raise offspring.

society has characteristics not found in any one member of the group: The whole is more than the sum of its parts. However, if cooperation and division of labor are to occur, there must be communication among the individuals, as well as coordination of effort.

Animal societies exhibit many levels of complexity, and the types of social organization differ from species to species. Some societies show little specialization of individuals other than that determined by sexual differences and differences in physical size and endurance. The African wild dog illustrates such a flexible social organization. These animals are nomadic and hunt in packs. Although an individual wild dog can kill prey about its own size, groups are able to kill fairly large animals if they cooperate in the chase and kill. Young pups are unable to follow the pack. When adults return from a successful hunt, they regurgitate food if the proper begging signal is presented to them (figure 18.22). Therefore, the young and the adults that remained behind to guard the young are fed by the returning hunters. The young are the responsibility of the entire pack, which cooperates in their feeding and protection. While the young are at the den site, the pack must give up its nomadic way of life. Therefore, the young are born during the time of year when prey are most abundant. Only the dominant female in the pack has young each year. If every female had young, the pack couldn't feed them all. At about 2 months of age, the young begin to travel with the pack, and it can return to its nomadic way of life.

Honeybees have a social organization with a high degree of specialization. A hive may contain thousands of individuals, but under normal conditions only the queen bee and the

male drones are capable of reproduction. None of the thousands of workers who are also females reproduce. The large number of sterile female worker honeybees collect food, defend the hive, and care for the larvae. As they age, the worker honeybees move through a series of tasks over a period of weeks. When they first emerge from their wax cells, they clean the cells. Several days later, their job is to feed the larvae. Next, they build cells. Later, they become guards, challenging all insects that land near the entrance to the hive. Finally, they become foragers, finding and taking back nectar and pollen to the hive to feed the other bees and to be stored for the winter. Foraging is usually the workers' last job before they die. Although this progression of tasks is the normal order, workers can shift from their main task to others if there is a need. Both the tasks performed and the progression of the tasks are instinctively (genetically) determined (figure 18.23).

FIGURE 18.23 Honeybee Society
Within the hive, the queen lays eggs, which the sterile workers care for. The workers also clean and repair the hive and forage for food.

Altruism

Altruism is behavior in which an individual animal gives up an advantage or puts itself in danger to aids others. In honeybee societies, the workers give up their right to reproduce and help raise their sisters. Is this a kind of self-sacrifice by the workers, or is there another explanation? In general, the workers are the daughters or sisters of the queen and, therefore, share a large number of her genes. This means that they are helping a portion of their genes get to the next generation by helping raise their own sisters, some of whom will become new queens. This argument has been used to partially explain behaviors in societies that might be bad for the individual but advantageous for the society as a whole.

In other cases, it is not clear that there is any advantage to altruistic behavior. Alarm calls may alert others to a danger but do not benefit the one who gives the alarm. In fact, the one giving the alarm may call attention to itself.

Culture

Culture often develops among social organisms. Extensive contact between parents and offspring allows the offspring to learn certain behavior patterns from their parents. Thus, there are behavioral differences among groups within species. This is obvious in humans, who have various languages, patterns of dress, and many other cultural characteristics. But it is even possible to see similar differences in other social animals. Different groups of chimpanzees use different kinds of tools to get food. Some groups of lions climb trees; others do not (figure 18.24).

Sociobiology

In many ways, honeybee and African wild dog societies are similar. Not all the females reproduce, raising the young is a shared responsibility, and there is some specialization of roles. The analysis and comparison of animal societies has led to the thought that there may be fundamental processes that shape all societies. **Sociobiology** is the systematic study of all forms of social behavior, both human and nonhuman.

How did various types of societies develop? What selective advantage does a member of a social group have? In what ways are social groups better adapted to their environment than nonsocial organisms? How does social organization affect the way populations grow and change? These are difficult questions because, although evolution occurs at the population level, it is individual organisms that are selected. Thus, new ways of looking at evolutionary processes are needed when describing the evolution of social structures.

The ultimate step is to analyze human societies according to sociobiological principles. Such an analysis is difficult and controversial, however, because humans have a much greater ability to modify behavior than do other animals. However, there are some clear parallels between human and nonhuman social behaviors. This implies that there are certain fundamental similarities among social organisms, regardless of their species. Do we see territorial behavior in humans? "No trespassing" signs and fences between neighboring houses seem to be clear indications of territorial behavior in our social species. Do groups of humans have dominance hierarchies? Most business, government, and social organizations have clear dominance hierarchies, in which those at the top get more resources (money, prestige) than those lower in the organization. Do human societies show division of labor? Our societies clearly benefit from the specialized skills of certain individuals. Do humans treat their own children differently than other children? Studies of child abuse indicate that abuse is more common between stepparents and their nongenetic stepchildren than between parents and their biological children. Although these few examples do not prove that human societies follow certain rules typical of other animal societies, it bears further investigation. Sociobiology will continue to explore the basis of social organization and behavior and will continue to be an interesting and controversial area of study.

FIGURE 18.24 Tree-Climbing Lions
In most of Africa, lions do not climb trees; however, in some areas, they do. The difference is cultural.

Summary

Behavior is how an organism acts, what it does, and how it does it. The kinds of responses organisms make to environmental changes (stimuli) include simple reflexes, very complex instinctive behavior patterns, or learned responses.

From an evolutionary viewpoint, behaviors represent adaptations to the environment. They increase in complexity and variety the more highly specialized and developed the organism is. All organisms have inborn, or instinctive, behavior, but higher animals also have one or more ways of learning. These include habituation, association, exploratory learning, imprinting, and insight. Communication for purposes of courtship and mating is accomplished through sounds, visual displays, touch, and chemicals, such as pheromones. Many animals have behavior patterns that are useful in the care and raising of their young.

Territorial behavior is used to obtain exclusive use of an area and its resources. Both dominance hierarchies and territorial behavior are involved in the allocation of scarce resources. To escape from seasonal stress, some animals estivate or hibernate, others store food, and others migrate. Migration to avoid seasonal extremes requires a timing sense and a way of determining direction. Animals navigate by means of sound, celestial light cues, and magnetic fields.

Societies consist of groups of animals in which individuals specialize and cooperate. Sociobiology attempts to analyze all social behavior in terms of evolutionary principles, ecological principles, and population dynamics.

Key Terms

*Use the interactive flash cards on the **Concepts in Biology,** 12/e website to help you learn the meaning of these terms.*

altruism 396	learning 382
anthropomorphism 378	observational learning (imitation) 384
association 382	
behavior 378	operant (instrumental) conditioning 383
classical conditioning 382	
communication 389	pheromones 389
conditioned response 382	photoperiod 395
critical period 384	response 379
dominance hierarchy 393	society 395
ethology 380	sociobiology 397
habituation 382	stimulus 379
imprinting 384	territorial behaviors 392
insight 385	territory 392
instinctive behavior 379	thinking 389
learned behavior 379	

Basic Review

1. Instinctive behavior differs from learned behavior in that instinctive behavior
   a. is inherited.
   b. is flexible.
   c. is found only in simple animals.
   d. is less valuable than learned behavior.

2. The thought that your dog is happy to see you is an example of _____.

3. Imprinting is different from other kinds of learning in that imprinting
   a. is of little value to an organism.
   b. is not reversible.
   c. can be changed easily.
   d. can occur at any time during the life of an individual.

4. Learning is a change in behavior as a result of experience. (T/F)

5. All of the following are typical of territorial behavior EXCEPT
   a. territorial behavior reserves resources for particular individuals or groups.
   b. territorial behavior involves the use of signals to denote territorial boundaries.
   c. territorial behavior is found only in higher animals, such as birds and mammals.
   d. territorial behavior reduces conflict after territories are established.

6. Social behavior typically involves individuals assuming specialized roles. (T/F)

7. Most methods of communication used by animals are learned. (T/F)

8. A social system in which each animal has a particular ranking in the group is a (n) _____.

9. Most kinds of animals provide no care for their offspring. (T/F)

10. Which of the following provide navigational clues to migrating animals?
    a. landmarks, such as rivers and shorelines
    b. the magnetic fields of the Earth
    c. the stars
    d. All of the above are correct.

Answers
1. a 2. anthropomorphism 3. b 4. T 5. c 6. T 7. F 8. dominance hierarchy 9. T 10. d

Concept Review

*Answers to the Concept Review questions can be found on the **Concepts in Biology**, 12/e website.*

18.1 Interpreting Behavior

1. List two ways in which learned and instinctive behaviors differ.
2. Why is it more difficult to study the evolution of behavior than the evolution of structural characteristics?
3. List two animal behaviors and state their significance to the animal's reproductive success.

18.2 The Problem of Anthropomorphism

4. Why do students of animal behavior reject the idea that a singing bird is happy?
5. Define the terms *ethology* and *anthropomorphism*.

18.3 Instinctive and Learned Behavior

6. Briefly describe an example of unlearned behavior in a particular animal. Explain why you know it is unlearned.
7. Briefly describe an example of learned behavior in a particular animal. Explain why you know it is learned.

18.4 Kinds of Learning

8. Give an example of a conditioned response. Describe one that is not mentioned in this chapter.
9. How are classical conditioning and operant conditioning different?
10. What is imprinting, and what value does it have to an organism?
11. Give an example of habituation in a wild animal.
12. In what kinds of animals is observational learning common?

13. Why is insight difficult to demonstrate in animals?

18.5 Instinct and Learning in the Same Animal

14. Which one of the following animals—goose, shark, or grasshopper—is likely to have the highest proportion of learning in its behavior? List two reasons for your answer.

18.6 Human Behavior

15. Give an example of instinctive behavior in humans.
16. Give examples of habituation, association, and imprinting in humans.

18.7 Selected Topics in Behavioral Ecology

17. Describe why communication is important to successful reproduction.
18. Describe two alternative strategies for assuring reproductive success.
19. How do territorial behavior and dominance hierarchies provide certain individuals with an advantage?
20. What distinguishes societies from simple aggregations of individuals?
21. How do animals use chemicals, light, and sound to communicate?
22. What is sociobiology?

Thinking Critically

*For guidelines to answer this question, visit the **Concepts in Biology**, 12/e website.*

If you were going to teach an animal to communicate a message new to that animal, what message would you select? How would you teach the animal to communicate the message at the appropriate time?

CHAPTER

19

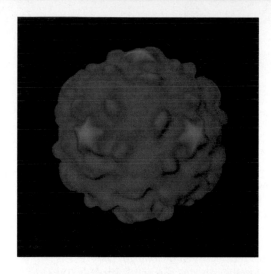

The Origin of Life and the Evolution of Cells

In 2004, scientists used off-the-shelf chemicals to synthesize the poliovirus. They had to use several highly technical techniques, but they were able to create a poliovirus from scratch. The process involved several steps. The genetic code of the poliovirus is in the form of RNA and the sequence of RNA nucleotides was known. First of all, the researchers assembled a DNA molecule that, when transcribed, would produce the correct RNA molecule. The DNA was manufactured from individual nucleotides and used to make the poliovirus RNA. Because polioviruses are normally parasites in human cells, the researchers then placed the RNA into a culture made from human cells. The cells were ground up and large structures, such as nuclei, were removed. When the RNA was placed in this juice, it resulted in the synthesis of the proteins that make the covering for the virus. The scientists started with individual chemical units and ended up with a functional poliovirus.

Although viruses are not cells, they show some similarities to living things. Does this experiment mean that we will be able to create living things from scratch? What conditions on Earth billions of years ago could have allowed for the creation of the first living thing? This chapter will explore the question of how life originated.

CHAPTER OUTLINE

19.1 Early Thoughts About the Origin of Life

For centuries, people have asked the question "How did life originate?" Scientists are still trying to answer that question today. One hypothesis for the origin of life is *spontaneous generation*. **Spontaneous generation** is the concept that living things arise from nonliving material. Aristotle proposed this concept (384–322 B.C.) and it was widely accepted until the seventeenth century. People believed that maggots arose from decaying meat; mice developed from wheat stored in dark, damp places; lice formed from sweat; and frogs originated from damp mud. However, as time passed, people began to question this long-held belief and proposed an alternative idea, called *biogenesis*. **Biogenesis** is the concept that life originates only from preexisting life. For several hundred years, proponents of these two alternative concepts about how life originates argued and performed scientific experiments to test their ideas.

In 1668, Francesco Redi, an Italian physician, performed an experiment that challenged the concept of spontaneous generation. He set up a controlled experiment to test the hypothesis that maggots arose spontaneously from rotting meat (figure 19.1). He used two sets of jars that were identical except for one aspect. Both sets of jars contained decaying meat, and both were exposed to the atmosphere; however, only one set of jars was covered by gauze. The other was uncovered. Redi observed that flies settled on the meat in the open jar, but the gauze blocked their access to the covered jars. When maggots appeared on the meat in the uncovered jars but not on the meat in the covered ones, Redi concluded that the maggots arose from the eggs of the flies (biogenesis), not from spontaneous generation in the meat. experiments, p. 6

In 1748, John T. Needham, an English priest, performed an experiment that led him to conclude that spontaneous generation occurred. He placed a solution of boiled mutton broth in containers, which he sealed with corks. Needham reasoned that boiling would kill any organisms in the broth and that the corks would prevent anything from entering the broth that could generate life. When, after several days, the broth became cloudy with a large population of microorganisms, he concluded that life in the broth was the result of spontaneous generation.

In 1767, another Italian scientist, Abbe Lazzaro Spallanzani, challenged Needham's findings. Spallanzani boiled a meat and vegetable broth, placed this mixture in clean glass containers, and sealed the openings by melting the glass over a flame. He placed the sealed containers in boiling water to kill any living things in the broth. As a control, he set up the same conditions but did not seal the necks, allowing air to enter the flasks (figure 19.2). Two days later, the open containers had a large population of microorganisms, but there were none in the sealed containers. He concluded that spontaneous generation did not occur and that something had entered the unsealed flasks from the air that caused the growth in the broth.

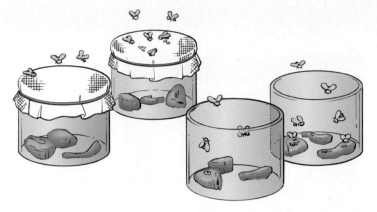

FIGURE 19.1 Redi's Experiment
The two sets of jars are identical in every way except the gauze covering. The set on the right has had nothing special done to it. It is called the control group. The set on the left has been manipulated by placing a gauze covering over the jars. It is the experimental group. Any differences seen between the control and the experimental groups are the result of this single variable (the gauze coverings). From this experiment, Redi concluded that the maggots in the meat arose from eggs laid by flies (biogenesis), not spontaneous generation.

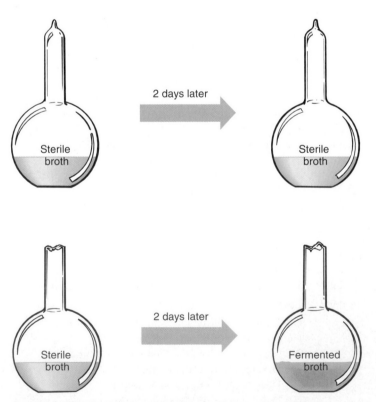

FIGURE 19.2 Spallanzani's Experiment
Spallanzani boiled a meat and vegetable broth and placed this mixture into clean flasks. He sealed one and put it in boiling water. As a control, he subjected another flask to the same conditions, except he left it open. Within 2 days, the open flask had a population of microorganisms. Spallanzani maintained that this demonstrated that spontaneous generation did not occur and that something from the air was responsible for the growth in the broth.

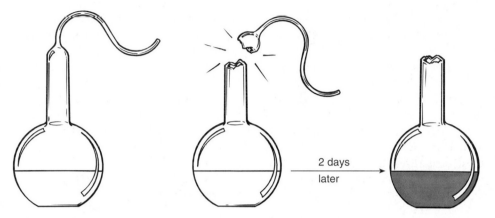

FIGURE 19.3 Pasteur's Experiment

Pasteur's experiment was designed to test the idea that a "vital element" from the air was necessary to produce life. He boiled a mixture of sugar and yeast in swan-neck flasks that allowed oxygen, but not airborne organisms, to enter them. He left some flasks intact and broke the neck off others. Within 2 days, there was growth in the flasks that had their swan necks removed. Pasteur concluded that oxygen in the air was not responsible for the growth of microorganisms in the broth because the broth was exposed to the air through the swan neck. However, microorganisms could enter the flasks from which the swan necks were removed. This supported the concept of biogenesis and argued against the concept of spontaneous generation.

Spallanzani's experiment did not completely disprove spontaneous generation to everyone's satisfaction, however. The supporters of spontaneous generation attacked Spallanzani by stating that he excluded air from his sealed flasks, a factor believed necessary for spontaneous generation. Supporters also argued that boiling had destroyed a "vital element." In 1774, when Joseph Priestly discovered oxygen, the proponents of spontaneous generation claimed that oxygen was the "vital element" that Spallanzani had excluded from his sealed containers.

In 1861, French chemist Louis Pasteur convinced most scientists that spontaneous generation could not occur. He placed a fermentable sugar solution and yeast mixture in a flask that had a long swan neck. The mixture and the flask were boiled for a long time. The flask was left open to allow oxygen, the "vital element," to enter, but no organisms developed in the mixture. The organisms that did enter the flask settled on the bottom of the curved portion of the neck and could not reach the sugar-water mixture. As a control, he cut off the swan neck (figure 19.3). This allowed microorganisms from the air to fall into the flask, and within 2 days the fermentable solution was supporting a population of microorganisms. In his address to the French Academy, Pasteur stated, "Never will the doctrine of spontaneous generation arise from this mortal blow."

19.2 Current Thinking About the Origin of Life

Today, it is clear that current living things come from other living things. They are the result of biogenesis. But that does not answer the question of how the *first* living thing developed. Because the evidence is strong that the universe had a time of origin, life must have originated spontaneously from nonliving material at least once. It is clear that we will never really know how the first living thing developed, because we cannot repeat the "original" experiment that led to the first living thing. However, that does not prevent scientists from speculating on the origin of life and evaluating relevant scientific information. There are many different theories with supporting evidence.

An Extraterrestrial Origin for Life on Earth

Early in the 1900s, Swedish scientist Svante Arrhenius popularized the idea of *panspermia*. **Panspermia** is the concept that life arose outside the Earth and that living things were transported to Earth to seed the planet with life. However, this idea does not explain *how* life arose originally. It sees the

Earth as similar to one of Spallanzani's or Pasteur's open flasks. Although Arrhenius's ideas had little scientific support at that time, his basic concept has since been revived and modified as a result of new evidence gained from examinations of meteorites and space explorations.

Evidence from Meteorites

Many meteorites contain organic molecules, and this suggests that life may have existed elsewhere in the solar system. In 1996, a meteorite from Antarctica was analyzed. Its chemical makeup suggests that it had been a portion of the planet Mars and had been ejected from that planet as a result of a collision with an asteroid. An analysis of the meteorite shows the presence of complex organic molecules and small globules, which resemble those found on Earth that are thought to be the result of the activity of ancient microorganisms. While scientists no longer think these structures in meteorites are from microorganisms, many still think Mars may have supported living things in the past.

Evidence from Mars Exploration

Although previous explorations of Mars's surface has failed to find evidence for life, many researchers are not satisfied with those results. In June and July 2003, two spacecrafts were launched by the National Aeronautics and Space Administration to explore the surface of Mars. One of the important goals of these missions is to search for signs of present or past life. The robotic rover vehicles from these two spacecrafts have gathered much information about the surface of Mars. One important piece of information is that it is highly likely that, in the past, liquid water existed in large enough quantities to form rivers, lakes, and perhaps salty oceans. Although no evidence of life on Mars has yet been found, these bodies of water could have supported life. Future explorations will continue to search for evidence of life on Mars. Thus, those who support the idea that life could have originated elsewhere in the universe argue that, if life originated on Mars, it could have been transported to Earth by meteorites.

An Earth Origin for Life on Earth

A popular alternative explanation for the origin of life on Earth focuses on chemical evolution. This theory proposes that inorganic matter changed into organic matter composed of complex carbon-containing molecules and that these, in turn, combined to form the first living cell. Furthermore, it is possible to perform experiments that test scientists' hypotheses about these processes. Several pieces of evidence support the idea that life could have arisen on the Earth.

1. The temperature range on Earth allows for water to exist as a liquid on its surface. This is important, because water is the most common compound of living things.
2. Analysis of the atmospheres of the other planets in our solar system shows that they all lack oxygen. The oxygen in the Earth's current atmosphere is the result of photosynthesis by living organisms. Therefore, before life on Earth, the atmosphere probably lacked oxygen.
3. Experiments demonstrate that organic molecules can be generated in an atmosphere that lacks oxygen.
4. Because it is assumed that all of the planets have been cooling off as they age, it is very likely that the Earth was much hotter in the past. The large portions of the Earth's surface that are of volcanic origin strongly suggest a hotter past.
5. The discovery of specialized prokaryotic organisms that live in extreme environments of high temperature, high salinity, low pH, or the absence of oxygen suggests that they may have been adapted to life in a world that is very different from today's Earth. These kinds of organisms are found today in unusual locations, such as hot springs and around thermal vents in the ocean floor, and may be descendants of the first organisms formed on the primitive Earth.

19.3 The "Big Bang" and the Origin of the Earth

An understanding of how Earth formed and what conditions were like on early Earth will help us think about how life may have originated. Astronomers note that the current stars and galaxies are moving apart from one another. This and other evidence has led to the concept that our universe began as a very dense mass of matter, which had a great deal of energy. This dense mass of matter exploded in a "big bang" about 13 billion years ago, resulting in the formation of simple atoms, such as hydrogen and helium. When huge numbers of these atoms collected in one place, stars formed. Stars consist primarily of hydrogen and helium atoms. The light from stars is the result of nuclear fusion, in which these atoms combine to form larger atoms.

Many scientists believe that Earth—along with other planets, meteors, asteroids, and comets—was formed at least 4.6 billion years ago. The *protoplanet nebular model* proposes that the solar system was formed from a large cloud of gases and elements produced by previously existing stars (figure 19.4). The simplest and most abundant gases were hydrogen and helium, but other, heavier elements were present as well. A gravitational force was created by the collection of particles within this cloud, which caused other particles to be pulled from the outer edges to the center. As the particles collected into larger bodies, gravity increased and more particles were attracted to the bodies. Ultimately, a central body (the Sun) was formed, with several other bodies (planets) moving around it (How Science Works 19.1). Like other stars, the Sun consists primarily of hydrogen and helium atoms, which are being fused together to form larger atoms, with the release of large amounts of thermonuclear energy.

Thermonuclear reactions also occurred as the particles became concentrated to form Earth. Thus, the early Earth

HOW SCIENCE WORKS 19.1

Gathering Information About the Planets

Exploration of our solar system has become quite sophisticated with the development of space travel. Only the planet Pluto has not been visited by a spacecraft. Spacecrafts have flown by the distant planets of Jupiter, Saturn, and Neptune and have sent back pictures. The *Galileo* spacecraft released a probe that entered Jupiter's atmosphere in 1996. In addition to the Earth's Moon, Mars and Venus have had spacecraft land on their surfaces and send back pictures and data about their temperature, their atmospheric composition, and the geologic nature of their surfaces.

Several characteristics of planets affect how likely it is that life exists or did exist on a planet. A planet's distance from the Sun determines the amount of energy it receives from the Sun. Planets near the Sun receive more solar energy. The larger the mass of a planet, the greater its force of gravity. The force of gravity influences how many other bodies it can capture (moons) and how much atmosphere it can hold. Some of the planets are gases, whereas others have solid surfaces. In order for life as we know it on Earth to exist, a planet must have liquid water, an appropriate atmosphere, and a solid surface. The only planet that may have had these conditions at one time is Mars.

Sizes of planets
(diameter in kilometers)

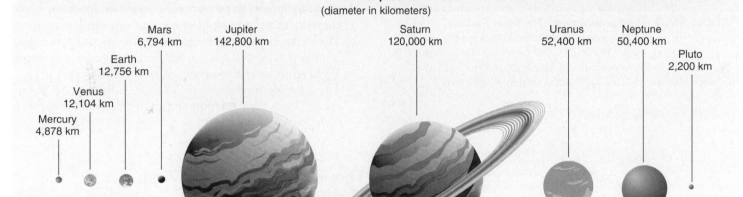

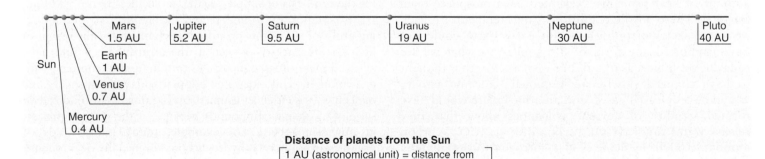

Distance of planets from the Sun

1 AU (astronomical unit) = distance from
Earth to Sun, approximately 150 million km

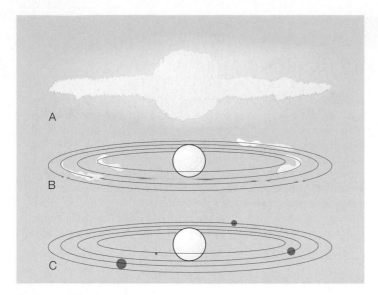

FIGURE 19.4 The Formation of Our Solar System
(a) As gravity pulled gases and other elements toward the center, the accumulating matter resulted in the formation of the Sun. *(b)* In other regions, smaller gravitational forces caused the formation of the Sun's planets. *(c)* Finally, the solar system was formed.

would have been a very hot place. Geologically, this period in the history of the Earth is called the "Hadean Eon." The term *Hadean* means "hellish." Although not as hot as the Sun, the material of Earth formed a molten core, which became encased by a thin outer crust as it cooled. In the Earth's early stages of formation, about 4 billion years ago, there probably was a considerable amount of volcanic activity.

Because the surface was hot, there was no water on the surface or in the atmosphere. In fact, the tremendous amount of heat probably prevented an atmosphere from forming. These hostile conditions (high temperature, lack of water, lack of atmosphere) on early Earth could not have supported any form of life similar to what we see today. Over hundreds of millions of years, Earth is thought to have changed slowly. As it cooled, volcanic activity would have released gases, and an atmosphere would have formed. Studies of current volcanoes show that they release water vapor (H_2O), carbon dioxide (CO_2), carbon monoxide (CO), methane (CH_4), nitrogen (N_2), ammonia (NH_3), hydrochloric acid (HCl), and various sulfur compounds. These gases would have formed a **reducing atmosphere**—an atmosphere that did not contain molecules of oxygen (O_2). Any oxygen would have quickly combined with other atoms to form compounds. Further cooling enabled the water vapor in the atmosphere to condense into droplets of rain, which ran over the land and collected to form oceans.

19.4 The Chemical Evolution of Life on Earth

When we consider the nature of the simplest forms of life today, we find that living things consist of an outer membrane, which separates the cell from its surroundings; genetic material in the form of nucleic acids; and many kinds of enzymes, which control the activities of the cell. Therefore, when we speculate about the origin of life from inorganic material, it seems logical that several events or steps were necessary:

1. Organic molecules must first have been formed from inorganic molecules.
2. Simple organic molecules must have combined to form larger organic molecules, such as RNA, proteins, carbohydrates, and lipids.
3. A molecule must have served as genetic material.
4. Some molecules must have functioned as enzymes.
5. Genetic material must have become self-replicating.
6. The molecules serving as genetic material and other large organic molecules must have been collected and segregated from their surroundings by a membrane.
7. The first life-forms would have needed a way to obtain energy from their surroundings in order to maintain their complex structure.

The Formation of the First Organic Molecules

In the 1920s, a Russian biochemist, Alexander I. Oparin, and a British biologist, J. B. S. Haldane, working independently, proposed that the first organic molecules were formed spontaneously in the reducing atmosphere thought to be present on the early Earth. The molecules in the primitive atmosphere supplied the atoms of carbon, hydrogen, oxygen, and nitrogen needed to build organic molecules. Lightning, heat from volcanoes, and ultraviolet radiation furnished the energy needed for the synthesis of simple organic molecules (figure 19.5).

It is important to understand the significance of a reducing atmosphere to this proposed mechanism for the origin of life. The absence of oxygen in the atmosphere would have allowed these organic molecules to remain and combine with one another. This does not happen today because organic molecules are either consumed by organisms or oxidized to simpler inorganic compounds because of the oxygen present in our atmosphere. For example, today many kinds of organic air pollutants (hydrocarbons) eventually degrade into smaller molecules in the atmosphere. Unfortunately, they participate in the formation of photochemical smog as they are broken down. organic molecules, p. 46

After simple organic molecules were formed in the atmosphere, they probably would have been washed from the air and carried into the newly formed oceans by rain and runoff from the land. There, the molecules could have reacted with one another to form the more complex molecules of simple

FIGURE 19.5 The Formation of Organic Molecules in the Atmosphere

The environment of the primitive Earth was harsh and lifeless. But many scientists believe that it contained the necessary molecules and sources of energy to fashion the first living cell. The energy furnished by volcanoes, lighting, and ultraviolet light could have broken the bonds in the simple inorganic molecules in the atmosphere. New bonds could have formed as the atoms from the smaller molecules were rearranged and bonded to form simple organic compounds in the atmosphere. Rain and runoff from the land would have carried these chemicals into the oceans, where they could have reacted with each other to form more complex organic molecules.

sugars, amino acids, and nucleic acids. This accumulation is thought to have occurred over half a billion years, resulting in oceans that were a dilute organic soup. These simple organic molecules in the ocean served as the building materials for more complex organic macromolecules, such as complex carbohydrates, proteins, lipids, and nucleic acids. Recognize that all the ideas discussed so far in this section cannot be confirmed by direct observation, because we cannot go back in time. However, several assumptions central to this model for the origin of life have been laboratory tested.

In 1953, Stanley L. Miller conducted an experiment to test the idea that organic molecules can be synthesized in a reducing environment. He constructed a simple model of the early Earth's atmosphere (figure 19.6). In a glass apparatus, he placed distilled water to represent the early oceans. Adding hydrogen (H_2), methane (CH_4), and ammonia (NH_3) to the water (H_2O) simulated the reducing atmosphere. Electrical sparks provided the energy needed to produce organic compounds. By heating parts of the apparatus and cooling others, he simulated the rains that are thought to have fallen into the early oceans. After a week of operation, he removed some of the water from the apparatus. When he analyzed this water, he found that it contained many simple organic compounds. Although Miller demonstrated the nonbiological synthesis of simple organic molecules, such as amino acids and simple sugars, his results did not account for complex organic molecules, such as proteins and nucleic acids (e.g., DNA). However, other researchers produced some of the components of nucleic acid under similar primitive conditions.

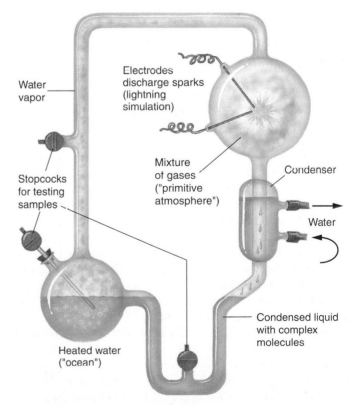

FIGURE 19.6 Miller's Apparatus

Stanley Miller developed this apparatus to demonstrate that the spontaneous formation of complex organic molecules could take place in a reducing atmosphere.

The idea that organic molecules could have formed on Earth is further supported by the discovery of many kinds of organic molecules in interstellar clouds and in the structure of meteorites. These pieces of evidence show that organic molecules form without the presence of living things.

The Formation of Macromolecules

Several ideas have been proposed to explain how simple organic molecules could have been concentrated and caused to combine to form larger macromolecules. One hypothesis suggests that a portion of the early ocean could have been separated from the main ocean by geologic changes. The evaporation of water from this pool could have concentrated the molecules, which might have led to the formation of macromolecules by dehydration synthesis. Second, it has been proposed that freezing may have been the means of concentration. When a mixture of alcohol and water is placed in a freezer, the water freezes solid and the alcohol becomes concentrated into a small portion of liquid. A similar process could have occurred on Earth's early surface, resulting in the concentration of simple organic molecules. In this concentrated solution, dehydration synthesis in a reducing atmosphere could have occurred, resulting in the formation of macromolecules. A third theory proposes that clay particles may have been a factor in concentrating simple organic molecules. Small particles of clay have electrical charges, which can attract and concentrate organic molecules, such as proteins, from a watery solution. Once the molecules became concentrated, it would have been easier for them to interact to form larger macromolecules.

RNA May Have Been the First Genetic Material

The reproduction of most current organisms involves the replication of DNA and the distribution of the copied DNA to subsequent cells (those that do not use DNA use RNA as their genetic material). DNA is responsible for the manufacture of RNA, which subsequently leads to the manufacture of proteins. Some of the proteins produced serve as enzymes that control chemical reactions. However, it is difficult to see how this complicated sequence of events, which involves many steps and the assistance of several enzymes, could have been generated spontaneously, so scientists have looked for simpler systems that could have led to the current DNA system.

Scientists involved in studying the structure and function of viruses discovered that many viruses do not contain DNA but, rather, store their genetic information in the structure of RNA. In order for these RNA-viruses to reproduce, they must enter a cell. The cell makes copies of the RNA and the proteins necessary to make more of the virus. Certain plant diseases are caused by viroids that consist of naked RNA, which enters cells and causes the cells to make copies of the RNA. In other words, in both of these cases, RNA serves as genetic material.

Other scientists who study viral diseases find that it is difficult to develop vaccines for many viral diseases because their genetic material mutates easily. Because of this, researchers have been studying the nature of viral DNA or RNA to see what causes the high rate of mutation. This has led others to explore the RNA viruses and ask the question "Can RNA replicate itself without DNA?" This is an important question, because, if RNA could replicate itself, it would have all of the properties it needed to serve as genetic material. It could store information, mutate, and make copies of itself. Other research about the nature of RNA provides interesting food for thought. RNA can be assembled from simpler subunits that could have been present on the early Earth. Scientists have also shown that RNA molecules are able to make copies of themselves without the need for enzymes, and they can do so without being inside cells. Furthermore, some RNA molecules serve as enzymes for specific reactions. Such molecules are called *ribozymes*. This evidence suggests that RNA may have been the first genetic material and helps solve one of the problems associated with the origin of life: how genetic information was stored. Because RNA is a much simpler molecule than DNA and can make copies of itself without the aid of enzymes, perhaps it was the first genetic material. Once a primitive life-form had the ability to copy its genetic material, it would be able to reproduce. Reproduction is one of the most fundamental characteristics of living things.

The Development of Membranes

One of the defining features of living things is the presence of a membrane surrounding its cells which regulates what enters and leaves them. Consider the formation of bubbles in water. Bubbles are particularly common when organic molecules are present in water. Perhaps the first membranes were formed in this way. There are several theories about what the first membranes were like. Some suggest that they were made of proteins, others that they were lipids or other organic molecules. Several kinds of experiments have sought to clarify what the first membrane might have been like. Alexander I. Oparin mentioned earlier, speculated that a structure could have formed consisting of a collection of organic macromolecules surrounded by a film of water molecules. This arrangement of water molecules, although not a membrane, could have functioned as a physical barrier between the organic molecules and their surroundings. He called these structures *coacervates*.

Coacervates have been synthesized in the laboratory (figure 19.7). They can selectively absorb molecules from the surrounding water and incorporate them into their structure. Also, the chemicals within coacervates have a specific arrangement—they are not random collections of molecules. Some coacervates contain proteins (enzymes) that direct a specific type of chemical reaction. Because they lack a definite membrane, no one claims that coacervates are alive, but these structures do exhibit some lifelike traits: They can increase in size and can split into smaller particles if the right conditions exist.

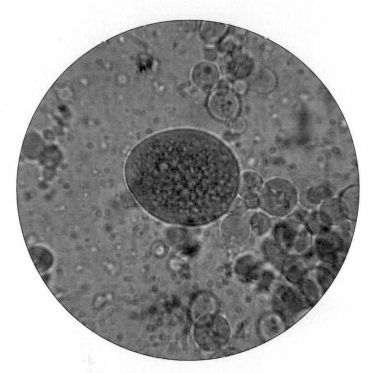

FIGURE 19.7 Coacervates
One hypothesis proposes that a film of water, which acted as a primitive cell membrane, could have surrounded organic molecules, forming a structure that resembles a living cell. Such a structure can easily be produced in the lab.

An alternative hypothesis is that the early, prebiotic cell structure could have consisted of a collection of organic macromolecules with a double-layered outer boundary. These structures have been called *microspheres*. Microspheres can be formed in the laboratory by heating simple, proteinlike compounds in boiling water and slowly cooling them. Some of the protein material produces a double-boundary structure enclosing the microsphere. Although these boundaries do not contain lipids, they exhibit some membranelike characteristics and suggest the structure of a cellular membrane.

Microspheres swell or shrink, depending on the osmotic potential in the surrounding solution. They also display a type of internal movement (streaming) similar to that exhibited by cells and contain some molecules that function as enzymes. Using ATP as a source of energy, microspheres can direct the formation of polypeptides and nucleic acids. They can absorb material from the surrounding medium and form buds, resulting in a second generation of microspheres.

A third possibility is that a membrane forms as a result of lipid materials interacting with water. Lipids do not dissolve in water and whenever lipids are mixed with water, spherical structures form, as in vinegar and oil salad dressing.

There is no way to know if any of these models represents what really happened in the origin of living things. However, some kind of structure was necessary to separate the complex, organized molecules from the watery environment in which they were dissolved. Once organic molecules were separated from their watery surroundings by a membrane, these structures were similar to primitive cells.

The Development of a Genetic System

The laboratory synthesis of coacervates and microspheres and the recognition that RNA can serve as enzymes and genetic material clarify how the first primitive living cells might have developed. However, it leaves a large gap in our understanding, because it does not explain how these first cells might have become the highly complex, living cells of today. Once the first primitive cell existed, several additional steps were required to reach our current state of complexity:

1. Proteins must have become the catalysts (enzymes) of the cells.
2. The control of protein synthesis must have been taken over by RNA.
3. DNA must have replaced RNA as the self-replicating genetic material of the cell.
4. Ultimately, the first cellular units must have been able to reproduce more of themselves.
5. The first cellular units must have had a way to obtain energy from their surroundings.

Research into these questions will continue. However, there has been a great deal of thought about the last of these questions: What kind of metabolism did the first living things have?

The Development of Metabolic Pathways

Fossil evidence indicates that there were primitive forms of life on Earth about 3.5 billion years ago. Regardless of how they developed, these first primitive cells would have needed a way to add new organic molecules to their structures as previously existing molecules were lost or destroyed. There are two ways to accomplish this. Heterotrophs capture organic molecules, such as sugars, amino acids, or organic acids, from their surroundings, which they use to make new molecules and provide themselves with energy. Autotrophs use an external energy source, such as sunlight or the energy from inorganic chemical reactions, to combine simple inorganic molecules, such as water and carbon dioxide, to make new organic molecules. These new organic molecules can then be used as building materials for new cells or can be broken down at a later date to provide energy.

The Heterotroph Hypothesis

Many scientists support the idea that the first living things on Earth were heterotrophs, which lived off organic molecules in the oceans. There is evidence to suggest that a wide variety of compounds were present in the early oceans, some of which could have been used, unchanged, by primitive cells. Because the earliest cells appear in the fossil record before any evidence of oxygen in the atmosphere, these early heterotrophs would have been anaerobic organisms. Therefore, they did not obtain the maximum amount of energy from the organic molecules they obtained from their environment. At first, this would not have been a problem; the organic molecules that had been accumulating in the oceans for hundreds of millions of years were a source of raw material for the heterotrophs. However, as their population increased through reproduction, they would have consumed the supply of organic material faster than it was being spontaneously produced in the atmosphere.

The compounds that could be used easily by these cells would have been the first to become depleted from the early environment. However, some of the heterotrophs may have contained a mutated form of nucleic acid, which allowed them to convert material that was not directly usable into a usable compound. Heterotrophic cells with such mutations could have survived, whereas those without such mutations would have become extinct as the compounds they used for food became scarce. It has been suggested that, through a series of mutations in the early heterotrophic cells, a more complex series of biochemical reactions originated within some of the cells. Such cells could use chemical reactions to convert unusable chemicals into usable organic compounds. Thus, additional steps would have been added to their metabolic processes, and new metabolic pathways would have evolved.

The Autotroph Hypothesis

Although the heterotroph hypothesis for the origin of living things was the prevailing theory for many years, recent discoveries have caused many scientists to consider an alternative—that the first organism was an autotroph. Several kinds of information support this theory. Many members of the Domain Archaea are autotrophic and live in extremely hostile environments resembling the conditions that may have existed on the early Earth. Today, these organisms are found in hot springs—such as those found in Yellowstone National Park; in Kamchatka, Russia (Siberia); and near thermal vents—areas where hot, mineral-rich water enters seawater from the deep ocean floor. They use energy released from inorganic chemical reactions to synthesize organic molecules from inorganic components. Because of this, they are called *chemoautotrophs*. If their ancient ancestors had similar characteristics, the first organisms may have been autotrophs.

There is much evidence that the Earth was a much hotter place in the past. The fact that today, many members of the Domain Archaea live in very hot environments suggests that they may have originated on an Earth that was much hotter than it is currently. If the first organisms were autotrophs, there would have been competition among different cells for the inorganic raw materials they needed for their metabolism, and there would have been changes in the metabolic processes as mutations occurred. If the first organisms were autotrophs, there could have been subsequent evolution of a variety of cells, both autotrophic and heterotrophic, which could have led to the diversity of prokaryotic cells seen today in the Domains Eubacteria and Archaea.

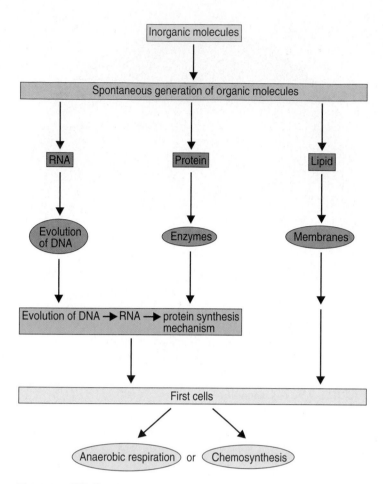

FIGURE 19.8 **The Chemical Evolution of Life**
This diagram summarizes the steps necessary for primitive cells to evolve from inorganic molecules.

Summary

As a result of this discussion you should understand that we do not know how life on Earth originated. Scientists look at many kinds of evidence and continue to explore new avenues of research. Thus, currently there are three competing theories of the origin of life on Earth:

1. Life arrived here from an extraterrestrial source.
2. Life originated on Earth as a heterotroph.
3. Life originated on Earth as an autotroph.

Figure 19.8 summarizes the steps that are thought to have been necessary for primitive cells to evolve from inorganic molecules.

19.5 Major Evolutionary Changes in Early Cellular Life

Once living things existed and had a genetic material that stored information but was changeable (mutational), they could evolve. Thus, living things could have proliferated into a variety of kinds that were adapted to specific environmental conditions. Remember that the Earth has not been static but has been changing as a result of its cooling, volcanic activity, and encounters with asteroids. In addition, the organisms have had an impact on the way in which the Earth has developed. Regardless of the way in which life originated on Earth, there have been several major events in the subsequent evolution of living things.

The Development of an Oxidizing Atmosphere

Since its formation, Earth has undergone constant change. In the beginning, it was too hot to support an atmosphere. Later, as it cooled and as gases escaped from volcanoes, a reducing atmosphere (lacking oxygen) was likely to have been formed. The early life-forms would have lived with a reducing atmosphere. However, today we have an **oxidizing atmosphere,** with 20 percent oxygen, and most organisms use the oxygen to extract energy from organic molecules through a process of aerobic respiration. But what caused the atmosphere to change? It is clear that the oxygen in our atmosphere is the result of the process of photosynthesis.

The Origin of Photosynthesis

Several kinds of Eubacteria perform photosynthesis. Several of these perform a type of photosynthesis that does not result in the release of oxygen. However, one major group, the cyanobacteria, use a form of photosynthesis that results in the release of oxygen. Therefore, it seems plausible that the first organisms, regardless of whether they were heterotrophs or chemoautotrophs, accumulated many mutations over time, resulting in photosynthetic autotrophs. Because oxygen is released from the most common form of photosynthesis, this would have resulted in the development of an oxidizing atmosphere.

The development of an oxidizing atmosphere created an environment *unsuitable* for the continued formation of organic molecules. Organic molecules tend to break down (oxidize) when oxygen is present. The presence of oxygen in the atmosphere would make it impossible for life to spontaneously originate in the manner described earlier in this chapter because an oxidizing atmosphere would not allow the accumulation of organic molecules in the seas. However, once living things existed, new life could be generated through reproduction, and new *kinds* of life could be generated through mutation and evolution. The presence of oxygen in the atmosphere had one other important outcome: It opened the door for the evolution of aerobic organisms.

An oxidizing atmosphere began to develop about 2.3 billion years ago. Although various chemical reactions released small amounts of molecular oxygen into the atmosphere, it was photosynthesis that generated most of the oxygen. The oxygen molecules also reacted with one another to form ozone (O_3). Ozone collected in the upper atmosphere and acted as a screen to prevent most of the ultraviolet light from reaching Earth's surface. The reduction of ultraviolet light

diminished the spontaneous formation of complex organic molecules. It also reduced the number of mutations in cells. In an oxidizing atmosphere, it was no longer possible for organic molecules to accumulate over millions of years to be later incorporated into living material.

The Origin of Aerobic Respiration

The appearance of oxygen in the atmosphere also allowed for the evolution of aerobic respiration. Because the first heterotrophs were, of necessity, anaerobic organisms, they did not derive large amounts of energy from the organic materials available as food. With the evolution of aerobic heterotrophs, there could be a much more efficient conversion of food into usable energy. Aerobic organisms would have a significant advantage over anaerobic organisms: They could use the newly generated oxygen as a final hydrogen acceptor and, therefore, generate many more adenosine triphosphates (ATPs) from the food molecules they consumed.

The Establishment of Three Major Domains of Life

Biologists have traditionally divided organisms into kingdoms, based on their structure and function. However, because of their small size, it is very difficult to do this with microscopic organisms. With the ability to decode the sequence of nucleic acids, it became possible to look at the genetic nature of organisms without being confused by their external structures. Biologist Carl Woese studied the sequences of nucleotides in the ribosomal RNA of many kinds of prokaryotic cells commonly known as bacteria and compared their similarities and differences. As a result of his studies and those of many others, a new concept of the relationships between various kinds of organisms has emerged. It is now clear that the bacteria that had been considered a group of similar organisms, are actually two very different kinds of organisms: the Eubacteria and the Archaea. Furthermore, the Archaea share some characteristics with eukaryotic organisms.

The three main kinds of living things—Eubacteria, Archaea, and Eucarya—are called Domains. The Domains Eubacteria and Archaea are different kinds of prokaryotic organisms. The Domain Eucarya contains organisms that have eukaryotic cells. Within each domain are several kingdoms. In the Domain Eucarya, there are four kingdoms: Animalia, Plantae, Fungi, and Protista. The process of identifying kingdoms within the Eubacteria and Archaea is currently ongoing.

Cell types:	Prokaryote		Eukaryote
Domains:	Eubacteria	Archaea	Eucarya

Kingdoms
1. Protista
2. Fungi
3. Plantae
4. Animalia

The oldest organisms gave rise to two major types of prokaryotic organisms (Eubacteria and Archaea). Strangely, the Domains Archaea and Eucarya share many characteristics, suggesting that they are more closely related to each other than either is to the Domain Eubacteria.

It appears that each domain developed specific abilities. The Archaea are primarily organisms that use inorganic chemical reactions to generate the energy they need to make organic matter. Often, these reactions result in the production of methane (CH_4), and these organisms are known as methanogens. Others use sulfur and produce hydrogen sulfide (H_2S); most of these organisms are found in extreme environments, such as hot springs, or in extremely salty or acidic environments.

The Eubacteria developed many different metabolic abilities. Today, many Eubacteria are heterotrophic and use organic molecules as a source of energy. Some of these heterotrophs use anaerobic respiration, whereas others use aerobic respiration. Other Eubacteria are autotrophic. Some, such as the cyanobacteria, carry on photosynthesis, whereas others are chemosynthetic and get energy from inorganic chemical reactions similar to Archaea.

The Eucarya are the most familiar and appear to have exploited the metabolic abilities of other organisms by incorporating them into their own structure. Chloroplasts and mitochondria are both bacteria-like structures found inside eukaryotic cells. Table 19.1 summarizes the major characteristics of these three domains.

The Origin of Eukaryotic Cells

Biologists generally believe that eukaryotes evolved from prokaryotes. Two major characteristics distinguish eukaryotic cells from prokaryotic cells. Eukaryotic cells have their DNA in a nucleus surrounded by a membrane and have many kinds of membranous organelles. The most widely accepted theory of how eukaryotic cells originated is the *endosymbiotic theory*. The **endosymbiotic theory** states that present-day eukaryotic cells evolved from the uniting of several types of primitive prokaryotic cells. It is thought that some organelles found in eukaryotic cells may have originated as free-living prokaryotes. For example, mitochondria and chloroplasts contain DNA and ribosomes that resemble those of bacteria. They also reproduce on their own and synthesize their own enzymes. Therefore, it has been suggested that mitochondria were originally free-living prokaryotes that carried on aerobic respiration and chloroplasts were free-living photosynthetic prokaryotes. If the combination of two different cells in this symbiotic relationship were mutually beneficial, the relationship could have become permanent (figure 19.9).

Although endosymbiosis explains how many of the membranous organelles may have arisen in eukaryotic cells, the origin of the nucleus is less clear. There are currently two ideas about how the nucleus came to be. One suggests that the nucleus formed in the same way as other organelles. An invading cell with a membrane around it became the nucleus when it took over the running of the cell from the cell's orig-

---

TABLE 19.1		
Major Domains of Life		
Eubacteria	**Archaea**	**Eucarya**
There is no nuclear membrane.	There is no nuclear membrane.	A nuclear membrane is present.
Chlorophyll-based photosynthesis is a bacterial invention but most are anaerobic heterotrophs.	None have been identified as pathogenic.	Many kinds of membranous organelles are present in cells.
Oxygen-generating photosynthesis was an invention of the cyanobacteria.	They probably have a common ancestor with Eucarya.	Chloroplasts are probably derived from cyanobacteria.
A large number of heterotrophs oxidize organic molecules for energy.	Many obtain energy from inorganic reactions to make organic matter.	Mitochondria are probably derived from certain aerobic bacteria.
Some use energy from inorganic chemical reactions to produce organic molecules.	They are typically found in extreme environments.	They probably have a common ancestor with Archaea.
Some live at high temperatures and may be ancestral to Archaea.	There are few heterotrophs.	The common evolutionary theme is the development of complex cells through endosymbiosis with other organisms.
There is much metabolic diversity among closely related organisms.	An example is *Pyrolobus fumarii*, which lives near deep-sea hydrothermal vents at temperatures up to 113°C	An Example is *Homo sapiens*, all cells of the human body are eukaryotic.
An example is *Streptococcus pneumoniae*, one cause of pneumonia.		

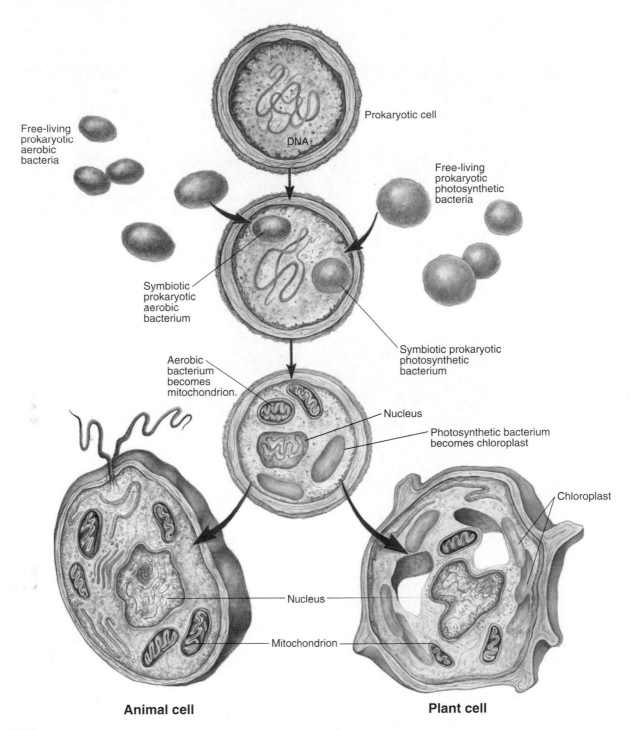

Prokaryotic cell

DNA

Free-living prokaryotic aerobic bacteria

Free-living prokaryotic photosynthetic bacteria

Symbiotic prokaryotic aerobic bacterium

Symbiotic prokaryotic photosynthetic bacterium

Aerobic bacterium becomes mitochondrion.

Nucleus

Photosynthetic bacterium becomes chloroplast

Chloroplast

Nucleus

Mitochondrion

Animal cell

Plant cell

FIGURE 19.9 The Endosymbiotic Theory

This theory proposes that some free-living prokaryotic bacteria entered a host cell and a symbiotic relationship developed. Mitochondria appear to have developed from certain aerobic bacteria and chloroplasts from photosynthetic cyanobacteria. Once eukaryotic cells were present, the subsequent evolution of more complex protozoa, algae, fungi, plants, and animals could take place.

inal DNA. The alternative hypothesis is that prokaryotic cells developed a nuclear membrane on their own from membranes in the cell. In other words, an increase in the number of membranes within prokaryotic cells could have produced an envelope that enclosed the DNA of the cell.

When the endosymbiotic theory was first suggested, it met with a great deal of criticism. However, continuing research has uncovered several other instances of the probable joining of two different prokaryotic cells to form one. In addition, it appears that endosymbiosis occurred among eukaryotic organisms as well. Several kinds of eukaryotic red and brown algae contain chloroplast-like structures, which appear to have originated as free-living eukaryotic cells (figure 19.10).

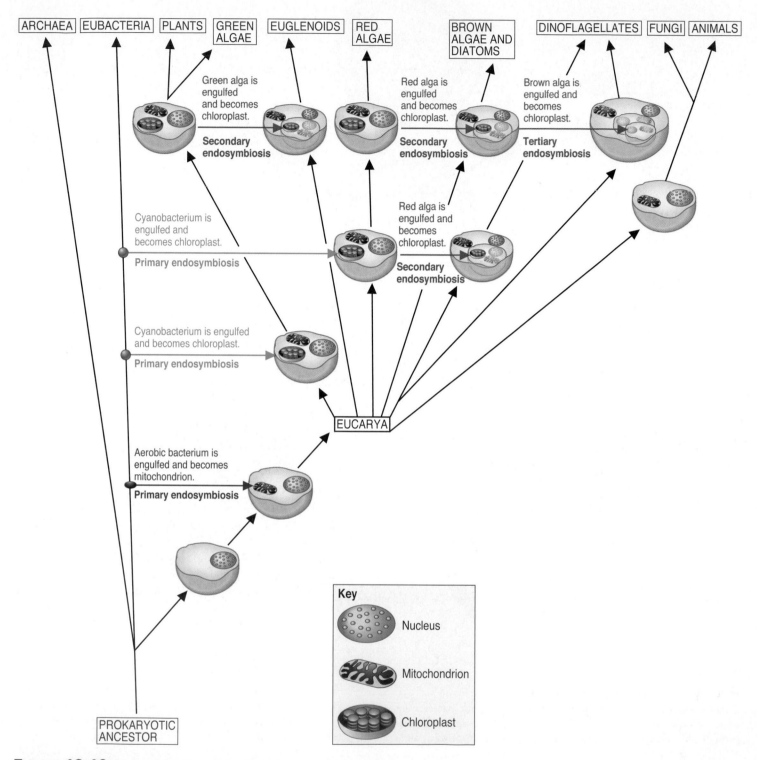

FIGURE 19.10 Endosymbiosis and the Evolution of Eucarya
Primary endosymbiosis involves the development of a symbiotic relationship between a prokaryotic cell and another cell. Mitochondria and some chloroplasts are thought to have developed in this way. As scientists looked closely at the chloroplasts in different kinds of photosynthetic organisms, it appears that some are the result of endosymbiosis between eukaryotic organisms. For example, it appears that ancestors of *Euglena* gained their chloroplasts by engulfing the cells of a green alga. Once engulfed, the nucleus and mitochondria of the green alga became nonfunctional, but the chloroplast remained as the functioning chloroplast in the *Euglena* cell. Because the endosymbiotic relationship developed between two eukaryotic cells, it is called secondary endosymbiosis. The illustration shows two other examples of secondary endosymbiosis and one of tertiary endosymbiosis.

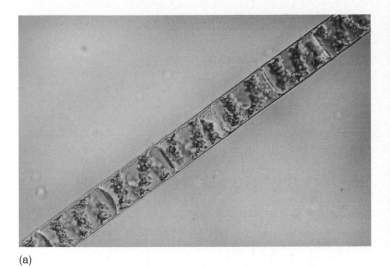

(a)

(b)

FIGURE 19.11 Simple and Complex Algae

(a) Spirogyra is a simple alga, which forms long, hairlike strands, in which all the cells are identical. *(b) Fucus* is a much more complex alga, with specialized structures, such as gas-filled bladders. It is much larger than *Spirogyra*.

The endosymbiotic theory is further supported by DNA studies. It is becoming clear that the genes within any one organism may have arisen from several sources and genetic information has moved about from one organism to another. We know that viruses carry genes from one organism to another, and parasitic organisms use their DNA to manipulate their hosts. Another possible source of the various kinds of genes in eukaryotic organisms is the combining of eukaryotic cells through endosymbiosis. Figure 19.10 illustrates several other examples of endosymbiosis thought to have been involved in the evolution of living things.

The Development of Multicellular Organisms

Following the development of eukaryotic cells, there was a long period in which single-celled organisms (both prokaryotic and eukaryotic) were the only ones on Earth. Eventually, organisms developed that consisted of collections of cells. At first, these collections may have been very similar to modern algae in which there was very little specialization of cells (figure 19.11a). However, eventually some cells within organisms became specialized for specific tasks, and the many kinds of multicellular algae, fungi, plants, and animals developed (figure 19.11b).

19.6 The Geologic Time Line and the Evolution of Life

A geologic time line shows a chronological history of living organisms based on the fossil record. The largest geologic time units are called *eons*. From earliest to most recent, the geologic eons are the Hadean, Archaean, Proterozoic, and

Phanerozoic. The Phanerozoic eon is divided into the eras Paleozoic, Mesozoic, and Cenozoic. Each of these *eras* is subdivided into smaller time units called *periods*. For example, Jurassic is the period of the Mesozoic era that began 208 million years ago (figure 19.12).

Recent evidence suggests that the first living thing most likely came into existence approximately 3.8 billion years ago during the Archaean eon. Prokaryotic cell types (Domain Eubacteria) appear about 3.5 billion years ago in the fossil record. One of the common fossils is of stromatolites, photosynthetic Eubacteria that grew in layers and formed columns in shallow oceans. Modern-day stromatolites still exist in western Australia (figure 19.13). Also about this time, the Archaea diverged from the Eubacteria. For approximately 2 billion years, the only organisms on Earth were Eubacteria and Archaea. For most of its existence the Earth was dominated by prokaryotic organisms. The photosynthetic eubacterial cyanobacteria are thought to have been responsible for the production of the molecular oxygen (O_2) that began to accumulate in the atmosphere, making conditions favorable for the evolution of other types of cells. The first members of the Domain Eucarya, the eukaryotic organisms, appeared approximately 1.8 billion years ago.

There is fossil evidence of multicellular algae at about 1 billion years ago and multicellular animals at about 0.6 billion years ago. During the Cambrian period of the Paleozoic era, an evolutionary explosion of multicellular animals occurred. Examples of most of the present-day phyla of marine invertebrate animals (e.g., echinoderms, arthropods, mollusks) are found in the fossil record at this time.

Several other "evolutionary explosions," or *adaptive radiations,* followed. Note on the geological time line that its divisions are associated with the major events in the evolutionary history of living things. For example, the Paleozoic era was

Eon	Era	Period	Epoch	Million Years Ago	Major Events and Characteristics
Phanerozoic	Cenozoic	Quaternary	Holocene (Recent)	0.01 to present	Dominance of modern humans Much of Earth modified by humans
			Pleistocene	1.8–0.01	Extinction of many large mammals Most recent period of glaciation
		Tertiary	Pliocene	5–1.8	Cool, dry period Grasslands and grazers widespread Origin of ancestors of humans
			Miocene	23–5	Warm, moist period Grasslands and grazers common
			Oligocene	38–23	Cool period—tropics diminish Woodlands and grasslands expand
			Eocene	54–38	All modern forms of flowering plants present All major groups of mammals present
			Paleocene	65–54	Warm period Many new kinds of birds and mammals
	Mesozoic	Cretaceous		144–65	Meteorite impact causes mass extinction Dinosaurs dominant Many new kinds of flowering plants and insects
		Jurassic		208–144	Giant continent begins to split up Dinosaurs dominant First flowering plants and birds
		Triassic		245–208	Giant continent—warm and dry Explosion in reptile and cone-bearing plant diversity First dinosaurs and first mammals
	Paleozoic	Permian		286–245	90% of species go extinct at end of Permian New giant continent forms Modern levels of oxygen in atmosphere Gymnosperms, insects, amphibians, and reptiles dominant
		Carboniferous		360–286	Gymnosperms and reptiles present by end Vast swamps of primitive plants—formed coal Amphibians and insects common
		Devonian		408–360	Glaciation the probable cause of extinction of many warm-water marine organisms Abundant fish, insects, forests, coral reefs
		Silurian		436–408	Melting of glaciers caused rise in sea level Numerous coral reefs First land animals (arthropods), jawed fish, and vascular plants
		Ordovician		505–436	Sea level drops causing major extinction of marine animals at end of Ordovician Primitive plants present—jawless fish common
		Cambrian		540–505	Major extinction of organisms at end of Cambrian All major phyla of animals present Continent splits into several parts and drift apart
Proterozoic	This period of time is also known as the Precambrian			2,500–540	First multicellular organisms about 1 billion years ago First eukaryotes about 1.8 billion years ago Increasing oxygen in atmosphere Single large continent about 1.1 billion years ago
Archaean				3,800–2,500	Continents form—no oxygen in atmosphere Fossil prokaryotes—about 3.5 billion years ago
Hadean				4,500–3,800	Crust of Earth in process of solidifying Origin of the Earth

FIGURE 19.12 A Geologic Time Chart

This chart shows the various geologic time designations, their time periods, and the major events and characteristics of each time period.

FIGURE 19.13 Stromatolites in Australia
This photo of stromatolites was taken at Hamelin Pool, Western Australia, a marine nature preserve. The dome-shaped structures are composed of cyanobacteria and materials they secrete, they grow to 60 centimeters tall. Similar structures are known from the fossil record. Taking samples from fossil stromatolites and cutting them into thin slices produces microscopic images that show some of the world's oldest cells.

dominated by nonvascular and primitive vascular plants; the Mesozoic era by cone-bearing evergreens; and the Cenozoic era by the flowering plants with which we are most familiar. Likewise, many periods of time are associated with specific animal groups. Among vertebrates, the Devonian period is considered the Age of Fishes and the Carboniferous period the Age of Amphibians. The Mesozoic era is considered the Age of Reptiles, and the Cenozoic era is considered the Age of Mammals. In each instance, the dominance of a particular animal group resulted from adaptive radiation events. It is also important to note that the ends of many geologic time designations are associated with major extinctions. These extinctions appear to be related to major changes in climate or sea level or asteroid impacts. Following each major extinction, there was an adaptive radiation, which led to a new dominant group.

For about 90% of Earth's history, life was confined to the sea. Primitive land plants probably arose about 430 million years ago and the first land animals (millipedes) at about 420 million years ago. In order to live on land, organisms needed several characteristics: an ability to breathe air, a way to prevent dehydration, some sort of skeleton for support, and an ability to reproduce out of water. The first major group of land animals to become abundant was the insects.

Among the vertebrates, the first land animals most likely evolved from a lobe-finned fish of the Devonian period. They possessed two important adaptations: lungs and paired, lobed fins, which allowed the organisms to pull themselves onto land and travel to new water holes during times of drought. Selective pressures resulted in fins evolving into legs, and the first amphibian came into being. The first vertebrates to spend part of their lives on land found a variety of unexploited niches, resulting in the rapid evolution of new amphibian species and their dominance during the Carboniferous period. During this time, abundant swamps allowed amphibians to return to water to reproduce. However, to exploit the land environment totally, a modification of the method of reproduction was required.

During this time, mutations continued to occur, and valuable modifications were passed on to future generations, eventually leading to the development of reptiles. One change allowed males to deposit sperm directly within females. Because the sperm could directly enter females and remain in a moist interior, it was no longer necessary for the animals to return to the water to mate, as the amphibians still had to do. However, the developing young still required a moist environment for early growth. A second modification, the amniotic egg, solved this problem. An amniotic egg, such as a chicken egg, protected the developing young from injury and dehydration while allowing for the exchange of gases with the external environment. A third adaptation, the development of protective scales and relatively impermeable skin, protected reptiles from dehydration. With these adaptations, the reptiles were able to outcompete the amphibians in most terrestrial environments. The amphibians that did survive were the ancestors of present day frogs, toads, and salamanders. With extensive adaptive radiation, the reptiles took to the land, sea, and air. A particularly successful group of reptiles was the dinosaurs, which were the dominant terrestrial vertebrates for more than 100 million years.

As reptiles diversified, some developed characteristics common to other classes of vertebrates found today, such as warm-bloodedness, feathers, and hair. Warm-blooded reptiles with scales modified as feathers for insulation eventually evolved into organisms capable of flight. Through natural selection, reptilian characteristics were slowly eliminated, and the characteristics typical of modern birds (multiple adaptations to flight, keen senses, and complex behavioral instincts) developed. *Archaeopteryx,* the first bird, had characteristics typical of both birds and reptiles.

Also evolving from reptiles were the mammals. The first reptiles with mammalian characteristics appeared in the Permian period, although the first true mammals did not appear until the Triassic period. These organisms remained relatively small in number and size until after the mass extinction of the reptiles. Extinctions opened many niches and allowed for the subsequent adaptive radiation of mammals. As with the other adaptive radiations, the mammals possessed unique characteristics that made them better adapted to the changing environment; the characteristics included insulating hair, constant body temperature, and internal development of the young. Figure 19.14 summarizes some of the major events in the evolutionary history of the Earth.

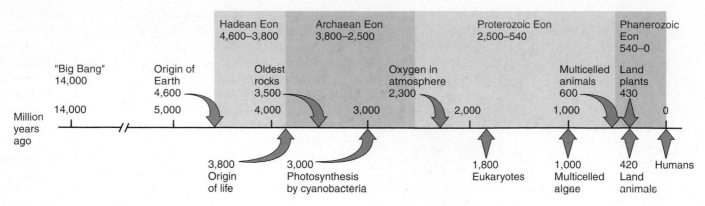

FIGURE 19.14 An Evolutionary Time Line
This chart displays how science sees the order of major, probable events in the origin and evolution of life from the "Big Bang" to the present day.

Summary

Current theories on the origin of life speculate that either the primitive Earth's environment led to the spontaneous organization of organic chemicals into primitive cells or primitive forms of life arrived on Earth from space. Regardless of how the first living things came to be on Earth, these basic units of life were probably similar to present-day prokaryotes. The primitive cells could have changed through time as a result of mutation and in response to a changing environment. The recognition that many prokaryotic organisms have characteristics that clearly differentiate them from the rest of the bacteria has led to the development of the concept that there are three major domains of life: the Eubacteria, the Archaea, and the Eucarya. The Eubacteria and Archaea are similar in structure, but the Archaea have metabolic processes that are distinctly different from those of the Eubacteria. Some people consider the Archaea, many of which can live in very extreme environments, good candidates for the first organisms to inhabit Earth. The origin of the Eucarya is less contentious. Similarities between cyanobacteria and chloroplasts and between aerobic bacteria and mitochondria suggest that eukaryotic cells may actually be a combination of ancient cell ancestors that lived together symbiotically. The likelihood of these occurrences is supported by experiments that have simulated primitive Earth environments and investigations of the cellular structure of simple organisms. Despite volumes of information, the question of how life began remains unanswered.

Key Terms

Use the interactive flash cards on the **Concepts in Biology,** *12/e website to help you learn the meaning of these terms.*

biogenesis 402
endosymbiotic theory 412
oxidizing atmosphere 411

panspermia 403
reducing atmosphere 406
spontaneous generation 402

Basic Review

1. The reproduction of an apple tree by seeds is an example of
   a. spontaneous generation.
   b. biogenesis.
   c. endosymbiosis.
   d. None of the above is correct.
2. The first organisms on Earth would have carried on aerobic respiration. (T/F)
3. Endosymbiosis involves one cell invading and living inside another cell. (T/F)

4. Which of the following suggests that organic molecules may have formed on the Earth from inorganic molecules?

   a. Experiments that simulated the early Earth's atmosphere have produced organic molecules.

   b. Organic molecules have been detected in interstellar gases in space.

   c. Meteorites contain organic molecules.

   d. All of the above are correct.

5. Oxygen in today's Earth atmosphere is the result of the process of _____.

6. The oldest fossils of living things are about _____ years old.

   a. 3.5 million

   b. 3.5 billion

   c. 4.5 billion

   d. 4.5 million

7. The first genetic material was probably

   a. protein.

   b. DNA.

   c. RNA.

   d. enzymes.

8. An organism that requires organic molecules for food is known as a(n) _____.

9. A prokaryotic organism that is a chemoautotroph and lives in a very hot environment is probably in the Domain

   a. Archaea.

   b. Eubacteria.

   c. Eucarya.

   d. Any of these are correct.

10. Which one of the following evolved before all of the others?

    a. terrestrial plants and animals

    b. prokaryotic cells

    c. eukaryotic cells

    d. multicellular organisms

Answers

1. b   2. F   3. T   4. d   5. photosynthesis   6. b   7. c
8. heterotroph   9. a   10. b

Concept Review

Answers to the Concept Review questions can be found on the Concepts in Biology, 12/e website.

19.1 Early Thoughts About the Origin of Life

1. What is meant by the term *spontaneous generation*?
2. What is meant by the term *biogenesis*?
3. Describe the contribution of each of the following scientists to the biogenesis/spontaneous generation debate: Spallanzani, Needham, Pasteur, and Redi.

19.2 Current Thinking About the Origin of Life

4. Provide two kinds of evidence that support the idea that life could have originated on Earth.
5. Provide two kinds of evidence that support the idea that life could have arrived on Earth from space.

19.3 The "Big Bang" and the Origin of the Earth

6. How did the atmosphere, the temperature, and the surface of the newly formed Earth differ from what exists today?
7. What is the approximate age of the Earth?

19.4 The Chemical Evolution of Life on Earth

8. The current theory of the origin of life as a result of the nonbiological manufacture of organic molecules depends on scientific knowledge about Earth's history. Why is this so?
9. List two kinds of evidence that suggest that RNA was the first genetic material.
10. Describe two models that suggest how collections of organic molecules could have been segregated from other molecules.

19.5 Major Evolutionary Changes in Early Cellular Life

11. Why must the first organism on Earth have been anaerobic?
12. What evidence supports the theory that eukaryotic cells arose from the development of an endosymbiotic relationship between primitive prokaryotic cells?
13. Can spontaneous generation occur today? Explain.
14. List two important effects caused by the increase of oxygen in Earth's atmosphere.
15. In what sequence did the following things happen: living cell, oxidizing atmosphere, respiration, photosynthesis, reducing atmosphere, first organic molecule?
16. Why do scientists believe life originated in the seas?

19.6 The Geologic Time Line and the Evolution of Life

17. For each of the following pairs of terms, select the one that is the earliest in geologic time.
    a. eukaryote—prokaryote
    b. marine—terrestrial
    c. vertebrate—invertebrate
    d. flowering plant—cone-bearing plant
    e. aerobic respiration—photosynthesis
    f. plants—animals

Thinking Critically

*For guidelines to answer this question, visit the **Concepts in Biology**, 12/e website.*

Imagine that there is life on another planet, in our galaxy (Planet X). Using the following data concerning the nature of this life, determine what additional information is necessary and how you would verify the additional data needed to develop a theory of the origin of life on Planet X.

1. The age of the planet is 5 billion years.
2. Water is present in the atmosphere.
3. The planet is farther from its Sun than our Earth is from our Sun.
4. The molecules of various gases in the atmosphere are constantly being removed.
5. Chemical reactions on this planet occur at approximately half the rate at which they occur on Earth.

20

American Robin

European Blackbird

European Robin

The Classification and Evolution of Organisms

Sing a song of sixpence a pocket full of rye,
Four and twenty blackbirds baked in a pie.
When the pie was opened the birds began to sing,
Oh, wasn't that a dainty dish to set before the king?

When the British colonists came to America, they discovered many plants and animals that were new to them. If they saw a resemblance between an American animal and one they knew from Britain, they applied the same name to the two animals. The blackbird described in the British nursery rhyme is very different from the blackbirds found in the Americas. In fact, the European blackbird, *Turdus merula,* is more closely related to the American robin, *Turdus migratorius,* than to what Americans call blackbirds. The European robin, *Erithacus rubecula,* is a completely different bird from the American robin, although it does have a reddish breast.

CHAPTER OUTLINE

20.1 The Classification of Organisms

In order to talk about items in our world, we must have names for them. As new items come into being or are discovered, we devise new words to describe them. For example, the words *laptop, palm pilot,* and *Internet* describe technology that did not exist 30 years ago. Similarly, in the biological world, people have identified newly discovered organisms by giving them names.

The Problem with Common Names

The common names people use vary from culture to culture. For example, *dog* in English is *chien* in French, *perro* in Spanish, and *cane* in Italian. Often different names are used in different regions within a country to identify the same organism. For example, the common garter snake is called a garden snake or gardner snake, depending on where you live (figure 20.1). Actually, there are several species of garter snakes that have been identified as distinct from one another. Thus, common names can be confusing, so scientists sought a more acceptable way to name organisms, one that all scientists would use that would eliminate confusion.

The naming of organisms is a technical process, but it is extremely important. When biologists are describing their research, common names, such as robin or blackbird, or garter snake are not good enough. They must be able to identify the organisms involved accurately, so that everyone who reads the report, wherever they live in the world, knows what organism is being discussed. The scientific identification of organisms involves two different but related activities. One, *taxonomy,* is the naming of organisms; the other, *phylogeny,* involves showing how organisms are related evolutionarily.

Taxonomy

Taxonomy is the science of naming organisms and grouping them into logical categories. The root of the word *taxonomy* is the Greek word *taxis,* which means arrangement.

During the Middle Ages, Latin was widely used as the scientific language. As new species were identified, they were given Latin names, often using as many as 15 words to describe a single organism. Although using Latin meant that most biologists, regardless of their native language, could understand a species name, it did not completely do away with duplicate names. Because many of the organisms were found over wide geographic areas and communication was slow, there could still be two or more Latin names for a species. To make the situation even more confusing, ordinary people still used common local names.

The Binomial System of Nomenclature

The modern system of classification began in 1758, when Carolus Linnaeus (1707–1778), a Swedish doctor and botanist, published his tenth edition of *Systema Naturae* (figure 20.2). (Linnaeus's original name was Carl von Linné, which he "latinized" to Carolus Linnaeus.) In the previous editions of his book, Linnaeus had used the polynomial Latin system of classification, which required many words to identify a species. However, in the tenth edition, he introduced the *binomial system of nomenclature.* The **binomial system of nomenclature,** uses only two Latin names—the *genus* name

FIGURE 20.1 Common Garter Snake
Depending on where you live, you may call this organism a garter snake, a garden snake, or a gardner snake. These common names can lead to confusion.

FIGURE 20.2 Carolus Linnaeus (1707–1778)
Carolus Linnaeus, a Swedish doctor and botanist, originated the modern system of taxonomy known as binomial nomenclature.

and the *specific epithet* (*epithet* = descriptive word) for each species of organism. Recall that a species is a population of organisms capable of interbreeding and producing fertile offspring. Individual organisms are members of a species. A **genus** (plural, *genera*) is a group of closely related organisms. The **specific epithet** is a word added to the genus name to identify which one of several species within the genus is being referred to.

This is similar to the naming system we use with people. When we look in the phone book, we look for a last name (surname), the correct general category. Then, we look for a first name (given name) to identify the specific individual we wish to call. The unique name given to an organism is its species name, or scientific name. In order to clearly distinguish the scientific name from other words, binomial names are either *italicized* or underlined. The first letter of the genus name is capitalized. The specific epithet is always written in lowercase. For example, *Thamnophis sirtalis* is the binomial name for the common garter snake. species, p. 240

When biologists adopted Linnaeus's binomial method, they eliminated the confusion of using common local names. Since the adoption of Linnaeus's system, international rules have been established to assure that an orderly system is maintained. The three primary sets of rules are the International Rules for Botanical Nomenclature, the International Rules for Zoological Nomenclature, and the International Bacteriological Code of Nomenclature. Although approximately 1.5 million species have been named, no one knows how many species of organisms live on Earth, but most biologists estimate that several million are yet to be identified.

The Organization of Species into Logical Groups

In addition to assigning a specific name to each species, Linnaeus recognized a need for placing organisms into groups. He originally divided all organisms into two broad groups, which he called the plant and animal kingdoms, and subdivided each kingdom into smaller units. Since Linneaus's initial attempts to place all organisms into categories, there have been many changes. One of the most fundamental is the recent recognition that there are three major categories of organisms, called *domains*.

Recall that a domain is the largest category into which organisms are classified, and there are three domains: Eubacteria, Archaea, and Eucarya (figure 20.3). Organisms are separated into these three domains based on the specific structural and biochemical features of their cells. The Eubacteria and Archaea are prokaryotic and the Eucarya are eukaryotic.

A **kingdom** is a subdivision of a domain. There are several kingdoms within the Eubacteria and Archaea based primarily on differences in the metabolism and genetic composition of the organisms. Within the Domain Eucarya, there are four kingdoms: Plantae, Animalia, Fungi, and Protista (protozoa and algae) (figure 20.4).

A **phylum** is a subdivision of a kingdom. However, microbiologists and botanists often use the term *division* rather than *phylum*. All kingdoms have more than one phylum. For example, the kingdom Plantae contains several phyla: flowering plants, conifer trees, mosses, ferns, and several other subgroups. Organisms are placed in phyla based on careful investigation of the specific nature of their structure, metabolism, and biochemistry. An attempt is made to identify

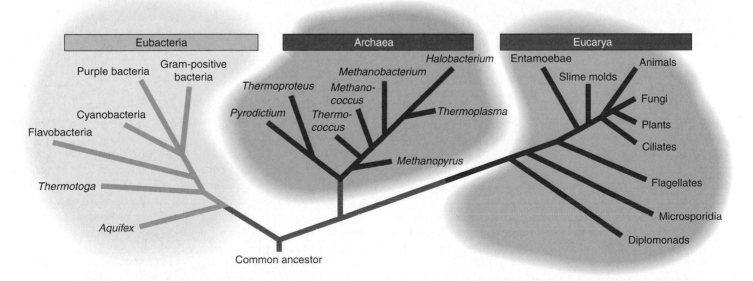

FIGURE 20.3 **The Three Domains of Life**
The three domains of living things are related to one another evolutionarily. The Domain Eubacteria is the oldest group. The Domains Archaea and Eucarya are derived from the Eubacteria.

(a)

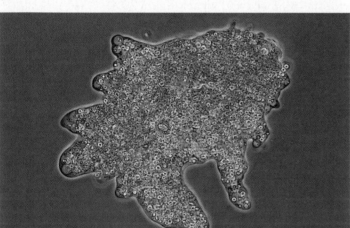

(b)

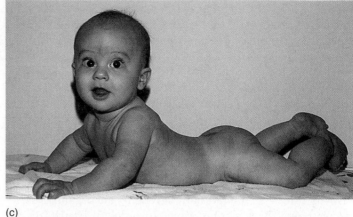

(c)

(d)

FIGURE 20.4 **Representatives of the Domain Eucarya**
There are four kingdoms in the Domain Eucarya, represented by the following organisms: *(a) Morchella esculenta*, kingdom Fungi; *(b) Amoeba proteus*, kingdom Protista; *(c) Homo sapiens*, kingdom Animalia; and *(d) Cypripedium calceolus*, kingdom Plantae.

natural groups, rather than artificial or haphazard arrangements. For example, although nearly all plants are green and carry on photosynthesis, only flowering plants have flowers and produce seeds; conifers lack flowers but have seeds in cones; ferns lack flowers, cones, and seeds; and mosses lack tissues for transporting water.

A **class** is a subdivision within a phylum. For example, within the phylum Chordata within the kingdom Animalia, there are seven classes: mammals, birds, reptiles, amphibians, and three classes of fishes. An **order** is a category within a class. The order Carnivora is an order of meat-eating animals within the class Mammalia. There are other orders of mammals, including horses and their relatives, cattle and their rel-

atives, rodents, rabbits, bats, seals, whales, humans, and many others.

A **family** is a subdivision of an order that consists of a group of closely related genera, which in turn are composed of groups of closely related species. The cat family, Felidae, is a subgroup of the order Carnivora; it includes many species in several genera, including the Canada lynx and bobcat (genus *Lynx*); the cougar (genus *Puma*); the leopard, tiger, jaguar, and lion (genus *Panthera*); the house cat (genus *Felis*); and several other genera. Thus, in the present-day science of taxonomy, each organism that has been classified has a unique binomial name. In turn, it is assigned to larger groupings that are thought to have a common evolutionary history.

TABLE 20.1

Classification of Humans

Taxonomic Category	Human Classification	Characteristics	Other Representatives
Domain	Eucarya	Cells containing a nucleus and many other kinds of organelles	Plants, animals, fungi, protozoa, algae
Kingdom	Animalia	Eukaryotic heterotrophs that are usually motile and have specialized tissues	Sponges, jellyfish, worms, clams, insects, snakes, cats
Phylum	Chordata	Animals with a stiffening rod down their back	Fish, amphibians, reptiles, birds, mammals
Class	Mammalia	Animals with hair and mammary glands	Kangaroos, mice, whales, skunks, monkeys
Order	Primates	Mammals with relatively a large brain and opposable thumbs	Monkeys, gorillas, chimpanzees, baboons
Family	Hominidae	Primates that lack a tail and have upright posture	Humans and extinct relatives (*Australopithecus, Paranthropus, Homo*)
Genus	*Homo*	Hominids with large brains	Humans are the only surviving member of the genus. Other species existed in the past (*Homo erectus, Homo neanderthalensis*)
Species	*Homo sapiens*	Humans	

Table 20.1 classifies humans to show how the various categories are used.

Phylogeny

Phylogeny is the science that explores the evolutionary relationships among organisms, seeking to reconstruct evolutionary history. Taxonomists and phylogenists work together, so that the products of their work are compatible. A taxonomic ranking should reflect the phylogenetic (evolutionary) relationships among the organisms being classified. New organisms and new information about organisms are discovered constantly. Therefore, taxonomic and phylogenetic relationships are constantly being revised. During this revision process, scientists often have differences of opinion about the significance of new information.

The Evidence Used to Establish Phylogenetic Relationships

Phylogenists use several lines of evidence to develop evolutionary histories: fossils, comparative anatomy, life cycle information, and biochemical and molecular evidence.

Fossils are physical evidence of previously existing life, and they are found in several forms. Some fossils are preserved whole and relatively undamaged. For example, mammoths and humans have been found frozen in glaciers, and bacteria and insects have been preserved after becoming embedded in plant resins. Other fossils are only parts of once living organisms. The outlines or shapes of extinct plant leaves are often found in coal deposits, and individual animal bones that have been chemically altered over time are often dug up. Animal tracks have also been discovered in the dried mud of ancient riverbeds (figure 20.5).

It is important to understand that some organisms are more easily fossilized than others. Those that have hard parts, such as cell walls, skeletons, and shells, are more likely to be preserved than are tiny, soft-bodied organisms. Aquatic organisms are much more likely to be buried in the sediments at the bottom of the oceans or lakes than are their terrestrial counterparts. Later, when these sediments are pushed up by geologic forces, aquatic fossils are found in layers of sediments on dry land.

Evidence obtained from the discovery and study of fossils allows biologists to place organisms in a time sequence. This can be accomplished by comparing one type of fossil with another. As geologic time passes and new layers of sediment are laid down, the older organisms should be in deeper layers, assuming that the sequence of layers has not been disturbed (figure 20.6). In addition, it is possible to age-date certain kinds of rocks by comparing the amounts of certain radioactive isotopes they contain. Older rocks have less of these specific radioactive isotopes than do younger rocks. Fossils associated with rocks of a known age are usually of a similar age.

It is also possible to compare subtle changes in particular kinds of fossils over time. For example, in studies of a certain kind of fossil plant, the size of the leaf changed extensively through long geologic periods. If one only looked at the two extremes, they would be classified into different categories. However, because there are fossil links that show intermediate stages between the extremes, scientists conclude that the younger plant is a descendant of the older.

(a)　　(b)

(c)

FIGURE 20.5 Fossil Evidence

A fossil is any evidence of previously existing life. Fossils can be the intact, preserved remains of organisms, as in *(a)* the remains of an ancient fly preserved in amber or the preserved parts of an organism, as in *(b)* the fossilized skeleton and body outline of a bony fish. It is even possible to have evidence of previously existing living things that are not the remains of organisms, as in *(c)* a dinosaur footprint.

Comparative anatomy studies of fossils or currently living organisms can be very useful in developing a phylogeny. Because the structures of an organism are determined by its genes, organisms having similar structures are thought to be related. For example, plants can be divided into several categories: All plants that have flowers are thought to be more closely related to one another than they are to plants that do not have flowers, such as ferns. In the animal kingdom, all organisms that have hair and mammary glands are grouped together, and all animals in the bird category have feathers, wings, and beaks.

FIGURE 20.6 Rock Layers and the Age of Fossils

Because new layers of sedimentary rock are formed on top of older layers of sedimentary rock, it is possible to determine the relative ages of fossils found in various layers. The layers of rock shown here represent millions of years of formation. The fossils of the lower layers are millions of years older than the fossils in the upper layers.

Life cycle information is another line of evidence useful to phylogenists and taxonomists. Many organisms have complex life cycles, which include many completely different stages. After fertilization, some organisms grow into free-living developmental stages that do not resemble the adults of their species. These are called *larvae* (singular, *larva*). Larval stages often provide clues to the relatedness of organisms. For example, adult barnacles live attached to rocks and other solid marine objects and look like small, hard cones. Their outward appearance does not suggest that they are related to shrimp; however, the larval stages of barnacles and shrimp are very similar. Detailed anatomical studies of barnacles confirm that they share many structures with shrimp; their outward appearance tends to be misleading (figure 20.7).

Both birds and reptiles lay eggs with shells. However, reptiles lack feathers and have scales covering their bodies. The fact that these two groups share this fundamental eggshell characteristic implies that they are more closely related to each other than they are to other groups, but they can be divided into two groups based on their anatomical differences.

This kind of evidence also applies to the plant kingdom. Many kinds of plants, such as peas, peanuts, and lima beans, produce large, two-parted seeds in pods. Even though peas grow as vines, lima beans grow as bushes, and peanuts have their seeds underground, all these plants are considered to be related.

Biochemical and molecular studies are recent additions to the toolbox of phylogenists. Like all aspects of biology, the science of phylogeny is constantly changing as new tech-

(a) Barnacle

(b) Shrimp

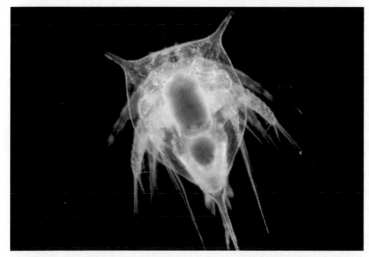

(c) Nauplius larva of barnacle

FIGURE 20.7 Developmental Stages and Phylogeny

The adult barnacle *(a)* and shrimp *(b)* are very different from each other, but their early larval stages *(c)* look very much alike.

niques develop. Recent advances in DNA analysis are being used to determine genetic similarities among species. In the field of ornithology, the study of birds, there are those who believe that storks and flamingos are closely related; others believe that flamingos are more closely related to geese. An analysis of the DNA points to a higher degree of affinity between flamingos and storks than between flamingos and geese. This is interpreted to mean that the closest relationship is between flamingos and storks.

There are five kinds of chlorophyll found in algae and plants: chlorophyll *a, b, c, d,* and *e*. Most photosynthetic organisms contain a combination of two of these chlorophyll molecules. Members of the kingdom Plantae have chlorophyll *a* and *b*. The large seaweeds, such as kelp, superficially resemble terrestrial plants, such as trees and shrubs. However, a comparison of their chlorophylls shows that kelp has chlorophyll *a* and *d*. Another group of algae, called the *green algae*, has chlorophyll *a* and *b*. Along with other anatomical and developmental evidence, this biochemical information has helped establish an evolutionary link between the green algae and plants. All the kinds of evidence (fossils, comparative anatomy, life cycle information, and biochemical evidence) have been used to develop phylogenetic relationships and taxonomic categories.

A Current Phylogenetic Tree

Given all the sources of evidence, biologists have developed a picture of how they think all organisms are related (figure 20.8). The three Domains—Eubacteria, Archaea, and Eucarya—diverged early in the history of life. Subsequently, many new kinds of organisms have evolved. It is important to remember that this diagram is a work in progress. As new information is discovered, there will be changes in the way biologists think organisms are related. Biologists have also developed new techniques that help in determining phylogenies. One such technique is cladistics (How Science Works 20.1).

20.2 A Brief Survey of the Domains of Life

Members of the Domains Eubacteria and Archaea are commonly known as *bacteria*. They are all tiny, prokaryotic cells that are difficult to distinguish from one another. Because of this, in the past it was assumed that the members of these groups were closely related. However, they show a wide variety of metabolic abilities. Evidence gained from studying DNA and RNA nucleotide sequences and a comparison of the amino acid sequences of proteins indicate that there are major differences between the Eubacteria and Archaea. The Eubacteria evolved first and then gave rise to the Archaea; finally, the Eucarya evolved. domains, p. 411

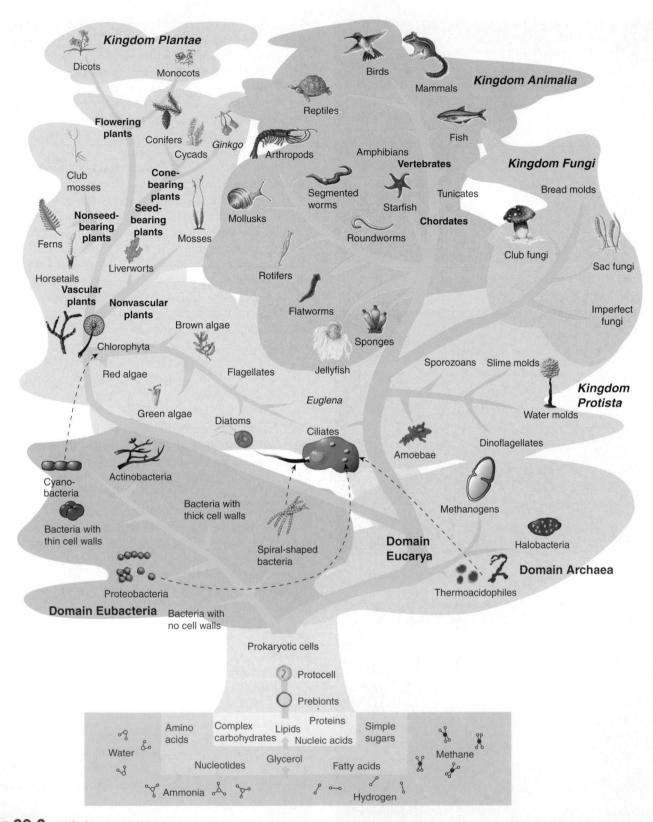

FIGURE 20.8 A Phylogenetic Tree

The first living cells probably came into being as a result of chemical evolution. The oldest fossils appear to be similar to the members of the Domain Eubacteria. It appears that the Archaea are derived from the Eubacteria. Some prokaryotic cells probably gave rise to eukaryotic cells through the process of endosymbiosis. The organisms formed from these early eukaryotic cells were probably similar to single-celled members of the kingdom Protista. Members of this kingdom are thought to have given rise to the kingdoms Animalia, Plantae, and Fungi. Thus, all present-day organisms evolved from simple prokaryotic cells.

HOW SCIENCE WORKS 20.1

Cladistics: A Tool for Taxonomy and Phylogeny

Classification, or taxonomy, is one part of the much larger field of phylogenetic systematics. Classification involves placing organisms into logical categories and assigning names to those categories. Phylogeny, or *systematics,* is an effort to understand the evolutionary relationships of living things in order to interpret the way in which life has diversified and changed over billions of years of biological history. Phylogeny attempts to understand how organisms have changed over time. *Cladistics* (*klados* = branch) is a method biologists use to evaluate the degree of relatedness among organisms within a group, based on how similar they are genetically. The basic assumptions behind cladistics are that

1. Groups of organisms are related by descent from a common ancestor.
2. The relationships among groups can be represented by a branching pattern, with new evolutionary groups arising from a common ancestor.
3. Changes in characteristics occur in organisms over time.

Several steps are involved in applying cladistics to a particular group of organisms. First, you must select characteristics that vary and collect information on the characteristics displayed by the group of organisms you are studying. The second step is to determine which expression of a characteristic is ancestral and which is more recently derived. Usually, this involves comparing the group in which you are interested with an *outgroup* that is related to, but not a part of, the group you are studying. The characteristics of the outgroup are then considered to be ancestral. Finally, you must compare the characteristics displayed by the group you are studying and construct a diagram known as a *cladogram.* For example, if you were interested in studying how various kinds of terrestrial vertebrates are related, you could look at the following characteristics:

In this example, the shark is the outgroup, and the ancestral conditions are lungs absent, skin not dry, cold-blooded, and hair absent. Using this information, you could construct the following cladogram.

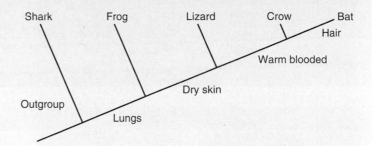

All of the organisms, except sharks, have lungs. Lizards, crows, and bats have dry skin, as well as lungs and so on. Crows and bats share the following characteristics; they have lungs, they have dry skin, and they are warm-blooded. Because they share more characteristics with each other than with the other groups, they are considered to be more closely related.

The choice of characteristics with which to make comparisons is important. Two organisms may share many characteristics but not be members of the same evolutionary group if the characteristics being compared are not from the same genetic background. For example, if you were to compare butterflies, birds, and squirrels using the presence or absence of wings and the presence or absence of bright colors as your characteristics, you would conclude that butterflies and birds are more closely related than birds and squirrels. However, this is not a valid comparison, because the wings of birds and butterflies are not of the same evolutionary origin.

Characteristic Organism	Lungs Present	Skin Dry	Warm- Blooded	Hair Present
Shark	0	0	0	0
Frog	+	0	0	0
Lizard	+	+	0	0
Crow	+	+	+	0
Bat	+	+	+	+

The Domain Eubacteria

The "true bacteria" (*eu* = true) are small, prokaryotic, single-celled organisms ranging in size from 1 to 10 micrometers (μm). Their cell walls contain a complex organic molecule known as peptidoglycan. Peptidoglycan is found only in the Eubacteria and is composed of two kinds of sugars linked together by amino acids. One of these sugars, muramic acid, is found only in the Eubacteria. The most common shapes of the cells are spheres, rods, and spirals.

Because they are prokaryotic, Eubacteria have no nucleus. Their genetic material consists of a single loop of

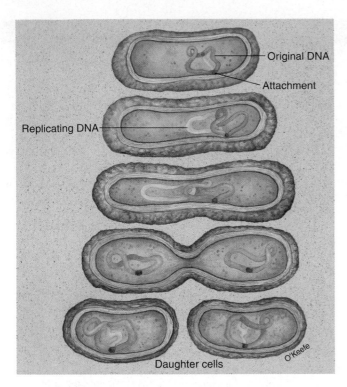

Original DNA
Attachment
Replicating DNA
Daughter cells
O'Keefe

FIGURE 20.9 **Asexual Reproduction in Eubacteria**
When Eubacteria reproduce, the loop of DNA replicates and the cell divides, with each cell having one of the two loops of DNA. This is a form of asexual reproduction.

DNA. When Eubacteria reproduce, the loop of DNA replicates to form two loops, and the cell divides into two, each having one copy of the DNA loop (figure 20.9). Since the two cells have the same DNA, this is a form of asexual reproduction.

Some cells move by secreting a slime that glides over their surface, causing them to move through the environment. Others move by means of a kind of flagellum. The structure of the flagellum is different from the flagellum found in eukaryotic organisms.

Because the early atmosphere is thought to have been a reducing atmosphere, the first Eubacteria were probably anaerobic organisms. Today, there are both anaerobic and aerobic Eubacteria. Many heterotrophic Eubacteria are **saprophytes,** organisms that obtain energy by the decomposition of dead organic material; others are parasites that obtain energy and nutrients from living hosts and cause disease; still others are mutualistic and have a mutually beneficial relationship with their host; finally, some are commensalistic and derive benefit from a host without harming it. Several kinds of Eubacteria are autotrophic. Many are called cyanobacteria because they contain a blue-green pigment, which allows them to capture sunlight and carry on photosynthesis. They can become extremely numerous in some polluted waters where nutrients are abundant. Others use inorganic chemical reactions for their energy sources and are called chemosynthetic.

The Domain Archaea

The term *archaea* comes from the Greek word *archaios,* meaning ancient. This is a little misleading, because the Eubacteria preceded the Archaea, and the Archaea are thought to have branched off from the Eubacteria between 2 and 3 billion years ago.

The Archaea are similar to the Eubacteria in that they both have a prokaryotic cell structure. However, the Archaea differ from the Eubacteria in several fundamental ways. The Archaea do not have peptidoglycan in their cell walls, and the structure of their DNA is different from that of the Eubacteria. Although the DNA of the Archaea is a loop, like that of the Eubacteria, the DNA of the Archaea appears to have a large proportion of genes that are different from either the Eubacteria or the Eucarya. Also, the cell membranes of the Archaea have a unique chemical structure, found in neither the Eubacteria nor the Eucarya. Members of the Archaea exist in many shapes, including rods, spheres, spirals, filaments, and flat plates.

Because they are found in many kinds of extreme environments, they have become known as *extremophiles*. Based on the particular extreme habitats they occupy and the kind of metabolism they display, the Archaea are divided into three groups:

1. *Methanogens* are anaerobic, methane-producing organisms. They can be found in sewage, swamps, and the intestinal tracts of ruminant animals, such as cows, sheep, and goats. They are even found in the intestines of humans.
2. *Halobacteria* (*halo* = salt) live in very salty environments, such as the Great Salt Lake (Utah), salt ponds, and brine solutions. Some contain a special kind of chlorophyll and are therefore capable of generating their ATP by photosynthesis.
3. The *thermophilic* Archaea live in environments that normally have very high temperatures and high concentrations of sulfur (e.g., hot sulfur springs and around deep-sea hydrothermal vents). Over 500 species of thermophiles have been identified at the openings of hydrothermal vents in the open oceans. One such thermophile, *Pyrolobus fumarii,* grows in a hot spring in Yellowstone National Park (figure 20.10). Its maximum growth temperature is 113°C, its optimum is 106°C, and its minimum is 90°C.

The Domain Eucarya

It is thought that eukaryotic cells evolved from prokaryotic cells through endosymbiosis. This hypothesis proposes that structures such as mitochondria, chloroplasts, and other membranous organelles originated from separate cells, which were ingested by larger cells. Once inside the larger cells, these structures and their functions became integrated with the host cells and ultimately became essential to their survival. This new type of cell was the forerunner of present-day eukaryotic cells.

FIGURE 20.10 Habitat for Thermophilic Archaea
The hot springs of Yellowstone National Park are the habitat for several kinds of Archaea, including *Pyrolobus fumarii.*

Eukaryotic cells are usually much larger than prokaryotic cells, typically having more than a thousand times the volume of prokaryotic cells. Their larger size was made possible by the presence of specialized membranous organelles, such as mitochondria, the endoplasmic reticulum, chloroplasts, and nuclei. endosymbiosis, p. 412

Kingdom Protista

The changes in cell structure that led to eukaryotic organisms probably gave rise to single-celled organisms similar to those currently grouped in the kingdom Protista. Most members of this kingdom are one-celled organisms, although there are some colonial forms.

There is a great deal of diversity among the approximately 60,000 known species of Protista. Many species live in freshwater; others are found in marine or terrestrial habitats, and some are parasitic, commensalistic, or mutualistic. All species can undergo mitosis, resulting in asexual reproduction. Some species can also undergo meiosis and reproduce sexually. Many contain chlorophyll in chloroplasts and are autotrophic; others require organic molecules as sources of energy and are heterotrophic. Both autotrophs and heterotrophs have mitochondria and respire aerobically.

Because members of this kingdom are so diverse with respect to form, metabolism, and reproductive methods, most biologists do not think that the Protista form a valid phylogenetic unit. However, it is still a convenient taxonomic grouping. By placing these organisms together in this group, it is possible to gain a useful perspective on how they relate to other kinds of organisms. After the origin of eukaryotic organisms, evolution proceeded along several pathways. Three major lines of evolution within the Protista can be seen. There are plantlike autotrophs (algae), animal-like heterotrophs (protozoa), and funguslike heterotrophs (slime molds). *Amoeba* and *Paramecium* are commonly encountered examples of protozoa. Many seaweeds and pond scums are collections of large numbers of algal cells. Slime molds are seen less frequently, because they live in and on the soil in moist habitats; they are most often encountered as slimy masses on decaying logs. Figure 20.11 shows some examples of this diverse group of organisms.

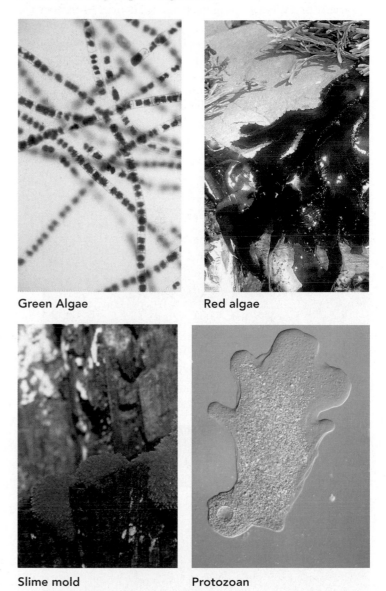

Green Algae

Red algae

Slime mold

Protozoan

FIGURE 20.11 A Diversity of Protista
The kingdom Protista includes a wide variety of organisms that are simple in structure. They are not a phylogenetic group.

Kingdom Fungi

Fungus is the common name for members of the kingdom Fungi. They probably arose from members of the kingdom Protista. Most fungi are nonmotile. They have a rigid, thin cell wall, which in most species is composed of chitin, a complex carbohydrate containing nitrogen. The members of the kingdom Fungi are nonphotosynthetic, eukaryotic organisms. Most of the approximately 70,000 species (mushrooms and molds) are multicellular, but a few, such as yeasts, are single-celled. In the multicellular fungi, the basic structural unit is a network of multicellular filaments.

Because all of these organisms are heterotrophs, they must obtain nutrients from organic sources. Most are saprophytes and secrete enzymes that digest large molecules into smaller units, which are absorbed as food for the organism. They are very important as decomposers in all ecosystems. They feed on a variety of nutrients ranging from dead organisms to such products as shoes, foodstuffs, and clothing. Most synthetic organic molecules are not attacked as readily by fungi; this is why plastic bags and foam cups are slow to decompose.

Some fungi are parasitic, whereas others are mutualistic. Many of the parasitic fungi are important plant pests. Some (e.g., chestnut blight, Dutch elm disease) attack and kill plants; others injure the fruit, leaves, roots, or stems. In crop plants, fungi are important pests that kill or weaken plants and reduce yields. The fungi that are human parasites are responsible for athlete's foot, vaginal yeast infections, valley fever, "ringworm," and other diseases. Mutualistic fungi are important in lichens and in combination with the roots of certain kinds of plants. Figure 20.12 shows some examples of this group of organisms.

Kingdom Plantae

The kingdom Plantae is another major group thought to be derived from the kingdom Protista; they are the green, photosynthetic plants. The ancestors of plants were most likely the green algae. The members of the kingdom Plantae are nonmotile, terrestrial, multicellular organisms that contain chlorophyll and produce their own organic compounds. All plant cells have a cellulose cell wall.

Most biologists believe that the evolution of this kingdom began nearly 500 million years ago, when the green algae of the kingdom Protista gave rise to nonvascular plants, such as mosses. The development of vascular plants was a major step in the evolution of plants, because it allowed plants to increase in size and to live in drier environmental conditions. Some of the vascular plants evolved into seed-producing plants, which today are the cone-bearing and flowering plants, whereas the ferns never developed seeds (figure 20.13). Over 300,000 species of

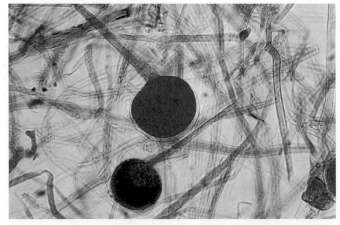

Mold

Puffball

Mushroom

FIGURE 20.12 Examples of Fungi
Molds, mushrooms, and puffballs are common examples of this kingdom.

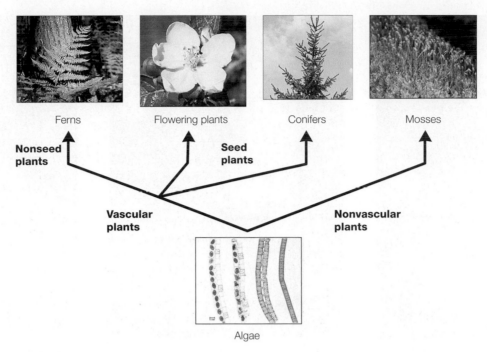

FIGURE 20.13 Plant Evolution

Two lines of plants are thought to have evolved from the green algae of the kingdom Protista. The nonvascular mosses evolved as one type of plant. The second type, the vascular plants, evolved into more complex plants. The early vascular plants did not produce seeds, like present-day ferns. Eventually, the evolution of the ability to produce seeds resulted in two distinct lines: the cone-bearing plants and the flowering plants.

plants have been classified; about 85% are flowering plants, 14% are mosses and ferns, and the remaining 1% are cone-bearers and several other small groups within the kingdom.

Plants have a unique life cycle. They have a haploid **gametophyte generation,** which produces haploid sex cells by mitosis. They also have a diploid **sporophyte generation** which produces haploid spores by meiosis. This **alternation of generations** is a unifying theme of all the members of this kingdom. In addition to sexual reproduction, plants are able to reproduce asexually.

Kingdom Animalia

Like the fungi and plants, the animals are thought to have evolved from the Protista. Over a million species of animals have been classified, ranging from microscopic types, such as mites and the aquatic larvae of marine animals, to huge animals, such as elephants and whales. All animals have some common traits. All are composed of eukaryotic cells and all species are heterotrophic and multicellular. All animals are motile, at least during some portion of their lives, although some, such as the sponges, barnacles, mussels, and corals, are sessile (i.e., nonmotile) as adults. All animals are capable of sexual reproduction, but many of the less complex animals also reproduce asexually.

It is thought that animals originated from certain kinds of Protista that had flagella. This idea proposes that colonies of flagellated Protista gave rise to simple, multicellular forms of animals such as the ancestors of present-day sponges. These first animals lacked specialized tissues and organs. As cells became more specialized, organisms developed special organs and systems of organs, and the variety of kinds of animals increased (figure 20.14).

20.3 Acellular Infectious Particles

All of the groups discussed so far are cellular forms of life. They all have at least the following features in common. They have (1) cell membranes, (2) nucleic acids as their genetic material, (3) cytoplasm, (4) enzymes and coenzymes, (5) and ribosomes, and they use ATP as their source of chemical-bond energy.

However, some particles show some characteristics of life and cause disease but do not have a cell structure. Because they lack a cell structure, they are referred to as *acellular* (*a* = lacking). Because they enter cells, cause disease, and can be passed from one organism to another, they are often called infectious particles. In causing disease, they make copies of themselves. There is no clear explanation of how these particles came to be. Therefore, they are not included in the classification system for cellular organisms. There are three kinds of acellular infectious particles: *viruses, viroids,* and *prions.*

Sea anemone

Sponge

Earthworm

Insect

Mammal and bird

FIGURE 20.14 Animal Diversity
Animals range in complexity from simple, marine sponges and sea anemones to complex, terrestrial insects, birds, and mammals.

Viruses

A **virus** is an acellular infectious particle consisting of a nucleic acid core surrounded by a coat of protein (figure 20.15). Viruses are often called *obligate intracellular parasites,* which means they are infectious particles that can function only when inside a living cell. Viruses are not considered to be living because they are not capable of living and reproducing by themselves and because they show some characteristics of life only when inside living cells.

Viruses vary in size and shape, which helps in classifying them. Some are rod-shaped, others are spherical, and still others are in the shape of a coil or helix. Most are too small to be visible through a light microscope and require an electron microscope to make them visible. Most viruses are identified

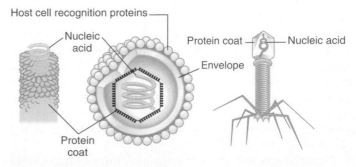

FIGURE 20.15 Typical Viruses
Viruses consist of a core of nucleic acid, either DNA or RNA, surrounded by a protein coat. Some have an additional layer, called an envelope, surrounding the protein coat.

TABLE 20.2

Viral Diseases

Type of Virus	Disease
Papovaviruses	Warts in humans
Paramyxoviruses	Mumps and measles in humans; distemper in dogs
Adenoviruses	Respiratory infections in most mammals
Poxviruses	Smallpox
Wound-tumor viruses	Diseases in corn and rice
Potexviruses	Diseases in potato
Bacteriophage	Infections in many types of bacteria

by the disease symptoms they cause when they infect cells. It is very likely that all species on Earth serve as hosts to viruses (table 20.2).

How Did Viruses Originate?

Soon after viruses were discovered in the late nineteenth century, biologists began to speculate on how they originated. One early hypothesis was that they were descendents of the first precells that did not evolve into cells. This idea was discarded as biologists learned more about the complex relationship between viruses and host cells. A second hypothesis was that viruses developed from intracellular parasites that became so specialized that they needed only the nucleic acid to continue their existence. Once inside a cell, this nucleic acid can take over and direct the host cell to provide for all of the virus's needs. A third hypothesis is that viruses are runaway genes that have escaped from cells and must return to a host cell to replicate. Regardless of how the viruses came into being, today they are important as parasites in all forms of life.

How Do Viruses Cause Disease?

Viruses are typically host-specific, which means that they usually attack only one kind of cell. The host is a specific kind of cell that provides what the virus needs to function. Viruses can infect only the cells that have the proper receptor sites to which the virus can attach. This site is usually a glycoprotein molecule on the surface of the cell membrane. For example, the virus responsible for measles attaches to the membranes of skin cells, hepatitis viruses attach to liver cells, and mumps viruses attach to cells in the salivary glands. Host cells for the human immunodeficiency virus (HIV) include some types of human brain cells and several types belonging to the immune system (Outlooks 20.1).

Once it has attached to the host cell, the virus either enters the cell intact or injects its nucleic acid into the cell. If it enters the cell, the virus loses its protein coat, releasing the nucleic acid. Once released into the cell, the virus's nucleic acid may remain free in the cytoplasm, or it may link with the host's

genetic material. Some viruses contain as few as 3 genes; others contain as many as 500. A typical eukaryotic cell contains tens of thousands of genes. Most viruses need only a small number of genes, because they rely on the host to perform most of the activities necessary for viral multiplication.

Some viruses have DNA as their genetic material but many have RNA. Many RNA viruses can be replicated directly, but others must have their RNA reverse-transcribed to DNA before they can reproduce. Reverse transcriptase, the enzyme that accomplishes this has become very important in the new field of molecular genetics, because its use allows scientists to make large numbers of copies of a specific molecule of DNA.

Viruses do not divide like true cells; they are *replicated*. Virus particles are replicated by using a set of genetic instructions from the virus and new building materials and enzymes from the host cell. Viral genes take command of the host's metabolic pathways and use the host's available enzymes and ATP to make copies of the virus. When enough new viral nucleic acid and protein coat are produced, complete virus particles are assembled and released from the host (figure 20.16). In many cases, this process results in the death of the host cell. When the virus particles are release, they can infect adjacent cells and the infection spreads. The number of viruses released ranges from 10 to thousands. The virus that causes polio affects nerve cells and releases about 10,000 new virus particles from each human host cell. Some viruses remain in cells and

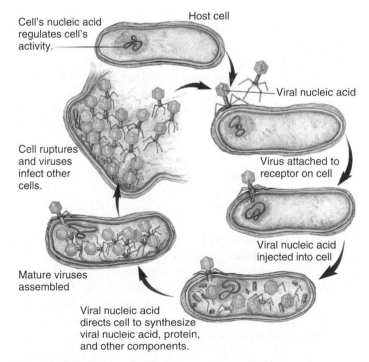

FIGURE 20.16 Viral Invasion of a Bacterial Cell

The viral nucleic acid takes control of the activities of the host cell. Because the virus has no functional organelles of its own, it can become metabolically active only while it is within a host cell. The viral genetic material causes the host cell to make copies of the virus, ultimately resulting in the destruction of the host cell.

are only occasionally triggered to reproduce, causing symptoms of disease. Herpes viruses, which cause cold sores, genital herpes, and shingles, reside in nerve cells and occasionally become active.

Viroids: Infectious RNA

A **viroid** is an infectious particle composed solely of a small, single strand of RNA in the form of a loop. To date, no viroids have been found to parasitize animals. The hosts in which they have been found are cultivated crop plants, such as potatoes, tomatoes, and cucumbers. Viroid infections result in stunted or distorted growth and may cause the plant to die. Pollen, seeds, and farm machinery can transmit viroids from one plant to another. Some scientists believe that viroids are parts of normal RNA that have gone wrong.

Prions: Infectious Proteins

Prions are proteins that can be passed from one organism to another and cause disease. All the diseases of this type cause changes in the brain that result in a spongy appearance called spongiform encephalopathies. Because these diseases can be transmitted from one animal to another, they are often called transmissible spongiform encephalopathies. The symptoms typically involve abnormal behavior and eventually death.

Examples of Prion Diseases

In animals, the most common examples are scrapie in sheep and goats and mad cow disease in cattle. Scrapie got its name because one of the symptoms of the disease is an itching of the skin associated with nerve damage that causes the animals to rub against objects and scrape their hair off.

In humans, there are several similar diseases. Kuru is a disease known to have occurred in the Fore people of the highlands of Papua New Guinea. The disease was apparently spread because the people ate small amounts of the brain tissue of dead relatives. (This ritual was performed as an act of love and respect for their relatives.) When the Fore people were encouraged to discontinue this ritual, the incidence of the disease declined. Creutzfeldt-Jakob disease (CJD) is found throughout the world. Its spread is associated with medical treatment; contaminated surgical instruments and tissue transplants, such as corneal transplants, are the most likely causes of transfer from infected to uninfected persons.

The occurrence of mad cow disease (bovine spongiform encephalopathy—BSE) in Great Britain in the 1980s and 1990s was apparently caused by the spread of prions from sheep to cattle. This occurred because of the practice of processing unusable parts of sheep carcasses into a protein supplement that was fed to cattle. It now appears that the original form of BSE has changed to a variety that is able to infect humans. This new form is called vCJD; scientists believe that BSE and CJD are, in fact, caused by the same prion.

A form of transmissible spongiform encephalopathy called *chronic wasting disease* is present in elk and deer in parts of the United States and Canada. It is called chronic wasting disease because the animals lose muscle mass and weight as a result of the prion infection. Similar diseases occur in mink, cats, and dogs.

How Prions Cause Disease

How are prions formed and how do they multiply? Prion multiplication appears to result from the disease-causing prion protein coming in contact with a normal body protein and converting it into the disease-causing form, a process called *conversion*. Because this normal protein is produced as a result of translating a DNA message, scientists looked for the genes that make the protein and have found it in a wide variety of mammals. The normal allele produces a protein that does not cause disease but is able to be changed by the invading prion protein into the prion form. Prions do not reproduce or replicate as do viruses and viroids. A prion protein (pathogen) presses up against a normal (not harmful) body protein and may cause it to change shape to that of the dangerous protein. When this conversion happens to a number of proteins, they stack up and interlock, as do the individual pieces of a Lego toy. When enough link together, they have a damaging effect—they form plaques (patches) of protein on the surface of nerve cells, disrupting the flow of the nerve impulses and eventually causing nerve cell death. Brain tissues taken from animals that have died of such diseases appear to be full of holes, thus the name spongiform encephalopathy.

A person's susceptibility to acquiring a prion disease, such as CJD, depends on many factors, such as his or her genetic makeup. If a person produces a functional protein with a particular amino acid sequence, the prion may not be able to convert it to its dangerous form. Other people may produce a protein with a slightly different amino acid sequence that can be converted to the prion form. Once formed, these abnormal proteins resist being destroyed by enzymes and most other agents used to control infectious diseases. Therefore, individuals with the disease-causing form of the protein can serve as the source of the infectious prions. There is still much to learn about the function of the prion protein and how the abnormal, infectious protein can cause copies of itself to be made. A better understanding of the alleles that produce proteins that can be transformed by prions will eventually lead to the prevention and effective treatment of these serious diseases in humans and other animals.

OUTLOOKS 20.1

The AIDS Pandemic

Epidemiology is the study of the transmission of diseases through a population. Diseases that occur throughout the world population and affect large numbers of people are called *pandemics*. Influenza, the first great pandemic of the first part of the twentieth century, killed millions of people. HIV/AIDS has become the greatest pandemic of recent times, claiming over 20 million victims so far.

The Magnitude of the Pandemic

AIDS is an acronym for acquired *immunodeficiency* syndrome; it is caused by the *human immunodeficiency* viruses (HIV-1 and HIV-2). This viral disease has been reported in all the countries of the world. The United Nations joint HIV/AIDS program (UNAIDS) has reported that each year about 3 million people die of AIDS and about 5 million became newly infected with HIV.

According to the World Health Organization (WHO), the incidence of HIV is lowest in the economically developed countries and highest in the developing countries. In economically developed countries, people can be diagnosed and treated. In the less-developed world, there is little medical care to treat AIDS and a lack of resources to identify those who have HIV. Many people do not know they are infected and continue to pass the disease to others. AIDS has become the fourth-leading cause of death in the world. In some countries in

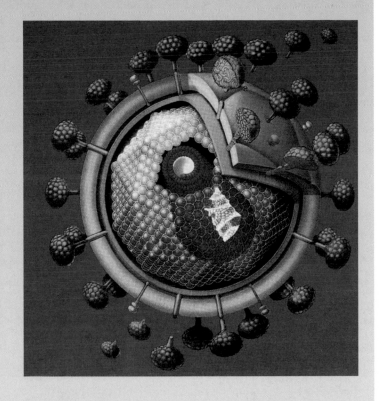

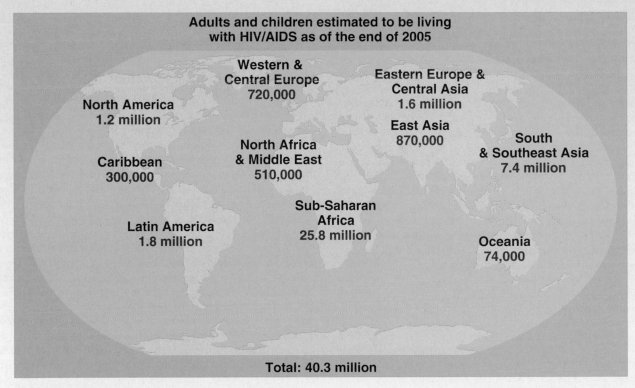

Adults and children estimated to be living with HIV/AIDS as of the end of 2005

North America
1.2 million

Western &
Central Europe
720,000

Eastern Europe &
Central Asia
1.6 million

East Asia
870,000

Caribbean
300,000

North Africa
& Middle East
510,000

South
& Southeast Asia
7.4 million

Latin America
1.8 million

Sub-Saharan
Africa
25.8 million

Oceania
74,000

Total: 40.3 million

Source: World Health Organization of the United Nations.

(continued)

southern Africa, AIDS is the leading cause of death resulting from disease. Over 2 million African children lose one parent to AIDS in a typical year.

The Nature and Life Cycle of the Virus

Evidence strongly supports the belief that HIV originated through many mutations of an African monkey virus during the late 1950s or early 1960s. The virus probably moved from its original monkey host to humans as a result of an accidental scratch or bite. Not until the late 1970s was the virus identified in human populations. The first reported case of AIDS was diagnosed in the United States in 1981 at the UCLA Medical Center. The virus had been present in Africa, however, for some time prior to 1981 without being recognized and diagnosed.

HIV is a spherical virus containing an RNA genome, a protein shell, and a lipid-protein envelope. Before the virus can be replicated by its host cell, its RNA must be used to make DNA. This is the reverse of the normal transcription process, in which a DNA template is used to manufacture an RNA molecule. Therefore, HIV and similar viruses are called retroviruses (*retro* = reverse). To accomplish reverse transcription, HIV has a gene for the enzyme reverse transcriptase. The virus gains entry into a suitable host cell through a very complex series of events involving the virus envelope and the host-cell membrane. Certain types of human cells can serve as hosts because they have a specific viral receptor site on their surface, identified as *CD4* (*CD* stands for "clusters of differentiation," a group of molecules on the surface of cells; *CD4* refers to a specific group of CD molecules). Several kinds of cells have CD4 receptors on their surface. These include some types of brain cells and several types of cells belonging to the immune system—namely, monocytes, macrophages, and T4-helper/inducer lymphocytes.

Once inside the host cell, the RNA of the HIV is used to make a DNA copy with the help of reverse transcriptase. When reverse transcriptase has completed its job, this piece of viral DNA is spliced into the host cell's DNA. In this integrated form, the virus is called a *provirus*. As a provirus, it can remain inside some host cells for an extended period without causing any harm. Some researchers estimate that this dormant period can last more than 30 years. Eventually, something triggers the host cell to replicate the virus. Then, multiple copies of the virus are produced, the host cell dies, and new viruses are released into the surrounding body fluids, where they can be transmitted to other cells in the body or to other individuals. Because of the dormant period, it is possible to differentiate two stages of an HIV infection. Persons can be HIV positive (HIV+) without showing the symptoms of AIDS. However, once replication of the virus begins and CD4 cells begin to die, the symptoms of AIDS become evident.

The Nature of the AIDS

A decrease in one type of CD4 cell—T4 lymphocytes—is an important diagnostic indicator of HIV infection and the onset of AIDS symptoms. The initial symptoms of the disease have

been referred to as *AIDS-related complex (ARC) or pre-AIDS*. HIV kills the immune system cells that are responsible for several important defense mechanisms against disease. The immune system cells

1. assist in the production of antibodies.
2. encourage the killing of tumor cells, microorganisms, and cells infected by microorganisms.
3. encourage disease-fighting cells to reproduce and increase their number.

When HIV destroys T4 lymphocytes, all of these defensive efforts are depressed. This leaves the body vulnerable to being invaded by many types of infecting microbes or to being overtaken by body cells that have changed into tumor cells. This means that HIV infections normally do not directly cause the death of the infected individual. Instead, interference with the immune system allows the development of a variety of secondary infections and cancers, which produce the symptoms known as AIDS. Most of the disease conditions associated with AIDS are rare in people with healthy immune systems. The following are some of the more common diseases associated with AIDS:

1. Kaposi's sarcoma, a form of skin cancer that shows up as purple-red bruises; this is one of the most common forms of cancer found among AIDS patients
2. A rare lung infection, *Pneumocystis carinii* pneumonia (PCP), caused by a fungus
3. Gastroenteritis (severe diarrhea) caused by the protozoan *Isospora*
4. Cytomegalovirus (CMV) infections of the retina of the eye

Prevention and Treatment

Because lymphocyte host cells are found in the blood and other body fluids, these fluids serve as carriers for the transmission of HIV. The virus is transmitted through direct contact with contaminated blood, semen, mucous secretions, serum, breast milk, and blood-contaminated hypodermic needles. If these body fluids contain the free viruses or infected cells (monocytes, macrophages, T4-helper/inducer lymphocytes) in sufficient quantity, they can be a source of infection.

To control the spread of the virus, there must be widespread public awareness of its nature and how it is transmitted. People must be able to recognize high-risk behavior and take action to avoid or change it. The most important risk factor is promiscuous sexual behavior (i.e., sex with many partners). This behavior increases the probability that eventually a person will be exposed to a partner who is HIV+. Other high-risk behaviors include intravenous drug use with shared needles, contact with blood-contaminated articles, and failure to use a condom during intercourse (vaginal, anal, oral) with persons whose HIV status is unknown. Babies born to women known to be HIV+ are at high risk of becoming infected.

Although it appears that the virus first entered the United States through the homosexual population, AIDS is not a

OUTLOOKS 20.1 (continued)
The AIDS Pandemic

disease unique to homosexuals. The transmission of HIV can occur in homosexual and heterosexual individuals. Today in all parts of the world, HIV is being spread primarily through heterosexual contact, transmission from infected mothers to their infants, and intravenous drug use. There is no evidence that the virus can be transmitted through the air; on toilet seats; by mosquitoes; by casual contact, such as shaking hands, hugging, touching, or closed-mouth kissing; by utensils, such as silverware or glasses; or by caring for AIDS patients. The virus is too fragile to survive transmission by these routes.

The progress of the infection can be slowed (but not stopped) by using drugs that can kill infected cells, improve the body's immune system, or selectively interfere with the life cycle of the virus. The life cycle may be disrupted when the virus enters the cell and the reverse transcriptase converts the RNA to DNA. If this enzyme does not operate, the virus is unable to function. Several drugs have been developed that block this reverse transcriptase step necessary for the replication of the virus. In addition, protease inhibitors block the protease enzyme that the virus needs. Usually, two or more drugs are given simultaneously or in sequence. This reduces the chance that the virus will develop resistance to the drugs being used.

Can a vaccine be developed to prevent the virus from infecting the body? Experimental vaccines have been created based on the body's ability to produce antibodies against the virus. Although some vaccines have been effective in monkeys, none have prevented infection in humans. One of the problems associated with developing a vaccine is HIV's high mutation rate. Studies have indicated that over 100 mutant strains of the virus have developed from a single infection in one individual. Such an astronomical mutation rate makes it very difficult to develop a vaccine, because any vaccine would have to stimulate an immune response against all possible mutant forms.

Blood tests can indicate whether a person has been exposed to HIV. The tests should be taken on a voluntary basis, absolutely anonymously, and with intensive counseling before and after. People who test positively (HIV+) should make lifestyle changes to prevent others from being exposed. They should also take measures to prevent reinfection, because new varieties of the virus may be more dangerous than those to which they were previously exposed. HIV+ individuals also should do everything they can to maintain good health—exercise regularly, eat a balanced diet, get plenty of rest, and reduce stress. We cannot stop this pandemic in its tracks, but it can be slowed.

Summary

To facilitate accurate communication, biologists assign a specific name to each species that is cataloged. The various species are cataloged into larger groups on the basis of similar traits.

Taxonomy is the science of classifying and naming organisms. Phylogeny is the science of trying to figure out the evolutionary history of a particular organism. The taxonomic ranking of organisms reflects their evolutionary relationships. Fossil evidence, comparative anatomy, developmental stages, and biochemical evidence are used in taxonomy and phylogeny.

The first organisms thought to have evolved were prokaryotic, single-celled organisms of the Domain Eubacteria. Current thinking is that the Domain Archaea developed from the Eubacteria and, ultimately, more complex, eukaryotic, many-celled organisms evolved. These organisms have been classified into the kingdoms Protista, Fungi, Plantae, and Animalia within the Domain Eucarya.

There are three kinds of acellular infectious particles: viruses, viroids, and prions. Viruses consist of pieces of genetic material surrounded by a protein coat. Viroids consist of naked pieces of RNA. Prions are proteins. Viruses and viroids can be replicated within the cells they invade. Prions cause already existing proteins within an organism to deform, resulting in disease.

Key Terms

*Use the interactive flash cards on the **Concepts in Biology, 12/e** website to help you learn the meaning of these terms.*

alternation of generations 433

binomial system of nomenclature 422

class 424

family 424

fungus 432

gametophyte generation 433

genus 423

kingdom 423

order 424

phylogeny 425

phylum 423

prions 436

saprophytes 430

specific epithet 423

sporophyte generation 433

taxonomy 422

viroid 436

virus 434

Basic Review

1. In the binomial system for naming organisms, an organism is given two names: the _____ and the specific epithet.

2. The most inclusive group into which an organism can be classified is the
   a. phylum.
   b. genus.
   c. domain.
   d. kingdom.

3. Phylogeny is the study that attempts to
   a. name organisms.
   b. organize organisms into groups based on how they evolved.
   c. decide on the names of phylums.
   d. classify organisms.

4. Closely related organisms should have very similar reproductive stages. (T/F)

5. Fossils
   a. provide information about when organisms lived.
   b. are found in sediments that form rock.
   c. of soft-bodied animals are rare, compared with those with hard body parts.
   d. All of the above are true.

6. The Eubacteria and Archaea are prokaryotic. (T/F)

7. Which one of the following organisms does not belong in the kingdom Protista?
   a. protozoa
   b. algae
   c. yeast
   d. slime mold

8. Which one of the following kinds of plants does not have seeds?
   a ferns
   b. pine trees
   c. roses
   d. apple trees

9. All viruses have DNA. (T/F)

10. Prions are proteins. (T/F)

Answers
1. genus 2. c 3. b 4. T 5. d 6. T 7. c 8. a 9. F 10. T

Concept Review

*Answers to the Concept Review questions can be found on the **Concepts in Biology**, 12/e website.*

20.1 The Classification of Organisms

1. List two ways that scientific names are different from common names for organisms.
2. Who designed the present-day system of classification? How does this system differ from the previous system?
3. What is the goal of phylogeny?
4. Name the categories of the classification system.

20.2 A Brief Survey of the Domains of Life

5. List two ways that Eubacteria and Archaea differ.
6. Describe typical modes of nutrition for Eubacteria.
7. Describe typical environmental conditions under which Archaea live.
8. List the four kingdoms of the domain Eucarya and give distinguishing characteristics for each.
9. Which of the following groups of organisms contain members that are autotrophic: Eubacteria, Archaea, Protista, Fungi, Plantae, Animalia?

20.3 Acellular Infectious Particles

10. Why do viruses invade only specific types of cells?
11. How do viruses reproduce?
12. Describe how viruses and viroids differ in structure.

Thinking Critically

*For guidelines to answer this question, visit the **Concepts in Biology**, 12/e website.*

A minimum estimate of the number of species of insects in the world is 750,000. Perhaps, then, it would not surprise you to see a fly with eyes on stalks as long as its wings, a dragonfly with a wingspread greater than 1 meter, an insect that can revive after being frozen at $-35°C$, and a wasp that can push its long, hairlike, egglaying tool directly into a tree. Only the dragonfly is not presently living, but it once was. What other curious features of this fascinating group can you discover? Have you looked at a common beetle under magnification? It will hold still if you chill it.

CHAPTER

21

The Nature of Microorganisms

Most bread is made by using yeast to ferment sugar to make carbon dioxide and alcohol. However, sourdough bread is produced in a different manner. The organisms that break down the sugars are a mixture of several kinds of microorganisms, including yeasts and bacteria. Each person's sourdough culture contains a slightly different combination of microbes. Cultures of sourdough are maintained by saving some of each new batch of bread dough and placing it in a refrigerator. This saved portion is called the starter. In order to maintain a sourdough culture, it must be used regularly, which involves adding flour and water to the starter and allowing the mixture to ferment whenever a new batch of bread is made. In this way, a sourdough culture can be kept "alive" for years. Some have been going for a century. The slightly sour flavor of the bread is the result of the organic acids produced by the bacteria present in the starter. It is said that miners in Alaska and the Yukon carried their sourdough starter in a pouch tied to their waist under their clothing. Their body heat kept the starter from freezing in the cold climate. Therefore, these miners were called sourdoughs.

CHAPTER OUTLINE

21.1 What Are Microorganisms?

A **microorganism,** or **microbe,** is a tiny organism that usually cannot be seen without the aid of a microscope. These are terms of convenience for a wide variety of organisms, including the Domains Eubacteria and Archaea and the kingdoms Protista and Fungi in the Domain Eucarya. Often, tiny animals, particularly those that cause disease, are considered microorganisms as well. However, we will not consider animals in this chapter.

In a very general sense, microorganisms share several characteristics. These organisms generally consist of cells, which function independently of the others. Many are single-celled organisms, although some single-celled microbes form loose aggregations, called colonies. Others are multicellular and have some specialization of cells for certain functions. Their primary method of reproduction is asexual reproduction, in which one cell divides to become two cells, although most kinds are also capable of sexual reproduction. Many have special structures involved in the production of gametes.

Microbes are extremely common organisms. Although no one can count them, it is estimated that there are about 10^{30} prokaryotes on the Earth at any one time. Your skin, mouth, and gut each contains trillions of microbes. Another estimate suggests that the total biomass of microbes is larger than the biomass of all other kinds of organisms combined. They live in any aquatic or moist environment and occur in huge numbers in the oceans and other bodies of water. However, because they are small, their moist habitat does not need to be large. Microbes can maintain huge populations in places such as human skin or intestine, temporary puddles, and soil. Most die if they dry out, but some have the special ability to become dormant and survive long periods without water. When moistened, they become actively growing cells again.

21.2 The Domains Eubacteria and Archaea

At one time, all prokaryotic organisms were lumped into one group of microorganisms called bacteria. Today, scientists recognize that there are two, very different kinds of prokaryotic organisms: the Domains Eubacteria and Archaea. The Eubacteria and Archaea differ in several ways: Eubacteria have a compound, called peptidoglycan, in their cell walls, which Archaea do not have. The chemical structure of the cell membranes of Archaea is different from that of all other kinds of organisms. When the DNA of Archaea are compared with that of other organisms, it is found that a large proportion of their genes are unique.

Today, most scientist still use the terms *bacterium* and *bacteria.* However, they are used in a restricted sense to refer to members of the Domain Eubacteria. The term *archeon* is frequently used to refer to members of the Domain Archaea.

The Domain Eubacteria

The Eubacteria are an extremely diverse group of organisms. Although only about 2,000 species of Eubacteria have been named, most biologists feel that there are many hundreds of thousands still to be identified. They are typically spherical, rod-shaped, or spiral-shaped. They are often identified by the characteristics of their metabolism or the chemistry of their cell walls (How Science Works 21.1). Many have a kind of flagellum, which rotates and allows for movement. Figure 21.1 shows the general structure of a bacterium. Some form resistant spores, which can withstand dry or other harsh conditions. Bacteria play several important ecological roles and interact with other organisms in many ways.

HOW SCIENCE WORKS 21.1

Gram Staining

Gram staining was developed in 1884 by Danish bacteriologist Christian Gram, who discovered that most bacteria can be divided into two main groups, based on their staining reactions. Such a technique is called differential staining, because it allows microbiologists to highlight the differences between cell types. Bacteria not easily decolorized with 95% ethyl alcohol after staining with crystal violet and iodine are said to be Gram-positive. Bacteria decolorized are Gram-negative and, thus, very difficult to see through the microscope. Another stain, called a counter-stain, is added to make Gram-negative cells more visible. A number of different stains can be used as a counterstain, but red is preferred, because it provides the greatest contrast. Knowing how some pathogenic bacteria react to Gram staining is of great value in determining how to handle those microbes in cases of infection. The Gram stain is probably the most widely performed diagnostic test in microbiology and can provide guidance in such matters as selecting the right antibiotic for treatment and predicting the kinds of symptoms a patient will show.

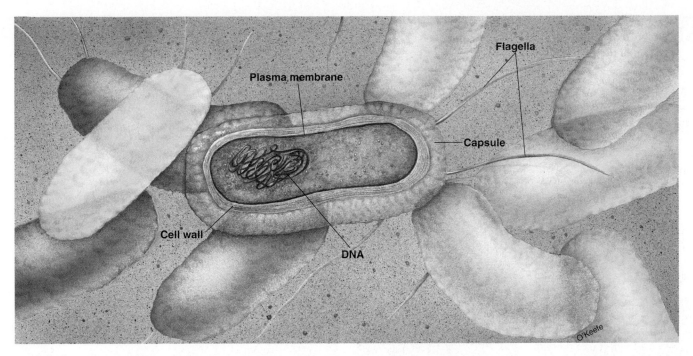

FIGURE 21.1 Bacteria Cell Structure

The plasma membrane regulates the movement of material between the cell and its environment. A rigid cell wall protects the cell and determines its shape. Some bacteria, usually pathogens, have a capsule to protect them from the host's immune system. The genetic material consists of a loop of DNA.

Decomposers

Many kinds of bacteria are heterotrophs that must break down organic matter to provide themselves with energy and raw materials for growth. Therefore, they function as decomposers in all ecosystems. Decomposers are often referred to as saprophytes, because they live on dead organic matter. Decomposers are a diverse group and use a wide variety of metabolic processes. Some are anaerobic and break down organic matter to simpler organic compounds. Others are aerobic and degrade organic matter to carbon dioxide and water. In nature, this decomposition process is important in the recycling of carbon, nitrogen, phosphorus, and many other elements.

The actions of decomposers have been harnessed for human purposes. Sewage treatment plants rely on bacteria and other organisms to degrade organic matter (figure 21.2).

The food industry uses controlled decomposition by certain bacteria to produce cheeses, yogurt, sauerkraut, and many other foods. Alcohols, acetones, acids, and other chemicals are produced by bacterial cultures. Some bacteria can even metabolize oil and are used to clean up oil spills.

Unfortunately, decomposer bacteria do not distinguish between items that we want to decompose and those that we don't want to decompose. Thus, it is often necessary to control the populations of some decomposer bacteria, so that foods and other valuable materials are not destroyed by rotting or spoiling.

FIGURE 21.2 Decomposers in Sewage

A sewage treatment plant is designed to encourage the growth of bacteria and other microorganisms that break down organic matter. The tank in the foreground contains a mixture of sewage and microorganisms, which is being agitated to assure the optimal growth of microbes.

TABLE 21.1

Common Bacteria in or on Humans

Skin	*Corynebacterium* sp., *Staphylococcus* sp., *Streptococcus* sp., *Escherichia coli*, *Mycobacterium* sp.
Eye	*Corynebacterium* sp., *Neisseria* sp., *Bacillus* sp., *Staphylococcus* sp., *Streptococcus* sp.
Ear	*Staphylococcus* sp., *Streptococcus* sp., *Corynebacterium* sp., *Bacillus* sp.
Mouth	*Streptococcus* sp., *Staphylococcus* sp., *Lactobacillus* sp., *Corynebacterium* sp., *Fusobacterium* sp., *Vibrio* sp., *Haemophilus* sp.
Nose	*Corynebacterium* sp., *Staphylococcus* sp., *Streptococcus* sp.
Intestinal tract	*Lactobacillus* sp., *Escherichia coli*, *Bacillus* sp., *Clostridium* sp., *Pseudomonas* sp., *Bacteroides* sp., *Streptococcus* sp.
Genital tract	*Lactobacillus* sp., *Staphylococcus* sp., *Streptococcus* sp., *Clostridium* sp., *Peptostreptococcus* sp., *Escherichia coli*

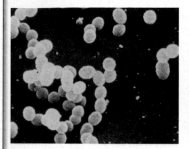

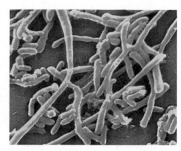

Streptococcus   *Escherichia coli*   *Lactobacillus*   *Corynebacterium*

Commensal Bacteria

Many kinds of bacteria have commensal relationships with other organisms. They live on the surface or within other organisms and cause them no harm, but neither do they perform any valuable functions. Most organisms are lined and covered by populations of bacteria called *normal flora* (table 21.1). In fact, if an organism lacks bacteria, it is considered abnormal. The bacterium *Escherichia coli* (commonly called *E. coli*) is common in the intestinal tract of humans, other mammals, and birds. A large proportion of human feces is composed of *E. coli* and other bacteria. Many of the odors humans produce from the skin and gut are the result of commensal bacteria.

Photosynthetic Bacteria

Several kinds of Eubacteria carry on a form of photosynthesis. A group called the cyanobacteria carries out a form of photosynthesis that is essentially the same as that in plants and photosynthetic protists. They use carbon dioxide and water as a raw material and release oxygen. In fact, the chloroplasts of eukaryotic organisms are assumed to be cyanobacteria that, in the past, formed an endosymbiotic relationship with other cells. Cyanobacteria are thought to be the first oxygen-releasing organisms; thus, their activities led to the presence of oxygen in the atmosphere and the subsequent evolution of aerobic respiration. Cyanobacteria are extremely common and are found in fresh and marine waters and soil and other moist environments. When conditions are favorable, asexual reproduction can result in what is called an algal **bloom**—a rapid increase in the population of microorganisms in a body of water (figure 21.3). Many

cyanobacteria form filaments or other kinds of colonies, which produce large masses when a bloom occurs. Some species of cyanobacteria produce toxins. When blooms occur, the levels of toxins in the water may be high enough to poison humans and other animals.

Within the filaments of many cyanobacteria are specialized, larger cells capable of nitrogen fixation which converts atmospheric nitrogen, N_2, to ammonia, NH_3. This provides a form of nitrogen usable to other cells in the colony—an example of division of labor.

FIGURE 21.3 Bloom of Cyanobacteria
Many kinds of cyanobacteria reproduce rapidly in nutrient-rich waters and produce masses of organisms known as a bloom.

Two kinds of Eubacteria, known as purple and green bacteria, carry on different forms of photosynthesis that do not release oxygen. Many of these organisms release sulfur as a result of their photosynthesis.

Mutualistic Bacteria

Mutualistic relationships occur between bacteria and other organisms. Some intestinal bacteria benefit humans by producing antibiotics, which inhibit the development of disease-causing bacteria. They also compete with disease-causing bacteria for nutrients, thereby helping keep them in check. They aid digestion by releasing various nutrients. They produce and release vitamin K. Mutualistic bacteria establish this symbiotic relationship when humans ingest them along with food or drink. When people travel, they consume local bacteria with their food and drink and may have problems establishing a new symbiotic relationship with these foreign bacteria. Both the host and the symbionts must adjust to their new environment, which can result in a very uncomfortable situation for both. Some people develop traveler's diarrhea as a result.

There are many other examples of mutualistic relationships between bacteria and other organisms. Many kinds of plants have nitrogen-fixing bacteria in their roots in a symbiotic relationship. Some fish and other aquatic animals have bioluminescent bacteria in their bodies, allowing them to produce light.

Bacteria and Mineral Cycles

Many different bacteria are involved in the nitrogen cycle. In addition to symbiotic nitrogen-fixing bacteria, free-living nitrogen-fixing bacteria in the soil convert N_2 to NH_3. Other bacteria convert ammonia to nitrite and nitrate. All of these bacteria are extremely important ecologically, because they are ultimately the source of nitrogen for plant growth. Finally, some bacteria convert nitrite to atmospheric nitrogen.

In addition to nitrogen; iron, sulfur, manganese, and many other inorganic materials are cycled by bacteria with specialized metabolic abilities. Some of these are important ecologically, because they produce acid mine drainage or convert metallic mercury to methylmercury, which can enter animals and cause health problems.

Disease-Causing Bacteria

Pathogens are organisms that cause disease. Only a small minority of bacteria fall into this category; however, because historically they have been responsible for huge numbers of deaths and continue to be a serious problem, they have been studied intensively and many pathogens are well understood.

Pathogenic bacteria can cause disease in several ways. Many are normally harmless commensals but become pathogenic when their populations increase to excessively high numbers so that they cause illness. For example, *Streptococcus pneumoniae* can grow in the throats of healthy people without any pathogenic effects. But if a person's resistance is lowered, as after a bout with viral flu, *Streptococcus pneumoniae* can invade the lungs and reproduce rapidly, causing pneumonia. The relationship changes from commensalistic to parasitic.

Other bacteria invade the healthy tissue of their host and cause disease by altering the tissue's normal physiology. Bacteria living in the host release a variety of enzymes that cause the destruction of tissue. The disease ends when the pathogens are killed by the body's defenses or an outside agent, such as an antibiotic. Examples are the infectious diseases strep throat, syphilis, pneumonia, tuberculosis, and leprosy.

Many other illnesses are caused by toxins or poisons produced by bacteria. Some of these bacteria release toxins that may be consumed with food or drink. In this case, disease can be caused even though the pathogens never enter the host. For example, botulism is a deadly disease caused by bacterial toxins in food or drink. Other bacterial diseases are the result of toxins released from bacteria growing inside the host tissue; tetanus and diphtheria are examples. In general, toxins cause tissue damage, fever, and aches and pains.

Bacterial pathogens are also important factors in certain plant diseases. Bacteria cause many types of plant blights, wilts, and soft rots. Apples and other fruit trees are susceptible to fire blight, a disease that lowers the fruit yield because it kills the tree's branches. Citrus canker, a disease of citrus fruits that causes cancerlike growths on stems and lesions on leaves and fruit, can generate widespread damage. Federal and state governments have spent billions of dollars controlling this disease (figure 21.4).

Probably all species of organisms have bacterial pathogens. Plants and animals get sick and die all the time. However, scientists are not likely to spend time and money studying these diseases unless the organisms have economic value to us. Therefore, scientists know much about bacterial

FIGURE 21.4 A Bacterial Plant Disease
Citrus canker is a disease of citrus trees caused by the bacterium *Xanthomonas axonopodis.* This photograph shows the typical lesions on the fruit and leaves of an orange tree.

diseases in humans, domesticated animals, and crop plants but know very little about the diseases of jellyfish, squid, or most plants.

Control of Bacterial Populations

The diseases and many kinds of environmental problems caused by bacteria are actually population control problems. Small numbers of bacteria cause little harm. However, when the population increases, their negative effects are multiplied. Despite large investments of time and money, scientists have found it difficult to control bacterial populations. Three factors operate in favor of the bacteria: and their reproductive rate, their ability to form resistant stages, and their ability to mutate and produce strains that resist antibiotics and other control agents.

Under ideal conditions, some bacteria can grow and divide every 20 minutes. If one bacterial cell and all its offspring were to reproduce at this ideal rate, in 48 hours there would be 2.2×10^{43} cells. In reality, bacteria cannot achieve such incredibly large populations, because they would eventually run out of food and be unable to dispose of their wastes. However, many of the methods used to control pathogenic bacteria are those that control their numbers by interfering with their ability to reproduce. Many antibiotics interfere with a certain aspect of bacterial physiology so that the bacteria are killed or become unable to divide and reproduce. This allows the host's immune system to gain control and destroy the disease-causing organism. Without the antibiotic, the immune system may be overwhelmed and the person may die.

Although antibiotics are life-saving discoveries, they don't always have the desired effect, because bacteria mutate and produce antibiotic-resistant strains. Because bacteria reproduce so rapidly, a few antibiotic-resistant cells in a bacterial population can increase to dangerous levels in a very short time. This requires the use of stronger doses or new types of antibiotics to bring the bacteria under control. Furthermore, these resistant strains can be transferred from one host to another, making it difficult to control the spread of disease. For example, sulfa drugs and penicillin, once widely used to fight infections, are now ineffective against many strains of pathogenic bacteria. As new antibiotics are developed, natural selection encourages the development of resistant bacterial strains. Therefore, humans are constantly waging battles against new strains of resistant bacteria.

In addition to antibiotics, various kinds of *antiseptics* are used to control the numbers of pathogenic bacteria. Antiseptics are chemicals able to kill or inhibit the growth of microbes. They can be used on objects or surfaces that have colonies of potentially harmful bacteria. Certain antiseptics are used on the skin or other tissues of people who are receiving injections or undergoing surgery. Reducing the numbers of bacteria lessens the likelihood that the microbes on the skin will be carried into the body, causing disease. We all are constantly in contact with pathogenic bacteria; however, as long as their numbers are controlled, they do not become a problem.

Another factor that enables some bacteria to survive a hostile environment is their ability to form *endospores*. An

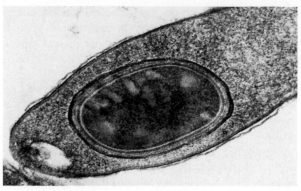

(a)

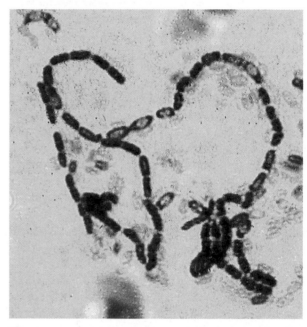

(b)

FIGURE 21.5 Bacterial Endospore

(a) The darker area in the cell is the endospore. It contains the bacterial DNA, as well as a concentration of cytoplasmic material surrounded and protected by a thick wall. *(b)* The photo shows a *Bacillus* bacterium that has formed endospores in some cells.

endospore is a unique bacterial structure with a low metabolic rate that can withstand hostile environmental conditions and germinate later, when there are favorable conditions to form a new, actively growing cell (figure 21.5). Endospores thought to be *Bacillus sphaericus* and estimated to be 25 million to 40 million years old have been isolated from the intestinal tract of a bee fossilized in amber. When placed in an optimum growth environment, they have germinated and grown into numerous colonies.

People who preserve food by canning often boil the food in the canning jars to kill the bacteria, but not all are killed by boiling, because some form endospores. The endospores of *Clostridium botulinum*, the bacterium that causes botulism, can withstand boiling and remain for years in the endospore state. However, endospores do not germinate and produce botulism toxin if the pH of the canned goods is in the acid range; in that case, the food remains preserved and edible. If conditions become favorable for *Clostridium* endospores to germinate, they become actively growing cells and produce toxin. Using a pressure cooker and heating the food to temperatures higher than 121°C for 15 to 20 minutes destroys both the botulism toxin and the endospores.

Anthrax is an acute infectious disease caused by the spore-forming bacterium *Bacillus anthracis*. Anthrax spores can live in the soil as spores for long periods and cause disease when they are inhaled, are swallowed, or invade the skin. Because anthrax spores can survive dry conditions, they have been used to contaminate mail as an agent of bioterrorism (Outlooks 21.1).

The Domain Archaea

The Archaea are distinct from the Eubacteria; they constitute a diverse group of organisms that generally live in extreme conditions. Therefore, they are often known as extremophiles (lovers of extremes). In addition to the spherical, rod-shaped, and spiral-shaped forms, as found in the Eubacteria, some Archaea are lobed, platelike, or irregular in shape.

Methanogens

Methanogens are members of the Archaea that are strict anaerobes (do not live where there is oxygen) and release methane as a waste product of cellular metabolism. Most produce methane by transferring hydrogen to carbon dioxide ($4H_2 + CO_2 \rightarrow CH_4 + 2H_2O$). Others are able to break down simple organic molecules, such as acetate, to produce methane ($CH_3COOH \rightarrow CO_2 + CH_4$). They live in a variety of environments where oxygen is absent. Many live in mud at the bottom of lakes and swamps, and some live in the intestinal tracts of animals, including humans, where they generate methane gas. The digestive system of cattle and some other organisms involves a complex mixture of microorganisms. Some are Eubacteria that break down cellulose to simpler molecules, such as acetate, and release hydrogen. Others are methanogens that convert the breakdown products of the Eubacteria to methane (Outlooks 21.2). Methanogens are also present in certain kinds of waste treatment systems used to manage animal and human waste. Anaerobic digesters containing methanogens can be used to produce methane from human or animal waste.

Extreme Halophiles

Extreme halophiles (salt lovers) are Archaea that can live only in extremely salty environments—such as the Great Salt Lake in Utah and the Dead Sea, located between Israel and Jordan. They require a solution of at least 8% salt and grow best in solutions that are about 20% salt. The Atlantic Ocean is about 3.5% salt. The Dead Sea is about 15% salt. They are also live in artificial salt ponds used to evaporate seawater to produce salt. Because they contain the reddish pigment carotene, they color these salt ponds pinkish or red.

Most of these organisms are aerobic heterotrophs. They use organic matter from their environment as a source of food. Some have been found growing on food products, such as salted fish, causing spoilage. However, some of them are also photosynthetic autotrophs that have a carotene-containing pigment, called bacteriorhodopsin, which absorbs sunlight and allows the cells to make ATP.

Thermophiles

The thermophiles (heat lovers) are a diverse group of the Archaea that live in extremely hot environments, such as the hot springs found in Yellowstone National Park and hydrothermal vents on the ocean floor (figure 21.6). All require high environmental temperatures—typically, above 50°C (122°F)—and some grow well at temperatures above 100°C (212°F). They are diverse metabolically; some are aerobic

FIGURE 21.6 Hydrothermal Vents
Extremely hot, mineral-rich water enters the ocean from hydrothermal vents on the ocean floor. Many kinds of specialized Archaea live in these places, where they use sulfur as a source of energy. These archeons are, in turn, eaten by other organisms that live in the vicinity.

OUTLOOKS 21.1

Microbes and Biological Warfare

Because bacteria and other microbes can spread through a population and cause disease, they have been used as biological warfare agents. However, only a limited number can be used effectively. To be an effective biological warfare agent, the agents should

1. Be human pathogens
2. Be easily cultured in a lab
3. Survive outside the lab long enough to reach the target
4. Enter a person in large enough numbers and in a way that causes disease
5. Cause a large number of soldiers and civilians to die or become severely ill
6. Spread easily from person to person
7. Cause panic and disruption in the infected population
8. Be difficult to diagnose
9. Have few, if any, antidotes
10. Require special action for public health preparedness

A serious drawback to the use of biological warfare agents is the problem of protecting the soldiers and civilian population of the country using the agent. Unlike explosives or chemical warfare agents, biological agents can reproduce, spread through the population, and cause epidemics, even after a war has ended. This is why only a few kinds of microbes or their toxins can be used as agents of warfare. As the historical chronology suggests, very few deaths have been due to biological warfare agents. Nevertheless, the history of biological warfare is long and with the development of genetic engineering and technology, the development of bio-weapons is an increasing concern. Equally if not more important, the potential and actual release of biological agents can be used to frighten people. Because one of the main goals of terrorism is to create panic and disrupt normal life, even a small number of infected persons and limited numbers of deaths due to biological agents can be effective.

History of Biological Warfare

Date	Event
Throughout history	Spear tips, swords, and knives were jabbed into the soil to inoculate them with pathogens before battle. Wounded soldiers later became incapacitated or died from their infections. In many conflicts, sharpened sticks were dipped into human and animal excrement and used as booby trap weapons.
6th century B.C.	Wells were poisoned with fungal rye ergot toxins. The purgative herb black hellebore was used to poison the well water during the siege of the Greek city Krissa.
1346	Corpses of persons who had died of plague were hurled by catapult over the city walls during the siege of the Genoese city of Kaffa, Italy.
1754–1767	Englishman Sir Jeffery Amherst gave smallpox-contaminated blankets to Indians fighting with the French against the English during the French and Indian war at Fort Carillon, renamed Fort Ticonderoga (New York).
World War I	German secret agents inoculated horses and mules awaiting shipment from the United States to Europe with the bacterium responsible for Glanders disease.

whereas others are anaerobic. Some can reduce sulfur or sulfur-containing compounds by attaching hydrogen to sulfur $(S + 2H \rightarrow H_2S)$. Thus, they release hydrogen sulfide gas (H_2S). Some live in extremely acidic conditions, with a pH of 1–2 or even less.

21.3 The Kingdom Protista

The kingdom Protista is a taxonomic category of convenience. Scientists do not actually think that the organisms in this group are closely related to one another. The only char-

acteristic they share is that they are the simplest eukaryotic organisms. Many of them are single-celled, but others are multicellular and show a degree of cellular specialization. The ancestors of this group are thought to have formed about 2 billion years ago.

There is great diversity among the more than 100,000 known species. Because of this diversity, it is a constant challenge to separate the kingdom Protista into meaningful subgroups. Furthermore, research continually reveals new evidence about the members of this group and their evolutionary relationships, requiring changes in taxonomy. Usually, the species are divided into three general groups,

OUTLOOKS 21.1 *(continued)*

History of Biological Warfare—*continued*

Date	Event
1937–1945	Japan began the biological warfare program known as Unit 731, in which prisoners of war were used as research subjects to investigate anthrax, plague, and other agents.
1943–1969	The United States began biological warfare agent research for offensive purposes at Fort Detrick, Maryland. All stockpiles of agents were destroyed between 1971 and 1972. The USSR produced and stored anthrax endospores for use as biological weapons (BW).
1953–present	The United States began a medical defensive program called the U.S. Army Medical Research Institute of Infectious Diseases (USAMRIID).
1972	The United States, United Kingdom, and USSR signed the *Convention on the Prohibition of the Development, Production and Stockpiling of Bacteriological (Biological) and Toxic Weapons and on Their Destruction,* commonly called the Biological Weapons Convention.
1978	A Bulgarian exile was assassinated in London by a Bulgarian secret service agent using a modified umbrella to inject lethal biotoxin.
1979	Sixty-six fatalities resulted from a Soviet military experiment in Sverdlovsk (now Yekaterinburg) in which *Bacillus anthracis* endospores were released into the air from the Soviet military microbiology facility known as Compound 19. The nature of the incident was denied until 1992.
1980	Members of the German Red Army Brigade were arrested in Paris, France, for growing *Clostridium botulinum* for use as biological weapons.
1984	Members of the religious cult Bhagwan Shree Rajneesh sprayed the salad bars of four restaurants in The Dalles, Oregon, with a solution containing *Salmonella* sp., 751 people were sickened, no fatalities.
1991	The United Nations carried out its first inspection of Iraq's biological warfare capabilities.
1992	A biological weapons program manager who defected from the USSR in 1992 revealed the existence of a biological warfare program.
1995	The religious cult Aum Shinriky released *Botulinum*–containing aerosols in downtown Tokyo, Japan, and near U.S. military installations in Japan.
1996	Twelve workers at a medical center in Texas contracted a rare form of dysentery. The bacterium in their stool was identical to one that was missing from the lab's stock of samples: No arrests were made, but the infections were clearly the result of an intentional act.
2001	Letters containing anthrax endospores were mailed from New Jersey. Five persons died and 22 were infected.

based on their mode of life: *algae,* autotrophic unicellular organisms; *protozoa,* heterotrophic unicellular organisms; and funguslike protists. These categories are helpful in discussing the major roles of these organisms but do not reflect how the organisms have evolved. For example, many kinds of flagellated protists are photosynthetic, but other, closely related organisms are not. Thus, two closely related organisms would be placed in the categories protozoa or algae, depending on their ability to perform photosynthesis. Figure 21.7 shows some current ideas about how some of these groups are related

Algae

Algae are protists that contain chlorophyll in chloroplasts and therefore carry on photosynthesis. Many of these organisms are single-celled, but some groups are multicellular. Although most live in the ocean and bodies of freshwater, they can also be found in other moist places, such as soil and the surface of other organisms in rainforests and other moist habitats. **Plankton** is a collection of small, floating or weakly swimming organisms. Algae are the major components of the **phytoplankton** that consists of the photosynthetic plankton

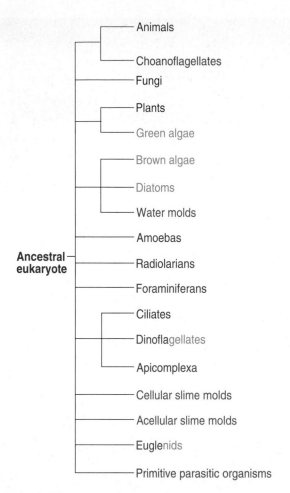

FIGURE 21.7 **Relationships Among Members of the Protista**
In this figure, organisms that are photosynthetic are show in green and are generally known as algae. Those that are shown in red are generally known as protozoa. Those shown in blue are funguslike. Some groups, such as the Euglenids and Dinoflagellates, have some members that are photosynthetic and others that are not.

that is the basis of most aquatic food chains. Cyanobacteria are the other major component of phytoplankton. **Zooplankton** consists of nonphotosynthetic plankton, including aquatic protozoa and tiny animals. **Benthic** organisms live attached to the bottom or to objects in the water. Many benthic algae form denses "forests" of large seaweeds in shallow sea water. The large number of benthic and planktonic algae makes them an important source of atmospheric oxygen (O_2). It is estimated that over 50% of the oxygen in the atmosphere is produced by marine algae. Furthermore, because algae are important producers in marine food chains, disruptions to the marine algal community can have serious implications for the production of fish and shellfish.

Because algae require light, they are found only near the surface of the water. Even in the clearest water, photosynthesis does not usually occur any deeper than 100 meters. Benthic forms are found in shallow water and are common along the ocean shoreline. Some phytoplankton have flagella or other methods of loco-

motion, which assist them in remaining near the surface. Others maintain their position by storing food as oil, which is less dense than water and enables the cells to float near the surface.

The various kinds of algae reproduce both sexually and asexually. However, their primary method of reproduction is asexual cell division. Like the cyanobacteria, in warm, nutrient-rich waters, various kinds of algae can reproduce rapidly by asexual reproduction and cause an algal bloom. The population can become so large that clumps of algae float on the surface or in the case of single-celled algae, the water may become colored or murky.

Single-Celled Algae

There are several common kinds of single-celled algae. *Euglenids* are single-celled algae that move by flagella. They have an outer covering, called a pellicle, which gives them shape but is flexible. *Euglenids* vary in terms of their metabolism. Some lack chloroplasts and are heterotrophs. Others have chloroplasts and are autotrophs. However, even among those that have chloroplasts, many are able to consume food and behave as heterotrophs, particularly when light levels are low. Most of them, like the common *Euglena,* are found in freshwater. They are widely studied because they are easy to culture.

Diatoms are extremely common single-celled algae found in freshwater, marine, and soil environments. They are a major component of the phytoplankton of the oceans and serve as a food source for zooplankton and many kinds of filter-feeding organisms, such as whales, clams, and barnacles. A few species are heterotrophs or parasites. They are typically brownish in color. Although they do not have cilia or flagella, they are able to move with a sort of gliding motion. They are unique because their cell walls contain silicon dioxide (silica). The walls fit together like the lid and bottom of a shoe box; the lid overlaps the bottom. The silica-containing walls have many pores, which form interesting patterns. Because their cell walls contain silicon dioxide, they readily form fossils. The fossil cell walls have many tiny holes and can be used in a number of commercial processes. They are used as filters for liquids and as abrasives in specialty soaps, toothpastes, and scouring powders.

Dinoflagellates, along with diatoms, are the most important food producers in the ocean's ecosystem. They are also very common is freshwater and brackish water. All members of this group of algae have two flagella, which is the reason for their name (*di* = two), and an outer cellulose covering made up of plates. Although many dinoflagellates are photosynthetic, some are heterotrophs and others are parasites. Some change from autotroph to heterotroph, depending on environmental conditions.

Some species of dinoflagellates have symbiotic relationships with marine animals, such as the reef corals; the dinoflagellates provide a source of nutrients for the reef-building coral. Corals that live in the light and contain dinoflagellates grow 10 times faster than corals without this symbiont. Thus, in coral reef ecosystems, dinoflagellates form the foundation of the food chain.

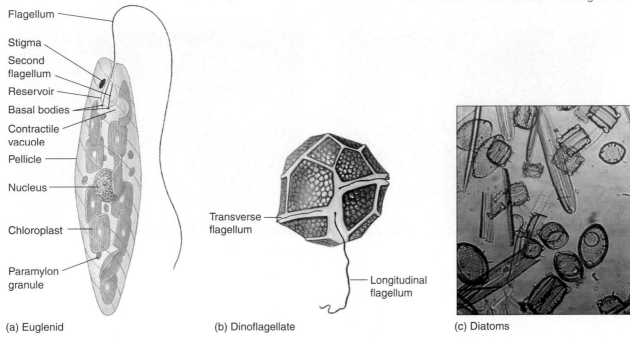

Flagellum

Stigma

Second
flagellum

Reservoir

Basal bodies

Contractile
vacuole

Pellicle

Nucleus

Chloroplast

Paramylon
granule

(a) Euglenid

Transverse
flagellum

Longitudinal
flagellum

(b) Dinoflagellate

(c) Diatoms

FIGURE 21.8 Single-Celled Algae
Three very common kinds of single-celled algae are the euglenids, dinoflagellates, and diatoms.

Some forms of dinoflagellates produce toxins. Because many of the toxin-producing dinoflagellates are reddish in color, a bloom of these organisms is called a *red tide*. Often, fish and other vertebrates such as birds and mammals are killed by exposure to the toxins. Although the toxins do not seem to harm shellfish, such as oysters, consuming the toxin along with the shellfish can cause sickness or death. Red tides usually occur in the warm months and are more common is tropical and semitropical waters. During red tide episodes, people are warned not to swim in areas that have a red tide or harvest fish or shellfish for food. Commercially available shellfish are tested for toxin content; if they are toxic, they are not marketed.

In recent years, a new problem has surfaced caused by the dinoflagellate *Pfiesteria piscidia*. It has been responsible for the death of millions of fish in estuaries of the eastern United States. These dinoflagellates release toxins that paralyze fish. The dinoflagellates then feed on the fish. They have also been responsible for human and wildlife poisoning. It appears that blooms of these organisms may be triggered by high amounts of nutrients in the water as a result of runoff from feedlots and agricultural land.

Many marine forms are bioluminescent; they are responsible for the twinkling lights seen at night in ocean waves or in a boat's wake. Figure 21.8 shows examples of euglenids, diatoms, and dinoflagellates.

Multicellular Algae

Many kinds of algae are multicellular and can be quite large, with some specialization of cells and body parts. These algae are commonly known as *seaweed*. They are found in shallow water attached to objects. Two types, red algae and brown algae, are mainly marine forms. The green algae are primarily freshwater species.

Red algae live in warm oceans and attach to the ocean floor by means of a holdfast structure. They are found from the splash zone, the area where waves are breaking to depths of 100 meters. Some red algae become encrusted with calcium carbonate and are important in reef building. Other species are commercially important, because they produce agar and carrageenin. *Agar* is widely used as a jelling agent for growth media in microbiology. *Carrageenin* is a gelatinous material used in paints, cosmetics, and baking. It is also used to make gelatin desserts harden faster and ice cream smoother. In Asia and Europe, some red algae are harvested and used as food.

Brown algae are found in cooler marine environments. Most species of brown algae have a holdfast organ. Colonies of these algae can reach 100 meters in length. Brown algae produce *alginates*, which are widely used as stabilizers in frozen desserts, as emulsifiers in salad dressings, and as thickeners to give body to foods such as chocolate milk and cream cheeses; they are also used to form gels in such products as fruit jellies.

The Sargasso Sea is a large mat of free-floating brown algae between the Bahamas and the Azores. It is thought that this huge mass (as large as the European continent) is the result of brown algae that have become detached from the ocean bottom, have been carried by ocean currents, and have accumulated in this calm region of the Atlantic Ocean. This large mass of floating algae provides a habitat for a large number of marine animals, such as marine turtles, eels, jellyfish, and innumerable crustaceans. Figure 21.9 shows examples of red and brown algae.

Red algae

Brown algae

FIGURE 21.9 Red and Brown Algae
Red and brown algae are primarily marine organisms. Most of them grow attached to the ocean bottom or other organisms in their environment.

Green algae are found primarily in freshwater ecosystems, although a few kinds live in oceans. Some are single-celled and have flagella; some lack flagella and form strings, which either float in the water or grow on surfaces. The members of this group can also be found growing on trees, in the soil, and even on snowfields in the mountains. Like land plants, green algae have cellulose cell walls and store food as starch. Green algae also have the same types of chlorophyll as do plants. Biologists believe that land plants evolved from the green algae. Figure 21.10 shows a variety of green algae.

Protozoa

Protozoa are members of the kingdom Protista; they are eukaryotic, heterotrophic, single-celled organisms that lack cell walls. Generally, protozoa lack all types of chlorophyll, but some organisms may contain chloroplasts at some times

in their lives and lack them at others. One common way to classify the protozoa into subgroups is by their method of locomotion. Although this is a convenient way to subdivide the organisms for the purposes of discussion, it is clearly not a valid phylogenic grouping.

Flagellates

Flagellates are an extremely diverse group of organisms that have flagella and lack cell walls and chloroplasts. They live in any moist environment, including marine waters and freshwater, moist soil, and as parasites or symbionts. Some flagellates have an extremely simple structure, suggesting that they may be the most primitive of all eukaryotic organisms. Some feed by absorbing simple organic molecules through their cell membranes; others engulf food particles or other organisms.

Many kinds of flagellates are mutualistic or parasitic. Termites are insects that eat wood but cannot digest it. Their

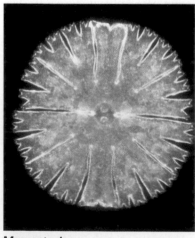

Macrasterias

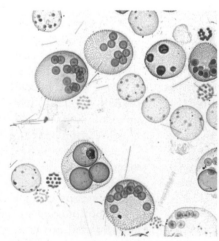

Volvox

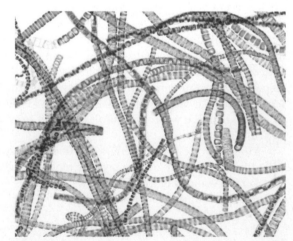

Ulothrix

FIGURE 21.10 Green Algae
Some green algae are single-celled; others form colonies.

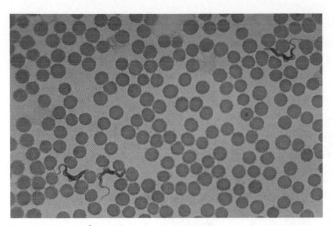

Trypanosoma brucei

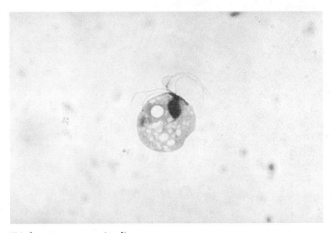

Trichomonas vaginalis

FIGURE 21.11 Flagellates

Several flagellated protozoa are parasites. *Trypanosoma brucei* cause sleeping sickness. It is shown here among red blood cells. *Trichomonas vaginalis* is the cause of a common sexually transmitted disease.

guts contain mutualistic flagellated protozoa capable of digesting cellulose. Thus, the termite benefits from a food source and the flagellate benefits from a good place to live and a continuous supply of food.

There are many examples of parasitic flagellates (figure 21.11). One is *Trichomonas vaginalis,* that can live in the reproductive tract of both men and women and is the cause of a common sexually transmitted disease. Often, it doesn't cause any symptoms but sometimes causes itching and a discharge. The symptoms are more common in women than men. Trypanosomes, which cause sleeping sickness in humans and domestic cattle, primarily in Africa are another example of a parasitic flagellate. The parasite develops in the circulatory system and moves to the cerebrospinal fluid surrounding the brain. When this occurs, the infected person develops the "sleeping" condition, which, if untreated, is eventually fatal.

Giardia lamblia is a flagellated protozoan that contaminates freshwater throughout the world. Because *Giardia* is a common intestinal parasite of deer, beaver, and many other animals, even "pure" mountain streams in wilderness areas are likely to be contaminated. Infection usually causes diarrhea, intestinal gas, and nausea, although it does not usually cause life-threatening illness. The most effective way to eliminate the spores formed by this protozoan is to filter out particles as small as 1 micrometer from the water or boil it for at least 5 minutes before drinking.

Choanoflagellates are colonial flagellates that many biologists believe are ancestral to all multicellular animals, because the simplest animals, sponges, contain cells that are extremely similar in structure to free-living choanoflagellates.

Amoeboid Protozoans

Amoeboid protozoans have extensions of their cell surface called pseudopods in which the cytoplasm flows. They range from the well-known *Amoeba,* with its constantly changing, lobelike pseudopods to species with thin, fiberlike pseudopods (figure 21.12). Most amoeboid protozoans are free-living and feed on bacteria, algae, or even small,

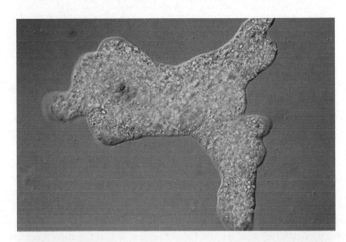

FIGURE 21.12 Amoeboid Protozoa

Amoeboid protozoa have extensions of their cell surface called pseudopods. Pseudopods contain moving cytoplasm. Some, such as *Amoeba,* have large, lobelike pseudopods, which change shape as the cell moves and feeds. Others have long, filamentous pseudopods that transport food molecules to the central cell from the objects they feed on.

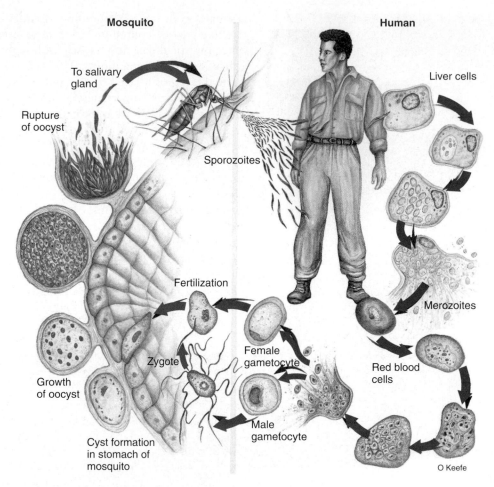

Mosquito

To salivary gland

Rupture of oocyst

Sporozoites

Fertilization

Zygote

Female gametocyte

Growth of oocyst

Male gametocyte

Cyst formation in stomach of mosquito

Human

Liver cells

Merozoites

Red blood cells

O Keefe

FIGURE 21.13 The Life Cycle of *Plasmodium vivax*

Plasmodium vivax is one of the members of the Apicomplexa that causes malaria. The life cycle requires two hosts, the *Anopheles* mosquito and the human. Humans get malaria when they are bitten by a mosquito carrying the larval stage of *Plasmodium*. The larva undergoes asexual reproduction and releases thousands of individuals, which invade the red blood cell. Their release from massive numbers of infected red blood cells causes the chills, fever, and headache associated with malaria. Inside the red blood cell, more reproduction occurs to form male gametocytes and female gametocytes. When the mosquito bites a person with malaria, it ingests some gametocytes. Fertilization occurs and zygotes develop in the stomach of the mosquito. The resulting larvae are housed in the mosquito's salivary gland. Then, when the mosquito bites someone, some saliva containing the larvae is released into the person's blood and the cycle begins again.

multicellular organisms. *Amoeba* uses pseudopods to move about and to engulf food.

Some forms are parasitic. *Entamoeba histolytica* is responsible for the disease known as amoebic dysentery. People become infected with this protozoan when they travel to parts of the world that have poor sewage and water treatment facilities and often have contaminated water.

Radiolarians and foraminiferans are two specialized groups of amoeboid protozoans that are extremely common in the oceans. Both kinds have long, thin pseudopods and float in the ocean, feeding on organic material and other living organisms. However, the radiolarians have a kind of skeleton composed of silicon dioxide, and the foraminiferans have a skeleton of calcium carbonate. When these organisms die, their cells disintegrate but their skeletons remain and sink to

the bottom of the sea. Extensive limestone deposits were formed from the accumulated skeletons of ancient foraminiferans. The cliffs of Dover, England, were formed from such shells.

Apicomplexa

All members of the Apicomplexa are nonmotile parasites with a sporelike stage in their life cycles. The disease malaria, one of the leading causes of disability and death in the world, is caused by members of the Apicomplexa. Two billion people live in malaria-prone regions of the world. There are 150 to 300 million new cases of malaria each year, and the disease kills 2 to 4 million people annually.

The organisms that cause malaria have a complex life cycle involving transmission by a mosquito vector (figure 21.13).

While in the mosquito vector, the parasite goes through the sexual stages of its life cycle. One of the best ways to control this disease is to eliminate the vector, which usually involves using a pesticide. Many of us are concerned about the harmful effects of pesticides in the environment. However, in the parts of the world where malaria is common, the harmful effects of pesticides are of less concern than the harm generated by the disease. Many diseases of insects, birds, and mammals are also caused by the members of this group.

Ciliates

Ciliates are a group of protozoans with a complex cellular structure and numerous short, flexible extensions from the cell called *cilia* (figure 21.14). The cilia move in an organized, rhythmic manner and propel the cell through the water. Some types of ciliates, such as *Paramecium*, have nearly 15,000 cilia per cell and move at a rapid speed of 1 millimeter per second. Most ciliates are free-living cells found in fresh water and salt water or damp soil, where they feed on bacteria and other small organisms. Ruminant animals have large numbers of ciliates in their digestive systems, where they are part of the complex ecology of the ruminant gut (Outlooks 21.2).

Ciliates have a complex cellular structure with two kinds of nuclei. Most have a macronucleus and one or more micronuclei. The macronucleus is involved in the day-to-day running of the cell, whereas the micronuclei are involved in sexual reproduction. Sexual reproduction involves a process called conjugation, in which two cells go through a series of nuclear divisions equivalent to meiosis and exchange some of their nuclear material. Although the exchange does not result in additional cells, it does result in cells that have a changed genetic mixture.

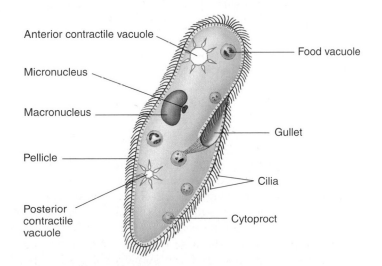

FIGURE 21.14 Ciliates
Ciliates, such as *Paramecium,* have a complex cell structure and a large number of cilia on their surface, which propel them through the water. They feed on a variety of organisms.

FIGURE 21.15 Slime Mold
Slime molds grow in moist conditions and are important decomposers. As slime molds grow, additional nuclei are produced by mitosis, but there is no cytoplasmic division. Thus, at this stage, a slime mold is a single mass of cytoplasm with many nuclei.

Funguslike Protists

Funguslike protists have a motile, reproductive stage, which differentiates them from true fungi. There are two kinds of funguslike protists: slime molds and water molds.

Slime Molds

Slime molds are amoeba-like organisms that crawl about and digest dead organic matter. Some slime molds look like giant amoebae. They are essentially a large mass several centimeters across, in which the nucleus and other organelles have divided repeatedly within a single large cell (figure 21.15). No cell membranes partition this mass into separate segments. They vary in color from white to bright red or yellow, and they can reach relatively large sizes (45 centimeters in length) when in an optimum environment.

Other kinds of slime mold exist as large numbers of individual, amoeba-like cells. These haploid cells get food by engulfing microorganisms. They reproduce by mitosis. When their environment becomes dry or otherwise unfavorable, the cells come together into an irregular mass. This mass glides along rather like an ordinary garden slug and is labeled the sluglike stage. This sluglike form may flow about for hours before it forms spores. When the mass gets ready to produce spores, it creates a stalk with cells that have cell walls. At the top of this specialized structure, cells are modified to become haploid spores. When released, these spores may be carried by the wind and, if they land in a favorable place, may develop into new amoeba-like cells.

Water Molds

Water molds were once thought to be fungi. However, they differ from fungi in two fundamental ways. Their cell walls are made of cellulose, not chitin, and water molds have a

OUTLOOKS 21.2

The Microbial Ecology of a Cow

Ruminants are animals, such as cattle, deer, bison, sheep, and goats, that have a special design to their digestive system. These animals have a large, pouchlike portion of the gut, called a rumen, connected to the esophagus. Ruminants eat plant materials that are often dry and consist of large amounts of cellulose from the plants' cell walls. They do not have enzymes (cellulases) that allow them to break down the cellulose. The rumen is essentially a fermentation chamber for a variety of anaerobic microorganisms, including members of the Domains Eubacteria and Archaea and fungi, ciliates, and other protozoa from the kingdom Protista.

Cows also chew their cud. This is a process in which the animal eat grasses and other plant materials, which go into the rumen. Later, they regurgitate the food and chew it again, which further reduces the size of the food particles and thoroughly mixes the food with liquids containing the mixture of microorganisms.

Some of the microorganisms in the rumen produce enzymes that can break down cellulose to short-chain fatty acids. These fatty acids are absorbed into the cow's bloodstream and are used by its cells to provide energy. But the story does not end there. Methanogens are common in the gut of ruminants. They metabolize some of the fatty acids to methane. A cow typically releases 200 to 500 liters of methane gas per day. Methane production by ruminants and termites that have a similar gut metabolism contributes significantly to the level of methane in the atmosphere; because methane is a greenhouse gas, it is a factor in global warming.

Agricultural researchers look at methane production by cows as an opportunity to increase the food efficiency of cattle. If the researchers could prevent the methanogens from using some of the fatty acids to make methane, there would be more for the cows to turn into meat or milk. They have experimented with substances that inhibit the growth of methanogens.

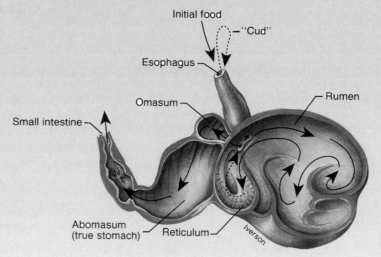

The digestive system of ruminants encourages the growth of microorganisms that assist in the breakdown of cellulose.

flagellated reproductive stage. Thus, they are considered to be more closely related to the diatoms and brown algae than to fungi. Although called water molds, they live in many moist environments, not just in bodies of water (figure 21.16).

Water molds are important saprophytes and parasites in aquatic ecosystems. They are often seen as fluffy growths on dead fish or other organic matter floating in water. A parasitic form of water mold is well known to people who rear tropical fish; it causes a cottonlike growth on the fish. Although these organisms are usually found in aquatic habitats, they are not limited to this environment. Some species

cause downy mildew on plants such as grapes. In the 1880s, this mildew almost ruined the French wine industry when it spread throughout the vineyards. A copper-based fungicide called *Bordeaux mixture*—the first chemical used against plant diseases—was used to save the vineyards. A water mold was also responsible for the Irish potato blight. In the nineteenth century, potatoes were the staple of the Irish diet. Cool, wet weather in 1845 and 1847 damaged much of the potato crop, and more than a million people died of starvation. Nearly one-third of the survivors left Ireland and moved to Canada or the United States.

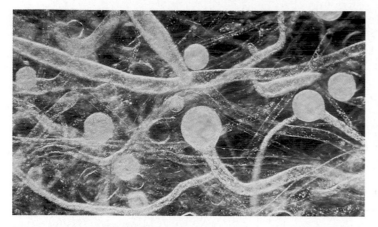

FIGURE 21.16 Water Mold
Rapidly reproducing water molds quickly produce a large mass of filaments. These filaments cause the fuzzy growth often seen on dead fish and other dead material in the water.

21.4 Multicellularity in the Protista

The three major types of organisms in the kingdom Protista (algae, protozoa, and funguslike protists) include both single-celled and multicellular forms. Biologists believe that there has been a similar type of evolution in all three of these groups. The most primitive organisms in each group are thought to have been single-celled and to have given rise to the more advanced, multicellular forms. Most protozoan organisms are single-celled however, some ciliates are colonial. The multicel-

lular forms of funguslike protists are the slime molds, which have both single-celled and multicellular stages.

Perhaps the most widely known example of this trend from a single-celled to a multicellular condition is found in the green algae. A very common single-celled green alga is *Chlamydomonas*, which has a cell wall and two flagella. It looks just like the individual cells of the colonial green algae *Volvox*. Some species of *Volvox* have as many as 50,000 cells (figure 21.17). All the flagella of each cell in the colony move in unison, allowing the colony to move in one direction. In some *Volvox* species, certain cells have even specialized to produce sperm or eggs. Biologists believe that the division of labor seen in colonial protists represents the beginning of the specialization that led to the development of true multicellular organisms with many kinds of specialized cells. Three types of multicellular organisms—fungi, plants, and animals—eventually developed.

21.5 The Kingdom Fungi

The members of the kingdom Fungi are nonphotosynthetic, eukaryotic organisms with rigid cell walls containing chitin. Most are multicellular, but a few are single-celled. The typical fungus consists of a filaments composed of many cells. Each filament is known as a hypha. The hyphae form a network known as a mycelium (figure 21.18).

Even though fungi are nonmotile, they are easily dispersed, because they form huge numbers of spores. A **spore** is a cell with a tough, protective cell wall that can resist extreme conditions. Some spores are produced by sexual reproduction,

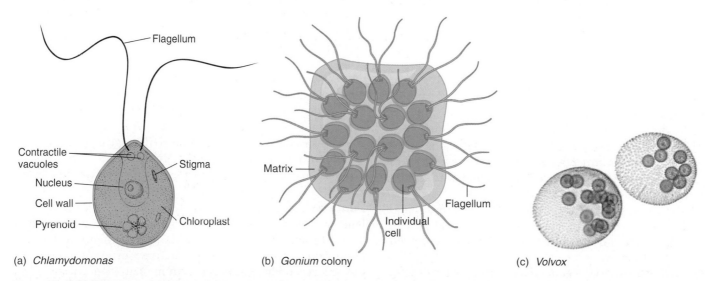

(a) *Chlamydomonas* (b) *Gonium* colony (c) *Volvox*

FIGURE 21.17 The Development of Multicellular Green Algae
(a) *Chlamydomonas* is a green, single-celled alga containing the same type of cholorophyll as that found in green plants. (b) *Gonium*, a green alga similar to *Chlamydomonas*, forms colonies composed of 4 to 32 cells that are essentially the same as *Chlamydomonas*. (c) *Volvox* is a colonial green alga that produces daughter colonies and also has specialized cells that produce eggs and sperm.

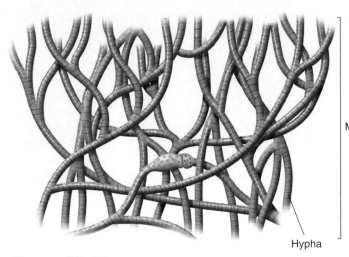

Mycelium

Hypha

FIGURE 21.18 Mycelium
The basic structure of a fungus is a multicellular filament known as a hypha. A mass of hyphae is collectively known as a mycelium.

others by asexual reproduction. An average-sized mushroom can produce over 20 billion spores; a good-sized puffball can produce as many as 8 trillion spores. When released, the spores are transported by wind or water. Because of their small size, spores can remain in the atmosphere a long time and travel thousands of kilometers. Fungal spores have been collected as high as 50 kilometers above the Earth.

All of these organisms are heterotrophs; that is, they must obtain nutrients from organic sources. Most secrete enzymes that digest large molecules into smaller units, which the fungi absorb. Fungi are either free-living or parasitic. Free-living fungi, such as mushrooms, decompose dead organisms as they absorb nutrients. Parasitic fungi are responsible for many plant diseases and human infections, such as athlete's foot, vaginal yeast infections, and ringworm.

The Taxonomy of Fungi

Fungi are divided into subgroups, based on their methods of reproduction (figure 21.19).

The *Chytridiomycota* are fungi that are aquatic or live in moist soil. They are distinct from other fungi because they have a flagellated spore stage. Infections caused by one member of this group, *Batrachochytrium dendrobatidis,* are thought to be one of the reasons for the decline in frogs and other amphibians throughout the world.

The *Zygomycota* are distinguished from other fungi because, although their hyphae are made of many cells, there are no walls separating the individual cells. One common example is bread mold.

The *Ascomycota* are distinguished from other fungi because they produce spores by sexual reproduction in a saclike structure. They also produce spores asexually. About 75% of all fungi are in this group. Molds, mildews, yeasts, morel mushrooms, and truffles are examples.

The *Basidiomycota* are unique in having sexual spores produced in sets of four on a club-shaped structure. Bracket fungi, most mushrooms, and puffballs are common examples. Many are parasites on plants; rusts and smuts are examples.

The Significance of Fungi

Fungi play many significant roles in ecosystems. They are also economically important to humans in many ways.

Decomposers

Because fungi produce so many spores, any dead organism is likely be colonized by a fungus. Because all fungi are capable of breaking down organic matter, fungi along with bacteria are the major decomposers of organic matter in ecosystems. This decomposer function recycles elements such as carbon, nitrogen, and phosphorus. Fungi that decompose organic matter in the soil often form rings of mushrooms called "Fairy rings" (Outlooks 21.3). However, fungi also colonize and break down organic matter that we do not want to have recycled and cause billions of dollars of damage each year. Clothing, wood, leather, and all types of food are susceptible to damage by fungi. One of the best ways to protect against such damage is to keep the material dry, because fungi grow best in a moist environment.

Fungi as Food

Fungi and their by-products have been used as sources of food for centuries. When we think of fungi and food, mushrooms usually come to mind. The common mushroom found in the grocer's vegetable section is grown in many countries and has an annual market value in the billions of dollars. But there are other uses for fungi as food. *Shoyu* (soy sauce) was originally made by fermenting a mixture of wheat, soybeans, and an ascomycote fungus for a year. Most of the soy sauce used

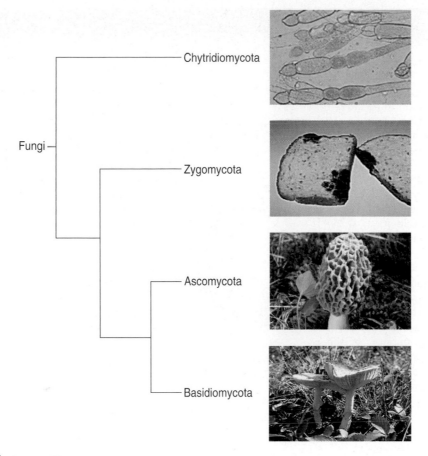

FIGURE 21.19 Fungus Taxonomy
There are four subgroups within the kingdom Fungi. They are distinguished from one another primarily by their methods of reproduction.

OUTLOOKS 21.3

Fairy Rings

Fairy rings are one of the most interesting formations caused by the grow of a mushroom colony. These formations result from the mushrooms' expanding growth. The colony originally begins at the center but grows out from there, because it exhausts the soil nutrients necessary for fungal growth. As the microscopic hyphae grow outward from the center, they stunt the growth of the grass, forming a ring. Just to the outside of this growth ring, the grass grows well, because the hyphae excrete enzymes, decomposing soil material into rich nutrients to support the growth of the grass. Often rings of mushrooms are associated with the outer edge of the colony. The name *fairy ring* comes from an old superstition that such rings were formed by fairies tramping down the grass while dancing in a circle.

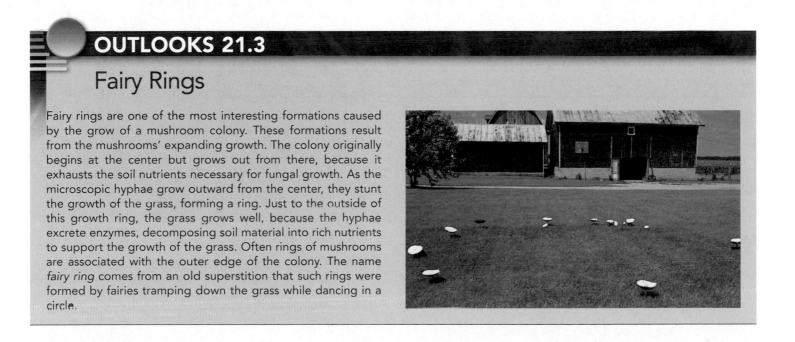

HOW SCIENCE WORKS 21.2

Penicillin

In 1928, Dr. Alexander Fleming was working at St. Mary's Hospital in London. As he sorted through some old petri dishes on his bench, he noticed something unusual. The mold *Penicillium notatum* was growing on some of the petri dishes. Apparently, the mold had found its way through an open window and onto a bacterial culture of *Staphylococcus aureus*. The bacterial colonies that were growing at a distance from the fungus were typical, but there was no growth close to the mold. Fleming isolated the agent responsible for this destruction of the bacteria and named it *penicillin*.

Through Fleming's research efforts and those of several colleagues, the chemical was identified and used for about 10 years in microbiological work in the laboratory. Many suspected that penicillin could be used as a drug, but the fungus could not produce enough of the chemical to make it worthwhile. When World War II began, and England was being firebombed, there was an urgent need for a drug that would control bacterial infections in burn wounds.

Two scientists from England were sent to the United States to begin research into the mass production of penicillin.

Their research in isolating new forms of *Penicillium* and purifying the drug were so successful that cultures of the mold now produce over 100 times more of the drug than the original mold discovered by Fleming. In addition, the price of the drug dropped considerably—from a 1944 price of $20,000 per kilogram to a current price of less than $250.00. The species of *Penicillium* used to produce penicillin today is *P. chrysogenum*, which was first isolated in Peoria, Illinois, from a mixture of molds found growing on a cantaloupe. The species name, *chrysogenum*, means golden and refers to the golden-yellow droplets of antibiotic that the mold produces on the surface of its hyphae. The spores of this mold were isolated and irradiated with high dosages of ultraviolet light, which caused mutations to occur in the genes. When some of these mutant spores were germinated, the new hyphae were found to produce much greater amounts of the antibiotic.

---

today is made by a cheaper method of processing soybeans with hydrochloric acid. True connoisseurs still prefer soy sauce made the original way. Another mold is important to the soft-drink industry. The citric acid that gives a soft drink its sharp taste was originally produced by squeezing juice from lemons and purifying the acid. Today, however, a mold is grown on a nutrient medium with table sugar (sucrose) to produce great quantities of citric acid at a low cost.

The primary flavors of blue cheeses, such as Danish, American, and the original Roquefort, are produced because the cheeses have been aged with the mold *Penicillium roquefortii*. Another species of *Penicillium* produces the antibiotic penicillin (How Science Works 21.2).

Yeasts are important organisms in the production of alcoholic beverages and in the making of bread. It is difficult to imagine a world in which these two food materials are not important.

Mycorrhizae

Mycorrhizae are associations between certain fungi and the roots of plants. There are two kinds of mycorrhizae. In one kind, the cells of fungi actually penetrate the cells of the plant roots. In the other kind, the fungal cells surround the root cells but do not invade them (figure 21.20). Mycorrhizal

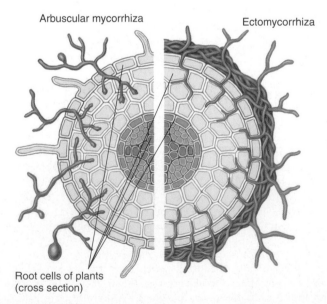

FIGURE 21.20 Mycorrhizae

Mycorrhizae are associations between fungi and the roots of plants. Both organisms benefit. The plants have increased absorption of water and minerals, and the fungi receive organic molecules as food from the plants.

(a)

(b)

FIGURE 21.21 Lichens

Lichens grow in a variety of habitats. (*a*) The shrubby British soldier lichen is growing on soil. (*b*) There are several kinds of lichen growing on the bark of this tree. The different colors are due to the different species of algae or cyanobacteria in the lichens.

fungi are found in 80–90% of all plants, and many plants cannot live without their mycorrhizal fungi. The addition of these fungi to the roots of plants increases the roots' surface area for absorption. Plants with mycorrhizal fungi can absorb as much as 10 times more minerals than those without the fungi. Some types of fungi also supply plants with growth hormones, whereas the plants supply carbohydrates and other organic compounds to the fungi.

Lichens

Lichens are organisms that consist of a symbiotic relationship between a fungus and either an alga or a cyanobacterium. The alga or cyanobacterium does photosynthesis and provides the fungus with organic molecules for food, while the fungus provides the moist environment required by the alga or cyanobacterium.

Because the fungi provide a damp environment and the algae produce the food, lichens require no soil for growth. For this reason, many kinds of lichens are commonly found growing on bare rock and are the pioneer organisms in the process of succession. Some lichens grow as fluffy growths on trees or as fleshy structures on the soil (figure 21.21). Lichens are important in the process of soil formation. They secrete an acid, which weathers the rock and makes minerals available for use by plants. When lichens die, they are a source of humus—decomposing organic material—which mixes with rock particles to form soil.

Lichens are found in a wide variety of environments, ranging from the frigid Arctic to the scorching desert. Reindeer moss of cold northern regions is actually a lichen. One reason for this success is their ability to withstand drought conditions. Some lichens can survive with only 2% water by weight. In this condition, they stop photosynthesis and go into a dormant stage until water becomes available and photosynthesis begins again.

Another factor in the success of lichens is their ability to absorb minerals. However, because air pollution increases the amounts of minerals in the air, many lichens are damaged. Some forms of lichen absorb concentrations of sulfur 1,000 times greater than those found in the atmosphere. This increases the amount of sulfuric acid in the lichen, resulting in damage or death. For this reason, areas with heavy air pollution are "lichen deserts." Because they can absorb minerals, certain forms of lichen have been used to monitor the amount of various pollutants in the atmosphere.

Pathogenic Fungi

Many kinds of fungi are important pathogens of plants, and a few are human pathogens. Chestnut blight and Dutch elm disease almost caused these two species of trees to become extinct. Wheat rust gets its common name because infected wheat plants look as if they are covered with rust. Corn smut is also due to a fungal pathogen (figure 21.22).

More recently, a fungal disease known as *sudden oak death* was identified in Germany, the Netherlands, California, and Oregon. It is caused by the fungus *Phytophthora ramorum,* which also infects rhododendrons, redwoods, and fir trees. Once a tree is infected, an oozing canker develops on the lower trunk; a few weeks later, its leaves turn yellow to brown. Control of the disease will require spending millions of dollars for fungicides. It will also be expensive and difficult to dispose of the millions of infected plants to prevent the spread of the fungus.

Pathogenic fungi that affect domestic crops cost billions of dollars yearly. Farmers and fruit growers must use large amounts of fungicides to control the spread of fungal disease in their fields and orchards.

The most common sites of human fungal infections are the skin, mouth, and lungs. For example, the organism *Pneumocystis* is present in the lungs of most people.

FIGURE 21.22 Corn Smut
Most people who raise corn have seen corn smut. Besides being unsightly, it decreases the corn yield.

FIGURE 21.23 Poisonous *Amanita* Mushroom
The *Amanita* mushroom is deadly poisonous. It destroys the liver cells of those who eat it, causing death.

However, in people with impaired immune systems, such as AIDS patients, the populations of this organism can increase and cause a form of pneumonia that is often fatal.

Toxic Fungi

A number of fungi produce deadly poisons called **mycotoxins.** The most deadly of these fungi is *Amanita verna*, known as "the destroying angel"; it can be found in woodlands during the summer (figure 21.23). Mushroom hunters must learn to recognize this deadly species. It is believed to be so dangerous that food accidentally contaminated by its spores can cause illness and possibly death.

The mushroom *Psilocybe mexicana* has been used for centuries in religious ceremonies by certain Mexican tribes because of the hallucinogenic chemical it produces. These mushrooms have been grown in culture, and the drug psilocybin has been isolated. In the past, it was used experimentally to study schizophrenia.

Claviceps purpurea is a parasite on rye and other grains. The metabolic activity of *C. purpurea* produces a toxin that can cause hallucinations, muscle spasms, insanity, and even death. However, it is also used to treat high blood pressure, to stop bleeding after childbirth, and to treat migraine headaches.

Summary

Organisms in the Domains Archaea and Eubacteria and in the kingdoms Protista and Fungi rely mainly on asexual reproduction, and each cell usually satisfies its own nutritional needs. The Eubacteria are prokaryotic and perform a wide variety of functions in ecosystems. Cyanobacteria are major photosynthesizers. Other functions include decomposer, mineral cycling, parasitism, and mutualism. Only a few species are pathogenic.

The Archaea are prokaryotic organisms that live in extreme environments. Some live at high temperatures, some in high-salt environments. Many produce methane.

The members of the kingdom Protista are one-celled eukaryotic organisms. In some species, there is minimal cooperation between cells. The protists include algae, cells that have a cell wall and carry on photosynthesis; protozoa, which lack cell walls and cannot carry on photosynthesis; and funguslike protists, whose motile, reproductive stage distinguishes them from true fungi. Some species of Protista developed a primitive type of specialization, and from these evolved the multicellular fungi, plants, and animals.

The kingdom Fungi consists of nonphotosynthetic, eukaryotic organisms with cell walls that contain chitin. Most species are multicellular. Fungi are nonmotile organisms that disperse by producing spores. Lichens are organisms that consist of a combination of organisms involving a mutualistic relationship between a fungus and an algal protist or a cyanobacterium.

Key Terms

*Use the interactive flash cards on the **Concepts in Biology**, 12/e website to help you learn the meaning of these terms.*

algae 449
benthic 450
bloom 444
endospore 446
lichens 461
microorganism (microbe) 442
mycorrhizae 460

mycotoxins 462
pathogens 445
phytoplankton 449
plankton 449
protozoa 452
spore 457
zooplankton 450

Basic Review

1. The Domains Eubacteria and Archaea differ from the Domain Eucarya in that
   a. the Eucarya are eukaryotic and the Eubacteria and Archaea are prokaryotic.
   b. the Eucarya are single-celled but the others are not.
   c. the Eucarya carry on photosynthesis but the others do not.
   d. the Eucarya are simpler organisms than the others.
2. The Archaea are often found living in extremely salty or hot environments. (T/F)
3. The Apicomplexa are parasitic species of protozoa. (T/F)
4. Which one of the following groups does not have members that carry on photosynthesis?
   a. the kingdom Protista
   b. the Domain Eubacteria
   c. the kingdom Fungi
   d. diatoms
5. The primary organisms that make up phytoplankton are dinoflagellates and _____.
6. Fungi are primarily dispersed by tiny structures called _____.

7. Methanogens are members of the Domain Eubacteria. (T/F)
8. The Chytridiomycota differ from other fungi in that they
   a. do not have chitin in their cell walls.
   b. have spores with flagella.
   c. do not cause disease.
   d. form mushrooms.
9. Benthic algae live attached to objects. (T/F)
10. Lichens and mycorrhizae are similar in that they both involve symbiotic relationships with fungi. (T/F)

Answers
1. a 2. T 3. T 4. c 5. diatoms 6. spores 7. F 8. b 9. T 10. T

Concept Review

*Answers to the Concept Review questions can be found on the **Concepts in Biology**, 12/e website.*

21.1 What Are Microorganisms?

1. What taxonomic groups are included in the category known as microorganisms?

21.2 The Domains Eubacteria and Archaea

2. What is meant by the term *bloom?*
3. What is a pathogen? Give two examples.
4. Define the term *saprophyte.*
5. Give an example of a symbiotic relationship involving Eubacteria.
6. What is a bacterial endospore?
7. What is meant by the term *pathogen?*

21.3 The Kingdom Protista

8. Why are the protozoa and the algae in different subgroups of the kingdom Protista?
9. What is phytoplankton?
10. Name three commercial uses of algae.
11. What is the best method to prevent the spread of malaria?

21.4 Multicellularity in the Protista

12. What is the difference between a tissue and the multicellularity found in some protists?
13. Give an example of a multicellular alga.

21.5 The Kingdom Fungi

14. Name two beneficial results of fungal growth and activity.
15. What types of spores do fungi produce?
16. What is sudden oak death, and why is it significant?

Thinking Critically

*For guidelines to answer this question, visit the **Concepts in Biology**, 12/e website.*

Throughout much of Europe, there has been a severe decline in the mushroom population. On study plots in Holland, data collected since 1912 indicate that the number of mushroom species has dropped from 37 to 12 per plot in recent years. Along with the reduction in the number of species, there has been a parallel decline in the number of individual plants; moreover, the surviving plants are smaller. The phenomenon of the disappearing mushrooms is also evident in England. One study noted that, in 60 fungus species, 20 exhibited declining populations. Mycologists are also concerned about a decline in the United States; however, there are no long-term studies, such as those in Europe, to provide evidence for such a decline. Consider the niche of fungi in the ecosystem. How would an ecosystem be affected by a decline in fungi numbers?

The Plant Kingdom

The mustard family has many plants in the genus *Brassica* that supply a wide variety of food materials. Several kinds of mustards are grown for their seeds, which are ground up and used as flavorings. Another mustard, called rape, is grown commercially for the seed from which rapeseed oil is extracted. Rapeseed oil is used for industrial purposes. A modified version is used as the cooking oil canola. Rutabagas and turnips, whose roots are eaten, are also members of this group. But probably the most interesting members are the various cabbagelike varieties of the species *Brassica oleracea*. Nearly every part of the above-ground portion of the plant can be eaten. Cabbage and kale are grown for their leaves, which are eaten raw or cooked. Purple cabbages are grown for their colorful leaves. Two kinds of *Brassica oleracea* are grown for their flowers: broccoli and cauliflower are immature flowers that have become large and fleshy. Brussels sprouts are grown for the buds that grow on the stem, and kohlrabi is grown for its thickened stems. The mustard family is just one of many families of plants that provide us with food.

CHAPTER OUTLINE

22.1 What Is a Plant?

Plants are eukaryotic, multicellular organisms that have chlorophyll *a* and chlorophyll *b* and carry on photosynthesis. Their cells have cellulose cell walls. Plants also have specializations for life on land. Plants are, for the most part, terrestrial organisms that can live in just about any environment, including deserts, arctic regions, and swamps (figure 22.1). Many plants live in shallow freshwater (e.g., water lilies, cattails), and a few live in shallow oceans (e.g., eelgrass, mangroves). Within the more than 250,000 species, plants show a remarkable variety of form, function, and activity. Plants range in size from tiny, floating duckweed the size of a pencil eraser to redwood trees over 100 meters tall. There are plants that lack chlorophyll and are parasites, plants that digest animals, and plants that have mutualistic relationships with animals.

22.2 Alternation of Generations

All plants have a life cycle that involves two distinctly different generations: the *sporophyte generation* and the *gametophyte generation*. The **sporophyte generation** is diploid (*2n*) and has plant parts in which meiosis takes place to produce haploid (*n*) spores. The word *spore* is used several different ways. There are structures called spores in bacteria, algae, protozoa, fungi, and plants. Each is distinct from the others. In our discussion of plants, the word *spore* refers to a haploid cell produced by meiosis that germinates to give rise to a multicellular haploid generation known as the *gametophyte generation*. The **gametophyte generation** is haploid and develops structures that produce gametes: eggs and sperm. Because the gametophyte is already haploid, eggs and sperm are produced by mitosis. When haploid gametes unite, a diploid zygote is formed. The zygote is the first cell in a new sporophyte generation. The zygote divides by mitosis, and a new multicellular sporophyte generation results. Recall that the term *alternation of generations* is used to describe this kind of life cycle, in which plants cycle between two stages in their life—the diploid sporophyte and the haploid gametophyte (figure 22.2).

22.3 The Evolution of Plants

At one time, all life was aquatic. Therefore, it is logical to look for the ancestors of plants among the aquatic, photosynthetic algae. Most scientists feel that the ancestor of plants was a freshwater member of the green algae. The strongest

Desert plants

Arctic plants

Aquatic plants

FIGURE 22.1 Plant Diversity

Plants are adapted to a wide range of terrestrial habitats, including such extremes as deserts and arctic regions. Many live in freshwater and a few live in shallow marine environments.

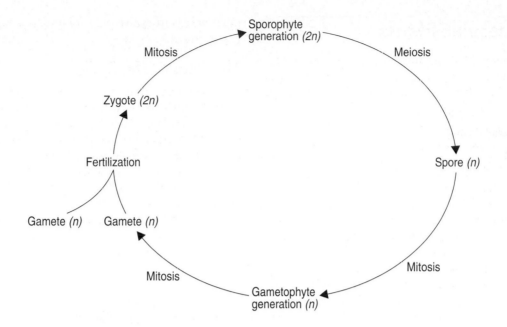

FIGURE 22.2 Alternation of Generations
Plants go through two distinctly different generations during their life cycle. The sporophyte generation is diploid and the gametophyte generation is haploid.

evidence for this is that green algae have the same kinds of chlorophyll (chlorophyll *a* and chlorophyll *b*) and the same kinds of chloroplasts as plants. Comparison of the DNA of plants and green algae also supports this conclusion. The evolution of plants shows two general trends. One is toward greater specialization for living in a dry environment; the other is toward a more prominent role for the sporophyte generation in the life cycle.

The most primitive plants lack vascular tissue to help carry water. Thus, most nonvascular plants are limited to moist habitats. The more advanced vascular plants have specialized cells that help transport water and other materials throughout the plant. A second major development that allowed advanced plants to exploit terrestrial habitats was the evolution of seeds that could resist drying.

In the more primitive plants, the gametophyte generation is the dominant generation, whereas, in more advanced groups, the sporophyte generation is dominant. The taxonomy of plants is based on these trends (figure 22.3). Subsequent sections will discuss the major groups of plants and show how life cycles and the capacity to live in a dry environment evolved.

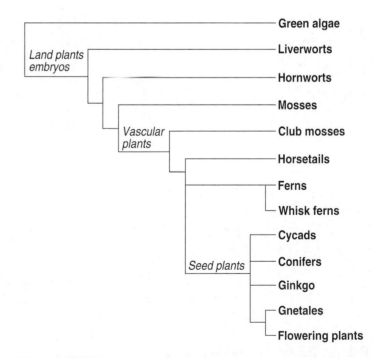

FIGURE 22.3 The Evolution of Plants
The evolution of plants involves specializations for living on land and changes in the way they reproduce.

22.4 Nonvascular Plants

The nonvascular plants, which include the mosses, hornworts, and liverworts, and are commonly known as *bryophytes*. These three kinds of plants share the following characteristics.

1. They lack vascular tissue.
2. They do not have true roots or leaves.
3. The gametophyte generation is the most prominent part of the life cycle.
4. Sperm swim to the egg.

Because they lack the vascular tissue and roots for the absorption and transportation of water, they must rely on the physical processes of diffusion and osmosis to move dissolved materials through their bodies. They also tend to lose water by evaporation from their surfaces, because they do not have a waterproof layer on their outer surfaces. They have sperm that swim to the egg, so they are similar to their algal ancestors and must have water to reproduce sexually. Thus, most of these organisms are small and live in moist habitats. In

other words, they are only minimally adapted to a terrestrial environment.

The Moss Life Cycle

The moss plant that you commonly recognize is the gametophyte generation. Recall that the gametophyte generation is haploid and is the gamete-producing stage in the plant life cycle. Although all the cells of the gametophyte have the haploid number of chromosomes (the same as gametes), not all of them function as gametes. At the top of the moss gametophyte are two kinds of structures that produce gametes: the *antheridium* and the *archegonium*.

The **antheridium** is made up of a jacket of cells surrounding the developing sperm. The **archegonium** is a flask-shaped structure that produces the egg; it has a tubular channel leading to the egg at its base (figure 22.4). There is usually only one egg cell in each archegonium. When the sperm are mature, the outer jacket of the antheridium splits open, releasing the flagellated sperm. The sperm swim through a film of dew or rainwater to the archegonium and continue down the channel of the archegonium to fertilize the

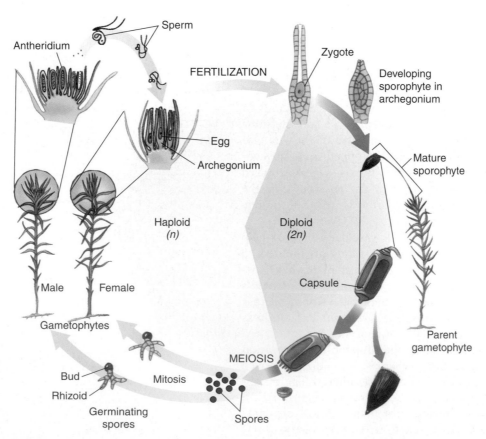

FIGURE 22.4 **The Life Cycle of a Moss**
The haploid gametophyte generation produces eggs in a structure called an archegonium and sperm in a structure called an antheridium. Sperm swim to the egg and fertilization occurs within the archegonium. The zygote is diploid and is the first stage of the sporophyte generation. The sporophyte grows and protrudes from the top of the gametophyte. Meiosis takes place in the capsule of the sporophyte, producing haploid spores. The spores are released and give rise to new gametophyte plants.

Moss

Liverwort

Hornwort

FIGURE 22.5 Nonvascular Plants

There are three kinds of nonvascular plants: mosses, liverworts, and hornworts. Because they lack vascular tissue, they are generally small and are usually found in moist environments.

egg. When the sperm and egg nuclei fuse, a diploid zygote is produced. The zygote is the first cell of the sporophyte generation. The zygote begins to grow within the archegonium and eventually develops into a mature sporophyte, which grows out of the top of the gametophyte.

The sporophyte consists of a long stalk with a capsule on the end of it. Within the capsule, meiosis takes place, producing haploid spores. These spores are released; when they germinate, they give rise to a multicellular filament known as a protonema. The protonema develops into a mature gametophyte plant. Thus, in mosses and other nonvascular plants, the gametophyte generation is the dominant stage and the sporophyte is dependent on the gametophyte.

Kinds of Nonvascular Plants

The mosses are the most common nonvascular plants. Mosses grow as a carpet composed of many parts. Each moss plant is composed of a central stalk less than 5 centimeters tall, with short, leaflike structures that are the sites of photosynthesis. There are over 10,000 species of mosses, and they are found anywhere there is adequate moisture.

The gametophytes of liverworts and hornworts are flat sheets only a few layers of cells thick. The name *liverwort* comes from the fact that these plants resemble the moist surface of a liver. There are about 8,000 species of liverworts. Hornworts derive their name from the presence of a long, slender sporophyte, which protrudes from the flat gametophyte plants. Their cells are unusual among plants, because they contain only one, large chloroplast in each cell, whereas other plants have many chloroplasts per cell. There are about 400 species of hornworts found throughout the world. Figure 22.5 shows examples of nonvascular plants.

22.5 The Significance of Vascular Tissue

A major step in the evolution of plants was the development of vascular tissue. Plants like ferns, pines, flowering plants and many others have vascular tissue. **Vascular tissue** consists of tube-like cells that allow plants to efficiently transport water and nutrients about the plant. The presence of vascular tissue is associated with the development of roots, leaves, and stems. **Roots** are underground structures that anchor the plant and absorb water and minerals. **Leaves** are structures specialized for carrying out the process of photosynthesis. **Stems** are structures that connect the roots with the leaves and position the leaves so that they receive sunlight.

There are two kinds of vascular tissue: *xylem* and *phloem*. **Xylem** consists of a series of dead, hollow cells arranged end to end to form a tube. The walls of these "cells" are strengthened with extra deposits of cellulose and a complex material called lignin. The plant's roots absorb water and minerals from the soil into the roots. The xylem carries water and minerals up from the roots through the stem to the leaves. There are two kinds of xylem cells: vessel elements and tracheids. Vessel elements are essentially dead, hollow cells, up to 0.7 mm in diameter, in which the endwalls are missing. Thus, vessel elements form long tubes similar to a series of pieces of pipe hooked together. Tracheids are smaller in diameter and consist of cells with overlapping, tapered ends. Holes in the walls allow water and minerals to move from one tracheid to the next (figure 22.6).

Phloem carries the organic molecules (primarily, sugars and amino acids) produced in the leaves to other parts of the plant where growth or storage takes place. Growth takes place at the tips of roots and stems and in the production of reproductive structures (cones, flowers, fruits). The roots are

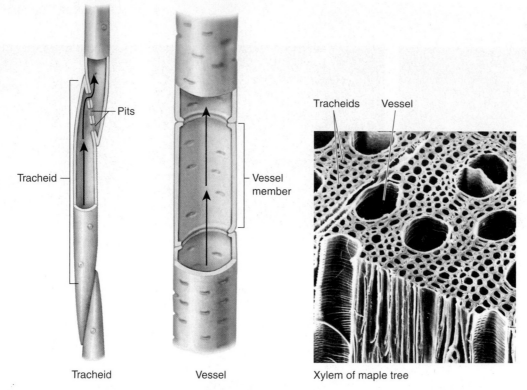

FIGURE 22.6 Xylem

Xylem consists of hollow, thick-walled cells. Vessel elements are connected end to end and have no endwalls. Tracheids have overlapping portions and pores that allow water to flow from one tracheid to the next.

typically the place where food is stored, but some plants store food in their stems. The phloem consists of two kinds of cells: sieve tube elements and companion cells. The sieve tube elements lack a nucleus and most organelles but retain a kind of cytoplasm. In addition, the sieve tube elements have holes in the endwalls that allow the flow of water and dissolved nutrients. The companion cells have direct connections to the sieve tube elements and assist in the movement of sugars and amino acids by active transport from cells in the leaves into the sieve tube elements (figure 22.7).

To be well adapted to a dry environment, plants need a waterproof layer on their surface in addition to vascular tissue. This layer reduces the amount of water they lose, and the presence of vascular tissue allows for the easy replacement of the water that is lost. Because vascular tissue allows for the more efficient transport of materials throughout the plant, it allows for an increase in plant size. Although not all vascular plants are large, many are able to become large, because vascular tissue allows them to transport water and nutrients efficiently (Outlooks 22.1).

22.6 The Development of Roots, Stems, and Leaves

Another factor associated with the presence of vascular tissue is the development of plant parts that are specialized for particular functions. A transport system allows for certain plant parts to specialize for certain functions (figure 22.8). Nearly all vascular plants have roots, stems, and leaves.

Roots

Roots are the underground parts of plants that anchor the plants in the soil and absorb water and nutrients, such as nitrogen, phosphorus, potassium, and other inorganic molecules from the soil. Vascular tissue allows these materials to be distributed for use by other parts of the plant. The vascular tissue in the roots transports water and nutrients to the stem and receives materials from the stem.

Roots grow from their tips. By growing constantly, roots explore new territory for available nutrients and water. As a plant becomes larger, it needs more root surface to absorb water and nutrients and to hold the plant in place. The actively growing portions of the root near the tips have many small, fuzzy, hairlike cell extensions called **root hairs,** which provide a large surface area for the absorption of nutrients and water.

Most roots are important storage places for the food produced by the above-ground parts of the plant. Many kinds of plants store food in their roots during the growing season and use this food to stay alive during the winter. The food also provides the raw materials necessary for growth for the next growing season. Although humans do not eat the roots of plants such as maple trees, rhubarb, and grasses, their roots are as important to them in food storage as are those of carrots, turnips, and radishes (figure 22.9).

Stems

Stems are, in most cases, the above-ground structures of plants that support the light-catching leaves in which photosynthesis occurs. However, stems can vary considerably. Trees

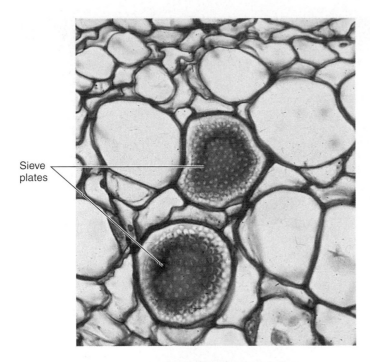

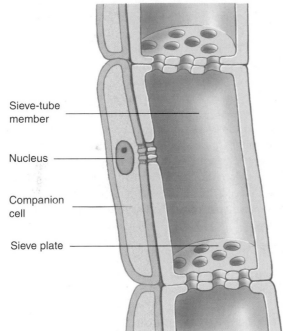

FIGURE 22.7 Phloem

Phloem consists of cells called sieve tube elements and companion cells. The endwalls of the sieve tube elements have holes that allow the movement of materials from one cell to the next. The cells lack a nucleus but have a form of cytoplasm. Companion cells have nuclei and help sieve tube elements transport materials.

have stems that support large numbers of branches; vines have stems that require support; some plants, such as dandelions, have very short stems, with all their leaves flat against the ground; and some stems are actually underground. Stems have two main functions:

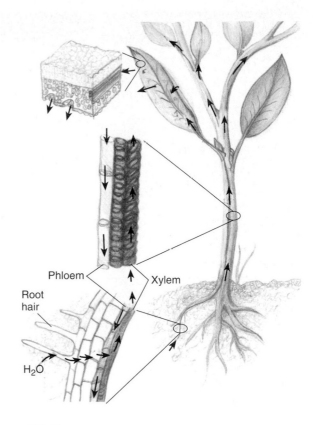

FIGURE 22.8 Vascular Tissue and the Specialization of Plant Parts

Vascular tissue allows for the specialization of leaves for photosynthesis, roots for the absorption of water and minerals, and stems for the positioning of the leaves. The xylem carries water and minerals from the roots to the leaves. The phloem transports organic molecules to places they are needed for growth or storage.

1. They support the leaves.
2. They transport raw materials from the roots to the leaves and manufactured food from the leaves to the roots.

The support that stems provide is possible because of the nature of plant cell walls. First, all plant cells are surrounded by a cell wall made of cellulose fibers interwoven to form a box, within which the plant cell is contained. Because the cell wall consists of fibers, it is like a wicker basket. There are spaces between cellulose fibers through which materials pass relatively easily. However, the cellulose fibers do not stretch; if the cell is full of water and other cellular materials, it becomes quite rigid. Remember that the process of osmosis results in cells having an internal pressure. It is this pressure against the nonstretchable cell wall that makes the cell rigid. Many kinds of small plants, called *herbaceous plants*, rely primarily on this mechanism for support.

The second way stems provide support involves the thickening of cell walls. *Woody plants* have especially thick cell walls, which provide additional support. The xylem of woody plants has thick cell walls and lignin deposited in the

OUTLOOKS 22.1

Plant Terminology

There are several sets of terminology used to discuss plants that describe their nature but do not have any relationship to their taxonomy.

Herbaceous and woody
Herbaceous plants have few cells with thick cell walls and, thus, are generally small and easily damaged. Mosses, grasses, and many garden plants are herbaceous. *Woody* plants have a great deal of tissue with thick cell walls. The xylem and associated cells are important in strengthening woody plants. There are ferns, gymnosperms, and angiosperms that are woody.

Annual and perennial
Annual plants live one year. They germinate from seeds, grow, produce flowers and seeds, and die within 1 year. The seeds produced germinate the following year and the cycle contin-

ues . Most annual plants are angiosperms. *Perennial* plants live many years. Most kinds of plants are perennials, including mosses, ferns, horsetails, pines, and most flowering plants. Many perennial plants are woody, such as various kinds of trees. Other perennial plants produce above-ground parts that die back at the end of a growing season. The plant regrows above-ground parts from the roots each year, as do tulips, daffodils, rhubarb, and ferns.

Trees, Shrubs, and Herbs
A *tree* is a large, woody plant that usually has a single main stem with branches. There are tree ferns, most gymnosperms are trees, and many flowering plants are trees. A *shrub* is a perennial woody plant that generally has several main stems and is relatively short. *Herbs* are generally nonwoody and small.

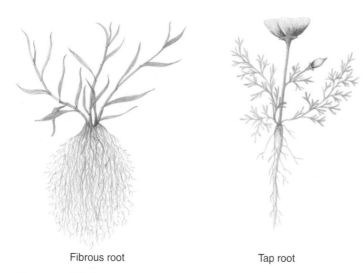

Fibrous root   Tap root

FIGURE 22.9 Kinds of Roots
The roots of grasses are often in the upper layers of the soil and form a dense network. This form of root is called a fibrous root and is an adaptation to relatively dry conditions. In grasslands, rainfall is infrequent and rarely penetrates very deeply into the soil. The network of roots near the surface efficiently captures the water when it is available. The roots of trees and many other plants typically extend deep into the soil, where they obtain moisture and serve as anchors to hold large plants upright. This kind of root is called a tap root.

cell walls that provides strength and binds cell walls to one another. This combination of thick cell walls with strengthening lignin is such an effective support mechanism that large trees and bushes are supported against the pull of gravity and can withstand strong winds for centuries. Some of

the oldest trees on Earth have been growing for several thousand years.

Another important function of the stem is to transport materials between the roots and the leaves (figure 22.10). A cross section of a stem examined under a microscope reveals that a large proportion of the stem consists of vascular tissue.

In addition to support and transport, the stems of some plants have additional functions. Some stems store food. This is true of sugar cane, yams, and potatoes. In addition, many plant stems are green and, therefore, are involved in photosynthesis. Stems also have a waterproof layer on the outside. In the case of herbaceous plants, it is usually a waxy layer. In the case of woody plants, there is a tough outer waterproof bark.

Leaves

Leaves are the specialized parts of plants that are the major sites of photosynthesis. (See chapter 7 for a discussion of photosynthesis.) To carry out photosynthesis, leaves must have certain characteristics (figure 22.11). Because it is a solar collector, a leaf should have a large surface area. Also, most leaves are relatively thin, compared with other plant parts. Thick leaves would not allow the penetration of light to the maximum number of photosynthetic cells. Throughout the leaf are bundles of vascular tissue, which transport water and minerals to the photosynthesizing cells and sugars and other molecules from these cells. The thick walls of the cells of vascular tissue also provide support for the leaf. In addition, the leaves of most plants are arranged so that they do not shade one another. This assures that the maximum number of cells in the leaf will be exposed to sunlight.

A drawback to having large, flat, thin leaves is an increase in water loss through evaporation. To help slow water loss, the outermost layer of cells, known as the epidermal layer, has

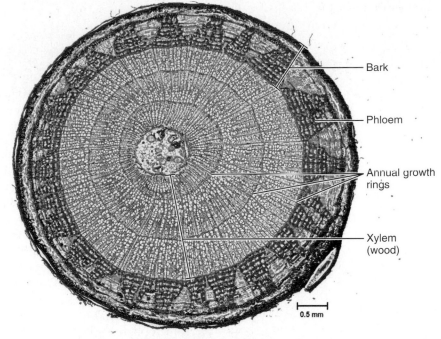

Bark

Phloem

Annual growth rings

Xylem (wood)

0.5 mm

Cross section of basswood stem

FIGURE 22.10 Cross Section of a Stem

This photo shows the cross section of a 3-year-old basswood tree. On the outside is the bark, which contains the phloem. Inside the bark are three layers of xylem tissue. Each layer of xylem constitutes 1 year's growth.

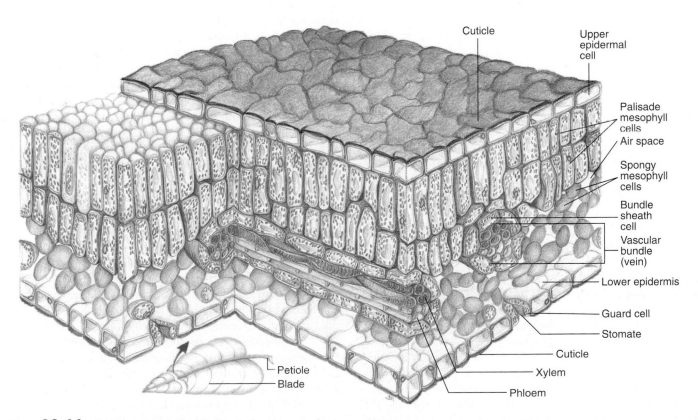

Cuticle

Upper epidermal cell

Palisade mesophyll cells

Air space

Spongy mesophyll cells

Bundle sheath cell

Vascular bundle (vein)

Lower epidermis

Guard cell

Stomate

Cuticle

Xylem

Phloem

Petiole

Blade

FIGURE 22.11 The Structure of a Leaf

Although a leaf is thin, it consists of several specialized layers. An outer epidermis lacks chloroplasts and has a waxy cuticle on its surface. In addition, the epidermis has openings, called stomates, that can open and close to regulate the movement of gases into and out of the leaf. The internal layers have many cells with chloroplasts, air spaces, and bundles of vascular tissue all organized so that photosynthetic cells can acquire necessary nutrients and transport metabolic products to other locations in the plant.

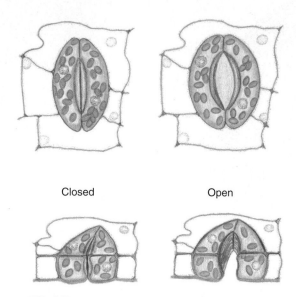

FIGURE 22.12 Stomates

The stomates are located in the covering layer (epidermis) on the outside of leaves. When these two elongated guard cells are swollen, the space between them is open and leaves lose water and readily exchange oxygen and carbon dioxide. In their less rigid and relaxed state, the two stomatal cells close. In this condition, the leaf conserves water but is not able to exchange oxygen and carbon dioxide with the outside air.

a waxy, waterproof coat on its outer surface. However, some exchange of gases and water must take place through the leaf. When water evaporates from the leaf, it creates a negative pressure, which tends to pull additional water and dissolved minerals through the xylem into the leaf, a process called *transpiration*. Because too much water loss can be deadly, the leaf must regulate transpiration. The amount of water, carbon dioxide, and oxygen moving into and out of the leaves of most plants is regulated by many tiny openings in the epidermis, called *stomates* (figure 22.12). The stomates can close or open to control the rate at which water is lost and gases are exchanged. Often during periods of drought or during the hottest, driest part of the day, the stomates are closed, reducing the rate at which the plant loses water.

22.7 Seedless Vascular Plants

Seedless vascular plants include the whisk ferns, horsetails, club mosses, and ferns. All of these plants have vascular tissue, but they do not produce seeds. These organisms are evolutionary links between the nonvascular plants, such as mosses, and the highly successful seed-producing plants.

With fully developed vascular tissues, these plants are not limited to wet areas, as are the mosses. They can absorb water through their roots and distribute it to leaves many meters above the surface of the soil. However, because they have swimming sperm, they must have moist conditions at least for a part of their life cycle.

The Fern Life Cycle

Although there are differences in the details of the life cycles of seedless vascular plants, we will use the life cycle of ferns as a general model. The conspicuous part of the life cycle is the diploid sporophyte generation. The leaves of most ferns are complex, branched structures commonly called fronds. The sporophyte produces haploid spores by meiosis in special structures of the leaves. The spore-producing parts are typically on the underside of the leaves in structures called sori. However, some species have specialized structures whose sole function is the production of spores. The spores give rise to a haploid heart-shaped gametophyte, which has archegonia and antheridia. The sperm swims to the archegonium and fertilizes the egg. Typically, the sperm fertilizes an egg of a different gametophyte plant. Following fertilization of the egg, the diploid zygote derives nourishment from the gametophyte and begins to grow. Eventually, a new sporophyte grows from the gametophyte to begin the life cycle again. Figure 22.13 illustrates the life cycle of the fern.

Kinds of Seedless Vascular Plants

Ferns are the most common of the seedless vascular plants. There are about 12,000 species of ferns in the world. They range from the common bracken fern, found throughout the world, to tree ferns, seen today in some tropical areas (figure 22.14). Along with horsetails and club mosses, tree ferns were important members of the forests of the Carboniferous period.

Whisk ferns are odd plants that lack roots and leaves. They appear to be related to ferns. They are anchored in the soil by an underground stem, which has filaments growing from it that absorb water and soil nutrients. It has flattened structures similar to leaves, but they are not considered leaves because they lack vascular tissue. Whisk ferns are most common in warm, moist environments. There are about 12 species. The spores are produced at the ends of branches; the spores produce a gametophyte in the soil.

Horsetails are low-growing plants with jointed stems. The leaves are tiny and encircle the stem at the joints. Much of the photosynthesis actually occurs in the green stem. There are about 30 species. They store silicon dioxide in their cells and, so, are often rough and called scouring rushes. The spores are produced in structures at the end of the stems. The spores give rise to gametophytes, which produce either eggs or sperm. Tree-sized members of this group are well known from the fossil record of the Carboniferous period 354 to 290 million years ago.

Club mosses are usually evergreen, bright green, low-growing, branching plants. Because they are evergreen and have a slight resemblance to small pine trees, some kinds are called ground pines. Their leaves are small and have only a single bundle of vascular tissue. Over 1,000 species are alive today. Like the whisk ferns and horsetails, they produce spores in structures at the end of stems. The spores give rise to gametophytes. Like the horsetails, these plants were prominent trees-like plants in ancient Carboniferous forests. Figure 22.15 shows examples of whisk ferns, club mosses, and horsetails.

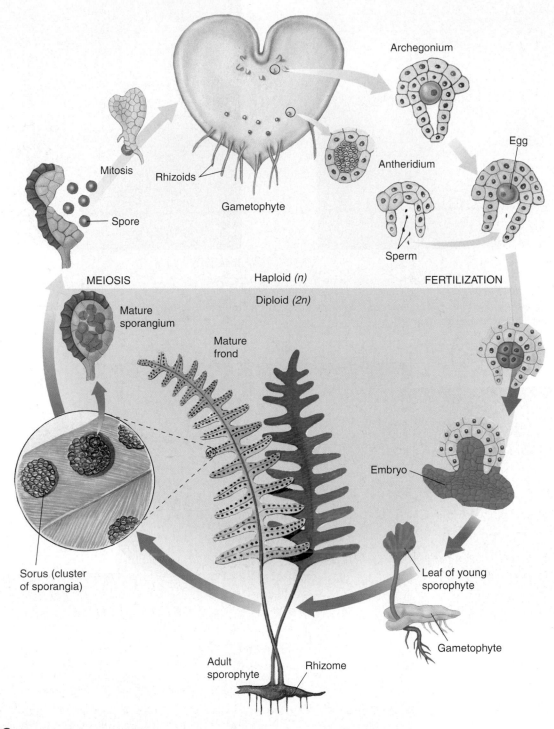

FIGURE 22.13 The Life Cycle of a Fern

The sporophyte of a fern consists of an underground stem (rhizome) and roots and above-ground leaves. The leaves are known as fronds. On the back of some fern leaves are small dots called sori. Each sorus contains structures that produce spores by meiosis. These haploid spores develop into the heart-shaped gametophyte. The gametophyte has archegonia, which produce eggs, and antheridia, which produce sperm. The sperm swim to the egg. The fertilized egg (zygote) is diploid and grows into a new sporophyte plant.

Bracken fern

Tree fern

FIGURE 22.14 Typical Ferns

There are many different kinds of ferns. The bracken fern is one of the most common. In some places, it is a weed that people try to eliminate from fields. The largest ferns are tree ferns.

Whisk fern

Horsetail

Club moss

FIGURE 22.15 Whisk Ferns, Club Mosses, and Horsetails

In addition to ferns, whisk ferns, club mosses, and horsetails are plants that have vascular tissue but do not have seeds. Whisk ferns are primarily tropical. They are odd plants that lack leaves and roots. Part of the stem is underground; the above-ground part is green and carries on photosynthesis. Club mosses and horsetails are common throughout North America.

22.8 Seed-Producing Vascular Plants

The most successful plants on Earth today are those that produce *seeds*. A **seed** is a specialized structure that contain an embryo, along with stored food, enclosed in a protective covering called the seed coat. The seed coat allows the embryo to resist drying. Thus, the seed is an adaptation that allows plants to live in dry terrestrial settings. Seeds are also important as dispersal devices for plants. Two major groups of plants produce seeds: the **gymnosperms** (conifers and their relatives) and the **angiosperms** (flowering plants). There were also extinct fernlike plants that had a kind of seed.

Another important innovation of seed-producing plants is the development of a life cycle in which *pollen* is produced. **Pollen** is the miniaturized male gametophyte generation. Pollen can be carried from one plant to another by wind or animals which makes the presence of water for fertilization unnecessary.

Gymnosperms

Several kinds of woody plants produce seeds that are not enclosed. Typically, their seeds are produced in woody structures called **cones**. The seeds are produced on the surface of flat, woody parts of the cone (figure 22.16). Because the seeds are not enclosed, they are said to be *naked*. Thus, these cone-producing plants, such as conifers, are called *gymnosperms*, which means naked seed plants. However, there are several distinct kinds of plants in this group: the cycads, ginkgos, and conifers. All are woody perennial plants. We will examine the life cycle of the pine tree to illustrate the life cycle of a gymnosperm.

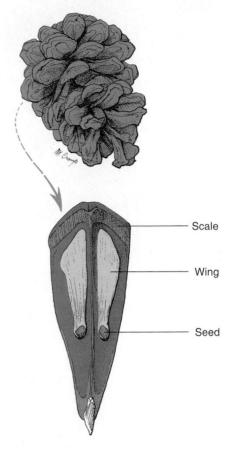

FIGURE 22.16   A Pine Cone with Seeds
On the scaly, leaflike portions of the cone are the seeds. Because these cones produce seeds "out in the open," they are aptly named the naked seed plants—gymnosperms. Pine tree seeds can be harvested by simply shaking a dried, open female cone. The seeds appear as small, brown, papery, winglike structures.

The Pine Life Cycle

The dominant portion of the pine life cycle is the large, diploid sporophyte generation. Pines produce two kinds of cones, which are part of the sporophyte generation but produce separate haploid male and female gametophytes. The conspicuous cones on pines are the female cones, which produce the female gametophyte generation. The female gametophyte consists of several thousand cells and produces several archegonia, each of which contains an egg.

The small male cones are located on the ends of branches. The male cones produce pollen. Each of these small, dustlike pollen grains contains haploid nuclei, which function as sperm. Pollen is released in such large quantities that clouds of pollen can be seen in the air when sudden gusts of wind shake the branches of the trees. Pollen is carried by wind to the female cone. The process of getting the pollen from the male cone to the female cone is called **pollination.** The pollen grain (male gametophyte) begins to grow a tubelike structure, which enters the archegonium and releases a sperm nucleus, which fertilizes the egg. The processes of pollination and fertilization are separate events. Pollination occurs with the transfer of pollen to the female cone. Fertilization occurs when the sperm cell from the pollen unites with the egg cell in the archegonium. Fertilization in gymnosperms may occur months or even years following pollination.

The fertilized egg (zygote) is diploid and develops into an embryo within a seed. Many kinds of birds and mammals use the seeds of pines and related trees for food. In most pines, the seeds are released from the cone when the scales of the cone fold back. The winged seeds are carried by wind and fall to the ground. The seed germinates and gives rise to a new sporophyte plant, and the life cycle continues. Figure 22.17 reviews the life cycle of the pine.

Thus, in addition to vascular tissue, the pine life cycle has two significant innovations that reduce the plant's dependence on water. The seed contains an embryo sporophyte plant along with some stored food, surrounded by a seed coat that reduces water loss. Thus, the seed can withstand dry periods. The production of pollen and the process of pollination also reduce the need for water. The sperm do not need to swim to the egg. The entire male gametophyte (pollen) is transported to the female gametophyte by wind. Then, the male gametophyte transfers the sperm nucleus to the egg,

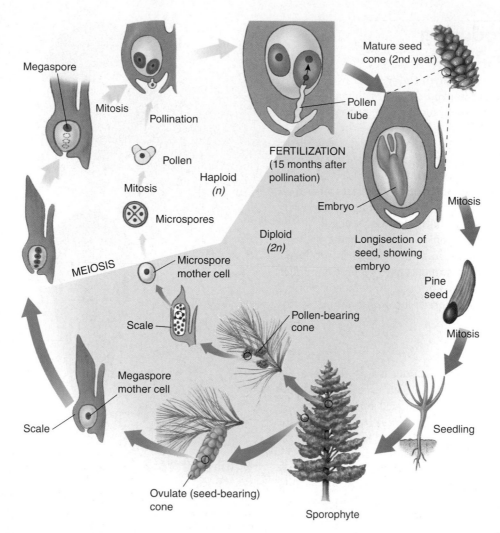

FIGURE 22.17 The Life Cycle of a Pine
The pine tree is the sporophyte plant. It produces two kinds of cones: small, male cones, which produce pollen, and larger, female cones, which contain the eggs and ultimately produce the seeds. The pollen is carried to the female cone by wind. After pollination, fertilization of the egg occurs when a pollen tube grows to the vicinity of the egg. The fertilized egg develops into an embryo enclosed within a seed.

when the pollen grain grows a tube that releases a sperm nucleus to fertilize the egg.

Kinds of Gymnosperms

Cycads are stout, woody gymnosperms that have a ring of fernlike leaves at the top. They live in tropical regions and are minor parts of the landscape. During the Jurassic period (144 to 213 million years ago), cycads were major organisms in the forests. The seeds are produced in tough, woody, conelike structures. There are about 300 species, and half of them are in danger of extinction.

Ginkgo trees were common in the Jurassic, but today there is only one species (*Ginkgo biloba*). It is a tree with fan-shaped leaves. Humans eat the seeds, and the leaves are used in many herbal medicines.

Conifers are the common trees and shrubs that bear seeds in cones. Most conifers have needle-shaped leaves. Because they retain their leaves throughout the year, they are often called evergreens. Although they have needles throughout they year, they do shed needles a few at a time throughout the year, as evidenced by the mat of needles under a conifer. A few conifers—for example, *Larix* (tamarack) and *Taxodium* (bald cypress), lose their leaves all at once in the fall. There are over 600 species of conifers, and about half are considered threatened. Many kinds of conifers are important in the production of lumber. Figure 22.18 shows the three types of gymnosperms.

Angiosperms

Angiosperms are plants that produce flowers and have their seeds enclosed in a *fruit*. A **fruit** is a modification of the ovary wall into a special structure that contains the seeds. Like the gymnosperms, the flowering plants are well adapted to terrestrial life with vascular tissue, a seed, and pollen.

Conifer (pine)

Cycad

Ginkgo

FIGURE 22.18 Several Gymnosperms
There are three common types of gymnosperms: conifers, such as pines, cedars, spruces, and firs; cycads, which are tropical and have fernlike leaves; and a single species of ginkgo.

Flower Structure

A **flower** is the structure, composed of highly modified leaves, that is responsible for sexual reproduction. At the center of a typical flower is the **pistil,** which is composed of the *stigma, style,* and *ovary.* The ovary produces the female gametophyte plant. Surrounding the pistil are the **stamens,** which consist of a long *filament* with *anthers* at the top. The anther produces the pollen (male gametophyte plant). **Petals** are a whorl of modified leaves surrounding the stamens and pistil. In many flowering plants, the petals are large and showy. Outside the petals is another whorl of modified leaves, known as **sepals.** Many kinds of flowers produce odors and a sugary liquid, known as nectar. Figure 22.19 shows the arrangement of parts in a typical flower.

There is a great degree of specialization of flowers. Some are large and showy, such as the flowers of roses and magnolias. Others are small and inconspicuous, such as the flowers of grasses and birch trees. In addition, some plants have two kinds of flowers. One kind has a pistil, whereas a different flower has the stamens. Any flower that has both pistil and stamens is called a **perfect flower;** a flower containing just a pistil or just stamens is called an **imperfect flower.**

The Life Cycle of a Flowering Plant

The life cycle of a flowering plant has both sporophyte and gametophyte generations. The sporophyte is the dominant stage of the life cycle, and the male and female gametophytes are produced within the flower. As in gymnosperms, the male gametophyte is the pollen. The female gametophyte is found within the ovary of the pistil and consists of only eight cells. One of the cells is an egg cell.

Pollination occurs when pollen is transferred from an anther to the stigma of the pistil. In some cases, pollination must be between flowers of different plants of the same species; this is called cross-pollination. However, some species of plants are able to pollinate themselves; this is called self-pollination.

The pollen grain germinates and produces a pollen tube, which grows down through the tissue of the style to the ovary, where the female gametophyte is located. The pollen tube releases two sperm. One fertilizes the egg nucleus and gives rise to the zygote. The other sperm nucleus combines with two other nuclei in the female gametophyte and produces a triploid nucleus, which develops into endosperm. The endosperm is the stored food of the seed. Both the embryo and the endosperm grow, and a seed coat develops around them to produce the mature seed. The wall of the ovary develops into a specialized seed-containing structure known as a fruit (figure 22.20).

Pollination Strategies

Plants use several strategies to ensure pollination. Most plants that have inconspicuous flowers are wind pollinated. They produce large numbers of flowers and huge amounts of pollen. This is necessary because the distribution of pollen by wind is a random process. By producing large amounts of pollen, the plant increases the chances that pollen will be transferred successfully. Grasses, sedges, and some other herbaceous plants are wind pollinated. Many trees, such as aspens, birches, and oaks, are also wind pollinated.

It is the pollen from wind-pollinated plants that is responsible for "hay fever," an allergic reaction to the presence of air borne pollen. Most tree species produce pollen in the spring, so people who have symptoms in the spring are usually reacting to the pollen of certain trees. Grasses and other herbaceous plants typically produce their pollen in late

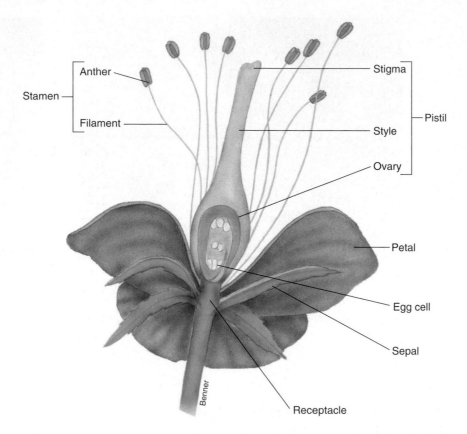

FIGURE 22.19 Flower Structure

The flower is the structure in angiosperms that produces sex cells. The female gametophyte is produced within the ovary. The pollen (male gametophyte) is produced within the anther. Following fertilization of the egg, the ovary modifies to form the fruit.

summer or fall. Ragweed is a common plant that produces pollen late in the growing season.

Plants with large, showy flowers are typically pollinated by animals. Although many showy flowers are pollinated by insects, such as bees, others are constructed to be pollinated by birds or small mammals. The animals feed on the nectar produced by the flower; in the process, they are dusted with pollen and carry it to another flower of the same species. Many flowers produce odors to help the animals find the plants. This is a mutualistic relationship between the flowers, which are pollinated, and the animals, which receive food in the form of nectar or pollen (figure 22.21).

Fruit

Because plants cannot move, they must have a method of dispersing their seeds to new locations. Most fruits are involved in the dispersal of plants.

Some fruits—for example, apples, watermelon, tomatoes, and raspberries—contain large amounts of nutritious materials. The nutrients are not for the plant's use but, rather, attract animals that eat the seeds along with the fruit. The seeds pass through the animals' digestive tract unharmed, dispersing the plant's offspring. Other fruits—such as cottonwoods, milkweeds, and dandelions—release fluffy seeds, which are carried by the wind. Many trees—such as maples and ashes—produce fruits with wings, which aid their dispersal by wind. Some have hooks or sticky surfaces, which become attached to the fur or feathers of passing animals. Figure 22.22 shows several kinds of fruits.

Angiosperm Diversity

Flowering plants are the most diverse group of plants, with about 260,000 species. Botanists classify all angiosperms into two groups: *dicots* or *monocots*. The names *dicot* and *monocot* refer to structures in the seeds of these plants called cotyledons. **Cotyledons,** also known as seed leaves, are embryonic leaves that have food stored in them. They are the first leaves that emerge when a seed germinates. A **monocot** has one cotyledon, and a **dicot** has two cotyledons (figure 22.23). A peanut is a dicot, as are lima beans and apples. Grasses, lilies, palms, and orchids are all monocots. Figure 22.24 lists the differences between monocots and dicots.

Even with this separation into monocots and dicots, the diversity is staggering. Although some monocots are woody species, such as yuccas and palms, most monocots are herbaceous. Many of these plants grow from underground structures, in which they store food. Many important food plants

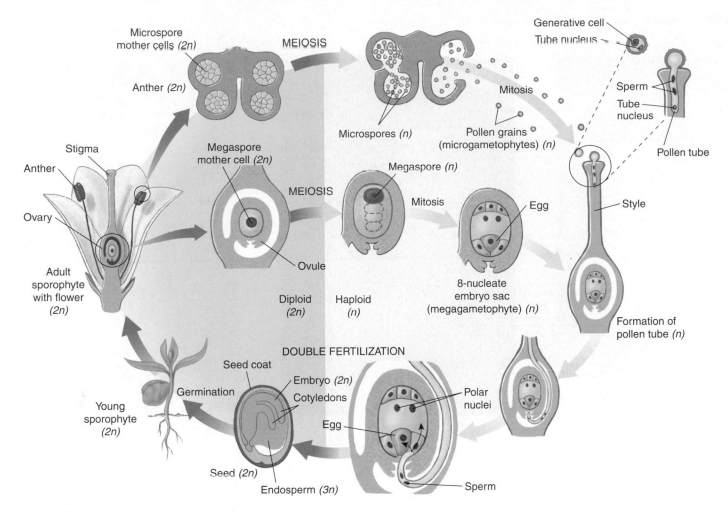

FIGURE 22.20 Life Cycle of an Angiosperm

Angiosperms have flowers, which produce eggs in the ovary and pollen in the anther. Pollen is carried to the pistil by wind or animals. A pollen tube releases haploid nuclei, fertilizing the egg. The fertilized egg develops into the embryo, and the remaining cells associated with the egg develop into a food storage material called endosperm. A seed coat develops around the embryo and endosperm. The ovary enlarges and becomes the fruit, which contains the seeds. The seed germinates to produce the next generation of flowering plant.

Wind-pollinated

Insect-pollinated

FIGURE 22.21 Wind- and Insect-Pollinated Flowers

The flowers of wind-pollinated plants, such like this oak, are small and inconspicuous. The flowers of plants that are pollinated by animals are typically large and showy, as are these California poppies and lupines.

481

Burdock

Milkweed

Raspberry

FIGURE 22.22 Types of Fruits
From a plant's point of view, fruits are useful as methods of dispersal. Each of these kinds of fruits uses a different method of dispersal. Burdock has hooks, which stick to passing animals. Milkweed fruits open to release seeds, which are carried by the wind. Raspberries are eaten by birds or other animals. The raspberry seeds pass through the gut and are distributed in the feces.

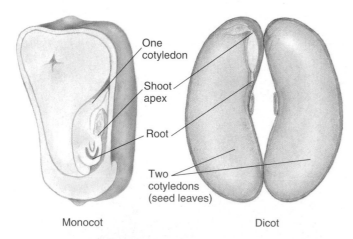

One cotyledon

Shoot apex

Root

Two cotyledons (seed leaves)

Monocot

Dicot

FIGURE 22.23 Embryos in Dicots and Monocots
The embryos of plants have embryonic leaves, called seed leaves or cotyledons, attached to the embryonic stem. The monocots have one cotyledon and the dicots have two.

are monocots, including wheat, rice, corn, sweet potatoes, yams, onions, and bananas. (How Science Works 22.1)

Common herbaceous dicots, such as mints, carrots, cabbages, mustards, tomatoes, potatoes, and peppers, are important food plants. About half of the dicots are woody trees and

shrubs, such as aspen trees and sagebrush. These plants add layers of new xylem and phloem each year, as the stem and root of the plant get larger in diameter. The phloem is in the outermost part of the stem called the bark. Dicot trees and shrubs generally produce broad, flat leaves. In colder parts of the world, most lose all their leaves during the fall. Such trees are said to be **deciduous** (figure 22.25). However, there are exceptions. Some are **nondeciduous,** keeping their leaves and staying green throughout the winter—for example, American holly (*Ilex opaca*). In tropical and semitropical regions, many of these trees shed leaves one at a time, so they retain their leaves throughout the year. In areas that have pronounced changes in rainfall, many trees and shrubs lose their leaves during the dry part of the year.

22.9 The Growth of Woody Plants

All gymnosperms and a large proportion of the dicots are woody plants. Although these two groups of plants have different evolutionary histories, they share the ability to grow continuously for many years. The trees get taller and larger in diameter each year. Both the stems and the roots grow from their tips and add to their length. In addition, they increase in diameter by adding new xylem and phloem to the outside of the stem.

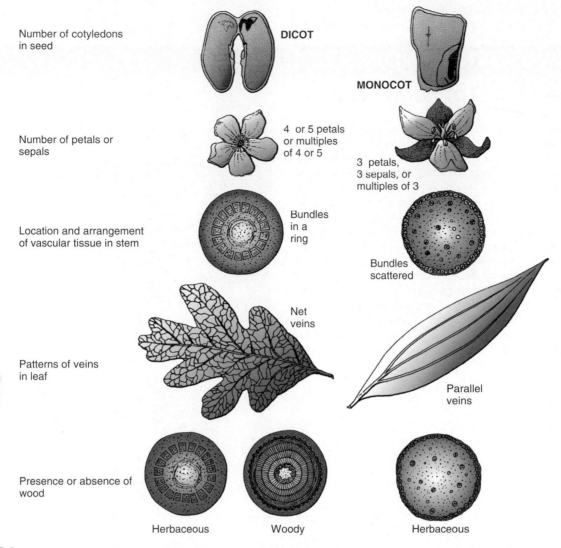

Number of cotyledons in seed

DICOT

MONOCOT

Number of petals or sepals

4 or 5 petals or multiples of 4 or 5

3 petals, 3 sepals, or multiples of 3

Location and arrangement of vascular tissue in stem

Bundles in a ring

Bundles scattered

Patterns of veins in leaf

Net veins

Parallel veins

Presence or absence of wood

Herbaceous   Woody   Herbaceous

FIGURE 22.24 A Comparison of Structures in Dicots and Monocots

Botanists classify all angiosperms into the monocots and the dicots. Several general characteristics help differentiate these two groups.

FIGURE 22.25 Deciduous Trees and Fall Colors

In the fall, a layer of waterproof tissue forms at the base of each leaf, cutting off the flow of water and other nutrients. The cells of the leaf die and their chlorophyll disintegrates. The color change seen in leaves in the fall in certain parts of the world is the result of the breakdown of the green chlorophyll. Other pigments (red, yellow, orange, brown) are always present but are masked by the green chlorophyll pigments. When the chlorophyll disintegrates, the reds, oranges, yellows, and browns are revealed.

HOW SCIENCE WORKS 22.1

The Big 10 of Food Crops

Modern agriculture has modified crops genetically, investigated the conditions needed for their optimal growth, and provided inputs in the form of fertilizers and herbicides to improve yields. All of these activities have required major research efforts around the world. Although there are about 300,000 species of plants, about 10 of them constitute the majority of the plant foods consumed by humans.

Wheat, rice, maize (corn), and barley are grasses that produce seeds containing large amounts of starch. In many parts of the world, most of the calories people consume come from one of these grains. Sugarcane is also a grass; its stems are crushed to extract sugar. The sugar beet is also used to produce sugar. Potato, sweet potato, and manioc (cassava) are grown for their starchy underground storage structures. The potato is actually an underground stem. Soybeans are grown for their nutritious seeds. These 10 crops account for about 80% of the food produced in the world.

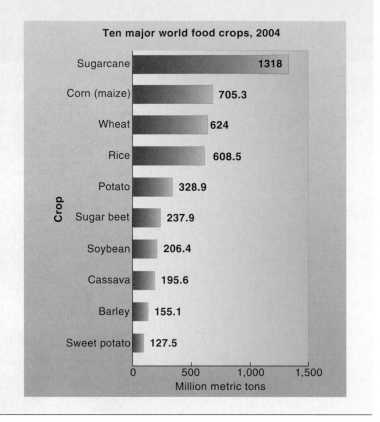

As a tree becomes larger, the strengthening tissue in the stem becomes more and more important. A layer of cells in the stem, called the **vascular cambium,** is responsible for this increase in diameter. The vascular cambium is between the xylem and the phloem. Xylem tissue is the innermost part of the tree trunk or limb, and phloem is outside. The bark of the tree, containing the phloem, is important in protecting the underlying tissues. Cambium cells go through a mitotic cell division, and two cells form. One cell remains cambium tissue, and the other specializes to form vascular tissue. If the cell is on the inside of the cambium ring, it becomes xylem; if it is on the outside of the cambium ring, it becomes phloem. As cambium cells divide again and again, one cell always remains cambium, and the other becomes vascular tissue. Thus, the tree constantly increases in diameter (figure 22.26).

The accumulation of the xylem in the trunk of gymnosperms and woody angiosperms is called **wood.** Wood is one of the most valuable biological resources of the world. We get lumber, paper products, turpentine, and many other valuable materials from the wood of gymnosperms and angiosperms.

As you can see from this discussion, both angiosperms and gymnosperms have members that are woody. Therefore the term *woody* cuts across taxonomic lines. Outlooks 22.1 discusses other common terms that apply across taxonomic lines.

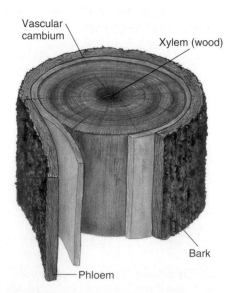

FIGURE 22.26 A Cross Section of Woody Stem
Notice that the xylem makes up most of what we call wood. The approximate age of a tree can be determined by counting the growth rings seen on the cut surface. It is also possible to learn something about the environment from these rings. Wide rings indicate good growth years, whereas narrow rings indicate poor growth.

22.10 Plant Responses to Their Environment

Our casual impression of plants is that they are unchanging objects. However, on closer examination we recognize that plants change over time. They grow, produce flowers and fruits at certain times of the year, and grow toward a source of light. Furthermore, they respond to organisms that harm them and may even mount an attack against competitors.

Tropisms

A **tropism** is a growth movement toward or away from a stimulus. For example, plants orient themselves toward light. This is known as *phototropism*. The value of this response is obvious, because plants need light to survive. The mechanism that allows this response involves a hormone known as auxin. The growing tip of the stem produces auxin, which is transported down the stems. Auxin stimulates cells to divide, grow, and elongate. If the growing tip of a plant receives more light on one side, the shaded side produces more auxin than does the lighted side. The larger amount of auxin on the shaded side causes greater growth in that area, and the tip of the stem bends toward the light. When all the sides of the stem are equally illuminated, the stem grows equally on all sides and grows straight. House plants near a window must be turned regularly, or they will grow more on one side than the other (figure 22.27).

Many kinds of climbing vines are able to wrap stringlike *tendrils* around sturdy objects in a matter of minutes. This kind of growth response is called *thigmotropism*. As the tendrils grow, they slowly wave about. When they encounter an object, their tropic response is to wrap around it and anchor the vine.

FIGURE 22.28 Thigmotropism—Clinging Stems
Some stems are modified to wrap themselves around objects and give support. The tendrils of this grapevine are a good example.

Once attached, the tendrils change into hard structures that bind the vine to its attachment. Sweet peas, grapevines, and ivy attach in this manner. Ivy can cause great damage as it grows, and its tendrils loosen siding and serve as a haven for the growth of other destructive organisms (figure 22.28).

Plants also have growth responses to gravity and the presence of water. Stems grow up; roots grow down and toward water in the soil.

Seasonal Responses

Plants live in a seasonal world, and they respond accordingly. Plants that live in temperate regions experience changes in temperature and the amount of sunlight. The days are longest in the summer and shortest in the winter. Plants are able to measure day length and manufacture hormones that cause changes in the growth and development of specific parts of the plant. Some plants produce flowers only when the days are getting longer, some only when the days are getting shorter, and some only after the days have reached a specific length. In deciduous trees, the shortening day length in the fall triggers a response that causes the tree to produce a wall of waterproof tissue between the leaves and the stem. Because the leaves no longer receive water, they die and fall to the ground.

Other plants store food materials in underground structures such as bulbs and tubers. The above-ground part of the plant dies back, but the underground portion remains. In the spring, as the soil warms, these structures send up new growth. In both of these cases, the plant goes into a dormant state. A portion of the plant is still alive, but it is not actively photosynthesizing.

FIGURE 22.27 Phototropism
These tomato plants were growing in a window and have oriented themselves to the light coming through the window.

Many parts of the world experience seasonal rainfall. Rain is abundant for several months and totally absent during others. Plants' responses to these conditions are similar to those of plants that respond to seasonal changes in temperature. During the dry season, the leaves fall from trees and many smaller plants become dormant. When it rains, the plants respond by growing, flowering, and fruiting.

Responses to Injury

Plants are attacked by a variety of disease-causing organisms and by plant-eating herbivores. Plants respond to disease in much the same way as animals do. When infected with a disease organism, they have an innate ability to fight the infection. Plants also repair wounds by growing a form of scar tissue over the wounded area.

Attack by herbivores is handled in a different way. Many plants produce toxic materials, which interfere with the metabolism of animals that eat plants. When the leaves of plants are eaten by animals, the new leaves produced to replace those lost often contain higher amounts of toxic materials than the original leaves.

Plants may even have the ability to communicate with one another. An experiment carried out in a greenhouse produced some interesting results. Some of the plants had their leaves mechanically "eaten" by an experimenter, whereas nearby plants were not harmed. Not only did the cut plants produce new leaves with more toxins, but the new growth on neighboring, nonmutilated plants had increased toxin levels as well. This raises the possibility that plants communicate in some way, perhaps by the release of molecules that cause changes in the receiving plant.

Many of the toxic chemicals produced by plants have been used as medications. Quinine has been used to fight malaria. Digitalis is used to strengthen heartbeat. Taxol is an anticancer drug.

22.11 The Coevolution of Plants and Animals

The first terrestrial organisms were plants. Shortly after the plants became established on land, animals, such as insects and amphibians, arrived. Thus, terrestrial plants and animals have a long history of interaction, which has had an influence on the evolution of each group. There are many examples of their coevolution.

Most flowering plants are pollinated by insects or other animals. Insect-pollinated plants produce flowers that are showy, have nectar, and produce odors. Many flowers that are pollinated by birds are red and produce much nectar. Some flowers bloom only at night and are pollinated by moths or bats.

Grasses and grazers have coevolved. Grasses have silica in their cell walls. This is a very hard material, and it tends to wear down the teeth of grazers. Most grazing animals have very long teeth, which can accommodate a lifetime of wear. Grasses also differ from most other plants in that their leaves and stems grow from the base of the plant rather than from the tip. Thus, they can withstand regularly having the tips of their leaves chewed off.

Many kinds of flowering plants produce large, nutritious fruits, which animals use for food, in the process distributing the seeds. There are even seeds that will not germinate unless they have passed through the gut of an animal. Birds eat small fruits and their seeds, which are dispersed when the birds defecate. In tropical forests, many trees have very large fruits, which are eaten by monkeys. They eat the fleshy part of the fruit and drop the seeds.

Browsers are animals that eat the leaves and small twigs of woody plants. Many kinds of woody plants have thorns, making this task more difficult, but browsers have techniques or structures that allow them to put up with the prickly deterrents.

Plants produce a variety of chemicals. Some of these chemicals are toxic or irritating and deter certain animals from eating portions of the plant. Others chemicals produce odors that help animals locate flowers or fruits. These attractive odors aid the plant by assuring that pollination and seed dispersal will take place. Humans use many of these plants and the chemicals they produce as spices and flavorings (Outlooks 22.2).

OUTLOOKS 22.2

Spices and Flavorings

Think about all the plant materials we use to season our foods. Black pepper comes from the hard, dried berries of a tropical plant, *Piper nigrum.* Cayenne pepper is made from the ground-up fruits of *Capsicum annuum,* and the hot, spicy chemical in the fruit and seeds is known as capsaicin. The seeds of the dill plant, *Anethum graveolens,* are used to flavor pickles and many other foods. The dried or fresh leaves of many herbs such as thyme, rosemary, chives, and parsley, are also used as flavorings. Many kitchen cabinets also contain cinnamon from the bark of a tree found in India; cloves, which are the dried flower buds of a tropical tree; ginger from the root of a tropical plant of Africa and China; nut-

meg from the seed of a tropical tree of Asia; and saffron from the dried, fragrant stigmas of the crocus flower (*Crocus sativus*) from Spain. Saffron's small size and difficulty in harvesting the stigmas make it the most expensive spice on Earth.

Centuries ago, spices such as these were so highly prized that fortunes were made in the spice trade. Beginning in the early 1600s, ships from Europe regularly visited the tropical regions of Asia and Africa, returning with cargoes of spices and other rare commodities that could be sold at great profit. Consequently, India has been greatly influenced by Britain, Indonesia has been greatly influenced by the Netherlands, and the development of various portions of Africa has been influenced by Britain and France.

Summary

The plant kingdom is composed of organisms that can manufacture their own food through photosynthesis. They have specialized structures for producing the male sex cell (sperm) and the female sex cell (egg). The relative importance of the haploid gametophyte and the diploid sporophyte that alternate in plant life cycles is a major characteristic used to determine an evolutionary sequence. The extent and complexity of the vascular tissue and the degree to which plants rely on water for fertilization are also used to classify plants as primitive (ancestral) or complex. Among the gymnosperms and the angiosperms, the methods of production, protection, and dispersal of pollen are used to name and classify the organisms into an evolutionary sequence. Based on the information available, mosses are the most primitive plants. Ferns, seed-producing gymnosperms, and angiosperms are the most advanced and show the development of roots, stems, and leaves. Plants respond to their environment in complex and interesting ways. They respond to the position of the sun, day length, contact with other objects, and injury.

Key Terms

*Use the interactive flash cards on the **Concepts in Biology,** 12/e website to help you learn the meaning of these terms.*

angiosperms 477	phloem 469
antheridium 468	pistil 479
archegonium 468	pollen 477
cones 477	pollination 477
cotyledons 480	root 469
deciduous 482	root hairs 470
dicot 480	seed 477
flower 479	sepals 479
fruit 478	sporophyte generation 466
gametophyte generation 466	stamens 479
gymnosperms 477	stem 469
imperfect flower 479	tropism 485
leaves 469	vascular cambium 484
monocot 480	vascular tissue 469
nondeciduous 482	wood 484
perfect flower 479	xylem 469
petals 479	

Basic Review

1. The sporophyte generation
   a. is diploid.
   b. produces spores.
   c. is the dominant stage in angiosperms.
   d. All of the above are correct.

2. Gymnosperms produce fruits. (T/F)

3. All of the following are vascular plants except
   a. ferns.
   b. mosses.
   c. horsetails.
   d. gymnosperms.

4. There are two kinds of vascular tissue: xylem and _____.

5. The central part of a woody stem consists of
   a. xylem.
   b. phloem.
   c. silica.
   d. bark.

6. Nearly all gymnosperms are woody perennials. (T/F)

7. Plants are able to sense changes in their environment. (T/F)

8. From a plant's point of view, a fruit
   a. is used to feed seeds.
   b. is used to disperse seeds.
   c. requires little energy to produce.
   d. All of the above are correct.

9. The pollination of plants by insects is an example of _____.

10. Which one of the following plant groups has the largest number of species?
    a. conifers
    b. ferns
    c. angiosperms
    d. mosses

Answers

1. d 2. F 3. b 4. phloem 5. a 6. T 7. T 8. b 9. coevolution
10. c

Concept Review

Answers to the Concept Review questions can be found on the Concepts in Biology, 12/e website.

22.1 What Is a Plant?

1. What do all plants have in common?

22.2 Alternation of Generations

2. Is the sporophyte generation haploid or diploid?
3. What reproductive cells do sporophytes produce? Are these structures haploid or diploid?

22.3 The Evolution of Plants

4. Why are green algae considered to be the ancestors of plants?
5. How does the significance of the sporophyte generation change with the evolution of advanced plants?

22.4 Nonvascular Plants

6. What sex cells are associated with antheridia? Archegonia?
7. What are the three major kinds of nonvascular plants?

22.5 The Significance of Vascular Tissue

8. What are the two kinds of vascular tissue, and what do they do?

22.6 The Development of Roots, Stems, and Leaves

9. How are the structures of roots, stems, and leaves related to their functions?

22.7 Seedless Vascular Plants

10. What kinds of vascular plants do not produce seeds?

22.8 Seed-Producing Vascular Plants

11. How is a seed different from pollen, and how do both of these differ from a spore?
12. Describe how wind- and insect-pollinated flowers differ.
13. Describe the major kinds of gymnosperms.
14. Describe how monocots and dicots differ.

22.9 The Growth of Woody Plants

15. What is the significance of the vascular cambium tissue in woody perennials?

22.10   Plant Responses to Their Environment

16   What is auxin? What role does it play in plant metabolism?

17.   What advantage does a plant with tendrils have?

22.11   The Coevolution of Plants and Animals

18.   Describe how the coevolution of grasses and grazing animals affected both kinds of organisms.

19.   In what way do some flowering plants encourage insects to visit them?

Thinking Critically

For guidelines to answer this question, visit the **Concepts in Biology,** *12/e website.*

Some people say the ordinary "Irish" potato is poisonous when the skin is green, and they are at least partly correct. A potato develops a green skin if the potato tuber grows so close to the surface of the soil that it is exposed to light. An alkaloid called *solanine* develops under this condition and may be present in toxic amounts. Eating such a potato raw can be dangerous. However, cooking breaks down the solanine molecules and makes the potato as edible and tasty as any other. The so-called Irish potato is of interest historically. Its country of origin is only part of the story. Check your local library to find out about this potato and its relatives. Are all related organisms edible? Where did this group of plants develop? Why is it called the Irish potato?

23

The Animal Kingdom

We humans are vertebrate mammals. We unconsciously view all other animals by comparing them with ourselves. If other animals are like us, we feel we understand them and can empathize with them. If they are not mammals like us, we often consider them to be weird. On the other hand, mammals are a tiny minority of all the animals that exist; 99.9% of all species of animals are invertebrates. Mammals are a minor ingredient in the whole collection of animals in the world.

Our parochial view of the world is evidenced in subtle ways. For example, most of the endangered species listed are vertebrate animals like us. About 7% of the known species of vertebrates are considered endangered. On the other hand, only about 0.17% of invertebrate species are considered endangered. We have organizations to save the whales, wild horses, and pandas but do not have organizations to save the poison arrow frog or endangered native crayfish and clams. This is more a reflection of our ignorance of these animal groups than their true importance.

This chapter presents an overview of the animal kingdom, with all of its variety and complexity.

CHAPTER OUTLINE

23.1 What Is an Animal?

Animals are eukaryotic, multicellular organisms whose cells lack cell walls. Like plants, animals have cells that are specialized for specific purposes. Functions such as ingesting food, exchanging gases, and removing wastes are more complicated in animals than in one-celled protists. In one-celled organisms, any exchange between the organism and the external environment occurs through the plasma membrane. However, because animals are multicellular, most of their cells are not on the body surface and, therefore, not in direct contact with the external environment. Thus, animals must have specialized ways to exchange materials between their internal and external environments.

Some simple animals, such as sponges and jellyfish, have a few kinds of specialized cells, whereas other, more complex animals, such as insects and mollusks, have bodies composed of groups of cells organized into tissues, organs, and organ systems. The more complex animals have gills, lungs, or other structures used to exchange gases; various kinds of structures are involved in capturing and digesting food; and there are usually special structures for getting rid of waste products. Like plants, many of the larger animals have a method of transporting materials throughout the body.

Animals are heterotrophs that eat other organisms to obtain organic molecules, and many are specialized for consuming certain kinds of food organisms. Although there are many ways of capturing and consuming food, all of them involve movement.

Movement is a characteristic associated with all animals; it involves specialized muscle cells that shorten. Many animals have appendages (e.g., legs; wings; tentacles; spines; or soft, muscular organs) that bend or change shape. Often, these structures are involved in moving the animal from place to place, but they are also used to capture food, to clean the animals' surfaces, and to move things in the animals' environment.

Sexual reproduction is important in all groups of animals. The methods vary greatly—from the release of sperm and eggs into the water, for many aquatic organisms, to fertilization inside the body, for most terrestrial organisms. A few kinds, such as birds and mammals, provide a great deal of care for their offspring. Many of the more primitive animals also reproduce asexually. Those that reproduce asexually also reproduce sexually at other times in their life cycle. Some animals reproduce asexually by a process called parthenogenesis. (Outlooks 23.1).

Animals also have sensory structures on their surface that sense changes in the environment. A system of nerve cells integrates information about the environment and coordinates movements, so that movements occur at appropriate times and with a proper orientation. Figure 23.1 shows a variety of animals.

23.2 The Evolution of Animals

Scientists estimate that the Earth is at least 4.5 billion years old and that life originated in the ocean about 3.8 to 3.7 billion years ago. The earliest animal-like fossils date to about

Frog

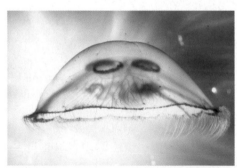
Jellyfish

Praying mantis

Snail

Starfish

FIGURE 23.1 Variety of Animals
Animals show an amazing variety of body forms and life styles.

600 million years ago. Most of the earliest animals were probably either small, planktonic organisms or wormlike organisms that crawled on the bottom or through the sediment on the ocean floor.

In the ocean, animals did not have a problem with dehydration. Because the ion content of the early ocean approximated that of the animals' cells, the animals needed little energy to keep their cells in osmotic balance. The temperature range in the ocean was not as great as that on land, and the rate of temperature change was lower. Therefore, animals in the ocean did not require mechanisms to deal with rapid or extreme changes in the environment.

Scientists have found examples of all the major groups of animals as far back as the Cambrian period, 540 to 490 million years ago. They were all marine animals, and many groups, such as sponges, jellyfish, mollusks, crustaceans, starfish, and many kinds of fish—are still primarily marine animals. Others have adapted to freshwater and terrestrial environments, but these are more recent developments. The first terrestrial animals (arthropods) are found in the fossils of the Silurian period, 443 to 417 million years ago. Major changes in the kinds of terrestrial animals occurred after that time. Two groups—arthropods (particularly the insects) and vertebrates (especially reptiles, birds, and mammals)—were extremely successful in terrestrial habitats.

It is common to divide animals into two categories: invertebrates and vertebrates. Animals with backbones made of vertebrae are called **vertebrates** and include various kinds of fishes, amphibians, reptiles, birds, and mammals. Those without backbones are called **invertebrates.** All early animals lacked backbones, and invertebrates still constitute 99.9% of all animal species in existence today.

The subsequent evolution of animals led to great diversity, and today animals are adapted to live in just about any environment. They live in all the oceans, on the shores of oceans, and in shallow freshwater. They live in the bitter cold of the Arctic, the dryness of the desert, and the driving rains of tropical forests. Some animals eat only other animals, some are parasites, some eat both plants and other animals, and some feed only on plants. They show a remarkable variety of form, function, and activity. They can be found in an amazing variety of sizes, colors, and body shapes.

The evolution of animals is complex and often difficult to interpret. Figure 23.2 shows the current knowledge of how various groups of animals are related.

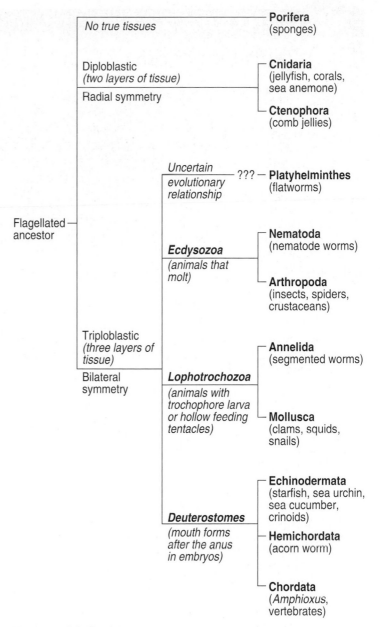

FIGURE 23.2 Animal Evolution
This diagram shows the current thinking about how the major groups of animals are related to one another. There are about 30 recognized phyla. Only the major phyla are shown here.

23.3 Temperature Regulation

The temperature of an organism affects the rate at which its metabolic reactions take place. Some animals do little to regulate their temperature, whereas others expend considerable energy to maintain a constant temperature. **Poikilotherms** are organisms whose body temperature varies with the environmental temperature. When the water or air temperature changes, so does the temperature of the organism. All microbes and plants and most animals, including insects, worms, and reptiles, are poikilotherms. This is significant because, at colder body temperatures, poikilotherms have lower metabolic rates; at higher body temperatures, they have higher metabolic rates.

Homeotherms are animals that maintain a constant body temperature that is generally higher than the environmental temperature, regardless of the external temperature (figure 23.3). These animals—birds and mammals—have high metabolic rates, because they use metabolic energy to maintain their

OUTLOOKS 23.1

Parthenogenesis

Parthenogenesis is an unusual method of reproduction common in certain species of insects, crustaceans, and rotifers. In these cases, an "egg" cell is produced but not fertilized. These cells develop into exact copies of the female parent. In most of these species, parthenogenesis is a major method of reproduction, but at other times these species also engage

in sexual reproduction, often when environmental conditions become challenging. In some species, however, true sexual reproduction does not occur. Because the offspring have only one parent, they are genetically identical to that parent. Certain fishes, amphibians, and reptiles also reproduce by parthenogenesis. Populations of certain whiptail lizards are entirely female.

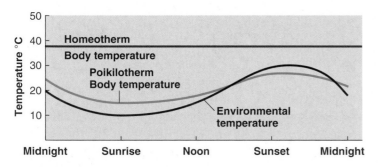

FIGURE 23.3 Regulating Body Temperature
The body temperature of a homeotherm remains relatively constant, regardless of changes in environmental temperature. The body temperature of a poikilotherm varies considerably, depending on the environmental temperature.

constant body temperature. This means that homeotherms have higher food demands than poikilotherms but are able to remain active at low environmental temperatures. Among homeotherms, the smaller animals have higher body temperatures and higher metabolic rates than the large animals. This is in part because small animals lose heat faster than large animals, because small animals have a larger surface area compared to their volume.

Sometimes, the terms *ectotherm* and *endotherm* are used to describe the same relationship from a slightly different point of view. **Ectotherms** are organisms whose body temperatures depend on the external temperature, and **endotherms** are animals that have internal heat-generating mechanisms and can maintain a relatively constant body temperature in spite of wide variations in the temperature of their environment.

However, poikilotherms or ectotherms are not necessarily totally at the mercy of their environment. Many poikilotherms use behavioral means to regulate body temperature.

One simple method is to position the body so that it absorbs heat from its surroundings. Many kinds of reptiles and insects "sun" themselves. By placing themselves in the Sun or on rocks or other surfaces that have been heated by the Sun, they can raise their body temperature above that of their environment. Similarly, many poikilotherms can generate heat by contracting their muscles. In cold weather, bumblebees often beat their wings prior to flying, raising their body temperature. Body temperature is important, because the rate at which the wings can beat is determined by body temperature.
heat and temperature, p. 29

23.4 Body Plans

Although animals come in a variety of sizes and shapes, their bodies do conform to a few basic body plans.

Symmetry

Symmetrical objects have similar parts that are arranged in a particular pattern. For example, the parts of a daisy flower and a bicycle are arranged symmetrically. **Asymmetry** is a condition in which there is no pattern to the individual parts. Asymmetrical body forms are rare and occur only in certain species of sponges, which are the simplest kinds of animals. **Radial symmetry** occurs when a body is constructed around a central axis. Any division of the body along this axis results in two similar halves. Although many animals with radial symmetry are capable of movement, they do not always lead with the same portion of the body; that is, there is no anterior, or head, end. Starfish and jellyfish are examples of organisms with radial symmetry.

Bilateral symmetry exists when an animal is constructed with equivalent parts on both sides of a plane. Animals with

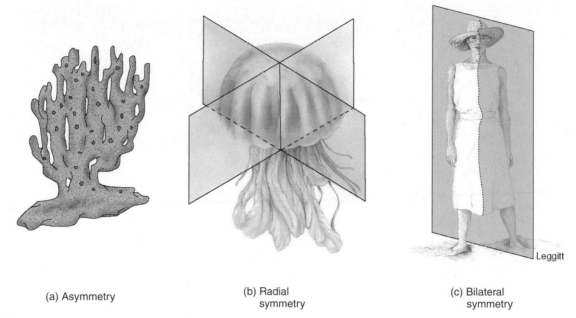

(a) Asymmetry

(b) Radial symmetry

(c) Bilateral symmetry

Leggitt

FIGURE 23.4 Radial and Bilateral Symmetry

(a) This sponge has a body that cannot be divided into symmetrical parts and is therefore asymmetrical. *(b)* In animals such as this jellyfish with radial symmetry, any cut along the central body axis results in similar halves. *(c)* In animals with bilateral symmetry, only one cut along one plane results in similar halves.

bilateral symmetry have a head and a tail region. There is only one way to divide bilateral animals into two mirrored halves. Animals with bilateral symmetry move head first, and the head typically has sense organs and a mouth. The feature of having an anterior head end is called *cephalization.* Most animals have bilateral symmetry (figure 23.4).

Embryonic Cell Layers

Animals differ in the number of layers of cells of which they are composed. When we look at the development of embryos we find that the embryos of the simplest animals (sponges) do not form distinct, tissuelike layers. However, jellyfishes and their relatives have embryos that consist of two layers. The **ectoderm** is the outer layer and the **endoderm** is the inner layer. Because their embryos are composed of two layers, these animals are said to be **diploblastic.** In adults, these embryonic cell layers give rise to an outer, protective layer and an inner layer that forms a pouch and is involved in processing food.

All the other major groups of animals are *triploblastic.* **Triploblastic** animals have three layers of cells in their embryos. Sandwiched between the ectoderm and endoderm is a third layer, the **mesoderm.** In the adult body, the ectoderm gives rise to the skin or other surface covering, the endoderm gives rise to the lining of the digestive system, and the mesoderm gives rise to muscles, connective tissue, and other organ systems involved in the excretion of waste, the circulation of material, the exchange of gases, and body support (figure 23.5).

Body Cavities

A **coelom** is a body cavity that separates the outer body wall of the organism from the gut and internal organs. Simple animals, such as jellyfish and flatworms, are **acoelomate,** which means that they have no space separating their outer surface from their internal organs and the internal organs are packed closely together.

More advanced animals, such as earthworms, insects, reptiles, birds, and mammals, have a coelom. The coelom in a turkey is the cavity where you stuff the dressing. In the living bird this cavity contains a number of organs, including those of the digestive, excretory, and circulatory systems. The development of the coelom was significant in the evolution of animals. In coelomates, there is less crowding of organs and less interference among them. Organs such as the heart, lungs, stomach, liver, and intestines have ample room to grow, move, and function. The coelom allows for the separation of the inner organs from the body-wall musculature; thus, the internal organs move independently of the outer body wall. This results in organ systems that are more highly specialized than those in acoelomate organisms. Organs are not loose in the coelom; they are held in place by sheets of connective tissue called **mesenteries.** Mesenteries also support the blood vessels connecting the various organs.

Some animals have a kind of coelom called a *pseudocoelom.* A **pseudocoelom** differs from a coelom in that it is located between the lining of the gut and the outer body wall. In other words, animals with a pseudocoelom do not have

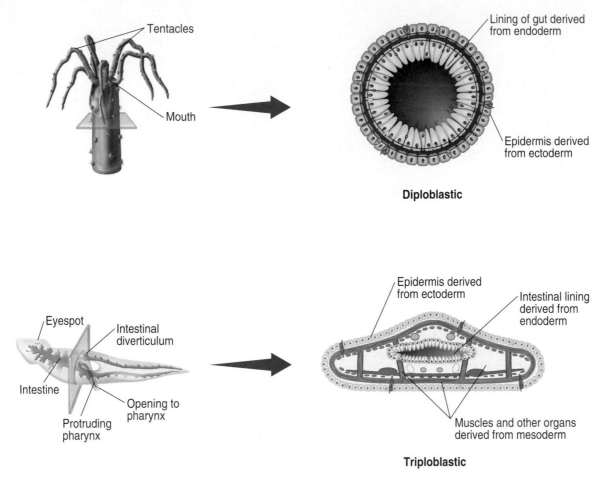

FIGURE 23.5 Embryonic Cell Layers

Diploblastic organisms have two embryonic cell layers. The outer ectoderm becomes the epidermis and the inner endoderm becomes the lining of the gut. Triploblastic organisms have three embryonic cell layers: the ectoderm, endoderm, and mesoderm. The mesoderm forms most of the tissues and organs of the body.

muscles around their digestive system. Nematode worms and several related groups of animals have a pseudocoelom (figure 23.6).

Segmentation

Many kinds of bilaterally symmetrical organisms have segmented bodies. **Segmentation** is the separation of an animal's body into a number of recognizable units from its anterior to its posterior end. Segmentation is associated with the specialization of certain parts of the body. Three common groups of animals show segmentation: annelid worms, arthropods, and chordates. Annelid worms have a series of very similar segments with minor differences between them. Segmentation in arthropods is modified so that several segments are specialized as a head region and more posterior segments are less specialized. Many of the posterior segments have legs and other appendages. Among the arthropods, insects show a

great deal of specialization of segments. In chordates, the segmentation is of a different sort but is obvious in the arrangement of muscles and the vertebral column (figure 23.7).

Skeletons

A **skeleton** is the part of an organism that provides structural support. Most animals have a skeleton. It serves as strong scaffolding, to which other organs can be attached. In particular, the skeleton provides places for muscle attachment and, if the skeleton has joints, the muscles can move one part of the skeleton with respect to others. Some aquatic organisms, such as sea anemones and many kinds of worms, are generally supported by the dense medium in which they live and lack well-developed skeletons. However, most aquatic animals have a skeleton, and terrestrial animals must have a strong supportive structure in the thin medium of the atmosphere.

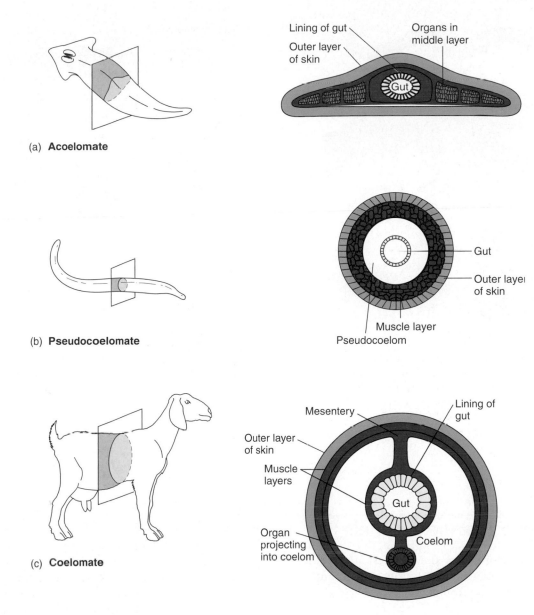

Lining of gut
Outer layer of skin
Organs in middle layer
Gut

(a) **Acoelomate**

Gut
Outer layer of skin
Muscle layer
Pseudocoelom

(b) **Pseudocoelomate**

Mesentery
Lining of gut
Outer layer of skin
Muscle layers
Gut
Organ projecting into coelom
Coelom

(c) **Coelomate**

FIGURE 23.6 Body Cavities

(a) Some animals, such as flatworms, have no open space between the gut and outer body layer. *(b)* Roundworms, commonly found in soil, have body cavities that have some cells and are not completely lined with thin sheets of connective tissue. *(c)* Other animals, including all vertebrates, have a coelom, an open area within the middle layer. Organs form from this middle layer, projecting into the coelom. The mesenteries are thin sheets of connective tissue that hold the organs in place within the coelom.

There are two major types of skeletons: internal skeletons (**endoskeletons**) and external skeletons (**exoskeletons**) (figure 23.8). The vertebrates (fish, amphibians, reptiles, birds, mammals) have internal skeletons. The various organs are attached to and surround the skeleton, which grows in size as the animal grows. Arthropods (crustaceans, spiders, insects, millipedes, centipedes) have an external skeleton, which surrounds all the organs. It is generally hard and has joints. These animals accommodate growth by shedding the old skeleton and producing a new, larger one. This period in the life of an arthropod is dangerous, because for a short period it is without its hard, protective outer layer. Many other animals have structures that have a supportive or protective function (such as clams, snails, and corals) and these are sometimes called skeletons, but they do not have joints.

Some organisms use water as a kind of supportive skeleton. Annelid worms and some other animals have fluid-filled coeloms. Because water is not compressible, but it is movable, compressive forces by muscles can cause the animal's shape to change. This is similar to what happens with a water-filled balloon; compression in one place causes it to bulge out somewhere else.

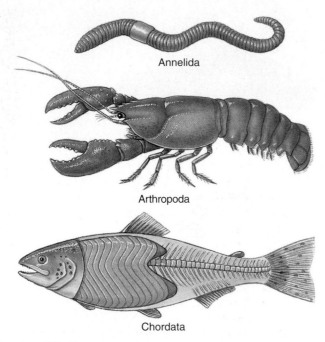

FIGURE 23.7 Segmentation

Segmentation is associated with the specialization of certain parts of the body. Annelid worms show many segments with little specialization. Arthropods show a highly developed head region, with the more posterior segments less specialized. Chordates show the segmentation of muscles and skeletal structures.

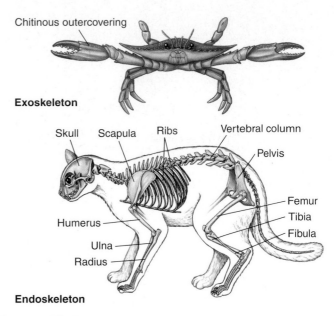

FIGURE 23.8 Skeletons

There are two major types of skeletons; endoskeletons and exoskeletons. Endoskeletons are typical of vertebrates and echinoderms, and exoskeletons are typical of arthropods and their relatives.

23.5 Primitive Marine Animals

Sponges, jellyfish, and corals are the simplest multicellular animals. They evolved about 600 million years ago and are usually found in saltwater environments.

Porifera—Sponges

Even though sponges are classified as multicellular, in many ways they are similar to colonial protozoa. There are two layers of cells, but most of the cells are in direct contact with the environment. All adult sponges are **sessile** (permanently attached) **filter feeders** with ciliated cells that cause a current of water to circulate through the organism. The individual cells obtain their nutrients directly from the water (figure 23.9). Although some sponges are radially symmetrical, most sponges are asymmetrical. Sponges have structures, called spicules, in a jellylike material between the two layers of cells. Some species have spicules of calcium carbonate, others of silicon dioxide, and some of a protein material. It is the skeletons composed of protein spicules that are used as washing sponges. (Actually, most sponges we buy are manufactured from various forms of plastic.)

Reproduction in sponges can occur by fragmentation. Wave action may tear off a part of a sponge, which eventu-ally settles down, attaches itself, and begins to grow. Sponges also reproduce by **budding**, a type of asexual reproduction in which the new organism is an outgrowth of the parent. They also reproduce sexually, and external fertilization results in a free-swimming, ciliated larval stage. The larva swims in the plankton and eventually settles to the bottom, attaches, and grows into an adult sponge.

Cnidaria—Jellyfish, Corals, and Sea Anemones

Cnidarians include the jellyfish, corals, and sea anemones. Like many sponges, they have two layers of cells. There is a jellylike material between the two layers of cells; it is very thick in jellyfishes. Cnidarians show radial symmetry, and all species have a single opening leading into a saclike digestive cavity. Surrounding the opening is a series of tentacles (figure 23.10). These long, flexible, armlike tentacles have specialized cells that produce structures called *nematocysts,* which can sting and paralyze small organisms. Nematocysts are unique to the Cnidaria. Even though they are primitive organisms, cnidarians are carnivorous.

Many species of Cnidaria exhibit alternation of generations and have both sexual and asexual stages of reproduction.

(a) **Cellular structure**   (b) **Vase-shaped sponge**   (c) **Asymmetrical sponge**

FIGURE 23.9 Sponge Structure and Function

(a) Sponges consist of an outer epidermal layer and an inner layer of flagellated cells called choanocytes. These flagella create a current, which brings water in through the openings formed by the porocyte cells, and it flows out through the osculum. The current brings food and oxygen to the inner layer of the cells. The food is filtered from the water as it passes through the animal. *(b)* Some sponges have a "vase shape"; however, *(c)* most sponges are asymmetrical and have no distinct form.

Jellyfish   **Sea anemone**   **Brain coral**

FIGURE 23.10 Phylum Cnidaria

There are three major types of cnidarians: jellyfish, sea anemones, and corals.

The **medusa** is a free-swimming adult stage that reproduces sexually. The **polyp** is a sessile larval stage that reproduces asexually (figure 23.11). Sea anemones and corals are polyps and some species lack a medusa stage. Jellyfish are medusas; they often produce small polyp stages during their life cycle.

Ctenophora—Comb Jellies

Ctenophorans are similar to cnidarians in that they are diploblastic and have radial symmetry. They swim by rows of cilia, and many have two long, armlike appendages, which help them gather food.

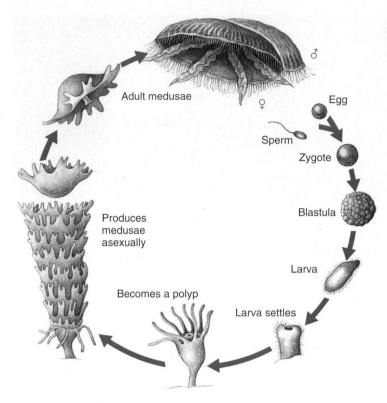

FIGURE 23.11 Phylum Cnidaria

The life cycle of *Aurelia* is typical of the alternation of generations seen in most species of Cnidaria. The free-swimming adult medusa (jellyfish) reproduces sexually, and the resulting larva develops into a polyp. The polyp undergoes asexual reproduction, which produces the free-swimming medusa stage.

23.6 Platyhelminthes—Flatworms

The simplest bilaterally symmetrical animals are the flatworms. Because they are also triploblastic, they represent a major increase in complexity over the sponges and cnidarians. They are considered more primitive than other bilaterally symmetrical, triploblastic animals, because they lack a coelom and have only one opening to the gut. They have no circulatory or respiratory system. Their flat structure allows for the diffusion of gases between their environment and the internal cells.

There are three basic types of flatworms—free-living flatworms (often called *planarians*), and the parasitic flukes and tapeworms (figure 23.12). Most free-living flatworms are bottom dwellers in marine waters or freshwater. A few species are found in moist terrestrial habitats. They are carnivores or scavengers that feed on dead organisms. The food that enters the gut is partly broken down; then, the food particles are engulfed by the cells that line the gut.

Parasitism is a very common mode of life for flatworms. All flukes and tapeworms are parasites, and they constitute the majority of species of flatworms. We may consider this form of nutrition rather unusual, but, of all the kinds of animals in the world, more are parasites than are not. Some of the flukes are external parasites on the gills and scales of fish, but most are internal parasites in the gut, liver, or lungs of vertebrate hosts. Most flukes have a complex life cycle involving more than one host. Usually, the larval stage infects an invertebrate host, whereas the adult parasite infects a vertebrate host.

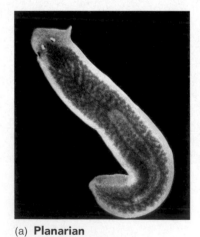

(a) **Planarian**

(b) **Tapeworm**

(c) **Fluke**

FIGURE 23.12 Flatworms

(a) Planarians are free-swimming, nonparasitic flatworms that inhabit freshwater. *(b)* Adult tapeworms are parasites found in the intestines of many carnivores. *(c)* Some flukes live attached to the outside surface of animals, but most are internal parasites of the lungs, liver, and intestine of vertebrate animals.

For example, the disease schistosomiasis is caused by adult *Schistosoma mansoni* flukes, which live in the blood vessels of the human digestive system. This disease is common in tropical countries; the presence of the worms in the body causes diarrhea, liver damage, anemia, and a lowering of the body's resistance to other diseases. Fertilized eggs pass out with the feces. Eggs released into the water hatch into free-swimming larvae. If a larva infects a snail, it undergoes additional asexual reproduction and produces a second larval stage. A single infected snail may be the source of thousands of larvae. These new larvae swim freely in the water. If they encounter a human, the larvae bore through the skin and enter the circulatory system, which carries them to the blood vessels of the intestine (figure 23.13).

In North America, people often encounter this kind of parasite as swimmer's itch. Swimmer's itch is caused by the larval stages of a fluke burrowing into the skin. However, the normal hosts for the worm are species of ducks. Humans are not ducks, so the larvae that have burrowed into the skin die and a rash often develops. In sensitive individuals, the reaction can be extreme and require medical attention.

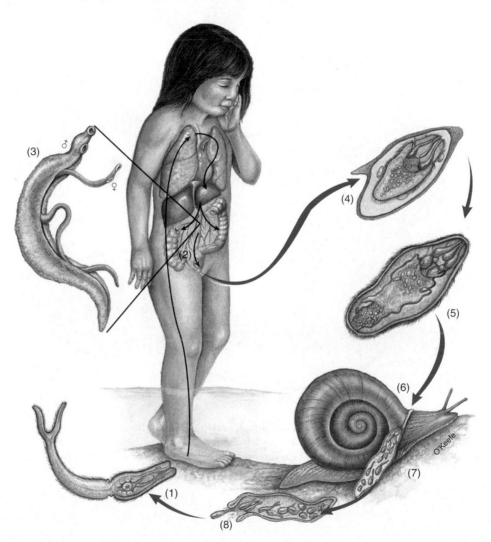

FIGURE 23.13 The Life History of *Schistosoma mansoni*

(1) Cercaria larvae in water penetrate human skin and are carried through the circulatory system to the veins of the intestine. They develop into *(2)* adult worms, which live in the blood vessels of the intestine. *(3)* Copulating worms are shown. The female produces eggs, which enter the intestine and leave with the feces. *(4)* A miracidium larva within an eggshell is shown. *(5)* The miracidium larva hatches in water and burrows into a snail *(6)*, where it develops into a mother sporocyst *(7)*. The mother sporocyst produces many daughter sporocysts *(8)*, each of which produces many cercaria larvae. These leave the snail's body and enter the water, thus completing the life cycle.

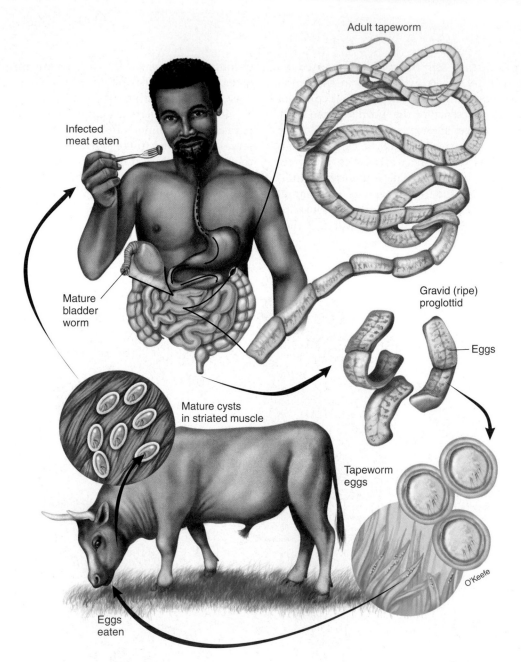

FIGURE 23.14 The Life Cycle of a Tapeworm

The adult beef tapeworm lives in the human small intestine. Proglottids, individual segments containing male and female sex organs, are the site of egg production. When the eggs are ripe, the proglottids drop off the tapeworm and pass out in the feces. If a cow eats an egg, the egg will develop into a cyst in the cow's muscles. When humans eat the cyst in the meat, the cyst develops into an adult tapeworm.

Tapeworms have a completely different way of life, involving two hosts. One of the hosts is an herbivore and the other is an omnivore or a carnivore. Both hosts are usually vertebrate animals. An herbivore eats tapeworm eggs that have been passed from an infected carnivore host through its feces. The eggs are eaten along with the vegetation the herbivore uses for food. An egg develops into a larval stage, which migrates to the herbivore's muscle and forms a cyst. When the herbivore is eaten by a carnivore, the tapeworm cyst develops into the adult form in the carnivore's intestine. The adult worm is highly specialized. It has no digestive system, but it has hooks and suckers for holding on to the intestine of its host. It produces a series of sections, called proglottids, by asexual reproduction. Each proglottid has a complete set of reproductive organs. Mating takes place between proglottids within the intestine, and the proglottids become filled with fertilized eggs. The end proglottids break off and pass out with the feces. Thus, the eggs can be easily dispersed in the feces (figure 23.14).

23.7 Nematoda—Roundworms

Nematode worms are extremely common animals, often called roundworms or nematodes. They have a simple, unsegmented, elongate body structure, which consist of an outer epidermis covered by a thick, flexible cuticle. The cuticle is shed periodically as the animal grows. Nematodes are triploblastic and have a pseudocoelom (figure 23.15). Their gut is continuous, with a mouth at one end and an anus at the other. Few animals are found in as many diverse habitats or in such numbers as the roundworms. They live in water, moist soil, and many live within other organisms as parasites. Soil nematodes are extremely common.

Although most roundworms are free-living, many are economically important parasites. Some are parasitic on plants, whereas others infect animals; collectively, they do billions of dollars worth of damage to crops and livestock. Roundworm parasites of animals range from the relatively harmless human intestinal pinworms *Enterobius*, which may cause irritation but do no serious harm, to *Dirofilaria*, which can cause heartworm disease in dogs. If untreated, this infection can be fatal. Often, the amount of damage inflicted by roundworms is directly proportional to their number. For example, hookworms (figure 23.16) feed on the host's blood. A slight infestation often results in mild anemia, but a heavy infestation of hookworms can result in mental or physical retardation in children because of the severe anemia it causes.

23.8 Marine Lifestyles

All groups of animals originated in the ocean, and most still call the oceans home. Marine animals exhibit several kinds of lifestyles.

Zooplankton

Zooplankton is a mixture of different kinds of small animals that drift with currents and feed on phytoplankton and other zooplankton. Crustaceans make up approximately 70% of the zooplankton. These include copepods, krill (small, shrimplike organisms), and other kinds of crustaceans. In addition, jellyfish and comb jellies are common. Other kinds of zooplankton include protozoa, several kinds of worms, a group of mollusks, and the larvae of many animals, including fish and primitive chordate relatives.

Nekton

Nekton includes many kinds of aquatic animals that are large enough and strong enough to be able to swim against currents and tides and go where they want to. They are carnivores that feed either on plankton or other nekton. Like the zooplankton, this is a diverse group of organisms, including some actively swimming jellyfish, squid and cuttlefish, shrimp, sharks, bony fish, turtles, sea snakes, birds (penguins,

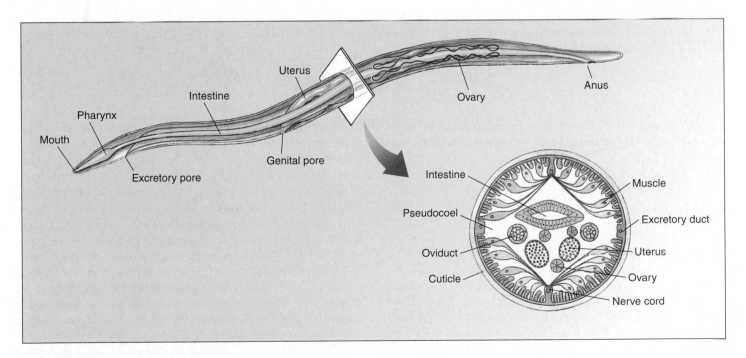

FIGURE 23.15 Nematode Worms
Nematode worms have a simple structure but have a complete gut with a mouth and an anus. Their body cavity is a pseudocoelom.

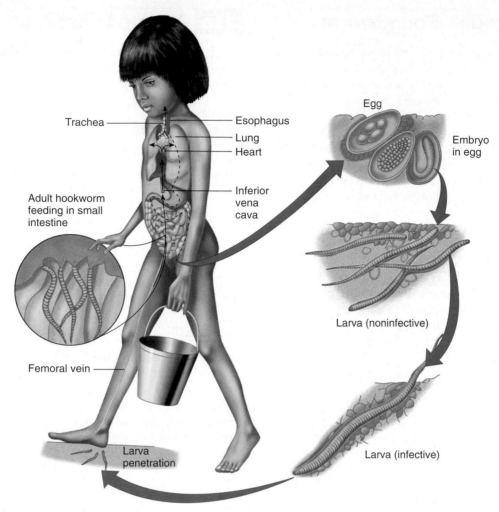

Trachea

Esophagus

Lung

Heart

Inferior vena cava

Adult hookworm feeding in small intestine

Femoral vein

Larva penetration

Egg

Embryo in egg

Larva (noninfective)

Larva (infective)

FIGURE 23.16 The Life Cycle of a Hookworm
Fertilized hookworm eggs pass out of the body in the feces and develop into larvae. In the soil, the larvae feed on bacteria. The larvae bore through the skin of humans and develop into adults. The adult hookworms live in the digestive system, where they suck blood from the host. The loss of blood can result in anemia, mental and physical retardation, and a loss of energy.

cormorants, and other diving birds), and several kinds of mammals (whales, porpoises, seals, sea otters, walruses, and so on).

Benthic Animals

A major ecological niche in the oceans includes large numbers of bottom-dwelling (benthic) organisms. Among the benthic organisms are such animals as the segmented worms; clams and snails; lobsters, crabs, and shrimp; starfish and sea urchins; and many kinds of fish. Benthic marine organisms fall into two general categories: those that move about in search of food and those that are filter feeders. Filter feeders are sessile. They feed by creating water currents with cilia or appendages that bring food to them. Sponges, corals, tube worms, clams, barnacles, and crinoids are examples of filter feeders.

Reproduction presents special problems for sessile animals, because they cannot move to find mates. However, because they are in an aquatic environment, the sperm can swim to the egg and fertilize it. The fertilized egg develops into a larval stage—the juvenile stage (figure 23.17). The larvae are usually ciliated or have appendages that enable them to move, even though the adults are sessile. The free-swimming larval stages allow the animal to disperse through its environment.

The larva differs from the sessile adult not only because it is free-swimming but also because it usually uses a different source of food and often becomes part of the plankton community. The larval stages of most organisms are subjected to predation, and the mortality rate is high. The larvae move to new locations, settle down, and develop into adults.

Even many marine animals that do not have sessile stages produce free-swimming larvae. For example, crabs, starfish, and eels move about freely and produce free-swimming larvae.

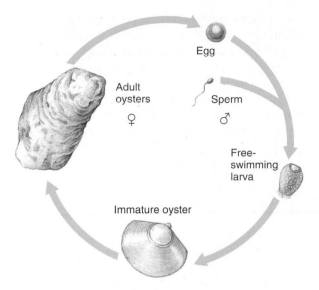

FIGURE 23.17 **The Life Cycle of an Oyster**
Each oyster can be either male or female at different times in its life. Sperm are released and swim to the egg. Fertilization results in the formation of a free-swimming larva. This larva undergoes several changes during the first 12 to 14 days and eventually develops into an immature oyster, which becomes attached and develops into the adult oyster.

Polychaete

Filter-feeding polychaete

Leech

Earthworm

FIGURE 23.18 **Annelids**
There are three major types of annelid worms. Polychaete worms are primarily marine worms. Some live in tubes and obtain food by filtering water. Others are free to move and are omnivores or carnivores. Oligochaetes (earthworms) burrow through soil and eat organic matter. Leeches either are carnivores or suck the blood of other animals.

23.9 Annelida—Segmented Worms

The annelid worms are bilaterally symmetrical and have a body structure consisting of repeating segments. Annelids are the first evolutionary group to display segmentation. Most of the segments are very similar to one another, although there is usually some specialization of anterior segments into a head, and there may be specialization related to reproduction and digestion. Each segment has a coelomic cavity. They have well-developed muscular, circulatory, digestive, excretory, and nervous systems. There are three major types of annelid worms (figure 23.18).

Polychaetes are primarily benthic, marine worms that have paddlelike appendages on each segment. They have a well-developed head region with sense organs and a mouth. Some live in tubes and are filter feeders. Others burrow in mud or sand; some swim in search of food. Most polychaetes have separate sexes and release eggs and sperm into the ocean, where fertilization takes place. The fertilized egg develops into a ciliated larva known as a trochophore larva; it is planktonic and eventually metamorphoses into the adult form of the worm.

Oligochaetes live in moist soil or freshwater. The most commonly seen members of this group are the various kinds of earthworms. They differ from polychaetes in that they do not have appendages and a well-developed head region. They are also hermaphroditic (have both sets of sex organs in the same individual). Mating between two earthworms results in each receiving sperm from the other. Earthworms are extremely important soil organisms. They eat organic matter in the soil or come to the surface to eat dead leaves and other organic matter. As they burrow through the soil, they create spaces, which allow water and air to penetrate the soil. They are also important food organisms to many other animals.

Leeches live in freshwater or moist terrestrial environments. They have suckers, which allow them to hold on to objects. Some are free-swimming carnivores, but many feed on the blood of various vertebrates. They attach to their victims, rasp a hole through the skin, and suck blood. They produce an anticoagulant, which aids in their blood-feeding lifestyle.

23.10 Mollusca

Like most other forms of animal life, the mollusks originated in the ocean, and, even though some forms have made the move to freshwater and terrestrial environments, most still live in the oceans. The members of this phylum display a true body cavity, a coelom. Reproduction is generally sexual; some species have separate sexes and others are hermaphroditic. They range from microscopic organisms to the giant squid, which is up to 18 meters long. A primary characteristic of mollusks is the presence of a soft body enclosed by a hard shell. They are not segmented but have three distinct body regions: the mantle, the foot, and the visceral mass. The mantle produces a shell and is involved in gas exchange, the foot is involved in movement, and the visceral mass contains the organs of digestion, circulation, and reproduction. Most of the mollusks have a tonguelike structure known as a radula, which has teeth on it. They use the radula like a rasp to scrap away at food materials and tear them to tiny particles that are ingested.

There are several kinds of mollusks; the most commonly observed are the chitons, bivalves, snails, and octopuses and squids (figure 23.19). The chitons have a series of shells along their back; they live attached to rocks along the seashore, where they feed on algae on the rocks. Various kinds of snails have a coiled shell. Most snails are marine, but some live in freshwater and moist terrestrial habitats. Slugs are a kind of snail that does not have a shell. The bivalves (clams, oysters, and mussels) have two shells. They also differ from other mollusks in that they are filter feeders and lack a radula. In the squids and octopuses, there is no external shell that serves

as a form of support structure, although a related organism—the nautilus—has a shell.

Except for the squids and octopuses, mollusks are slow-moving benthic animals. Some are herbivores and feed on marine algae; others are scavengers and feed on dead organic matter. A few are even predators of other slow-moving or sessile neighbors. As with most other marine animals, the mollusks produce free-swimming larval stages, which aid in dispersal.

23.11 Arthropoda

The arthropods are the most successful group of animals on Earth. The earliest arthropods were marine, and today arthropods are found as plankton, nekton, and benthic animals. However, the most successful group of animals is the terrestrial insects. Over three-fourths of the species of animals are arthropods, and over three-fourths of the arthropods are insects.

Arthropods have an external skeleton composed of chitin. Their bodies are segmented, but the segmentation is highly modified, compared with that of the annelids. There is generally a head region, and the segments that follow may be modified into regions as well. Many of the segments have paired appendages. Both the body and the appendages are jointed.

Because they have an exoskeleton, in order to grow they must shed their skeleton and manufacture a new, larger one at intervals. They have well-developed nervous, muscular, respiratory, circulatory, and reproductive systems.

There are many types of arthropods: crustaceans, millipedes, centipedes, arachnids, and insects (figure 23.20). Nearly all crustaceans are aquatic (both marine and freshwater) but range in size from large crabs and lobsters to tiny, planktonic organisms and sessile barnacles. Crustaceans are typically omnivores, which feed on a variety of living and dead materials.

Millipedes and centipedes have long bodies with many legs, and all the posterior segments are similar to one another. The earliest fossils of terrestrial animals are of this type.

The arachnids are the scorpions, spiders, mites, and ticks. They are very successful as land animals. Mites are extremely common as soil organisms. Spiders are carnivores, and most ticks are blood feeders.

Insects are terrestrial animals with a body divided into three regions: head, thorax, and abdomen. They have three pairs of legs, and most species have wings.

Chiton

Snail

Clam

Octopus

FIGURE 23.19 Mollusks
Mollusks range in complexity from a small, slow-moving, grazing animals, such as chitons and snails, to clams, which are filter feeders, to intelligent, rapidly moving carnivores, such as octopuses.

23.12 Echinodermata

The echinoderms represent a completely different line of evolution from other invertebrates. They are actually more closely related to the vertebrates than to other invertebrates.

Insect (grasshopper)

Millipede

Centipede

Crustacean
(mantis shrimp)

Arachnid
(scorpion)

FIGURE 23.20 Arthropods
Arthropods are the most successful group of organisms on Earth.

Echinoderms, along with chordates, have an embryological development, in which the anus is the first opening of the gut to form. Such animals are called deuterostomes. In all other triploblastic organisms, the mouth is the first opening of the gut to form; these are called protostomes. All echinoderms are marine benthic animals and are found in all regions, from the shoreline to the deep portions of the ocean. Echinoderms are the most common type of animal on much of the ocean floor. Most species are free-moving and either are carnivores or feed on detritus.

They are unique among the more advanced invertebrates in that they display radial symmetry. They have a five-part radial symmetry; thus, they have five arms or regions of the body that project from a central axis. However, the larval stage has bilateral symmetry, leading many biologists to believe that the echinoderm ancestors were bilaterally symmetrical.

Another unique characteristic of this group is their water vascular system (figure 23.21). In this system, water is taken in through a structure on the top side of the animal and then moves through a series of canals. The passage of water through this water vascular system is involved in the organism's locomotion.

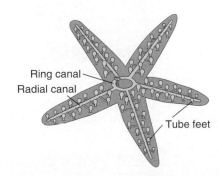

Ring canal
Radial canal
Tube feet

FIGURE 23.21 Water Vascular System
Echinoderms move by means of a water vascular system. Water enters the system through an opening, travels to the radial canals, and is forced into the tube feet.

They have a hard, jointed internal skeleton covered by a thin skin. Many of them have spines that are part of the skeleton and project from the animal.

There are five major types of echinoderms: starfish, brittle stars, sea urchins, sea cucumbers, and crinoids (figure 23.22). Starfish, brittle stars, and sea urchins move slowly along the ocean bottom and feed primarily on detritus. Some starfish are carnivores that eat clams. Crinoids are sessile, and many of them are stalked. Because they have five arms, that are often feather-like, they are often called sea lilies or feather stars. They are filter feeders. Sea cucumbers are sausage-shaped organisms that lie on the bottom or burrow in the mud. Some are detritus feeders and some are filter feeders.

23.13 Chordata

The chordates are a diverse group of animals. They have a hollow nerve cord down the back of the body; they have a flexible rod just below the nerve cord, known as the notochord; they have a tail, which extends beyond the anus; and they have a region posterior to the mouth, known as a pharynx, that has pouches in it. The most primitive chordates are small marine organisms that have a notochord but no backbone. They are *Branchiostoma* (formerly named *Amphioxus*) and tunicates. *Branchiostoma* has a notochord as an adult, but the tunicates have only a notochord in the tail region of the larval stage (figure 23.23). However, most chordates are vertebrates (animals with a backbone) and only have a notochord during their embryonic stages. In the aquatic chordates the pouches in the pharynx region form gill slits. In terrestrial vertebrates, these pouches are obvious in embryos but are highly modified in the adults.

There are several kinds of vertebrates that have always been aquatic organisms and are commonly called fishes. Hagfish and lampreys lack jaws and are the most primitive of the fish. Hagfish are strictly marine forms and are scavengers; lampreys are mainly marine but can also be found in freshwater (figure 23.24). Adult lampreys suck blood from their larger fish hosts. Lampreys reproduce in freshwater streams,

Starfish

Brittle star

Sea urchin

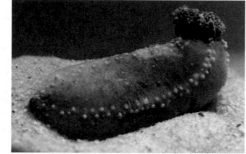

Sea cucumber

Feather star

FIGURE 23.22 Echinoderms
Echinoderms have a five-part radial symmetry and an internal skeleton. Many of them have spines, which protrude from the surface. There are five major types of echinoderms: starfish, brittle stars, sea urchins, sea cucumbers, and crinoids (feather stars).

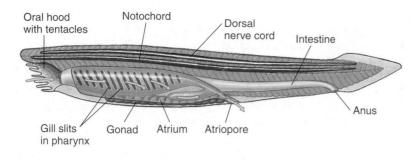

Branchiostoma

Oral hood with tentacles — Notochord — Dorsal nerve cord — Intestine — Anus — Atriopore — Atrium — Gonad — Gill slits in pharynx

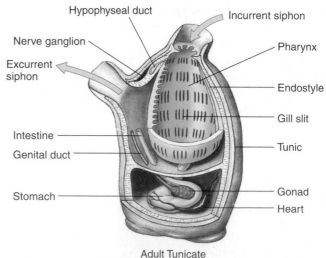

Adult Tunicate

Hypophyseal duct — Incurrent siphon — Nerve ganglion — Pharynx — Excurrent siphon — Endostyle — Gill slit — Tunic — Intestine — Genital duct — Stomach — Gonad — Heart

FIGURE 23.23 Invertebrate Chordates
Branchiostoma and tunicates are invertebrate chordates. They have a notochord but no backbone of vertebrae. Tunicates have a larval stage, which has a notochord in the tail. Most tunicates change into a saclike body form that is sessile. *Branchiostoma* lives in the sand on the ocean floor. Both organisms are filter feeders.

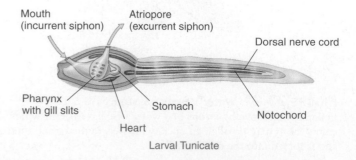

Larval Tunicate

Mouth (incurrent siphon) — Atriopore (excurrent siphon) — Dorsal nerve cord — Pharynx with gill slits — Stomach — Heart — Notochord

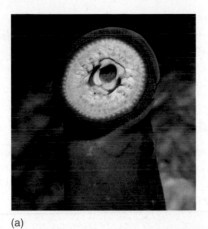

(a)

(b)

FIGURE 23.24 The Lamprey
The lamprey uses its round mouth *(a)* to attach to a fish and then *(b)* to suck blood from the fish.

(a) **Silvertip shark**

(b) **Blue-spotted stingray**

FIGURE 23.25 Cartilaginous Fish
Although *(a)* sharks and *(b)* rays are large animals, they do not have bones; their skeletal system is made entirely of cartilage.

where the eggs develop into filter-feeding larvae. After several years, the larvae change to adults and migrate to open water.

Sharks and rays are marine animals that have an internal skeleton made entirely of cartilage (figure 23.25). These animals have no swim bladder to adjust their body density in order to maintain their position in the water; therefore, they must constantly swim or they will sink. Rays feed by gliding along the bottom and dredging up food, usually invertebrates. Sharks are predatory and feed primarily on other fish. They travel great distances in search of food. Of the 40 species of sharks, only 7 are known to attack people. Most sharks grow no longer than a meter. The whale shark, the largest, grows to 16 meters, but it is strictly a filter feeder.

The bony fish are the class most familiar to us (figure 23.26). Their skeleton is composed of bone. Most species have a swim bladder and can regulate the amount of gas in the bladder to control their density. Thus, the fish can remain at a given level in the water without expending large amounts of energy. Bony fish are found in marine and freshwater habitats, and some, such as salmon, can live in both. There are many kinds of bony fish. Some are bottom-dwelling, whereas others are wide-ranging in the open water. Some fish are highly territorial and remain in a small area their entire lives.

Of the many kinds, some feed primarily on algae and detritus. However, many are predators.

The remaining groups of vertebrates (amphibians, reptiles, birds, and mammals) all are terrestrial animals. The amphibians are transitional organisms, which must return to water to reproduce and have aquatic larval stages. Reptiles, birds, and mammals have become very successful terrestrial animals. They are discussed in section 23.14.

23.14 Adaptations to Terrestrial Life

Plants began to colonize land over 400 million years ago, during the Silurian period, and they were well established on land before the animals. Thus, they served as a source of food and shelter for the animals. All animals that live on land must overcome certain common problems. Terrestrial animals must have (1) a moist membrane that allows for an adequate gas exchange between the atmosphere and the organism, (2) a means of support and locomotion suitable for land travel, (3) methods to conserve internal water, (4) a means of reproduction and early embryonic development in which large

Lined sweet lips

Sea horse

FIGURE 23.26 Bony Fish

Bony fish have a skeleton composed of bone. Most fish have a streamlined body and are strong swimmers. A few, such as the sea horse, have unusual body shapes and are relatively sedentary.

amounts of water are not required, and (5) methods to survive the rapid and extreme climatic changes that characterize many terrestrial habitats.

When the first terrestrial animals evolved, there were many unfilled niches; therefore, much adaptive radiation occurred, resulting in a large number of different animal species. Of all the many phyla of animals in the ocean, only a few made the transition from the ocean to the extremely variable environments found on the land. The annelids (earthworms and leeches) and the mollusks (land snails) have terrestrial species but are confined to moist habitats. Many of the arthropods (centipedes, millipedes, scorpions, spiders, mites, ticks, and insects) and vertebrates (reptiles, birds, and mammals) adapted to a wide variety of drier terrestrial habitats.

Terrestrial Arthropods

The first terrestrial animals were scorpion-like organisms, which are known from the fossil record from over 400 million years ago. The exoskeleton of arthropods was important in allowing them to adapt to land. It provides the support needed in the less buoyant air and serves as a surface for muscle attachment that permits rapid movement. The exoskeleton of terrestrial arthropods is waterproof, which reduces water loss.

Terrestrial arthropods have an internal respiratory system that prevents the loss of water from their respiratory surface. Insects and spiders have a tracheal system of thin-walled tubes extending into all regions of the body, thus providing a large surface area for gas exchange (figure 23.27a). These tubes have small openings to the outside, which reduce the amount of water lost to the environment. Another important method of conserving water in insects and spiders is the presence of Malpighian tubules, thin-walled tubes that surround the gut (figure 23.27b). If the insect is living in a dry environment, most of the water in waste materials is reabsorbed into the body by the Malpighian tubules and conserved.

Internal fertilization is typical of terrestrial arthropods. It involves copulation in which the sperm is placed into the reproductive tract of the female (insects) or the production of special sperm containing sacs that are picked up by the female (spiders). This is important because both the sperm and egg are protected from drying. Terrestrial arthropods have evolved a number of means of survival under hostile environmental conditions. Their rapid reproductive rate is one means. Most of a population may be lost because of an unsuitable environmental change, but, when favorable conditions return, the remaining individuals can quickly increase in number. Others survive unfavorable conditions in the egg or larval form and develop into adults when conditions become suitable (figure 23.28).

The terrestrial arthropods occupy an incredible variety of niches. Many are herbivores that compete directly with humans for food. They are capable of decimating plant populations that serve as human food. Many farming practices, including the use of pesticides, are directed at controlling insect populations. Other kinds of terrestrial arthropods are carnivores that feed primarily on herbivorous insects. Insects have evolved in concert with the flowering plants; their role in pollination is well understood. Bees, butterflies, and beetles transfer pollen from one flower to another as they visit the flowers in search of food. Many kinds of crops rely on bees for pollination, and farmers even rent beehives to ensure adequate pollination for fruit production.

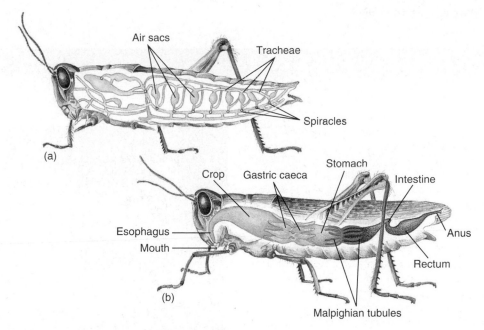

FIGURE 23.27 Insect Respiratory and Waste-Removal Systems

(a) Spiracles are openings in the exoskeleton of an insect. These openings connect to a series of tubes (tracheae), which allow for the transportation of gases in the insect's body. *(b)* Malpighian tubules are used in the elimination of waste materials and the reabsorption of water into the insect's body. Both systems are means of conserving body water.

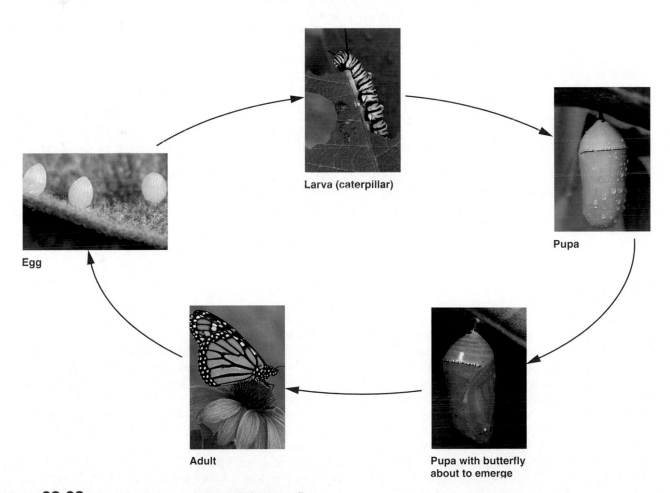

FIGURE 23.28 The Life History of a Monarch Butterfly

The adult female lays eggs on a milkweed plant. The fertilized egg hatches into the larval stage, known as a caterpillar, which feeds on the leaves on the milkweed plant. The caterpillar grows and eventually metamorphoses into a pupa. After emerging from the pupa, the adult's wings expand to their full size. The adults feed on the nectar of flowers. After mating, the female lays eggs and the life cycle starts over again.

HOW SCIENCE WORKS 23.1

Coelacanth Discoveries

Sometimes, scientific discoveries are made because the right person is in the right place at the right time. In 1938, a coelacanth—a fish that was thought to have been extinct for about 80 million years—was discovered near the mouth of the Chalumna River on the east coast of South Africa. The fishing boat captain who caught the fish contacted Marjorie Courtney-Latimer, the director of the local museum in the town of East London. She immediately contacted a fish biologist, J. L. B. Smith, and the coelacanth became a scientific celebrity. The coelacanth was named *Latimeria chalumnae*, incorporating both Courtney-Latimer's name and the name of the Chalumna River in its name. Subsequently, it was discovered that the fish was probably a stray and that the center of the population was farther north in the waters between Mozambique and the island of Madagascar.

In 1997, a biologist, Mark Erdmann, saw a coelacanth in a market in Indonesia. He photographed the fish and questioned the fishers. He then returned to Indonesia and, over a 5-month period, interviewed over 200 fishers. Eventually, he discovered two fishermen who occasionally caught coelacanths. Eventually, he was rewarded with a live specimen, and it was eventually described as a new species, *Latimeria menadoensis*. Scientists used structural differences and DNA analysis as evidence that the Indonesian coelacanth was a different species from the African species.

What was once thought to have been extinct for 80 million years can now be studied in the flesh. Over 200 specimens have been captured, and diving submarines have been used to observe coelacanths in the wild. They live in deep water and typically stay in caves during the day.

The coelacanth has several characteristics that make it of interest to biologists. Primary among them is the presence of

lobe fins. The fins are on short, limblike appendages that resemble stubby legs. Because of this, many biologists thought that coelacanths may have been the ancestors of terrestrial, four-legged vertebrates. Although most no longer feel this is the case, coelacanths do have several interesting characteristics. When they swim, they move their fins alternately in a manner similar to the way four-legged animals walk. They do not have a vertebral column but have a notochord that serves the same purpose as a spine. The have internal fertilization and the young develop from large eggs that are retained within the body of the mother and are born when they have completed development.

Sometimes accidental discoveries lead to a new body of scientific information.

Terrestrial Vertebrates

The first vertebrates on land were probably the ancestors of present-day amphibians (frogs, toads, and salamanders). The endoskeleton of vertebrates is an important prerequisite for life on land. It provides the support in the air and provides the places for muscle attachment necessary to movement. However, appendages are needed to move about. Certain of the bony fishes have lobe-fins, which can serve as primitive legs. It is likely that the amphibians evolved from a fish with modified fins (How Science Works 23.1). The first amphibians made the transition to land about 360 million years ago during the Devonian period. This was 50 million years after plants and arthropods had become established on land. Thus, when the first vertebrates developed the ability to live on land, shelter and food for herbivorous as well as carnivorous animals were available. But vertebrates faced the same prob-

lems that the insects, spiders and other invertebrates faced in their transition to life on land.

Amphibians

In amphibians (frogs, toads, and salamaders), the development of lungs was an adaptation that provided a means for land animals to exchange oxygen and carbon dioxide with the atmosphere. However, amphibians do not have an efficient method of breathing; they swallow air to fill the lungs, and most gas exchange between amphibians and the atmosphere must occur through their moist skin. In addition to needing water to keep their skin moist, amphibians must reproduce in water. When they mate, the female releases eggs into the water, and the male releases sperm amid the eggs. External fertilization occurs in the water, and the fertilized eggs must remain in water or they will dehydrate. Thus, amphibians live on "dry" land but are not found far from water, because they lose water through their moist skin and reproduction must occur in water. The most common present-day amphibians are frogs, toads, and salamanders (figure 23.29).

Reptiles

For 40 million years, amphibians were the only vertebrate animals on land. However, eventually they were replaced by reptiles, which were better adapted to life on land. In addition to having internal lungs, reptiles have a waterproof skin and water-conserving kidneys to reduce water loss. Furthermore, their reproduction involves internal fertilization, which protects the egg and sperm from drying. However, the young still require a moist environment in which to develop.

Reptiles became completely independent of an aquatic environment with the development of the amniotic egg, which protects the developing young from injury and dehydration (figure 23.30). The covering on the egg retains moisture and protects the developing young from dehydration while allowing for the exchange of gases. The reptiles were the first animals to develop such an egg.

The development of a means of internal fertilization and the amniotic egg allowed the reptiles to spread over much of the Earth and occupy a large number of previously unfilled niches. For about 200 million years, they were the only large vertebrate animals on land. The evolution of reptiles increased competition with the amphibians for food and space. The amphibians generally lost in this competition; consequently, most became extinct. Some evolved into present-day frogs, toads, and salamanders.

In a similar fashion, as birds and mammals evolved, many kinds of reptiles became extinct. However, there are still many kinds of reptiles present today. They include turtles, lizards, snakes, crocodiles, and alligators (figure 23.31).

Birds

The reptiles gave rise to two other groups of vertebrates: birds and mammals. About 65 million years ago, a mass extinction of many kinds of reptiles occurred. At that time, birds and mammals began to diversify and became the dominant forms of vertebrates on land. Like the structures of their

Frog tadpole

Frog

Salamander

Toad

FIGURE 23.29 Amphibians

Amphibian larvae (tadpoles) are aquatic organisms that have external gills and feed on vegetation. Most adult salamanders, frogs, and toads feed on insects, worms, and other small animals.

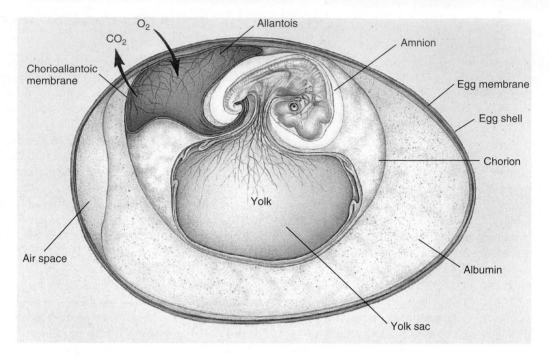

FIGURE 23.30 The Amniotic Egg

An amniotic egg has a shell and a membrane, which prevent the egg from dehydrating and allow for the exchange of gases between the egg and the environment. The egg yolk provides a source of nourishment for the developing young. The embryo grows three extraembryonic membranes: the amnion is a fluid-filled sac that allows the embryo to develop in a liquid medium, the allantois collects the embryo's metabolic waste material and exchanges gases, and the chorion is a membrane that encloses the embryo and the other two membranes.

Turtle

Snake

Alligator

Lizard

FIGURE 23.31 Reptiles

Present-day reptiles include turtles, crocodiles and alligators, snakes, and lizards.

reptile ancestors, the skin, lungs, and kidneys of birds reduce water loss, and reproduction involves internal fertilization and the shelled amniotic egg.

They also have adaptations that allow them to be successful as land animals. Birds are homeothermic and have feathers. As homeotherms, they have a high body temperature and a more rapid metabolic rate than reptiles. Feathers serve two primary functions in birds. They form an insulating layer, which helps prevent heat loss, and they provide structural surfaces enabling the birds to fly. There are several values to flight. Animals that fly are able to travel long distances in a short time and use less energy than animals that must walk or run. They are able to cross barriers, such as streams, lakes, oceans, bogs, ravines, or mountains, that other animals cannot easily cross. They can also escape many kinds of predators by taking flight quickly.

The exploitation of flight has shaped the entire structure and function of birds. The forelimbs are modified into wings for flight, although it is actually the flight feathers that provide most of the flight surface. Large breast muscles provide the power for flight. Feathers help provide a streamlined shape. Homeothermism enables a high constant temperature, which allows for the rapid wing beats typical of most birds. In addition, the skeleton is reduced in weight and the jaws lack teeth and are modified into a beak. Birds have successfully occupied many niches. Some are nectar feeders, some are carnivores, some are seedeaters, some are aquatic, and some have even lost their ability to fly (figure 23.32).

Because they are homeotherms, birds' eggs must be kept at a warm temperature. Thus, birds build nests, incubate their eggs, and care for their young.

Mammals

The first mammals appeared about 200 million years ago, when reptiles were at their most diverse. Like birds, mammals are homeotherms, with a high constant body temperature. Like reptiles and birds, they have a waterproof skin and water-conserving lungs and kidneys. However, they differ from birds in several respects. Their insulating covering is hair, rather than feathers, and they provide nourishment to their young with milk produced by special glands.

There are three categories of mammals. Monotremes (platypus and echidna) are egg-laying mammals whose young still develop in an external egg. Following hatching, the mother provides nourishment in the form of milk. The milk glands are widely distributed on her underside, and the young lap up the milk from her fur.

Marsupials (pouched mammals) have internal development of the young. However, the young are born in a very immature state and develop further in an external pouch in the belly region of the female. In the pouch, the young attach

Hummingbird

Eagle

Ostrich

FIGURE 23.32 Birds
Birds range in size from the small hummingbird with a rapid wingbeat to the large, flightless ostrich. In food habits, they range from the nectar-feeding hummingbird to the omnivorous ostrich and the carnivorous eagle.

to a nipple. They continue their development and eventually begin to leave the pouch for periods of time until they are able to forage for themselves.

Placental mammals have a form of development in which the young remain within the female much longer. The embryo is attached to the wall of the uterus by an organ known as a placenta. In the placenta, capillaries from the circulatory systems of the mother and embryo are adjacent to one another. This allows for the exchange of materials between the developing young and the mother, and the young are born in a more advanced stage of development than is typical for marsupials. The young rely on milk as a source of food for some time before they begin to feed on their own (figure 23.33).

Monotreme (echidna)

Placental mammal (zebra)

Marsupial (kangaroo)

FIGURE 23.33 Mammals
All mammals have hair and produce milk. However, there are three, very different kinds of mammals. Monotremes lay eggs. Marsupials have pouches; the young are born in an immature stage and are carried in the pouch while they finish their development. Placental mammals have a placenta, which connects the mother and embryo; they retain the young in the uterus for a longer period of time.

Summary

The 4 million known species of animals, which inhabit widely diverse habitats, are all multicellular and heterotrophic. Animal body shape is asymmetrical, radial, or bilateral. All animals with bilateral symmetry have a body structure composed of three layers.

Animal life originated in the ocean about 600 million years ago; for the first 200 million years, all animal life remained in the ocean. Many simple marine animals have life cycles that involve alternation of generations.

Many kinds of flatworms and nematodes have a parasitic lifestyle with complicated life cycles. A major ecological niche for many marine animals is the ocean bottom—the benthic zone. Many marine animals are planktonic. These zooplankton feed on phytoplankton and, in turn, are fed upon by larger, free-swimming marine animals—the nekton.

Animals that adapted to a terrestrial environment had to have (1) a moist membrane for gas exchange, (2) support and locomotion suitable for land, (3) a means of conserving body water, (4) a means of reproducing and providing for early embryonic development out of water, and (5) a means of surviving in rapid and extreme climatic changes.

Key Terms

*Use the interactive flash cards on the **Concepts in Biology,** 12/e website to help you learn the meaning of these terms.*

acoelomate 495

asymmetry 494

bilateral symmetry 494

budding 498

coelom 495

diploblastic 495

ectoderm 495

ectotherms 494

endoderm 495

endoskeletons 497

endotherms 494

exoskeletons 497

filter feeders 498

homeotherms 493

invertebrates 493

medusa 499

mesenteries 495

mesoderm 495

nekton 503

poikilotherms 493

polyp 499

pseudocoelom 495

radial symmetry 494

segmentation 496

sessile 498

skeleton 496

triploblastic 495

vertebrates 493

Basic Review

1. The most abundant group of terrestrial animals are the
   a. mammals.
   b. birds.
   c. earthworms.
   d. insects.

2. The sponges and cnidarians
   a. reproduce sexually.
   b. are primarily marine.
   c. are not bilaterally symmetrical.
   d. All of the above are correct.

3. Which of the following animal groups shows radial symmetry?
   a. annelids
   b. echinoderms
   c. vertebrates
   d. None of the above is correct.

4. The most successful group of animals are the _____.

5. The three general parts of the mollusk body are the visceral mass, mantle, and _____

6. Nematode worms are extremely common in soil. (T/F)

7. Which of the following organisms has a body consisting of a linear arrangement of similar segments?
   a. nematodes
   b. mollusks
   c. annelids
   d. cnidarians

8. All of the following animals have an amniotic egg except
   a. monotremes.
   b. marsupials.
   c. reptiles.
   d. birds.

9. Terrestrial animals have
   a. internal fertilization.
   b. a water-proof skin.
   c. a strong skeleton.
   d. All of the above are correct.

10. The two groups of animals that are homeotherms are the _____.

Answers
1. d 2. d 3. b 4. arthropods 5. foot 6. T 7. c 8. b 9. d 10. birds and mammals

Concept Review

*Answers to the Concept Review questions can be found on the **Concepts in Biology,** 12/e website.*

23.1 What Is an Animal?

1. List three characteristics shared by all animals.

23.2 The Evolution of Animals

2. Where did all the various groups of animal life originate?

3. How are invertebrates and vertebrates different?

23.3 Temperature Regulation

4. What is the value in being able to maintain a constant body temperature?

5. Define the term *ectotherm*.

23.4 Body Plans

6. Describe body forms that show asymmetry, radial symmetry, and bilateral symmetry.

7. Give an example of an animal that has a coelom.

8. Give an example of an animal with an exoskeleton, an endoskeleton.

9. How does an animal with an exoskeleton grow?
10. How do diploblastic and triploblastic animals differ?

23.5 Primitive Marine Animals

11. How do sponges feed?
12. What are the differences between a polyp and a medusa?

23.6 Platyhelminthes—Flatworms

13. Explain the tapeworm's life cycle.
14. What animal causes the disease schistosomiasis?

23.7 Nematoda—Roundworms

15. Describe the general body structure of a nematode.

23.8 Marine Lifestyles

16. What is a sessile filter feeder?
17. How do plankton and nekton organisms differ?

23.9 Annelida—Segmented Worms

18. What are the three kinds of segmented worms?

23.10 Mollusca

19. Describe the general body plan of mollusks.
20. How does the lifestyle of a cephalopod differ from that of a chiton?

23.11 Arthropoda

21. List three characteristics typical of all arthropods.
22. How do crustaceans and insects differ structurally?

23.12 Echinodermata

23. Describe the general body plan of an echinoderm.
24. What is a water vascular system, and what does it do?

23.13 Chordata

25. List three characteristics shared by all chordates.
26. How does a shark differ from most freshwater fish?

23.14 Adaptations to Terrestrial Life

27. List the problems animals had to overcome to adapt to a terrestrial environment.
28. Why can't amphibians live in all types of terrestrial habitats?
29. What is the importance of the amniotic egg?
30. How does a marsupial differ from a placental mammal?

Thinking Critically

*For guidelines to answer this question, visit the **Concepts in Biology,** 12/e website.*

Animals have been used routinely as models for the development of medical techniques and strategies. They have also been used in the development of pharmaceuticals and other biomedical products, such as heart valves, artificial joints, and monitors. The techniques necessary to perform heart, kidney, and other organ transplants were first refined using chimpanzees, rats, and calves. Antibiotics, hormones, and chemotherapeutic drugs have been tested for their effectiveness and for possible side effects using laboratory animals that are very sensitive and responsive to such agents. Biologists throughout the world have bred research animals that readily produce certain types of cancers that resemble the cancers found in humans. By using these animals instead of humans to screen potential drugs, the risk to humans is greatly reduced. The emerging field of biotechnology is producing techniques that enable researchers to manipulate the genetic makeup of organisms. Research animals are used to perfect these techniques and highlight possible problems.

Animal rights activists are very concerned about using animals for these purposes. They are concerned about research that seems to have little value in relation to the suffering these animals are forced to endure. Members of the American Liberation Front (ALF), an animal rights organization, vandalized a laboratory at Michigan State University where mink were used in research to assess the toxicity of certain chemicals. Members of this group poured acid on tables and in drawers containing data, smashed equipment, and set fires in the laboratory. This attack destroyed 32 years of research records, including data used for developing water-quality standards. In 1 year, 80 similar actions were carried out by groups advocating animal rights. What type of restrictions or controls should be put on such research? Where do you draw the line between "essential" and "nonessential" studies? Do you support the use of live animals in experiments that may alleviate human suffering?

CHAPTER

24

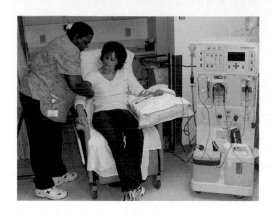

Materials Exchange in the Body

Born in 1916, he grew up the son of a farmer. The meals were "healthy" by the standards of the time. His diet consisted of meat from the farm animals, lots of eggs and animal fats, bacon, sausage, and gravy. When the family was hungry and supplies were scarce, potatoes and other starches filled empty stomachs. Smoking was very popular, and he smoked unfiltered cigarettes for 50 years. He was a moderate drinker of alcohol. He drank caffeinated coffee and lots of it. He was a thin man and a hard worker—a blue-collar worker for General Motors. He also owned a small farm, where he, like his family before him, raised cows and pigs for consumption.

As the years went by, he developed a cough. Every morning, he coughed up yellowish tan phlegm from his lungs. At the age of 50, he had surgery to repair the blocked arteries in his legs. The doctor told him he could not work anymore. One evening, he had several episodes in which he could not talk, and he lost the use of his left arm. He was admitted to the hospital and had surgery to clean out the plaque that clogged the artery in his neck. He was told to change his diet, no animal fat and less salt, but he didn't. He finally quit smoking, but he continued to become increasingly short of breath. His legs became weaker and his feet were always cold; his thin face and neck were puffy. He became confused and forgetful and often complained of back pain. He had settled into a sedentary life. He didn't drive anymore, and getting out, even with help, was more work than he thought it was worth. He suffered from emphysema and acute renal failure. During his last hospitalization, his mental status continued to deteriorate due to the impaired gas exchange in his lungs, and his kidneys were no longer able to remove the fluid and waste from his body. The chemicals in his bloodstream were no longer being regulated by either system. His blood sugar could not be controlled easily. He became more confused; when he saw steam rising from a roof vent outside his hospital window, he thought the hospital was on fire and became very agitated. Pain medication, sedatives, and oxygen kept him comfortable until his death. For the body to function properly nutrients and wastes must be exchanged between cells and between the body and the environment. This chapter focuses on the systems and mechanisms used to accomplish this task.

CHAPTER OUTLINE

24.1   The Basic Principles of Materials Exchange

Large, multicellular organisms, such as humans, consist of trillions of cells. Because many of these cells are buried within the organism far from the body surface, the body needs ways to solve its materials-exchange problems. Cells are highly organized units that require a constant flow of energy to maintain themselves. The energy they require is provided in the form of nutrient molecules. Oxygen is required for the efficient release of energy from the large, organic molecules that serve as fuel. Inevitably, as oxidation takes place, waste products form that are useless or toxic, and they must be removed. The five organ systems engaged in these efforts are the cardiovascular, lymphatic, respiratory, digestive, and excretory systems. All these organ systems are integrated, affecting one another in many ways. The interactions of these systems requires regulation and specialization for them to maintain life.

24.2   Circulation: The Cardiovascular System

The **cardiovascular system** of all vertebrates, including humans, is the organ system that pumps blood around the body; and consists of the blood, heart, and blood vessels. **Blood** is the fluid tissue that assists in the transport of materials and heat. The **heart** is a muscular pump, which forces the blood from one part of the body to another. The heart pumps blood into blood vessels. **Arteries** are the vessels that carry blood away from the heart and distribute it to the organs. The blood flows through successively smaller arteries until it reaches tiny vessels called *capillaries*. **Capillaries** are the thinnest blood vessels that exchange materials between the blood and tissues that surround the vessels. The blood flows from the capillaries into **veins**, the vessels that return blood to the heart.

The Nature of Blood

Blood is a fluid tissue consisting of several kinds of cells and platelets, called **formed elements**, suspended in a watery matrix, called **plasma** (table 24.1). This fluid plasma also contains many kinds of dissolved molecules, including nutrients, wastes, salts and proteins. The blood's primary function is to transport molecules, cells, and heat from one part of the body to another. The major kinds of molecules distributed by the blood are respiratory gases (oxygen and carbon dioxide), nutrients of various kinds, waste products, disease-fighting cells and antibodies, and chemical messengers (hormones). Also, substances in the blood are capable of forming clots or plugs to prevent blood loss.

Heat is generated by metabolic activities and must be lost from the body. To handle excess body heat, blood is moved through the vessels to the surface of the body, where it can be radiated away. In addition, humans and some other animals have the ability to sweat. The evaporation of sweat from the body surface also gets rid of excess heat. If the body is losing heat too rapidly, blood is moved away from the skin, and metabolic heat is conserved. Vigorous exercise produces an excess of heat, so that, even in cold weather, blood is moved to the skin and the skin feels hot.

Formed Elements

Red blood cells (rbcs) are small, disk-shaped cells that lack a nucleus. They are formed elements, their characteristics allow the blood to distribute respiratory gases efficiently. RBCs contain the iron-containing pigment **hemoglobin,** the protein molecule to which oxygen molecules bind readily. Red blood cells are the primary carriers used to transport oxygen around the body since little oxygen is carried as free, dissolved oxygen in the plasma. Because hemoglobin is inside red blood cells, some health problems can be diagnosed by counting the number of red blood cells a person has. If the number is low, the person's blood cannot carry oxygen efficiently and he or she tires easily. *Anemia* is a condition in which a person has reduced oxygen-carrying capacity. Anemia can also result when a person does not consume enough iron. Because iron is a central atom in hemoglobin molecules, people with an iron deficiency cannot manufacture sufficient hemoglobin; they can be anemic even if their number of red blood cells is normal.

Red blood cells are also important in the transport of carbon dioxide. Carbon dioxide is a result of the normal aerobic cellular respiration of food materials in the cells of the body. If carbon dioxide is not eliminated, it causes the blood to become more acidic (lowers its pH), eventually resulting in death. Carbon dioxide can be carried in the blood in three forms: about 7% is dissolved in the plasma, about 23% is carried attached to hemoglobin molecules, and 70% is carried as bicarbonate ions. The enzyme carbonic anhydrase in red blood cells helps convert carbon dioxide into bicarbonate ions (HCO_3^-), which can be carried as dissolved ions in the plasma of the blood. The following reversible chemical equation shows the changes that occur. pH, p. 38

$$CO_2 + H_2O \longleftrightarrow H_2CO_3 \longleftrightarrow H^+ + HCO_3^-$$

When the blood reaches the lungs, dissolved carbon dioxide is lost from the plasma, and carbon dioxide is released from the hemoglobin molecules. In addition, the bicarbonate ions reenter the red blood cells and can be converted back into molecular carbon dioxide by the same enzyme-assisted process that converts carbon dioxide to bicarbonate ions. The importance of this mechanism will be discussed in the section "Gas Exchange: The Respiratory System."

White blood cells (wbcs), also called leukocytes, are formed elements in the blood that lack hemoglobin; they are larger than RBCs and have a nucleus. Under the microscope, they are clear, and some appear to have granules in their cytoplasm, whereas others do not. The granule-containing

TABLE 24.1

Components of Blood

Formed Elements	Function and Description	Source		Plasma	Function	Source
Red blood cells (erythrocytes) 4 million–6 million per mm³ blood	Transport O_2 and help transport CO_2 7–8 μm in diameter Bright red to dark purple, biconcave disks without nuclei	Red bone marrow		Water (90–92% of plasma)	Maintains blood volume; transports molecules	Absorbed from intestine
White blood cells (leukocytes) 4,000–11,000 per mm³ blood	Fight infection	Red bone marrow		Plasma proteins (7–8% of plasma)	Maintain blood osmotic pressure and pH	Liver
Granular leukocytes				Albumin	Maintains blood volume and pressure	
• Basophil 20–50 per mm³ blood	10–12 μm in diameter Spherical cells with lobed nuclei; large, irregularly shaped, deep blue granules in cytoplasm			Globulins	Transport; fight infection	
• Eosinophil 100–400 per mm³ blood	10–14 μm in diameter Spherical cells with bilobed nuclei; coarse, deep red, uniformly sized granules in cytoplasm			Fibrinogen	Clotting	
• Neutrophil 3,000–7,000 per mm³ blood	10–14 μm in diameter Spherical cells with multilobed nuclei; fine, pink granules in cytoplasm			Salts (less than 1% of plasma)	Maintain blood osmotic pressure and pH; aid metabolism	Absorbed from intestine
Agranular leukocytes				Gases		
• Lymphocyte 1,500–3,000 per mm³ blood	5–17 μm in diameter (average 9–10 μm) Spherical cells with large, round nuclei			Oxygen Carbon dioxide	Cellular respiration End product of metabolism	Lungs Tissues
• Monocyte 100–700 per mm³ blood	10–24 μm in diameter Large, spherical cells with kidney-shaped, round, or lobed nuclei			Nutrients Fats Glucose Amino acids	Food for cells	Absorbed from intestine
Platelets (thrombocytes) 150,000–300,000 per mm³ blood	Aid clotting 2–4 μm in diameter Disk-shaped cell fragments with no nuclei; purple granules in cytoplasm	Red bone marrow		Nitrogenous waste Urea Uric acid	Excretion by kidneys	Liver
				Other Hormones, vitamins, etc.	Aid metabolism	Varied

Plasma 55%

Formed elements 45%

leukocytes are basophils, eosinophils, and neutrophils. Those without granules are lymphocytes and monocytes. All of these cells play vital roles in defending the body against potentially dangerous agents, such as microbes, chemicals, and cancer.

Platelets are another kind of formed element in the blood; they are fragments of specific kinds of white blood cells; they are important in blood clotting. About 200 billion are formed each day. They collect at the site of a wound, where they break down, releasing molecules called clotting factors. There are at least 12 known clotting factors. Their release begins a series of reactions, resulting in the formation of fibers that trap blood cells and form a plug in the opening of the wound. Blood clots, sometimes called scabs, appear red because they contain the leftovers of destroyed rbcs. Scabs are short-lived and are replaced by healthy tissue as the wound heals.

Plasma

Recall that plasma is the liquid part of the blood, consisting of water and dissolved materials, such as a variety of salts. Many of these serve as buffers, helping maintain the blood's pH. The salts also play a role in maintaining the blood's osmotic pressure. This helps keep the fluid that surrounds the cells of the body's tissues, called **tissue fluid,** at the right concentration, so that it tends to flow into the capillaries, maintaining blood pressure. The tissue fluid is the liquid that baths the body's cells; it contains the same chemicals as plasma but smaller amounts of the blood protein albumin. Tissue fluid is produced when the blood pressure forces water and some small, dissolved molecules through the capillary walls. It bathes the cells but must eventually be returned to the circulatory system, or swelling will occur.

Proteins are other important plasma molecules. Antibodies and other chemicals involved in defending the body against harm circulate freely in the plasma. The most plentiful plasma protein, albumin, is the primary molecule involved in maintaining osmotic balance. Salts can flow through capillary walls, but albumin cannot; and therefore plays the more important role in regulating osmotic pressure. Albumin transports a breakdown product of hemoglobin from rbcs called bilirubin. Normally, bilirubin is transported to the liver where it is destroyed. The broken-down products are passed into the bile and excreted through the intestine. However, some newborns become jaundiced (a yellowing of the skin and whites of the eyes) because their systems have not matured enough to do this important job. If bilirubin is allowed to accumulate in their blood, jaundiced babies can develop deafness, cerebral palsy, or brain damage. Placing babies under a "bili" light helps relieve the problem, because the wavelength of the light helps breaks down the bilirubin.

Plasma carries nutrient molecules from the gut to other locations, where they are modified, metabolized, or incorporated into cell structures. Amino acids and simple sugars are carried as dissolved molecules in the blood. Lipids, which are not water-soluble, are combined with proteins and carried as

suspended particles called lipoproteins. Some organs manufacture or modify molecules for use elsewhere; therefore, they must constantly receive raw materials and distribute their products to the cells that need them through the blood. In addition, many kinds of hormones are produced by the brain, reproductive organs, digestive organs, and glands. These are secreted into the bloodstream and transported throughout the body. Tissues with appropriate receptors bind to these molecules and respond to these chemical messengers.

The Heart

Blood can perform its transportation function only if it moves. The heart is responsible for providing the energy to pump the blood throughout the body. For a fluid to flow through a tube, there must be a pressure difference between the two ends of the tube; water flows through pipes because it is under pressure. Because the pressure is higher behind a faucet than at the spout, water flows from the spout when the faucet is opened. The circulatory system can be analyzed from the same point of view. The heart is the muscular pump that provides the pressure necessary to propel the blood throughout the body. It must continue its cycle of contraction and relaxation or the blood stops flowing and the body's cells cannot obtain nutrients or get rid of wastes. Some cells, such as brain cells, are extremely sensitive to interruptions in the flow of blood, because they require a constant supply of glucose and oxygen. Others, such as muscle cells and skin cells, are much better able to withstand temporary interruptions of blood flow.

The hearts of humans, other mammals, and birds consist of four chambers and four sets of valves, which work together to ensure that blood flows in one direction only. Two of these chambers, the right and left **atria** (singular, *atrium*), are relatively thin-walled structures that collect blood from the major veins and empty it into the larger, more muscular **ventricles** (figure 24.1). Most of the blood flow from the atria to the ventricles is caused by the lowered pressure produced within the ventricles as they relax. The contraction of the thin-walled atria helps empty them more completely.

The right and left ventricles are chambers with powerful muscular walls, whose contractions force blood to flow through the arteries to all parts of the body. The valves between the atria and the ventricles, the **atrioventricular valves,** are important one-way valves, which allow blood to flow from the atria to the ventricles but prevent flow in the

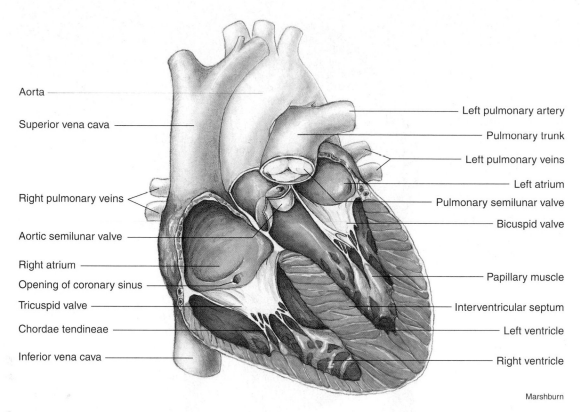

Aorta

Superior vena cava

Right pulmonary veins

Aortic semilunar valve

Right atrium

Opening of coronary sinus

Tricuspid valve

Chordae tendineae

Inferior vena cava

Left pulmonary artery

Pulmonary trunk

Left pulmonary veins

Left atrium

Pulmonary semilunar valve

Bicuspid valve

Papillary muscle

Interventricular septum

Left ventricle

Right ventricle

Marshburn

FIGURE 24.1 **The Anatomy of the Heart**
The heart consists of two thin-walled chambers, called atria, which contract to force blood into the two ventricles. When the ventricles contract, the atrioventricular valves (bicuspid and tricuspid) close, and blood is forced into the aorta and pulmonary artery. Semilunar valves in the aorta and pulmonary artery prevent the blood from flowing back into the ventricles when they relax.

opposite direction. Similarly, there are valves in the aorta and pulmonary artery, the **semilunar valves.** The **aorta** is the large artery that carries blood from the left ventricle to the body, and the **pulmonary artery** carries blood from the right ventricle to the lungs. The semilunar valves prevent blood from flowing back into the ventricles. If the atrioventricular or semilunar valves are damaged or function improperly, the heart's efficiency as a pump is diminished, and the person may develop an enlarged heart or other symptoms. Malfunctioning heart valves are often diagnosed because they cause abnormal sounds—called heart murmurs—as the blood passes through them. Similarly, if the muscular walls of the ventricles are weakened because of infection, damage from a heart attack, or lack of exercise, the heart's pumping efficiency is reduced and the person develops symptoms such as chest pain, shortness of breath, and fatigue. The pain is caused by an increase in the amount of lactic acid in the heart muscle, because the muscle is not getting sufficient blood to satisfy its needs. The heart receives blood from coronary arteries that are branches of the aorta; it is not nourished by the blood that flows through its chambers. If heart muscle does not get sufficient oxygen for a period of time, the portion of the heart muscle not receiving blood dies. Shortness of breath and fatigue result because the heart is not able to pump blood efficiently to the lungs, muscles, and other parts of the body.

The right and left sides of the heart have slightly different jobs, because they pump blood to different parts of the body. The right side of the heart receives blood from the general body and pumps it through the pulmonary arteries to the lungs, where the exchange of oxygen and carbon dioxide takes place and the blood returns from the lungs to the left atrium. This is called **pulmonary circulation.** The larger, more powerful left side of the heart receives blood from the lungs, delivers it through the aorta to all parts of the body, and returns it to the right atrium by way of the veins. This is known as **systemic circulation.** Both circulatory pathways are shown in figure 24.2. The systemic circulation is responsible for gas, nutrient, and waste exchange in all parts of the body except the lungs.

Blood Vessels: Arteries, Veins, and Capillaries

Recall that arteries and veins are the tubes that transport blood from one place to another within the body. Arteries carry blood away from the heart, because it is under considerable pressure from the contraction of the ventricles. The contraction of the ventricle walls increases the pressure in the arteries. A typical pressure recorded in a large artery while the heart is contracting is about 120 millimeters of mercury

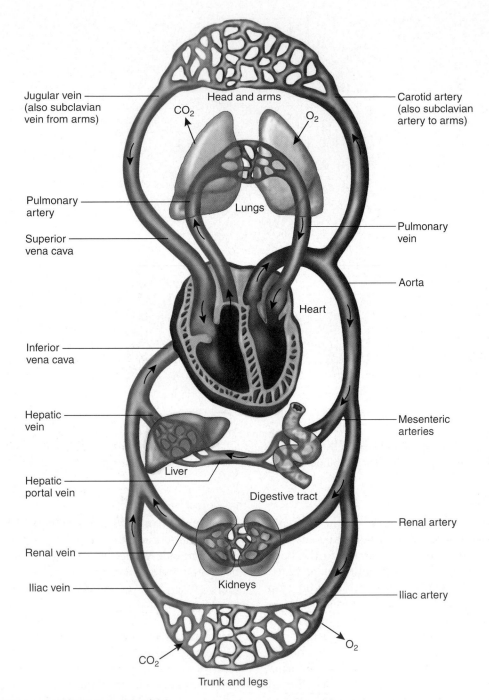

Jugular vein (also subclavian vein from arms)

Head and arms

CO_2

O_2

Carotid artery (also subclavian artery to arms)

Pulmonary artery

Lungs

Pulmonary vein

Superior vena cava

Aorta

Heart

Inferior vena cava

Mesenteric arteries

Hepatic vein

Liver

Hepatic portal vein

Digestive tract

Renal artery

Renal vein

Iliac vein

Kidneys

Iliac artery

CO_2

O_2

Trunk and legs

FIGURE 24.2 Pulmonary and Systemic Circulation
The right ventricle pumps blood that is poor in oxygen to the lungs by way of the pulmonary arteries, where it receives oxygen and turns bright red. The blood is then returned to the left atrium by way of four pulmonary veins. This part of the circulatory system is known as pulmonary circulation. The left ventricle pumps oxygen-rich blood by way of the aorta to all parts of the body except the lungs. This blood returns to the right atrium, depleted of its oxygen, by way of the superior vena cava from the head region and the inferior vena cava from the rest of the body. This portion of the circulatory system is known as systemic circulation.

(mmHg). This is known as the **systolic blood pressure.** The typical pressure recorded while the heart is relaxing is about 80 millimeters of mercury. This is known as the **diastolic blood pressure.** A blood pressure reading includes both numbers—for example, 120/80. (Originally, blood pressure was measured by how high the pressure of the blood caused

a column of mercury [Hg] to rise in a tube. Although the devices used today have dials or digital readouts and contain no mercury, they are still calibrated in mmHg.)

The walls of arteries are relatively thick and muscular yet elastic. Healthy arteries have the ability to expand as blood is pumped into them and return to normal as the pressure

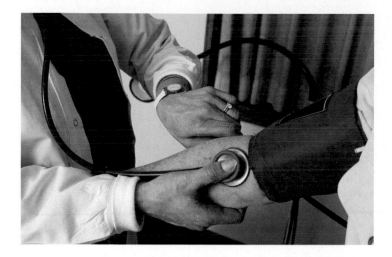

allowing a large volume of blood to flow to the capillaries of the skin. Because the blood is red, their skin reddens. Similarly, when people exercise, blood flow to the muscles increases to accommodate their higher metabolic needs—the muscles need to take in more oxygen and glucose and to get rid of wastes. Exercise also results in an increased blood flow to the skin, which allows for heat loss. At the same time, the amount of blood flowing to the digestive system is reduced. Athletes do not eat a full meal before exercising, because the additional blood flow to the digestive system reduces the amount of blood available to the muscles and lungs. Muscular cramps can result if insufficient blood flows to the muscles.

Veins collect blood from the capillaries and return it to the heart. The pressure in veins is very low. Some of the largest veins may have a blood pressure of 0.0 mmHg for brief periods. The walls of veins are not as muscular as those of arteries. Because of the low pressure, veins must have valves to prevent the blood from flowing backwards, away from the heart. Veins are often found on the surface of the body and are seen as blue lines. *Varicose veins* result when veins contain faulty valves, which do not allow the efficient return of blood to the heart. Therefore, blood pools in these veins, and they become swollen, bluish networks.

Because the pressure in veins is so low, muscular movements of the body are important in helping return blood to the heart. When the muscles of the body contract, they compress nearby veins, and this pressure pushes blood along in the veins. Because the valves allow blood to flow only toward the heart, this activity acts as an additional pump to help return blood to the heart. People who sit or stand for long periods without using their leg muscles tend to have a considerable amount of blood pool in the veins of their legs and lower body. Thus, less blood may be available to go to the

drops. This ability to expand absorbs some of the pressure and reduces the peak pressure within the arteries, thus reducing the likelihood that they will burst. If arteries become hardened and less resilient, peak systolic blood pressure rises and the arteries are more likely to rupture. The elastic nature of the arteries is also helps in blood flow. When the arteries return to normal from their stretched condition, they give a little push to the blood flowing through them.

Blood is distributed from the large aorta through smaller and smaller blood vessels to millions of tiny capillaries (figure 24.3). Some of the smaller arteries, called **arterioles**, contract or relax to regulate the flow of blood to specific parts of the body. The major parts of the body that receive differing amounts of blood, depending on need, are the digestive system, muscles, and skin. When light-skinned people blush, it is because many arterioles in the skin have expanded,

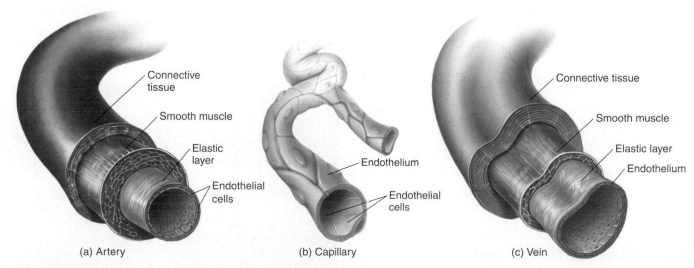

(a) Artery   (b) Capillary   (c) Vein

FIGURE 24.3 The Structure of Arteries, Veins, and Capillaries
The walls of *(a)* arteries are much thicker than the walls of *(c)* veins. (The pressure in arteries is much higher than the pressure in veins.) The pressure generated by the ventricles of the heart forces blood through the arteries. Veins often have very low pressure. The valves in the veins prevent the blood from flowing backward, away from the heart. *(b)* Capillaries have only a single layer of cells.

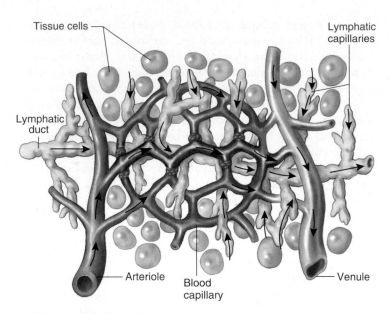

FIGURE 24.4 Capillaries

Capillaries are tiny blood vessels. The exchange of cells and molecules can occur between blood and tissues through their thin walls. Molecules diffuse in and out of the blood, and cells, such as white blood cells, can move from the blood through the thin walls into the surrounding tissue.

brain, causing people to faint. Although the arteries are responsible for distributing blood to various parts of the body and arterioles regulate where blood goes, it is the function of capillaries to assist in the exchange of materials between the blood and cells.

Capillaries are tiny, thin-walled tubes that receive blood from arterioles. They are so small that red blood cells must go through them in single file. They are so numerous that each cell in the body has a capillary located near it. It is estimated that there are about 1,000 square meters of surface area of capillary surface in a typical human. Each capillary wall consists of a single layer of cells and, therefore, presents only a thin barrier to the diffusion of materials between the blood and cells. It is also possible for liquid to flow through tiny spaces between the cells of most capillaries (figure 24.4). The flow of blood through these smallest blood vessels is relatively slow. This allows time for the diffusion of such materials as oxygen, glucose, and water from the blood to surrounding cells, and for the movement of such materials as carbon dioxide, lactic acid, and ammonia from the cells into the blood.

24.3 The Lymphatic System

The human **lymphatic system** is a collection of thin-walled tubes (lymph vessels) that branch throughout the body and lymph organs. The lymphatic system plays three important roles: (1) It moves fat from the intestinal tract to the bloodstream, (2) it transports excess tissue fluid back to the cardiovascular system, and (3) it defends against harmful agents, such as bacteria and viruses. **Lymph** is tissue fluid that moves through this system.

Lymph vessels collect lymph and ultimately empty it into two large veins near the heart. Lymph is moved one-way through the vessels as a result of the contraction of the body's muscles; the lymph vessels have valves to prevent backflow. For this system to function properly, the person must be active, or the tissue fluid will not move into the lymph vessels and will build up, causing the tissue to swell, a condition called edema. Edema is common in circulatory disorders. Another cause of edema results from an increase in capillary permeability as a result of injury. Fluid moves from insidethe blood vessels into the damaged tissue and is seen as swelling. A swollen sprained ankle or smashed thumb are examples you may have experienced. Figure 24.5 shows the structure of the lymphatic system.

There are five types of lymph organs: (1) lymph nodes, (2) tonsils, (3) spleen, (4) thymus, and (5) red bone marrow. **Lymph nodes** are small encapsulated bodies found along the lymph vessels that contain large numbers of white blood cells (wbc), macrophages and lymphocytes that remove microorganisms and foreign particles from the lymph. As the lymph makes its way back to the circulatory system, it is filtered by lymph nodes. When nodes are actively engaged in this cleanup, they swell as the wbc population increases. During an examination, physicians palpate or feel the nodes (usually in the neck) to help determine if a person has an infection.

The tonsils are partially encapsulated lymph organs located around the throat. There are two parts, the portion that most people refer to as their 'tonsils,' the palatine tonsil and the pharyngeal adenoids. Their job, like that of lymph nodes, is to clean the lymph of pathogens. Because they are so close to the mouth and nose, tonsils are likely to be the first to filter pathogens. They can become swollen as a result of their antimicrobial activity. In the past, if tonsils repeatedly became "infected" and unable to manage pathogens, they

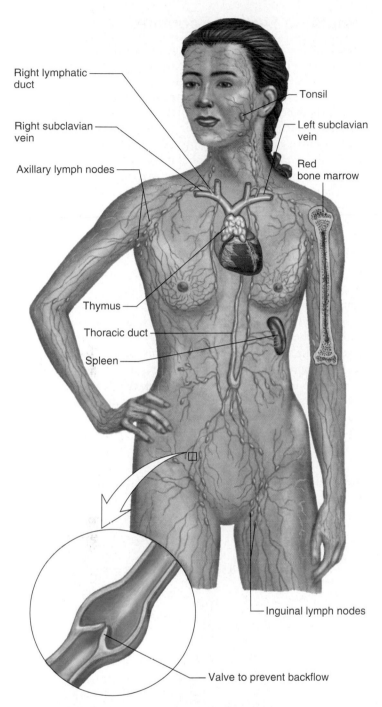

Right lymphatic duct

Right subclavian vein

Axillary lymph nodes

Thymus

Thoracic duct

Spleen

Tonsil

Left subclavian vein

Red bone marrow

Inguinal lymph nodes

Valve to prevent backflow

FIGURE 24.5 The Lymphatic System

Lymphatic vessels drain the fluid that surrounds cells, tissue fluid, and return it to the cardiovascular system near the heart. The enlargement shows that lymphatic vessels have valves to prevent backflow. The tonsils, spleen, thymus gland, and red bone marrow are lymph organs.

were removed by a surgical procedure called a tonsillectomy. This procedure was performed on children who displayed chronic sore throats or infections. Today, tonsillectomies are rarely performed.

The spleen also contains large numbers of white blood cells and filters the blood. It is about the size of a small pickle, located in the upper left side of the body just below the diaphragm. Its job is to clean the blood of pathogens and worn-out or damaged rbcs. Should the spleen be ruptured as a result of an accident or a wound, it may have to be removed. In such cases, the person loses an important antimicrobial organ.

The thymus gland, located over the breastbone, is large and active in children. As a person ages, the thymus shrinks and may actually disappear. Its primary function is to produce wbcs, which are vital to the functioning of the immune system. These cells, called T-lymphocytes, are distributed from the thymus throughout the body and establish themselves in lymph nodes. Even though the thymus shrinks in size into adulthood, the descendants of the T-lymphocytes it produced earlier in life are still active throughout the lymphatic system.

The red bone marrow produces red and white blood cells and platelets. Stem cells found inside these bones continue to divide throughout a person's life, providing a continuous supply of rbcs and wbcs. Red bone marrow is found in most of a child's bones. In adults, the red bone marrow is only found in the pelvis (hipbone), sternum (breastbone), skull, clavicle (collarbone), and vertebrae (backbone).

24.4 Gas Exchange: The Respiratory System

The **respiratory system** is the organ system that moves air into and out of the body; it consists of the *lungs, trachea, air-transport pathway,* and *diaphragm.* The **lungs** are organs of the body that allow gas exchange to take place between the air and blood. Associated with the lungs is a set of tubes, part of the air-transport pathway, that conducts air from outside the body to the lungs. The single, large-diameter **trachea** is supported by rings of cartilage that prevent its collapse. It branches into two major **bronchi** (singular, *bronchus*), which deliver air to smaller and smaller branches. Bronchi are also supported by cartilage. The smallest tubes, **bronchioles**, contain smooth muscle and are therefore capable of constricting. Finally, the bronchioles deliver air to clusters of tiny sacs, known as **alveoli** (singular, *alveolus*), where the exchange of gases between the air and the blood takes place.

The nose, mouth, and throat are also important parts of the air-transport pathway, because they modify the humidity and temperature of the air and clean the air as it passes. The lining of the trachea contains cells with cilia, which beat in one direction to move mucus and foreign materials from the lungs. The foreign matter can then be expelled by swallowing or coughing. Figure 24.6 illustrates the various parts of the respiratory system.

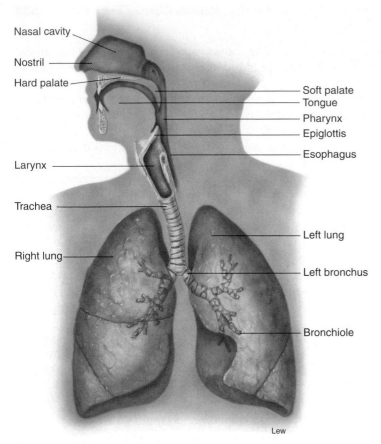

Nasal cavity

Nostril

Hard palate

Soft palate

Tongue

Pharynx

Epiglottis

Esophagus

Larynx

Trachea

Left lung

Right lung

Left bronchus

Bronchiole

Lew

FIGURE 24.6 Respiratory Anatomy

Although gas exchange takes place in the alveoli of the lungs, there are many other important parts of the respiratory system. The nasal cavity cleans, warms, and humidifies the air entering the lungs. The trachea is also important in cleaning the air going to the lungs.

Breathing System Regulation

Breathing is the process of moving air in and out of the lungs. It is accomplished by the movement of a muscular organ known as the **diaphragm,** which separates the chest cavity and the lungs from the abdominal cavity. In addition, muscles located between the ribs (*intercostal* muscles) are attached to the ribs in such a way that their contraction causes the chest wall to move outward and upward, which increases the size of the chest cavity. During inhalation, the diaphragm moves downward and the external intercostal muscles of the chest wall contract, causing the volume of the chest cavity to increase. This results in a lower pressure in the chest cavity compared with the outside air pressure. Thus, air flows from the outside high-pressure area through the trachea, bronchi, and bronchioles to the alveoli. During normal, relaxed breathing, exhalation is accomplished by the chest wall and diaphragm simply returning to their normal, relaxed positions. Muscular contraction is not involved (figure 24.7).

When the body's demand for oxygen increases during exercise, the breathing system responds by exchanging the gases in the lungs more rapidly. This can be accomplished both by increasing the breathing rate and by increasing the volume of air exchanged with each breath. An increase in volume exchanged per breath is accomplished in two ways. First, the muscles of inhalation can contract more forcefully, resulting in a greater change in the volume of the chest cavity. Second, the lungs can be emptied more completely by contracting the muscles of the abdomen, which forces the abdominal contents upward against the diaphragm and com-

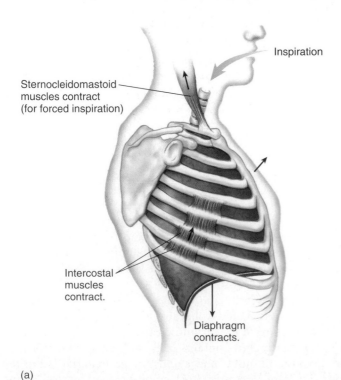

Inspiration

Sternocleidomastoid muscles contract (for forced inspiration)

Intercostal muscles contract.

Diaphragm contracts.

(a)

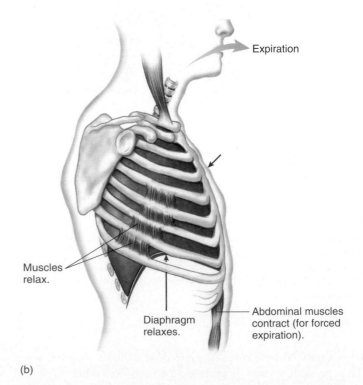

Expiration

Muscles relax.

Diaphragm relaxes.

Abdominal muscles contract (for forced expiration).

(b)

FIGURE 24.7 Breathing Movements

During (a) inhalation, the diaphragm and external intercostal muscles between the ribs contract, causing the volume of the chest cavity to increase. During a normal (b) exhalation, these muscles relax, and the chest volume returns to normal.

presses the lungs. A set of internal intercostal muscles also helps compress the chest.

Several mechanisms can cause changes in the rate and depth of breathing, but the primary mechanism is the amount of carbon dioxide present in the blood. Carbon dioxide, a waste product of aerobic cellular respiration, becomes toxic in high quantities because it combines with water to form carbonic acid:

$$CO_2 + H_2O \rightarrow H_2CO_3$$

As mentioned previously, if carbon dioxide cannot be eliminated, the pH of the blood is lowered. Eventually, this can cause death.

Exercising causes an increase in the amount of carbon dioxide in the blood, because the muscles are oxidizing glucose more rapidly. This lowers the blood's pH. Certain brain cells and specialized cells in the aortic arch and carotid arteries are sensitive to changes in blood pH. When these brain cells sense a lower blood pH, they send nerve impulses more frequently to the diaphragm and intercostal muscles. These muscles contract more rapidly and more forcefully, resulting in more rapid, deeper breathing. Because more air is being exchanged per minute, carbon dioxide is lost from the lungs more rapidly. When exercise stops, blood pH rises, and breathing eventually returns to normal (figure 24.8). However, moving air in and out of the lungs is of no value

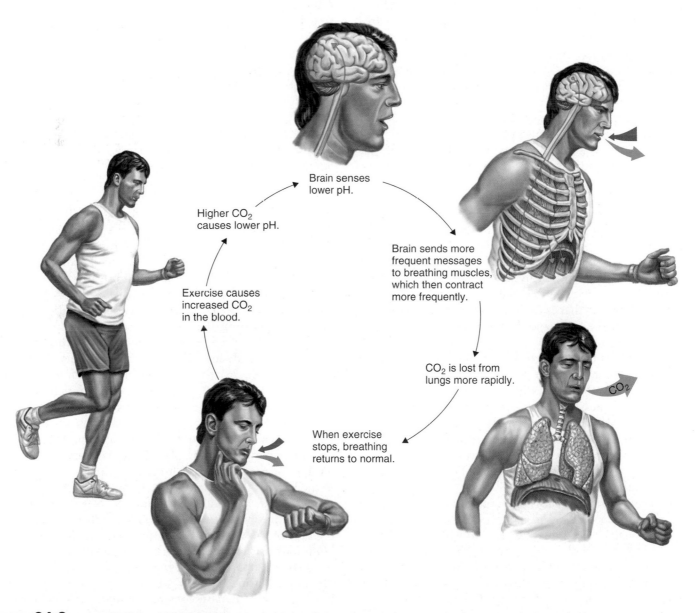

FIGURE 24.8 Control of Breathing Rate

The rate of breathing is controlled by cells in the brain that sense the pH of the blood. When the amount of CO_2 increases, the pH drops (the blood becomes more acidic) and the brain sends more frequent messages to the diaphragm and intercostal muscles, causing the breathing rate to increase. More rapid breathing increases the rate at which CO_2 is lost from the blood; thus, the blood pH rises (it becomes less acidic) and the breathing rate decreases.

unless oxygen is diffusing into the blood and carbon dioxide is diffusing out.

Lung Function

The lungs allow blood and air to come into close contact with each other. Air flows in and out of the lungs during breathing. The blood flows through capillaries in the lungs and is in close contact with the air in the alveoli of the lungs. For oxygen to enter the body or carbon dioxide to exit the body, the molecules must pass through a surface. The efficiency of exchange is limited by the surface area available. This problem is solved in the lungs by the large number of tiny sacs, the alveoli. Each alveolus is about 0.25 to 0.5 millimeters across. However, alveoli are so numerous that the total surface area of all these sacs is about 70 square meters—comparable to the floor space of many standard-sized classrooms. Associated with these alveoli are large numbers of capillaries (figure 24.9). The walls of both the capillaries and the alveoli are very thin, and the close association of alveoli and capillaries in the lungs allows the easy diffusion of oxygen and carbon dioxide across these membranes.

Another factor that increases the efficiency of gas exchange is that both the blood and the air are moving. Because blood is flowing through capillaries in the lungs, the capillaries continually receive new blood that is poor in oxygen and high in carbon dioxide. As blood passes by the alveoli, it is briefly exposed to the gases in the alveoli, where it gains oxygen and loses carbon dioxide. Thus, blood that leaves the lungs is high in oxygen and low in carbon dioxide. Although the movement of air in the lungs is not in one direction, as is the case with blood, the cycle of inhalation and exhalation allows air that is high in carbon dioxide and low in oxygen to exit the body and brings in new air that is rich in oxygen and low in carbon dioxide. This oxygen-rich, bright red blood is then sent to the left side of the heart and pumped throughout the body.

Any factor that interferes with the flow of blood or air or alters the effectiveness of gas exchange in the lungs reduces the efficiency of the organism. A poorly pumping heart sends less blood to the lungs, and the person experiences shortness of breath. Similarly, diseases such as asthma, which cause constriction of the bronchioles, reduce the flow of air into the lungs and inhibit gas exchange.

Any process that reduces the number of alveoli also reduces the efficiency of gas exchange in the lungs. For example, *emphysema* is a progressive disease in which some of the alveoli are lost. As the disease progresses, those afflicted have less and less respiratory surface area and experience greater and greater difficulty getting adequate oxygen, even though they may be breathing more rapidly. Often, emphysema is accompanied by an increase in the amount of connective tissue and the lungs do not stretch as easily, further reducing their ability to exchange gases.

24.5 Obtaining Nutrients: The Digestive System

The **digestive system** is the organ system responsible for the processing and distribution of nutrients; it consists of a muscular tube and glands that secrete digestive juices into the tube. Four kinds of activities are involved in getting nutrients to the cells that need them: mechanical processing, chemical processing, nutrient uptake, and chemical alteration.

Mechanical and Chemical Processing

The purpose of the digestive system is to break down large chunks of food into smaller molecules, which the circulatory system can distribute to the body's cells. The first step in this process is mechanical processing.

It is important to grind large particles into small pieces by chewing to increase their surface areas and allow for more efficient chemical reactions. It is also important to add water to the food, which further disperses the particles and provides the watery environment needed for these chemical reactions. Materials also must be mixed so that all the molecules that need to interact with one another have a good chance of doing so. The oral cavity and the stomach are the major body regions involved in reducing the size of food particles. The teeth are involved in cutting and grinding food to increase its surface area. The watery mixture that is added to the food in the oral cavity is called *saliva*, and the three pairs of glands that produce saliva are known as **salivary glands.** Saliva contains the enzyme salivary amylase, which begins the chemical breakdown of starch. Saliva also lubricates the oral cavity and helps bind food before swallowing.

In addition to having taste buds that help identify foods, the tongue performs the important service of helping position the food between the teeth and pushing it to the back of the throat for swallowing. The oral cavity is very much like a food processor, in which mixing and grinding take place. Figure 24.10 describes and summarizes the functions of these and other structures of the digestive system.

Once the food has been chewed, it is swallowed and passes down the esophagus to the stomach. The process of swallowing involves a complex series of events. First, a ball of food, known as a *bolus*, is formed by the tongue and moved to the back of the mouth cavity. There, it stimulates the walls of the throat, also called the **pharynx.** Nerve endings in the lining of the pharynx are stimulated, causing a reflex contraction of the walls of the esophagus, which transports the bolus to the stomach. Because both food and air pass through the pharynx, it is important to prevent food from getting into the lungs. During swallowing, the larynx is pulled upward, causing a flap of tissue, called the *epiglottis,* to cover the opening to the trachea and preventing food from entering the trachea. In the stomach, additional liquid, called **gastric juice,** is added to the food. Gastric juice contains enzymes and

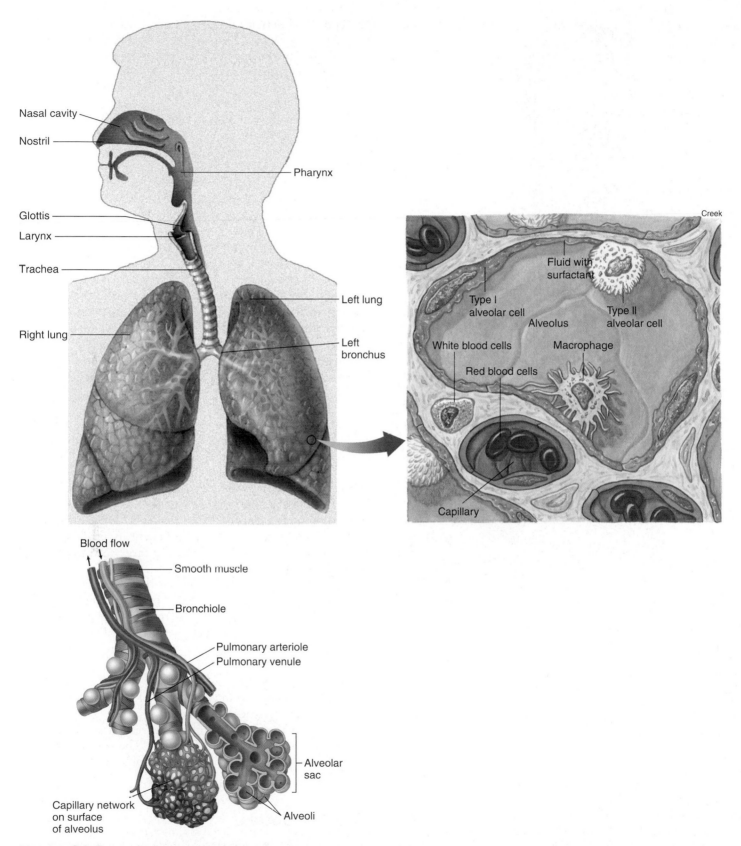

FIGURE 24.9 The Association of Capillaries with Alveoli

The exchange of gases takes place between the air-filled alveolus and the blood-filled capillary. The capillaries form a network around the saclike alveoli. The thin walls of the alveolus and capillary are in direct contact with one another; their combined thickness is usually less than 1 micrometer (a thousandth of a millimeter).

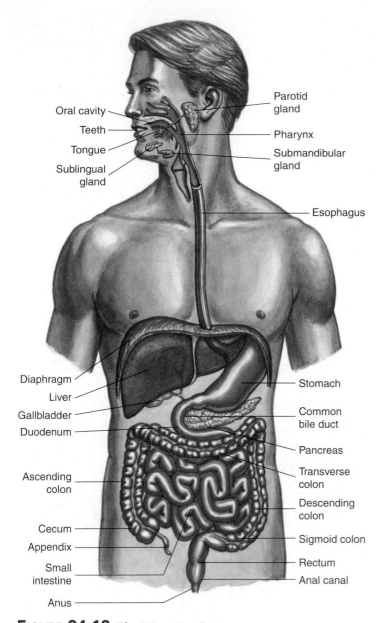

Oral cavity

Teeth

Tongue

Sublingual gland

Parotid gland

Pharynx

Submandibular gland

Esophagus

Diaphragm

Liver

Gallbladder

Duodenum

Ascending colon

Cecum

Appendix

Small intestine

Anus

Stomach

Common bile duct

Pancreas

Transverse colon

Descending colon

Sigmoid colon

Rectum

Anal canal

FIGURE 24.10 The Digestive System

The teeth, tongue, and enzymes from the salivary glands modify the food before it is swallowed. The stomach adds acid and enzymes and further changes the food's texture. The food is eventually emptied into the duodenum, where the liver and pancreas add their secretions. The small intestine also adds enzymes and is involved in absorbing nutrients. The large intestine is primarily involved in removing water.

hydrochloric acid. The major enzyme of the stomach is **pepsin,** which initiates the chemical breakdown of protein. The pH of gastric juice is very low, generally around pH 2. Consequently, very few kinds of bacteria or protozoa emerge from the stomach alive. Those that do survive have characteristics that protect them from the gastric juice as they pass through the stomach. The entire mixture is churned by the contractions of the three layers of muscle in the stomach wall.

The combined activities of enzymatic breakdown, chemical breakdown by hydrochloric acid, and mechanical processing by muscular movement result in a thoroughly mixed liquid. This liquid eventually leaves the stomach through a valve known as the **pyloric sphincter** and enters the **small intestine** (How Science Works 24.1).

The first part of the small intestine is called the **duodenum.** In addition to producing enzymes, the duodenum secretes several kinds of hormones, which regulate the release of food from the stomach and the release of secretions from the pancreas and liver. The **pancreas** produces a number of digestive enzymes and secretes large amounts of bicarbonate ions, which neutralize the acids that enter from the stomach so that the pH of the duodenum is about pH 8. The liver, a large organ in the upper abdomen, performs several functions, one of which is the secretion of **bile.** When bile leaves the liver, it is stored in the **gallbladder** prior to being released into the duodenum. When bile is released from the gallbladder, it assists mechanical processing by breaking large fat globules into smaller particles, much as soap breaks up fat particles into smaller globules that are suspended in water and washed away. This process is called *emulsification.* Emulsification is important because fats are not soluble in water, yet the reactions of digestion must take place in a water solution.

Along the length of the small intestine, additional watery juices are added until the mixture reaches the **large intestine.** The large intestine is primarily involved in reabsorbing the water that has been added to the food tube along with saliva, gastric juice, bile, pancreatic secretions, and intestinal juices. The large intestine is also home to a variety of bacteria. Most live on the undigested food that makes it through the small intestine. Some provide additional benefit by producing vitamins that can be absorbed from the large intestine. A few cause disease.

Several kinds of enzymes have been mentioned in this section. Each is produced by a specific organ and has a specific function. Chapter 5 introduced the topic of enzymes and how they work. Some enzymes, such as those involved in glycolysis, the Krebs cycle, and protein synthesis, are produced and used inside cells; others, such as the digestive enzymes, are produced by cells and secreted into the digestive tract. Digestive enzymes are simply a class of enzymes; they have the same characteristics as all other enzymes. They are protein molecules that speed up chemical reactions and are sensitive to changes in temperature or pH. The various digestive enzymes, the sites of their production, and their functions are listed in table 24.2.

Nutrient Uptake

Digestion results in a variety of simple organic molecules that are available for absorption from the tube of the gut into the circulatory system. As simple sugars, amino acids, glycerol, and fatty acids move into the circulatory system, surface area again becomes important. The amount of material that can be

HOW SCIENCE WORKS 24.1

An Accident and an Opportunity

On the morning of June 6, 1822, on Mackinac Island in northern Lake Huron, a 19-year-old French-Canadian fur trapper named Alexis St. Martin was shot in the stomach by the accidental discharge from a shotgun. The army surgeon at Fort Mackinac, Michigan, Dr. William Beaumont, was called to attend the wounded man. Part of St. Martin's stomach and body wall had been shot away, and parts of his clothing were embedded in the wound. Although Dr. Beaumont did not expect St. Martin to live, he quickly cleaned the wound, pushed portions of the lung and stomach that were protruding back into the cavity, and dressed the wound. Finding St. Martin alive the next day, Beaumont was surprised and encouraged to do what he could to extend his life. In fact, Dr. Beaumont cared for St. Martin for 2 years. When the wound was completely healed, the stomach had fused to the body wall and a hole allowed Dr. Beaumont to look into the stomach. Fortunately for St. Martin, a flap of tissue from the lining of the stomach closed off the opening, so that what he ate did not leak out.

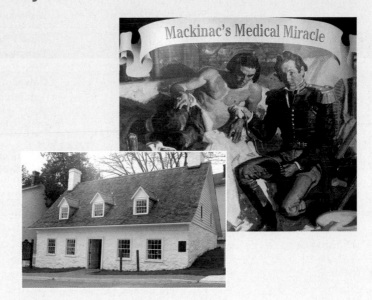

Dr. Beaumont found that he could look through the opening and observe the activities in the stomach and recognized that this presented an opportunity to study the stomach's function in a way that had not been done before. He gathered gastric juice, had its components identified, introduced food into the hole with a string attached so that he could retrieve the food particles that were partially digested for examination, and observed the effect of emotion on digestion. He discovered many things that were new to science and contrary to the teachings of the time. He recounted many of his observations and experiments in his journal: "I consider myself but a humble inquirer after truth—a simple experimenter. And if I have been led to conclusions opposite to the opinions of many who have been considered luminaries of physiology, and, in some instances, from all the professors of this science, I hope the claim of sincerity will be conceded to me, when I say that such difference of opinion has been forced upon me by the convictions of experiment, and the fair deductions of reasoning."

Following are some of his important discoveries.

1. He measured the temperature of the stomach and found that it does not heat up when food is introduced, as was thought at the time: "But from the result of a great number of experiments and examinations, made with a view to asserting the truth of this opinion, in the empty and full state of the organ, . . . I am convinced that there is no alteration of temperature. . . ."

2. He found that pure gastric juice contains large amounts of hydrochloric acid. This was contrary to the prevailing opinion that gastric juice was simply water: "I think I am warranted, from the result of all the experiments, in saying, that the gastric juice, so far from being "inert as water," as some authors assert, is the most general solvent in nature, of alimentary matter—even the hardest bone cannot withstand its action."

3. He observed that gastric juice is not stored in the stomach but, rather, is secreted when food is eaten: "The gastric juice does not accumulate in the cavity of the stomach, until alimentary matter is received, and excite its vessels to discharge their contents, for the immediate purpose of digestion."

4. He realized that digestion begins immediately when food enters the stomach. The prevailing opinion of the day was that nothing happened for an hour or more: "At 2 o'clock P.M.—twenty minutes after having eaten an ordinary dinner of boiled, salted beef, bread, potatoes, and turnips, and drank [sic] a gill of water, I took from his stomach, through the artificial opening, a gill of the contents. . . . Digestion had evidently commenced, and was perceptually progressing, at the time."

5. He discovered that food in the stomach satisfies hunger even though it is not eaten: "To ascertain whether the sense of hunger would be allayed without food being passed through the oesophagus [sic], he fasted from breakfast time, til 4 o'clock, P.M., and became quite hungry. I then put in at the aperture, three and a half drachms of lean, boiled beef. The sense of hunger immediately subsided, and stopped the borborygmus, or croaking noise, caused by the motion of the air in the stomach and intestines, peculiar to him since the wound, and almost always observed when the stomach is empty."

St. Martin did not take kindly to these probings and twice ran away from Dr. Beaumont's care back to Canada, where he married, had two children, and resumed his former life as a voyager and fur trapper. He did not die until the age of 83, having lived over 60 years with a hole in his stomach.

TABLE 24.2

Digestive Enzymes and Their Functions

Enzyme	Site of Production	Molecules Altered	Molecules Produced
Salivary amylase	Salivary glands	Starch	Smaller polysaccharides
Pepsin	Stomach lining	Proteins	Peptides
Gastric lipase	Stomach lining	Fats	Fatty acids and glycerol
Chymotrypsin	Pancreas	Polypeptides (long chains of amino acids)	Peptides
Trypsin	Pancreas	Polypeptides	Peptides
Carboxypeptidase	Pancreas	Peptides (several amino acids)	Smaller peptides and amino acids
Pancreatic amylase	Pancreas	Polysaccharides (many sugar molecules attached together)	Disaccharides
Pancreatic lipase	Pancreas	Fats	Fatty acids and glycerol
Nuclease	Pancreas	Nucleic acids	Nucleotides
Aminopeptidase	Intestinal lining	Peptides	Smaller peptides and amino acids
Dipeptidase	Intestinal lining	Dipeptides	Amino acids
Lactase	Intestinal lining	Lactose	Glucose and galactose
Maltase	Intestinal lining	Maltose	Glucose
Sucrase	Intestinal lining	Sucrose	Glucose and fructose
Nuclease	Intestinal lining	Nucleic acids	Nucleotides

taken up is limited by the surface area available. This problem is solved by increasing the surface area of the intestinal tract in several ways. First, the small intestine is a very long tube; the longer the tube, the greater the internal surface area. In a typical adult human, it is about 3 meters long. In addition to length, the lining of the small intestine consists of millions of fingerlike projections, called **villi,** which increase the surface area. The cells that make up the villi also have folds in their surface membranes. All of these characteristics increase the surface area available for the transport of materials from the gut into the circulatory system. Scientists estimate that the cumulative effect of all these features produces a total intestinal surface area of about 250 square meters. That is equivalent to about the area of a basketball court.

The surface area by itself would be of little value if it were not for the intimate contact of the circulatory system with the intestinal lining. Each villus contains several capillaries and a branch of the lymphatic system called a **lacteal.** The close association between the intestinal surface and the circulatory and lymphatic systems allows for the efficient uptake of nutrients from the cavity of the gut into the circulatory system (figure 24.11).

Several processes are involved in the transport of materials from the small intestine to the circulatory system. Some molecules, such as water and many ions, simply diffuse through the wall of the small intestine into the circulatory system. Other materials, such as amino acids and simple sugars, are assisted across the membrane by carrier molecules. Fatty acids and glycerol are absorbed into the intestinal lining cells, where they are resynthesized into fats and enter lacteals in the villi. Because the lacteals are part of the lymphatic system, which eventually empties its contents into the circulatory system, fats are also transported by the blood. They just reach the blood by a different route.

Chemical Alteration: The Role of the Liver

When the blood leaves the small intestine, it flows directly to the liver through the **hepatic portal vein.** Portal veins are blood vessels that collect blood from capillaries in one part of the body and deliver it to a second set of capillaries in another part of the body without passing through the heart. Thus, *hepatic portal veins* collect nutrient-rich blood from the small intestine and delivers it directly to the liver. As the blood flows through the liver, enzymes in the liver cells modify many of the molecules and particles that enter them. One of the functions of the liver is to filter any foreign organisms from the blood that might have entered through the intestinal cells. It also detoxifies many dangerous molecules that might have entered with the food.

Many foods contain toxic substances that could be harmful if not destroyed by the liver. Ethyl alcohol is one obvious example. Many plants contain various kinds of toxic molecules that

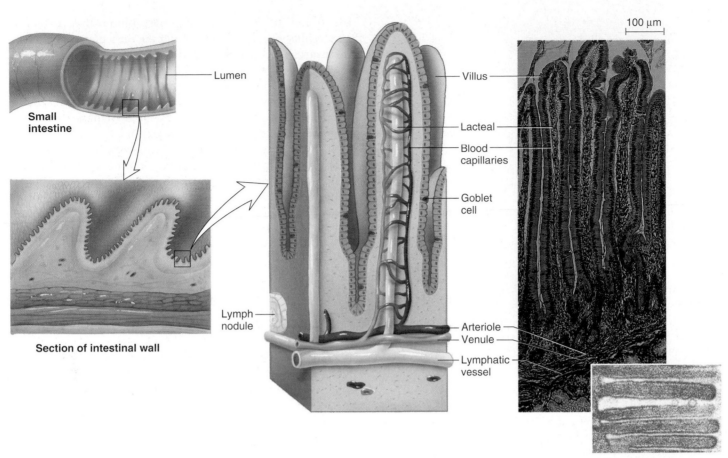

Small intestine

Lumen

Section of intestinal wall

Lymph nodule

Villus

Lacteal

Blood capillaries

Goblet cell

Arteriole

Venule

Lymphatic vessel

100 μm

Electron micrograph of microvilli

FIGURE 24.11 **The Exchange Surface of the Intestinal Tract**

The surface area of the intestinal lining is increased by the many fingerlike projections known as villi. Within each villus are capillaries and lacteals. Most kinds of materials enter the capillaries, but most fat-soluble substances enter the lacteals, part of the lymphatic system, giving them a milky appearance. Because the lymphatic system empties into the circulatory system, fat-soluble materials also eventually enter the circulatory system. The close relationship between the vessels and the epithelial lining of the villi allows for the efficient exchange of materials from the intestinal cavity to the circulatory system.

are present in small quantities and could accumulate to dangerous levels if the liver did not perform its role of detoxification.

In addition, the liver is responsible for modifying nutrient molecules. The liver collects glucose molecules and synthesizes glycogen, which can be stored in the liver for later use. When glucose is in short supply, the liver can convert some of its stored glycogen back into glucose. Although amino acids are not stored, the liver can change the relative numbers of various amino acids circulating in the blood. It can remove the amino group from one kind of amino acid and attach it to a different carbon skeleton, generating a different amino acid. The liver can also take the amino group off amino acids, so that what remains of the amino acid can be used in aerobic cellular respiration. The toxic amino groups are then converted to urea by the liver. Urea is secreted back into the bloodstream for disposal in urine.

24.6 Waste Disposal: The Excretory System

The **excretory system** is the organ system responsible for the processing and elimination of metabolic waste products; it consists of the kidneys, ureters, urinary bladder, and urethra (figure 24.12). Because cells modify molecules during metabolic processes, harmful waste products are constantly being formed. Urea is a common waste; many other toxic materials must be eliminated as well. Among these are large numbers of hydrogen ions produced by metabolism. This excess of hydrogen ions must be removed from the bloodstream. Other molecules, such as water and salts, may be consumed in excessive amounts and must also be removed.

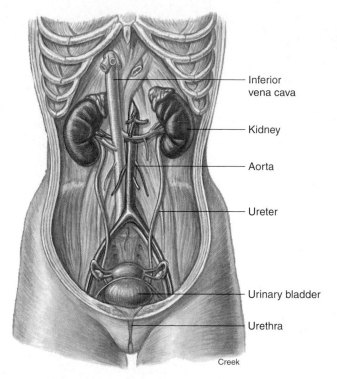

Inferior
vena cava

Kidney

Aorta

Ureter

Urinary bladder

Urethra

Creek

FIGURE 24.12 The Excretory System
The primary organs involved in removing materials from the blood
are the kidneys. The urine produced by the kidneys is transported
by the ureters to the urinary bladder. From the bladder, the urine is
emptied to the outside of the body by way of the urethra.

Kidney Structure

The **kidneys** are the primary organs involved in regulating the
level of toxic or unnecessary molecules in the body. The kidneys consist of about 2.4 million tiny units called **nephrons.**
At one end of a nephron is a cup-shaped structure called a
Bowman's capsule, which surrounds a knot of capillaries
known as a **glomerulus** (figure 24.13). In addition to
Bowman's capsule, a nephron consists of three distinctly different regions: the **proximal convoluted tubule,** the **loop of
Henle,** and the **distal convoluted tubule.** The distal convoluted tubule of a nephron is connected to a collecting duct,
which transports fluid to the ureters and ultimately to the urinary bladder, where it is stored until it can be eliminated
through the urethra.

Kidney Function

As in the other systems discussed in this chapter, the excretory
system involves a close connection between the circulatory
system and a surface. In this case, the large surface is provided by the walls of the millions of nephrons, which are surrounded by capillaries. Three major activities occur at these
surfaces: filtration, reabsorption, and secretion. The glomerulus presents a large surface for the filtering of material from

the blood to Bowman's capsule. Blood that enters the
glomerulus is under pressure from the muscular contraction of
the heart. The capillaries of the glomerulus are quite porous
and provide a large surface area for the movement of water
and small dissolved molecules from the blood into Bowman's
capsule. Normally, only the smaller molecules, such as glucose, amino acids, and ions, are able to pass through the
glomerulus into the Bowman's capsule at the end of the
nephron. The various kinds of blood cells and larger molecules, such as proteins, do not pass out of the blood into the
nephron. This physical filtration process allows many kinds of
molecules to leave the blood and enter the nephron. The volume of material filtered in this way through the approximately
2.4 million nephrons of the kidneys is about 7.5 liters per
hour. Because the entire blood supply is about 5 to 6 liters,
there must be a method for recovering much of this fluid.

Surrounding the various portions of the nephron are capillaries that passively accept or release molecules on the basis
of diffusion gradients. The walls of the nephron are made of
cells that actively assist in the transport of materials. Some
molecules are reabsorbed from the nephron and picked up by
the capillaries that surround them, whereas other molecules
are actively secreted into the nephron from the capillaries.
Each portion of the nephron has cells with specific secretory
abilities.

The proximal convoluted tubule is primarily responsible
for reabsorbing valuable materials from the fluid moving
through it. Molecules such as glucose, amino acids, and
sodium ions are actively transported across the membrane of
the proximal convoluted tubule and returned to the blood. In
addition, water moves across the membrane because it follows the absorbed molecules and diffuses to the area where
water molecules are less common. By the time the fluid has
reached the end of the proximal convoluted tubule, about
65% of the fluid has been reabsorbed into the capillaries surrounding this region.

The next portion of the tubule, the loop of Henle, is primarily involved in removing additional water from the
nephron. Although the details of the mechanism are complicated, the principles are rather simple. The cells of the ascending loop of Henle actively transport sodium ions from the
nephron into the space between nephrons where sodium ions
accumulate in the fluid that surrounds the loop of Henle. The
collecting ducts pass through this region as they carry urine
to the ureters. Because the area these collecting ducts pass
through is high in sodium ions, water within the collecting
ducts diffuses from the ducts and is picked up by surrounding
capillaries. However, the ability of water to pass through the
wall of the collecting duct is regulated by hormones. Thus, it
is possible to control water loss from the body by regulating
the amount of water lost from the collecting ducts. For example, if you drank a liter of water or some other liquid, the
excess water would not be allowed to leave the collecting
duct and would exit the body as part of the urine. However,
if you were dehydrated, most of the water passing through
the collecting ducts would be reabsorbed, and very little urine

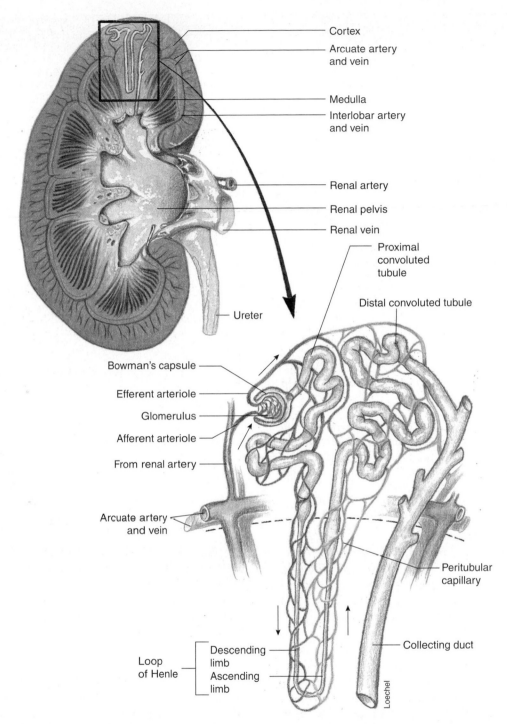

FIGURE 24.13 The Structure of the Nephron
The nephron and the closely associated blood vessels create a system that allows for the passage of materials from the circulatory system to the nephron by way of the glomerulus and Bowman's capsule. Materials are added to and removed from the fluid in the nephron via the tubular portions of the nephron and their associated capillaries.

would be produced. The primary hormone involved in regulating water loss is the antidiuretic hormone (ADH). When the body has excess water, cells in the hypothalamus of the brain respond and send a signal to the pituitary, and only a small amount of ADH is released and water is lost. When you are dehydrated, the same brain cells cause more ADH to be released and water leaves the collecting duct and is returned to the blood.

The distal convoluted tubule is primarily involved in fine tuning the amounts of various kinds of molecules that are lost

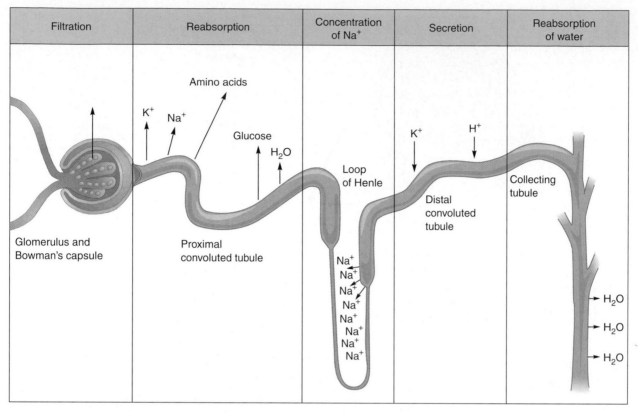

Filtration	Reabsorption	Concentration of Na$^+$	Secretion	Reabsorption of water

FIGURE 24.14 Specific Functions of the Nephron

Each portion of the nephron has specific functions. The glomerulus and Bowman's capsule accomplish the filtration of fluid from the bloodstream into the nephron. The proximal convoluted tubule reabsorbs most of the material filtered. The loop of Henle concentrates Na$^+$, so that water will move from the collecting tubule. The distal convoluted tubule regulates pH and ion concentration by differential secretion of K$^+$ and H$^+$.

in the urine. Hydrogen ions (H$^+$), sodium ions (Na$^+$), chloride ions (Cl$^-$), potassium ions (K$^+$), and ammonium ions (NH$_4^+$) are regulated in this way. Some molecules that pass through the nephron are relatively unaffected by the various activities going on in the kidneys. One of these is urea, which is filtered through the glomerulus into Bowman's capsule. As it passes through the nephron, much of it stays in the tubule and is eliminated in the urine. Many other kinds of molecules, such as minor metabolic waste products and some drugs, are also treated in this manner. Figure 24.14 summarizes the major functions of the various portions of the kidney tubule system.

Summary

The body's systems must be integrated in such a way that the internal environment stays relatively constant. This chapter surveys five systems of the body—the cardiovascular, lymphatic, respiratory, digestive, and excretory systems—and describes how they are integrated. All of these systems are involved in the exchange of materials across cell membranes.

The cardiovascular system consists of a pump, the heart, and blood vessels that distribute the blood to all parts of the body. The blood is a carrier fluid that transports molecules and heat. The exchange of materials between the blood and body cells takes place through the walls of the capillaries. Because the flow of blood can be regulated by the contraction of arterioles, blood can be sent to different parts of the body at different times. Hemoglobin in red blood cells is very important in the transport of oxygen. Carbonic anhydrase is an enzyme in red blood cells that converts carbon dioxide into bicarbonate ions, which can be easily carried by the blood.

The lymphatic system is a collection of thin-walled tubes (lymph vessels), which branch throughout the body and lymph organs. Lymph moves fat from the intestinal tract to the bloodstream through lacteals, transports excess tissue fluid back to the cardiovascular system, and defends against harmful agents, such as bacteria and viruses. The lymph organs include the lymph nodes, tonsils, spleen, thymus gland, and red bone marrow.

The respiratory system consists of the lungs and the associated tubes that allow air to enter and leave the lungs. The diaphragm and muscles of the chest wall are important in the process of breathing. In the lungs, alveoli provide a large surface area in association with capillaries, which allows for the rapid exchange of oxygen and carbon dioxide.

The digestive system disassembles food molecules. This involves several processes: grinding by the teeth and stomach, the emulsification of fats by bile from the liver, the addition of water to dissolve molecules, and enzymatic action to break complex molecules into simpler molecules for absorption. The small intestine provides a large surface area for the absorption of nutrients, because it is long and its wall contains many tiny projections that increase surface area. Once absorbed, the materials are carried to the liver, where molecules can be modified.

The excretory system is a filtering system of the body. The kidneys consist of nephrons, into which the circulatory system filters fluid. Most of this fluid is useful and is reclaimed by the cells that make up the walls of these tubules. Materials that are present in excess and those that are harmful are allowed to escape. Some molecules may be secreted into the tubules before being eliminated from the body.

Key Terms

Use the interactive flash cards on the **Concepts in Biology,** *12/e website to help you learn the meaning of these terms.*

alveoli 527
aorta 523
arteries 520
arterioles 525
atria 522
atrioventricular valves 522
bile 532
blood 520
Bowman's capsule 536
breathing 528
bronchi 527
bronchioles 527
capillaries 520
cardiovascular system 520
diaphragm 528
diastolic blood pressure 524
digestive system 530
distal convoluted tubule 536
duodenum 532
excretory system 535
formed elements 520
gallbladder 532
gastric juice 530
glomerulus 536
heart 520

hemoglobin 520
hepatic portal vein 534
kidneys 536
lacteal 534
large intestine 532
loop of Henle 536
lungs 527
lymph 526
lymph nodes 526
lymphatic system 526
nephrons 536
pancreas 532
pepsin 532
pharynx 530
plasma 520
platelets 522
proximal convoluted tubule 536
pulmonary artery 523
pulmonary circulation 523
pyloric sphincter 532
red blood cells (RBCs) 520
respiratory system 527
salivary glands 530
semilunar valves 523

small intestine 532
systemic circulation 523
systolic blood pressure 524
tissue fluid 522
trachea 527

veins 520
ventricles 522
villi 534
white blood cells (wbcs) 520

Basic Review

1. The vessels that carry blood away from the heart are
   a. veins.
   b. arteries.
   c. capillaries.
   d. lacteals.
2. Which of the following is not a formed element?
   a. rbc
   b. wbc
   c. platelets
   d. plasma
3. _____ is the liquid that baths the body's cells and contains the same chemicals as plasma but smaller amounts of albumin.
4. Which of the following is not a function of the lymphatic system?
   a. It moves fat from the intestinal tract to the bloodstream.
   b. It transports excess tissue fluid back to the cardiovascular system.
   c. It defends against harmful agents, such as bacteria and viruses.
   d. It carries oxygen to cells deep in the body.
5. The first part of the small intestine is known as the
   a. duodenum.
   b. ileum.
   c. pancreas.
   d. glomerulus.
6. Saliva contains the enzyme _____, which begins the chemical breakdown of starch.
   a. salivary amylase
   b. pepsin
   c. trypsin
   d. pylorase
7. In the respiratory system, the small sacs in the lungs where gas exchange takes place are called
   a. nephrons.
   b. alveoli.
   c. nodes.
   d. platelets.

8. The lining of the small intestine consists of millions of fingerlike projections, called
   a. nodes.
   b. alveoli.
   c. villi.
   d. glomeruli.

9. The distal convoluted tubule is primarily involved in fine-tuning the amounts of various kinds of molecules that are lost in the waste material known as
   a. urine.
   b. epinephrine.
   c. pepsin.
   d. cytokine.

10. About 200 billion _____ are formed each day and are important in blood clotting.
   a. nephrons
   b. lacteals
   c. platelets
   d. formed elements

Answers
1. b 2. d 3. Tissue fluid 4. d 5. a 6. a 7. b 8. c 9. a 10. c

Concept Review

Answers to the Concept Review questions can be found on the Concepts in Biology, 12/e website.

24.1 The Basic Principles of Materials Exchange

1. Why do multicellular organisms have materials-exchange problems?
2. In general, what kinds of methods can be used to overcome materials-exchange problems?

24.2 Circulation: The Cardiovascular System

3. What are the functions of the heart, arteries, veins, arterioles, blood, and capillaries?
4. How do red blood cells assist in the transportation of oxygen and carbon dioxide?

24.3 The Lymphatic System

5. What purposes are served by the lymphatic system?

6. List the organs of the lymphatic system, and describe the function of each.

24.4 Gas Exchange: The Respiratory System

7. Describe the mechanics of breathing.
8. How are blood pH and breathing interrelated?
9. Why does a poorly functioning heart result in a person experiencing shortness of breath?

24.5 Obtaining Nutrients: The Digestive System

10. Describe three ways in which the digestive system increases its ability to absorb nutrients.
11. List three functions of the liver.
12. Name five digestive enzymes and describe their functions.
13 What is the role of bile in digestion?
14. How is fat absorption different from the absorption of carbohydrate and protein?

24.6 Waste Disposal: The Excretory System

15. What are the functions of the glomerulus, proximal convoluted tubule, loop of Henle, and distal convoluted tubule?
16. Describe how the kidney regulates the amount of water in the body.

Thinking Critically

For guidelines to answer this question, visit the Concepts in Biology, 12/e website.

It is possible to keep a human being alive even if the heart, lungs, kidneys, and digestive tract are not functioning by using heart-lung machines in conjunction with kidney dialysis and intravenous feeding. This implies that the basic physical principles involved in the functioning of these systems is well understood because the natural functions can be duplicated with mechanical devices. However, these machines are expensive and require considerable maintenance. Should society be spending money to develop smaller, more efficient mechanisms that could be used to replace diseased or damaged hearts, lungs, and kidneys?

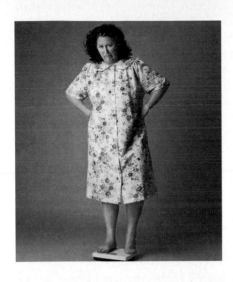

25

Nutrition

Food and Diet

There is an epidemic of obesity in the United States. In 2001, 20 states had obesity rates of 15–19%, 29 states had rates of 20–24%, and 1 state had a rate of more than 25%. Even states such as California, Hawaii, Oregon, and Washington, considered weight-conscious, saw a doubling in their overweight and obesity rates during the 1990s. The increases have occurred no matter the age, ethnicity, educational level, social status, or income level.

Public and private health agencies are very worried about the increasing number of obesity-related health problems. There has been an increase in fad dieting, an unwillingness to engage in regular exercise, and a higher demand for bariatric surgery, an operation to reduce the size of the stomach and/or intestines. Obesity is so pervasive in our society that some health care providers now cover bariatric surgery, an irreversible major surgical procedure. Insurance companies recognize that it is in their best interest to cover the costs of the surgery in hopes of reducing the future costs associated with obesity-related problems.

People appear to be unwilling to make the changes necessary to lose weight and become fit. Most are interested only in the "quick fix" approaches that go along with fad diets and infrequent exercise.

As our poor eating and exercise habits spread to other countries, they, too, are seeing an increase in obesity rates. Globally, over 300 million people are obese and an additional 225 million are overweight. Even countries such as Japan and Italy, which have histories of healthy diets, are now seeing their populations becoming overweight and obese. The cost to society is taking its toll financially, emotionally, and physically.

Getting and maintaining a healthy weight demands lifestyle changes that many find difficult to achieve. It also requires a good understanding of the basic principles of nutrition.

CHAPTER OUTLINE

25.1 Living Things as Chemical Factories: Matter and Energy Manipulators

Organisms maintain themselves by constantly processing molecules to obtain energy and building blocks for new living material. Autotrophs can manufacture organic molecules from inorganic molecules, but heterotrophs must consume organic molecules to get what they need. **Nutrients** are all the molecules required to support living things. Some nutrients are elements, such as calcium, iron, and potassium; others are organic molecules, such as carbohydrates, proteins, fats, and vitamins. All heterotrophs obtain the nutrients they need from food, and each kind of heterotroph has particular nutritional requirements. This chapter examines the nutritional requirements of humans. evolution of heterotrophs, p. 409

The word *nutrition* is used in two contexts. First, **nutrition** is the branch of science that seeks to understand food, its nutrients, how the body uses nutrients, and how inappropriate combinations or quantities of nutrients lead to ill health. The word *nutrition* also refers to all the processes by which we take in food and use it, including *ingestion, digestion, absorption,* and *assimilation*. **Ingestion** is the process of taking food into the body through eating. **Digestion** is the breakdown of complex food molecules to simpler molecules. **Absorption** is the movement of simple molecules from the digestive system to the circulatory system for dispersal throughout the body. **Assimilation** is the modification and incorporation of absorbed molecules into the structure of the organism.

Many of the nutrients that enter living cells undergo chemical changes before they are incorporated into the body. These interconversion processes are ultimately under the control of the cell's genetic material. It is DNA that codes the information necessary to manufacture the enzymes required to extract energy from chemical bonds and to convert raw materials (nutrients) into the structure (anatomy) of the organism.

Diet is the food and drink consumed by a person from day to day. It must contain the minimal nutrients necessary to manufacture and maintain the body's structure (e.g., the bones, skin, tendon, muscle) and regulatory molecules (enzymes and hormones) and to supply the energy (ATP) needed to run the body's machinery. If the diet is deficient in nutrients, or if a person's body cannot process nutrients efficiently, a dietary deficiency and ill health may result. A good understanding of nutrition can promote good health; it requires an understanding of the energy and nutrient content in various foods.

The *kilocalorie (kcal)* is the unit used to measure the amount of energy in foods. One **kilocalorie** is the amount of energy needed to raise the temperature of 1 *kilogram* of water 1°C. Remember that the prefix *kilo-* means "1,000 times" the value listed. Therefore, a kilocalorie is 1,000 times more heat energy than a *calorie*. A **calorie** is the amount of heat energy needed to raise the temperature of 1 *gram* of water 1°C. However, the amount of energy contained in food is usually called a Calorie with a capital C. This is unfortunate because it is easy to confuse a dietary Calorie, which is really a kilocalorie, with a calorie. Most books on nutrition and dieting use the term Calorie to refer to *food calories* (How Science Works 25.1).

25.2 The Kinds of Nutrients and Their Function

Nutritionists have divided nutrients into six major classes: carbohydrates, lipids, proteins, vitamins, minerals, and water. Chapters 2 and 3 presented the chemical makeup of these

HOW SCIENCE WORKS 25.1

Measuring the Caloric Value of Foods

A *bomb calorimeter* is an instrument used to determine the heat of combustion or caloric value of foods. To operate the instrument, a small food sample is formed into a pellet and sealed inside a strong container called a bomb. The bomb is filled with 30 atmospheres of oxygen and is then placed in a surrounding jacket filled with water. The sample is electrically ignited; as it aerobically reacts with the oxygen, the bomb releases heat to the surrounding water jacket. As the heat is released, the temperature of the water rises. Recall that a calorie is the amount of energy (or heat) needed to increase the temperature of 1 gram of water 1°C. Therefore, if 1 gram of food results in a water temperature increase of 13°C for each gram of water, that food has the equivalent of 13 calories.

types of molecules, and chapter 6 explored the nature of cellular respiration. A look at each of these classes of nutrients from a nutritionist's point of view should reveal how the human body works and how its nutritional needs can be met.

Carbohydrates

When the word *carbohydrate* is mentioned, many people think of table sugar, pasta, or potatoes. The term *sugar* is usually used to refer to mono- or disaccharides, but the carbohydrate group also includes more complex polysaccharides, such as starch, glycogen, and cellulose. Each of these has a different structural formula, has different chemical properties, and plays a different role in the body (figure 25.1). Many simple carbohydrates taste sweet and stimulate the appetite. Starch is the primary form in which we obtain carbohydrates. When complex carbohydrates, such as starch or glycogen, are broken down to monosaccharides, they can be used in cellular respiration to provide energy in the form of ATP molecules. Simple sugars are also used as building blocks in the manufacture of molecules such as nucleic acids. Complex carbohydrates can also be a source of **fiber,** which slows the absorption of nutrients and stimulates peristalsis (rhythmic contractions) in the intestinal tract.

A diet deficient in carbohydrates results in fats being oxidized and converted to ATP. Unfortunately, protein is also metabolized to provide cells with the glucose they need for survival. In situations in which carbohydrates are absent, most of the fats are metabolized to *keto acids*. Large numbers of keto acids may be produced in extreme cases of fasting, resulting in a potentially dangerous change in the body's pH. If a person does not have stored fat to metabolize, a carbohydrate deficiency will result in an even greater use of the body's proteins as a source of energy. This is usually only encountered in starvation, extreme cases of fasting, or eating disorders. In extreme cases, this can be fatal, because the oxidation of protein results in an increase in toxic, nitrogen-containing compounds. A more typical situation for humans is the consumption of too much food. As with other nutrients, if there is an excess of carbohydrates in the diet, the body converts them to lipids and stores them in fat cells.

FIGURE 25.1 Sources of Carbohydrates
The diet includes a wide variety of carbohydrates. Some are monosaccharides (simple sugars); others are more complex disaccharides, trisaccharides, and polysaccharides. Notice that the complex carbohydrates shown are primarily from plants. With the exception of milk, animal products are not a good nutritional source of carbohydrates, because animals do not store them in great quantities or use large amounts of them as structural materials.

Lipids

The class of nutrients technically known as *lipids* is often called *fat* by many people. This may lead to some confusion, because fats are only one of three subclasses of lipids. Each subclass of lipids—phospholipids, steroids, and true fats—plays an important role in human nutrition. Although various kinds of phospholipids are sold as dietary supplements, they are unnecessary, because all food composed of cells contains phospholipids. Many steroids are hormones that help regulate a variety of body processes. With the exception of vitamin D, it is not necessary to include steroids as a part of the diet. Cholesterol is a steroid, commonly found in certain foods, that causes health problems in some people. The *true fats* (also called *triglycerides*) are an excellent source of energy. They are able to release 9 kilocalories of energy per gram compared to 4 kilocalories per gram for carbohydrate or protein. Some fats contain the **essential fatty acids,** linoleic acid and linolenic acid, which cannot be synthesized by the human body and, therefore, must be a part of the diet. The body requires these essential fatty acids for normal growth, blood clotting, and healthy skin. Most diets that incorporate a variety of foods, including meats and vegetable oils, have enough of these essential fatty acids. A diet high in linoleic acid has also been shown to help reduce the amount of cholesterol in the blood. Some vitamins, such as A, D, E, and K, do not dissolve in water but dissolve in fat and, therefore, require fat for their absorption from the gut. omega fatty acids, p. 60

Fat is an insulator against outside cold and internal heat loss. Cold is the absence of heat and insulation slows heat loss. Fat is also an excellent shock absorber. Also, fat deposits in the back of the eyes serve as cushions when the head suffers a severe blow. During starvation, these deposits are lost, and the eyes become deep-set in the eye sockets, giving the person a ghostly appearance.

The pleasant taste and "mouth feel" of many foods is the result of fats. Their ingestion provides a full feeling after a meal because they leave the stomach more slowly than do other nutrients. Excess kilocalories obtained directly from fats are stored more efficiently than excess kilocalories obtained from carbohydrates and proteins, because the body does not need to expend energy to convert the molecules to fat. The fat molecules can simply be disassembled in the gut and reassembled in the cells.

Proteins

Proteins are composed of amino acids linked together by peptide bonds; however, not all proteins contain the same amino acids. From a nutritional point of view, proteins can be divided into two main groups: *complete proteins* and *incomplete proteins*. **Complete proteins** contain all the amino acids necessary for good health, whereas **incomplete proteins** lack certain amino acids that the body must have to function efficiently. The human body can manufacture some amino acids but is unable to manufacture others. Those the body cannot manufacture are called **essential amino acids** (table 25.1). Without adequate amounts of these essential amino acids in the diet, a person may develop a protein-deficiency disease. Proteins also provide a last-ditch source of energy during starvation when carbohydrate and fat reserves are used up. three categories of protein, p. 56

Unlike carbohydrates and fats, proteins cannot be stored for later use. Because they are not stored and because they have many important functions in the body, adequate amounts of protein must be present in the daily diet. However, a high-protein diet is not necessary. Only small amounts of protein—20 to 30 grams—are metabolized and lost from the body each day and must be replaced. Any protein in excess of that needed to rebuild lost molecules is metabolized to provide the body with energy. Protein is the most expensive but least valuable energy source. Carbohydrates and fats are much better sources.

TABLE 25.1	
Sources of Essential Amino Acids	
Essential Amino Acids*	**Food Sources**
Threonine	Dairy products, nuts, soybeans, turkey
Lysine	Dairy products, nuts, soybeans, green peas, beef, turkey
Methionine	Dairy products, fish, oatmeal, wheat
Arginine (essential to infants only)	Dairy products, beef, peanuts, ham, shredded wheat, poultry
Valine	Dairy products, liverwurst, peanuts, oats
Phenylalanine	Dairy products, peanuts, calves' liver
Histidine (essential to infants only)	Human and cow's milk and standard infant formulas
Leucine	Dairy products, beef, poultry, fish, soybeans, peanuts
Tryptophan	Dairy products, sesame seeds, sunflower seeds, lamb, poultry, peanuts
Isoleucine	Dairy products, fish, peanuts, oats, lima beans

*The essential amino acids are required in the human diet for protein building and, along with the nonessential amino acids, allow the body to metabolize all nutrients at an optimum rate. Combinations of plant foods can provide essential amino acids even if complete protein foods (e.g., meal, fish, and milk) are not in the diet.

The body has a mechanism that protects protein from being metabolized to provide cells with energy. **Protein-sparing** is the body's conservation of proteins by first oxidizing carbohydrates and fats as a source of ATP energy. During fasting or starvation, many of the body's cells can use fat as their primary source of energy, thus protecting the more valuable protein. However, red blood cells and nervous tissue must have glucose, which can be supplied by small amounts of carbohydrates. Because very little glucose is stored, after a day or two of fasting the body begins to convert some of the amino acids from protein into glucose to supply these vital cells. Although fat can supply energy for many cells during fasting or starvation, the fat cannot completely protect the proteins if there are no carbohydrates in the diet. With prolonged starvation, the fat stores are eventually depleted and structural proteins are used for all the body's energy needs. When starvation reaches this point, it is usually fatal.

Most people have a misconception about the amount of protein necessary in the diet. The total amount necessary is actually quite small (about 50 grams/day) and can be obtained easily. One-quarter pound of hamburger, a half chicken breast, or a fish sandwich contains the daily amount of protein needed by most people.

Vitamins

Vitamins are organic molecules needed in minute amounts to maintain essential metabolic activities. Like essential amino acids and fatty acids, vitamins cannot be manufactured by the body. Table 25.2 lists vitamins for which there are recommended daily intakes. Contrary to popular belief, vitamins are not a source of energy; rather, they help specific enzymes bring about biochemical changes. Some vitamins are actually incorporated into the structure of the enzymes. Such vitamins are called *coenzymes*. For example, the B-complex vitamin niacin helps enzymes in the respiration of carbohydrates. cofactors, vitamins, and coenzymes, p. 99

Most vitamins are acquired from food; however, vitamin D can be formed when the ultraviolet light in sunshine strikes a cholesterol molecule in the skin, converting this cholesterol to vitamin D. This means that vitamin D is not really a vitamin at all. It came to be known as a vitamin because of the mistaken idea that it was acquired only through food, rather than formed in the skin on exposure to sunshine. It would be more correct to call vitamin D a hormone, but most people do not.

Because many vitamins are inexpensive and their functions are poorly understood, many advocate large doses (megadoses) of vitamins to prevent a wide range of diseases. Often, the benefits advertised are based on fragmentary evidence and lack a clearly defined mechanism of action. The consumption of high doses of vitamins is unwise, because megadoses of many vitamins are toxic. For example, fat-soluble vitamins, such as vitamins A and D, are stored in body fat and the liver. Excess vitamin A causes joint pain, hair

loss, and jaundice. Excess vitamin D results in calcium deposits in the kidneys, high amounts of calcium in the blood, and bone pain. High doses of some of the water-soluble vitamins also have toxic effects; vitamin B_6 (pyridoxine) in high concentrations causes nervous symptoms such as unsteady gait and numbness in the hands. However, inexpensive multivitamins that provide 100% of the recommended daily allowance can prevent or correct deficiencies caused by a poor diet without the danger of toxic consequences. Most people do not need vitamin supplements *if they eat a well-balanced diet* (How Science Works 25.2).

Minerals

Minerals are elements found in nature that cannot be synthesized by the body. Table 25.3 lists the sources and functions of several common minerals. Because minerals are elements, they cannot be broken down or destroyed by metabolism or cooking. They commonly occur in many foods and in water. Minerals retain their characteristics whether they are in foods or in the body, and each plays a different role in metabolism. Minerals can function as regulators, activators, transmitters, and controllers of various enzymatic reactions. For example, sodium (Na^+) and potassium ions (K^+) are important in the transmission of nerve impulses, whereas magnesium ions (Mg^{++}) facilitate energy release during reactions involving ATP. Without iron, not enough hemoglobin would be formed to transport oxygen, a condition called *anemia*, and a lack of calcium may result in *osteoporosis*. **Osteoporosis** is a condition that results from calcium loss leading to painful, weakened bones. Many minerals are important in the diet. In addition to those just mentioned, we need chlorine, cobalt, copper, iodine, phosphorus, sulfur, and zinc to remain healthy. With few exceptions, adequate amounts of minerals are obtained in a normal diet. Calcium and iron supplements may be necessary, however, particularly in women.

Water

Water is crucial to all life and plays many essential roles. Humans can survive weeks without food but would die in a matter of days without water. It is known as the universal solvent, because so many types of molecules are soluble in it. The human body is about 65% water. Even dense bone tissue consists of 33% water. All the chemical reactions in living things take place in water. It is the primary component of blood, lymph, and body tissue fluids. Inorganic and organic nutrients and waste molecules are also dissolved in water. Dissolved inorganic ions, such as sodium (Na^+), potassium (K^+), and chloride (Cl^-), are electrolytes because they form a solution capable of conducting electricity. The concentration of these ions in the body's water must be regulated to prevent electrolyte imbalances. electrolytes, p. 32

Excesses of many types of waste are eliminated from the body dissolved in water; that is, they are excreted from the kidneys as urine or in small amounts from the lungs or skin

TABLE 25.2

Sources and Functions of Vitamins

Name	Recommended Daily Intake for Adults	Physiological Value	Readily Available Sources	Other Information
Vitamin A	800–1,000 µg	Important in vision; maintains skin and intestinal lining	Orange, red, and dark green vegetables	Fat-soluble, stored in liver; children have little stored
Vitamin B$_1$ (thiamin)	1.1–1.2 mg	Maintains nerves and heart; involved in carbohydrate metabolism	Whole-grain foods, legumes, and pork	Larger amounts needed during pregnancy and lactation
Vitamin B$_2$ (riboflavin)	1.1–1.3 mg	Central role in energy metabolism; maintains skin and mucous membranes	Dairy products, green vegetables, whole-grain foods, and meat	
Vitamin B$_3$ (niacin)	14–16 mg	Energy metabolism	Whole-grain foods and meat	
Vitamin B$_6$ (pyridoxine pyridoxol, pyradoxamine)	1.3 mg	Forms red blood cells; maintains nervous system	Whole-grain foods; milk; green, leafy vegetables; meats; legumes; and nuts	
Vitamin B$_{12}$ (cobalamin)	2.4 µg	Protein and fat metabolism; forms red blood cells; maintains nervous system	Animal products only: dairy products, meats, and seafood	Stored in liver; pregnant and lactating women and vegetarians need larger amounts
Vitamin C	60 mg	Maintains connective tissue, bones, and skin	Citrus fruits, leafy green vegetables, tomatoes, and potatoes	Toxic in high doses Fat-soluble
Vitamin D	5 µg	Needed to absorb calcium for strong bones and teeth	Vitamin D–fortified milk; exposure of skin to sunlight	
Vitamin E	8–10 µg	Antioxidant; protects cell membranes	Whole-grain foods, seeds and nuts, vegetables, and vegetable oils	Fat-soluble; only two cases of deficiency ever recorded
Vitamin K	60–70 µg	Blood clotting	Dark green vegetables	Fat-soluble; small amount of fat needed for absorption
Folate (folic acid)	400 µg	Coenzyme in metabolism	Most foods: beans and fortified cereals	Important in pregnancy
Pantothenic acid	5 mg	Involved in many metabolic reactions; formation of hormones, normal growth	All foods	Typical diet provides adequate amounts
Biotin	30 µg	Maintains skin and nervous system	All foods	Typical diet provides adequate amounts; added to intravenous feedings
Choline (lecithin)	425–550 mg	Component of cell membranes	All foods	Only important in people unable to consume food normally

HOW SCIENCE WORKS 25.2

Preventing Scurvy

Scurvy is a nutritional disease caused by a lack of vitamin C in the diet. Vitamin C is essential to the formation of collagen, a protein important in most tissues. The symptoms of scurvy include the poor healing of wounds; fragile blood vessels, resulting in bleeding; a lack of bone growth; and a loosening of the teeth.

Scurvy is not common today; however, in the past, many people on long sea voyages developed scurvy because their diets lacked fresh fruits and vegetables. This was such a common problem that the disease was often called sea scurvy. Excerpts from a letter by a Dr. Harness to the First Lord of the Admiralty of the British navy describe the practice of using lemons to prevent scurvy on

British ships: "During the blockade of Toulon in the summer of 1793, many of the ships' companies were afflicted with symptoms of scurvy; . . . I was induced to propose . . . the sending a vessel into port for the express purpose of obtaining lemons for the fleet; . . . and the good effects of its use were so evident . . . that an order was soon obtained from the commander in chief, that no ship under his lordship's command should leave port without being previously furnished with an ample supply of lemons. And to this circumstance becoming generally known may the use of lemon juice, the effectual means of subduing scurvy, while at sea, be traced."

A common term applied to British seamen during this time was *limey*.

TABLE 25.3

Sources and Functions of Minerals

Name	Recommended Daily Intake for Adults	Physiological Value	Readily Available Sources	Other Information
Calcium	1,000 mg	Builds and maintains bones and teeth	Dairy products	Children need 1,300 mg; vitamin D needed for absorption
Fluoride	3.1–3.8 mg	Maintains bones and teeth; reduces tooth decay	Fluoridated drinking water and seafood	
Iodine	150 μg	Necessary to make the hormone thyroxine	Iodized table salt and seafood	The soils in some parts of the world are low in iodine, so iodized salt is very important
Iron	15–10 mg	Necessary to make hemoglobin	Grains, meat, seafood, poultry, legumes, and dried fruits	Women need more than men; pregnant women need two times the normal dose
Magnesium	320–420 mg	Bone mineralization; muscle and nerve function	Dark green vegetables, whole grains, and legumes	
Phosphorus	700 mg	Acid/base balance; enzyme cofactor	All foods	Children need 1,250 mg; most people get more than recommended
Selenium	55–70 μg	Involved in many enzymatic reactions	Meat, grains, and seafood	
Zinc	12–15 mg	Wound healing; fetal development; involved in many enzymatic and hormonal activities	Meat, fish, and poultry	

through evaporation. In a similar manner, the evaporation of water from the skin cools the body. Water molecules are also essential reactants in all the various hydrolytic reactions of metabolism. Without it, the breakdown of molecules such as starch, proteins, and lipids would be impossible. With all these important roles played by water, it's no wonder that nutritionists recommend that we drink the equivalent of at least eight glasses each day. This amount of water can be obtained from tap water, soft drinks, juices, and numerous food items, such as lettuce, cucumbers, tomatoes, and applesauce.

25.3 Dietary Reference Intakes

Over the years, the U.S. Department of Agriculture (USDA) has published various guidelines for maintaining good nutritional health. The current guidelines, the **Dietary Reference Intakes,** provide information on the amounts of certain nutrients various members of the public should receive. These guidelines are very detailed, and they are different for males and females of various ages and for pregnant and nursing mothers. They also provide information about the maximum amount of certain nutrients that people should get.

A *kilocalorie (kcal)* is the unit used to measure the amount of energy in foods. One **kilocalorie** is the amount of energy needed to raise the temperature of 1 *kilogram* of water 1°C. Remember that the prefix *kilo-* means 1,000 times the value listed. Therefore, a kilocalorie is 1,000 times more heat energy than a *calorie.* Recall from chapter 2 that a calorie is the amount of heat energy needed to raise the temperature of 1 *gram* of water 1°C. However, the amount of energy contained in food is usually called a Calorie, with a capital C. A dietary Calorie is actually a kilocalorie. temperature, heat, and calories, p. 24

Dietary Reference Intakes are used when preparing product labels. By law, labels must list ingredients from the greatest to the least in quantity. In addition to carbohydrates fats, proteins, and fiber, about 25 vitamins and minerals have Dietary Reference Intakes. Table 25.4 gives examples of some of the more common nutrients and their reference amounts for young adults.

25.4 The Food Guide Pyramid

Using Dietary Reference Intakes and product labels or counting kilocalories is a complicated way to plan a diet. Planning a diet around basic food groups is generally easier. The USDA's **Food Guide Pyramid** is a useful tool for planning a nutritious diet. Over the years, as new nutritional information has become available, its suggestions have been modified (figure 25.2). The current Food Guide Pyramid is a modification of an earlier version and as new discoveries are made it will be changed.

Grains

Grains include vitamin-enriched or whole-grain cereals and products such as breads, bagels, buns, crackers, dry and cooked cereals, pancakes, pasta, and tortillas. Whole-grain foods contain many more nutrients. Most of the items in this group are dry and seldom need refrigeration.

The Food Guide Pyramid recommends that women consume 6 oz a day of grains, 8 oz for men; 1 oz is equal to about one slice of bread or half a bagel. Grains should provide most of the kilocalorie requirements in the form of complex carbohydrates, such as starch, which is the main ingredient in most grain products. These foods help satisfy the appetite, and many are very low in fat. They also provide fiber, which assists in the proper functioning of the digestive tract. Foods high in fiber are also a source of several vitamins and minerals.

Cereal group

Significant nutritional components of grains are carbohydrate, fiber, several B vitamins, vitamin E, and the minerals iron and magnesium.

Fruits

Is a tomato a vegetable or a fruit? The confusion arises from the fact that the term *vegetable* is not scientifically precise but for nutritional purposes generally means a plant material that is not sweet and is eaten during the main part of a meal. *Fruit,* on the other hand, is a botanical term for the structure produced from the female part of the flower that contains the seeds. Although botanically green beans, peas, and corn are all fruits, nutritionally speaking they are placed in the vegetable category, because they are generally eaten during the main part of a meal. From a

Fruits group

nutritional point of view, fruits include such sweet plant products as melons, berries, apples, oranges, and bananas. The Food Guide Pyramid recommends 1 to 2 cups per day, depending on age. However, because fruits tend to be high in natural sugars, the consumption of large amounts of fruits can add significant kilocalories to the diet.

Significant nutritional components of fruits are carbohydrate, fiber, water, and vitamin C.

Vegetables

The Food Guide Pyramid suggests 41/2 cups be eaten from this group each day for those who need 2,000 calories to maintain their weight and health. The items in this group include nonsweet plant materials, such as broccoli, carrots,

TABLE 25.4

Dietary Reference Intakes for Some Common Nutrients

Nutrient	Women, 19–30 Years Old	Men, 19–30 Years Old	Maximum, Persons 19–30 Years Old	Value of Nutrient
Carbohydrates	130 g/day (45–65% of kilocalories)	130 g/day (45–65% of kilocalories)	No maximum set but refined sugars should not exceed 25% of total kilocalories	A source of energy
Proteins	46 g/day (10–35% of kilocalories)	56 g/day (10–35% of kilocalories)	No maximum but high-protein diets stress kidneys	Proteins are structural components of all cells; there are 10 essential amino acids that must be obtained in the diet
Fats	20–35% of kilocalories	20–35% of kilocalories	Up to 35% of total kilocalories	Energy source and building blocks for many molecules needed
Saturated and *trans* fatty acids	No reference amount determined	No reference amount determined	As low as possible	
Linoleic acid	12 g/day	17 g/day	No maximum set	Essential fatty acid needed for enzyme function and maintenance of epithelial cells
Linolenic acid	1.1 g/day	1.6 g/day	No maximum set	Essential fatty acid needed to reduce coronary heart disease
Cholesterol	No reference amount determined	No reference amount determined	As low as possible	None needed, because the liver makes cholesterol
Total fiber	25 g/day	38 g/day	No maximum set	Improve gut function
Calcium	1 g/day	1 g/day	2.5 g/day	Needed for the structure of bones and many other functions
Iron	18 mg/day	8 mg/day	45 mg/day	Needed to build the hemoglobin of red blood cells
Sodium	1.5 g/day	1.5 g/day	5.8 g/day (most people exceed this limit)	Needed for normal cell function
Vitamin A	700 ug/day	900 ug/day	3,000 ug/day	Maintains skin and intestinal lining
Vitamin C	75 mg/day	90 mg/day	2,000 mg/day	Maintains connective tissue and skin
Vitamin D	5 ug/day	5 ug/day	50 ug/day	Needed to absorb calcium for bones

Vegetables group

cabbage, corn, green beans, tomatoes, potatoes, lettuce, and spinach. A cup is considered 1 cup of raw, leafy vegetables or 1/2 cup of other types. There is increasing evidence that cabbage, broccoli, and cauliflower can provide some protection from certain types of cancers. This is a good reason to include these foods in the diet. Foods in this group are the primary source of many vitamins. Because different vegetables contain different amounts of vitamins, it is wise to include as much variety as possible in this group. They also provide fiber, which assists in the proper functioning of the digestive tract.

Significant nutritional components of vegetables are carbohydrate, fiber; several B vitamins; vitamins A, C, E, and K; and the minerals iron and magnesium.

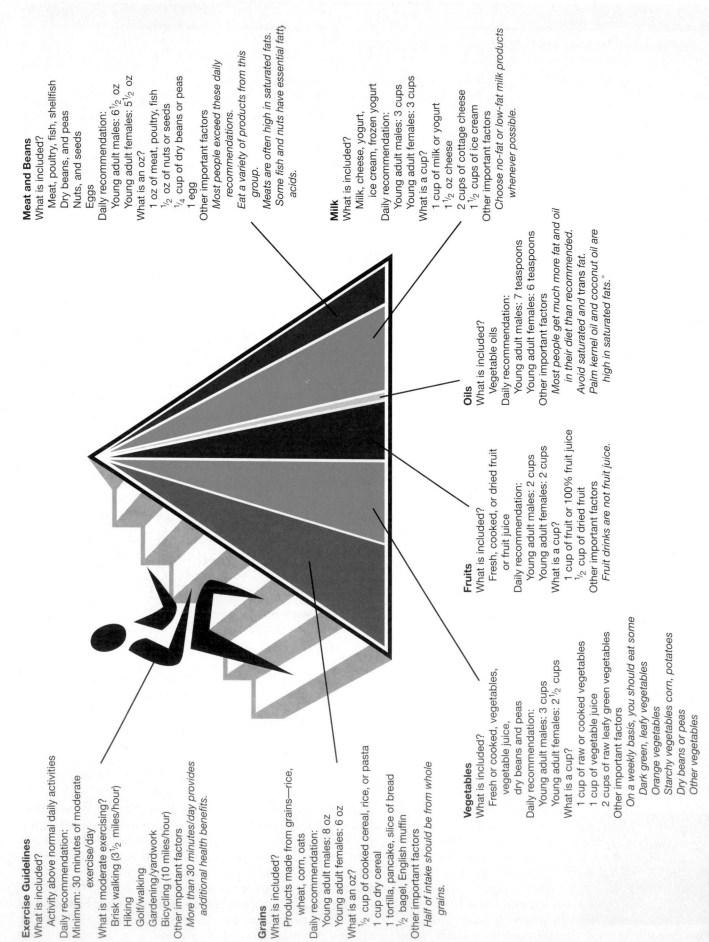

Exercise Guidelines
What is included?
    Activity above normal daily activities
Daily recommendation:
    Minimum: 30 minutes of moderate
        exercise/day
What is moderate exercising?
    Brisk walking (3½ miles/hour)
    Hiking
    Golf/walking
    Gardening/yardwork
    Bicycling (10 miles/hour)
Other important factors
    More than 30 minutes/day provides
        additional health benefits.

Grains
What is included?
    Products made from grains—rice,
        wheat, corn, oats
Daily recommendation:
    Young adult males: 8 oz
    Young adult females: 6 oz
What is an oz?
    ½ cup of cooked cereal, rice, or pasta
    1 cup dry cereal
    1 tortilla, pancake, slice of bread
    ½ bagel, English muffin
Other important factors
    Half of intake should be from whole
        grains.

Vegetables
What is included?
    Fresh or cooked, vegetables,
        vegetable juice,
        dry beans and peas
Daily recommendation:
    Young adult males: 3 cups
    Young adult females: 2½ cups
What is a cup?
    1 cup of raw or cooked vegetables
    1 cup of vegetable juice
    2 cups of raw leafy green vegetables
Other important factors
    On a weekly basis, you should eat some
    Dark green, leafy vegetables
    Orange vegetables
    Starchy vegetables corn, potatoes
    Dry beans or peas
    Other vegetables

Fruits
What is included?
    Fresh, cooked, or dried fruit
        or fruit juice
Daily recommendation:
    Young adult males: 2 cups
    Young adult females: 2 cups
What is a cup?
    1 cup of fruit or 100% fruit juice
    ½ cup of dried fruit
Other important factors
    Fruit drinks are not fruit juice.

Oils
What is included?
    Vegetable oils
Daily recommendation:
    Young adult males: 7 teaspoons
    Young adult females: 6 teaspoons
Other important factors
    Most people get much more fat and oil
        in their diet than recommended.
    Avoid saturated and trans fat.
    Palm kernel oil and coconut oil are
        high in saturated fats.

Meat and Beans
What is included?
    Meat, poultry, fish, shellfish
    Dry beans, and peas
    Nuts, and seeds
    Eggs
Daily recommendation:
    Young adult males: 6½ oz
    Young adult females: 5½ oz
What is an oz?
    1 oz of meat, poultry, fish
    ½ oz of nuts or seeds
    ¼ cup of dry beans or peas
    1 egg
Other important factors
    Most people exceed these daily
        recommendations.
    Eat a variety of products from this
        group.
    Meats are often high in saturated fats.
    Some fish and nuts have essential fatty
        acids.

Milk
What is included?
    Milk, cheese, yogurt,
        ice cream, frozen yogurt
Daily recommendation:
    Young adult males: 3 cups
    Young adult females: 3 cups
What is a cup?
    1 cup of milk or yogurt
    1½ oz cheese
    2 cups of cottage cheese
    1½ cups of ice cream
Other important factors
    Choose no-fat or low-fat milk products
        whenever possible.

FIGURE 25.2 The Food Guide Pyramid
The Food Guide Pyramid suggests that we eat certain amounts of five food groups while decreasing our intake of fats and sugars. This guide simplifies menu planning and helps us ensure that we get all the recommended amounts of basic nutrients. In order to be healthy, exercising on a regular basis is also essential.

Milk

All of the cheeses, ice cream, yogurt, and milk are in this group. Two to 3 cups, depending on age, are recommended each day. One and one-half ounces of hard cheese is equivalent to a cup. Product labels state the appropriate serving size of individual items. Vitamin D–fortified dairy products are the primary dietary source of vitamin D. Remember that many cheeses contain large amounts of cholesterol and fat per serving. Low-fat dairy products are now recommended in the pyramid and are becoming increasingly common as manufacturers seek to match their products with the public's desire for less fat in the diet.

Dairy group

Significant nutritional components of milk are protein, carbohydrate, fat, several B vitamins, vitamin D, and the mineral calcium.

Meat and Beans

This group contains most of the things we eat as a source of protein—for example, beef, chicken, fish, nuts, peas, tofu, and eggs. The Food Guide Pyramid recommends 5.5–6.5 oz of protein per day. This means that one hamburger exceeds this recommendation for daily intake, and most people eat many times what they need. Recall that daily protein intake is essential, because protein is not stored in the body, as are fats and carbohydrates, and that the body cannot manufacture the 10 essential amino acids, so they must be included in the diet. Animal proteins are complete proteins.

Meat group

Because many sources of protein also contain significant amounts of fat, and health recommendations suggest reducing our fat intake, more attention is being paid to the quantity of the protein-rich foods in the diet. Beans (except for the oil-rich soybean) are excellent sources of protein without unwanted fat. Food selection and preparation are also important in reducing fat consumption. Selecting foods that have less fat, broiling rather than frying, and removing the fat before cooking all reduce fat in the diet. For example, most of the fat in chicken and turkey is attached to the skin, so removing the skin removes most of the fat.

Eating excessive amounts of protein, however, can stress the kidneys by causing higher concentrations of calcium in the urine, can increase the demand for water to remove toxic keto acids produced from the breakdown of amino acids, and can lead to weight gain because of the intake of fat normally associated with many sources of protein. However, vegetarians must pay particular attention to acquiring adequate sources of protein, because they have eliminated a major source from their diet. They can get all the essential amino acids if they eat proper combinations of plant materials. Although nuts and soybeans are high in protein, they should not be consumed in large quantities because they are also high in fats.

Significant nutritional components of meat and beans are protein, fat, several B vitamins, vitamin E from seeds and nuts, and the mineral iron

Oils

Small amounts of oils are important in the diet, because certain essential fatty acids cannot be manufactured by the body and must be obtained in the diet. However, because oils are fats, they have a high Caloric content. The Food Guide Pyramid recommends that less than 10% of total Calories be from saturated fatty acids. This is about 3–7 teaspoons per day, depending on age and sex. The diet should also contain less than 300 mg/day of cholesterol, and trans fatty acid consumption should be as low as possible. For an adult, total fat intake should be between 20 and 35% of total Calories consumed daily. Most fats should be polyunsaturated and monounsaturated fatty acids, such as fish, nuts, and vegetable oils.

Oils group

The oils group includes canola, corn, olive, and sunflower oils, which are used in cooking. Some oils, such as olive, sesame, and walnut, are used to flavor foods. Mayonnaise, some salad dressings, and soft margarine with no *trans* fats are almost entirely made of oils. Most oils are high in monounsaturated or polyunsaturated fats and low in saturated fats. Plant oils do not contain cholesterol. A few plant oils, such as coconut and palm oil, are high in saturated fats and have health effects similar to those of solid fats.

Solid fats come from animals and include butter, beef fat (tallow, suet), chicken fat, pork fat (lard), stick margarine, and shortening. Fats can also be made from vegetable oils by adding more hydrogens to the molecules, a process called hydrogenation.

Significant nutritional components of oils are essential fatty acids and vitamin E.

Exercise

Since over 65 percent of adult Americans are overweight and over 30 percent are obese, exercise is important to improving nutritional health. Therefore, exercise is included in the Food Guide Pyramid; although it is not directly related to nutrition,

TABLE 25.5

Typical Energy Requirements for Common Activities

Light Activities (120–150 kcal/hr)	Light to Moderate Activities (150–300 kcal/hr)	Moderate Activities (300–400 kcal/hr)	Heavy Activities (420–600 kcal/hr)
Dressing	Sweeping floors	Pulling weeds	Chopping wood
Typing	Painting	Walking behind a lawnmower	Shoveling snow
Slow walking	Walking 2–3 mi/hr	Walking 3.5–4 mi/hr on a level surface	Walking or jogging 5 mi/hr
Standing	Bowling	Golf (no cart)	Walking up hills
Studying	Store clerking	Doubles tennis	Cross-country skiing
Sitting activities	Canoeing 2.5–3 mi/hr	Canoeing 4 mi/hr	Swimming
	Bicycling on a level surface at 5.5 mi/hr	Volleyball	Bicycling 11–12 mi/hr or up and down hills

the amount of exercise people get affects the number of Calories they can consume on a daily basis without gaining weight. Exercise has other health benefits as well. The pyramid recommends at least 30 minutes of moderate exercise per day; longer periods and more vigorous exercise have additional health benefits. Moderate exercise elevates the heart rate significantly, although activities such as doing household chores or walking while shopping do not elevate heart rate and, therefore, do not count as moderate exercise. The energy requirements in kilocalories (Kcal) for a variety of activities are listed in table 25.5.

25.5 Basal Metabolic Rate, Diet, and Weight Control

Basal Metabolic Rate (BMR)

Significant energy expenditure is required for muscular activity. However, even when the body is at rest, energy is required to maintain breathing, heart rate, and other normal body functions. The **basal metabolic rate (BMR)** is the rate at which the body uses energy when it is at rest. The basal metabolic rate of most people requires more energy than their voluntary muscular activity. Much of this energy is used to keep the body temperature constant. A true measurement of basal metabolic rate requires a measurement of oxygen used over a specific period under controlled conditions (How Science Works 25.3).

Several factors affect an individual's basal metabolic rate, including age, gender, height, and weight. Children have higher basal metabolic rates, and they decline throughout life. Elderly people have the lowest basal metabolic rates. In general, men have higher metabolic rates than women. The larger

a person, the higher his or her metabolic rate. With all of these factors taken into account, most young adults fall into the range of 1,200 to 2,200 kilocalories per day for a basal metabolic rate.

Because few of us rest 24 hours a day, we normally require more than the energy needed for basal metabolism. One of these requirements is **specific dynamic action (SDA)**, the amount of energy needed to process the food we eat. It is equal to approximately 10% of the total daily kilocalorie intake.

In addition to basal metabolic rate and specific dynamic action, the activity level of a person determines the number of kilocalories needed. A good general indicator of the number of kilocalories needed above basal metabolic rate is the type of occupation a person has (table 25.6). Because most adults are relatively sedentary, they would receive adequate amounts of energy if women consumed 2,200 kilocalories and men consumed 2,900 kilocalories per day.

TABLE 25.6

Additional Kilocalories as Determined by Occupation

Occupation	Kilocalories Needed per Day Above Basal Metabolic Rate*
Sedentary (student)	500–700
Light work (businessperson)	750–1,200
Moderate work (laborer)	1,250–1,500
Heavy work (professional athlete)	1,550–5,000 and up

*These are general figures and will vary from person to person, depending on the specific activities performed in the job.

TABLE 25.7

Are You Overweight?

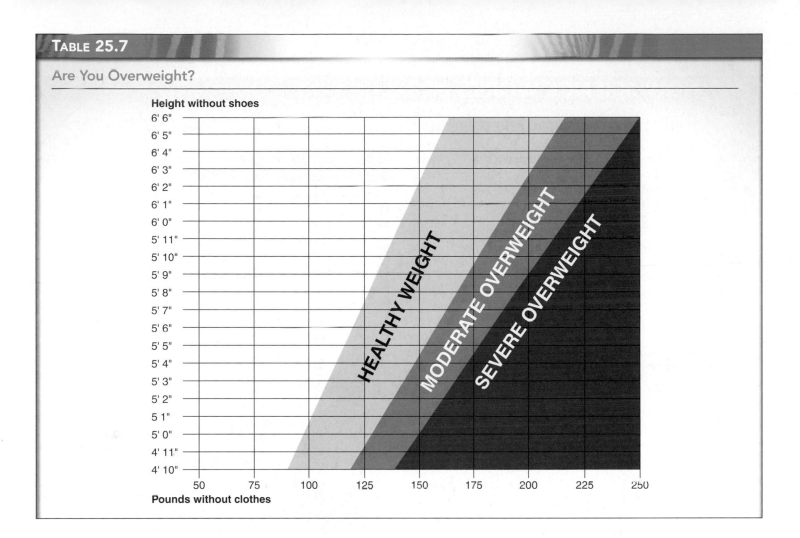

Body Mass Index (BMI)

Table 25.7 shows guidelines for determining whether one is overweight or obese. It is based on a specific method for determining *body mass index*. **Body mass index (BMI)**, body weight compared with height. BMI is calculated by determining a person's weight (without clothing) in kilograms and barefoot height in meters. The BMI is the weight in kilograms divided by height in meters squared:

$$BMI = \frac{\text{weight in kilograms}}{(\text{height in meters})^2}$$

(Appendix 1 of this book gives conversions to the metric system of measurements.) For example, a person with a height of 5 feet 6 inches (1.68 meters) who weighs 165 pounds (75 kilograms) has a body mass index of 26.6 kg/m²:

$$BMI = \frac{kg}{m^2} = \frac{75 \text{ kg}}{(1.68 \text{ m})^2} = \frac{75 \text{ kg}}{2.82 \text{ m}^2} = 26.6 \text{ kg/m}^2$$

A person is considered **obese** if his or her body mass index is 30 kg/m² or more. An obese person has a significantly increased risk for many diseases. In general, the higher the body mass index, the more significant the risk. An adult who has a BMI between 25 and 29.9 is considered overweight. An adult who has a BMI of 30 or higher is considered obese. The health risks associated with being overweight or even obese can be lessened with increasing one's *fitness*. **Fitness** is a measure of how efficiently a person can function both physically and mentally. A person with a BMI of 30 may be more fit than someone with a BMI of 23 if he or she is involved in a regular program of physical and mental exercise. As fitness increases, so does metabolism, strength, mental acuity, and coordination.

Weight Control

Why is weight control a problem for such a large portion of the population? There are several metabolic pathways that convert carbohydrates (glucose) or proteins to fat. Stored body fat was very important for our prehistoric ancestors, because it allowed them to survive periods of food scarcity. In periods of food scarcity, stored body fat can be used to supply energy. Today, however, for most of us in the United States food scarcity is not a problem, and even small amounts of excess food consumed daily add to our fat stores.

553

HOW SCIENCE WORKS 25.3

Estimating Basal Metabolic Rate

A person's basal metabolic rate (also known as resting metabolic rate) is influenced by their genetic makeup, height, weight, age, skin surface area, and level of activity. For example, a 20-year-old female who is 160 centimeters tall (5'3") who weighs 50 kilograms (110 pounds) has 1.5 square meters of skin-surface area. The heat production in kilocalories released per square meter of skin varies with a person's age and sex (see box table 25.1). This woman uses about 886 kilocalories per day for each square meter of skin and her basal metabolic rate is estimated to be 1,329 kilocalories per day (1.5 × 886 = 11,329 kcal).

To estimate your basal metabolic rate,

1. For women,
   655.1 + (9.563 × kg) + (1.850 × cm) − (4.676 × age)
2. For men,
   66.5 + (13.75 × kg) + (5.003 × cm) − (6.775 × age)

(To convert pounds to kilograms, divide your weight in pounds by 2.2. To convert height in inches to centimeters, multiply inches by 2.54.) Please note that this is only a rough estimate of a person's basal metabolic rate.

Box Table 25.1

Kilocalories per Day per Square Meter of Skin

Age	Male	Female	Age	Male	Female
6	1,265	1,217	20–24	984	886
8	1,229	1,154	25–29	967	878
10	1,188	1,099	30–40	948	876
12	1,147	1,042	40–50	912	847
14	1,109	984	50–60	886	826
16	1,073	924	60–70	859	806
18	1,030	895	70–80	828	782

Although energy doesn't weigh anything, the nutrients that contain the energy do. Weight control is a matter of balancing the kilocalories ingested with the kilocalories of energy expended by normal daily activities and exercise. There is a limit to the rate at which a moderately active human body can use fat as an energy source. At most, 0.45 to 0.90 kilogram (1 or 2 pounds) of fat tissue per week are lost by an average person when dieting. Because 0.45 kilogram (1 pound) of fatty tissue contains about 3,500 kilocalories, decreasing the kilocalorie intake by 500 to 1,000 kilocalories per day while maintaining a balanced diet (including proteins, carbohydrates, and fats) will result in fat loss of 0.45 to 0.90 kilogram (1 to 2 pounds) per week. (A pound of pure fat contains about 4,100 kilocalories, but fat tissue contains other materials besides fat, such as water.)

Many diets promise large, rapid weight loss but, in fact, result only in temporary water loss. They may encourage eating and drinking foods that are diuretics, which increase the amount of urine produced and thus increase water loss. Or they may encourage exercise or other activities that cause people to lose water through sweating. Or they may simply encourage people to eat less, which results in a temporary weight loss because there is less food in the gut.

For most people who need to gain weight, increasing kilocalorie intake by 500 to 1,000 kilocalories per day will result in a gain of 1 or 2 pounds per week, provided the low weight is not the result of a health problem.

If you have calculated your body mass index and wish to modify your body weight, what are the steps you should take? You should first check with your physician before making any drastic change in your eating habits. You also need to determine the number of kilocalories you are consuming. That means keeping an accurate diet record for at least a week. Record everything you eat and drink and determine the number of kilocalories in those nutrients. This can be done by estimating the amounts of protein, fat, and carbohydrate (including alcohol) in your diet. Roughly speaking, 1 gram of carbohydrate is the equivalent of 4 kilocalories, 1 gram of fat is the equivalent of 9 kilocalories, 1 gram of protein is the equivalent of 4 kilocalories, and 1 gram of alcohol is 7 kilocalories. Most nutrition books have food-composition tables that tell you how much protein, fat, and carbohydrate are in a particular food.

Packaged foods also have serving sizes and nutritive content printed on the package. Do the arithmetic and determine your total kilocalorie intake for the week. If your intake (from your diet) in kilocalories equals your output (from your basal metabolic rate plus voluntary activity plus SDA), you should not have gained any weight. You can double-check this by weighing yourself before and after your week of recordkeeping. If your weight is constant and you want to lose weight, reduce the amount of food in your diet. To lose 1 pound each week, reduce your kilocalorie intake by 500 kilocalories per day. Be careful not to eat less than 600 kilocalories of carbohydrates or reduce your total daily intake below 1,200 kilocalories unless you are under the care of a

physician. It is important to have some carbohydrate in your diet, because a lack of carbohydrate leads to a breakdown of the protein that provides the cells with the energy they need. Also you may not be getting all the vitamins required for efficient metabolism and you could cause yourself harm. To gain 1 pound, increase your intake by 500 kilocalories per day.

A second ingredient valuable in a weight-loss plan is an increase in exercise while keeping food intake constant. This can involve organized exercise in sports or fitness programs. It can also include simple things, such as walking up stairs rather than taking an elevator, parking at the back of the parking lot so you walk farther, riding a bike for short errands, and walking down the hall to someone's office rather than using the phone. Many people who initiate exercise plans as a way of reducing weight are frustrated because they initially gain weight rather than lose it. This is because muscle weighs more than fat. Typically, they are "out of shape" and have low muscle mass. If they gain a pound of muscle at the same time they lose a pound of fat, they do not lose weight. However, if the fitness program continues, they will eventually stop increasing muscle mass and will lose weight. Even so, weight as muscle is more healthy than weight as fat.

If you, like millions of others, find that you are overweight, you have probably tried numerous diet plans. Not all of these plans are the same, and not all are suitable to your situation. If a diet plan is to be valuable in promoting good health, it must satisfy your needs in several ways. It must provide kilocalories, proteins, fats, and carbohydrates. It should also contain readily available foods from all the basic food groups, and it should provide enough variety to prevent you from becoming bored with the plan and going off the diet too soon. A diet should not be something you follow only for a while, then abandon and regain the lost weight.

25.6 | Eating Disorders

The three most common eating disorders are obesity, bulimia, and anorexia nervosa. All three disorders are related to the prevailing perceptions and values of the culture in which we live. In many cases, there is a strong psychological component as well.

Obesity

The health and life span of people who are obese are adversely affected by their obesity. The U.S. government blames 300,000 deaths a year on weight-related diseases. Because the children of this generation are more physically inactive than their ancestors, they may be the first generation in American history to live a shorter life than their parents. The early death associated with this problem has been termed *sedentary death syndrome (SDS)*.

Obesity occurs when people consistently take in more food energy than is necessary to meet their daily requirements. On the surface, obesity appears to be a simple problem to solve. To lose weight, all people must do is eat fewer kilocalories, exercise more, or both. On average, people in the United States gain 2 pounds a year, which equals 50 extra Calories stored each day. Because the body can store half of the Calories consumed, preventing a 2-pound weight gain might simply require eating 100 fewer Calories a day.

Although all obese people have an imbalance between their need for kilocalories and the amount of food they eat, the reasons for this imbalance are complex and varied. It is clear that the prevailing culture has much to do with the incidence of obesity. For example, obesity rates have increased over time, which strongly suggests that most cases of obesity are due to changes in lifestyle, not inherent biological factors. Furthermore, immigrants from countries with low rates of obesity show increased rates of obesity when they integrate into the American culture.

Overeating to solve problems is encouraged by our culture. Furthermore, food consumption is central to most kinds of celebrations. Social gatherings of almost every type are considered incomplete without food and drink. If snacks (usually high-calorie foods) are not made available by the host, many people feel uneasy or even unwelcome. It is also true that Americans and people of other cultures show love and friendship by sharing

OUTLOOKS 25.1

The Genetic Basis of Obesity

Advances have been made in identifying the genetic and biochemical bases that regulate body weight. While not yet fully understood, a person's ability to regulate food intake and body weight is controlled by how several genes function and how their proteins interact. This has turned out to be a very complex interaction, and simple solutions to obesity may not be so simple until we learn how body weight is regulated.

The molecules involved are messengers that control the flow of information between the stomach and nerve cells in a portion of the brain called the *hypothalamus*. This flow of information controls the appetite—whether one is hungry or not depends on how these cells are stimulated. Several molecules are involved in this control:

1. *Leptin* is a hormone produced by white fat tissue; it is the product of the *ob* gene (formerly called the "obesity" gene). Leptin acts on nerve cells in the brain to regulate food intake. The *absence* of leptin causes severe obesity; the presence of leptin suppresses the appetite center of the brain. As a person's weight increases, the additional fat cells increase the production of leptin. In animals, the increase in leptin signals the brain to eat less and the body to do more. In humans, however, this does not appear to

be the case, because many obese people have high levels of leptin and no suppression of appetite.

2. *Neuropeptide Y (NPY)* is a small protein that increases appetite. Leptin's appetite-suppressing effects may be due to the fact that leptin can inhibit NPY.

3. *Alpha-melanocyte-stimulating hormone (α-MSH),* the brown pigmentation of the skin, suppresses the appetite by acting on the hypothalamus. Leptin is believed to stimulate the production of α-MSH, which suppresses the appetite.

4. *Melanin-concentrating hormone (MCH)* is another neuropeptide. High levels of MCH in the brain increase food consumption and low levels decrease food consumption. Some believe that the smell of food may stimulate the production of MCH and enhance appetite. The gene for MCH is significantly more active in the brains of the obese.

5. *Ghrelin* is a growth hormone peptide secreted by the stomach. It has the opposite effect of leptin—it appears to be a powerful appetite stimulant.

---

a meal. Many photographs in family albums have been taken at mealtime. In addition, less than half the meals consumed in the United States are prepared in the home. Under these conditions, the consumer has reduced choices in the kind of food available, no control over the way foods are prepared, and little control over serving size. Meals prepared in restaurants and fast-food outlets emphasize meat and minimize the fruit, vegetable, and cereal portions, in direct contrast to the Food Guide Pyramid. The methods of preparation also typically involve cooking with oils and serving with dressings or fat-containing condiments. In addition, portion sizes are generally much larger than recommended.

Recent discoveries of genes in mice suggest that there are genetic components that contribute to obesity. In one study, mice without a crucial gene gained an extraordinary amount of weight. There may be similar genes in humans. It is clear also that some people have much lower metabolic rates than most of the population and, therefore, need much less food

than is typical. Still other obese individuals have a chemical imbalance of the nervous system, which prevents them from feeling "full" until they have eaten an excessive amount of food. This imbalance prevents the brain from turning off the desire to eat after a reasonable amount of food has been eaten. Research into the nature and action of this brain chemical indicates that, if obese people lacking this chemical receive it in pill form, they can feel "full" even when their food intake is decreased by 25% (Outlooks 25.1).

Health practitioners are changing their view of obesity from one of blaming the obese person for lack of self-control to one of treating the condition as a chronic disease that requires a varied approach to control. For most people, dietary counseling and increased exercise is all that is needed, but some need psychological counseling, drug therapy, or surgery. Regardless, controlling obesity can be very difficult, because it requires basic changes in a person's eating habits, lifestyle, and value system.

Bulimia

Bulimia ("hunger of an ox") is a disease condition in which the person has a cycle of eating binges followed by purging the body of the food by inducing vomiting or using laxatives. Many bulimics also use diuretics, which cause the body to lose water, to reduce their weight. It is often called the silent killer, because it is difficult to detect. Bulimics are usually of normal body weight or are overweight. Although women are more likely than men to become bulimic, there has been an increasing number of men with this disorder. The cause is thought to be psychological, stemming from depression or other psychological problems.

A bulimic person may induce vomiting physically or by using nonprescription drugs. Case studies have shown that some bulimics take 40 to 60 laxatives a day to rid themselves of food. For some, the laxatives become addictive. The binge-purge cycle and the associated use of diuretics result in a variety of symptoms that can be deadly. The following is a list of the major symptoms observed in many bulimics:

Excessive water loss
Diminished blood volume
Extreme mineral deficiencies
Kidney malfunction
Increase in heart rate
Loss of rhythmic heartbeat
Lethargy
Diarrhea
Severe stomach cramps
Damage to teeth and gums
Loss of body proteins
Migraine headaches
Fainting spells
Increased susceptibility to infections

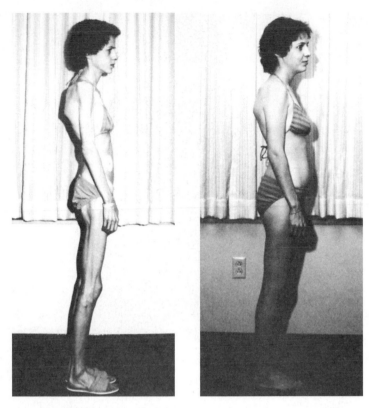

FIGURE 25.3 Anorexia Nervosa
Anorexia nervosa is a psychological eating disorder afflicting many Americans. These photographs were taken of an individual before and after treatment. Restoring a person with this disorder requires both medical and psychological efforts.

Anorexia Nervosa

Anorexia nervosa (figure 25.3) is a nutritional deficiency disease characterized by severe, prolonged weight loss as a result of a voluntary severe restriction in food intake. It is most common among adolescent and preadolescent women. An anorexic person's fear of becoming overweight is so intense that, even though weight loss occurs, it does not lessen the fear of obesity, and the person continues to diet, even refusing to maintain the optimum body weight for his or her age, sex, and height. Persons who have this disorder have a distorted perception of their bodies. They see themselves as fat when, in fact, they are starving to death. Society's preoccupation with body image, particularly among young people, may contribute to the incidence of this disease. The following are some of the symptoms of anorexia nervosa:

Thin, dry, brittle hair
Degradation of fingernails
Constipation
Amenorrhea (lack of menstrual periods)
Decreased heart rate
Loss of body proteins
Weaker than normal heartbeat
Calcium deficiency
Osteoporosis
Hypothermia (low body temperature)
Hypotension (low blood pressure)
Increased skin pigmentation
Reduction in the size of the uterus
Inflammatory bowel disease
Slowed reflexes
Fainting
Weakened muscles

FIGURE 25.4 Kwashiorkor
This starving child shows the symptoms of kwashiorkor, a protein-deficiency disease. If this child were treated with a proper diet containing all the amino acids, the symptoms could be reduced.

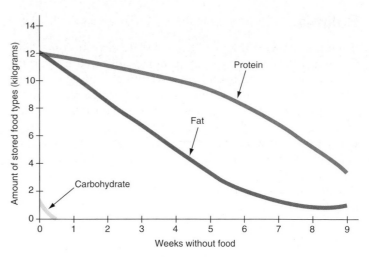

FIGURE 25.5 Starvation and Stored Foods
Fasting results in a very selective loss of the kinds of nutrients stored in the body. Notice how the protein level in the body has the slowest decrease of the three nutrients. This protein-conservation mechanism enables the body to preserve essential amounts of enzymes and other vital proteins.

25.7 Deficiency Diseases

Without minimal levels of the essential amino acids in the diet, a person can develop health problems that could ultimately lead to death. In many parts of the world, people live on diets that are very high in carbohydrates and fats but low in complete protein, because carbohydrates and fats are inexpensive to grow and process, in comparison with proteins. For example, corn, rice, wheat, and barley are all high-carbohydrate foods. Corn and its products (e.g., meal, flour) contain protein, but it is an incomplete protein that has very low amounts of the amino acids tryptophan and lysine. Without enough of these amino acids, the body cannot make many of the necessary enzymes in sufficient amounts to keep a person healthy. One protein-deficiency disease is **kwashiorkor;** its symptoms are easily seen (figure 25.4). A person with this deficiency has a distended belly, slow growth, and slow movement and is emotionally depressed. If the disease is caught in time, brain damage can be prevented and death averted. This requires a change in diet, including expensive protein, such as poultry, fish, beef, shrimp, or milk. As the world food problem increases, these expensive foods will be in even shorter supply and will become more and more costly.

Starvation is a common problem in many parts of the world. Very little carbohydrate is stored in the body; during starvation this small amount lasts as a stored form of energy for only about 2 days. Even after a few hours of fasting, the body begins to use its stored fat deposits as a source of

energy; as soon as the carbohydrates are gone, the body begins to use proteins as a source of glucose. Some of the keto acids produced during the breakdown of fats and amino acids are released in the breath and can be detected as an unusual odor. People who are fasting, anorexic, or diabetic or have other metabolic problems often have this "ketone breath." During the early stages of starvation, the amount of fat in the body steadily decreases, but the amount of protein drops only slightly—20–30 grams per day (figure 25.5). This can continue only up to a certain point. After several weeks of fasting, so much fat has been lost from the body that proteins are no longer protected, and cells begin to use them as a primary source of energy (as much as 125 grams per day). This results in a loss of proteins from the cells, which prevents them from carrying out their normal functions. When starvation gets to this point, it is usually fatal. People who are chronically undernourished do not have the protective effect of fat and experience the effects of starvation much more quickly than those who have stored fat. Children are particularly at risk, because they also need nutrients as building blocks for growth.

The lack of a particular vitamin in the diet can result in a **vitamin-deficiency disease.** Vitamin-deficiency diseases that show recognizable symptoms are extremely rare, except in cases of extremely poor nutrition. Some people claim that dietary vitamin supplements are essential; others claim that a well-balanced diet provides adequate amounts of vitamins and minerals. Supporters of vitamin supplements have even claimed that extremely high doses of certain vitamins can prevent poor health or even create "superhumans." It is very difficult to substantiate many of these claims.

Osteoporosis is a calcium-deficiency disease in older adults that is tied to diet. Persons with this disease lose bone mass; their bones become more brittle and subject to fracture. The body needs calcium to maintain bone, so many adults take calcium supplements. However, calcium alone does not prevent osteoporosis. Bone strength is directly related to the amount of stress placed on the bone. Therefore, exercise is extremely important to assure that calcium will be incorporated into bones and improve their strength. Folic acid and other B vitamins may also help in preventing osteoporosis.

25.8 Nutrition Through the Life Cycle

Nutritional needs vary throughout life and are related to many factors, including age, sex, reproductive status, and level of physical activity. Infants, children, adolescents, adults, and the elderly all require essentially the same types of nutrients but have special nutritional needs related to their stage of life, requiring slight adjustments in the kinds and amounts of nutrients they consume.

Infancy

A person's total energy requirements per kilogram are highest during the first 12 months of life: 100 kilocalories per kilogram of body weight per day; 50% of this energy is required for an infant's basal metabolic rate. Infants triple their weight and increase their length by 50% during their first year; this is their so-called first growth spurt. Because they are growing so rapidly, they require food that contains adequate proteins, vitamins, minerals, and water. They also need food that is high in kilocalories. For many reasons, the food that most easily meets these needs is human breast milk (table 25.8). Even with breast milk's many nutrients, many physicians strongly recommend multivitamin supplements as part of an infant's diet.

Childhood

As infants reach childhood, their dietary needs change. The rate of growth generally slows between 1 year of age and puberty, and girls increase in height and weight slightly faster than boys. During childhood, the body becomes leaner and the bones elongate; the brain reaches 100% of its adult size between the ages of 6 and 10. To meet children's growth and energy needs adequately, protein intake should be high enough to take care of the development of new tissues. Minerals, such as calcium, zinc, and iron, as well as vitamins are necessary to support growth and prevent anemia. Many parents continue to give their children multivitamin supplements, but this should be done only after a careful evaluation of their children's diets; three groups of children are at risk for vitamin deficiency and should receive such supplements:

TABLE 25.8

Comparison of Human Breast Milk and Cow's Milk*

Nutrient	Human Milk	Cow's Milk
Energy (kilocalories/1,000 grams)	690	660 (whole milk)
Protein (grams per liter)	9	35
Fat (grams per liter)	40	38
Lactose (grams per liter)	68	49
Vitamins		
A (international units)	1,898	1,025
C (micrograms)	44	17
D (activity units)	40	14
E (international units)	3.2	0.4
K (micrograms)	34	170
Thiamin (B_1) (micrograms)	150	370
Riboflavin (B_2) (micrograms)	380	1,700
Niacin (B_3) (milligrams)	1.7	0.9
Pyridoxine (B_6) (micrograms)	130	460
Cobalamin (B_{12}) (micrograms)	0.5	4
Folic acid (micrograms)	41–84.6	2.9–68
Minerals (All in Milligrams)		
Calcium	241–340	1,200
Phosphorus	150	920
Sodium	160	560
Potassium	530	1,570
Iron	0.3–0.56	0.5
Iodine	200	80

*All milks are not alike. Each milk is unique to the species that produces it for its young, and each infant has its own growth rate. Humans have one of the slowest infant growth rates, and human milk contains the least amount of protein. Because cow's milk is so different, many pediatricians recommend that human infants be fed either human breast milk or formulas developed to be comparable to breast milk during the first 12 months of life. The use of cow's milk is discouraged. This table lists the relative amounts of the nutrients in human breast milk and cow's milk.

1. Children from deprived families and those suffering from neglect or abuse
2. Children who have anorexia nervosa or poor eating habits or who are obese
3. Children who are strict vegetarians

During childhood, eating habits are very erratic and often cause parental concern. Children often limit their intake of milk, meat, and vegetables while increasing their intake of sweets. To get around these problems, parents can provide calcium by serving cheeses, yogurt, and soups as alternatives to milk. Meats can be made more acceptable if they are in easy-to-chew, bite-sized pieces, and vegetables might be more readily accepted if smaller portions are offered on a more frequent basis. Steering children away from high-fat foods (e.g., french fries, potato chips) and sugar (e.g., cookies, soft drinks) by offering healthy alternatives can lower their risk for potential health problems. For example, sweets in the form of fruits can help reduce dental caries (*carrion* = rotten). Parents can better meet the dietary needs of children by making food available on a more frequent basis, such as every 3 to 4 hours. Obesity is an increasing problem among children. Parents sometimes encourage this by insisting that children eat everything served to them. Before the age of 2, most children automatically regulate the food they eat to an appropriate amount. After that age, parents should be concerned about both the kinds and the amounts of food children eat.

Adolescence

The nutrition of an adolescent is extremely important because, during this period, the body changes from nonreproductive to reproductive. Puberty is usually considered to last between 5 and 7 years. Before puberty, males and females have similar proportions of body fat and muscle. Both body fat and muscle make up between 15% and 19% of the body's weight. Lean body mass, primarily muscle, is about equal in males and females. During puberty, female body fat increases to about 23%, and in males it decreases to about 12%. Males double their muscle mass in comparison with females. The changes in body form that take place during puberty constitute the second growth spurt. Because of their more rapid rate of growth and unique growth patterns, males require more of certain nutrients than females (protein, vitamin A, magnesium, and zinc). During adolescence, youngsters gain as much as 20% of their adult height and 50% of their adult weight, and many body organs double in size. Nutritionists have taken these growth patterns and spurts into account by establishing Dietary Reference Intakes for males and females 10 to 20 years old, including requirements at the peaks of growth spurts. Dietary Reference Intakes at the peak of the growth spurt are much higher than they are for adults and children.

Adulthood

People who have completed the changes associated with adolescence are considered to have entered adulthood. During adulthood, the body enters a plateau phase, and diet and nutrition focus on maintenance and disease prevention. Nutrients are used primarily for tissue replacement and repair, and changes such as weight loss occur slowly. Because the BMR slows, as does physical activity, the need for food energy decreases from about 2,900 kilocalories in average young adult males (ages 20 to 40) to about 2,300 for elderly men. For women, the corresponding numbers decrease from 2,200 to 1,900 kilocalories. Protein intake for most U.S. citizens is usually in excess of the recommended amount. The Dietary Reference Intakes standard for protein is about 63 grams for men and 50 grams for women each day. About 25–50%

should come from animal foods to ensure intake of the essential amino acids. The rest should be from plant-protein foods, such as whole grains, legumes, nuts, and vegetables.

Old Age

As people move into their sixties and seventies, the digestion and absorption of all nutrients through the intestinal tract is not impaired but does slow down. The number of cells undergoing mitosis is reduced, resulting in an overall loss in the number of body cells. With age, complex organs, such as the kidneys and brain, function less efficiently, and protein synthesis becomes inefficient. With regard to nutrition, energy requirements for the elderly decrease as the BMR slows, physical activity decreases, and eating habits change.

The change in eating habits is significant, because it can result in dietary deficiencies. For example, linoleic acid, an essential fatty acid, may fall below required levels as an older person reduces the amount of food he or she eats. The same is true for some vitamins and minerals. Therefore, it may be necessary to supplement the diet daily with 1 tablespoon of vegetable oil. Vitamin E, multiple vitamins, or a mineral supplement may also be necessary. The loss of body protein means that people must be certain to meet their daily Dietary Reference Intakes for protein and participate in regular exercise to prevent muscle loss. As with all stages of the life cycle, regular exercise is important in maintaining a healthy, efficiently functioning body.

The two minerals that demand special attention are calcium and iron, especially for women. A daily intake of 1,200 milligrams of calcium should prevent calcium loss from bones (figure 25.6), and a daily intake of 15 milligrams of iron should allow adequate amounts of hemoglobin to be manufactured to prevent anemia in women over 50 and men over 60. In order to reduce the risk for chronic diseases such as heart attack and stroke, adults should eat a balanced diet, participate in regular exercise programs, control their weight, avoid tobacco and alcohol, and practice stress management.

Pregnancy and Lactation

The period of pregnancy and milk production (lactation) requires that special attention be paid to the diet to ensure proper fetal development, a safe delivery, and a healthy milk supply. Studies have shown that an inadequate supply of essential nutrients can result in infertility, spontaneous abortion, and abnormal fetal development.

The daily amount of essential nutrients must be increased, as should kilocalorie intake. Kilocalories must be increased by 300 per day to meet the needs of an increased BMR; the development of the uterus, breasts, and placenta; and the work required for fetal growth. Some of these kilocalories can be obtained by drinking milk, which simultaneously supplies the calcium needed for fetal bone development. Women who cannot tolerate milk should consume supplementary sources of calcium. In addition, their protein intake

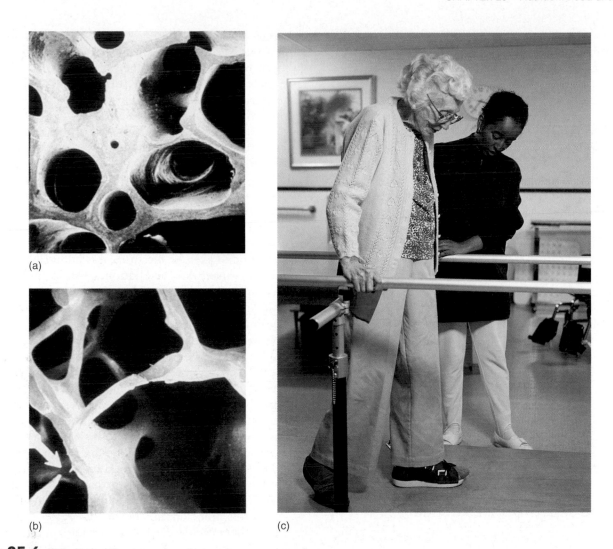

(a)

(b)

(c)

FIGURE 25.6 Osteoporosis

(a) Healthy bone and *(b)* a section of bone from a person with osteoporosis. This calcium-deficiency disease results in a change in the density of the bones as a result of the loss of bone mass. Bones that have undergone this change look "lacy" or like Swiss cheese, with larger than normal holes. *(c)* A few risk factors associated with this disease are being female and fair-skinned; having a sedentary lifestyle; using alcohol, caffeine, and tobacco; and having reached menopause.

should be at least 65 grams per day; however, as mentioned earlier, most people in developed countries consume much more than this per day. Two essential nutrients, folic acid and iron, should be obtained through prenatal supplements, because they are essential to cell division and the development of the fetal blood supply.

The mother's nutritional status affects the developing baby in several ways. If she is under 15 years of age or has had three or more pregnancies in a 2-year period, her nutritional stores are inadequate to support a successful pregnancy. The use of drugs, such as alcohol, caffeine, nicotine, and "hard" drugs (e.g., heroin), can result in decreased nutrient exchange between the mother and fetus. In particular, heavy smoking can result in low birth weight, and alcohol abuse is responsible for *fetal alcohol syndrome (FAS)*. Children with FAS may show the characteristics such as: small size for their

age, facial abnormalities, poor coordination, hyperactive behavior; learning disabilities; developmental disabilities (e.g., speech and language delays); mental retardation or low IQ.

Nutrition for Fitness and Sports

There has been a heightened interest in fitness and sports. Along with this, an interest has developed in the role nutrition plays in providing fuel for activities, controlling weight, and building muscle. The cellular respiration process described in chapter 6 is the source of the energy needed to take a leisurely walk or run a marathon. However, the specific molecules used to get energy depend on the length of the period of exercise, whether or not one warms up before exercise, and how much effort one exerts during exercise. The molecules respired by muscle cells to produce ATP may be glucose, fatty acids, or amino acids. Glucose is stored as glycogen in the muscles, liver, and some other organs. Fatty acids are stored as triglycerides in fat cells. Amino acids are found in the circulatory system. Which molecules are respired depends on the duration and intensity of exercise. Glucose from glycogen and fatty acids from triglycerides are typically the primary fuels. Amino acids provide 10% or less of a person's energy needs, even in highly trained athletes. **aerobic cellular respiration, p. 112**

Aerobic and Anaerobic Exercise

Aerobic exercise occurs when the level of exertion allows the heart and lungs to keep up with the oxygen needs of the muscles. **Anaerobic exercise** involves bouts of exercise that are so intense that the muscles cannot get oxygen as fast as they need it; therefore, they must rely on the anaerobic respiration

of glucose to provide the energy needed. During a long, brisk walk, the heart and lungs of most people should be able to keep up with the muscle cells' requirement for oxygen. The oxygen is used by mitochondria to run the aerobic Krebs cycle and electron-transport system. (They are exercising aerobically.) During the first 20 minutes of moderate exercise, glucose from the blood and glycogen from the muscles are used as the fuel for aerobic respiration. When exercise continues for more than 20 minutes, triglycerides stored in cells and fatty acids from the circulatory system are used. Many exercise programs encourage a warm-up period before the exercise begins. During the warm-up period, the body is using glucose and glycogen but will switch to the burning of triglycerides and fatty acids later in the exercise period. This is why moderate, longer periods of exercise are most beneficial in weight loss.

But what about extreme exertion or vigorous exercise for longer periods? Extreme bouts of activity that last only several seconds, such as a 100-meter dash, rely on the ATP already present in the muscle cells. Because the activity period is so short and the ATP is already in the muscles, oxygen is not needed. Intense exercise that lasts for a few minutes, such as a 400-meter run or gymnastic events, requires additional ATP—more than can be stored in the muscles. This is supplied primarily by the anaerobic respiration of glycogen and glucose. As the period of exercise increases, it is impossible to continue to rely on anaerobic respiration. Aerobic respiration of triglycerides and fatty acids becomes more important. Without a warm-up period, muscle cells begin to respire mostly muscle-stored glycogen. At first, the heart and lungs cannot supply all the oxygen needed by the mitochondria. Glycolysis will provide ATP for muscle contraction, but there will be an increase in the amount of pyruvic and lactic acids. About 20 minutes into vigorous exercise, 20% of muscle glycogen is gone and little triglyceride has been respired. At this time, the heart and lungs are "warmed up" and are able to provide the oxygen necessary to respire larger amounts of triglycerides and fatty acids along with the glycogen and glucose. Fatty acid levels in the blood increase greatly and ATP output increases dramatically. A person who has not warmed up experiences this metabolic shift as a "second wind."

From this point on, all sources of energy—glucose from glycogen, fatty acids from triglycerides, and even small amounts of amino acids from proteins—are used, but the balance shifts during the period of exercise. The glycogen continues to be used, but at approximately 1 hour into exercise glycogen stores have largely been depleted, lowering blood glucose. The athlete then experiences debilitating fatigue, known as "hitting the wall."

Metabolic shifts also occur as one stops exercising. If exercise is stopped suddenly while there are high concentrations of fatty acids in the blood, they will be converted to keto acids, potentially negatively affecting kidney function; causing a loss of sodium, calcium, and other minerals; and changing the pH of the blood. For this reason, many exercise physiologists suggest a cool-down period of light exercise to allow the body to shift back slowly to a more normal fuel balance of glucose and fatty acids.

Diet and Training

Diet is an important aspect of an athlete's training program. The diet needs to be tailored to the kind of performance expected from the athlete. For example, to avoid or postpone "hitting the wall," many marathon athletes practice *carbohydrate loading*. This should be done only by those engaged in periods of hard exercise or competition lasting 90 minutes or more. It requires a week-long diet and exercise program. On

Myths or Misunderstandings About Diet and Nutrition

Myth or Misunderstanding	Scientific Basis
1. Exercise burns calories.	1. Calories are not molecules, so they cannot be burned in the physical sense, but we do oxidize (burn) the fuels to provide the energy needed to perform various activities.
2. Active people who are increasing their fitness need more protein.	2. The amount of protein needed is very small—about 50 grams. Most people get many times the amount of protein required.
3. Vitamins supply energy.	3. Most vitamins assist enzymes in bringing about chemical reactions, some of which may be energy yielding, but they are not sources of energy.
4. Large amounts of protein are needed to build muscle.	4. A person can build only a few grams of new muscle per day. Therefore, consuming large amounts of protein will not increase the rate of muscle growth.
5. Large quantities (megadoses) of vitamins will fight disease, build strength, and increase the lifespan.	5. Quantities of vitamins that greatly exceed recommendations have not been shown to be beneficial. Large doses of some vitamins (e.g., vitamins A, D, and B_3) are toxic.
6. Protein supplements are more quickly absorbed than dietary protein and can build muscle faster.	6. There is adequate protein in nearly all diets. The supplements may be absorbed faster, but that does not mean that they are incorporated into muscle mass faster.
7. Vitamins prevent cancer, heart disease, and other health problems.	7. Vitamins are important to health. However, it is a gross oversimplification to suggest that the consumption of excess amounts of specific vitamins will prevent certain diseases. Many factors contribute to the causes of disease.

the first day, the muscles needed for the event are exercised for 90 minutes; carbohydrates, such as fruits, vegetables, or pasta, provide a source of dietary kilocalories. On the second and third days of carbohydrate loading, the person continues the 50% carbohydrate diet, but the exercise period is reduced to 40 minutes. On the fourth and fifth days, the workout period is reduced to 20 minutes and the carbohydrates are increased to 70% of total kilocalorie intake. On the sixth day, the day before competition, the person rests and continues the 70% carbohydrate diet. Following this program increases muscle glycogen levels and makes it possible to postpone "hitting the wall."

Conditioning includes many interrelated body adjustments in addition to energy considerations. Training increases the strength of muscles, including the heart, and increases the efficiency of their operation. Practicing a movement allows for the development of a smooth action, which is more energy-efficient than a poorly trained motion. As the body is conditioned, the number of mitochondria per cell increases, the Krebs cycle and the ETS run more efficiently, the number of capillaries increases, fats are respired more efficiently and for longer periods, and weight control becomes easier.

The amount of protein in an athlete's diet has also been investigated. An increase in dietary protein does not automatically increase strength, endurance, or speed. In fact, most Americans eat the 10% additional protein that athletes require as a part of their normal diets. The body uses the additional percentage for many things, including muscle growth; however, increasing protein intake does not automatically increase muscle size. Only when there is a need is the protein used to increase muscle mass. Exercise provides that need. The body will build the muscle it needs to meet the demands placed on it. Vitamins and minerals operate in much the same way. No supplements should be required as long as the diet is balanced and complex. (Outlooks 25.2).

Athletes must monitor their water intake, because dehydration can cripple an athlete very quickly. A water loss of only 5% of body weight can decrease muscular activity by as much as 30%. One way to replace water is to drink 1 to 11/2 cups of water 15 minutes before exercising and 1/2 cup during exercise. In addition, drinking 16 ounces of cool tap water for each pound (16 ounces) of body weight lost during exercise will prevent dehydration. Another method is to drink diluted orange juice (one part juice to five parts water). This is an excellent way to replace water and resupply a small amount of lost glucose and salt. Salt pills (so-called electrolyte pills) are not recommended, because salt added to the digestive tract tends to reduce the rate at which water is absorbed from the gut. When the salt is absorbed into the bloodstream, additional water is needed to dilute the blood to the proper level.

Summary

To maintain good health, people must ingest nutrient molecules that can enter the cells and function in the metabolic processes. The proper quantity and quality of nutrients are essential to good health. Nutritionists have classified nutrients into six groups: carbohydrates, proteins, lipids, minerals, vitamins, and water. Energy for metabolic processes can be obtained from carbohydrates, lipids, and proteins and is measured in kilocalories. An important measure of the amount of energy required to sustain a human at rest is the basal metabolic rate. To meet this and all additional requirements, the U.S. Department of Agriculture (USDA) has established the Dietary Reference Intakes, recommended dietary allowances for each nutrient. The USDA released an updated version of the Food Guide Pyramid in 2005, the first since it was created over a decade ago. The new version makes it easier to choose the foods and amounts that are right for you.

Should there be metabolic or psychological problems associated with a person's normal metabolism, a variety of disorders can occur, including obesity, anorexia nervosa, bulimia, kwashiorkor, and vitamin-deficiency diseases. As people move through the life cycle, their nutritional needs change, requiring a reexamination of their eating habits to maintain good health.

Key Terms

*Use the interactive flash cards on the **Concepts in Biology**, 12/e website to help you learn the meaning of these terms.*

absorption 542
aerobic exercise 562
anaerobic exercise 562
anorexia nervosa 557
assimilation 542
basal metabolic rate (BMR) 552
body mass index (BMI) 553
bulimia 557
calorie 542
complete proteins 544
diet 542
Dietary Reference Intakes 548
digestion 542
essential amino acids 544
essential fatty acids 544
fiber 543

fitness 553
Food Guide Pyramid 548
incomplete proteins 544
ingestion 542
kilocalorie (kcal) 542
kwashiorkor 558
minerals 545
nutrients 542
nutrition 542
obese 553
osteoporosis 545
protein-sparing 545
specific dynamic action (SDA) 552
vitamin-deficiency disease 558
vitamins 545

Basic Review

1. _____ is the modification and incorporation of absorbed molecules into the structure of an organism.
   a. Nutrition
   b. Assimilation
   c. Dieting
   d. Absorption

2. The rate at which the body uses energy when at rest is called its
   a. basal metabolic rate.
   b. basal metabolic index.
   c. specific dynamic action.
   d. work.

3. A body mass index between 25 and 30 kg/m^2 indicates that a person is
   a. within a normal range.
   b. obese.
   c. overweight.
   d. extremely active.

4. Some complex carbohydrates are a source of _____, which slows the absorption of nutrients and stimulate peristalsis (rhythmic contractions) in the intestinal tract.
   a. fiber
   b. vitamins
   c. minerals
   d. lipids

5. Linoleic acid and linolenic acid are called _____ because they cannot be synthesized by the human body and, therefore, must be a part of the diet.
   a. amino acids
   b. fatty acids
   c. essential fatty acids
   d. B-complex vitamins

6. Inorganic elements, found throughout nature, that cannot be synthesized by the body are called
   a. vitamins.
   b. essential amino acids.
   c. electrolytes.
   d. minerals.

7. When proteins are conserved and carbohydrates and fats are oxidized first as a source of ATP energy, the body is involved in
   a. protein-sparing.
   b. specific dynamic action.
   c. absorption.
   d. dieting.

8. Information on the amounts of certain nutrients various member of the public should receive is contained in
   a. the Food Guide Pyramid.
   b. the Protein-Sparing Guide.
   c. kwashiorkor.
   d. the Dietary Reference Intakes.

9. A person with _____ has a cycle of eating binges followed by purging of the body of the food by inducing vomiting or using laxatives.
   a. bulimia
   b. anorexia nervosa
   c. obesity
   d. kwashiorkor

10. The use of alcohol during pregnancy can result in decreased nutrient exchange between the mother and fetus, as well as other developmental abnormalities called
    a. osteoporosis.
    b. fetal alcohol syndrome (FAS).
    c. bulimia.
    d. fetal deficiency disease.

Answers
1. b 2. a 3. c 4. a 5. c 6. d 7. a 8. d 9. a 10. b

Concept Review

Answers to the Concept Review questions can be found on the **Concepts in Biology,** *12/e website.*

25.1 Living Things as Chemical Factories: Matter and Energy Manipulators

1. What is a nutrient?
2. What is the difference between digestion and assimilation?

25.2 The Kinds of Nutrients and Their Function

3. Why are some nutrients referred to as essential? Name them.
4. List the six main classes of nutrients as described by nutritionists.

25.3 Dietary Reference Intakes

5. How many essential amino acids are required in the human diet?

6. Women 19–30 years old should have how much calcium in their diet?

25.4 The Food Guide Pyramid

7. Name the six basic food groups and give two examples of each.
8. Why has exercise been included in the latest revision of the Food Guide Pyramid?

25.5 Basal Metabolic Rate, Diet and Weight Control

9. Define the terms *basal metabolic rate, specific dynamic action,* and *voluntary muscular activity.*
10. What is BMI? How is it calculated?

25.6 Eating Disorders

11. What are the three primary eating disorders?
12. What is the role of leptin in controlling appetite?

25.7 Deficiency Diseases

13. What type of deficiency is associated with 'ketone breath'?
14. How common are vitamin-deficiency diseases in the United States?

25.8 Nutrition Through the Life Cycle

15. During which phase of the life cycle is a person's demand for kilocalories per unit of body weight the highest?
16. What changes need to be made to the diets of the elderly?

25.9 Nutrition for Fitness and Sports

17. What is 'carbohydrate loading'?
18. How do you recognize when a person shifts from aerobic to anaerobic exercise?

Thinking Critically

For guidelines to answer this question, visit the **Concepts in Biology,** *12/e website.*

Imagine that you're a 21-year-old woman who has never been involved in any kind of sport, but suddenly you have become interested in playing rugby, a very demanding contact sport. If you are to play well with only minor injuries, you must get into condition. Describe the changes you should make in your daily diet and exercise program to prepare for this new experience.

26

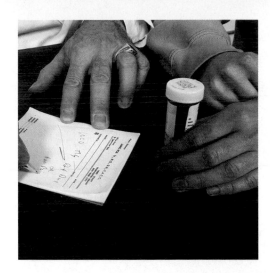

The Body's Control Mechanisms and Immunity

In the U.S., Preventive Services Task Force (USPSTF) was set up by the United States Public Health Service to determine the merits of hormone replacement therapy (HRT). In 1996, the USPSTF recommended that physicians advise all women about the potential benefits and problems of HRT, while stating that the evidence was insufficient to recommend for or against it for all women after menopause. The task force concluded in 1996 that HRT might have long-term benefits for women with osteoporosis and heart disease, but it might increase the risk for breast cancer.

Although recent evidence has confirmed that hormone replacement therapy reduces the risk for bone fractures and may reduce the likelihood of developing colorectal cancer, the evidence has also confirmed that hormone replacement therapy increases the risk for blood clots, breast cancer, and stroke. New evidence also indicates that HRT does not reduce the risk for coronary heart disease and that combined estrogen and progestin may actually increase the risk. The USPSTF now recommends against the routine use of combined estrogen and progestin for the prevention of chronic conditions (i.e., thinning of the bones, which can lead to fractures; heart disease; colon cancer; and dementia) in postmenopausal women because the problems are likely to outweigh the benefits for most women. In 2004, two studies found that, for women using hormone replacement therapy for 5 years or longer to prevent chronic conditions, the harms could exceed benefits. Its problems include an increased risk for blood clots and stroke, an increased risk for breast cancer with 5 or more years of use, and a probable increase in gallbladder disease. This most recent evidence suggests that HRT does not reduce the risk for heart disease and may modestly increase it. So far, the evidence isn't strong enough to recommend for or against the use of estrogen alone for the prevention of chronic conditions in postmenopausal women who have had a hysterectomy.

CHAPTER OUTLINE

26.1 Coordination in Multicellular Animals

A large, multicellular organism, which consists of many kinds of systems, must have a way of integrating various functions so that it can survive. Its systems must be coordinated to maintain a reasonably constant internal environment. Recall from chapter 1 that **homeostasis** is the maintenance of a constant internal environment as a result of monitoring and modifying the functioning of various systems. Homeostatic mechanisms, are involved in maintaining blood pressure, body temperature, body fluid levels, and blood pH. When we run up a hill, our leg and arm muscles move in a coordinated way to provide power. They burn fuel (glucose) for energy and produce carbon dioxide and lactic acid as waste products, which tend to lower the pH of the blood. The heart beats faster to provide oxygen and nutrients to the muscles, we breathe faster to supply the muscles with oxygen and get rid of carbon dioxide, and the blood vessels in the muscles dilate to allow more blood to flow to them. Running generates excess heat. As a result, more blood flows to the skin to get rid of the heat, and sweat glands begin to secrete fluid, thus cooling the skin. All these automatic internal adjustments help the body maintain a constant level of oxygen, carbon dioxide, and glucose in the blood; constant pH; and constant body temperature.

One common homeostatic mechanism is **negative-feedback inhibition,** a mechanism in which an increase in the stimulus causes a decrease in the response (figure 26.1a). A common negative-feedback inhibition mechanism is a household furnace thermostat. When the heat (stimulus) in a room increases to the set temperature, the thermostat sends a signal (response) to the furnace to shut off. When the heat in the room drops too low, the furnace turns on again and the temperature in the room rises to the set point. Negative feed-back systems work in your body in a similar manner. Messenger molecules move throughout the body conveying information about the kinds of responses that are happening. When a response reaches a set point, a message is sent back to a control center telling it to stop or cut back that activity. For example, your body requires a certain amount of insulin to process blood glucose. When too much insulin is produced by cells in the pancreas, a message is sent back to the pancreas to stop producing so much insulin and the amount of insulin in the blood decreases. When the insulin level reaches a low point, the cells begin to produce insulin again and the amount increases in the blood. Negative feed-back results in a fluctuation around the set point, thus maintaining a relatively constant concentration of insulin.

The alternative to negative feedback is *positive feedback regulation*. In a case of **positive feedback,** an increase in the stimulus causes an increase of the response and does not result in homeostasis. While this is rarer than negative feedback, it can play an important role in homeostasis. For example, positive feedback occurs during childbirth. Each time the uterus contracts, signals are sent to the brain where more

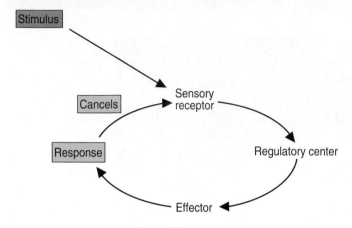

(a) Negative-feedback control

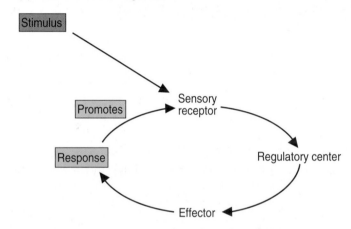

(b) Positive-feedback regulation

FIGURE 26.1 **Negative and Positive Feedback Controls** *(a)* Negative-feedback control occurs when an increase in a stimulus reduces a response. *(b)* Positive-feedback happens when a response to a stimulus causes an increase in the stimulus which further increases the response.

oxytocin is produced causing the uterine muscles to contract again. In this positive feed-back example, the activity (uterine muscle contraction) is stimulated to occur again, not to be stopped as in negative feed-back. This repeated stimulation continues until the baby is born (figure 26.1b).

The nervous, endocrine, and immune systems are the three major systems of the body that play key roles by integrating stimuli and generating the appropriate responses necessary to maintain homeostasis.

26.2 Nervous System Function

The nervous system is well suited to managing the rapid adjustments that must take place within the body. It functions very much as a computer. A message is sent very rapidly along established pathways from a specific initiating point to a specific end point. The following sections describe the major structural and functional characteristics of the nervous system.

The Structure of the Nervous System

The **nervous system** consists of a network of cells, with fibrous extensions, that carry information along specific pathways from one part of the body to another. A **neuron**, or **nerve cell**, is the basic unit of the nervous system. A neuron consists of a central body, called the **soma** or *nerve cell body*, which contains the nucleus. It has several long extensions, called nerve fibers. There are two kinds of nerve fibers: **axons**, which carry information away from the cell body, and **dendrites**, which carry information toward the cell body (figure 26.2). Most neurons have one axon and several dendrites.

Neurons are arranged into two major systems. The **central nervous system**, which consists of the brain and spinal cord, is surrounded by the skull and the vertebrae of the spinal column. The **spinal cord** is a collection of nerve fibers surrounded by the vertebrae that conveys information to and from the brain. It receives input from sense organs, interprets information, and generates responses. The **peripheral nervous system** is located outside the skull and spinal column; it consists of bundles of long axons and dendrites called **nerves.** In the peripheral nervous system, cranial nerves connect to the brain, whereas spinal nerves connect to the spinal cord. **Motor neurons** carry messages from the central nervous system to muscles and glands. Some motor nerves constituting the *somatic nervous system,* control the skeletal (voluntary) muscles. Other motor nerves, collectively called the *autonomic nervous system,* control the smooth (involuntary) muscles, the heart, and glands. There are two sets of neurons in the peripheral nervous system. **Sensory neurons** carry input from sense organs to the central nervous system. Motor neurons have one long axon, which runs from the spinal cord to a muscle or gland; sensory neurons have long dendrites that carry input from the sense organs to the central nervous system (figure 26.3).

The Nature of the Nerve Impulse

Because most neurons have long, fibrous extensions, information can be passed along the nerve cell from one end to the other. The **nerve impulse** is the message that travels along a neuron. A nerve impulse is not like an electric current; instead, it involves a specific sequence of chemical events at the cell membrane.

Because all cell membranes are differentially permeable, it is difficult for some ions to pass through the membrane, and the combination of ions inside the membrane is different from that outside it. Cell membranes also contain proteins that actively transport specific ions from one side of the membrane to the other. Active transport involves the cell's use of adenosine triphosphate (ATP) to move materials from one side of the cell membrane to the other. Because ATP is

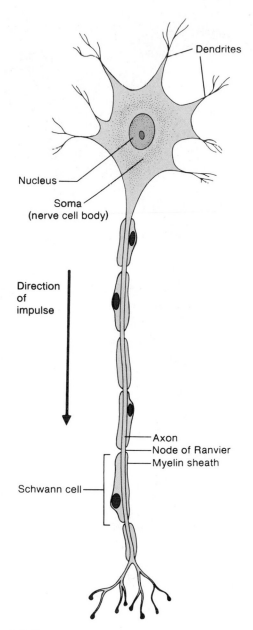

FIGURE 26.2 Structure of a Neuron
Neurons consist of a nerve cell body, which contains the nucleus and several fibrous extensions. The shorter, more numerous fibers that carry impulses to the nerve cell body are dendrites. The long fiber that carries the impulse away from the cell body is the axon.

required, cells lose this ability when they die. One of the ions that is actively transported from cells is the sodium ion (Na^+). At the same time sodium ions are being transported out of cells, potassium ions (K^+) are being transported into the normal resting cells. However, there are more sodium ions transported out than potassium ions transported in. ions, p. 31

Because a normal resting cell has more positively charged Na^+ ions on the outside of the cell than on the inside, a small but measurable voltage exists across the membrane of the cell. (**Voltage** is a measure of the electrical charge difference

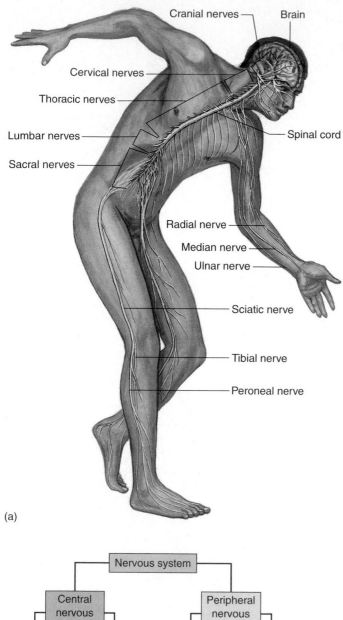

Cranial nerves

Brain

Cervical nerves

Thoracic nerves

Lumbar nerves

Sacral nerves

Spinal cord

Radial nerve

Median nerve

Ulnar nerve

Sciatic nerve

Tibial nerve

Peroneal nerve

(a)

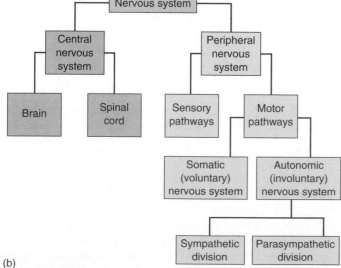

(b)

FIGURE 26.3 Organization of the Nervous System
(a) A generalized view of the central nervous system and some of the nerves of the peripheral nervous system. (b) The chart shows how the various functional portions of the nervous system are related to one another.

that exists between two points or objects.) The voltage difference between the inside and the outside of a cell membrane is about 70 millivolts (0.07 volt). The two sides of the cell membrane are, therefore, polarized in the same sense that a battery is polarized, with a positive and a negative pole. A resting neuron has its positive pole on the outside of the cell membrane and its negative pole on the inside of the membrane (figure 26.4).

When a cell is stimulated at a specific point on the cell membrane, the cell membrane changes its permeability and lets sodium ions (Na^+) pass through it from the outside to the inside. The membrane is thus **depolarized;** it loses its difference in charge as sodium ions diffuse into the cell from the outside. Sodium ions diffuse into the cell because, initially, they are in greater concentration outside the cell than inside. When the membrane becomes more permeable, they are able to diffuse into the cell, toward the area of lower concentration. The depolarization of one point on the cell membrane causes the adjacent portion of the cell membrane to change its permeability as well, and it also depolarizes. Thus, a wave of depolarization passes along the length of the neuron from one end to the other (figure 26.5). The depolarization and passage of an impulse along any portion of the neuron is a momentary event. As soon as a section of the membrane has been depolarized, potassium ions diffuse out of the cell. This reestablishes the original polarized state, and the membrane is said to be *repolarized*. Subsequently, the continuous active transport of sodium ions out of the cell and potassium ions into the cell restores the original concentration of ions on both sides of the cell membrane. When the nerve impulse reaches the end of the axon, it stimulates the release of a molecule that stimulates depolarization of the next neuron in the chain.

Activities at the Synapse

The **synapse** is the space between the fibers of adjacent neurons in a chain. Many chemical events occur in the synapse that are important in the function of the nervous system. When a neuron is stimulated, an impulse passes along its length from one end to the other. When the impulse reaches a synapse, a molecule called a **neurotransmitter** is released into the synapse from the axon. It diffuses across the synapse and binds to specific receptor sites on the dendrite of the next neuron. When enough neurotransmitter molecules have been bound to the second neuron, an impulse is initiated in it as well. Several kinds of neurotransmitters are produced by specific neurons. These include dopamine, epinephrine, acetylcholine, and several other molecules. The first neurotransmitter to be identified was *acetylcholine*. **Acetylcholine** molecules are neurotransmitters manufactured in the soma; they migrate down the axon, where they are stored until needed.

As long as a neurotransmitter is attached to its receptor, it continues to stimulate the neuron. Thus, if acetylcholine continues to occupy receptors, the neuron continues to be stimulated again and again. **Acetylcholinesterase**, an enzyme, destroys acetylcholine and prevents this from happening.

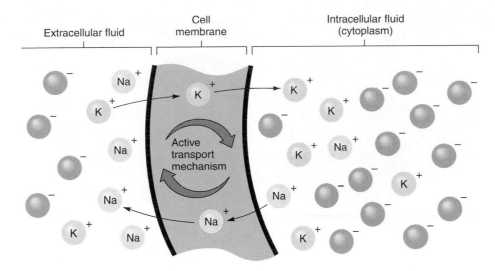

FIGURE 26.4 The Polarization of Cell Membranes

All cells, including neurons have an active transport mechanism that pumps Na^+ out of cells and simultaneously pumps K^+ into them. The end result is that there are more Na^+ ions outside the cell and more K^+ ions inside the cell. In addition, negative ions, such as Cl^-, are more numerous inside the cell. Consequently, the outside of the cell is positive ($+$) compared with the inside, which is negative ($-$).

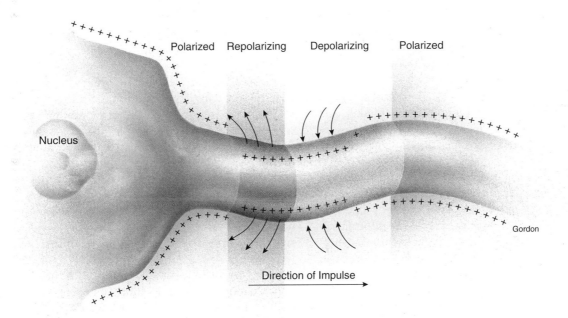

FIGURE 26.5 A Nerve Impulse

When a neuron is stimulated, a small portion of the cell membrane depolarizes as Na^+ flows into the cell through the membrane. This encourages the depolarization of an adjacent portion of the membrane, and it depolarizes a short time later. In this way, a wave of depolarization passes down the length of the neuron. Shortly after a portion of the membrane is depolarized, the ionic balance is reestablished. It is repolarized and ready to be stimulated again.

(The breakdown products of the acetylcholine can be used to remanufacture new acetylcholine molecules.) The destruction of acetylcholine allows the second neuron in the chain to return to normal (figure 26.6). Thus, it will be ready to accept another burst of acetylcholine from the first neuron a short time later. Neurons must also constantly manufacture new acetylcholine molecules, or they will exhaust their supply and be unable to conduct an impulse across a synapse.

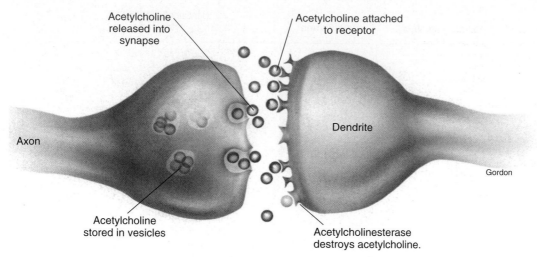

Acetylcholine released into synapse

Acetylcholine attached to receptor

Axon

Dendrite

Gordon

Acetylcholine stored in vesicles

Acetylcholinesterase destroys acetylcholine.

FIGURE 26.6 Activities at the Synapse

When a nerve impulse reaches the end of an axon, it releases a neurotransmitter into the synapse. In this illustration, the neurotransmitter is acetylcholine. When acetylcholine is released into the synapse, acetylcholine molecules diffuse across the synapse and bind to receptors on the dendrite, initiating an impulse in the next neuron. Acetylcholinesterase is an enzyme that destroys acetylcholine, preventing continuous stimulation of the dendrite.

Certain drugs, such as curare and strychnine, interfere with the activities at the synapse. Curare blocks the synapse and causes paralysis, whereas strychnine causes neurons to be stimulated continually. Many insecticides are nerve poisons and, therefore, are quite hazardous.

Because of the way the synapse works, impulses can go in only one direction: Only axons secrete acetylcholine, and only dendrites have receptors. This explains why there are sensory and motor neurons to carry messages to and from the central nervous system.

As mentioned earlier, the nervous system is organized in a fashion similar to a computer. Information from various input devices (sense organs) is delivered to the central processing unit (brain) by way of wires (sensory nerves). The information is interpreted in the central processing unit. Eventually, messages can be sent by way of cables (motor nerves) to drive external machinery (muscles and glands). This concept reveals how the functions of various portions of the nervous system have been identified. It is possible to stimulate specific portions of the nervous system electrically or to damage certain parts of the nervous system in experimental animals and determine the functions of different parts of the brain and other parts of the nervous system. For example, because peripheral nerves carry bundles of both sensory and motor fibers, damage to a nerve may result in both a lack of feeling because sensory messages cannot get through and an inability to move because the motor nerves have been damaged.

The Organization of the Central Nervous System

Certain parts of the brain are involved in controlling fundamental functions, such as breathing and heart rate. Others are involved in generating emotions, others decode sensory input,

and others coordinate motor activity. The human brain also has considerable capacity to store information and create new responses to environmental stimuli.

The brain consists of several regions, each with a specific function. The functions of the brain can be roughly divided into three major levels: automatic activities, basic decision making and emotions, and finally thinking and reasoning. If we begin with the spinal cord and work our way forward, we will proceed from the more fundamental, automatic activities of the brain to its more complex, thinking portions. At the base of the brain, where the spinal cord enters the skull, is the **medulla oblongata,** the region of the brain that controls fundamental activities, such as blood pressure, breathing, and heart rate. Most of the fibers of the spinal cord cross from one side of the body to the other in the medulla oblongata. This is why the left side of the brain affects the right side of the body.

The **cerebellum** is the large bulge at the base of the brain that is connected to the medulla oblongata. The primary function of the cerebellum is the coordination of muscle activity. It receives information from sense organs, such as the portions of the ear involved in balance, the eyes, and pressure sensors in muscles and tendons. The cerebellum uses this information to make adjustments in the strength and order of muscle contractions, so that the body moves in a coordinated fashion.

The **pons** is the region of the brain that is connected to the anterior end of the medulla oblongata. It also connects to the cerebellum and to higher levels of the brain. It is involved in controlling many sensory and motor functions of the sense organs of the head and face. The pons is also connected to a portion of the brain that forms two bulblike structures, which ultimately connect to the cerebrum. Although these regions of the brain control many automatic activities, some level of decision making occurs in this region. The primary regions are the *thalamus* and the *hypothalamus.* The **thalamus** is the

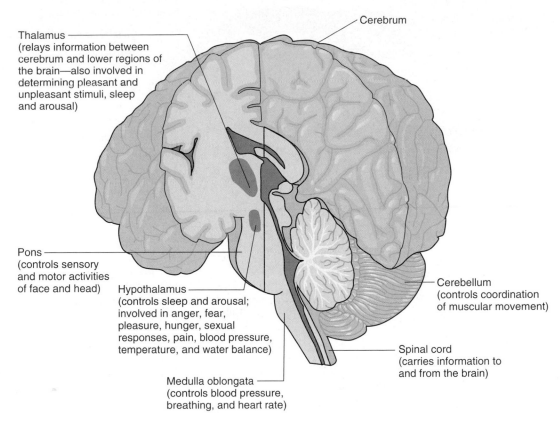

FIGURE 26.7 **Functions of the More Primitive Brain Regions**
The brain is organized into several levels of function. The more primitive regions of the brain connected to the spinal cord monitor and manage many essential functions automatically.

region of the brain that relays information between the cerebrum and lower portions of the brain. It also provides a level of awareness in that it determines pleasant and unpleasant stimuli and is involved in sleep and arousal. The **hypothalamus** is the region of the brain involved in sleep and arousal; it is important in emotions such as anger, fear, pleasure, and the sensations that accompany hunger, sexual response, and pain. Several other, more automatic functions are regulated in this region, such as body temperature, blood pressure, and water balance. The hypothalamus also is connected to the pituitary gland and influences the manufacture and release of its hormones. Figure 26.7 shows the relationship of the various, more primitive parts of the brain.

The **cerebrum** is the logical part of the brain. It is also the largest portion of the brain in humans. The two hemispheres of the cerebrum cover all other portions of the brain except the cerebellum (figure 26.7). The surface of the cerebrum has been extensively mapped as to the locations of many functions. Abilities such as memory, language, the control of movement, the interpretation of sensory input, and thought are associated with specific areas of the cerebrum. Figure 26.8 illustrates the cerebrum and the locations of specific functions.

The function of the brain is not determined by structure alone. Many parts of the brain have specialized neurons, which produce specific neurotransmitter molecules used only to stimulate particular neurons that have the proper receptor sites. As scientists learn more about the functioning of the brain, they are finding more kinds of specialized neurotransmitter molecules, allowing for the treatment of many types of mental and emotional diseases. Manipulating these neurotransmitter molecules can help correct inappropriate functioning of the brain. However, the brain is still not completely understood. Scientists are at an early stage in their search to comprehend this organ, which sets us apart from other animals (How Science Works 26.1).

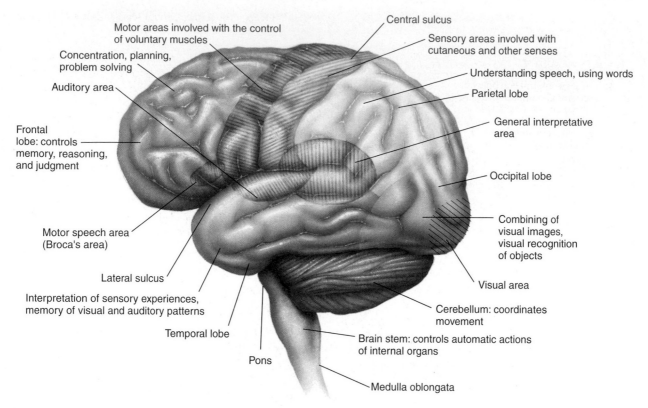

Motor areas involved with the control of voluntary muscles

Concentration, planning, problem solving

Auditory area

Frontal lobe: controls memory, reasoning, and judgment

Motor speech area (Broca's area)

Lateral sulcus

Interpretation of sensory experiences, memory of visual and auditory patterns

Temporal lobe

Pons

Central sulcus

Sensory areas involved with cutaneous and other senses

Understanding speech, using words

Parietal lobe

General interpretative area

Occipital lobe

Combining of visual images, visual recognition of objects

Visual area

Cerebellum: coordinates movement

Brain stem: controls automatic actions of internal organs

Medulla oblongata

FIGURE 26.8 Specialized Areas of the Cerebrum
Each portion of the cerebrum has particular functions.

HOW SCIENCE WORKS 26.1

How Do We Know What the Brain Does?

Scientists know a great deal about the function of the brain, although there is still much more to learn. Certain functions have been identified as residing in specific portions of the brain, as a result of many kinds of studies over the past century. For example, persons who have had specific portions of their brains altered by damage from accidents or strokes have been studied. Their changes in behavior or the way they perceive things can be directly correlated with the portion of the brain that was damaged. During surgeries that require the brain to be exposed, a local anesthetic can be given and the patient can be conscious while the surgery is taking place. (The brain perceives pain from pain receptors throughout the body; however, because the brain does not have many pain receptors within it, touching or manipulating the brain does not cause pain to be perceived.) Specific portions of the brain can be stimulated and the patient can be asked to describe his or her sensations and the patient's motor functions can be observed.

Many kinds of experiments have also been done with animals, in which specific portions of the brain have been destroyed and the animals' changes in behavior noted.

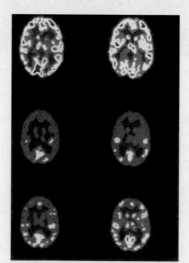

Electrodes have been inserted into the brains of animals to stimulate certain portions of the brain.

More recently, techniques have been developed to observe changes in the electrical activity of specific portions of the brain without electrodes or other invasive procedures. This allows researchers to present stimuli to human subjects and determine which parts of the brain alter their activity. In addition to localizing the part of the brain that responds, it is also possible to determine what parts of a complex stimulus are most important in changing brain activity. For example, these techniques have revealed that languages learned by adults are processed in different places in the brain than the languages they learned as children and that the brain has a built-in mechanism for recognizing unexpected words or musical notes.

Although much is known about the brain, there is still much to learn. Current experiments are seeking ways to regenerate neurons that have been damaged. A better understanding of the chemical events that take place in the brain would enable us to cure many kinds of debilitating mental illnesses.

26.3 Endocrine System Function

In the 1890s, many physicians began to describe the workings of chemicals in the body, suggesting that they were "internal secretions." Ernest Starling named these chemical messengers *hormones*. A **hormone** is a specific molecule that is released by one organ and transported to another organ, where it triggers a change in the other organ's activity. The **endocrine system** consists of a number of glands that communicate with one another and with other tissues through chemicals distributed throughout the organism. **Glands** are organs that manufacture molecules that either are secreted into surrounding tissue, where they are picked up by the circulatory system, or are secreted through ducts into the cavity of an organ or to the body surface. **Endocrine glands** have no ducts; they secrete their products—hormones—into the circulatory system (figure 26.9). Other glands, such as the digestive glands and sweat glands, empty their contents through ducts. These kinds of glands are called **exocrine glands.**

The endocrine system functions the way a radio broadcast system does. Radio stations send their signals in all directions, but only the radio receivers that are tuned to the correct frequency can receive the signals. Messenger molecules (hormones) are distributed throughout the body by the circulatory system, but only the cells that have the proper receptor sites can receive and respond to them. These cells are called *target cells;* they respond in one of three ways: (1) Some cells release products that have been previously manufactured; (2) some cells synthesize molecules or begin metabolic activities; and (3) some cells divide and grow. As a result of these different kinds of responses, some endocrine responses are relatively rapid, whereas others are very slow. For example, the release of the hormones **epinephrine** and **norepinephrine** (formerly called adrenalin and noradrenalin) from the adrenal medulla, located near the kidney, causes a rapid change in an organism's behavior. The heart rate increases, blood pressure rises, blood is shunted to the muscles, and the breathing rate increases. You have certainly experienced this reaction many times in your lifetime, such as when you nearly had an automobile accident or slipped and nearly fell.

Antidiuretic hormone (ADH) acts more slowly. It is released from the posterior pituitary gland at the base of the brain; it regulates the rate at which the body loses water through the kidneys. It does this by encouraging the reabsorption of water from the collecting ducts of the kidneys (see chapter 24). The effects of ADH can be noticed in a matter of minutes to hours.

Insulin is a hormone whose effects are quite rapid; it is produced by the pancreas, located near the stomach. Insulin stimulates cells—particularly muscle, liver, and fat cells—to take up glucose from the blood. After a high-carbohydrate meal, the glucose level in the blood begins to rise, stimulating the pancreas to release insulin. The increased insulin causes glucose levels to fall as the sugar is taken up by the cells. People with diabetes have insufficient or improperly acting insulin or lack the receptors to respond to the insulin; therefore, they have difficulty regulating glucose levels in their blood.

The responses that result from the growth of cells may take weeks or years to occur. For example, **growth-stimulating hormone (GSH)** is produced by the anterior pituitary gland over a period of years and results in typical human growth. After sexual maturity, the amount of this hormone generally drops to very low levels, and body growth stops. Sexual development is also largely the result of the growth of specific tissues and organs. The male sex hormone **testosterone,** produced by the testes, causes the growth of male sex organs and a change to the adult body form. The female counterpart, **estrogen,** results in the development of female sex organs and body form. In all of these cases, it is the release of hormones over long periods, continually stimulating the growth of sensitive tissues, that results in a normal

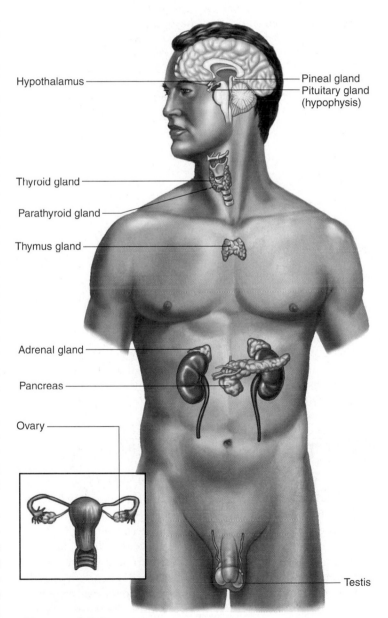

FIGURE 26.9 Endocrine Glands
The endocrine glands, located in various places in the body, secrete hormones.

OUTLOOKS 26.1

Negative-Feedback Inhibition and Hormones

Glands within the endocrine system interact with one another and control the production of hormones. In negative-feedback inhibition, an increased amount of one hormone interferes with the production of a different hormone in the chain of events. The production of *thyroxine* and *triiodothyronine* by the thyroid gland exemplifies this kind of control. The production of these two hormones is stimulated by the increased production of a hormone from the anterior pituitary called *thyroid-stimulating hormone (TSH)*. The thyroid gland's control lies in the quantity of TSH produced. When the anterior pituitary produces high levels of thyroid-stimulating hormone, the thyroid is stimulated to grow and secrete more thyroxine and triiodothyronine. But when increased amounts of thyroxine and triiodothyronine are produced, these hormones have a negative effect on the pituitary, so that it decreases its production of thyroid-stimulating hormone, leading to reduced production of thyroxine and triiodothyronine. If the amount of the thyroid hormones falls too low, the pituitary is no longer inhibited and releases additional thyroid-stimulating hormone. As a result of the interaction of these hormones, their concentrations are maintained within certain limits.

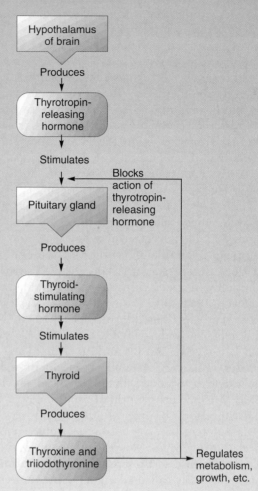

developmental pattern. The absence or inhibition of any of these hormones early in life changes the normal growth process (Outlooks 26.1).

26.4 The Integration of Nervous and Endocrine Function

The nervous and endocrine systems can interact (How Science Works 26.2). The pituitary gland, located at the base of the brain, is divided into two parts. The posterior pituitary, which is directly connected to the brain, develops from nervous tissue during embryology. The other part, the anterior pituitary, develops from the lining of the roof of the mouth in early fetal development. Certain pituitary hormones are produced in the brain and transported down axons to the posterior pituitary, where they are stored before being released. The anterior pituitary also receives a continuous input of messenger mole-

cules from the brain, but these are delivered by way of blood vessels, which pick up hormones produced by the hypothalamus and deliver them to the anterior pituitary.

The pituitary gland produces a variety of hormones that are responsible for causing other endocrine glands, such as the thyroid, ovaries and testes, and adrenals, to secrete their hormones. Pituitary hormones also influence milk production, skin pigmentation, body growth, mineral regulation, and blood glucose levels (figure 26.10).

Because the pituitary is constantly receiving information from the hypothalamus, many kinds of sensory stimuli to the body can affect the functioning of the endocrine system. One example is the way in which the nervous system and endocrine system interact to influence the menstrual cycle. At least three hormones are involved in the cycle of changes that affects the ovary and the lining of the uterus (see chapter 27 for details). It is well documented that stress caused by tension or worry can interfere with the normal cycle of hor-

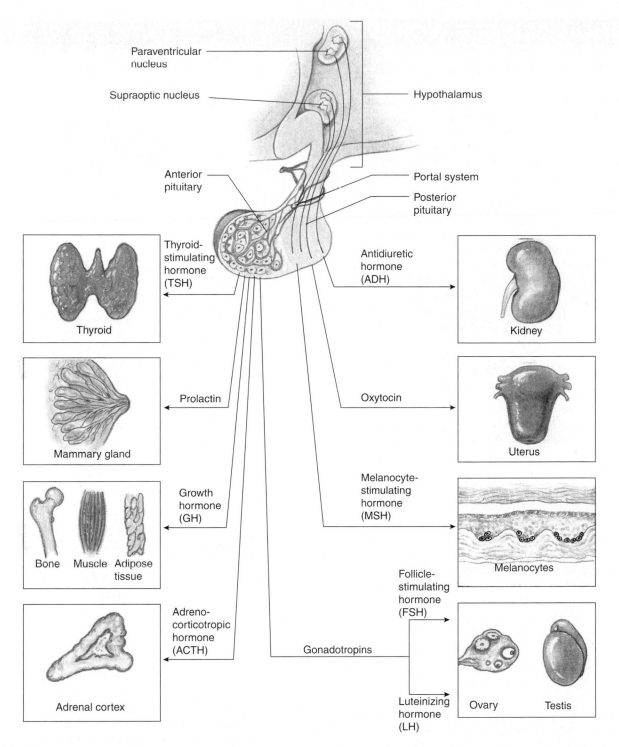

FIGURE 26.10 Hormones of the Pituitary and Their Target Organs
The anterior pituitary gland produces several hormones that regulate growth and the secretions of target tissues. The posterior pituitary produces hormones that change the behavior of the kidney and uterus but do not influence the growth of these organs.

HOW SCIENCE WORKS 26.2

Endorphins: Natural Pain Killers

The pituitary gland and the brain produce a group of small molecules that act as pain suppressors, called *endorphins*. It is thought that these molecules are released when excessive pain or stress occurs in the body. They attach to the same receptor molecules of brain cells associated with the feeling of pain. Endorphins work on the brain in the same manner as morphine and opiate drugs. Once endorphins are attached to the brain cells, the feeling of pain goes away, and a euphoric feeling takes over. Long-distance runners and other athletes talk about feeling good once they have "reached their stride," get their "second wind," or experience a "runner's high." These responses may be due to an increase in endorphin production. It is thought that endorphins are also released by mild electric stimulation and the use of acupuncture needles.

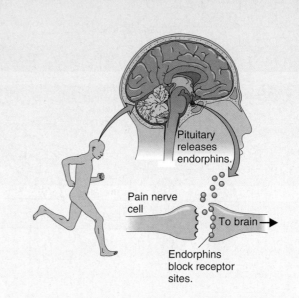

mones and delay or stop menstrual cycles. In addition, young women living in groups, such as in college dormitories, often find that their menstrual cycles become synchronized. Although the exact mechanism involved in this phenomenon is unknown, it is suspected that input from the nervous system causes this synchronization. (Odors and sympathetic feelings have been suggested as causes.)

In many animals, the changing length of the day causes hormonal changes related to reproduction. In the spring, birds respond to lengthening days and begin to produce hormones that gear up their reproductive systems for the summer breeding season. The pineal body, a portion of the brain, serves as the receiver of light stimuli and changes the amounts of hormones secreted by the pituitary, resulting in changes in the levels of reproductive hormones. These hormonal changes modify the birds' behavior. Courtship, mating, and nest-building behaviors increase in intensity. Therefore, it appears that a change in hormone level is affecting the animals' behavior; the endocrine system is influencing the nervous system (figure 26.11). It has been known for centuries that changes in the levels of sex hormones cause changes in animals' behavior. The castration (the removal of the testes) of male domesticated animals, such as cattle, horses, and pigs, is sometimes done in part to reduce their aggressive behavior and make them easier to control. The use of anabolic steroids by humans to increase muscle mass is known to cause behavioral changes and "moodiness."

Although we still tend to think of the nervous and endocrine systems as separate and different, it is becoming clear that they are interconnected. As scientists learn more about the molecules produced in the brain, it is becoming clear that the brain produces many molecules that act as hormones. Some of these molecules affect adjacent parts of the brain, others affect

the pituitary, and still others may have effects on more distant organs. In any case, these two systems cooperate to bring about appropriate responses to environmental challenges. The nervous system is specialized for receiving and sending short-term messages, whereas activities that require long-term, growth-related actions are handled by the endocrine system.

26.5 Sensory Input

The activities of the nervous and endocrine systems are often responses to input received from the sense organs. Sense organs of various types are located throughout the body. Many of them are located on the surface, where environmental changes can be detected easily. Hearing, sight, and touch are good examples. Other sense organs are located within the body and indicate to the organism how its various parts are changing. For example, pain and pressure are often used to monitor internal conditions. The sense organs detect changes, but the brain is responsible for **perception**—the recognition that a stimulus has been received. Sensory input is detected through many kinds of mechanisms, including chemical recognition, the detection of energy changes, and the monitoring of physical forces.

Chemical Detection

All cells have receptors on their surfaces that can bind selectively to molecules they encounter. This binding process can cause changes in the cells in several ways. In some cells, it causes depolarization. When this happens, the binding of

Light

Changing length of day stimulates growth of reproductive organs.

Reproductive organs secrete hormones that cause changed behavior.

FIGURE 26.11 Interaction Between the Nervous and Endocrine Systems
In birds and many other animals, the brain receives information about the changing length of day, which causes the pituitary to produce hormones that stimulate sexual development. The testes or ovaries grow and secrete their hormones in increased amounts. Increased levels of testosterone or estrogen result in changed behavior, with increased mating, aggression, and nest-building activity.

molecules to the cell can stimulate neurons and cause messages to be sent to the central nervous system, informing it of a change in the surroundings. Many internal sense organs respond to specific molecules. For example, the brain and aorta contain cells that respond to concentrations of hydrogen ions, carbon dioxide, and oxygen in the blood. In other cases, a molecule binding to the cell surface may cause certain genes to be expressed, and the cell responds by changing the molecules it produces. This is typical of the way the endocrine system receives and delivers messages. Taste and smell are two ways of detecting chemicals in the environment.

Taste

Most cells have specific binding sites for specific molecules. Some, such as the taste buds on the tongue, appear to respond to classes of molecules. Traditionally, four kinds of tastes have been identified: sweet, sour, salt, and bitter. However, recently, a fifth kind of taste, *umami* (meaty), has been identified. Umami receptors respond to the amino acid glutamate, which is present in many kinds of foods and is added as a flavor enhancer (monosodium glutamate) to many kinds of foods.

The taste buds that give us the sour sensation respond to the presence of hydrogen ions (H^+). (Acidic foods taste sour.)

The hydrogen ions stimulate the cells in two ways: They enter the cell directly or they alter the normal movement of sodium and potassium ions across the cell membrane. In either case, the cell depolarizes and stimulates a neuron. In similar fashion, sodium chloride stimulates the taste buds that give us the sensation of a salty taste by directly entering the cell, which causes the cell to depolarize.

However, the sensations of sweetness, bitterness, and umami occur when molecules bind to specific surface receptors on the cell. Sweetness can be stimulated by many kinds of organic molecules, including sugars and artificial sweeteners, as well as by inorganic lead compounds. When a molecule binds to a sweetness receptor, a molecule is split, and its splitting stimulates an enzyme, leading to the depolarization of the cell. The sweet taste of lead salts in old paints partly explains why children sometimes eat paint chips. Because the lead interferes with normal brain development, this behavior can have disastrous results. Many other kinds of compounds give the bitter sensation. The cells that respond to bitter sensations have a variety of receptor molecules on their surface. When a substance binds to one of the receptors, the cell depolarizes. In the case of umami, it is the glutamate molecule that binds to receptors on the cells of the taste buds.

Each of these tastes has a significance from an evolutionary point of view. Carbohydrates are a major food source, and many carbohydrates taste sweet; therefore, this sense is useful in identifying foods that have high food value. Similarly, proteins and salts are necessary in the diet. Therefore, being able to identify these items in potential foods is extremely valuable. This is true for salt, which must often be obtained from mineral sources. On the other hand, many bitter and sour materials are harmful. Many plants produce bitter-tasting toxic materials, and acids are often the result of the bacterial decomposition (spoiling) of foods. Being able to identify bitter and sour allows organisms to avoid potentially harmful foods.

Much of what we often refer to as *taste* involves such inputs as temperature, texture, and smell. Cold coffee has a different taste than hot coffee, even though they are chemically the same. Lumpy, cooked cereal and smooth cereal have different tastes. If we are unable to smell food, it doesn't taste as it should, which is why we sometimes lose our appetite when we have a stuffy nose. There still is much to learn about how the tongue detects chemicals and the role of associated senses in modifying taste.

Smell

The sense of smell is much more versatile than taste; it can detect thousands of different molecules at very low concentrations. The cells that make up the **olfactory epithelium,** the lining of the nasal cavity, which responds to smells, apparently bind molecules to receptors on their surfaces. Exactly how this can account for the large number of recognizably different odors is unknown, but the receptor cells are extremely sensitive. In some cases, a single molecule of a substance is sufficient to cause a receptor cell to send a message

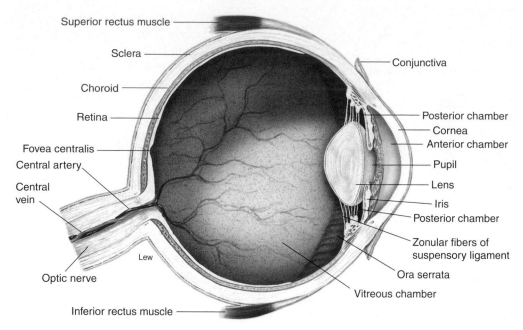

FIGURE 26.12 The Structure of the Eye

The eye contains a cornea and a lens, which focus light on the retina. The light causes pigments in the rods and cones of the retina to decompose. This leads to the depolarization of these cells and the stimulation of neurons, which send messages to the brain.

to the brain, where the sensation of odor is perceived. These sensory cells also fatigue rapidly. For instance, when we first walk into a room, we readily detect specific odors; however, after a few minutes, we are unable to detect them. Most perfumes and aftershaves are undetectable after 15 minutes of continuous exposure to them.

Vision

The eyes primarily respond to changes in the flow of light energy. The structure of the eye is designed to focus light on a light-sensitive layer of the back of the eye, known as the **retina** (figure 26.12). There are two kinds of receptors in the retina of the eye. The cells called **rods** respond to a broad range of wavelengths of light and are responsible for black-and-white vision. Because rods are very sensitive to light, they are useful in dim light. Rods are located over most of the retinal surface, except for the area of most acute vision, known as the **fovea centralis.** The other kind of receptor cells, called **cones,** are not as sensitive to light, but they can detect different wavelengths of light. They are found throughout the retina but are concentrated in the fovea centralis. This combination of receptors gives us the ability to detect color when light levels are high, but we rely on black-and-white vision at night. There are three varieties of cones: One type responds best to red light, one type responds best to green light, and the third type responds best to blue light. The stimulation of various combinations of these three kinds of cones allows us to detect different shades of color (figure 26.13).

Rods and the three kinds of cones each contain a pigment that decomposes when struck by light of the proper wave-

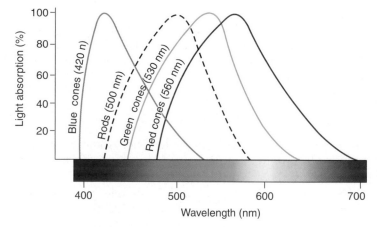

FIGURE 26.13 Light Reception by Cones

There are three different kinds of cones. Each responds differently to red, green, and blue wavelengths of light. Stimulation of combinations of these three kinds of cones gives us the ability to detect many different shades of color.

length and sufficient strength. **Rhodopsin** is the pigment found in rods. This change in the structure of rhodopsin causes the rod to depolarize. Cone cells have a similar mechanism of action, and each of the three kinds of cones has a different pigment. Because rods and cones synapse with neurons, they stimulate a neuron when depolarized and cause a message to be sent to the brain. Thus, the pattern of color and light intensity recorded on the retina is detected by rods and cones and converted into a series of nerve impulses, which the brain receives and interprets.

Hearing and Balance

Sound is produced by the vibration of molecules. The ears respond to changes in sound waves. Consequently, the ears detect changes in the quantity of energy and the quality of sound waves. Sound has several characteristics. Loudness, or volume, is a measure of the intensity of sound energy that arrives at the ear. Very loud sounds literally vibrate the body and can cause hearing loss if they are too intense. Pitch is a quality of sound that is determined by the frequency of the sound vibrations. High-pitched sounds have short wavelengths; low-pitched sounds have long wavelengths.

The sound that arrives at the ear is first funneled by the external ear to the **tympanum**, also known as the *eardrum*. The cone shape of the external ear focuses sound on the tympanum and causes it to vibrate at the same frequency as the sound waves reaching it. Attached to the tympanum are three tiny bones, the **malleus** (hammer), **incus** (anvil), and **stapes** (stirrup).

The malleus is attached to the tympanum, the incus is attached to the malleus and stapes, and the stapes is attached to a small, membrane-covered opening, called the **oval window,** in a snail-shaped, fluid-filled structure known as the **cochlea.** The vibration of the tympanum causes the tiny bones (malleus, incus, and stapes) to vibrate; in turn, they cause a corresponding vibration in the membrane of the oval window.

The cochlea detects sound. When the oval window vibrates, the fluid in the cochlea begins to move, causing the **basilar membrane** to vibrate. High-pitched, short-wavelength sounds cause the basilar membrane to vibrate at the base of the cochlea near the oval window. Low-pitched, long-wavelength sounds vibrate the basilar membrane far from the oval window. Loud sounds cause the basilar membrane to vibrate more vigorously than do faint sounds. Cells on this membrane depolarize when they are stimulated by its vibrations. Because they synapse with neurons, messages can be sent to the brain (figure 26.14).

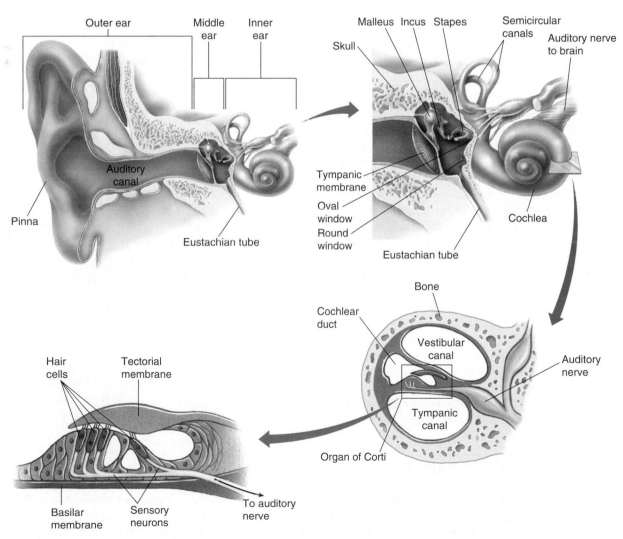

FIGURE 26.14 The Anatomy of the Ear
The ear consists of an external cone, which directs sound waves to the tympanum. Vibrations of the tympanum move the ear bones and vibrate the oval window of the cochlea, where the sound is detected. The semicircular canals monitor changes in the position of the head, helping us maintain balance.

Because sounds of different wavelengths stimulate different portions of the cochlea, the brain is able to determine the pitch of a sound. Most sounds consist of a mixture of pitches. Louder sounds stimulate the membrane more forcefully, causing the sensory cells in the cochlea to send more nerve impulses per second. Thus, the brain can perceive the loudness of various sounds as well as their pitch. Associated with the cochlea are two fluid-filled chambers and a set of fluid-filled tubes, called the *semicircular canals*. The **semicircular canals** are not involved in hearing but are involved in maintaining balance and posture. In the walls of these canals and chambers are cells similar to those found on the basilar membrane. These cells are stimulated by movements of the head and by the position of the head with respect to the force of gravity. The head's constantly changing position gives sensory input that is important in maintaining balance.

Touch

What we normally call the sense of *touch* consists of a variety of kinds of input, which are responded to by three kinds of receptors: Some receptors respond to pressure, others to temperature, and others, which we call *pain receptors*, to cell damage. When these receptors are appropriately stimulated, they send a message to the brain. Because receptors are stimulated in particular parts of the body, the brain can localize the sensation. However, not all parts of the body are equally supplied with these receptors. The finger tips, lips, and external genitals have the highest density of these nerve endings, whereas the back, legs, and arms have far fewer receptors.

Some internal receptors, such as pain and pressure receptors, are important in allowing us to monitor our internal activities. Many pains generated by the internal organs are often perceived as if they were somewhere else. For example, the pain associated with a heart attack is often perceived to be in the left arm or jaw. Pressure receptors in joints and muscles provide information about the degree of stress being placed on a portion of the body. This is also important information to send back to the brain so that adjustments can be made in movements to maintain posture. If you have ever had your foot "go to sleep" because the nerve stopped functioning, you have experienced what it is like to lose this constant input of nerve messages from the pressure sensors that assist in guiding the movements you make. Your movements become uncoordinated until the nerve function returns to normal.

26.6 | Output Coordination

The nervous system and endocrine system cause changes in several ways. Both systems can stimulate muscles to contract and glands to secrete. The endocrine system is also able to change the metabolism of cells and regulate the growth of tissues. The nervous system acts on two kinds of organs: muscles and glands. The actions of muscles and glands are simple and direct: Muscles contract and glands secrete.

Muscular Contraction

The ability to move is one of the fundamental characteristics of animals. Through the coordinated contraction of many muscles, the intricate, precise movements of a dancer, basketball player, or writer are accomplished. Muscles pull only by contracting; they are unable to push by lengthening. The work of any muscle is done during its contraction. Relaxation is the muscle's passive state. There must always be some force available to stretch a muscle after it has stopped contracting and relaxes. Therefore, the muscles that control the movements of the skeleton are present in antagonistic sets—for every muscle's action there is another muscle with the opposite action. For example, the biceps muscle causes the arm to flex (bend) as the muscle shortens. The contraction of its antagonist, the triceps muscle, causes the arm to extend (straighten) and simultaneously stretches the relaxed biceps muscle (figure 26.15).

What we recognize as a muscle is composed of many muscle cells, which in turn are made up of threadlike fibers, myofibrils, composed of two kinds of myofilamnets arranged in a regular pattern. Thin myofilaments composed of the proteins **actin, tropomyosin,** and **troponin** alternate with thick myofilaments composed primarily of the protein **myosin** (figure 26.16). The mechanism by which muscle contracts involves the movement of protein filaments past one another as adenosine triphosphate (ATP) is used.

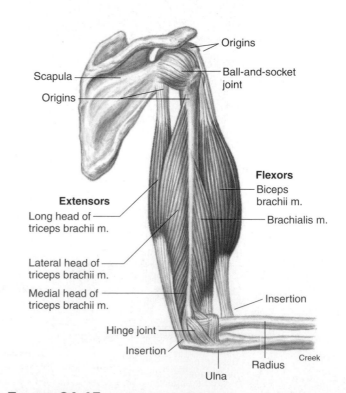

FIGURE 26.15 Antagonistic Muscles
Because muscles cannot actively lengthen, sets of muscles oppose one another. The contraction and shortening of one muscle causes the stretching of a relaxed muscle.

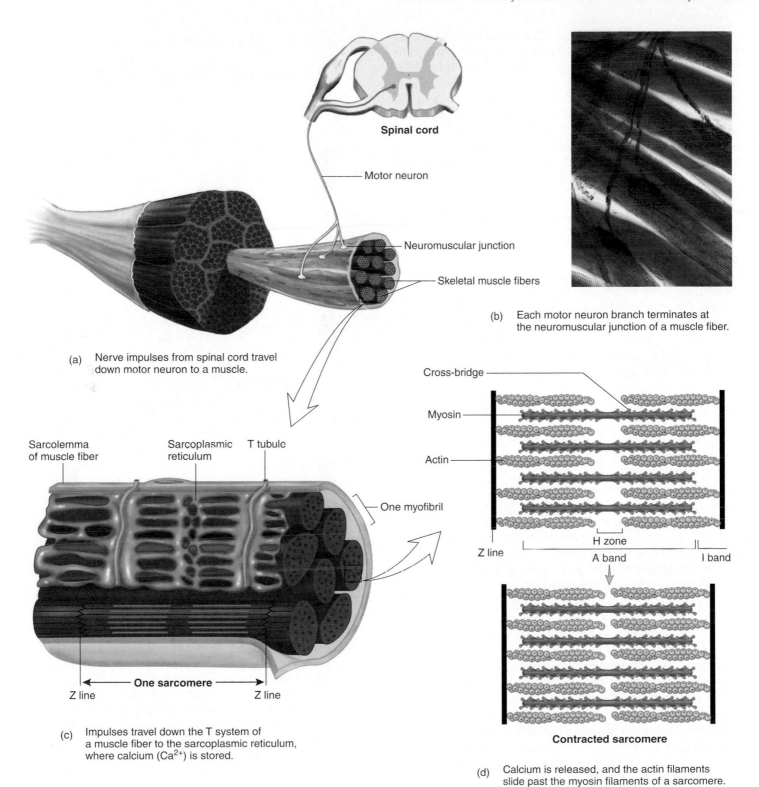

(a)   Nerve impulses from spinal cord travel down motor neuron to a muscle.

(b)   Each motor neuron branch terminates at the neuromuscular junction of a muscle fiber.

(c)   Impulses travel down the T system of a muscle fiber to the sarcoplasmic reticulum, where calcium (Ca^{2+}) is stored.

Contracted sarcomere

(d)   Calcium is released, and the actin filaments slide past the myosin filaments of a sarcomere.

FIGURE 26.16 The Microanatomy of a Muscle

(*a–c*) Muscles are made of cells that contain bundles known as myofibrils. The myofibrils are composed of two kinds of myofilaments: thick myofilaments composed of myosin, and thin myofilaments containing actin, troponin, and tropomyosin. (*d*) The actin- and myosin-containing myofilaments are arranged in a regular fashion into units called sarcomeres. Each sarcomere consists of two sets of actin-containing myofilaments inserted into either end of bundles of myosin-containing myofilaments. The actin-containing myofilaments slide past the myosin-containing myofilaments, shortening the sarcomere.

Myosin molecules are shaped like a golf club. The head of the club-shaped molecule sticks out from the thick myofilament and can combine with the actin of the thin myofilament. However, the troponin and tropomyosin proteins associated with the actin cover (i.e., a troponin-tropomyosin-actin complex) the actin in such a way that myosin cannot bind with it. When actin is uncovered, myosin can bind to it, and muscle contraction occurs when ATP is used.

The process of muscle-cell contraction involves several steps. The arrival of a nerve impulse at a muscle cell causes the muscle cell to depolarize. When muscle cells depolarize, calcium ions (Ca^{2+}) contained within membranes are released among the actin and myosin myofilaments. The calcium ions (Ca^{2+}) combine with the troponin molecules, causing the troponin-tropomyosin complex to expose actin, so that it can bind with myosin. While the actin and myosin molecules are attached, the head of the myosin molecule can flex as ATP is used and the actin molecule is pulled past the myosin molecule. Thus, a tiny section of the muscle cell shortens (figure 26.17). When one muscle contracts, thousands of such interactions take place within a tiny portion of a muscle cell, and many cells within a muscle contract at the same time.

The Types of Muscle

There are three major types of muscle: skeletal, smooth, and cardiac. These differ from one another in several ways. *Skeletal muscle* is voluntary muscle; it is under the control of the nervous system. The brain or spinal cord sends a message to skeletal muscles, and they contract to move the legs, fingers, and other parts of the body. This does not mean that we must make a conscious decision every time we want to move a muscle. Many of the movements we make are learned initially but become automatic as a result of practice. For example, walking, swimming, and riding a bicycle required a great amount of practice originally but become automatic for many people. They are, however, still considered voluntary actions.

Skeletal muscles are constantly bombarded with nerve impulses, which result in repeated contractions of differing strength. Many neurons end in each muscle, and each one stimulates a specific set of muscle cells, called a *motor unit* (figure 26.18). A **motor unit** is a single neuron and all the muscle fibers to which it connects. Because each muscle consists of many motor units, it is possible to have a wide variety of intensities of contraction within one muscle organ. This allows a single set of muscles to have a wide variety of functions. For example, the same muscles of the arms and shoulders that are used to play a piano can be used in other combinations to grip and throw a baseball. If the nerves going to a muscle are destroyed, the muscle becomes paralyzed and begins to shrink. The regular nervous stimulation of skeletal muscle is necessary for it to maintain its size and strength. Any kind of prolonged inactivity leads to the degeneration of muscles, known as atrophy. Muscle maintenance is one of the primary functions of physical therapy and a benefit of regular exercise.

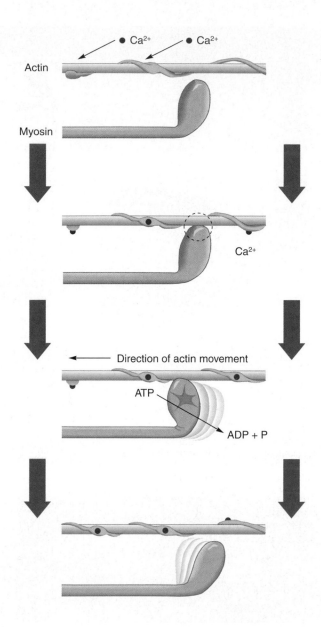

FIGURE 26.17 Interaction Between Actin and Myosin
When calcium ions (Ca^{2+}) enter the region of the muscle cell containing actin and myosin, they allow the actin and myosin to bind to each other. ATP is broken down to ADP and P with the release of energy. This energy allows the club-shaped head of the myosin to flex and move the actin along, causing the two molecules to slide past each other.

Skeletal muscles can contract quickly, but they cannot remain contracted for long periods. Even when we contract a muscle for a minute or so, the muscle is constantly shifting the individual motor units within it that are in a state of contraction. A single skeletal muscle cell cannot stay in a contracted state.

Smooth muscles make up the walls of muscular internal organs, such as the gut, blood vessels, and reproductive organs. They contract as a response to being stretched.

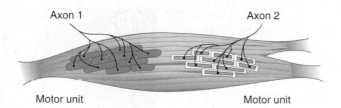

FIGURE 26.18 Motor Units
When a skeletal muscle contracts, individual groups of muscle cells are stimulated by specific neurons. They contract momentarily, then relax. The strength of the contraction is determined by the number of motor units stimulated. The sequence of movements is determined by the order in which the motor units are stimulated.

Because much of the digestive system is being stretched constantly, the responsive contractions contribute to the normal rhythmic movements associated with the digestive system. These are involuntary muscles; they can contract on their own without receiving direct messages from the nervous system. This can be demonstrated by removing portions of the gut or uterus from experimental animals. When these muscular organs are kept moist with special solutions, they go through cycles of contraction without any possible stimulation from neurons. However, they do receive nervous stimulation, which can modify the rate and strength of their contraction. This kind of muscle also has the ability to stay contracted for long periods without becoming fatigued. Many kinds of smooth muscle, such as the muscle of the uterus, also respond to the presence of hormones. Specifically, the hormone **oxytocin,** which is released from the posterior pituitary, causes strong contractions of the uterus during labor and birth. Similarly, several hormones produced by the duodenum influence certain muscles of the digestive system to either contract or relax.

Cardiac muscle makes up the heart. It can contract rapidly, like skeletal muscle, but does not require nervous stimulation to do so. Nervous stimulation can, however, cause the heart to speed or slow its rate of contraction. Hormones, such as epinephrine and norepinephrine, also influence the heart by increasing its rate and strength of contraction. Cardiac muscle also has the characteristic of being *unable* to stay contracted; it will contract quickly but must have a short period of relaxation before it is able to contract a second time. This makes sense in light of its continuous, rhythmic, pumping function. Table 26.1 summarizes the differences among skeletal, smooth, and cardiac muscles.

The Activities of Glands

Recall that there are two types of glands: endocrine glands, such as the pituitary, thyroid, ovary, and testis, and exocrine glands, such as the salivary glands, intestinal mucous glands, and sweat glands. Some of these glands, such as salivary glands and sweat glands, are under nervous control. When stimulated by the nervous system, they secrete their contents.

Russian physiologist Ivan Petrovich Pavlov showed that salivary glands are under the control of the nervous system, when he trained dogs to salivate in response to hearing a bell. You may recall from chapter 18 that, initially, the animals were presented with food at the same time the bell was rung. Eventually, they would salivate when the bell was rung even if food was not present. This demonstrated that saliva release is under the control of the central nervous system.

Many other exocrine glands are under hormonal control. Many of the digestive enzymes of the stomach and intestine are secreted in response to local hormones produced in the gut. These are circulated through the blood to the digestive glands, which respond by secreting the appropriate digestive enzymes and other molecules.

Growth Responses

The hormones produced by the endocrine system can have a variety of effects. Hormones can stimulate smooth muscle to contract and can influence the contraction of cardiac muscle. In addition, the hormones released by one gland can cause another gland to secrete its own hormone. However, the endocrine system has one major effect that is not equaled by the nervous system: Hormones regulate growth. Several examples of the many kinds of long-term growth changes that are caused by the endocrine system were given earlier in the chapter. Growth stimulating hormone (GSH) is produced over a period of years to bring about the increase in size of most of the structures of the body. A low level of this hormone results in a person with small body size. The amount of

TABLE 26.1			
Characteristics of the Three Kinds of Muscle			
Kind of Muscle	Stimulus	Length of Contraction	Rapidity of Response
Skeletal	Nervous system	Short; tires quickly	Most rapid
Smooth	Self-stimulated; also responds to nervous and endocrine systems	Long; doesn't tire quickly	Slow
Cardiac	Self-stimulated; also responds to nervous and endocrine systems	Short; cannot stay contracted	Rapid

growth-stimulating hormone (GSH) present varies from time to time; it is present in fairly high amounts throughout childhood and results in steady growth. It also appears to be present at higher levels at certain times, resulting in growth spurts. Finally, as adulthood is reached, the level of this hormone falls, and growth stops.

Similarly, testosterone produced during adolescence influences the growth of bone and muscle to provide men with larger, more muscular bodies than those of women. In addition, there is growth of the penis, of the larynx, and of hair on the face and body. The primary female hormone, estrogen, causes the growth of reproductive organs and the development of breast tissue. It is also involved, along with other hormones, in the cyclic growth and sloughing of the wall of the uterus.

26.7 The Body's Defense Mechanisms—Immunity

Immunity is the body's ability to maintain homeostasis by resisting or defending against potentially harmful agents, including microbes, toxins, and abnormal cells, such as tumor cells. The body's **immune system** comprises specialized cells and messenger molecules, which work together to defend against infection and disease, using methods that can be nonspecific or specific. *Nonspecific defense methods* are more general and able to protect the body against many agents. These threats can be chemical (e.g., a toxin), biological (e.g., a pathogen), or mechanical (e.g., a splinter). *Specific defense methods* are programmed to distinguish exact differences between types of threats based on their exact chemical makeup. Once programmed, they continue to protect only against that agent. After recognizing an enemy, the body selects one of these unique combinations of chemicals and cells to send into the battle to destroy only that enemy.

Nonspecific Defenses

The body has three nonspecific defenses—defensive barriers, chemicals, and certain kinds of cells. These defenses are able to function from the time you are born, in other words, you do not develop these defenses after coming in contact with potentially harmful agents. All three may work together to safeguard the body. For example, the skin and mucous membranes are barriers to bacteria, viruses, parasitic worms, and large toxic molecules; these harmful agents are not able to burrow into the skin to cause infection. The enzyme *lysozyme,* found in sweat, destroys bacterial cells on the skin. The cells of the upper respiratory system are ciliated and covered with mucus. Should bacteria or dust particles be caught in the fluid, the cilia's upward beating moves them toward the mouth and nose, where they can be eliminated by sneezing, coughing, or swallowing. If the threatening agent is swallowed, stomach acid (about pH 2) also acts as a nonspecific defense. A mixture of materials with a low pH forms a protective barrier in the vagina.

Certain cells of the body produce chemicals that help in defense. *Complement* is a group of proteins that circulates in the blood and helps other systems defend against threats of cellular pathogens, such as bacteria, fungi, and protozoa. Complement can help form holes in pathogens, allowing their content to leak out. *Interferons* are

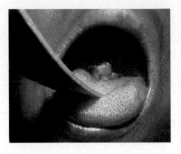

Inflamed tonsils

proteins that interfere with the attachment of viruses to the surface of human cells. If interferons do their job, viruses cannot enter the cells. If viruses cannot get inside, they cannot cause harm, and the viruses are destroyed by more specific methods. If a cell does become infected, it produces interferons, which act as messengers. These are sent to other, nearby cells, informing them that they should get ready for the battle by making their own protective interferons.

Cells and chemicals interact in a nonspecific defense mechanism called *inflammation.* **Inflammation** is a series of events that clear an area of harmful agents and damaged tissue. The symptoms of inflammation seen on the skin and mucous membranes are heat, redness, swelling, and pain. The events that occur during this process are show in table 26.2. Inflammation occurs frequently and is one of the body's most important nonspecific defense mechanisms.

Because of inflammation's symptoms, most people don't think of it as a good thing, and, under certain circumstances, the inflammatory process can get out of hand. If inflammation continues unchecked, the process can result in tissue damage. Under such circumstances, anti-inflammatory medications, such as nonsteroidal anti-inflammatory drugs (NSAIDs), can be used to interfere with the inflammatory process. Aspirin, ibuprofen, ketorolac, and naproxen are all NSAIDs.

Specific Defenses

A person is born with the ability to use specific defense mechanisms, but these defenses must be "turned on" in order to work. This system also has a memory, being able to call into action the same specific defenses when threatened at a later time. This mechanism is also called *acquired immunity,* because it begins to work only after certain cells of the immune system have come into contact with a specific agent, called an *antigen.* An **antigen** is a large organic molecule, usually a protein, that is able to stimulate the production of a specific defense response and becomes neutralized or destroyed by that response. Antigens can be individual molecules, such as botulism toxin. They can also be parts of cells, such as molecules found on the surface or inside bacteria, viruses, or eukaryotic cells. The cells that respond to antigens are primarily T-lymphocytes and B-lymphocytes. The way each kind of cell responds is very different, so specific defense methods are subdivided into *B-cell or antibody-mediated*

TABLE 26.2

Inflammation

What Happens on the Surface	What's Really Happening
The event that triggers inflammation	Damage to tissue by molecules, microbes, or other materials for example, a scratch, infecting bacteria releasing toxins, a splinter.
General events of inflammation	Inflammatory chemicals released by injured cells serve as signals calling neutrophils from nearby capillaries. White blood cells (lymphocytes, monocytes, macrophages) go through the blood vessel walls and move to the damaged area, where they attack the enemy and clean up damaged cells by phagocytosis. The pH of tissue fluid decreases, helping control pathogens.
Heat	The inflammatory chemicals cause increased blood flow and the dilation (an increase in the diameter) of capillaries.
Swelling	As the capillaries dilate, they also become more leaky or porous, allowing plasma and white blood cells to move from the vessel into the tissue.
Redness	Inflammatory chemicals cause increased blood flow in the area.
Pain	Tissue fluid pressure increases and inflammatory chemicals affect nerve endings.

immunity and *T-cell or cell-mediated immunity*. However, a great deal of interaction between these two types methods must occur to defend a person against a specific enemy.

B-Cells and Antibody Mediated Immunity

B-lymphocytes are formed in the bone marrow, mature in the bone marrow; they are hard to find in the blood and are found in large numbers in the lymph nodes and spleen. When a B-cell contacts an antigen, chemical messengers are sent from the cell surface to its nucleus and genes are activated. In these B-cells, certain portions of the DNA are cut and spliced to produce a new nucleotide sequence. This new gene is capable of directing the synthesis of a specific protein (antibody) that will be able to combine with that antigen. This new gene is capable of directing the synthesis of a specific protein (anti-

body) that will be able to combine with that antigen. These protein molecules are called immunoglobulins or *antibodies*. An **antibody** is a protein made by the B-cells in response to an antigen (figure 26.19). There are five classes of antibodies, differing in their structures and functions (table 26.3).

When B-cells become specialized for the production of a specific antibody, they remain specialized for the rest of their lives. These specialized B-cells undergo mitosis, and their descendents become either plasma cells, which produce antibodies, or memory B-cells. Memory B-cells are like reserve military troops. They have been taught how to fight (i.e., have the gene to produce the antibody), are not fighting at the moment (i.e., are circulating through the body), but could be called into the battle at a later time (i.e., will turn on the antibody gene). At this point in the antibody-mediated response,

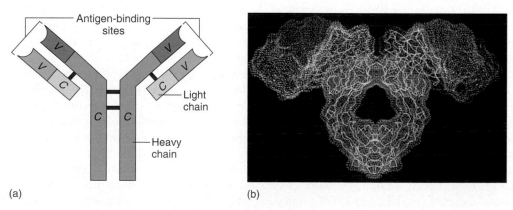

(a)   (b)

FIGURE 26.19 Structure of the most common antibody (IgG)
(a) An IgG antibody contains two heavy (long) polypeptide chains and two light (short) chains arranged so that there are two "variable regions" at the top of the Y-shaped molecule. This is where the antigen combines with the antibody (V = variable region, C = constant region). *(b)* A computer model of an antibody molecule.

TABLE 26.3

Classes of Antibodies

Class	Percent of Total Amount Circulating Antibody	Location	Function
Immunoglobulin gamma (IgG)	80–85	Blood and tissue fluids	Most abundant class, also called gammaglobulin: (1) binds with toxins, (2) activates complement proteins, (3) encourages phagocytosis, and (4) is the only class passed to the fetus through the placenta
Immunoglobulin mu (IgM)	5–10	Blood	Is the first to appear after infection: (1) activates complement and (2) causes cells to clump, encouraging phagocytosis
Immunoglobulin alpha (IgA)	5–15	Blood, mucous secretions	Are the only antibodies that are secreted onto the surface of mucosal membranes; prevents pathogens from attaching to the surface of the digestive and respiratory systems
Immunoglobulin delta (IgD)	1	Blood	Is found on B-cells; its function is not clear
Immunoglobulin epsilon (IgE)	0.002	Blood, attached to basophiles	(1) May help protect again parasitic worm infections and (2) Is responsible for many allergic reactions, such as asthma and hay fever

there are many more plasma cells than memory B-cells, and the plasma cells produce large quantities of antibodies, which enter the circulatory system. When an antibody molecule contacts the antigen, they combine. If the antigen is an isolated toxin molecule or a virus, it is prevented from causing harm or *neutralized*. If the antigen is part of a pathogen, the antigen-antibody combination causes holes to be formed in the cell, killing the pathogen. The antigen-antibody combination can also act as an attractants for phagocytes, such as macrophages and neutrophils, causing them to move in and destroy the pathogen. mitosis, p. 169

Once the trouble is brought under control, the number of plasma cells slowly drops. This leaves the memory B-cells to serve as a backup team, should that antigen appear in the future. Memory B-cells know how to make the correct antibody to defend against that specific antigen, they but don't make much of it. The antibodies they produce is on their surface. If the same antigens appears again, they attach to the antibodies causing the memory B-cells to undergo mitosis, changing into a large population of antibody-secreting plasma cells. The amount of antibody increases rapidly and again protects the body from harm.

Immunization is the technique used to induce the immune system to develop an acquired immunity to a specific disease by the use of a *vaccine*. **Vaccines** are antigens made so they can start an active immunity without causing disease. A vaccine can be a solution that contains whole, live pathogens in a form so changed that they are not harmful. It can also contain dead organisms or just the antigenic portion of the pathogen; some vaccines are synthetic. When the vaccine is first given, B-cells become activated. This is called the primary immune response (figure 26.20). The next time the person is exposed to the vaccine, there is a sharp increase in the amount of antibody produced. This is called the secondary immune response. As time passes, the amount of antibody will begin to fall. It will remain high only if the person is repeatedly exposed to the same antigen, either by accident or as the result of a booster shot. For example, if a person lives or works in an area where tetanus toxin can regularly stimulate a secondary immune response (e.g., through a puncture wound), a constant high level of immunity to tetanus toxin is naturally maintained. Each time the memory B-cells come in contact with tetanus toxin, the amount of antibody is "boosted" to its maximum protective level. If natural exposure does not occur, the amount of antibody falls slowly, sometimes reaching a low, unprotective level. There are no quick ways to determine how "naturally" protected someone is. If a physician suspects that a patient might develop a case of tetanus, a booster injection of tetanus vaccine is given to stimulate antibody production artificially to ensure a high level of protection.

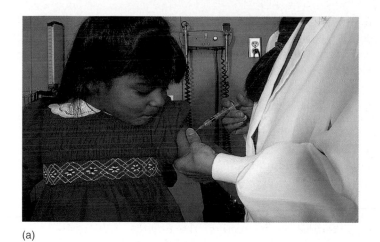

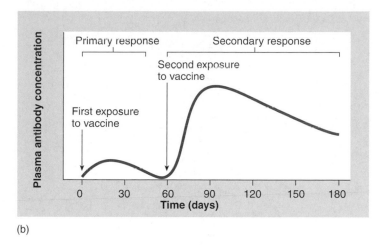

(a)   (b)

FIGURE 26.20 Active Immunity Due to Immunization

(a) Vaccines immunize children against various childhood diseases. *(b)* The primary immune response, after the first exposure to a vaccine, is minimal, but the secondary immune response, which may occur after the second exposure, shows a dramatic rise in the amount of antibody present in the plasma.

T-Cells and Cell-Mediate Immunity

T-lymphocytes are formed in the bone marrow, mature in the thymus gland, and are found in the blood, lymph, and lymph tissue. T-cells are responsible for regulating B-cell activity (i.e., turning their genes on an off), and rupturing cellular pathogens (i.e., bacteria), virus infected body cells, and cancer cells. Like B-cells, T-cells must be brought into battle through an activation process. To activate T-cells, the antigen must be "presented" to them by other cells, called antigen-presenting cells (APCs), macrophages that have phagocytosed the pathogen. Phagocytosis was covered in the cell chapter, chapter 4. Inside an APC, the pathogen is broken down, and each antigenic molecule is attached to a molecule already present in the APC. This new combination is inserted into the APCs cell membrane so that each molecule sticks out on the APCs surface like a microscopic hair. The APC activates the T-cells by presenting the antigen to T-cells when they attach to one another at receptor sites on the surface of the T-cells (figure 26.21). As with B-cells, chemical messengers are sent to the T-cell nucleus and its DNA is cut and spliced to form new genes. Depending on the genes produced, T-cell may be differentiated into one of three kinds of specialized cells: T-regulator cells, cytotoxic T-cells and T-memory cells.

T-regulator cells are able to communicate with B-cells, helping them control the amount of antibody they produce. If B-cells did not produce enough antibodies, an infection would go unchecked. If they produced too many antibodies, not only would this be a waste of metabolic energy but the excess antibodies could cause abnormal tissue damage, known as hypersensitivities or allergies. Two kinds of T-regulator cells have been identified: T-helper cells encourage B-cells to make antibodies, and T-suppressor cells discourage B-cells from making antibodies.

Cytotoxic T-cells are the specialized killers of cells that cause disease. After identifying the enemy cells, cytotoxic T-cells move toward their target cells, press up against their surface, and make holes in them. Cells that are recognized as cytotoxic T-cell targets are bacteria; foreign cells, such as cancer and cells from a transplant; and parasite-infected cells. T-cells are also responsible for a variety of chemicals called *cytokines,* which are generally responsible for attracting white blood cells to the site of infection, encouraging inflammation, stimulating blood cell production, and killing tumor cells.

T-memory cells are generated in much the same fashion as B-memory cells, and they play a similar role. They remember the specific antigen of their enemy and can multiply into T-regulator cells and cytotoxic T-cells when exposed to the same antigen at a later time (Outlooks 26.2).

Allergic Reactions

The immune system plays many roles in maintaining health, but it sometimes malfunctions. Under certain conditions, antibody-antigen reactions result in abnormal responses known as *allergies* or *type I hypersensitivities*. An **allergy** is an abnormal immune reaction to an antigen. Possibly the most familiar are allergies to foods, pollens, and drugs. All allergies involve the interaction of an antigen with a B-cell antibody. An antigen that causes an allergy and comes from outside the body, such as dust, pollen, tobacco, fungi, eggs, and penicillin, is called an *allergen.*

Type I hypersensitivities are associated with immunoglobulin E (IgE). B-cells capable of producing IgE are first exposed to the allergen through the skin, respiratory tract, or intestinal tract and become activated. Unlike most other immunoglobulins, IgE can bind to the surface of basophiles

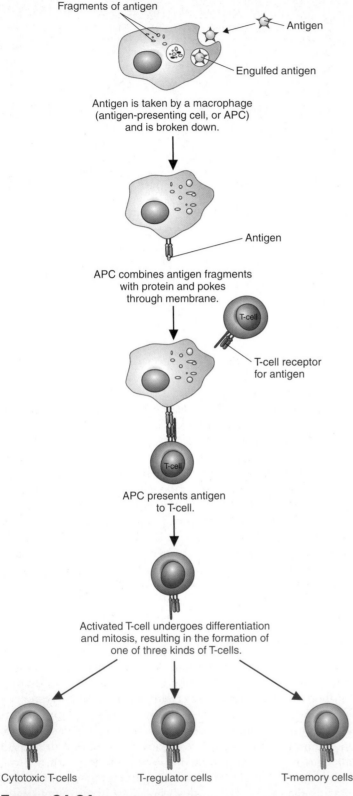

Fragments of antigen

Antigen

Engulfed antigen

Antigen is taken by a macrophage
(antigen-presenting cell, or APC)
and is broken down.

Antigen

APC combines antigen fragments
with protein and pokes
through membrane.

T-cell

T-cell receptor
for antigen

T-cell

APC presents antigen
to T-cell.

Activated T-cell undergoes differentiation
and mitosis, resulting in the formation of
one of three kinds of T-cells.

Cytotoxic T-cells

T-regulator cells

T-memory cells

FIGURE 26.21 Specialized T-Cells

T-cells learn to be T-regulator cells, cytotoxic T-cells, or T-memory cells by their teacher, an antigen-presenting cell (APC). APCs are macrophages that have engulfed an antigen and present it to their "students." Once differentiated, each kind of T-cell has a special job to perform in defending the body against harm.

(figure 26.22). When a person comes in contact with the allergen a second time, the allergen (antigen) attaches to the antibody, which is attached to the basophiles. This causes the release of large amounts of IgE and a secondary immune response. The symptoms are caused by the release of chemicals, including histamine, leukotrienes, and prostaglandins. Symptoms vary greatly in people and include skin rashes, hives, itching throat, asthma, eczema, migraine headaches, and bedwetting. The most severe reaction is anaphylactic shock; within minutes after exposure to the allergen, this shock reaction progresses from (1) extensive skin reddening and itching to (2) smooth muscle spasms to (3) capillary breakage in the skin, eyes, and internal organs to (4) a burning sensation in the rectum, the mouth, and possibly the vagina to (5) severe headache to (6) hot flashes to (7) a quick drop in blood pressure to (8) a constriction of bronchial smooth muscle, resulting in respiratory failure, to (9) death. People who are extremely allergic to certain antigens may go into anaphylactic shock. To counteract anaphylactic shock, people are advised to carry and use an EpiPen as soon as they have been exposed to the allergen. The medication in this single-use hypodermic syringe is epinephrine, which quickly counteracts the symptoms. However, after using the pen, the person should seek medical attention immediately, because the EpiPen may not totally control the reaction.

Autoimmune and Immunodeficiency Diseases

Autoimmune diseases are disorders that result from the immune system's turning against the normal chemicals and cells of the body. In other words, the immune response makes a mistake, sees healthy cells as harmful, and attacks them. Some examples of autoimmune diseases are rheumatoid arthritis, and type 1 (insulin-dependent) diabetes, systemic lupus, Crohn's disease, and psoriasis. In rheumatoid arthritis, B-cells produce antibodies that mistakenly identify proteins in the cartilage and bone as dangerous and destroy them. In type 1 diabetes, the insulin-producing cells of the pancreas are destroyed by T-cells.

Immunodeficiency diseases are disorders that result from the immune system's not having one or more component cells or chemicals. In this situation, the system cannot provide the protection needed to maintain health. People with such disorders are more susceptible in infections and cancers. People born with the genetic disorder severe combined immunodeficiency disease (SCIDS), or "boy-in-the-bubble" disease, lack the ability to synthesize the enzyme adenosine deaminase (ADA), which is crucial to the formation of T-cells. Acquired immunodeficiency syndrome (AIDS) is also an immunodeficiency disease but is not genetic. AIDS is the result of a virus

First Exposure

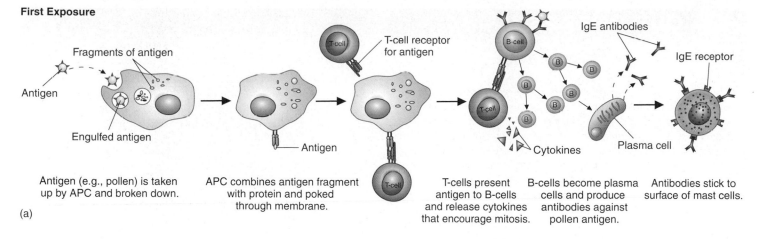

Antigen (e.g., pollen) is taken up by APC and broken down.

APC combines antigen fragment with protein and poked through membrane.

T-cells present antigen to B-cells and release cytokines that encourage mitosis.

B-cells become plasma cells and produce antibodies against pollen antigen.

Antibodies stick to surface of mast cells.

(a)

Second Exposure

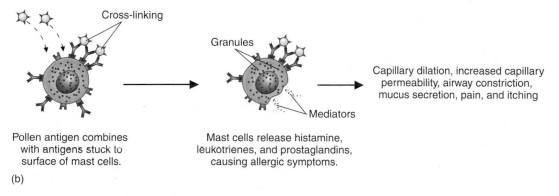

Pollen antigen combines with antigens stuck to surface of mast cells.

Mast cells release histamine, leukotrienes, and prostaglandins, causing allergic symptoms.

Capillary dilation, increased capillary permeability, airway constriction, mucus secretion, pain, and itching

(b)

FIGURE 26.22 How an IgE Allergy Works

(a) When a person is first exposed to an allergen, such as pollen, a complex series of events leads to the release of IgE antibodies, which stick to the surface of basophils. Memory B-cells are formed and are responsible for the ongoing allergic reaction. A few days later, when pollen is encountered again, the increased population of antibody-coated mast cells combines with the allergen, causing them to release the chemical responsible for the symptoms.

infection caused by the human immunodeficiency virus (HIV) (see chapter 20). HIV infects T- and B-cells and eventually destroys them. When the immune system's population of T-cells has been reduced significantly, it can no longer defend the body against microbial infections. It also loses its ability to recognize and destroy tumor cells.

Summary

A nerve impulse is caused by sodium ions entering the cell as a result of a change in the permeability of the cell membrane. Thus, a wave of depolarization passes down the length of a neuron to the synapse. The axon of a neuron secretes a neurotransmitter, such as acetylcholine, into the synapse, where these molecules bind to the dendrite of the next cell in the chain, resulting in an impulse in it as well. The acetylcholinesterase present in the synapse destroys acetylcholine, so that it does not repeatedly stimulate the dendrite. The brain is composed of several functional units. The lower portions of the brain control automatic activities, the middle portion of the brain controls the basic categorizing of sensory input, and the higher levels of the brain are involved in thinking and self-awareness.

Several kinds of sensory inputs are possible. Many kinds of chemicals can bind to cell surfaces and be recognized. This is how the senses of taste and smell function. Light energy can be detected because light causes certain molecules in the retina of the eye to decompose and stimulate neurons. Sound can be detected because fluid in the cochlea of the ear is caused to vibrate, and special cells detect this movement and stimulate neurons. The sense of touch consists of a variety of receptors that respond to pressure, cell damage, and temperature.

Muscles shorten because of the ability of actin and myosin to bind to one another. A portion of the myosin molecule is caused to bend when ATP is used, resulting in the sliding of actin and myosin molecules past each other. Skeletal muscle responds to nervous stimulation to cause movements of the skeleton. Smooth muscle and cardiac muscle have internally generated contractions, which can be modified by nervous stimulation or hormones.

OUTLOOKS 26.2

Organ Transplants

Physicians have long attempted to restore health by transplanting tissues from a donor to a recipient. In early attempts, whole blood, limbs, and internal organs from animals and humans were transferred to recipients, with the hope that the new and old tissues would "grow together." However, very little success was achieved, even though the tissues "looked alike." Within a matter of days, the transplanted organs withered and died, and in countless cases so did the recipient. The degree of resemblance of donor tissue to recipient tissue is very important to the success of transplantation operations but not on the superficial level most people think. For a donor tissue to be transplanted successfully, it must antigenically resemble the recipient as closely as possible. In addition, the recipient's immune system must be temporarily suppressed by medication to give the new tissue a chance to be accepted.

The immune responses that result in the destruction of transplanted tissues, known collectively as rejection responses, are triggered by naturally occurring antigens on the surface of transplanted cells. Cell-surface antigens of a different species are seen by the immune system of the human recipient as "nonself" and are rejected very quickly. Antigens on the surface of cells from a genetically related individual of the same species stimulate the rejection response less rapidly and, with help, may be successfully incorporated into the recipient's body. Glycoproteins within the same species that function as cell-surface antigens are known as isoantigens. Their structure is

This illustration (circa 1693) shows a primitive attempt at whole-blood transfusion. The patient's blood is being "let" from his right arm, while the blood from a dog is being transfused into his left arm.

genetically controlled, and they are found on the surfaces of blood cells (e.g., red cells, white cells, and platelets) and on tissue cells other than blood (e.g., skin, heart, kidney, etc.). Cells have a variety of isoantigens on their surfaces, but the most clinically important are those on red blood cells associated with the ABO and Rh blood systems and those histocompatibility antigens (refer to chapter 4) found on the surface of white blood cells and most fixed tissues.

There are two types of glands: exocrine glands, which secrete through ducts into the cavity of an organ or to the surface of the skin, and endocrine glands, which release their secretions into the circulatory system. Digestive glands and sweat glands are examples of exocrine glands. Endocrine glands, such as the ovaries, testes, and pituitary gland, change the activities of cells and often cause responses resulting in growth over a period of time. The endocrine system and the nervous system are interrelated. Actions of the endocrine system can change how the nervous system functions, and the reverse is also true.

The immune system defends the body against potentially harmful agents, including microbes, toxins, and abnormal cells. Nonspecific defense methods protect the body against many kinds of harmful agents. Specific defense methods can tell the difference between one kind of threat and another based on their exact chemical makeup. After recognizing the enemy, the body selects a specific combination of chemicals and cells to send into the battle to destroy only that enemy. Cells important to the system include B-lymphocytes, T-lymphocytes, and macrophages called antigen-presenting cells (APCs). Abnormally functioning immune systems can result in allergies, immunodeficiency, and autoimmune diseases.

Key Terms

Use the interactive flash cards on the **Concepts in Biology**, *12/e website to help you learn the meaning of these terms.*

Basic Review

1. The _____ is the space between the fibers of adjacent neurons in a chain.

2. In the _____, the somatic nervous system controls the skeletal (voluntary) muscles and the autonomic nervous system controls smooth (involuntary) muscles, the heart, and glands.
   a. peripheral nervous system
   b. central nervous system
   c. immune system
   d. medulla oblongata

3. The specific cells that a certain hormone affects are called
   a. target cells.
   b. neurons.
   c. motor units.
   d. T-cells.

4. _____ glands have no ducts and secrete their products into the circulatory system.
   a. Exocrine
   b. Endocrine
   c. Salivary
   d. Pituitary

5. The mechanism by which an increase in a stimulus causes a reduction in a response is called
   a. positive-feedback regulation.
   b. immune.
   c. negative-feedback control.
   d. absolute.

6. Molecules that are neurotransmitters manufactured in the soma and migrate down the axon, where they are stored until needed, are called
   a. somas.
   b. acetylcholine.
   c. cytokines.
   d. estrogen.

7. Which of the following is a resulting symptom of inflammatory chemicals that cause increased blood flow and dilation of capillaries?
   a. heat
   b. estrogens
   c. artherosclerosis
   d. nausea

8. Which cells of the immune system produce memory cells?
   a. H-cells
   b. APCs
   c. cytokines
   d. B-cells

9. Histamine, leukotrienes, and prostaglandins are compounds produced and released by which cells during an allergic reaction?
   a. WBCs
   b. RBCs
   c. basophiles
   d. APCs

10. In muscle, thin myofilaments composed of the proteins actin, tropomyosin, and troponin alternate with thick myofilaments composed primarily of the protein _____.

Answers
1. synapse 2. a 3. a 4. b 5. c 6. b 7. a 8. d 9. c 10. myosin

Concept Review

*Answers to the Concept Review questions can be found on the **Concepts in Biology**, 12/e website.*

26.1 Coordination in Multicellular Animals

1. Describe two ways in which the function of the nervous system differs from that of the endocrine system.
2. How do exocrine and endocrine glands differ?

26.2 Nervous System Function

3. Describe how the changing permeability of the cell membrane and the movement of sodium ions cause a nerve impulse.
4. What is the role of acetylcholine in a synapse? What is the role of acetylcholinesterase?
5. List the differences between the central and peripheral nervous systems and between the motor and sensory nervous systems.

26.3 Endocrine System Function

6. Give an example of negative-feedback control in the endocrine system.
7. List three hormones and give their functions.

26.4 The Integration of Nervous and Endocrine Function

8. Give an example of the interaction between the endocrine system and the nervous system.
9. How do the anterior pituitary and the posterior pituitary differ in the way they function?

26.5 Sensory Input

10. What is detected by the nasal epithelium, the cochlea of the ear, and the retina of the eye?
11. Name the five kinds of taste humans are able to detect. What other factors are involved in taste?

12. List three kinds of receptors associated with touch.

26.6 Output Coordination

13. How do skeletal, cardiac, and smooth muscles differ in (1) speed of contraction, (2) ability to stay contracted, and (3) cause of contraction?
14. What is the role of each of the following in muscle contraction: actin, myosin, ATP, troponin, and tropomyosin?
15. Why must muscles be in antagonistic pairs?

26.7 The Body's Defense Mechanisms—Immunity

16. What are the differences between nonspecific and specific immunity?
17. How do B-cell and T-cell systems work to defend against disease?

Thinking Critically

*For guidelines to answer this question, visit the **Concepts in Biology**, 12/e website.*

Humans are considered to have a poor sense of smell. However, when parents are presented with baby clothing, they are able to identify the clothing with which their own infant had been in contact with a high degree of accuracy. Specially trained individuals, such as wine and perfume testers, are able to identify large numbers of different kinds of molecules that the average person cannot identify. Birds rely primarily on sound and sight for information about their environment; they have a poor sense of smell. Most mammals are known to have a very well-developed sense of smell. Is it possible that we have evolved into sound-and-sight-dependent organisms, like birds, and have lost the keen sense of smell of our ancestors? Or is it that we just don't use our sense of smell to its full potential? Devise an experiment that would help shed light on this question.

Human Reproduction, Sex, and Sexuality

Recent advances in reproductive biology have led to the possibility of cloning in humans. Developing eggs can be harvested from the ovary, fertilized in dishes, and implanted in a female who is not the woman who donated the egg. In nonhuman animals, the nucleus of an egg has been removed and replaced with the nucleus from an adult cell. This manipulated cell begins development and can be implanted into the uterus. In this way, sheep, mice, and monkeys have been cloned that are genetically identical to their nuclear donors.

Can humans be cloned? In 2001, an Italian embryologist Dr. Severino Antinori claimed to have cloned a human embryo. In 2002, the United States-based company Clonaid claimed that it had created the first-ever cloned human baby, a girl nicknamed Eve. However, since no evidence—including DNA testing—confirming the event has been provided, scientists are skeptical. Chinese scientists have also claimed to have produced early human clones. More recently, one of the embryos was used as a source of stem cells for further research. Later in 2004, Britain granted its first license for human cloning, joining South Korea on the leading edge of stem cell research. In 2005, a team of British scientists cloned that country's first human embryo. The Newcastle University group used eggs from 11 women, removed the genetic material and replaced it with DNA from embryonic stem cells.

What scientific evidence would be needed to confirm such births? Should humans be cloned? To answer this question it is necessary to have a fundamental understanding of sex and sexuality.

CHAPTER OUTLINE

27.1 Sexuality from Various Points of View

Probably nothing interests people more than sex and sexuality. **Sex** is the nature of the biological differences between males and females. By **sexuality,** we mean all the factors that contribute to one's female or male nature. A persons sexuality includes the structure and function of the sex organs, sexual behavior, and the ways in which culture influences sexual behavior. Males and females have different behavior patterns for a variety of reasons. Some behavioral differences are learned (e.g., patterns of dress, the use of facial makeup), whereas others appear to be less dependent on culture (e.g., degree of aggressiveness, the frequency of sexual thoughts).

finding a mate, p. 390

There are several ways of looking at human sexuality. The behavioral sciences tend to focus on the behaviors associated with being male and female and what is considered appropriate or inappropriate sexual behavior. Psychologists consider sexual behavior to be a strong drive, appetite, or urge (figure 27.1). They describe the sex drive as a basic impulse to satisfy a biological, social, or psychological need. Other social scientists, such as sociologists and cultural anthropologists, are interested in sexual behavior as it occurs in various cultures. When a variety of cultures are examined, it becomes very difficult to classify various kinds of sexual behavior as normal or abnormal; What is considered abnormal in one culture may be normal in another. For example, public nudity is considered abnormal in many cultures, but not in others.

Biologists have studied the sexual behavior of nonhuman animals for centuries. They have long considered the function of sexuality in light of its value to the population or species. Sexual reproduction results in new combinations of genes, which are important in the process of natural selection. Many biologists are attempting to look at human sexual behavior from an evolutionary perspective and speculate on why certain sexual behaviors are common. The behaviors of courtship, mating, the raising of the young, and the division of labor between the sexes are complex in all social animals, including humans, as demonstrated in the elaborate social behaviors surrounding picking a mate and forming a family. It is difficult to draw the line between the biological development of sexuality and the social customs related to the sexual aspects of human life. However, the biological mechanism that determines whether an individual will develop into a female or a male has been well documented.

27.2 A Spectrum of Human Sexuality

Sexuality ranges from strongly heterosexual to strongly homosexual. Human sexual orientation is a complex trait, and evidence suggests that there is no one gene that deter-

FIGURE 27.1 The Study of Sex and Sexuality
Famous psychologist and anthropologist Margaret Mead (1901–1978) authored the best-seller *Coming of Age in Samoa* in 1925. The book was the first to focus on the idea that the individual experience of developmental stages might be shaped by cultural demands and expectations. Therefore, adolescence might be more or less stormy and sexual development more or less problematic in different cultures.

mines where a person falls on the sexuality spectrum. It is most likely a combination of various genes acting together and interacting with environmental factors.

In some organisms, the reproductive structures of both sexes are formed as a normal part of their reproductive strategy, these organisms are classified as *hermaphrodites.* In humans, incidences of partial development of the genitalia (sex organs) of both sexes in one individual may be more frequent than most people realize. These people are referred to as *pseudo-hermaphrodites,* because they do not have complete sets of male and female organs. Sometimes, this abnormal development occurs because the hormone levels are out of balance at critical times in the development of the embryo. This hormonal imbalance may be related to an abnormal number of sex-determining chromosomes, or it may be the result of abnormal functioning of the endocrine glands.

Some pseudo-hermaphrodites are primarily female, with a slightly enlarged clitoris, whereas some are primarily male, with an underdeveloped penis and a normal vagina and labia. When children with abnormal combinations of sex organs are born they are usually assessed by a physician in consultation with the parents to determine which sexual structures should be retained or surgically reconstructed. The physician might also decide that hormone therapy might be a more successful treatment. These decisions are not easily made, because they involve children who have not fully developed their sexual nature. An increasingly vocal group advocates that children who are diagnosed with this condition not be surgically "corrected" as infants, recommending that, if the parents can cope with the unusual genitalia, they allow the child to grow older without surgery. They believe that people should choose for themselves once they are more mature. However, few long-term studies have examined the stability of reconstructive surgery in children or the extent of the social adjustment required.

More and more frequently, we are becoming aware of individuals whose physical gender does not match their psychological gender. Gender is the sexual identity based on an person's anatomy. These people are sometimes labeled *gender-confused*. A male with normal external male genitals may "feel" like a female. The same situation may occur with structurally female individuals. In private, such people may dress as the other sex, behaving in social situations appropriately for their sex. Others completely change their public and private behavior to reflect their inner desire to function as the other sex. A male may dress as a female, work in a traditionally female occupation, and make social contacts as a female. Tremendous psychological and emotional pressures develop from this condition. Frequently, many of these individuals choose to undergo gender reassignment surgery, a sex-change operation. Because the most frequent behavior of gender-confused individuals is dressing as a member of the opposite sex, they are called *cross-dressers* or *transvestites*. However, psychological other than gender confusion can also lead to cross-dressing, so not all transvestites are gender-confused. The way people dress is not biological but, rather, is related to cultural customs. This surgery and the follow-up hormonal treatment can cost tens of thousands of dollars and take several years.

Individuals with an occasional or a consistent feeling of being another gender than that indicated by their genitalia experience *gender dysphoria*. People who feel gender dysphoria (also known as trans-gendered or transsexuals) most intensely may be candidates for gender reassignment surgery, including hormonal treatments, counseling, and even training on how to act as a member of the other gender. Their goal is to interact with society as a member of the other gender without being detected as having been the other gender at one time, called "passing" as a man or woman.

There have been some studies on brain structures and trans-gendered individuals. A small region of the hypothalamus is proportionally larger in normal males than in normal females. In one study, it was found that, in trans-gendered persons, the proportions of this region of the brain are switched: Males in a female body have a larger region than

expected for a normal female; females in a male body have a larger region than expected for a normal male. This study was originally done in rats, in which malelike behavior was induced in females who were exposed to testosterone during their fetal development. Brain structure correlations were first established in the

rat. Follow-ups on trans-gendered humans have shown similar results. Scientists don't know what those brain cells actually do. They may be only involved in activating certain hormones.

Similar studies were done to correlate brain structure with homosexual orientation. However, these studies were laden with controversy. Scientists have difficulty defining *homosexuality* in practical terms, so control groups were hard to establish. Many of the homosexual brain donors in this research were HIV+ and had taken medications such as AZT. The controls did not consider HIV+ status or time spent on the antiviral drugs. This is a significant issue, because HIV can infect and destroy certain brain cells, causing AIDS-associated dementia. AZT and its ability to affect neural cell function were not known at the time of the studies.

The sexual orientation of people seems to be unrelated to whether they are transvestites. Some males who are attracted to males ("gay") may also be gender-confused, whereas others are not. The reverse is also true. Some females who are attracted to females ("lesbians") are gender-confused, whereas others are not. Certain studies suggest that genetic regions on chromosomes 7, 8, and 10 may influence homosexuality. Some regions on chromosome 7 and 8 have been linked with male sexual orientation, regardless of whether the male receives the chromosomes from his mother or father. The regions on chromosome 10 appear to be linked with male sexual orientation only if they were inherited from the mother. Homosexuality is a complex behavioral pattern that appears to be separate and distinct from gender confusion.

27.3 The Chromosomal Determination of Sex

Recall from chapter 10 that, of the 46 human chromosomes, 44 contain genes that are not involved in determining the sex of an individual; they are called autosomes. Two of the 46 chromosomes are involved in determining sex and are called **sex determining chromosomes**. There are two kinds of sex determining chromosomes: the **X chromosome** and the **Y chromosome**. (figure 27.2). The two sex-determining chromosomes, do not carry equivalent amounts of information, nor do they have equal functions.

X chromosomes carry genetic information about the production of a variety of proteins, in addition to their function in determining sex. For example, the X chromosome carries information on blood clotting, color vision, and other characteristics. The Y chromosome, with about 80 genes, however, appears to be primarily concerned with determining male sexual differentiation.

When a human **sperm,** consisting of haploid sex cells produced by sexually mature males, is produced, it carries 22 autosomes and a sex-determining chromosome. Unlike eggs, which always carry an X chromosome, half the sperm cells carry an X chromosome and the other half carries a

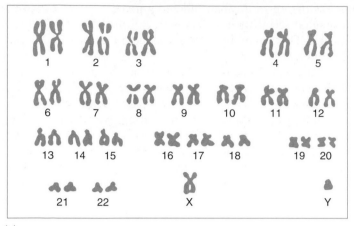

(a)

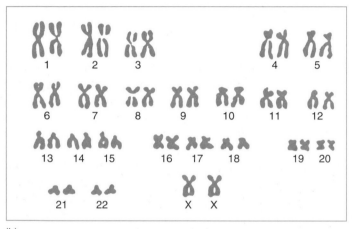

(b)

FIGURE 27.2 Human Male and Female Chromosomes
The chromosomes have been arranged into homologous pairs: *(a)* a male karyotype, with an X and a Y chromosome, and *(b)* a female karyotype, with two X chromosomes.

Y chromosome. If an X-carrying sperm cell fertilizes an X-containing egg cell, the resultant embryo will develop into a female. If a Y-carrying sperm cell fertilizes the egg, a male embryo will develop. It is the presence or absence of the *sex-determining region Y* (SRY) gene located on the short arm of the Y chromosome that determines the sex of the developing individual. The SRY gene produces a chemical, called *testes determining factor* (TDF), which acts as a master switch that triggers the events that converts the embryo into a male. Without this gene, the embryo would become female (Outlooks 27.1).

The early embryo resulting from fertilization and cell division is not recognizable as either male or female. Sexual development begins when certain cells become specialized, forming the embryonic *gonads* known as the female ovaries and the male testes. This specialization of embryonic cells is called differentiation. If the SRY gene is present and functioning, the embryonic gonads begin to differentiate into testes 5 to 7 weeks after conception (fertilization).

Evidence that the Y chromosome and its SRY gene control male development comes from many kinds of studies, including research on individuals who have an abnormal number of chromosomes. An abnormal meiotic division that results in sex cells with too many or too few chromosomes is a form of *nondisjunction* (see chapter 9). If nondisjunction affects the X and Y chromosomes, a gamete might be produced that has only 22 chromosomes and lacks a sex-determining chromosome. On the other hand, it might have 24, with 2 sex-determining chromosomes. If a cell with too few or too many sex chromosomes is fertilized, sexual development is usually affected. If a normal egg cell is fertilized by a sperm cell with no sex chromosome, the offspring will have only 1 X chromosome. These people, always women, are designated as XO. They develop a collection of characteristics known as *Turner's syndrome* (figure 27.3).

About 1 in 2,000 girls born has Turner's syndrome. A female with this condition is short for her age and fails to mature sexually, resulting in sterility. In addition, she may

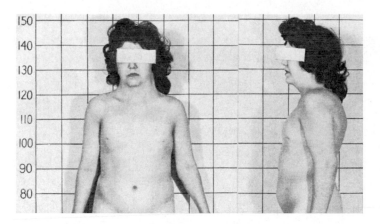

FIGURE 27.3 Turner's Syndrome

Individuals with Turner's syndrome have 45 chromosomes. They have only 1 of the sex chromosomes, and it is an X chromosome. Individuals with this condition are female, have delayed growth, and fail to develop sexually. This woman is less than 150 cm (5 ft) tall and lacks typical secondary sexual development for her age. She also has the "webbed neck" that is common among individuals with Turner's syndrome.

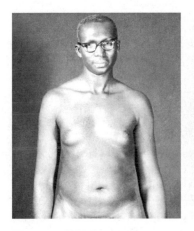

FIGURE 27.4 Klinefelter's Syndrome

Individuals with two X chromosomes and a Y chromosome are male, are sterile, and often show some degree of breast development and female body form. They are typically tall. The two photos show an individual with Klinefelter's syndrome before and after receiving testosterone hormone therapy.

have a thickened neck (termed webbing), hearing impairment, and some abnormalities in her cardiovascular system. When the condition is diagnosed, some of the physical conditions can be modified with treatment. Treatment involves the use of growth-stimulating hormone to increase her growth rate and female sex hormones to stimulate sexual develop-

ment, although sterility is not corrected. monosomy and trisomy, p. 185

An individual who has XXY chromosomes is basically male (figure 27.4). This genetic condition is termed *Klinefelter's syndrome.* It is one of the most common examples of abnormal chromosome number in humans. This condition is present in about 1 in 500 to 1,000 men. Most of these men lead healthy, normal lives and it is impossible to tell them apart from normal males. However, those with Klinefelter's syndrome may be sterile and show breast enlargement, incomplete masculine body form, lack of facial hair, and some minor learning problems. These traits vary greatly in degree, and many men are diagnosed only after they undergo testing to determine why they are infertile. Treatments include breast-reduction surgery and testosterone therapy.

27.4 Male and Female Fetal Development

The development of embryonic gonads begins very early during fetal growth. First, a group of cells begins to differentiate into primitive gonads at about week 5. By week 5 to 7, if a Y chromosome is present, the gene product (testes determining factor) from the chromosome begins the differentiation of these embryonic gonads into testes. They will develop into ovaries beginning about week 12 if 2 X chromosomes are present (and the Y chromosome is absent).

As soon as the gonad has differentiated into an embryonic testis at about week 8, it begins to produce testosterone. The presence of testosterone results in the differentiation of male sexual anatomy, and the absence of testosterone results in the differentiation into female sexual anatomy in the developing embryo.

At about the seventh month of pregnancy (gestation), in normal males each testis moves from a position in the abdominal cavity to an external sac, called the scrotum. They pass through an opening called the **inguinal canal** (figure 27.5). This canal closes off but continues to be a weak area in the abdominal wall, and it may rupture later in life. This can happen when strain (e.g., from improperly lifting heavy objects) causes a portion of the intestine to push through the inguinal canal into the scrotum, a condition known as an inguinal hernia.

Occasionally, the testes do not descend, resulting in a condition known as **cryptorchidism** (*crypt* = hidden; *orchidos* = testes). Sometimes, the descent occurs during puberty; if not, there is a 25 to 50 times increased risk for testicular cancer. Because of this increased risk, surgery can be done to allow the undescended testes to be moved into the scrotum. Sterility will result if the testes remain in the abdomen. This happens because normal sperm cell development cannot occur in a very warm environment. The temperature in the abdomen is higher than the temperature in the scrotum. Normally the temperature of the testes is very carefully

Evidence Through a Microscope—Barr Bodies

In 1949, M. L. Barr discovered a darkly staining body generally present in female cells but not present in male cells. The tightly coiled structure in the cells of female mammals is called a **Barr body,** after its discoverer. Now we know that it is an X chromosome that is largely nonfunctional. Therefore, female cells have only one dose of X-chromosome genetic information that is functional. The single X chromosome of the male functions as expected, and the Y

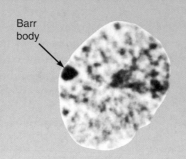

Barr body

chromosome primarily directs male sex-determining activities. The fact that only one of the X chromosomes is functional explains why identical twin females can have quite different patterns of gene expression, depending on which of their X chromosomes becomes the Barr body. Using this knowledge is also helpful in crime scene investigations. One can identify the blood at a crime scene as male or female based on the presence or absence of Barr bodies.

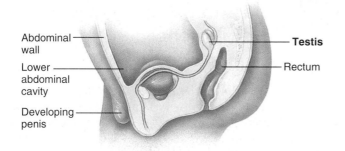

Abdominal wall
Lower abdominal cavity
Developing penis
Testis
Rectum

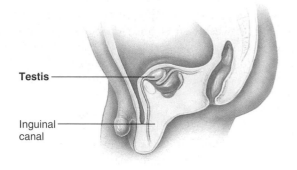

Testis
Inguinal canal

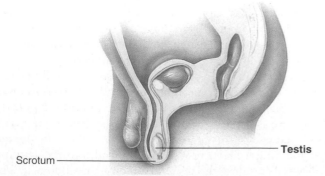

Scrotum
Testis

FIGURE 27.5 The Descent of Testes
During the development of the male fetus, the testes originate in the abdomen and eventually descend through the inguinal canal to the scrotum.

regulated by muscles that control their distance from the body. Physicians have even diagnosed cases of male infertility as being caused by tight-fitting pants that hold the testes so close to the body that the temperature increase interferes with normal sperm development. Recent evidence has also suggested that teenage boys and young men working with computers in the laptop position for extended periods may also be at risk for lowered sperm counts.

27.5 The Sexual Maturation of Young Adults

Following birth, sexuality plays only a small part in physical development for several years. However culture and environment shape the responses that the individual will come to recognize as normal behavior for their geades. **Puberty** is the developmental period when the body changes to the adult form and becomes able to reproduce. During puberty, normally between 11 and 14 years of age, an increased production of sex hormones causes major changes as the individual reaches sexual maturity. Generally, females reach puberty 6 months to 2 years before males.

The Maturation of Females

Girls typically begin to produce sex hormones from several glands, marking the onset of puberty. The hypothalamus, pituitary gland, ovaries, and adrenal glands begin to produce sex hormones at 9 to 12 years of age (figure 27.6).

The hypothalamus is an endocrine gland that controls the functioning of many glands throughout the body, including the **pituitary gland.** At puberty, the hypothalamus begins to release **gonadotropin-releasing hormone (GnRH),** which stimulates the pituitary to release **luteinizing hormone (LH)**

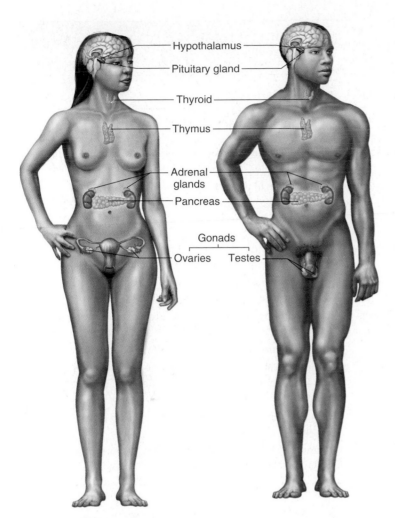

FIGURE 27.6 Hormones and Sexual Function
These are the primary hormone-producing glands that affect sexual development and behavior.

under the control of a number of hormones produced by the pituitary gland and ovaries. The ovaries are stimulated to release their hormones by the pituitary gland, which is in turn influenced by the ovarian hormones, both follicle-stimulating hormone (FSH) and luteinizing hormone (LH). FSH causes the maturation and development of the ovaries, and LH is important in causing **ovulation,** the periodic release of sex cells from the surface of the ovary (figure 27.7), and converting the ruptured follicle into a structure known as the **corpus luteum.** In addition to their role in bringing about sexual development LH and FSH are also involved in regulating the menstrual cycle.

The corpus luteum produces the hormone **progesterone,** which is important in maintaining the lining of the uterus. Changes in progesterone levels result in a periodic buildup and shedding of the lining of the uterus known as the menstrual cycle (table 27.1). Initially, as girls go through puberty, menstruation and ovulation may be irregular; however, in most women hormone production eventually becomes regulated, so that ovulation and menstruation take place on a monthly basis, although normal cycles vary from 21 to 45 days.

As girls progress through puberty, curiosity about the changing female body form and new feelings leads to self-investigation. Studies have shown that sexual activity, such as manipulation of the clitoris, which causes pleasurable sensations, is performed by a large percentage of young women. This stimulation is termed **masturbation,** and it is a normal part of sexual development. **Orgasm** is the pleasurable climax during sexual activity. This event has many other names, including "cuming," climaxing, and having a "big O." During orgasm, all the muscles that were tightened during sexual arousal relax, resulting in a frenzy of enjoyable sensations. During orgasm, the heart rate increases, breathing quickens, and blood pressure increase; muscles throughout the body spasm, but mainly those in the vagina, uterus, anus, and pelvic floor.

and **follicle-stimulating hormone (FSH).** Increased levels of FSH stimulate the development of **follicles,** saclike structures that produce eggs in the ovaries. The increased luteinizing hormone stimulates the ovary to produce larger quantities of estrogens. The increasing supply of estrogens is responsible for the many changes in sexual development. These changes include breast growth, changes in the walls of the uterus and vagina, increased blood supply to the clitoris, and changes in the pelvic bone structure.

Estrogens also stimulates the female adrenal gland to produce **androgens,** male sex hormones. The androgens are responsible for the production of pubic hair, and they seem to have an influence on the female sex drive. The features that are not primarily involved in sexual reproduction but are characteristic of a sex are called **secondary sexual characteristics.** In women, the distribution of body hair, the patterns of fat deposits, and a higher voice are secondary sexual characteristics.

A major development during this time is the establishment of the **menstrual cycle,** the periodic growth and shedding of the lining of the uterus, the *endometrium.* These changes are

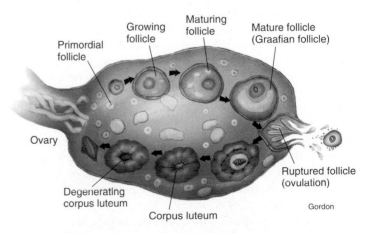

FIGURE 27.7 Ovulation
In the ovary, the egg begins development inside the follicle. Each month, one of these follicles develops and releases its product. This release through the wall of the ovary is ovulation. The ruptured follicle is converted into a corpus luteum which produces progesterone.

TABLE 27.1

Human Reproductive Hormones

Hormone	Production Site	Target Organ	Function
Prolactin, lactogenic, or luteotropic hormone	Anterior pituitary	Breast, ovaries	Stimulates milk production; also helps maintain normal ovarian cycle
Follicle-stimulating hormone (FSH)	Anterior pituitary	Ovaries, testes	Stimulates ovary and testis development; stimulates egg production in females and sperm production in males
Luteinizing hormone (LH) or interstitial cell-stimulating hormone (ICSH)	Anterior pituitary	Ovaries, testes	Stimulates ovulation in females and sex-hormone (estrogen and testosterone) production in both males and females
Estrogen	Ovaries	Entire body	Stimulates the development of the female reproductive tract and secondary sexual characteristics
Testosterone	Testes	Entire body	Stimulates the development of the male reproductive tract and secondary sexual characteristics
Progesterone	Corpus luteum of ovaries	Uterus, breasts	Causes uterine thickening and maturation; maintains pregnancy
Oxytocin	Posterior pituitary	Uterus, breasts	Causes the uterus to contract and breasts to release milk
Androgens	Testes, adrenal glands	Entire body	Stimulates the development of the male reproductive tract and secondary sexual characteristics in males and females
Gonadotropin-releasing hormone (GnRH)	Hypothalamus	Anterior pituitary	Stimulates the release of FSH and LH from the anterior pituitary
Human chorionic gonadotropin	Placenta	Corpus luteum	Maintains the corpus luteum, so that it continues to secrete progesterone and maintains pregnancy

The Maturation of Males

Males typically reach puberty about 2 years later (ages 11–14) than females, but puberty in males also begins with a change in hormone levels. At puberty, the hypothalamus releases increased amounts of gonadotropin-releasing hormone (GnRH), resulting in increased levels of follicle-stimulating hormone (FSH) and luteinizing hormone (LH). These are the same changes that occur in female development. Luteinizing hormone is often called interstitial cell-stimulating hormone (ICSH) in males. LH stimulates the testes to produce testosterone, the primary sex hormone in males. The testosterone produced by the embryonic testes causes the differentiation of internal and external genital anatomy in the male embryo. At puberty, the increase in testosterone is responsible for the development of male secondary sexual characteristics and is important in the maturation and production of sperm.

The major changes during puberty include growth of the testes and scrotum, pubic hair development, and increased penis size. Secondary sex characteristics also begin to become apparent; facial hair, underarm hair, and chest hair are some of the most obvious. The male voice changes as the larynx (voice box) begins to change shape. Body contours also change, and a growth spurt increases height. In addition, the proportion of the body that is muscle increases and the proportion of body fat decreases. At this time, a boy's body begins to take on the characteristic adult male shape, with broader shoulders and heavier muscles.

In addition to these external changes, increased testosterone causes the production of **semen,** also known as *seminal fluid,* a mixture of sperm and secretions from three *accessory glands*—the *seminal vesicles, prostate,* and *bulbourethral glands.* They produce secretions that nourish and activate the sperm. They also lubricate the tract and act as the vehicle to help propel the sperm.

Seminal vesicles secrete an alkaline fluid, containing fructose, and hormones. Its alkaline nature helps neutralize the acidic environment in the female reproductive tract, improving the sperm's chances of reaching the egg. The fructose provides energy for the sperm. Seminal vesicle secretions make up about 60% of the seminal fluid. The prostate gland produces a thin, milky fluid, which makes up about 25% of semen. It contains sperm-activating enzymes. Bulbourethral gland secretions make up the remaining 15%; they, too, are alkaline.

FSH stimulates the production of sperm cells. The release of sperm cells and seminal fluid begins during puberty and is termed **ejaculation.** This release is generally accompanied by the pleasurable sensations of orgasm. During sleep, males frequently have erections, sometimes resulting in ejaculation of seminal fluid. This is termed a *wet dream*. It is normal and related to the amount of seminal fluid produced. Wet dreams occur less often in men who engage in frequent sexual intercourse or masturbation. Masturbation is a common and normal activity as a boy goes through puberty. Studies of adult sexual behavior have shown that nearly all men masturbate at some time during their lives.

27.6 Spermatogenesis

The process of producing gametes is called **gametogenesis** (gamete formation), and it includes meiosis (figure 27.8). **Spermatogenesis** is gametogenesis that takes place in the testes of males, producing sperm. The two, olive-shaped testes are composed of many small, sperm-producing tubes, called **seminiferous tubules,** and collecting ducts that store sperm. These are held together by a thin covering membrane. The seminiferous tubules join together and eventually become the epididymis, a long, narrow, twisted tube in which sperm cells are stored and mature before ejaculation (figure 27.9).

Leading from the epididymis is the vas deferens, or sperm duct; this empties into the urethra, which conducts the sperm out of the body through the penis (figure 27.10). Before puberty, the seminiferous tubules are packed solid with diploid cells. These cells, which are found just inside the tubule wall, undergo *mitosis*, thus providing a continuous supply of cells. Beginning about age 11, some of these cells specialize and begin the process of *meiosis*, whereas others continue to divide by mitosis. Once spermatogenesis begins, the seminiferous tubules become hollow and can transport the mature sperm.

Spermatogenesis consists of several steps. It begins when some of the cells in the walls of the seminiferous tubules differentiate and enlarge. These diploid cells undergo the first meiotic division, which produces two haploid cells. These cells go through the second meiotic division, resulting in four haploid cells called spermatids. Spermatids then lose much of their cytoplasm, develop long tails, and mature into sperm (figure 27.11). Sperm have only a small amount of food reserves. Therefore, once they are released and become active swimmers, they live no more than 72 hours. However, if the sperm are placed in a protective solution, the temperature can be lowered drastically to −196°C. Under these conditions, the sperm become deactivated, freeze, and can survive for years outside the testes. This has led to the development of sperm banks. The artificial insemination (placing stored sperm into the reproductive tract of a female) of cattle, horses, and other domesticated animals with sperm from sperm banks is a common breeding practice. This technique is also used in humans who experience infertility.

Spermatogenesis takes place continuously throughout a male's reproductive life, although the number of sperm produced decreases as a man ages. Sperm counts can be taken and used to determine the probability of successful fertiliza-

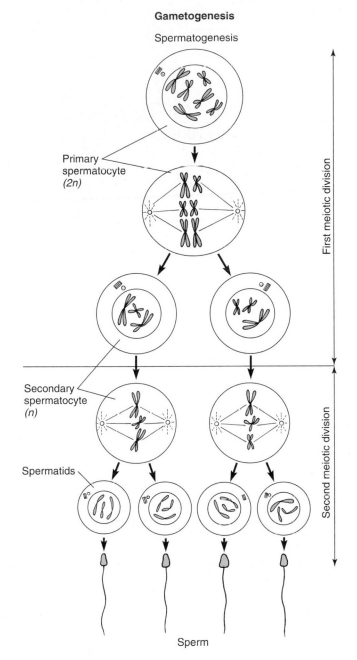

Gametogenesis

Spermatogenesis

Primary spermatocyte (2n)

First meiotic division

Secondary spermatocyte (n)

Second meiotic division

Spermatids

Sperm

FIGURE 27.8 **Spermatogenesis**
This diagram illustrates the process of spermatogenesis in human males. Not all of the 46 chromosomes are shown. Carefully follow the chromosomes as they segregate, recalling the details of the process of meiosis.

tion. A healthy male releases about 150 million sperm per milliliter with each ejaculation. For reasons not totally understood, a man must be able to release at least 100 million sperm per milliliter to be fertile. Many men with sperm counts under 50 million per milliliter are infertile; those with sperm counts below 20 million per milliliter are clinically infertile. These vast numbers of sperm are necessary because each contains enzymes in its cap that are able to digest

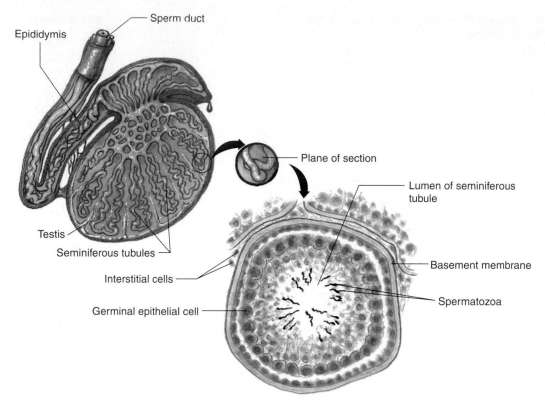

FIGURE 27.9 Sperm Production

The testis consists of many tiny tubes, called seminiferous tubules. The walls of the tubes consist of cells that continually divide, producing large numbers of sperm. The sperm leave the seminiferous tubules and enter the epididymis, where they are stored prior to ejaculation through the sperm duct.

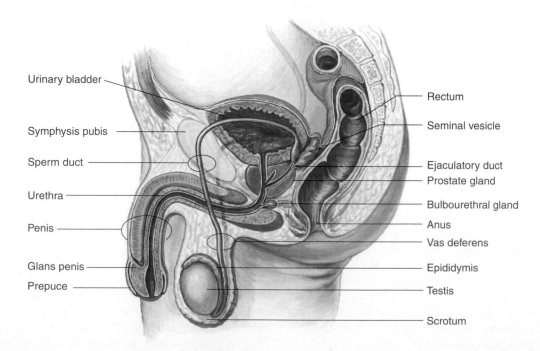

FIGURE 27.10 The Human Male Reproductive System

The male reproductive system consists of two testes, which produce sperm; ducts, which carry the sperm; and various glands. Muscular contractions propel the sperm through the vas deferens past the seminal vesicles, prostate gland, and bulbourethral gland, where most of the liquid of the semen is added. The semen passes through the urethra of the penis to the outside of the body.

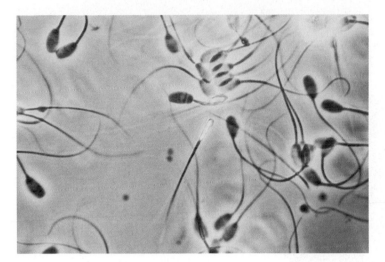

FIGURE 27.11 Human Sperm Cells

Sperm cells have a head region, with an enzyme-containing cap and the DNA. It also has a neck region, with ATP-generating mitochondria, and a tail flagellum, which propels the sperm.

through the mucus and protein found in the female reproductive tract. Millions of sperm contribute in this way to the process of fertilization, but only one fertilizes the egg.

27.7 Oogenesis

Oogenesis is the production of egg cells. This process starts before a girl is born, during prenatal development of the ovaries. This occurs when diploid cells in each ovary cease dividing by mitosis and enlarge in preparation to divide by meiosis. These are the potential egg cells that will become the eggs. All of these cells form in the embryonic ovaries before a female is born. At birth, they number approximately 2 million, but that number has been reduced by cell death to between 300,000 and 400,000 cells by puberty. These cells stop their development in an early stage of meiosis and remain just under the surface of the ovary inside a follicle.

At puberty and on a regular monthly basis thereafter, the sex hormones stimulate one of these cells to continue meiosis. However, in telophase I, the two cells that form receive unequal portions of cytoplasm, a kind of lopsided division (figure 27.12). The smaller of the two cells is called a **polar body,** and the larger haploid cell is commonly refered to as an *egg* or *ovum,* although technically it is not. Just prior to ovulation, the follicle of the soon-to-be-released egg grows and moves near the surface of the ovary. When this maturation is complete, ovulation occurs, when the follicle ruptures and the egg is released. It is swept into the **oviduct** (Fallopian tube) by ciliated cells and travels toward the uterus (figure 27.13). Because of the action of luteinizing hormone, the follicle from which the egg was ovulated develops into a glandlike structure, the corpus luteum. This gland produces hormones

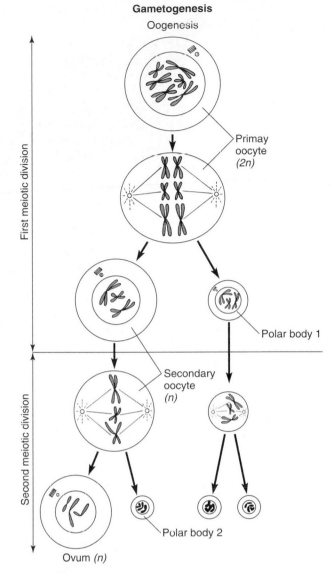

Gametogenesis
Oogenesis

First meiotic division

Primay oocyte *(2n)*

Polar body 1

Secondary oocyte *(n)*

Second meiotic division

Polar body 2

Ovum *(n)*

FIGURE 27.12 Oogenesis

This diagram illustrates the process of oogenesis in human females. Not all of the 46 chromosomes are shown. Carefully follow the chromosomes as they segregate, recalling the details of the process of meiosis.

(progesterone and estrogen), which prevent the release of other eggs. If the egg is fertilized, it completes meiosis with the sperm DNA inside and the haploid egg and sperm nuclei unite to form the zygote. If the egg is not fertilized, it passes through the vagina to the outside of the body during menstruation.

One of the characteristics to note here is the relative age of males' and females' sex cells. In males, sperm production continues throughout the life span. Sperm do not remain in the tubes of the male reproductive system for very long; They are either released shortly after they form or they die and are harmlessly absorbed. In females, meiosis begins before birth, but the oogenesis process is not completed, and an egg cell is not released for many years. An egg released when a woman

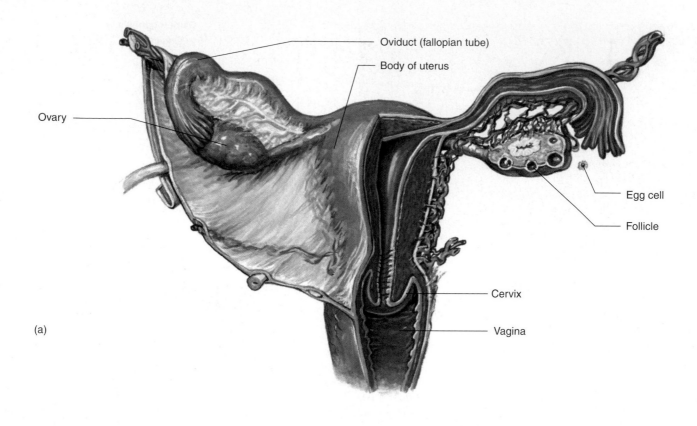

Oviduct (fallopian tube)

Body of uterus

Ovary

Egg cell

Follicle

Cervix

Vagina

(a)

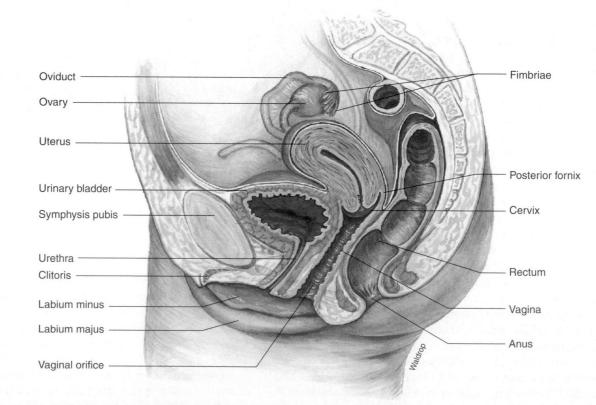

Oviduct

Ovary

Uterus

Urinary bladder

Symphysis pubis

Urethra

Clitoris

Labium minus

Labium majus

Vaginal orifice

Fimbriae

Posterior fornix

Cervix

Rectum

Vagina

Anus

Waldrop

(b)

FIGURE 27.13 The Human Female Reproductive System

(a) After ovulation, the cell travels down the oviduct to the uterus. If it is not fertilized, it is shed when the uterine lining is lost during menstruation. *(b)* The human female reproductive system, side view.

A Link Between Menstruation and Endometriosis?

Most of the endometrium, along with its blood and capillaries, is released from the uterus when a woman has her period. It passes out through the cervix and the vagina. However, medical exams often reveal that a small amount of blood and other tissues from the uterine lining has moved up through the oviducts. If these tissues come in contact with the ovary and top of the oviduct, they may cause inflammation and the tissues may grow together. Some physicians believe that this is a likely cause of a condition called *endometriosis*, in which the endometrial tissue enters the body cavity

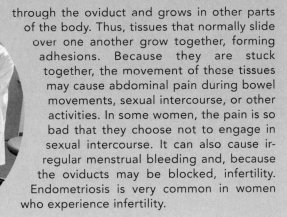

through the oviduct and grows in other parts of the body. Thus, tissues that normally slide over one another grow together, forming adhesions. Because they are stuck together, the movement of those tissues may cause abdominal pain during bowel movements, sexual intercourse, or other activities. In some women, the pain is so bad that they choose not to engage in sexual intercourse. It can also cause irregular menstrual bleeding and, because the oviducts may be blocked, infertility. Endometriosis is very common in women who experience infertility.

---

is 37 years old began meiosis 37 years earlier. During that time, the cell was exposed to many influences, a number of which may have damaged the DNA or interfered with the meiotic process. The increased risk for abnormal births in older mothers may be related to the age of their eggs. Such alterations are less likely to occur in males, because new gametes are being produced continuously. mutations, p. 158

Hormones control the cycle of changes in breast tissue, the ovaries, and the uterus. In particular, estrogen and progesterone stimulate milk production by the breasts and cause the uterine lining to become thicker and filled with blood vessels prior to ovulation. This ensures that, if fertilization occurs, the resultant embryo will be able to attach itself to the uterine wall and receive nourishment. If the cell is not fertilized, the lining of the endometrium is shed. This is known as *menstruation, menstrual flow,* the *menses,* or a *period.* Once the endometruim has been shed, it begins to build up again (figure 27.14) (Outlooks 27.2).

The activities of the ovulatory cycle and the menstrual cycle are coordinated. During the first part of the menstrual cycle, increased amounts of FSH cause the follicle to increase in size. Simultaneously, the follicle secretes increased amounts of estrogen, which cause the uterine lining to thicken. When ovulation occurs, the remains of the follicle are converted into a corpus luteum by LH. The corpus luteum begins to secrete progesterone and the nature of the uterine lining changes as a result of the development of many additional blood vessels. This is organized so that, if an embryo arrives in the uterus shortly after ovulation, the uterine lining is prepared to accept it. If pregnancy does not occur, the corpus luteum degenerates, resulting in a reduction in the amount of progesterone needed to maintain the uterine lining, and it is shed. At the same time that hormones are regulating ovulation and the

menstrual cycle, changes are taking place in the breasts. The same hormones that prepare the uterus to receive the embryo also prepare the breasts to produce milk. These changes in the breasts, however, are relatively minor unless pregnancy occurs.

27.8 The Hormonal Control of Fertility

An understanding of how various hormones influence the menstrual cycle, ovulation, milk production, and sexual behavior has led to the medical use of certain hormones. Some women are unable to have children because they do not release eggs from their ovaries, or they release them at the wrong time. Physicians can regulate the release of eggs using certain hormones, commonly called *fertility drugs.* These hormones can be used to stimulate the release of eggs for capture and use in what is called *in vitro* fertilization, also called *test-tube* or *IV* fertilization. The hormones can also be used to stimulate the release of eggs in women with irregular cycles to increase the probability of natural conception; that is, *in vivo* fertilization (*in-life* fertilization) (How Science Works 27.1).

The use of these techniques often results in multiple embryos being implanted in the uterus. This is likely to occur because the drugs may cause too many eggs to be released at one time. In the case of in vitro fertilization, because there is a high rate of failure and the process is expensive, typically several early-stage embryos are inserted into the uterus to increase the likelihood that one will implant. If several are successful, multiple embryos implant. The implantation of multiple embryos makes it difficult for

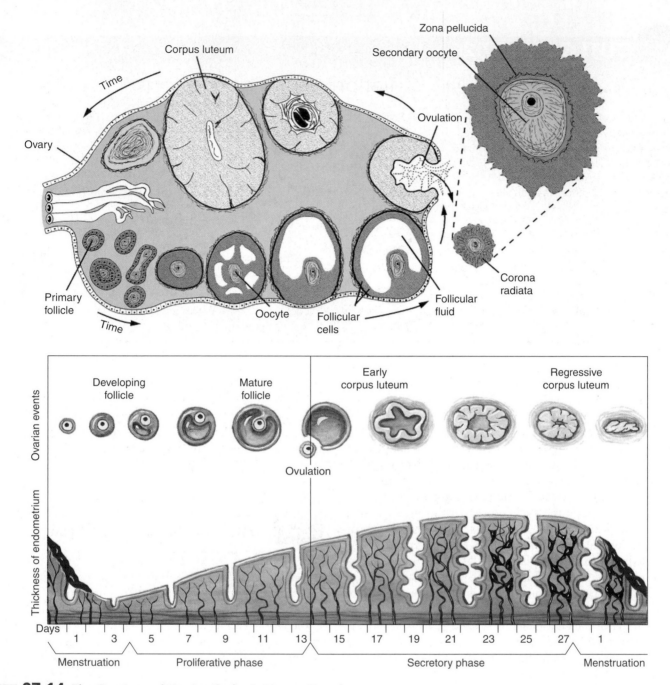

FIGURE 27.14 **The Ovarian and Uterine Cycles in Human Females**

The release of an egg (ovulation) is timed to coincide with the thickening of the lining of the uterus. The uterine cycle in humans involves the preparation of the uterine wall to receive the embryo if possible fertilization occurs. Knowing how these two cycles compare, it is possible to determine when pregnancy is most likely to occur.

one embryo to develop properly and be carried through the entire nine-month pregnancy. When scientists understand the action of hormones better, they may be able to control the effects of fertility drugs and eliminate the problem of multiple implantations.

Another medical use of hormones is the control of conception with birth-control pills—oral contraceptives (table 27.2). Birth-control pills have the opposite effect of fertility drugs.

They raise the levels of estrogen and progesterone, which slows the production of FSH and LH, preventing the release of eggs from the ovaries. They can also help relieve premenstrual syndrome (PMS), which causes irritability, emotional instability, depression, headache, and other aches and pains.

The hormonal control of fertility is not as easy to achieve in men, because there is no comparable cycle of gamete release. However a new, reversible male conception

Assisted Reproductive Technology

The Centers for Disease Control and Prevention defines assisted reproductive technology (ART) as "all fertility treatments in which both eggs and sperm are manipulated. In general, ART involves surgically removing eggs from a woman's ovaries, combining them with sperm in the laboratory, and returning them to the woman's body or donating them to another woman. It does NOT include procedures in which only sperm are manipulated (i.e., artificial insemination or intrauterine insemination) or procedures in which a woman takes drugs only to stimulate egg production, without the intention of having eggs retrieved." There are three types of ART: in vitro fertilization, (IVF) gamete intra-fallopian transfer, (GIFT), and zygote intra-fallopian transfer ZIFT.

In vitro fertilization is a method that uses hormones to stimulate egg production, removing the egg or eggs from the ovary, and fertilizing it with donated sperm. The fertilized egg is incubated to stimulate cell division in a laboratory dish and then placed in the uterus.

GIFT relies on the same hormonal treatment as IVF to stimulate ovulation. With ultrasound, the physician observes the transfer of unfertilized eggs and sperm into the woman's oviduct through small incisions in her abdomen. Once fertilized, the zygote moves down through the oviduct into the uterus and implants. GIFT is only an option for women with open fallopian tubes.

In the ZIFT procedure, a woman's mature eggs are collected and fertilized in the laboratory. Then, a scope is used to see where things are and other instruments are used to guide.

TABLE 27.2

Common Causes of Infertility

Female

Endometriosis	Endometrial tissue grows outside the uterus, forming adhesions between tissues.
Ovulation problems	Usually, a hormonal condition prevents the release of a mature egg from an ovary.
Poor egg quality	Eggs become damaged or develop chromosomal abnormalities, usually age-related.
Polycystic ovary syndrome	Ovaries contain many small cysts, causing hormonal imbalances, and therefore do not ovulate regularly.
Female tube blockages	Causes can include pelvic inflammatory disease (PID), sexually transmitted diseases, and previous sterilization surgery.

Male

Male tube blockages	Varicose veins in the testicles are the most common cause. Sexually transmitted diseases also cause blockage.
Sperm problems	Sperm problems include low or no sperm count, poor sperm motility (the ability to move), and abnormally shaped sperm.
Sperm allergy	This very rare condition is caused by an immune reaction to sperm, resulting in their being killed by antibodies produced by the woman.

control method for males has been developed. It relies on using an implant containing testosterone, and injections of a progestin, a hormone used in female contraceptive pills. The combination of the two hormones temporarily turns off the normal brain signals that stimulate sperm production. The extra injection of testosterone are given to maintain sex drive.

27.9 Fertilization, Pregnancy, and Birth

In most women, an egg is released from the ovary about 14 days before the next menstrual period. The menstrual cycle is usually said to begin on the first day of menstruation. Therefore, if a woman has a regular 28-day cycle, the cell is released approximately on day 14 (review figure 27.14). If a woman normally has a regular 21-day menstrual cycle, ovulation occurs about day 7 in the cycle. If a woman has a regular 40-day cycle, ovulation occurs about day 26. Some women, however, have very irregular menstrual cycles, and it is difficult to determine just when the egg will be released to become available for fertilization. Once the cell is released, it is swept into the oviduct and moved toward the uterus. If sperm are present, they swarm around the egg as it passes down the oviduct, but only one sperm penetrates the outer layer to fertilize it and cause it to complete meiosis. The other sperm contribute enzymes, which digest away the protein and mucous barrier between the egg and the successful sperm (How Science Works 27.2).

During the second meiotic division, the second polar body is pinched off and the larger of the two cells, the true ovum, is formed. Because chromosomes from the sperm are already inside, they simply intermingle with those of the

HOW SCIENCE WORKS 27.2

History of Pregnancy Testing

The answer to the question "Am I pregnant?" once demanded a combination of guesswork, "old wives' tales" and time. An ancient Egyptian papyrus described a test in which a woman who might be pregnant urinated on wheat and barley seeds over the course of several days: "If the barley grows, it means a male child. If the wheat grows, it means a female child. If both do not grow, she will not bear at all."

In the Middle Ages, so-called piss prophets said they could tell if a woman was pregnant by the color of her urine. Urine was described as a "clear pale lemon color leaning toward off-white, having a cloud on its surface" if she was pregnant. Some mixed wine with urine and watched for changes. In fact, alcohol reacts with certain proteins in urine, so they may have had a little help in guessing.

In 1927, a test was developed that required the injection of a women's urine into an immature female rat or mouse. If the woman was not pregnant, the animal showed no reaction. If she was pregnant, the animal went into heat (displayed behaviors associated with a desire to mate) despite its immaturity. This test implied that, during pregnancy, there was an increased production of certain hormones, and that they were excreted in the urine. In the 1930s, a similar test involved injecting urine into rabbits, frogs, toads, and rats. If pregnancy hormones were in the urine, they induced ovulation in the animals.

However, in 1976, advertisements proclaimed "a private little revolution," the first home pregnancy tests. The FDA granted approval to four tests: Early Pregnancy Test, Predictor,

ACU-TEST, and Answer. All the tests identified changes in hormone levels in the urine of pregnant women.

The next generation of home pregnancy tests arrived in 2003, when the FDA approved Clearblue Easy's digital pregnancy test. Instead of showing a thin blue line, the indicator screen now displayed the words "pregnant" or "not pregnant."

---

ovum, forming a diploid zygote. As the zygote continues to travel down the oviduct, it begins to divide by mitosis into smaller and smaller cells, without an increase in the mass of cells (figure 27.15). This division process is called *cleavage*.

Eventually, a solid ball of cells is produced, known as the morula stage of embryological development. The morula continues down the oviduct and continues to divide by mitosis. The result is called a *blastocyst*. The blastocyst becomes embedded or implanted in the uterine lining when it reaches the uterus. stages of mitosis, p. 169

The blastocyst stage of the embryo eventually develops a tube, which becomes the gut. The formation of the primitive digestive tract is just one of a series of changes that result in an embryo that is recognizable as a miniature human being.

Most of the time during its development, the embryo is enclosed in a water-filled membrane, the amnion, which protects it from blows and keeps it moist. Two other membranes, the chorion and allantois, fuse with the lining of the uterus to form the **placenta** (figure 27.16). A fourth sac, the yolk sac, is well developed in reptiles, fish, amphibians, and birds. The yolk sac in these animals contains a large amount of food used by the developing embryo. Although a yolk sac is present in mammals, it is small and does not contain yolk. The

embryo's nutritional needs are met through the placenta. The placenta also produces the hormone chorionic gonadotropin, which stimulates the corpus luteum to continue producing progesterone and thus prevents menstruation and ovulation during pregnancy.

As the embryo's cells divide and grow, some of them become differentiated into nerve cells, bone cells, blood cells, or other specialized cells. In order to divide, grow, and differentiate, cells must receive nourishment. This is provided by the mother through the placenta, in which both fetal and maternal blood vessels are abundant, allowing for the exchange of substances between the mother and embryo. The materials diffusing across the placenta include oxygen, carbon dioxide, nutrients, and a variety of waste products. The materials entering the embryo travel through blood vessels in the umbilical cord. The diet and behavior of the mother are extremely important. Any molecules consumed by the mother can affect the embryo. Cocaine; alcohol; heroin; the chemicals in cigarette smoke; and medications, such as Phenytoin (or Dilantin) and Accutane (acne treatment), can all cross the placenta and affect the development of the embryo. Infections by such microbes as rubella virus (German measles), Varicella (chicken pox), and the protozoan *Toxoplasma* can also result in fetal

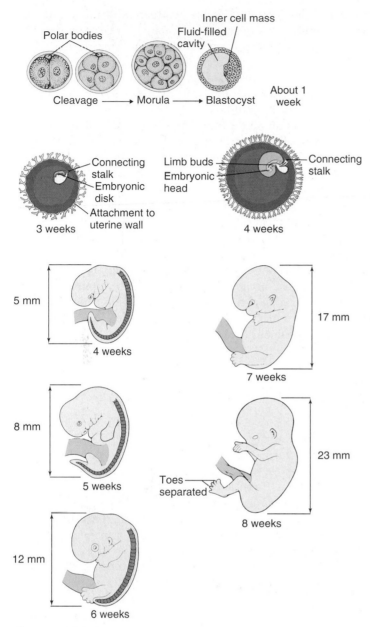

FIGURE 27.15

Human Embryonic Development
Between fertilization and birth, many changes take place in the embryo. These are some of the changes that take place during the first 8 weeks.

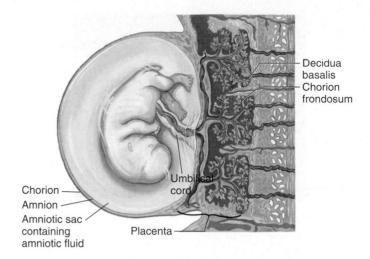

FIGURE 27.16 Placental Structure
The embryonic blood vessels that supply the developing embryo with nutrients and remove the metabolic wastes are separate from the blood vessels of the mother. Because of this separation, the placenta can selectively filter many types of incoming materials and microorganisms.

abnormalities or even death. The growth of the embryo results in the development of major parts of the body by the tenth week of pregnancy. After this time, the embryo continues to increase in size, and the structure of the body is refined.

Twins

In the United States, women giving birth have a 1 in 40 chance of delivering twins and a 1 in 650 chance of triplets or other multiple births. Twins happen in two ways. In the case of identical twins (approximately one-third of all twins), the embryo splits during cleavage into two separate groups of cells. Each group develops into an independent embryo. Because they come from the same single fertilized ovum, they have the same genes and are the same sex.

Fraternal twins result from the fertilization of two separate eggs by two different sperm. Therefore, they resemble each other no more than do regular brothers and sisters. They do not contain the same genetic information and may be different sexes.

Fraternal Twins

Identical Twins

Birth

The process of giving birth is also known as parturition, or birthing. At the end of about 9 months, hormonal changes in the mother's body stimulate contractions of the uterine muscles during a period prior to birth called *labor*. These contractions are stimulated by the hormone oxytocin, which is released from the posterior pituitary. The contractions normally move the baby headfirst and face up through the vagina, or birth canal. One of the first effects of these contractions may be the bursting of the amnion ("bag of water") surrounding the baby. Following

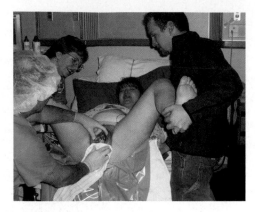

FIGURE 27.17 Childbirth

What many people refer to as "natural" childbirth is really called a vaginal delivery.

this, the uterine contractions become stronger, and shortly thereafter the baby is born. In some cases, the baby becomes turned in the uterus before labor. If this occurs, the feet or buttocks appear first. Such a birth is called a *breech birth*. This can be a dangerous situation, because the baby's source of oxygen may be cut off as the placenta begins to separate from the mother's body before the baby's head emerges (figure 27.17). If for any reason the baby does not begin to breathe on its own, it will not receive enough oxygen to prevent the death of nerve cells; thus, brain damage or death can result.

Occasionally, a baby cannot be born vaginally because of the position of the baby in the uterus. Other factors include the location of the placenta on the uterine wall, the size of the birth canal. A common procedure to resolve this problem is the surgical removal of the baby through the mother's abdomen. This procedure is known as a cesarean, or C-section. The procedure was apparently named after Roman Emperor Julius Caesar, who was said to have been the first child to be delivered by this method. During 2003, more than a quarter of the babies born in the United States, about 1.13 million were delivered by C-section, an all-time record. This rate reflects the fact that many women who are prone to problem pregnancies are having children rather than forgoing pregnancy. In addition, changes in surgical techniques have made the procedure much safer. Finally, many physicians who are faced with liability issues related to problem pregnancies may encourage cesarean section rather than vaginal birth.

Following the birth of the baby, the placenta, also called the *afterbirth,* is expelled. Once born, the baby begins to function on its own. The umbilical cord collapses and the baby's lungs, kidneys, and digestive system must now support all bodily needs. Birth is quite a shock, but the baby's loud protests fill the lungs with air and stimulate breathing.

Over the next few weeks, the mother's body returns to normal, with one major exception. The breasts, which underwent changes during pregnancy, are ready to produce milk to feed the baby. Following birth, prolactin, a hormone from the pituitary gland, stimulates the production of milk, and oxytocin stimulates its release. If the baby is breast-fed, the stimulus of the baby's sucking prolongs the time during which milk is produced. This response involves both the nervous and endocrine systems. The sucking stimulates nerves in the nipples and breasts, resulting in the release of prolactin and oxytocin from the pituitary.

Recent studies have found that breast-fed babies are 20% less likely to die between the ages of 1 and 12 months than are those who are not breast-fed. In addition, the longer babies are breast-fed, the lower is their risk for early death. The American Academy of Pediatrics has recommended exclusive breast-feeding for the first 6 months of life. In some cultures, breast-feeding continues for 2 to 3 years, and the continued production of milk often delays the reestablishment of the normal cycles of ovulation and menstruation. Many people believe that a woman cannot become pregnant while she is nursing a baby. However, because there is so much variation among women, relying on this as a natural conception-control method is not a good choice. Many women have been surprised to find themselves pregnant again a few months after delivery.

27.10 Contraception

Throughout history, people have tried various methods of conception control. In ancient times, conception control was encouraged during times of food shortage or when tribes were on the move from one area to another in search of a new home. Writings from as early as 1500 B.C. indicate that the Egyptians used a form of tampon medicated with the ground powder of a shrub to prevent fertilization. This may sound primitive, but we use the same basic principle today to destroy sperm in the vagina.

Chemical Methods

Contraceptive jellies and foams make the vaginal environ-

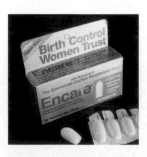

ment more acidic, which diminishes the sperm's chances of survival. Spermicidal (sperm-killing) foam or jelly is placed in the vagina before intercourse. When the sperm make contact with the acidic environment, they stop swimming and soon die. Aerosol foams are an effective method of conception control, but interfering with the hormonal regulation of ovulation is more effective.

Hormonal Control Methods

The first successful method of hormonal control was "the pill," or "birth control pill." However, the way the birth-control pill works is by preventing ovulation, not preventing

a birth. The quantity and balance of hormones (estrogen and progesterone) in the pill fools the ovaries into functioning as if the woman were already pregnant. Therefore, ovulation does not occur so conception is highly unlikely. The emergency contraceptive pill (ECP), or "morning-after pill," uses a high dose of the same hormones found in oral contraceptives, which prevents the woman from becoming pregnant in the first place. ECPs prevent conception in five ways: (1) by delaying or inhibiting ovulation; (2) by inhibiting the transport of the egg or sperm; (3) by preventing fertilization; (4) by inhibiting the implantation of a fertilized egg; or (5) by stimulating an autoimmune response, which kills the sperm. This method is effective if used within 72 hours after unprotected intercourse. One more form of oral contraceptive, the progesteron-only pill (POP or mini pill), is to be taken within the same three hour time period each day, and taken continuously without a break.

Another method of conception control also involves hormones. The hormones are contained within small rods, or

capsules, known as implants, which are placed under a woman's skin. These rods, when properly implanted, slowly release hormones and prevent the maturation and release of eggs from the follicle. The major advantage of the implant is its convenience. Once the implant has been inserted, the woman can forget about contraceptive protection for several years. If she wants to become pregnant, the implants are removed and her normal menstrual and ovulation cycles return over a period of weeks.

The hormone-infused contraceptive patch, an adhesive square that can be attached to the abdomen, buttocks, upper arm, or upper torso. The patch works by slowly releasing a combination of estrogen and progestin through the skin, preventing ovulation. The hormones also cause the cervical mucus to thicken, creating a barrier to prevent sperm from entering the uterus. When used correctly, it is about 99% effective. It does not protect against reproductive tract infections, STDs.

The vaginal ring also uses hormones. The ring is inserted into the vagina, where it releases a continuous low dose of estrogen and progestin for 21 days. At the end of the 21 days, the ring is removed for 7 days to allow the menstrual period to occur. A new ring is then inserted monthly.

The Timing Method

Killing sperm and preventing ovulation are not the only methods of preventing conception. Any method that prevents sperm from reaching the egg prevents conception. One method is to avoid intercourse during the times of the month when an egg may be present. This is known as the *rhythm method* of conception control. Although at first glance it appears to be the simplest and least expensive, determining when an egg is likely to be present can be very difficult. A woman with a regular 28-day menstrual cycle typically ovulates about 14 days before the onset of the next menstrual flow. To avoid pregnancy, couples need to abstain from intercourse a few days before and after this date. However, if a woman has an irregular menstrual cycle, there may be only a few days each month for intercourse without the possibility of pregnancy. In addition to calculating safe days based on the length of the menstrual cycle, a woman can better estimate the time of ovulation by keeping a record of changes in her body temperature and vaginal pH. Both changes are tied to the menstrual cycle and can therefore help a woman predict ovulation. In particular, at about the time of ovulation, a woman has a slight rise in body temperature—less than 1°C. Thus, she should use an extremely sensitive thermometer. A digital-readout thermometer on the market spells out the word *yes* or *no*.

Barrier Methods

Other methods of conception control that prevent sperm from reaching the egg include the diaphragm, vaginal cap, sponge, and condom. A diaphragm is a specially fitted membranous shield that is inserted into  the vagina before intercourse and positioned so that it covers the cervix, which contains the opening of the uterus. Because of anatomical differences among females, diaphragms must be fitted by a physician. The diaphragm's effectiveness is increased if spermicidal foam or jelly is also used. A vaginal cap functions in a similar way. A contraceptive sponge, as the name indicates, is a small amount of absorbent material soaked in a spermicide. The sponge is placed within the vagina, where it chemically and physically prevents sperm cells from reaching the egg.

The male condom is probably the most popular contraceptive device. It is a thin sheath placed over the erect penis before intercourse. In addition to preventing sperm from reaching the egg, this physical barrier also helps prevent the transmission of the microbes that cause sexually transmitted diseases (STDs), such as syphilis, gonorrhea, and AIDS, from being passed from one person to another during sexual intercourse (Outlooks 27.3). The most desirable condoms are

Sexually Transmitted Diseases

Sexually transmitted diseases (STDs) were formerly called *venereal diseases (VDs)*. The term *venereal* is derived from the name of the Roman goddess for love, Venus. Although these kinds of illnesses are most frequently transmitted by sexual activity, many can also be spread by other methods of direct contact, such as hypodermic needles, blood transfusions, and blood-contaminated materials. Currently, the Centers for Disease Control and Prevention (CDC) in Atlanta, Georgia, recognizes over 25 diseases as being sexually transmitted (table 27.A). The United States has the highest rate of sexually transmitted disease among the industrially developed countries—65 million people (nearly one-fifth of the population) in the United States have incurable sexually transmitted diseases. The CDC estimates that there are 15 million new cases of sexually transmitted diseases each year, nearly 4 million among teenagers. Table 27.B lists the most common STDs and estimates of the number of new cases each year. The portions of the public that are most at risk are teenagers, minorities, and women. Some of the most important STDs are described here because of their high incidence in the population and the inability to bring some of them under control. For example, there is no known cure for HIV, responsible for AIDS. Recently, the CDC has discovered that fewer U.S. teenagers and adults have genital herpes. However, there has also been an increase in the number of cases of syphilis, especially among gay and bisexual men. In addition, there has been a sharp rise in the number of gonorrhea cases in the United States caused by a form of the bacterium *Neisseria gonorrhoeae*. It appears that this bacterium has become resistant to penicillin by producing an enzyme that actually destroys the antibiotic. However, most infectious agents can be controlled if diagnosis occurs early and treatment programs are carefully followed by the patient.

The spread of STDs during sexual intercourse is significantly diminished by the use of condoms. Other types of sexual contact (e.g., hand, oral, anal) and transmission from a mother to the fetus during pregnancy help maintain some of these diseases in the population. Therefore, public health organizations, such as the U.S. Public Health Service, the CDC, and state and local public health agencies, regularly keep an eye on the number of cases of STDs. All of these agencies are involved in attempts to raise the general public health to a higher level. Their investigations have resulted in the successful control of many diseases and the identification of special problems, such as those associated with the STDs. Because the United States has an incidence rate of STDs that is 50 to 100 times higher than that of other industrially developed countries, there is still much that needs to be done.

The high-risk behaviors associated with contracting STDs include sex with multiple partners and the failure to use condoms. Whereas some STDs are simply inconvenient or annoying, others severely affect health and can result in death. As one health official stated, "We should be knowledgeable enough about our own sexuality and the STDs to answer the question, Is what I'm about to do worth dying for?"

TABLE 27.A

Sexually Transmitted Diseases

Disease	Agent
Genital herpes	Virus
Gonorrhea	Bacterium
Syphilis	Bacterium
Acquired immunodeficiency syndrome (AIDS)	Virus
Candidiasis	Yeast
Chancroid	Bacterium
Genital warts	Virus
Gardnerella vaginalis	Bacterium
Genital *Chlamydia* infection	Bacterium
Genital cytomegalovirus infection	Virus
Genital *Mycoplasma* infection	Bacterium
Group B *Streptococcus* infection	Bacterium
Nongonococcal urethritis	Bacterium
Pelvic inflammatory disease (PID)	Bacterium
Molluscum contagiosum	Virus
Crabs	Body lice
Scabies	Mite
Trichomoniasis	Protozoan
Hepatitis B	Virus
Gay bowel syndrome	Variety of agents

TABLE 27.B

Cases of Sexually Transmitted Diseases United States, 2003

Cases Reported by State Health Departments

Syphills	34,270
Chlamydia	877,478
Gonorrhea	335,104
Chancroid	54
HIV/AIDS	43,171

Cases Reported by Initial Visits to Physicians' Offices

Genital herpes	203,000
Genital warts	264,000
Hepatitis B	73,000
Vaginal trichomoniasis	179,000

made of a thin layer of latex that does not reduce the sensitivity of the penis. Latex condoms have also been determined to be the most effective in preventing transmission of HIV. The condom is most effective if it is prelubricated with a spermicidal material, such as nonoxynol-9. This lubricant also has the advantage of providing some protection against the spread of HIV.

Condoms for women are also available. One, called the Femidom, is a polyurethane sheath, which, once inserted,

lines the contours of the woman's vagina. It has an inner ring, which sits over the cervix, and an outer ring, which lies flat against the labia. Research shows that this device protects against STDs and is as effective a contraceptive as the condom used by men.

The IUD

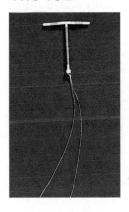

The intrauterine device (IUD) is not a physical barrier that prevents the gametes from uniting; in fact, the way in which this device works is not completely known. It might interfere with the implantation of the embryo. The IUD must be fitted and inserted into the uterus by a physician, who can also remove it if pregnancy is desired. IUDs are used successfully in many countries. The IUD can also be used for "emergency contraception"—in cases of unprotected sex, forced sex, or the failure of a conception-control method (e.g., a broken condom). The IUD must be inserted within 7 days of unprotected sex.

Surgical Methods

Two contraceptive methods that require surgery are tubal ligation and vasectomy (figure 27.18). Tubal ligation involves the cutting and tying off of the oviducts; it can be done on an outpatient basis, in most cases. An alternative to sterilization by tubal ligation involves the insertion of small, flexible devices, called Micro-inserts, into each oviduct (Fallopian tube). Once inserted, tissue grows into the inserts, blocking the tubes. Ovulation continues as usual, but the sperm and egg cannot unite.

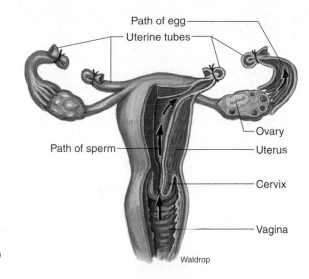

(a)

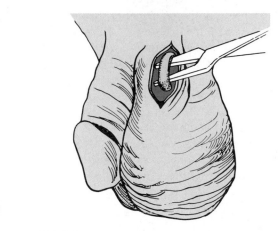

(b)

FIGURE 27.18 Tubal Ligation and Vasectomy
Two very effective contraceptive methods require surgery. *(a)* Tubal ligation involves severing the oviducts and suturing or sealing the cut ends. This prevents the sperm cell and the egg from meeting. This procedure is generally considered ambulatory surgery; at most, it requires a short hospitalization period. *(b)* Vasectomy requires minor surgery, usually in a clinic under local anesthesia. Following the procedure, minor discomfort may be experienced for several days. The severing and sealing of the vas deferens prevents the release of sperm cells from the body during ejaculation.

A vasectomy can be performed in a physician's office. A small opening is made above the scrotum, and the spermatic cord (vas deferens) is cut and tied. This prevents sperm from moving through the ducts to the outside. Because most of the semen is produced by the seminal vesicles, prostate gland, and bulbourethral glands, a vasectomy does not interfere with normal ejaculation. The sperm that are still being produced die and are reabsorbed in the testes.

Neither tubal ligation nor vasectomy interferes with normal sex drives. However, these medical procedures are generally not reversible and should not be considered by those who may want to have children at a future date.

27.11 Abortion

Another medical procedure often associated with birth control is abortion, which has been used throughout history. Abortion is the death and removal of a developing embryo through various medical procedures. Abortion is not a method of conception control; rather, it prevents the normal development of the embryo. Abortion is a highly charged subject: Some people believe that abortion should be prohibited by law in all cases; others think that abortion should be allowed in certain situations, such as in pregnancies that endanger the mother's life or in pregnancies that are the result of rape or incest. Still others think that abortion should be available to any woman under any circumstances. Regardless of the moral and ethical issues that surround abortion, it is still a common method of terminating unwanted pregnancies.

The three most common techniques are scraping the inside of the uterus with instruments, (called *dilation and curettage,* or *D and C);* injecting a saline solution into the uterine cavity; and using a suction device to remove the embryo from the uterus. In the future, abortion may be accomplished by a medication prescribed by a physician. One drug, RU-486, is currently used in about 15% of the elective abortions in France, and it has received approval for use in the United States. The medication is administered orally under the direction of a physician, and several days later, a hormone is administered. This usually results in the onset of contractions, which expel the fetus. A follow-up examination of the woman is made after several weeks to ensure that there are no serious side effects.

27.12 Changes in Sexual Function with Age

Although there is a great deal of variation, at about age 50, a woman's hormonal balance begins to change because of changes in the ovaries production of hormones. At this time, the menstrual cycle becomes less regular and ovulation is often unpredictable. Over several years, the changes in hormone levels cause many women to experience mood swings and physical symptoms, including cramps and hot flashes. **Menopause** is the period when a woman's body becomes nonreproductive, because reproductive hormones stop being produced. This causes the ovaries to stop producing eggs, and menstruation ends. Occasionally, the symptoms associated with menopause becomes to severe that they interfere with normal life and the enjoyment of sexual activity. A physician might recommend *hormone replacement therapy (HRT)* to augment the natural production of the hormones estrogen and progestin. Normally, the sexual enjoyment of a healthy woman continues during menopause and for many years thereafter.

Human males do not experience a relatively abrupt change in their reproductive or sexual lives. However, as men age, their production of sperm declines and they may experience a variety of problems related to their sexuality. The word *impotence* is used to describe problems that interfere with sexual intercourse and reproduction. These may include a lack of sexual desire, problems with ejaculation or orgasm, and erectile dysfunction (ED). Erectile dysfunction is the recurring inability to get or keep an erection firm enough for sexual intercourse. In older men, this is usually the result of injury, disease, or the side effects of medication. Damage to nerves, arteries, smooth muscles, and other tissues associated with the penis is the most common cause of ED. Diseases linked with ED include diabetes, kidney disease, chronic alcoholism, multiple sclerosis, atherosclerosis, vascular disease, and neurologic disease. Blood pressure drugs, antihistamines, antidepressants, tranquilizers, appetite suppressants, and certain ulcer drugs have been associated with ED. Other possible causes are smoking, which reduces blood flow in veins and arteries, and lowered amounts of testosterone. ED is frequently treated with psychotherapy, behavior modification, oral or locally injected drugs, vacuum devices, and surgically implanted devices. ED it is not an inevitable part of aging. Rather, sexual desires tend to wane slowly as men age. They produce fewer sperm cells and less seminal fluid. Nevertheless, healthy individuals can experience a satisfying sex life during aging. Recent evidence also indicates that men also experience hormonal and emotional changes similar to those seen as women go through menopause.

Human sexual behavior is quite variable. The same is true of older persons. The range of responses to sexual partners continues, but generally in a diminished form. People who were very active sexually when young continue to be active, but less so as they reach middle age. Those who were less active tend to decrease their sexual activity also. It is reasonable to state that one's sexuality continues from before birth until death.

Summary

The human sex drive is a powerful motivator for many activities in our lives. Although it provides for reproduction and the improvement of the gene pool, it also has a nonbiological, sociocultural dimension. Sexuality begins before birth, as sexual anatomy is determined by the sex-determining chromosome complement received at fertilization. Females receive 2 X chromosomes. Only 1 of these is functional; the other remains tightly coiled as a Barr body. A male receives 1 X and 1 Y sex-determining chromosome. It is the presence of the Y chromosome that causes male development and the absence of a Y chromosome that allows female development.

At puberty, hormones influence the development of secondary sex characteristics and the functioning of testes and ovaries. As the ovaries and testes begin to produce gametes, fertilization becomes possible. Sexual reproduction involves the production of gametes by meiosis in the ovaries and testes. The production and release of these gametes is controlled by the interaction of hormones. In males, each cell that undergoes spermatogenesis results in four sperm; in females, each cell that undergoes oogenesis results in one egg and two polar bodies. Humans have specialized structures for the support of the developing embryo, and many factors influence its development in the uterus. Sexual reproduction depends on proper hormone balance, proper meiotic division, fertilization, placenta formation, proper diet of the mother, and birth. Hormones regulate ovulation and menstruation and can be used to encourage or discourage ovulation. Fertility drugs and birth-control pills, for example, involve hormonal control. In addition to the pill, a number of contraceptive methods have been developed, including the diaphragm, condom, IUD, spermicidal jellies and foams, contraceptive implants, the sponge, tubal ligation, and vasectomy.

Hormones continue to direct our sexuality throughout our lives. Even after menopause, when fertilization and pregnancy are no longer possible for a female, normal sexual activity can continue in both men and women.

Key Terms

Use the interactive flash cards on the **Concepts in Biology,** *12/e website to help you learn the meaning of these terms.*

androgens 601
Barr body 599
corpus luteum 601
cryptorchidism 600
ejaculation 602

follicles 601
follicle-stimulating hormone (FSH) 601
gametogenesis 603

gonadotropin-releasing hormone (GnRH) 600
inguinal canal 600
luteinizing hormone (LH) 600
masturbation 601
menopause 616
menstrual cycle 601
oogenesis 605
orgasm 601
oviduct 605
ovulation 601
pituitary gland 600
placenta 610

polar body 605
progesterone 601
puberty 600
secondary sexual characteristics 601
semen 602
seminiferous tubules 603
sex-determining chromosomes 597
sexuality 596
sperm 597
spermatogenesis 603
X chromosome 597
Y chromosome 597

Basic Review

1. All the factors that contribute to one's female or male nature are referred to as _____.

2. Humans have a total of _____ chromosomes, _____ of which are considered sex-determining.
   a. 44, 2
   b. 23, 2
   c. 48, 4
   d. 46, 2

3. In humans, the _____ acts as a master switch, triggering the events that convert an embryo into a male.
   a. SRY gene
   b. TDF gene
   c. Y-gene
   d. Turner's gene

4. Men with Turner's syndrome may be sterile and show breast enlargement, an incomplete masculine body form, a lack of facial hair, and some minor learning problems. (T/F)

5. Teenage boys and young men working with computers in the laptop position for extended periods may be at risk for lowered sperm counts. (T/F)

6. Gonadotropin-releasing hormone (GnRH) is produced by the
   a. hypothalamus.
   b. pituitary.
   c. testes.
   d. corpus luteum.

7. Semen is a mixture of sperm and secretions from glands called

   a. testes.

   b. ovaries.

   c. accessory glands.

   d. adrenal glands.

8. The unequal division that occurs during oogenesis results in the formation of smaller cells that do not develop into true eggs, called _____.

9. If a woman normally has a regular 21-day menstrual cycle, ovulation occurs about day _____ in the cycle.

10. If an embryo splits during cleavage into two separate groups of cells, _____ twins may develop.

Answers

1. sexuality 2. d 3. a 4. F 5. T 6. a 7. c 8. polar bodies 9. 7 10. identical

Concept Review

*Answers to the Concept Review questions can be found on the **Concepts in Biology,** 12/e website..*

27.1 Sexuality from Various Points of View

1. Define the term *sexuality*.
2. Name one advantage of a sexually reproducing species.

27.2 A Spectrum of Human Sexuality

3. What does the term *hermaphrodite* mean?
4. Describe gender reassignment.

27.3 The Chromosomal Determination of Sex

5. Describe the processes that cause about 50% of babies to be born male and 50% to be born female.
6. Name two genetic abnormalities associated with nondisjunction of chromosomes.

27.4 Male and Female Fetal Development

7. List the events that occur as an embryo matures.
8. What is cryptorchidism?

27.5 The Sexual Maturation of Young Adults

9. What are the effects of the secretions of the pituitary, the gonads, and the adrenal glands at puberty in females?
10. What role does testosterone play in male sexual maturation?

27.6 Spermatogenesis

11. What structures are associated with the human male reproductive system? What are their functions?

12. Explain the path taken by sperm from their source until they exit the body.

27.7 Oogenesis

13. What structures are associated with the human female reproductive system? What are their functions?
14. What are the differences between oogenesis and spermatogenesis in humans?

27.8 The Hormonal Control of Fertility

15. What changes occur in ovulation and menstruation during pregnancy?
16. How are ovulation and menstruation related to each other?

27.9 Fertilization, Pregnancy, and Birth

17. What are the functions of the placenta?
18. What causes the genetic differences between fraternal twins?

27.10 Contraception

19. Describe hormonal methods of conception control.
20. Describe barrier methods of conception control.

27.11 Abortion

21. What is dilation and curettage?
22. How does RU-486 work?

27.12 Changes in Sexual Function with Age

23. What is menopause?
24. How does sexual function change with age?

Thinking Critically

*For guidelines to answer this question, visit the **Concepts in Biology,** 12/e website.*

The practice of medicine has increasingly moved in the direction of using medications containing steroids to control disease or regulate function, as in birth-control pills. To meet these demands, the pharmaceutical industry produces enormous amounts of these drugs. However, what happens to these drugs once they enter the body? Although some of the drug is destroyed in controlling disease or regulating a function, such as ovulation, a certain amount is not and is excreted. There is some concern about the amount of medical steroids entering the environment in this way. What effects might they have on the public, who unintentionally ingest these as environmental contaminants? Consider the topics of sexual reproduction, the regulation of hormonal cycles, and fetal development, and explain (1) how you would determine "acceptable levels" of such contaminants, (2) what might happen if these levels were exceeded, and (3) what steps might be taken to control such environmental contamination.

Appendix 1

The Metric System

Standard Metric Units

		Abbreviation
Standard unit of mass	gram	g
Standard unit of length	meter	m
Standard unit of volume	liter	L

Common Prefixes   **Examples**

kilo	1,000	A kilogram is 1,000 grams.
centi	0.01	A centimeter is 0.01 meter.
milli	0.001	A milliliter is 0.001 liter.
micro (μ)	one-millionth	A micrometer is 0.000001 (one-millionth) of a meter.
nano (n)	one-billionth	A nanogram is 10^{-9} (one-billionth) of a gram.
pico (p)	one-trillionth	A picogram is 10^{-12} (one-trillionth) of a gram.

Units of Length

Unit	Abbreviation	Equivalent
meter	m	approximately 39 in.
centimeter	cm	10^{-2} m
millimeter	mm	10^{-3} m
micrometer	μm	10^{-6} m
nanometer	nm	10^{-9} m
angstrom	Å	10^{-10} m

Length Conversions

1 in. = 2.5 cm	1 mm = 0.039 in
1 ft = 30 cm	1 cm = 0.39 in
1 yd = 0.9 m	1 m = 39 in
1 mi = 1.6 km	1 m = 1.094 yd
	1 km = 0.6 mi

To Convert	Multiply By	To Obtain
inches	2.54	centimeters
feet	30	centimeters
centimeters	0.39	inches
millimeters	0.039	inches

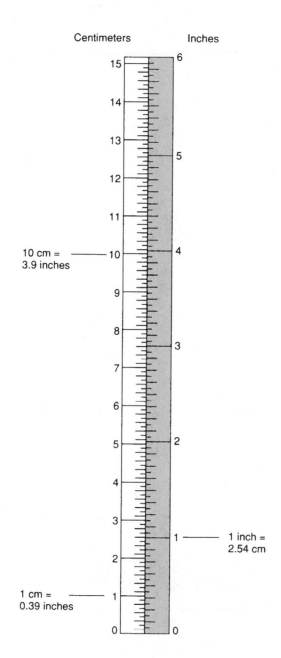

Centimeters   Inches

10 cm = 3.9 inches

1 cm = 0.39 inches

1 inch = 2.54 cm

Units of Volume

Unit	Abbreviation	Equivalent
liter	L	approximately 1.06 qt
milliliter	mL	10^{-3} L (1 mL = 1 cm^3 = 1 cc)
microliter	μL	10^{-6} L

Volume Conversions

1 tsp	=	5 mL	1 mL	=	0.03 fl oz
1 tbsp	=	15 mL	1 L	=	2.1 pt
1 fl oz	=	30 mL	1 L	=	1.06 qt
1 cup	=	0.24 L	1 L	=	0.26 gal
1 pt	=	0.47 L			
1 qt	=	0.95 L			
1 gal	=	3.8 L			

To Convert	Multiply By	To Obtain
fluid ounces	30	milliliters
quarts	0.95	liters
milliliters	0.03	fluid ounces
liters	1.06	quarts

Units of Weight

Unit	Abbreviation	Equivalent
kilogram	kg	10^3 g (approximately 2.2 lb)
gram	g	approximately 0.035 oz
milligram	mg	10^{-3} g
microgram	μg	10^{-6} g
nanogram	ng	10^{-9} g
picogram	pg	10^{-12} g

Weight Conversions

1 oz = 28.3 g   1 g = 0.035 oz
1 lb = 453.6 g   1 kg = 2.2 lb
1 lb = 0.45 kg

To Convert	Multiply By	To Obtain
ounces	28.3	grams
pounds	453.6	grams
pounds	0.45	kilograms
grams	0.035	ounces
kilograms	2.2	pounds

Apothecary System of Weight and Volume*

Metric Weight		Apothecary Weight	Metric Volume		Apothecary Volume
30 g	=	1 ounce	1,000 mL	=	1 quart
15 g	=	4 drams	500 mL	=	1 pint
10 g	=	2.5 drams	250 mL	=	8 fl ounces
4 g	=	60 grains	90 mL	=	3 fl ounces
		(= 1 dram)	30 mL	=	1 fl ounce
2 g	=	30 grains			
1 g	=	15 grains			

*Used by pharmacists in preparing medications.

Temperature

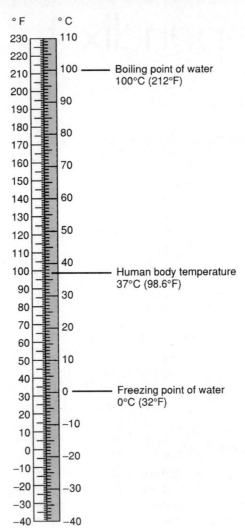

Boiling point of water 100°C (212°F)

Human body temperature 37°C (98.6°F)

Freezing point of water 0°C (32°F)

Temperature Conversions

$$°C = \frac{(°F - 32) \times 5}{9}$$

$$°F = \frac{(°C \times 9)}{5} + 32$$

Some Equivalents

0°C = 32°F
37°C = 98.6°F
100°C = 212°F

Appendix 2

Acronyms

ADH	antidiuretic hormone
AIDS	acquired immunodeficiency syndrome
AMU	atomic mass unit
ATP	adenosine triphosphate
BMI	body mass index
BMR	basal metabolic rate
CAM	crassulacean acid metabolism
DDT	dichlorodiphenyltrichloroethane
DNA	deoxyribonucleic acid
DRI	dietary reference intakes
EGF	epidermal growth factor
ER	endoplasmic reticulum
ETS	electron-transport system
FAS	fetal alcohol system
FAD	flavin adenine dinucleotide
FSH	follicle-stimulating hormone
GnRH	gonadotropin-releasing hormone
GSH	growth-stimulating hormone
HDL	high-density lipoprotein
HIV	human immunodeficiency virus
HLA	human leukocyte antigens
ICSI I	interstitial cell-stimulating hormone
IPCC	Intergovernmental Panel on Climate Change
LDL	low-density lipoprotein
LH	luteinizing hormone
Lp(a)	lipoprotein a
NAD	nicotinamide adenine dinucleotide
NADP	nicotinamide adenine dinucleotide phosphate
pH	hydrogen ion concentration
PCB	polychlorinated biphenyl
PKU	phenylketone urea
RBC	red blood cell
RNA	ribonucleic acid
RuBisCo	ribulose bisphosphate carboxylase
SARS	severe acute respiratory syndrome
SDA	specific dynamic action
STD	sexually transmitted disease
TCA	tricarboxylic acid cycle
TSH	thyroid-stimulating hormone
UNAIDS	United Nations Joint HIV/AIDS Program
USAMRIID	U.S. Army Medical Research Institute of Infectious Diseases
VLDL	very-low-density lipoprotein
WBC	white blood cell

Glossary

A

abiotic factors (a-bi-otik fak′tōrz) Nonliving parts of an organism's environment. **304**

absorption (ab-sorp′shun) The movement of simple molecules from the digestive system to the circulatory system for dispersal throughout the body. **542**

accessory pigments (ak-ses′o-re pig′ments) Photosynthetic pigments other than the chlorophylls that enable an organism to use more colors of the visible light spectrum for photosynthesis; e.g., carotinoids (yellow, red, and orange); phycoerythrins (red); and phycocyanin (blue). **132**

acetyl (ă-sēt′l) The 2-carbon remainder of the carbon skeleton of pyruvic acid that is able to enter the mitochondrion. **116**

acetylcholine (ă-sēt″l-kō′lēn) A neurotransmitter secreted into the synapse by many axons and received by dendrites. **570**

acetylcholinesterase (ă-sēt″l-kō″lĭ-nes′tĕ-rās) An enzyme present in the synapse that destroys acetylcholine. **570**

acid-base reactions (ăs′id bās re-ak′shunz) When the ions of one compound (acid) interact with the ions of another compound (base), forming a salt and water. **37**

acids (ăs′idz) Compounds that release a hydrogen ion in a solution. **38**

acoelomate (a-se′lă-mat) Without a coelom; the internal organs have no spaces between them. **495**

acquired characteristics (ă-kwīrd′ kar″ak-ter-iss′tiks) Characteristics of an organism gained during its lifetime, not determined genetically and therefore not transmitted to the offspring. **260**

actin (ak′tin) A protein found in the thin myofilaments of muscle fibers that binds to myosin. **582**

activation energy (ak′tĭ-va′shun en′ur-je) Energy required to start a reaction. **96**

active site (ak″tiv sīt) The place on the enzyme that causes the substrate to change. **97**

active transport (ak′tiv trans′port) The use of a carrier molecule to move molecules across a plasma membrane in a direction opposite that of the concentration gradient. The carrier requires an input of energy other than the kinetic energy of the molecules. **88**

adaptive radiation (uh-dap′tiv ra-de-a′shun) A specific evolutionary pattern in which there is a rapid increase in the number of kinds of closely related species. **288**

adenine (ad′ĕ-nēn) A double-ring nitrogenous-base molecule in DNA and RNA; the complementary base of thymine or uracil. **144**

adenosine triphosphate (ATP) (uh-den′o-sēn tri-fos′fāt) A molecule formed from the building blocks of adenine, ribose, and phosphates; it functions as the primary energy carrier in the cell. **106**

aerobic cellular respiration (aer-o′-bik sel′yu-lar res″ pĭ-ra′shun) The biochemical pathway that requires oxygen and converts food, such as carbohydrates, to carbon dioxide and water. During this conversion, it releases the chemical-bond energy as ATP molecules. **79**

aerobic exercise (aer-o′-bik ek-ser-sīz) The level of exercise at which the level of exertion allows the heart and lungs to keep up with the oxygen needs of the muscles. **562**

age distribution (āj dis″trĭ-biu′shun) The number of organisms of each age in the population. **360**

alcoholic fermentation (al-ko-hol′ik fur″men-ta′shun) The anaerobic respiration pathway in yeast cells; during this process, pyruvic acid from glycolysis is converted to ethanol and carbon dioxide. **122**

algae (al′je) Protists that have cell walls and chlorophyll and can therefore carry on photosynthesis. **449**

allele (a-lēl) An alternative form of a gene for a particular characteristic (e.g., attached earlobe and free earlobe are alternative alleles for ear shape). **182**

allele frequency (a-lēl′ fre′kwen-sē) A measure of how common a specific allele is, compared with other alleles for the same characteristic. **241**

allergy (al′er-g) An abnormal immune reaction to an antigen. Possibly the most familiar are allergies to foods, pollens, and drugs. **589**

alternation of generations (awl″tur-na′shun uv gen″uh-ra′shunz) The aspect of the life cycle in which there are two distinctly different forms of an organism; each form is involved in the production of the other and only one form is involved in producing gametes; the cycling of a diploid sporophyte generation and a haploid gametophyte generation in plants. **433**

alternative splicing (awl″·tur-na′tiv splis′ing) A process which selects which exons will be retained as part of the mature mRNA that will be used during translation. Alternative splicing allows for the possibility that a single gene can produce more than one type of protein. **156**

altruism (al′tru-ism) Behavior in which an individual animal gives up an advantage or puts itself in danger to aid others. **396**

alveoli (al-ve′o-li″) Tiny sacs found in the lungs; where gas exchange takes place. **527**

amino acid (ah-mēn′o ă′sid) A basic subunit of protein consisting of a short carbon skeleton that contains an amino group, a carboxylic acid group, and one of various side groups. 53

anabolism (ah-nab′o-lizm″) Metabolic pathways that result in the synthesis of new, larger compounds; e.g., protein synthesis. 105

anaerobic cellular respiration (an′uh-ro″bik sel′yu-lar res″pi-ra′shun) A biochemical pathway that does not require oxygen for the production of ATP and does not use O_2 as its ultimate hydrogen ion acceptor. 122

anaerobic exercise (anuh-ro″bik ek′ser-sīz) Bouts of exercise that are so intense that the muscles cannot get oxygen as fast as they need it. 562

analogous structures (a-nal′o-gous struk-churz) Structures that have the same function (ex. the wing of a butterfly and the wing of a bird) but different evolutionary backgrounds. 292

anaphase (an′ă-fāz) The third stage of mitosis, characterized by division of the centromeres and movement of the chromosomes to the poles. 167

androgens (an′dro-jenz) Male sex hormones, produced by the testes, that cause the differentiation of typical internal and external genital male anatomy. 601

angiosperms (an′je-o″spurmz) Plants that produce flowers, seeds, and fruits. 477

anorexia nervosa (an″o-rek′se-ah ner-vo′sah) A nutritional deficiency disease characterized by severe, prolonged weight loss for fear of becoming obese; this eating disorder is thought to stem from sociocultural factors. 557

anther (an′ther) The sex organ in plants that produces the pollen that contains the sperm. 174

antheridium (an″thur-id′e-um) The structure in lower plants that bears sperm. 468

anthropomorphism (an-thro-po-mōr′fizm) The ascribing of human feelings, emotions, or meanings to the behavior of animals. 378

antibiotics (an-te-bi-ot′iks) Drugs that selectively kill or inhibit the growth of a particular cell type. 90

antibody (an′ti-bod-y) A protein made by B-cells in response to a molecule known as the antigen. 587

anticodon (an″te-ko′don) The trio of bases in the tRNA that is involved in base-pairing. 149

antidiuretic hormone (ADH) (an-tĭ-di″u-rĕ′tik hōr′mōn) The hormone produced by the pituitary gland that stimulates the kidney to reabsorb water. 575

antigen (an′ti-jun) A large organic molecule, usually a protein, which is able to stimulate the production of a specific defense response and becomes neutralized or destroyed by that response. 586

aorta (a-or′tah) The large blood vessel that carries blood from the left ventricle to the majority of the body. 523

apoptosis (ap″op-to′sis) Death that has a genetic basis and not the result of injury. 171

archegonium (ar″ke-go′ne-um) The structure in lower plants that bears eggs. 468

arteries (ar′-tĕ-rēz) The blood vessels that carry blood away from the heart. 520

arterioles (ar-ter′e-ōlz) Small arteries, located just before capillaries, that can expand and contract to regulate the flow of blood to parts of the body. 525

asexual reproduction (sek′shoo-al re-pro-duk′shun) a form of duplication that requires only one parent and results in two organisms that are genetically identical to the parent. 164

assimilation (ă-si″mĭ-la′shun) The physiological process that takes place in a living cell as it converts nutrients in food into specific molecules required by the organism. 542

association (ă-sō″se-ā′-shun) An animal learns that a particular outcome is associated with a particular stimulus. 382

aster (as′ter) Microtubules that extend from the centrioles to the plasma membrane of an animal cell. 167

asymmetry (a-sĭm′ĭ-tre) The characteristic of animals with no particular body shape. 494

atomic mass unit (ă-tom′ik mas yu′nit) A unit of measure used to describe the mass of atoms and is equal to 1.67 × 10^{-24} grams, approximately the mass of 1 proton. 25

atomic nucleus (ă-tom′ik nu′kle-us) The central region of an atom. 25

atomic number (ă-tom′ik num′bur) The number of protons in an atom. 25

atomic weight (ă-tom′ik wāt) The weight of an atomic nucleus, expressed in atomic mass units (the sum of the protons and neutrons). 27

atoms (ă′tomz) The fundamental units of matter; the smallest parts of an element that still act like that element. 12

atria (a′trē-uh) Thin-walled sacs of the heart that receive blood from the veins of the body and empty into the ventricles. 522

atrioventricular valves (a″trē-o-ven-trĭk′u-lar valvz) Valves, located between the atria and ventricles of the heart, that prevent the blood from flowing backwards from the ventricles into the atria. 522

attachment site (uh-tatch′munt sīt) A specific point on the surface of the enzyme where it can physically attach itself to the substrate; also called binding site. 97

autoimmune diseases (aw′to- ĭ-myun di-z̄ez′) Disorders that result from the immune system turning against normal chemicals and cells of the body. 590

autosomes (aw′to-sōmz) Chromosomes that typically carry genetic information used by an organism for characteristics other than the primary determination of sex. 209

autotrophs (aw′to-trōfs) Organisms that are able to make their food molecules from inorganic raw materials by using basic energy sources, such as sunlight. 112

auxin (awk′sin) A plant hormone that stimulates cell elongation, cell division, and growth. 485

axon (ak′sahn) A neuronal fiber that carries information away from the nerve cell body. 569

B

bacteria (bak-tĭr′ē-ah) Prokaryotic, unicellular organisms of the Domains Archaea and Eubacteria. 442

Barr body (bar bod′ē) A tightly coiled X chromosome in the cells of female mammals. **599**

basal metabolic rate (BMR) (ba′sal mĕ-ta-bah′lik rāt) The amount of energy required to maintain normal body activity while at rest. **552**

bases (bāsez) Compounds that release hydroxide ions or accept hydrogen ions in a solution. **38**

basilar membrane (ba′sĭ-lar mem′brān) A membrane in the cochlea containing sensory cells that are stimulated by the vibrations caused by sound waves. **581**

behavior (be-hāv′yur) How an organism acts, what it does, and how it does it. **378**

behavioral isolating mechanisms (be-hāv′yu-ral eye′-so-lā′-ting mek′-ă-nizmz) Reproductive isolating mechanisms that prevent interbreeding between species because of differences in behavior. **286**

benign tumor (be-nīn′ too′mor) A cell mass that does not fragment and spread beyond its original area of growth. **172**

benthic (ben′thik) A term used to describe organisms that live in bodies of water, attached to the bottom or to objects in the water. **450**

bilateral symmetry (bi-lat′er-al sĭm′ĭ-tre) The characteristic of animals that are constructed along a plane running from a head to a tail region, so that only a cut along one plane of this axis results in two mirror halves. **494**

bile (bīl) The product of the liver, stored in the gallbladder, that is responsible for the emulsification of fats. **532**

binary fission (bi′ne-re fish′en) A method of asexual cell division used by prokaryotic cells. **164**

binding site (attachment site) (bin′ding sīt) A specific point on the surface of the enzyme where it can physically attach itself to the substrate. **97**

binomial system of nomenclature (bi-no′mi-al sis′tem uv no′men-kla-ture) A naming system that uses two Latin names, genus and specific epithet, for each species of organism. **422**

biochemical isolating mechanisms (bi′o-kem″ i-kal eye′-so-lā′-ting mek′-ă-nizmz) Differences in biochemical activities that prevent mating between individuals of different species. **287**

biochemical pathway (metabolic pathway) (bi′o-kem″ĭ-kal path′wā) A major series of enzyme-controlled reactions linked together. **105**

biochemistry (bi-o-kem′iss-tre) The chemistry of living things, often called biological chemistry. **46**

biogenesis (bi-o-jen′uh-sis) The concept that life originates only from preexisting life. **402**

biological species concept (bi′o-loj′ĭ-cal spe′shez kon′sept) The concept that species are distinguished from one another by their inability to interbreed. **240**

biology (bi-ol′o-je) The science that deals with the study of living things and how living entities interact with things around them. **2**

biomagnification (bi″o-mag′nĭf-fĭ-ka′shun) The accumulation of a compound in increasing concentrations in organisms at successively higher trophic levels. **349**

biomass (bi′o-mas) The dry weight of a collection of designated organisms. **309**

biomes (bi′ōmz) Large, regional communities primarily determined by climate. **334**

bioremediation (bi′o-ri-me′dea′shun) The use of living organism to remove toxic agents from the environment. **230**

biotechnology (bi-o-tek-nol′uh-je) The science of gene manipulation. **216**

biotic factors (bi-ah′tik fak′tōrz) Living parts of an organism's environment. **304**

blood (blud) The fluid medium, consisting of cells and plasma, that assists in the transport of materials and heat. **520**

bloom (bloom) A rapid increase in the number of microorganisms in a body of water. **444**

body mass index (BMI) (bah′-dee măs in′-deks) The weight of a person in kilograms divided by the person's height in meters squared. **533**

Bowman's capsule (bo′manz kap′sl) The saclike structure at the end of a nephron that surrounds the glomerulus. **536**

breathing (bre′thing) The process of pumping air into and out of the lungs. **528**

bronchi (brŏng′ki) Major branches of the trachea that ultimately deliver air to bronchioles in the lungs. **527**

bronchioles (brŏng′ke-ōlz) Small tubes that deliver air to the alveoli in the lung. **527**

budding (bud′ing) A type of asexual reproduction in which the new organism is an outgrowth of the parent. **498**

bulimia (bu-lēm′e-ah) A nutritional deficiency disease characterized by a binge-and-purge cycle of eating; it is thought to stem from psychological disorders. **557**

C

calorie (kal′or-ee) The amount of heat energy necessary to raise the temperature of 1 gram of water 1°C. **29**

Calvin cycle (kal′vin sī′kl) A cyclic sequence of reactions that make up the light-independent reactions stage of photosynthesis. **136**

cambium (kam′be-um) A tissue in higher plants that produces new xylem and phloem. **484**

capillary (cap′il-lar-y) The thinnest blood vessel that exchanges materials between the blood and tissues that surround these vessels. **520**

carbohydrates (kar-bo-hi′drāts) One class of organic molecules composed of carbon, hydrogen, and oxygen in a ratio of 1:2:1; the basic building block of carbohydrates is a simple sugar (monosaccharide). **49**

carbon skeleton (kar′bon skel′uh-ton) The central portion of an organic molecule composed of rings or chains of carbon atoms. **48**

carcinogens (kar-sin′o-jenz) Agents that cause cancer. **171**

cardiovascular system (kar″de-o- vas′kyu-lar) The organ system of all vertebrates including humans that pumps blood around the body and consists of blood, the heart, and vessels. **520**

carnivores (kar′nĭ-vōrz) Animals that eat other animals. **306**

carrier proteins (ka′re-er prō′tēnz) A category of proteins that pick up molecules at one place and transport them to another. **56**

carrying capacity (ka're-ing kuh-pas'i-te) The maximum sustainable population for an area. **368**

catabolism (kah-tab'o-lizm) Metabolic pathways that result in the breakdown of compounds; e.g., glycolysis. **105**

catalyst (cat'uh-list) A chemical that speeds up a reaction but is not used up in the reaction. **96**

cell division (sel di-vi'zhun) The process in which a cell becomes two new cells. **164**

cell plate (sel plāt) A plant-cell structure that begins to form in the center of the cell and proceeds to the cell membrane, resulting in cytokinesis. **168**

cell theory (sel the'o-re) The concept that all living things are made of cells. **68**

cell wall (sel wawl) An outer covering on some cells; may be composed of cellulose, chitin, or peptidoglycan, depending on the kind of organism. **68**

cells (selz) The basic structural units of all living things. The smallest units that display the characteristics of life. **14**

cellular membranes (sel'yu-lar mem'brānz) Thin sheets of material composed of phospholipids and proteins. Some of the proteins have attached carbohydrates or fats. **72**

cellular respiration (sel'yu-lar res"pi-ra'shun) A major biochemical pathway by which cells release the chemical-bond energy from food and convert it into a usable form (ATP). **112**

central nervous system (sen'trul ner'vus sis'tem) The portion of the nervous system consisting of the brain and spinal cord. **569**

centriole (sen'tre-ōl) Two sets of nine short microtubules; each set of tubules is arranged in a cylinder. **82**

cerebellum (ser"ĕ-bel'um) The region of the brain, connected to the medulla oblongata, that receives many kinds of sensory stimuli and coordinates muscle movement. **572**

cerebrum (ser'ĕ-brum) The region of the brain that surrounds most of the other parts of the brain and is involved in consciousness and thought. **573**

chemical bonds (kem'ĭ-kal bondz) Forces that combine atoms or ions and hold them together. **29**

chemical equation (kem'i-kal e-kwa'zhun) A way of describing what happens in a chemical reaction. **36**

chemical reaction (kem'ĭ-kal re-ak'shun) The formation or rearrangement of chemical bonds, usually indicated in an equation by an arrow from the reactants to the products. **30**

chemistry (kem'iss-tre) The science concerned with the study of the composition, structure, and properties of matter and the changes it undergoes. **24**

chemosynthesis (kem"o-sin'thuh-sis) The use of inorganic chemical reactions as a source of energy to make larger, organic molecules. **112**

chlorophyll (klo'ro-fil) The green pigment located in the chloroplasts of plant cells associated with trapping light energy. **79**

chloroplasts (klo'ro-plasts) Energy-converting, membranous, saclike organelles in plant cells containing the green pigment chlorophyll. **79**

chromatid (kro'mah-tid) One of two component parts of a chromosome formed by replication and attached at the centromere. **165**

chromatin (kro'mah-tin) An area or a structure within the nucleus of a cell composed of long molecules of deoxyribonucleic acid (DNA) in association with proteins. **83**

chromosomal aberration (kro-mo-sōm'al ab-e-rā-shen) A change in the structure of chromosomes, which can affect the expression of genes; e.g., translocation, duplication mutations. **160**

chromosomes (kro'mo-sōmz) Double-stranded DNA molecules with attached protein (nucleoprotein) coiled into a short, compact unit. **57**

cilia (sil'e-ah) Numerous short, hairlike structures, projecting from the cell surface, that enable locomotion. **82**

class (class) A group of closely related families within a phylum. **424**

classical conditioning (klas'ĭ kul kon-dĭ'shun-ing) Learning that occurs when an involuntary, natural, reflexive response to a natural stimulus is

transferred from the natural stimulus to a new stimulus. **382**

cleavage furrow (kle'vaj fuh'ro) An indentation of the cell membrane of an animal cell that pinches the cytoplasm into two parts during cell division. **168**

climax community (klī'maks ko-miu'nĭ-te) A relatively stable, long-lasting community. **340**

clone (klōn) Exact copies of biological entities such as genes, organisms, or cells. **229**

cochlea (kŏk'le-ah) The part of the ear that converts sound into nerve impulses. **581**

coding strand (kō-ding strand) One of the two DNA strands that serves as a template, or pattern, for the synthesis of RNA. **147**

codominance (ko-dom'ĭ-nans) A situation in which both alleles in a heterozygous organism express themselves. **202**

codon (ko'don) A sequence of three nucleotides of an mRNA molecule that directs the placement of a particular amino acid during translation. **148**

coelom (se'lōm) A body cavity in which internal organs are suspended. **495**

coenzyme (ko-en'zim) A molecule that works with an enzyme to enable the enzyme to function as a catalyst. **99**

cofactor (co'fak-tōr) Inorganic ions or organic molecules that serve as enzyme helpers. **99**

commensalism (ko-men'sal-izm) A relationship between two organisms in which one organism is helped and the other is not affected. **332**

communication (ko-mu'ni-ka-shun) The use of signals to convey information from one animal to another. **389**

community (ko-miu'nĭ-te) Populations of different kinds of organisms that interact with one another in a particular place. **304**

competition (com-pe-tĭ'shun) A relationship between two organisms in which both organisms are harmed. **328**

competitive exclusion principle (com-pe'-tĭ-tiv eks-klu'zhun prin'sĭ-pul) No two species can occupy the same niche at the same time. **328**

competitive inhibition (com-pĕ′-tĭ-tiv in″hĭ-bĭ′shun) The formation of a temporary enzyme-inhibitor complex that interferes with the normal formation of enzyme-substrate complexes, resulting in a decreased turnover. 103

complete proteins (kom-plēt′ prō′tēnz) Protein molecules that provide all the essential amino acids. 544

complex carbohydrates (kom′pleks kar-bo-hi′drāts) Macromolecules composed of simple sugars combined by dehydration synthesis to form a polymer. 51

compound (kom′pound) A kind of matter that consists of a specific number of atoms (or ions) joined to each other in a particular way and held together by chemical bonds. 29

concentration gradient (diffusion gradient) (kon″sen-tra′shun gra′de-ent) (dĭ-fiu′zhun gra′de-ent) The gradual change in the number of molecules per unit of volume over distance. 85

conditioned response (kon-dĭ′shund respons′) The modified behavior displayed in which a new response is associated with a natural stimulus. 382

cones (kōnz) Reproductive structures of gymnosperms that produce pollen in males or eggs in females. 477

consumers (kon-soom′urs) Organisms that must obtain energy in the form of organic matter. 306

control group (con-trōl′ grūp) The situation used as the basis for comparison in a controlled experiment. The group in which there are no manipulated variables. 6

control processes (con-trōl′ pro′ses-es) Mechanisms that ensure an organism will carry out all life activities in the proper sequence (coordination) and at the proper rate (regulation). 13

controlled experiment (con-trold′ ek-sper′ĭ-ment) An experiment that includes two groups, one in which the variable is manipulated in a particular way and one in which there is no manipulation. 6

convergent evolution (kon-vur′jent ĕv-o-lu′shun) An evolutionary pattern in which widely different organisms show similar characteristics. 290

corpus luteum (kōr′pus lu′te-um) The remainder of a follicle after the release of the secondary oocyte; it develops into a glandlike structure that produces hormones (progesterone and estrogen) that prevent the release of other eggs. 601

cotyledons (kot-ah-lee′don) Also know as seed leaves, are embryonic leaves that have food stored in them. 480

covalent bond (ko-va′lent bond) The attractive force formed between two atoms that share a pair of electrons. 32

cristae (krĭs′te) Folded surfaces of the inner membranes of mitochondria. 79

critical period (krit′ĭ-kl pir-ē-ed) The period of time during the life of an animal when imprinting can take place. 384

crossing-over (kro′sing o′ver) The exchange of a part of a chromatid from 1 chromosome with an equivalent part of a chromatid from a homologous chromosome. 176

cryptorchidism (krip-tor′kĭ-dizm) A developmental condition in which the testes do not migrate from the abdomen through the inguinal canal to the scrotum. 600

cytokinesis (si-to-kĭ-ne′sis) The division of the cytoplasm of one cell into two new cells. 166

cytoplasm (si′to-plazm) The portion of the protoplasm that surrounds the nucleus. 69

cytosine (si′to-sēn) A single-ring nitrogenous-base molecule in DNA and RNA; it is complementary to guanine. 144

cytoskeleton (si″ to-skel′ē-ton) The internal framework of eukaryotic cells composed of intermediate filaments, microtubules, and microfilaments; provides the cell with a flexible shape, and the ability to move through the environment, to move molecules internally, and to respond to environmental changes. 81

D

death phase (deth fāz) The portion of some population growth curves in which the size of the population declines. 367

deciduous (de-sid′yu-us) A type of trees that lose their leaves at the end of the growing season. 482

decomposers (de-kom-po′zurs) Organisms that use dead organic matter as a source of energy. 306

deductive reasoning (deduction) (de′-duk-tiv re′son-ing) (de′duk-shun) The mental process of using accepted generalizations to predict the outcome of specific events. From the general to the specific. 8

dehydration synthesis reaction (de-hī-dra′shun sin′thuh-sis re-ak′shun) A reaction that results in the formation of a macromolecule when water is removed from between the two smaller component parts. 37

deletion abberration (de-le′shun abe-rā-shen) A major change in DNA that can be observed at the level of the chromosome. 160

deletion mutation (de-le′shun miu-ta′shun) A change in the DNA that has resulted from the removal of one or more nucleotides. 159

denatured (de-nā′churd) Altered so that some of the protein's original properties are diminished or eliminated. 56

dendrites (den′drīts) Neuronal fibers that receive information from axons and carry it toward the nerve-cell body. 569

denitrifying bacteria (de-ni′trĭ-fi-ing bak-te′re-ah) Several kinds of bacteria capable of converting nitrite to nitrogen gas. 315

density-dependent limiting factors (den′sĭ-te de-pen′dent lim-ĭ-ting fak′torz) Population-limiting factors that become more effective as the size of the population increases. 366

density-independent limiting factors (den′sĭ-te in-de-pen′dent lim-ĭ-ting fak′torz) Population-controlling factors that are not related to the size of the population. 366

deoxyribonucleic acid (DNA) (de-ok″se-ri-bo-nu-kle′ik ă′sid) A polymer of nucleotides that serves as genetic information. In prokaryotic cells, it is a double-stranded loop and contains attached HU proteins. In eukaryotic cells, it is found in strands with attached histone proteins. When tightly coiled, the DNA and histone structure is known as a chromosome. 56

dependent variable (de-pen′dent var′ē-abul) A variable that changes in direct response to (depends on) how another variable (independent variable) is manipulated. 7

depolarized (de-po'lă-rīzd) Having lost the electrical difference existing between two points or objects. 570

determination (de-ter'mi-na-shun) The process a cell goes through to select which genes it will eventually express on a more or less permanent basis. 173

diaphragm (di'uh-fram) The muscle separating the lung cavity from the abdominal cavity; it is involved in exchanging the air in the lungs. 528

diastolic blood pressure (di"uh-stol'ik blud presh'yur) The pressure present in a large artery when the heart is not contracting. 524

dicot (di'kot) An angiosperm whose embryo has two seed leaves. 480

diet (di'et) The food and drink consumed by a person from day to day. 542

Dietary Reference Intakes Published by the USDA these guidelines provide information on the amounts of certain nutrients various members of the public should receive. 548

differentiation (dĭf-ĕ-ren"she-a'shun) The process of forming specialized cells within a multicellular organism. 173

diffusion (dĭ-fiu'zhun) The net movement of a kind of molecule from an area of higher concentration to an area of lesser concentration. 85

digestion (di-jest'yun) The breakdown of complex food molecules to simpler molecules; the chemical reaction of hydrolysis. 542

digestive system (di-jest·v sis'tum) The organ system responsible for the processing and distribution of nutrients and consists of a muscular tube and glands that secrete digestive juices into the tube. 530

diploblastic (dip-lo-blas'tik) A condition in which some simple animals only consist of two layers of cells. 495

diploid (dip'loid) Having two sets of chromosomes: one set from the maternal parent and one set from the paternal parent. 174

directional selection (dye-rek'-shun-l sĕ-lek'shun) Selection that occurs when individuals at one extreme of the range of a characteristic are consistently selected for. 271

disruptive selection (dis-rup'tive sĕ-lek'shun) Selection that occurs when both extremes of a range for a charcteristic are selected for and the intermediate condition is selected against. 271

distal convoluted tubule (dis'tul kon'vo-lu-ted tūb'yūl) The downstream end of the nephron of the kidney, primarily responsible for regulating the amount of hydrogen and potassium ions in the blood. 536

divergent evolution (di-vur'jent ĕv-o-lu'shun) A basic evolutionary pattern in which individual speciation events cause many branches in the evolution of a group of organisms. 287

DNA fingerprinting (DNA fing"er-prin'ting) A laboratory technique that detects differences in DNA to identify a unique individual; the differences in DNA are detected by using variable number tandem repeats, restriction enzymes, and electrophoresis. 216

DNA library (DNA li'bra-re) A collection of cloned DNA fragments that represent all the genetic information of an organism. 228

DNA replication (rep"lĭ-ka'shun) The process by which the genetic material (DNA) of the cell reproduces itself prior to its distribution to the next generation of cells. 145

domain (dō'mān) The first (broadest) classification unit of organisms; there are three domains: Eubacteria, Archaea, and Eucarya. 90

dominance hierarchy (dom'in-ants hi'ur-ar-ke) A relatively stable, mutually understood order of priority within a group. 393

dominant allele (dom'in-ant a-lēl') An allele that expresses itself and masks the effects of other alleles for the trait. 193

double bond (dub'l bond) A pair of covalent bonds formed between two atoms when they share two pairs of electrons. 47

double-factor cross (dub'l fak'tur kros) A genetic study in which two pairs of alleles are followed from the parental generation to the offspring. 194

Down syndrome (down sin'drom) A genetic disorder resulting from the presence of an extra chromosome 21. Symptoms include slightly slanted eyes, flattened facial features, a large tongue, and a tendency toward short stature and fingers. Individuals usually display mental retardation. 185

duodenum (dew"o-de'num) The first part of the small intestine, which receives food from the stomach and secretions from the liver and pancreas. 532

duplications (du"pli-ka'shunz) A form of chromosomal aberration in which a portion of a chromosome is replicate and attached to the original section in sequence. 160

dynamic equilibrium (di-nam'ik e-kwĭ-lib'reum) The condition in which molecules are equally dispersed; therefore, movement is equal in all directions. 85

E

ecology (e-kol'o-je) The branch of biology that studies the relationships between organisms and their environment. 304

ecosystem (e'ko-sis"tum) A unit consisting of a community of organisms (populations) and its interactions with the physical surroundings. 304

ectoderm (ek'to-derm) The outer embryonic layer. 495

ectotherms (ek"to-thermz') Animals that are unable to regulate their body temperature by automatic physiological processes but can regulate their temperature by moving to places where they can be most comfortable. 494

ejaculation (e-jak"u-lā'shun) The release of sperm cells and seminal fluid through the penis of a male. 602

electron (e-lek'tron) A negatively charged particle moving at a distance from the nucleus of an atom; balance it the positive charges of the protons. 25

electron-transport system (ETS) (e-lek'tron trans'port sis'tem) The series of oxidation-reduction reactions in aerobic cellular respiration in which the energy is removed from hydrogens and transferred to ATP. 115

electrophoresis (e-lek"tro-fo-re'sis) A technique that separates DNA fragments on the basis of size. 217

elements (el'ĕ-ments) Fundamental chemical substances that are made up of collections of only one kind of atom. **25**

endocrine glands (en′do-krĭn glandz) Glands that secrete into the circulatory system. **575**

endocrine system (en′do-krĭn sis′tem) A number of glands that communicate with one another and other tissues through chemical messengers transported throughout the body by the circulatory system. **575**

endocytosis (en″ dō-sĭ-tō′-sis) The process cells use to wrap membrane around a particle (usually food) and engulf it. **88**

endoderm (en′do-derm) The inner embryonic layer. **495**

endoplasmic reticulum (ER) (en″do-plaz′mik re-tĭk′yu-lum) Folded membranes and tubes throughout the eukaryotic cell that provide a large surface on which chemical activities take place. **75**

endoskeletons (en″do-skel′ĕ-tonz) Skeletons typical of vertebrates in which the skeleton is surrounded by muscles and other organs. **497**

endospore (en′do-spōr″) A unique bacterial structure with a low metabolic rate that germinates under favorable conditions to grow into a new cell. **446**

endosymbiotic theory (en″do-sim-be-ot′ik the′o-re) A theory suggesting that some organelles found in eukaryotic cells may have originated as free-living cells. **412**

endotherms (en″do-thermz′) Animals that have internal temperature-regulating mechanisms and can maintain a relatively constant body temperature in spite of wide variations in the temperature of their environment. **494**

energy (en′er-je) The ability to do work or cause things to move. **12**

energy level (en′er-je lev′el) A region surrounding an atomic nucleus that contains electrons moving at approximately the same speed and having approximately the same amount of kinetic energy. **25**

environment (en-vi′ron-ment) Anything that affects an organism during its lifetime. **304**

environmental resistance (en-vi-ron-men′tal re-zis′tans) The collective factors that limit population growth. **366**

enzymatic competition (en-zi-mă′tik com-pĭ tĭ′chun) Competition among several different available enzymes to combine with a given substrate material. **101**

enzymes (en′zīms) Molecules, produced by organisms, that are able to control the rate at which chemical reactions occur. **14**

enzyme-substrate complex (en′zīm sub′ strāt kom′pleks) A temporary molecule formed when an enzyme attaches itself to a substrate molecule. **97**

epinephrine (ĕ″pĭ-nef′rin) A hormone produced by the adrenal medulla and certain nerve cells that increases heart rate, blood pressure, and breathing rate. **575**

epiphytes (ep′e-fīts) Plants that live on the surface of other plants without doing harm. **332**

essential amino acids (ĕ-sen′shul ah-me′no ă′sids) Amino acids that cannot be synthesized by the human body and must be part of the diet (e.g., lysine, tryptophan, and valine). **544**

essential fatty acids (ĕ-sen′shul fă′te ă′sidz) The fatty acids linoleic and linolenic, which cannot be synthesized by the human body and must be part of the diet. **544**

estrogen (es′tro-jen) One of the female sex hormones responsible for the growth and development of female sexual anatomy. **575**

ethology (e-thol′uh-je) The scientific study of the nature of behavior and its ecological and evolutionary significance in its natural setting. **380**

eukaryotic cells (yu′ka-re-ah″tik sels) One of the two major types of cells; cells that have a true nucleus, as in plants, fungi, protists, and animals. **69**

evolution (ĕv-o-lu′shun) A change in the frequency of genetically determined characteristics within a population over time. **260**

excretory system (ex′cre-to-ry sis-tum) the organ system responsible for the processing and elimination of metabolic waste products and consists of the kidneys, ureters, urinary bladder, and urethra. **535**

exocrine glands (ek′so-krĭn glandz) Glands that secrete through ducts to the surface of the body or into hollow organs of the body. **575**

exocytosis (ek′sō-sĭ-to′-sis) The process cells use to wrap membrane around a particle (usually cell products or wastes) and eliminate it from a cell. **88**

exon (ek″s-on) Sequences of mRNA that are used to code for proteins **155**

exoskeletons (ek″sō-skel′ĕ-tonz) Skeletons typical of many invertebrates, in which the skeleton is on the outside of the animal. **497**

experiment (ek-sper′ĭ-ment) A re-creation of an event that enables a scientist to gain valid and reliable empirical evidence. **6**

experimental group (ek-sper′ĭ-men′tal grūp) The group in a controlled experiment that has a variable manipulated. **6**

exponential growth phase (ek-spo-nen′shul grōth fāz) The period of time during population growth when the population increases at an accelerating rate. **364**

expressivity (ek-spre′si″vi-te) A term used to describe situations in which the gene expresses itself but not equally in all individuals that have it. **267**

external parasites (ek-stur′nal pĕr′uh-sīts) Parasites that live on the outside of their hosts. **331**

extinction (eks-tingt′shun) The loss of a species. **288**

extrinsic limiting factors (eks-trin′sik lim-ĭ-ting fak′tōrz) Population-controlling factors that arise outside the population. **366**

F

facilitated diffusion (fah-sil′ĭ-ta″ted dĭ-fiu′zhun) Diffusion assisted by carrier-molecules. **87**

family (fam′ĭ-le) A group of closely related genera within an order. **424**

fats (fats) A class of water-insoluble macromolecules composed of a glycerol and fatty acids. **59**

fatty acid (fat′ē ă′sid) One of the building blocks of a fat, composed of a long-chain carbon skeleton with a carboxylic acid functional group. **60**

fermentation (fer-men-ta′shun) Pathways that oxidize glucose to generate ATP energy using something other than O_2 as the ultimate hydrogen acceptor. **122**

fertilization (fer″tĭ-lĭ-za′shun) The joining of haploid nuclei, usually from an egg and a sperm cell, resulting in a diploid cell called a zygote. **174**

fiber (fi'ber) Natural (plant) or industrially produced polysaccharides that are resistant to hydrolysis by human digestive enzymes. **543**

filter feeders (fil'ter fēd'erz) Animals that use cilia or other appendages to create water currents and filter food out of the water. **498**

fitness (fit'nes) The concept that those who are best adapted to their environment produce the most offspring. **264** Nutritionally, a measure of how efficiently a person can function both physically and mentally. **553**

flagella (flah-jel'luh) Long, hairlike structures, projecting from the cell surface, that enable locomotion. **82**

flavin adenine dinucleotide (FAD) A hydrogen carrier used in respiration. **99**

flower (flow'er) A complex plant reproductive structure made from modified stems and leaves that produces pollen and eggs. **479**

fluid-mosaic model (flu'id mo-za'ik mod'l) The concept that the cellular membrane is composed primarily of protein and phospholipid molecules that are able to shift and flow past one another. **72**

follicles (fol'ĭ-kulz) Saclike structures near the surface of the ovary, which encases the soon-to-be-released secondary oocyte. **601**

follicle-stimulating hormone (FSH) (fol'ĭ-kul stim'yu-lā-ting hōr'mōn) The pituitary secretion that causes the ovaries to begin to produce larger quantities of estrogen and to develop the follicle and prepare the egg for ovulation. **601**

food chain (food chān) A sequence of organisms that feed on one another, resulting in a flow of energy from a producer through a series of consumers. **305**

Food Guide Pyramid (food gīd pī'ra-mĭd) A tool developed by the U.S. Department of Agriculture to help the general public plan for good nutrition; it contains guidelines for required daily intake from each of the six food groups. **548**

food web (food web) A system of interlocking food chains. **324**

formed elements (form-d' el'ĕ-ments) Red, white blood cells and platelets suspended in a watery matrix called plasma. **520**

formula (for'mu-lah) Pertains to a chemical compound; describes what elements it contains (as indicated by a chemical symbol) and in what proportions they occur (as indicated by the subscript number). **29**

fossil (fŏ'-silz) Physical evidence of former life. **283**

founder effect (faùn'der e'fekt) The concept that small, newly established populations are likely to have reduced genetic diversity because of the small number of individuals in the founding population. **247**

fovea centralis (fo've-ah sen-tral'is) The area of sharpest vision on the retina, containing only cones, where light is sharply focused. **580**

frameshift (frām-shift) A form of mutation which occurs when insertions or deletions cause the ribosome to read the wrong sets of three nucleotides. **159**

free-living nitrogen-fixing bacteria (ni'tro-jen fik'sing bak-te're-ah) Soil bacteria that convert nitrogen gas molecules into nitrogen compounds plants can use. **314**

fruit (froot) The structure (mature ovary) in angiosperms that contains seeds. **478**

functional groups (fung'shun-al grūps) Specific combinations of atoms attached to the carbon skeleton that determine specific chemical properties. **48**

fungus (fung'us) The common name for members of the kingdom Fungi. **457**

G

gallbladder (gol'blad"er) The organ, attached to the liver, that stores bile. **532**

gametes (gam'ēts) Haploid sex cells. **174**

gametogenesis (gă-me"to-jen'ĕ-sis) The generating of gametes; the meiotic cell-division process that produces sex cells; oogenesis and spermatogenesis. **603**

gametophyte (gă-me'to-fīt) A haploid plant that produces gametes; it alter-

nates with the sporophyte through the life cycle. **433**

gametophyte generation (gă-me'to-fīt jen'uh-ra"shun) A life cycle stage in plants in which a haploid sex cell is produced by mitosis. **433**

gas (gas) The phase of matter in which the molecules are more energetic than the molecules of a liquid, resulting in only a slight attraction for each other. **30**

gastric juice (gas'trik jūs) The secretions of the stomach; they contain enzymes and hydrochloric acid. **530**

gene (jēn) Any molecule usually segments of DNA, able to (1) replicate by directing the manufacture of copies of themselves; (2) mutate, or chemically change, and transmit these changes to future generations; (3) store information that determines the characteristics of cells and organisms; and (4) use this information to direct the synthesis of structural and regulatory proteins. **54**

gene expression (jēn ek-spresh'un) The cellular process of transcribing and translating genetic information. **152**

gene flow (jēn flo) The movement of genes within a population because of migration or the movement of genes from one generation to the next by gene replication and reproduction. **282**

gene frequency (jēn fre'kwen-see) A measure of how often a specific gene shows up in the gametes of a population. **241**

gene pool (jēn pool) All the genes of all the individuals of a species. **240**

gene therapy (jēn ther'ah-pe) A technique that introduces new genetic material into an organism to correct a genetic deficiency. **231**

generative processes (jen'uh-ra"tiv pros'es-es) Actions that increase the size of an individual organism (growth) or increase the number of individuals in a population (reproduction). **12**

gene-regulator proteins (jēn reg'yu-la-tor prō'tēns) Chemical messengers within a cell that inform the genes as to whether protein-producing genes should be turned on or off or whether they should have their protein-producing activities increased or decreased; e.g., gene-repressor proteins and gene-activator proteins. **102**

genetic bottleneck (jĕ-ne′tik ba′tul-nek) The concept that, when populations are severely reduced in size, they may lose some of their genetic diversity. **247**

genetic diversity (jĕ-ne′tik dĭ-ver′sĭ-te) The degree to which individuals in a population possess alternative alleles for characteristics. **243**

genetic drift (jĕ-ne′tik drĭft) A change in gene frequency that is not the result of natural selection; this typically occurs in a small population. **245**

genetic recombination (jĕ-net′ik re-kom-bĭ-na′shun) The gene mixing that occurs during sexual reproduction. **266**

genetically modified (GM) (jĕ-ne′tik-le ma′di-fid) Engineered to contain genes from at least one other species. **229**

genetics (jĕ-net′iks) The study of genes, how genes produce characteristics, and how the characteristics are inherited. **192**

genome (jĕ′nōm) A set of all the genes necessary to specify an organism's complete list of characteristics. **192**

genomics (jĕ-nŏ′-miks) A new field of science that has developed since the sequencing of the human genome; the field looks at how genomes are organized and compares them with genomes of other organims. **227**

genotype (jē′no-tĭp) The catalog of genes of an organism, whether or not these genes are expressed. **193**

genus (jē′nus) A group of closely related species within a family. **423**

geographic isolation (jē-o-graf′ik i-so-la′shun) A condition in which part of the gene pool is separated by geographic barriers from the rest of the population. **283**

glands (glandz) Organs that manufacture and secrete a material either through ducts or directly into the circulatory system. **575**

glomerulus (glo-mer′u-lus) A cluster of blood vessels, surrounded by Bowman's capsule in the kidney. **536**

glyceraldehyde-3-phosphate (glĭ-ser-al′-dĕ-hĭd 3 fos′-fāt) A 3-carbon compound, produced during glycolysis and photosynthesis, that can be

converted to other organic molecules. **134**

glycerol (glis′er-ol) One of the building blocks of a fat, composed of a carbon skeleton that has three alcohol groups (OH) attached to it. **59**

glycolysis (gli-kol′ĭ-sis) The anaerobic first stage of cellular respiration, consisting of the enzymatic breakdown of a sugar into two molecules of pyruvic acid. **114**

Golgi apparatus (gōl′je ap″pah-rat′us) A stack of flattened, smooth, membranous sacs; the site of synthesis and packaging of certain molecules in eukaryotic cells. **75**

gonads (go′nads) Organs in which meiosis occurs to produce gametes; ovaries or testes. **174**

gonadotropin-releasing hormone (GnRH) (go″nad-o-tro′-pin re-le′sing hōr′mōn) A hormone released from the hypothalamus that stimulates the release of follicle-stimulating hormone (FSH) and luteinizing hormone (LH). **600**

gradualism (grad′u-al-ism) A model for evolutionary change that assumes that evolution occurred slowly by accumulating small changes over a long period of time. **292**

grana (gra′nuh) Areas of the chloroplast membrane where chlorophyll molecules are concentrated. **79**

granules (gran′yūls) Materials whose structure is not as well defined as that of other organelles. **82**

growth-stimulating hormone (GSH) (grōth stĭ′mu-la-ting hōr′mōn) The hormone produced by the anterior pituitary gland that stimulates tissues to grow. **575**

guanine (gwah′nēn) A double-ring nitrogenous-base molecule in DNA and RNA; it is the complementary base of cytosine. **144**

gymnosperms (jim′no-spurmz) Plants that produce their seeds in cones. **477**

H

habitat (hab′-ĭ-tat) The place or part of a community occupied by an organism. **326**

habitat preference (ecological isolating mechanisms) (hab′i-tat pref′ur-ents) Reproductive isolating mecha-

nisms that prevent interbreeding between species because they live in different areas. **285**

habituation (hah-bit′u-a′shun) A change in behavior in which an animal ignores a stimulus after repeated exposure to it. **382**

haploid (hap′loid) Having a single set of chromosomes, resulting from the reduction division of meiosis. **174**

Hardy-Weinberg concept (har′de wĭn′burg kon′sept) Populations of organisms will maintain constant gene frequencies from generation to generation as long as mating is random, the population is large, mutation does not occur, migration does not occur, and no genes provide more advantageous characteristics than others. **273**

heart (hart) The muscular pump that forces blood through the blood vessels of the body. **520**

heat (hēt) The total internal kinetic energy of molecules. **29**

hemoglobin (hēm′o-glo-bin) An iron-containing molecule found in red blood cells, to which oxygen molecules bind. **520**

hepatic portal vein (hĕ-pat′ik pōr′tul vān) A blood vessel that collects blood from capillaries in the intestine and delivers it to a second set of capillaries in the liver. **534**

herbivores (er′bĭ-vōrz) Animals that feed directly on plants. **306**

heterotrophs (hĕ′tur-o-trōfs) Organisms that require a source of organic material from their environment; they cannot produce food on their own. **112**

heterozygous (hĕ″ter-o-zi′gus) Describes a diploid organism that has 2 different alleles for a particular characteristic. **194**

high-energy phosphate bond (hi en′ur-je fos′fāt bond) The bond between two phosphates in an ADP or ATP molecule that readily releases its energy for cellular processes. **106**

homeostasis (ho″me-o-sta′sis) The maintenance of a constant internal environment. **14**

homeotherms (ho′me-o-thermz) Animals that maintain a constant body temperature. **493**

homologous chromosomes (ho-mol′o-gus kro′mo-sōmz) A pair of chromosomes in a diploid cell that contain similar genes at corresponding loci throughout their length. **174**

homologous structures (ho-mol'o-gus struk'chu-rz) similar structures in different species that have the same general function indicating that they been derived from a common ancestor. **291**

homozygous (ho"mo-zi'gus) Describes a diploid organism that has 2 identical alleles for a particular characteristic. **194**

hormone (hōr'mōn) A chemical messenger secreted by an endocrine gland to regulate other parts of the body. **575**

host (host) An organism that a parasite lives in or on. **330**

hybrid inviability (hī-brid in-vi-a-bil'i-ty) Mechanisms that prevent the offspring of two different species from continuing to reproduce. **287**

hydrogen bonds (hi'dro-jen bondz) Weak attractive forces between molecules; important in determining how groups of molecules are arranged. **34**

hydrolysis reactions (hi-drol'ī-sis re-ak' shuns) Processes that occur when large molecules are broken down into smaller parts by the addition of water. **37**

hydrophilic (hi-dro-fil'ik) Readily absorbing or dissolving in water. **72**

hydrophobic (hi'dro-fo'bik) Tending not to combine with, or incapable of dissolving in, water. **72**

hydroxide ions (hi-drok'sīd i'onz) Negatively charged particles (OH⁻) composed of oxygen and hydrogen atoms released from a base when dissolved in water. **38**

hypertonic (hi'pur-tŏn'ik) A comparative term describing one of two solutions; a hypertonic solution is one with higher amount of dissolved material. **86**

hypothalamus (hi"po-thal'ah-mus) The region of the brain located in the floor of the thalamus and connected to the pituitary gland; it is involved in sleep and arousal; emotions, such as anger, fear, pleasure, hunger, sexual response, and pain; and automatic functions, such as temperature, blood pressure, and water balance. **573**

hypothesis (hi-poth'e-sis) A possible answer to or explanation for a question that accounts for all the observed facts and that is testable. **6**

hypotonic (hi'po-tŏn'ik) A comparative term describing one of two solutions; a hypotonic solution is one with a lower amount of dissolved material. **86**

I

immune system (ĭ-myun' sis'tem) The system of white blood cells specialized to provide the body with resistance to disease. **586**

immunity (ĭ-myun-a-t) The ability to maintain homeostasis by resisting or defending against potentially harmful agents including microbes, toxins, and abnormal cells such as tumor cells. **586**

immunization (ĭ-myun·ā shun) The technique used to induce the immune system to develop an acquired immunity to a specific disease by the use of a vaccine. **588**

immunodeficiency diseases (im"i-no-de-fish'en-se di-z̄ ez') Disorders that result from the immune system not having one or more component cells or chemicals. **590**

imperfect flower (im"pur'fekt flow'er) A flower that contains either male (stamens) or female (pistil) reproductive structures, but not both. **479**

imprinting (im'prin-ting) A form of learning that occurs in a very young animal that is genetically primed to learn a specific behavior in a very short period. **384**

inclusions (in-klu'zhuns) Materials inside a cell that are usually not readily identifiable; stored materials. **82**

incomplete proteins (in-kom-plēt' prō'tēnz) Protein molecules that do not provide all the essential amino acids. **544**

incus (in'kus) The ear bone that is located between the malleus and the stapes. **581**

independent assortment (in"de-pen'dent ă-sort'ment) The segregation, or assortment, of one pair of homologous chromosomes independently of the segregation, or assortment, of any other pair of chromosomes. **183**

independent variable (in"de-pen'dent var'ē-a-bul) A variable that is purposely manipulated to determine how it will affect the outcome of an event. **7**

inductive reasoning (induction) (in-duk'tiv re'son-ing) (in'duk-shun) The mental process of examining many sets of facts and developing generalizations. From the specific to the general. **8**

inflammation (in"flah-ma'shun) A nonspecific defense method that is a series of events that clear an area of harmful agents and damaged tissue. **586**

ingestion (in-jes'chun) The process of taking food into the body through eating. **542**

inguinal canal (ing'gwĭ-nal că-nal') The opening in the floor of the abdominal cavity through which the testes in a human male fetus descend into the scrotum. **600**

inhibitor (in-hib'ĭ-tor) A molecule that temporarily attaches itself to an enzyme, thereby interfering with the enzyme's ability to form an enzyme-substrate complex. **102**

inorganic molecules (in-or-gan'ik mol'uh-kiuls) Molecules that do not contain carbon atoms in rings or chains. **46**

insertion mutation (in-ser'shun miu-ta'shun) A change in DNA resulting from the addition of one or more nucleotides to the normal DNA dequence. **159**

insight (in'sīt) Learning in which past experiences are reorganized to solve new problems. **385**

instinctive behavior (in-stink'tiv be-hāv'yur) Automatic, preprogrammed, genetically determined behavior. **379**

intermediate filaments (in"ter-me'de-it fil'ah-ments) Protein fibers that connect microtubules and microfilaments as part of the cytoskeleton. **81**

internal parasites (in-tur'nal pĕr'uh-sīt) Parasites that live inside their hosts. **331**

interphase (in'tur-fāz) The stage between cell divisions in which the cell is engaged in metabolic activities. **164**

interspecific competition (in-ter-spĕ-sī'-fik com-pe-tī'shun) Interaction between two members of *different* species that is harmful to both organisms. **328**

interspecific hybrids (in-ter-spĕ-sī'-fik hī'-bridz) Hybrids between two different species. **248**

intraspecific competition (in-tră-spĕ-sī'-fik com-pe-tī'shun) Interaction between two members of the *same* species that is harmful to both organisms. **328**

intraspecific hybrids (in-tră-spĕ-sĭ′-fik hĭ′-bridz) Organisms that are produced by the controlled breeding of separate varieties of the same species. **248**

intrinsic limiting factors (in-trin′sik lim-ĭ-ting fak′tōrz) Population-controlling factors that arise from within the population. **366**

intron (intron) Sequences of mRNA that do not code for protein. **155**

invertebrates (in-vur′tuh-brāts) Animals without backbones. **493**

ion (i′on) Electrically unbalanced or charged atoms. **31**

ionic bonds (i-on′ik bondz) The attractive forces between ions of opposite charge. **31**

isomers (i′so-meers) Molecules that have the same empirical formula but different structural formulas. **47**

isotonic (i′so-tŏn′ik) A term used to describe two solutions that have the same concentration of dissolved material. **86**

isotope (i′so-tōp) An atom of the same element that differs only in the number of neutrons. **27**

K

kidneys (kid′nēz) The primary organs involved in regulating blood levels of water, hydrogen ions, salts, and urea. **536**

kilocalorie (kcal) (kil″o-kal′o-re) A measure of heat energy 1,000 times larger than a calorie; food Calories are kilocalories. **542**

kinetic energy (ki-net′ik en′er-je) Energy of motion. **24**

kinetic molecular theory (ki-net′ik mol-·ĕ-kyūar the-o-re) states that all matter is made up of tiny particles that are in constant motion. **29**

kinetochore (ki-ne′to-kor) A protein attached to each chromatid at the centromere. **168**

kingdom (king′dom) A classification category larger than a phylum and smaller than a domain. **423**

Krebs cycle (krebs si′kl) The series of reactions in aerobic cellular respiration that results in the production of two carbon dioxide, the release of four

pairs of hydrogens, and the formation of an ATP molecule. **114**

kwashiorkor (kwa″she-or′kōr) A protein-deficiency disease, common in malnourished children, caused by prolonged protein starvation leading to reduced body size, lethargy, and low mental ability. **558**

L

lacteal (lak′tēl) A tiny lymphatic vessel located in a villus. **534**

lactic acid fermentation (lak′-tik ă′-sid fermen-tā′-shun) A process during which the pyruvic acid ($CH_3COCOOH$) that results from glycolysis is converted to lactic acid ($CH_3CHOHCOOH$) by the transfer of electrons that had been removed from the original glucose. **123**

lag phase (lag fāz) The period of time following colonization when a population remains small or increases slowly. **364**

large intestine (larj in-tes′tin) The last portion of the food tube; it is primarily involved in reabsorbing water. **532**

law of conservation of energy The law that states that energy is never created or destroyed. **24**

Law of Dominance (law uv dom′in-ans) When an organism has 2 different alleles for a trait, the allele that is expressed and overshadows the expression of the other allele is said to be dominant; the allele whose expression is overshadowed is said to be recessive. **198**

Law of Independent Assortment (law uv in″ depen′dent ă-sort′ment) Members of one allelic pair will separate from each other independently of the members of other allele pairs. **201**

Law of Segregation (law uv seg″rĕ-ga′shun) When haploid gametes are formed by a diploid organism, the 2 alleles that control a trait separate from one another into different gametes, retaining their individuality. **194**

leaf (lee-f) Plant structure specialized for carrying out the process of photosynthesis. **469**

learned behavior (lur′ned be-hāv′yur) A change in behavior as a result of experience. **379**

learning (lur′ning) A change in behavior as a result of experience. **382**

lichen (li·kĕn) An organism comprised of a fungus and an alga protist or cyanobacterium existing in a mutualistic relationship. **461**

light-capturing events (līt kap′chu-ring e-vents) The first stage in photosynthesis; involves photosynthetic pigments capturing light energy in the form of excited electrons. **130**

light-dependent reactions (līt de-pen′dent re-ak′shuns) The second stage in photosynthesis, during which excited electrons from the light-capturing events are used to make ATP, and water is broken down to hydrogen and oxygen. The hydrogens are transferred to electron carrier coenzymes, $NADP^+$. **131**

light-independent reactions (līt in′de-pen″dent re-ak′shuns) The third stage of photosynthesis; involves cells using ATP and NADPH from the light-dependent reactions to attach CO_2 to 5-carbon starter molecules to manufacture organic molecules; e.g., glucose ($C_6H_{12}O_6$). **131**

limiting factors (lim′ĭ-ting fak′tōrz) Environmental influences that limit population growth. **366**

linkage (lingk′ij) A situation in which the genes for different characteristics are inherited together more frequently than would be predicted by probability. **208**

linkage group (lingk′ij grūp) A group of genes located on the same chromosome that tend to be inherited together. **208**

lipids (lĭ′pidz) Large organic molecules that do not easily dissolve in water; classes include true (neutral) fats, phospholipids, and steroids. **59**

liquid (lik′wid) The phase of matter in which the molecules are strongly attracted to each other, but, because they have more energy and are farther apart than in a solid, they move past each other more freely. **30**

locus (lo′kus) The spot on a chromosome where an allele is located. **182**

loop of Henle (loop uv hen′le) The middle portion of the nephron, which is primarily involved in regulating the amount of water lost from the kidney. **536**

lung (lung) Organs of the body that allow gas exchange to take place between the air and blood. **527**

luteinizing hormone (LH) (loo'te-in-i"zing hōr'mōn) A hormone produced by the anterior pituitary gland that stimulates ovulation. **600**

lymph (limf) Liquid material that leaves the circulatory system to surround cells. **526**

lymphatic system (lim-fă'tik sis'tem) A collection of thin-walled tubes that collect, filter, and return lymph from the body to the circulatory system. **526**

lymph nodes (limf nōdz) Small encapsulated bodies found along the lymph vessels that contain large numbers of white blood cells (WBCs), macrophages and lymphocytes that remove microorganisms and foreign particles from the lymph. **526**

lysosomes (li'so-sōmz) Specialized, submicroscopic organelles that hold a mixture of hydrolytic enzymes. **77**

M

macromolecules (mă-krō-mol'-ĕ-kyūlz) Very large molecules, many of which are composed of many smaller, similar monomers that are chemically bonded together. **48**

malignant tumors (mah-lig'nant too'morz) Nonencapsulated growths of tumor cells that are harmful; they may spread to or invade other parts of the body. **172**

malleus (mă'le-us) The ear bone that is attached to the tympanum. **581**

mass number (mas num'ber) The weight of an atomic nucleus expressed in atomic mass units (the sum of the protons and neutrons). **27**

masturbation (măs"tur-ba'shun) Stimulation of one's own sex organs. **601**

matter (mat'er) Anything that has weight (mass) and takes up space (volume). **12**

mechanical (morphological) isolating mechanisms (mĕ-kan'i kal [mor"fo-loj-i-kal] i-so-la'ting me'kan-izm) Structural differences that prevent mating between members of different species. **286**

medulla oblongata (mĕ-dul'ah ob"long-ga' tah) The region of the more primitive portion of the brain, connected to the spinal cord, that controls such automatic functions as blood pressure, breathing, and heart rate. **572**

medusa (muh-du'sah) A free-swimming adult stage in the phylum Cnidaria that reproduces sexually. **499**

meiosis (mi-o'sis) The specialized pair of cell divisions that reduces the chromosome number from diploid ($2n$) to haploid (n). **164**

Mendelian genetics (men-dē'le-an jĕ-net'iks) The pattern of inheriting characteristics that follows the laws formulated by Gregor Mendel. **196**

menopause (mĕn'o-pawz) The period beginning at about age 50 when the ovaries stop producing viable secondary oocytes and ovarian hormones. **616**

menstrual cycle (men'stru-al si'kul) The repeated building up and shedding of the lining of the uterus. **601**

mesenteries (mes'en-ter"ez) Connective tissues that hold the organs in place and serve as support for blood vessels connecting the various organs. **495**

mesoderm (mes'o-derm) The middle embryonic layer. **495**

messenger RNA (mRNA) (mes'en-jer) A molecule composed of ribonucleotides that functions as a copy of the gene and is used in the cytoplasm of the cell during protein synthesis. **58**

metabolic processes (me-tah-bol'ik pros' es-es) The total of all chemical reactions within an organism—for example, nutrient uptake and processing and waste elimination. **12**

metabolism (mĕ-tab'o-lizm) The total of all the chemical reactions and energy changes that take place in an organism. **12**

metaphase (me'tah-fāz) The second stage in mitosis, characterized by alignment of the chromosomes at the equatorial plane. **167**

metastasize (me-tas'tah-sīz) The process by which cells of tumors move from the original site and establish new colonies in other regions of the body. **172**

microfilaments (mi"kro-fil'ah-ments) Long, fiberlike, submicroscopic structures made of protein and found in cells, often in close association with the microtubules; provide structural support and enable movement. **81**

microorganism (microbe) (mi"kro-or'guh-niz'm) A small organism that cannot be seen without magnification. **442**

microtubules (mi'kro-tū"byuls) Submicroscopic, hollow tubes of protein that function throughout the cytoplasm to provide structural support and enable movement. **81**

minerals (mĭn'er-alz) Inorganic elements that cannot be manufactured by the body but are required in low concentrations; essential to metabolism. **545**

missense mutation (mis-sens miu-ta'shun) A change in the DNA at a single point that causes the wrong amino acid to be used in making a protein. **158**

mitochondria (mi-to-kahn'dre-uh) Membranous organelles resembling small bags with a larger bag inside that is folded back on itself; serve as the site of aerobic cellular respiration. **79**

mitosis (mi-to'sis) A process that results in equal and identical distribution of replicated chromosomes into two newly formed nuclei. **164**

mixture (miks'chur) Matter that contains two or more substances *not* in set proportions. **34**

molecules (mol'ĕ-kūlz) The smallest particles of a chemical compound; the smallest naturally occurring parts of an element or a compound. **29**

monocot (mon'o-kot) An angiosperm whose embryo has one seed leaf (cotyledon). **480**

monoculture (mon"o-kul'chur) The agricultural practice of planting the same varieties of a species over large expanses of land. **249**

monohybrid cross (mon"o-hi'brid kros) A mating between two organisms that are both heterozygous for the one observed gene. **194**

monosomy (mon'o-so"me) A cell with only 1 of the 2 chromosomes of a homologous pair. **185**

mortality (mor-tal'ĭ-te) The number of individuals leaving the population by death per thousand individuals in the population. **364**

motor neurons (mo'tur noor'onz") Neurons that carry information from the central nervous system to muscles or glands. **569**

motor unit (mo′tur yoo′nit) All the muscle cells stimulated by a single neuron. **584**

multigene families (mul··ti-jēn fam·ĭ-lez) A type of variation in an organism's DNA sequence; this variation consists of several different genes that produce different proteins that are related in function. **226**

multiple alleles (mul′tĭ-pul a-lēlz′) Several different alleles for a particular characteristic within a population, not just 2. **205**

multiregional hypothesis (mul″ti-re′jun-al hi-poth′e-sis) The concept that *Homo erectus* migrated to Europe and Asia from Africa and evolved into *Homo sapiens*. **298**

mutation (miu-ta′shun) Any change in the genetic information of a cell. **171**

mutualism (miu′chu-al-izm) A relationship between two organisms in which both organisms benefit. **333**

mycorrhizae (mi″ko-ri′zah) Symbiotic relationships between fungi and plant roots. **460**

mycotoxins (mi″ko-tok′sinz) Deadly poisons produced by fungi. **462**

myosin (mi′o-sin) The protein molecule, found in the thick filaments of muscle fibers, that attaches to actin, bends, and moves actin molecules along its length, causing the muscle fiber to shorten. **582**

N

natality (na-tal′ĭ-te) The number of individuals entering the population by reproduction per thousand individuals in the population. **364**

negative-feedback inhibition (nĕg′ăh-tiv fēd′bak in″hĭ-bĭ′shun) A regulatory mechanism in which an increase in the stimulus causes a decrease of the response and results in homeostasis. **104**

nekton (nek′ton) Many kinds of aquatic animals that are large enough and strong enough to be able to swim against currents and tides and go where they want to. **503**

nephrons (nef′ronz) Tiny tubules that are the functional units of kidneys. **536**

nerve impulse (nerv im′puls) A series of changes that take place in the neuron resulting in a wave of depolar-

ization, which passes from one end of the neuron to the other. **569**

nerves (nervz) Bundles of neuronal fibers. **569**

nervous system (ner′vus sis′tem) A network of neurons that carry information from sense organs to the central nervous system and from the central nervous system to muscles and glands. **569**

net movement (net mūv′ment) Movement in one direction minus the movement in the other. **84**

neuron (nerve cell) (noor′on″) The cellular unit consisting of a cell body and fibers that makes up the nervous system. **569**

neurotransmitter (noor″o-trans′mĭt-er) A molecule released by the axons of neurons, stimulative other cells. **570**

neutron (nu′tron) A particle in the nucleus of an atom that has no electrical charge; named *neutron* to reflect this lack of electrical charge. **25**

niche (knee-sh) An organisms specific functional role in its community. **326**

nicotinamide adenine dinucleotide (NAD^+) An electron acceptor and hydrogen carrier used in respiration. **99**

nicotinamide adenine dinucleotide phosphate ($NADP^+$) An electron acceptor and hydrogen carrier used in photosynthesis. **107**

nitrifying bacteria (ni′trĭ-fi-ing bak-te′re-ah) Several kinds of bacteria capable of converting ammonia to nitrite, or nitrite to nitrate. **314**

noncoding strand (non-kō-ding strand) The strand of DNA that is not read directly by the enzymes. **147**

nondeciduous (non″de-sid′yu-us) A term used a type of trees that do not lose their leaves all at once. **482**

nondisjunction (non″dis-junk′shun) An abnormal meiotic division that results in sex cells with too many or too few chromosomes. **185**

nonhomologous chromosomes (non-ho-mol′o-gus kro′mo-sōmz) Chromosomes that have different genes on their DNA. **174**

nonsense mutation (non-sens miu-ta′shun) A type of point mutation that causes a rilbosome to stop protein synthesis by introducing a stop codon too early. **158**

norepinephrine (nor-ĕ″pĭ-nef′rin) The hormone produced by the adrenal medulla and certain nerve cells that increases heart rate, blood pressure, and breathing rate. **575**

nuclear membrane (nu′kle-ar mem′brān) The structure surrounding the nucleus that separates the nucleoplasm from the cytoplasm. **78**

nucleic acids (nu-kle′ik ă′sidz) Complex molecules that store and transfer information within a cell. They are constructed of fundamental monomers known as nucleotides. **56**

nucleolus (nu-kle′o-les) A nuclear structure composed of completed or partially completed ribosomes and the specific parts of chromosomes that contain the information for their construction. **83**

nucleoplasm (nu′kle-o-plazm) The liquid matrix of the nucleus, composed of a mixture of water and the molecules used in the construction of the rest of the nuclear structures. **84**

nucleoprotein (nu-kle′o-prō′tēn) DNA strands with attached proteins that become visible during cell division. **152**

nucleosome (nu-kle′o- sōm) Histone protein with their encircling DNA. **152**

nucleotides (nu″kle-o-tĭ d′z) Fundamental subunits of nucleic acid constructed of a phosphate group, a sugar, and an organic nitrogenous base. **56**

nucleus (nu′kle-us) The central body that contains the information system for the cell; also the central part of an atom, containing protons and neutrons. **69, 25**

nutrients (nu′tre-ents) Molecules required by organisms for growth, reproduction, or repair. **96**

nutrition (nu-trĭ′shun) Collectively, the processes involved in taking in, assimilating, and utilizing nutrients. **542**

O

obese (o-bēs) A person with a body mass index that is 30 kg/m² or more and has a significantly increased risk of many different kinds of diseases. **553**

observation (ob-sir-vā′shun) The process of using the senses or extensions of the senses to record events. **3**

observational learning (imitation) (ob-sir-vā' shun-uhl lur'ning) A form of association that involves a complex set of associations used in watching another animal being rewarded for performing a particular behavior and then performing the same behavior oneself. 384

offspring (of'spring) Descendants of a set of parents. 193

olfactory epithelium (ōl-fak'to-re ě"pǐ-thē'leum) The cells of the nasal cavity that respond to chemicals. 579

omnivores (om'nǐ-vōrz) Animals that are carnivores at some times and herbivores at others. 306

oogenesis (oh"ō-jen'ě-sis) The gametogenesis process that leads to the formation of eggs. 605

operant (instrumental) conditioning (op'ě-rant kon' dǐ'shun-ing) A change in behavior that results from associating a stimulus with a response by either rewarding or punishing the behavior after it has occurred. 383

order (or'der) A group of closely related classes within a phylum. 424

organ (or'gun) A structure composed of groups of tissues that perform particular functions. 16

organ system (or'gun sis'tem) A structure composed of groups of organs that perform particular functions. 16

organelles (or-gan-elz') Cellular structures that perform specific functions in the cell; the function of organelles is directly related to their structure. 69

organic molecules (or-gan'ik mol'uh-kiulz) Complex molecules whose basic building blocks are carbon atoms in chains or rings. 46

organism (or'gun-izm) An independent living unit. 14

orgasm (or'gaz-um) The complex series of responses to sexual stimulation that results in an intense frenzy of sexual excitement. 601

osmosis (os-mo'sis) The net movement of water molecules through a selectively permeable membrane. 86

osteoporosis (os"te-o-po-ro'sis) A disease condition resulting from the demineralization of the bone, resulting in pain, deformities, and fractures; related to a loss of calcium. 545

out-of-Africa hypothesis The concept that modern humans (*Homo sapiens*) originated in Africa and migrated from Africa to Europe and Asia and displaced existing hominins. 298

oval window (o'val win'do) The membrane-covered opening of the cochlea, to which the stapes is attached. 581

ovaries (o'vah-rez) Female sex organs, which produce haploid sex cells, called eggs. 174

oviduct (o'vǐ-dukt) The tube (*fallopian tube*) that carries the egg to the uterus. 605

ovulation (ov-yu-lā'shun) The release of a secondary oocyte from the surface of the ovary. 601

oxidation-reduction reaction (ok'sǐ-day-shun re-duk'shun re-ak'shun) An electron-transport reaction in which the molecules losing electrons become oxidized and those gaining electrons become reduced. 36

oxidizing atmosphere (ok'sǐ-di-zing at'mosfer) An atmosphere that contains molecular oxygen. 411

oxytocin (ok"sǐ-to'sin) The hormone, released from the posterior pituitary, that causes contraction of the uterus. 585

P

pancreas (pan'kre-as) The organ of the body that secretes many kinds of digestive enzymes into the duodenum. 532

panspermia (pan-sper'me-a) A hypothesis by Svante Arrhenius in the early 1900s that life arose outside the Earth and that living things were transported to Earth to seed the planet with life. 403

parasite (pěr'uh-sīt) An organism that lives in or on another organism and derives nourishment from it. 330

parasitism (pěr'uh-sīt-izm) A relationship between two organisms that involves one organism living in or on another organism and deriving nourishment from it. 330

pathogens (path'uh-jenz) Agents that cause specific diseases. 445

penetrance (pen'ě-trens) A term used to describe how often an allele expresses itself when present. 267

pepsin (pep'sin) The enzyme, produced by the stomach, that is responsible for beginning the digestion of proteins. 532

perception (per-sep'shun) Recognition by the brain that a stimulus has been received. 577

perfect flower (pur'fekt flow'er) A flower that contains both male (stamen) and female (pistil) reproductive structures. 479

peripheral nervous system (pǔ-rǐ'fě-ral ner'vus sis'tem) The fibers that communicate between the central nervous system and other parts of the body. 569

peroxisomes (pě-roks'ǐ-sōmz) Membrane-bound, submicroscopic organelles that hold enzymes capable of producing hydrogen peroxide that aids in the control of infections and other dangerous compounds. 77

petals (pě'tuls) Modified leaves of angiosperms; accessory structures of a flower. 479

pH A scale used to indicate the concentration of an acid or a base. 39

phagocytosis (fa"go-si-to'sis) The process by which the cell wraps around a particle and engulfs it. 88

pharynx (far'inks) The region at the back of the mouth cavity; the throat. 530

phases of matter (fāzez uv mat'er) Physical conditions of matter (solid, liquid, and gas) determined by the relative amounts of energy of the molecules. 30

phenotype (fēn'o-tīp) The physical, chemical, and behavioral expression of the genes possessed by an organism. 193

pheromones (fěr'uh-mōnz) Chemicals produced by an animal and released into the environment to trigger behavioral or developmental processes in another animal of the same species. 389

phloem (flo'em) One kind of vascular tissue found in higher plants; it transports food materials from the leaves to other parts of the plant. 469

phospholipids (fos"fo-lǐ'pidz) A class of water-insoluble molecules that resembles fats but contains a phosphate group (PO_4) in its structure. 61

phosphorylation reaction (fos"for lā shun re-ak shun) The reaction that takes place when a custer of atoms, known as a phosphate group is added to another molecule. 37

photoperiod (fo″to-pĭr′e-ud) The length of the light part of the day. 395

photosynthesis (fo-to-sin′thuh-sis) A series of reactions that take place in chloroplasts and result in the storage of sunlight energy in the form of chemical-bond energy. 79

photosystems (fo′to-sis″temz) Clusters of photosynthetic pigments (e.g., chlorophyll) that serve as energy-gathering or energy-concentrating mechanisms; used during the light-capturing events of photosynthesis. 133

phylogeny (fi-laj′uh-ne) The science that explores the evolutionary relationships among organisms and seeks to reconstruct evolutionary history. 425

phylum (fi′lum) A subdivision of a kingdom. 423

phytoplankton (fye-tuh-plank′tun) Microscopic, photosynthetic species that form the basis for most aquatic food chains. 449

pinocytosis (pĭ″no-si-to′sis) The process by which a cell engulfs some molecules dissolved in water. 88

pioneer community (pi″o-nēr′ ko-miu′nĭ-te) The first community of organisms in the successional process established in a previously uninhabited area. 341

pioneer organisms (pi″o-nēr′ or′gu-nizms) The first organisms in the successional process. 341

pistil (pis′til) The female reproductive structure in flowers, which contains the ovary, which produces eggs. 174

pituitary gland (pĭ-tu′ĭ-tĕ-re gland) The gland at the base of the brain that controls the functioning of other glands throughout the organism. 600

placenta (plah-sen′tah) An organ made up of tissues from the embryo and the uterus of the mother that allows for the exchange of materials between the mother's bloodstream and the embryo's bloodstream; it also produces hormones. 610

plankton (plank′tun) Small, floating or weakly swimming organisms. 449

plasma (plaz′muh) The watery matrix that contains the molecules and cells of the blood. 520

plasma membrane (cell membrane) (plaz′muh mem′brān) The outer boundary membrane of the cell. 73

plasmid (plaz′mid) A plasmid is a circular piece of DNA that is found free in the cytoplasm of some bacteria. 228

platelets (plat′letez) Fragments of specific kinds of white blood cells and are important in blood clotting. 522

pleiotropy (ple″o-tro′-pe) The multiple effects that a gene may have on the phenotype of an organism. 207

poikilotherms (poy-ki′luh-thermz) Animals with a variable body temperature, which changes with the external environment. 493

point mutation (point miu-ta′shun) A change in the DNA of a cell as a result of a loss or change in a nitrogenous-base sequence. 158

polar body (po′lar bod′e) The smaller of two cells formed by unequal meiotic division during oogenesis. 605

pollen (pol′en) The male gametophyte in gymnosperms and angiosperms. 477

pollination (pol″ĭ-na′shun) The transfer of pollen in gymnosperms and angiosperms. 477

polygenic inheritance (pol″e-jen′ik in-her′ĭ-tans) The concept that a number of different pairs of alleles may combine their efforts to determine a characteristic. 206

Polymers (po-lim′er) Combinations of many smaller, similar building blocks called monomers (mono = single) bonded together. 48

polymerase chain reaction (PCR) (po-lim′er-ās chān re-ak′shun) A laboratory technique that is able to generate useful quantities of DNA from very small amounts of DNA. 216

polyp (pol′ĭp) A sessile larval stage in the phylum Cnidaria that reproduces asexually. 499

polypeptide (pŏ″le-pep′tīd) A macromolecule composed of a specific sequence of amino acids. 54

polyploidy (pah″lĭ-ploy′de) A condition in which cells contain multiple sets of chromosomes. 284

pons (ponz) The region of the brain, immediately anterior to the medulla oblongata, that connects to the cerebellum and higher regions of the brain and controls several sensory

and motor functions of the head and face. 572

population (pop″u-la′shun) A group of organisms of the same species located in the same place at the same time. 240

population density (pop″u-la′shun den′sĭ-te) The number of organisms of a species per unit area. 362

population distribution (pop″u-la′shun dis··trĭ-biu·shun) The way individuals within a population are arranged with respect to one another. 362

population genetics (pop″u-la′shun jĕ-ne·tiks) The study of the kinds of genes within a population, their relative numbers, and how these numbers change over time. 240

population growth curve (pop″u-la′shun grōth kurv) A graph of the change in population size over time. 364

population pressure (pop″u-la′shun presh′yur) Intense competition as a result of high population density that leads to changes in the environment and the dispersal of organisms. 363

positive feedback (poz′i-tiv fēd′bak) A regulatory mechanism in which an increase in the stimulus causes an increase of the response and does not result in homeostasis. 568

potential energy (po-ten′shul en′er-je) The energy an object has because of its position. 24

predation (prĕ-da′shun) A relationship between two organisms that involves the capturing, killing, and eating of one by the other. 329

predator (pred′uh-tor) An organism that captures, kills, and eats another animal. 329

prey (prā) An organism captured, killed, and eaten by a predator. 329

primary carnivores (pri′mar-e kar′nĭ-vōrz) Carnivores that eat herbivores and are therefore on the third trophic level. 306

primary consumers (pri′mar-e kon-su′merz) Organisms that feed directly on plants—herbivores. 306

primary succession (pri′mar-e suk-sĕ′shun) The orderly series of changes that begins in a previously uninhabited area and leads to a climax community. **340**

prions (pri′onz) Infectious protein particles responsible for diseases such as Creutzfeldt-Jakob disease and bovine spongiform encephalitis. **436**

probability (prob″a-bil′ĭ-te) The chance that an event will happen, expressed as a percentage or fraction. **195**

producers (pro-du′surz) Organisms that produce new organic material from inorganic material with the aid of sunlight. **305**

productivity (pro-duk-tiv′ĭ-te) The rate at which an ecosystem can accumulate new organic matter. **316**

products (prŏ′dukts) New molecules resulting from a chemical reaction. **36**

progesterone (pro-jes′ter-ōn) A hormone produced by the corpus luteum that maintains the lining of the uterus. **601**

prokaryotic cells (pro′kār′e-ot″ik sels) One of the two major types of cells. They do not have a typical nucleus bound by a nuclear membrane and lack many of the other membranous cellular organelles—for example, members of Eubacteria and Archaea. **69**

prophase (pro′fāz) The first phase of mitosis, during which individual chromosomes become visible. **166**

proteins (prō′tēnz) Macromolecules made up of one or more polypeptides attached to each other by bonds. **53**

protein-sparing (prō′tēn spē′ring) The conservation of proteins by first oxidizing carbohydrates and fats as a source of ATP energy. **545**

proteomics (prō-tee-o′-miks) A new field of science that has developed since the sequencing of the human genome; it groups proteins by similarities to help explain their function and how they may have evolved. **227**

proton (pro′ton) The particle in the nucleus of an atom that has a positive electrical charge. **25**

proto-oncogene (pro′to-onko-gen) Genes that code for proteins that provide signals to the cell that encourage cell division. **170**

protoplasm (pro′to-plazm) The living portion of a cell, as distinguished from the nonliving cell wall. **69**

protozoa (pro″to-zo′ah) Heterotrophic, eukaryotic, unicellular organisms. **452**

proximal convoluted tubule (prok′sĭ-mal kon′vōl-lu-ted tŭb′yŭl) The upstream end of the nephron of the kidney, which is responsible for reabsorbing most of the valuable molecules filtered from the glomerulus into Bowman's capsule. **536**

pseudocoelom (su dō-se′lo-m) Or false body cavity. A body cavity located between the lining of the gut and the outer body wall and does not have muscles around the digestive system. **495**

pseudoscience (su-dō-si′ens) An activity that uses the appearance or language of science to convince or mislead people into thinking that something has scientific validity. **10**

puberty (pu′ber-te) A time in the life of a developing individual characterized by the increasing production of sex hormones, which cause it to reach sexual maturity. **600**

pulmonary artery (pul′muh-nă-rē ar′tuh-rē) The major blood vessel that carries blood from the right ventricle to the lungs. **523**

pulmonary circulation (pul′muh-nă-rē ser-kyu-lā′shun) The flow of blood through certain chambers of the heart and blood vessels to the lungs and back to the heart. **523**

punctuated equilibrium (pung′chu-a-ted e-kwĭ-lib′re-um) The theory stating that evolution occurs in spurts, between which there are long periods with little evolutionary change. **292**

Punnett square (pun′net sqwār) A method used to determine the probabilities of allele combinations in a zygote. **194**

pyloric sphincter (pi-lor′ik sfingk′ter) The valve located at the end of the stomach that regulates the flow of food from the stomach to the duodenum. **532**

R

radial symmetry (ra′de-ul sĭm ĭ-tre) The characteristic of an animal with a body constructed around a central axis; any division of the body along this axis results in two similar halves. **494**

reactants (re-ak′tants) Materials that will be changed in a chemical reaction. **36**

receptor mediated endocytosis (re-sep′tor me-de-it′d en″dō-si-tō′sis) The process in which molecules from the cell's surroundings bind to receptor molecules on the plasma membrane, followed by the membrane folding into the cell so that the cell engulfs these molecules. **88**

recessive allele (re-sĕ′siv a-lēl′) An allele that, when present with its homolog, does not express itself and is masked by the effect of the other allele. **193**

recombinant DNA (re-kom′bĭ-nant) DNA that has been constructed by inserting new pieces of DNA into the DNA of an organism. **229**

red blood cells (**rbcs**) (red blud sels) small disk-shaped cells that lack a nucleus and contain the iron-containing pigment hemoglobin. **520**

reducing atmosphere (re-du′sing at′mos-fēr) An atmosphere that does not contain molecular oxygen (O_2). **406**

reduction division (re-duk′shun dĭ-vĭ′zhun) A type of cell division, in which daughter cells get only half the chromosomes from the parent cell. **176**

regulator proteins (reg′yu-la-tōr prō′tēnz) Proteins that influence the activities that occur in an organism—for example, enzymes and some hormones. **56**

reproductive capacity (**biotic potential**) (re-pro-duk′tiv kuh-pas′ĭ-te) The theoretical maximum rate of reproduction. **364**

reproductive (genetic) isolating mechanisms (re-produk′tiv i-so-la′ting me′kan-izmz) Mechanisms that prevent interbreeding between species. **285**

respiratory system (res-pĭ-ra·tory sis-tum) The organ system that moves air into and out of the body and consists of lungs, trachea, the air-transport pathway, and diaphragm. **527**

response (re-spons′) The reaction of an organism to a stimulus. **379**

responsive processes (re-spon′siv pros′es-es) Abilities to react to external and internal changes in the environment—for example, irritability, individual adaptation, and evolution. **13**

restriction enzymes (re-strik′shun en-zīm) Proteins that catalyze the cutting of the DNA helix; these proteins cut the DNA helix into two pieces at specific DNA sequences. **216**

restriction fragments (re-strik′shun frag′ments) Pieces of DNA that are created by cutting DNA with restriction enzymes; restriction fragments are used in cloning and characterizing DNA. 216

restriction sites (re-strik′shun sits) The DNA sequence that is recognized by a restriction enzyme; the restriction enzyme cuts at this sequence, generating two DNA fragments. 216

retina (rĕ′tĭ-nah) The light-sensitive region of the eye. 580

rhodopsin (ro-dop′sin) A light-sensitive pigment found in the rods of the retina. 580

ribonucleic acid (RNA) (ri-bo-nu-kle′ik ă′sid) A polymer of nucleotides formed on the template surface of DNA by transcription. Three forms that have been identified are mRNA, rRNA, and tRNA. 58

ribosomal RNA (rRNA) (ri-bo-sōm′al) A globular form of RNA; a part of ribosomes. 58

ribosomes (ri′bo-sōmz) Small structures composed of two protein and ribonucleic acid subunits, involved in the assembly of proteins from amino acids. 75

ribulose (ri′bu-lōs) A 5-carbon sugar molecule used in photosynthesis. 134

ribulose biphosphate carboxylase (RuBisCo) (ri′bu-lōs bi-fos′fāt kar-bak′se-lāz) An enzyme found in the stroma of chloroplast that speeds the combining of the CO_2 with an already present 5-carbon carbohydrate, ribulose. 134

RNA polymerase (po-lim′er-ās) An enzyme that bonds RNA nucleotides together during transcription after they have aligned on the DNA. 147

rods (rahdz) Light-sensitive cells in the retina of the eye that respond to low-intensity light but do not respond to different colors of light. 580

root (root) Underground structures that anchor the plant and absorbe water and minerals. 469

root hairs (root hārs) Tiny cellular outgrowths of roots that improve the ability of plants to absorb water and minerals. 470

S

salivary glands (să′lĭ-vĕ′rē glanz) Glands that produce saliva. 530

salts (sŏlts) Ionic compounds formed from a reaction between an acid and a base. 39

saprophytes (sap′ruh-fīts) Organisms that obtain energy by the decomposition of dead organic material. 430

saturated (sat′yu-ra-ted) A term used to describe the carbon skeleton of a fatty acid that contains no double bonds between carbons. 60

science (si′ens) A process used to solve problems or develop an understanding of natural events. 2

scientific law (si-en-tif′ik law) A uniform or constant fact that describes what happens in nature. 8

scientific method (si-en-tif′ik meth′ud) A way of gaining information (facts) about the world around you that involves observation, hypothesis formation, testing of hypotheses, theory formation, and law formation. 2

seasonal isolating mechanisms (se′zun-al eye-sō-lā′ting mek-ă-nizmz) Reproductive isolating mechanisms that prevent interbreeding between species because they reproduce at different times of the year. 286

secondary carnivores (sĕk′on-dĕr-e kar′nĭ-vōrz) Carnivores that feed on primary carnivores and are therefore at the fourth trophic level. 306

secondary consumers (sek′on-dĕr-e konsu′merz) Animals that eat other animals—carnivores. 306

secondary sexual characteristics (sek′on-dĕr-e sek′shoo-al kăr-ak-tĕ-ris′tiks) Characteristics of the adult male or female, including the typical shape that develops at puberty: broader shoulders; heavier long-bone muscles; development of facial hair, axillary hair, and chest hair; and changes in the shape of the larynx in the male; rounding of the pelvis and breasts and changes in deposition of fat in the female. 601

secondary succession (sek′on-dĕr-e suk-sĕ′shun) The orderly series of changes that begins with the disturbance of an existing community and leads to a climax community. 341

seed (see-d) A specialized structure that contains an embryo along with stored food enclosed in a protective covering called the seed coat. 477

seed leaves (sēd lēvz) Cotyledons; embryonic leaves in seeds. 480

segmental duplications (seg′men-tal du″pli-ka-shunz) A type of variation in an organism's DNA sequence; this variation occurs when a segment of DNA, which may contain several genes, occurs twice in the genome. 226

segmentation (seg″men-ta′shun) The separation of the body of an animal into a number of recognizable units from the anterior to the posterior end of the animal. 177

segregation (seg″rĕ-ga′shun) The separation and movement of homologous chromosomes to the opposite poles of the cell. 183

selecting agents (se-lek′ting a′jents) Factors that affect the probability that a gene will be passed on to the next generation. 263

selectively permeable (se-lek′tiv-le per′me-uh-bul) The property of a membrane that allows certain molecules to pass through it but interferes with the passage of others. 85

semen (se′men) The sperm-carrying fluid produced by the seminal vesicles, prostate gland, and bulbourethral glands of males. 602

semicircular canals (sĕ-mi-ser′ku-lar că-nalz′) A set of tubular organs, associated with the cochlea, that sense changes in the movement or position of the head. 582

semilunar valves (sĕ-me-lu′ner valvz) Valves, located in the pulmonary artery and aorta, that prevent the flow of blood backwards, into the ventricles. 523

seminiferous tubules (sem″ĭ-nif′ur-us tub′yūlz) Sperm-producing tubes in the testes. 603

sensory neurons (sen′so-re noor′onz″) Neurons that send information from sense organs to the central nervous system. 569

sepals (se′pals) Accessory structures of flowers. 479

sessile (ses′il) Firmly attached. 498

sex (seks) The nature of the biological differences between males and females. 596

sex chromosomes (seks kro'mo-sōmz) Chromosomes that carry genes that determine the sex of an individual (X and Y in humans). **209**

sex-determining chromosomes (seks de-ter'mĭ-ning kro'mo-sōmz) The chromosomes X and Y, which are primarily responsible for determining if an individual will develop as a male or a female. **597**

sex linkage (seks lingk'ij) Refers to genes that are located on the chromosomes that determine the sex of an individual. **209**

sex ratio (seks ra'sho) The number of males in a population compared with the number of females. **362**

sexual reproduction (sek'shu-al re"pro-duk'shun) The propagation of organisms involving the union of gametes from two parents. **164**

sexual selection (sek'shu-al se-lek'shun) Selection that is the result of specific individuals being chosen by members of the opposite sex for mating purposes. **270**

sexuality (seks"u-al'i-te) All the factors that contribute to one's female or male nature. **596**

signal transduction (sig'-nal trans-duk'-shun) The process by which cells detect specific signals and transmit those signals to the cell's interior. **75**

silent mutation (si'lent miu-ta'shun) A change of a single nucleotide that does not cause a change in the amino acids used to build a protein. **158**

single-factor cross (sing'ul fak'tur kros) A genetic study in which a single characteristic is followed from the parental generation to the offspring. **194**

sister chromosomes (sis'ter kro'mo-sōmz) The 2 chromatids of a chromosome that are produced by DNA replication and that contain the same DNA. **165**

skeleton (skel'e-ton) The part of an organism that provides structural support. **496**

small intestine (smahl in-tes'ten) The portion of the digestive system immediately following the stomach; it is responsible for digestion and absorption. **532**

society (so-si'uh-te) Interacting groups of animals of the same species that show division of labor. **395**

sociobiology (so-se-o-bi-ol'o-je) The systematic study of all forms of social behavior, both human and nonhuman. **397**

solid (sol'id) The phase of matter in which the molecules are packed tightly together; they vibrate in place. **30**

solute (sol'lūt) The component that dissolves in a solvent. **36**

solution (sol'ū shun) A homogeneous mixture of ions or molecules of two or more substances. **34**

solvent (sol'vent) The component present in the larger amount. **36**

soma (so'mah) The cell body of a neuron, which contains the nucleus. **569**

somatic cell nuclear transfer (so-mat'ik sel nu'kle-ar trans'fer) A laboratory technique in which the nucleus of a cell is placed into an unfertilized egg cell; the cell may then be stimulated to grow; cells that are generated from this growth will have the same genetic information as the cell that donated the nucleus. **231**

speciation (spe-she-a'shun) The process of generating new species. **283**

species (spe'shēz) A population of organisms potentially capable of breeding naturally among themselves and having offspring that also interbreed. **240**

specific dynamic action (SDA) (spĕ-sĭ'fik dină'mik ak'shun) The amount of energy required to digest and assimilate food. SDA is equal to approximately 10% of total daily kilocalorie intake. **552**

specific epithet (spĕ-sĭ'fik ep"ĭ-the't) A word added to the genus name to identify which one of several species within the genus is being identified (i.e., *Homo sapiens*: *Homo* is the genus name and *sapiens* is the specific epithet). **423**

sperm (spurm) The haploid sex cells produced by sexually mature males. **597**

spermatogenesis (spur-mat-o-jen'uh-sis) The gametogenesis process that leads to the formation of sperm. **603**

spinal cord (spi'nal kōrd) A collection of nerve fibers surrounded by the verte-

brae that conveys information to and from the brain. **569**

spindle (spin'dul) An array of microtubules extending from pole to pole; used in the movement of chromosomes. **166**

spindle fiber (spin'dul fi'bers) Microtubules that are individual strands of the spindle. **166**

spontaneous generation (spon-ta'ne-us jen-uhra'shun) The idea that living organisms arose from nonliving material. **402**

spontaneous mutations (spon-ta'ne-us miu-ta'shunz) Natural changes in the DNA caused by unidentified environmental factors. **266**

spore (spōr) In the kingdom Fungi, a cell with a tough protective cell wall that can resist extreme conditions. **457**

sporophyte generation (or stage) (spōr·o-fit jen-uh-ra·shun) A stage in the life cycle of plants in which this diploid (2*n*) plant, which has special plant parts where meiosis takes place, produces haploid (n) spores. **433**

sporophyte stage (spōr'o-fi t stāj) A life cycle stage in plants in which a haploid spore is produced by meiosis. **433**

stabilizing selection (stā-bill-ĭ-zing se-lĕk'-shun) Selection that occurs when individuals at the extremes of the range of a characteristic are consistently selected against. **271**

stable equilibrium phase (stā'bul e-kwi-lib're-um fāz) The period of time during population growth when the number of individuals entering a population and the number leaving the population are equal, resulting in a stable population size. **364**

stamens (sta'menz) Male reproductive structures of a flower. **479**

stapes (sta'pēz) The ear bone that is attached to the oval window. **581**

stem (stem) Plant structures that connect the roots with the leaves and position the leaves so that they receive sunlight. **469**

stem cells (stem selz) Cells that can differentiate into any type of cell, including liver cells, skin cells, and brain cells; embryonic and hematopoietic cells. **232**

steroids (stēr′oidz) One of the three kinds of lipid molecules characterized by their arrangement of interlocking rings of carbon. **62**

stimulus (stim′yu-lus) Any change in the internal or external environment of an organism that it can detect. **379**

stroma (stro′muh) The region within a chloroplast that has no chlorophyll. **79**

structural proteins (struk′chu-ral prō′tēns) Proteins that are important for holding cells and organisms together, such as the proteins that make up the cell membrane, muscles, tendons, and blood. **56**

subspecies (**breeds, varieties, strains, races**) (sub′spe-shēz) Regional groups within a species that are significantly different structurally, physiologically, or behaviorally yet are capable of exchanging genes by interbreeding. **243**

substrate (sub′strāt) A reactant molecule with which the enzyme combines. **97**

succession (suk-sĕ′shun) The process of changing one type of community to another. **340**

successional (**stage**) (**successional community**) (suk-sĕ′shunal stāj) An intermediate stage in succession. **340**

symbiosis (sim-bi-o′sis) A close physical relationship between two kinds of organisms; parasitism, commensalism, and mutualism are examples of symbiosis. **330**

symbiotic nitrogen-fixing bacteria (sim-bi-ah′tik ni′tro-jen fik′sing bak-te′re-ah) Bacteria that live in the roots of certain kinds of plants, where they convert nitrogen gas molecules into compounds that plants can use. **314**

synapse (sin-aps) The space between the axon of one neuron and the dendrite of the next, where chemicals are secreted to cause an impulse to be initiated in the second neuron. **570**

synapsis (sin-ap′sis) The condition in which the two members of a pair of homologous chromosomes come to lie close to one another. **176**

systemic circulation (sis-tĕ′mik ser-kyu-la′shun) The flow of blood through certain chambers of the heart and blood vessels to the general body and back to the heart. **523**

systolic blood pressure (sis-tah′lik blud presh′yur) The pressure generated in a large artery when the ventricles of the heart are contracting. **524**

T

tandem clusters (tan′dem klus′ters) A type of variation in an organism's DNA sequence; this variation occurs when a single gene is duplicated several times in the same region of DNA. **226**

taxonomy (tak-son′uh-me) The science of classifying and naming organisms. **422**

telomere (tel′o-mēr) A chromosome cap composed of repeated, specific sequences of nucleotide pairs; its activity or inactivity is associated with cell aging and cancer. **157**

telophase (tel′uh-fāz) The last phase in mitosis, characterized by the formation of daughter nuclei. **168**

temperature (tem′per-̆a-chiur) A measure of molecular energy of motion. **29**

territorial behavior (ter″ĭ-tor′e-al be-hāv′yur) Behavior involved in establishing, defending, and maintaining a territory for food, mating, or other purposes. **392**

territory (ter′ĭ-tor-e) A space that an animal defends against others of the same species. **392**

testes (tes′tēz) The male sex organs, which produce haploid cells—sperm. **174**

testosterone (tes-tos-tur-ōn) The male sex hormone, produced in the testes, that controls male sexual development. **575**

thalamus (thal′ah-mus) The region of the brain that relays information between the cerebrum and lower portions of the brain; it also provides some level of awareness in that it determines pleasant and unpleasant stimuli and is involved in sleep and arousal. **572**

theory (the′o-re) A widely accepted, plausible generalization about fundamental concepts in science that is supported by many experiments and explains why things happen in nature. **7**

theory of natural selection (the′o-re uv nat′chu-ral se-lek′shun) In a species of genetically differing organisms, the organisms with the genes that enable them to survive better in the environment and thus reproduce more offspring than others will transmit more of their genes to the next generation. **261**

thinking (thingk′ing) A mental process that involves memory, a concept of self, and an ability to reorganize information. **389**

thylakoid (thī′la-koid) A thin, flat disk found in the chloroplast of plant cells that is the site of the light-capturing events and light-dependent reactions of photosynthesis. **79**

thymine (thi′mĕn) A single-ring nitrogenous-base molecule in DNA but not in RNA; it is complementary to adenine. **144**

tissue (tish′yu) A group of specialized cells that work together to perform a specific function. **16**

tissue fluid (tish′yu flu′id) The liquid that baths the body's cells and contains the same chemicals as plasma but smaller amounts of the blood protein albumin. **522**

trachea (trā′ke-uh) The major tube, supported by cartilage rings, that carries air to the bronchi; also known as the windpipe. **527**

transcription (tran-skrip′shun) The process of manufacturing RNA from the template surface of DNA; three forms of RNA that can be produced are mRNA, rRNA, and tRNA. **147**

transcription factors (tran-skrip′shun fak′torz) Proteins that help control the transcription process by binding DNA or other transcription factors and regulating when RNA polymerase begins transcription. **154**

transcriptomics (tran-skrip-tō′-miks) A new field of science developed since the sequencing of the human genome; it looks at when and how much of a particular transcript is made by an organism. **227**

transfer RNA (**tRNA**) (trans′fur) A molecule composed of ribonucleic acid. It is responsible for transporting a specific amino acid into a ribosome for assembly into a protein. **58**

transformation (trans″for-ma′shun) A technique or process in which an organism gains new genetic information from its environment; this is known to happen to bacteria and is used to introduce DNA fragments to bacteria during the DNA cloning process. **229**

translation (trans-la'shun) The process whereby tRNA uses mRNA as a guide to arrange the amino acids in their proper sequence according to the genetic information in the chemical code of DNA. 160

transpiration (tran"spĭ-ra'shun) In plants, the transportation of water from the soil by way of the roots to the leaves, where it evaporates. 313

triploblastic (trĭp"low-bla'stik) A condition typical of most animals in which their bodies consist of three layers of cells. 495

trisomy (trī s'oh-me) The presence of 3 chromosomes instead of the normal 2, resulting from the nondisjunction of homologous chromosomes during meiosis—as in Down syndrome. 185

trophic level (tro'fik lĕ'vel) A step in the flow of energy through an ecosystem. 305

tropism (tro'pizm) Any reaction to a particular stimulus in which the organism orients toward or away from the stimulus. 485

tropomyosin (tro"po-mi'o-sin) A molecule, found in thin myofilaments of muscle, that helps regulate when muscle cells contract. 582

troponin (tro'po-nin) A molecule, found in thin myofilaments of muscle, that helps regulate when muscle cells contract. 582

true (neutral) fats (troo fats) Important organic molecules composed of glycerol and fatty acids that are used to provide energy. 59

tumor (too'mor) A mass of undifferentiated cells not normally found in a certain portion of the body. 172

tumor suppressor gene (too'mor su-pre'or jēn) Code for proteins that provide signals that discourage cell division. 170

turnover number (turn'o-ver num'ber) The number of molecules of substrate with which a single molecule of enzyme can react in a given time. 100

tympanum (tim'pă-num) The eardrum. 581

U

unsaturated (un-sat'yu-ra-ted) A term used to describe the carbon skeleton of a

fatty acid containing carbons that are double-bonded to each other at one or more points. 60

uracil (yu'rah-sil) A single-ring nitrogenous-base molecule in RNA but not in DNA; it is complementary to adenine. 146

V

vaccines (vak'sen) Antigens made so they can start an active immunity without causing disease. 588

vacuoles (vak'yu-ōlz) Large sacs within the cytoplasm of a cell, composed of a single membrane. 77

variable number tandem repeats (va're-ah-b'l num'ber tan'dem re'petz) A type of variation in an organism's DNA sequence that is a repeated sequence; different individuals may have the sequence repeated a different number of times; this variation can occur in regions of the DNA that do not code for genes. 216

variables (var'e-ă-bulz) Factors in an experimental situation or other circumstance that are changeable. 6

vascular tissue (vas'ku-ler tish'yu) Consists of tube-like cells that allow plants to efficiently transport water and nutrients about the plant. 469

vector (vek'tor) An organism that carries a disease or parasite from one host to the next. 228

veins (vānz) The blood vessels that return blood to the heart. 520

ventricles (ven'trĭ-klz) The powerful muscular chambers of the heart whose contractions force blood to flow through the arteries to all parts of the body. 522

vertebrates (vur'tuh-brāts) Animals with backbones. 493

vesicles (vĕ'sĭ-kuls) Small, intracellular, membrane-bound sacs in which various substances are stored. 77

villi (vil'e) Tiny, fingerlike projections in the lining of the small intestine that increase the surface area for absorption. 534

viroid (vi'roid) An infectious particle composed solely of single-stranded RNA. 436

virus (vi'rus) A nucleic acid particle coated with protein that functions as an obligate intracellular parasite. 434

vitamin-deficiency disease (vi'tah-min defish'en-se di-zēz') Poor health caused by the lack of a certain vitamin in the diet—for example, scurvy from lack of vitamin C. 558

vitamins (vi'tah-minz) Organic molecules that cannot be manufactured by the body but are required in very low concentrations for good health. 99

voltage (vōl'tij) A measure of the electrical difference between two points or objects. 569

W

white blood cells (wbcs) (hwit blud sel) Also called leukocytes; formed elements in the blood that lack hemoglobin, are larger than RBCs, and have a nucleus. 520

wood (wood)The accumulation of the xylem in the trunk of gymnosperms. 484

X

X chromosome (eks kro'mo-sōm) The chromosome in a human female egg (and in one-half of sperm cells) that is associated with the determination of sexual characteristics. 597

X-linked genes (eks-lingt jēnz) Genes located on the sex-determining X chromosome. 209

xylem (zi'lem) A kind of vascular tissue that transports water from the roots to other parts of the plant. 469

Y

Y chromosome (wi kro'mo-sōm) The sex-determining chromosome in one-half the sperm cells of human males responsible for determining maleness. 597

Y-linked gene (Y lingk'd jēn) Genes found only on the Y chromosome. 209

Z

zooplankton (zo"o-plank'ton) Nonphotosynthetic aquatic protozoa and tiny animals. 450

zygote (zi'gōt) A diploid cell that results from the union of an egg and a sperm. 173

Credits

Line Art and Text

Chapter 1
1.19, pg 17. : Data from the U.S. Department of Agriculture, Natural Agriculture Statistics. **Box Figure 1:** Advisory Committee on Immunization Practices. American Academy of Pediatric and American Academy of Family Physicians. **How Science Works 1.1, pg. 18:** Advisory Committee on Immunization Practices. American Academy of Pediatric and American Academy of Family Physicians.

Chapter 12
12.4, pg. 244: The MapQuest logo is a registered trademark of MapQuest.com, Inc. Map content © 2005 by MapQuest, Inc. Used with permission.

Chapter 13
13.8, pg. 270: Source: Data from Ian Heap, "The International Survey of Herbicide Resistant Weeds." Available at *www. weedscience.com.*

Chapter 15
Pg. 312: Source: Scripps Institute of Oceanography. Pg. 317: Source: The Canada Centre for Inland Waters, World Wide Web Servers operated by Computing and Programming Services. National Water Research Institute, Environment Canada. Http://www.cciw.ca/green-lane/wildlife/gl-factsheet/herring

Chapter 16
16.2, pg. 325: A Food Web from Ralph D. Bird, "Biotic Communities of the Aspen Parkland of Central Canada," in Ecology, 11 (2): 410. Reprinted with permission of the Ecological Society of America. **Pg. 354:** Source: Mittermeier, R.A., Robles-Gil, P., Hoffmann, M., Pilgrim, J.D., Brooks, T.M., Mittermeier, C.G., Lamoreux, J.L. & Fonsesca, G. (eds). 2004. Hotspots Revisited: Earth's Biologically Richest and Most Endangered Ecoregions. Second Edition. Cemex, Mexico. (Information updated based on content available at *www.biodiversityhotspots.org*; downloaded October 25, 2005.

Chapter 17
17.3, pg. 361: Source: Data from Population Reference Bureau. Reprinted with permission. **17.11, pg. 368:** Gilg, O., Hanski, I., and Sittler, B. 2003. Cyclic dynamics in a simple vertebrate predator-prey community. Science 302, pp. 866-868. Reprinted with permission. **17.14, pg. 372:** Source: Data from Jean Van Der Tak, et al., "Our Population Predicament: A New Look," in Population Reference Bureau, Washington D.C. and updates from recent Population Data Sheets. Reprinted with permission.

Chapter 20
Pg. 438: Source: World Health Organization of the United Nations.

Chapter 22
Pg. 488: Data from: Food and Agriculture Organization of the United Nations.

Chapter 25
25.2, pg. 554: Source: U.S. Department of Agriculture. Table 25.7, pg. 558: Source: Report of the Dietary Guidelines Advisory Committee on the Dietary Guidelines for Americans, 2000.

Chapter 27
Tables 27.A & 27.B, pg. 614: Data from the Center for Disease Control and Prevention. Tracking the Hidden Epidemics: Trends in STD's in the United States 2000.

Photographs

Chapter 1
Opener: © Vol. 71/PhotoDisc/Getty Images; 1.1: © Eldon Enger & Fred Ross/Bob Coyle, photographer; 1.3: © Vol. 43/PhotoDisc/Getty Images; .4(top): © Vol. 86/CORBIS; 1.4(bottom): © CORBIS royalty-free; 1.5: Steve Maslowski/U.S. Fish & Wildlife Services; 1.6: © Barrington Brown/Photo Researchers; 1.7: © The McGraw-Hill Companies, Inc./John Thoeming, photographer; 1.8a: © David M. Phillips/Visuals Unlimited; 1.8b: © Hank Morgan/Photo Researchers; 1.9(left): © Bettmann/CORBIS; 1.9(right): © Bob Coyle/Red Diamond Photos; 1.10a: © Mark E. Gibson/Visuals Unlimited; 1.10b: © PhotoDisc/Getty Images; 1.11: © The McGraw-Hill Companies, Inc./Jim Shaffer, photographer; 1.13: © Vol. 86/CORBIS; 1.14: © Vol. 1/PhotoDisc/Getty Images; 1.15(left): © Vol. 44/PhotoDisc/Getty Images; 1.15(right): © Vol. 6/Photo Disc/Getty Images; 1.16: © Vol. 50/Photo Disc/Getty Images; 1.17(Euplotes): © A.M. Siegelman/Visuals Unlimited; 1.17(Protozoan): © Science VU/Visuals Unlimited; 1.17(Orchid): © Steven P. Lynch; 1.17(Toddlers): © CORBIS royalty-free; 1.20(top): © Stockbyte; 1.20(bottom): © Eldon Enger; p. 18: © Bettmann/CORBIS

Chapter 2
Opener: © REUTERS/Beawiharta; 2.1: © Vol. 85/PhotoDisc/Getty Images; 2.2a,b: © Vol. OS14/PhotoDisc/Getty Images; p. 30: © Vol. OS43/PhotoDisc/Getty Images; 2.6a: © Vol. 16/PhotoLink/Getty Images; 2.6b: © Vol. 24/PhotoLink/Getty Images; 2.6c: © CORBIS royalty-free; 2.7: © SS11/PhotoDisc/Getty Images; 2.10: © SS04/PhotoDisc/Getty Images; p. 35: © Vol. 44/Geostock/Getty Images; 2.13: © CORBIS royalty-free; p. 40(both): © Fred Ross; 2.15a,b: © 2003 Fred Ross.

Chapter 3
Opener: © Fred Ross; p. 45: © Eldon Enger; 3.1(All): © Eldon Enger & Fred Ross/Jim Shaffer, photographer; 3.2a: © Susan McCartney/Photo Researchers; 3.2b: © Will & Deni McIntyre/Photo Researchers; p. 51 (top): © Vol. 83/Photo Disc/Getty Images; p. 51 (bottom): © Vol. 30/Getty Images; 3.10(left, middle): © CORBIS royalty-free; 3.10(right): © Vol. 10/PhotoLink/Getty Images; p. 52(top): © Vol. 19/PhotoLink/Getty Images; p. 53: © Vol. 30/PhotoDisc/Getty Images; 3.14b: From C.E. Creutz, The annexins and exocytosis from *Science,* 258, Nov. 6, 1992 p. 927 fig 3b (photo provided by A. Burger & R. Huber).

Reprinted by permission of the author; p. 56: © Vol. 8/PhotoLink/Getty Images.

Chapter 4
Opener: © Vol. 40/PhotoDisc/Getty Images; 4.1a: © Stock Montage; 4.1b: © Dr. Jeremy Burgess/Science Photo Library/Photo Researchers; 4.2: © Kathy Talaro/Visuals Unlimited; 4.3b: © Ken Collins/Visuals Unlimited; 4.9 (Vacuole): © David M. Phillips/Visuals Unlimited; 4.9 (Golgi): © 2006 Richard Rodewald/Biological Photo Service; 4.9 (Endoplasmic): © 2006 Warren Rosenberg/Biological Photo Service; 4.9 (Lysosome): © K.G. Murti/Visuals Unlimited; 4.15: © Don Fawcett/Photo Researchers; 4.16b: M. Sameni and B.F. Sloane, Wayne State University School of Medicine, Detroit, MI; 4.18: © William Dentler; 4.19: From William Jensen and R.B. Park, *Cell Ultrastructure,* 1967, p. 57. © Brooks/Cole Publishing, a division of International Thompson Publishing, Inc.

Chapter 5
Opener: © Vol. 25/LifeFile/Getty Images; 5.3(both): © The McGraw-Hill Companies, Inc./Jim Shaffer, photographer; p. 98 (top): © Vol. 126/Getty Images; p. 102: © The McGraw-Hill Companies, Inc./Auburn University Photographic Services

Chapter 6
Opener: © Vol. 12/PhotoDisc/Getty Images; p. 122: © Vol. 76/PhotoDisc/Getty Images; p. 123(top): © Vol. 30127/Getty Images; p. 123 (bottom): © CORBIS royalty-free; p. 125: © Vol. 67/Getty Images; p. 126(top): © PhotoLink/Getty Images; p. 126(bottom): Courtesy Tampa Hyperbaric Enterprise, Inc., *www.oxytank.com.*

Chapter 7
Opener: © CORBIS royalty-free; p. 129(bottom): © Vol. 1/PhotoLink/Getty Images; 7.1a: © Steven P. Lynch; 7.1b: © Vol. 16/PhotoLink/Getty Images; p. 133: © Vol. IL063/PhotoDisc/Getty Images; p. 137(Algae): © Vol. 39/Getty Images; p. 137(Field): © Vol. 16/PhotoLink/Getty Images; p. 137(Jade): © Eldon Enger; 7.10: © Vol. 18/PhotoLink/Getty Images.

Chapter 8
Opener: © Julia Kamlish/Science Photo Library; p. 148: © Vol. 18/PhotoDisc/Getty Images; 8.12b: © Biophoto Associates/Photo Researchers p. 154: Centers for Disease Control; 8.17a,b: © Stanley Flegler/Visuals Unlimited; 8.19: © George Musil/Visuals Unlimited.

Chapter 9
Opener: © CORBIS royalty-free; p. 169(left 1-7 whitefish): © The McGraw-Hill Companies, Inc./Kingsley Stern, photographer; p. 169(left 8 - daughter cells): © Ed Reschke; p. 169(mitosis in onion root-all): © The McGraw-Hill Companies, Inc./Kingsley Stern, photographer; p. 171: © Vol. 25/PhotoDisc/Getty Images; 9.13: Courtesy of Fred Williams, U.S. Environmental Protection Agency; 9.14: © James Stevenson/Photo Researchers; 9.16: © Vol. 54/PhotoDisc/Getty Images; 9.38: Courtesy Darlene Schueller.

Chapter 10
Opener: © CORBIS royalty-free; 10.1a,b: © The McGraw-Hill Companies, Inc./Bob Coyle, photographer; p. 195: © Vol. OS23/Getty Images; p. 196: © Index Stock/Getty RF; 10.3a,b: Courtesy V. Orel, Mendelianum Mussei Moraviae, Brno; p. 199: © Steven P. Lynch; 10.5: © Fred Ross; 10.6: © John D. Cunningham/Visuals Unlimited; 10.8a,c,d: Courtesy of Jeanette Navia; 10.8b: © CORBIS royalty-free; p. 211: © Vol. 19/PhotoDisc/Getty Images; 10.11: © The McGraw-Hill Companies, Inc./Bob Coyle, photographer; 10.12(top): © Renee Lynn/Photo Researchers; 10.12(left & right): Courtesy of Mary Drapeau.

Chapter 11
Opener: © Vol. 95/BrandX/Getty Images; 12.1: © Vol. 69/PhotoDisc/Getty Images; 11.8a: © Monsanto; 11.8b: © Vol. 19/PhotoDisc/Getty Images; 11.8c: © Peter Beyer, University of Freiburg, Germany; 11.11a,b: Photo Courtesy: College of Veterinary Medicine, Texas A&M University.

Chapter 12
Opener: © Vol. OS42/PhotoDisc/Getty Images; 12.1: © Vol. 69/PhotoDisc/Getty Images; 12.3(Asia, Europe, Africa): © CORBIS royalty-free; 12.4(top): © John Cancalosi/Tom Stack & Associates; 12.4(bottom): © John Serrao/Visuals Unlimited; p. 244 (top): © Vol. 23/Photo Disc/Getty Images; p. 244(bottom): © Vol. 19/PhotoLink/Getty Images; 12.5: © CORBIS royalty-free; p. 246(left): Courtesy White Sands National Monument, U.S. Department of the Interior, National Park Service; 12.7: © Gary Milburn/Tom Stack & Associates; p. 248: © Smallhorn Land & Cattle, Huson, MT; 12.9(left): © Vol. 7/PhotoLink/Getty Images; 12.9(right): © Vol. 74/Getty Images; 12.10: Courtesy National Center for Genetic Resources Preservation (NCGRP), USDA; 12.11: © Fred Ross; 12.12a: © Vol. 66/PhotoDisc/Getty Images; 12.12b,c: © Fred Ross; 12.12d: © OS18/Getty Images RF; p. 252: Courtesy Virginia Living Museum, Newport News, Virginia; 12.14a,b: © Stanley L. Flegler/Visuals Unlimited; 12.15: © Vol. 2/SS45/Getty Images RF; p. 254: © The McGraw-Hill Companies, Inc./Christopher Kerrigan, photographer; p. 255: © CORBIS royalty-free.

Chapter 13
Opener: Courtesy Centers for Disease Control, Susan Lindsey; p. 261(Darwin): Courtesy Library of Congress; p. 261(Wallace): National Library of Medicine, Historical Images Collection; p. 262: © Christopher Ralling; p. 263 (left): CORBIS royalty-free; p. 263 (right): © Vol. 44/PhotoDisc/Getty Images; p. 264: © Vol. 44/PhotoDisc/Getty Images; 13.3: © Ted Levin/Animals Animals/Earth Scenes; 13.4: © Vol. 46/Getty Images; p. 266: © Vol. 9/PhotoLink/Getty Images; p. 267: © Siebert/Custom Medical Stock Photo; 13.5: Joe McDonald/Visuals Unlimited; 13.6(top left): © Eldon Enger; 13.6(bottom left): © Vol. 6/PhotoLink/Getty Images; 13.6(right): © Vol. 44/PhotoDisc/Getty Images; p. 270: © CORBIS royalty-free; 13.9: © Punchstock RF/

Comstock; 13.11: Courtesy U.S. Fish & Wildlife Service; p. 274 (top): © Vol. 76/ PhotoDisc/Getty Images; p. 274 (bottom): © Getty RF; p. 275: © CORBIS royalty-free; 13.12: © Mitch Reardon/Photo Researchers.

Chapter 14

Opener: © AP/Wide World Photos; p. 282(left): Courtesy John and Karen Hollingsworth, U.S. Fish and Wildlife Service; 14.1a,c: © Fred Ross; 14.1b: © Eldon Enger; p. 283 (left): © Native Organic Products; 14.3: © BrandX/ Punchstock; 14.4: © Vol. 19/PhotoDisc/ Getty Images; 14.5: © Mike Blair; 14.6a: © Vol. 44/PhotoDisc/Getty Images; 14.6b: Courtesy C. Nelson; p. 289(shar pei): Courtesy Marilyn R. Vinson; p. 289(wolf): © PunchStock RF; p. 289(basenji): Courtesy M.C. Felix; p. 289(husky); p. 289(pharoah): Courtesy Sarah and Emily Kerridge; p. 289(Ibizan): Courtesy Lisa Puskas; p. 296: © BrandX Pictures/PunchStock; p. 299: Courtesy of the Field Museum.

Chapter 15

Opener: © CORBIS royalty-free; 15.1a: © Getty RF; 15.1b: © Eldon Enger; 15.3 (#1,2,3, and 5): © Getty RF; 15.3 (#4): © Digital Vision RF; 5.3 (#6): © Image 100 Ltd/RF; 15.13: © CORBIS royalty-free; 15.14(top): © Getty RF; 15.14(bottom): © CORBIS royalty-free.

Chapter 16

Opener: © CORBIS royalty-free; 16.4a: © W. Perry Conway/CORBIS; 16.4b: © John D. Cunningham/Visuals Unlimited; 16.5a: © Kennan Ward/CORBIS; 16.5b: © Getty RF; 16.7a: © Vol. 44/ PhotoDisc/Getty Images; 16.7b: © Stephen Dalton/Photo Researchers; 16.7c: © Vol. 44/ PhotoDisc/Getty Images; 16.8a: © J.H. Robinson/Photo Researchers; 16.8b: © Vol. 54/PhotoDisc/Getty Images; 16.8c: © Malcolm Kitto, Papillo/CORBIS; 16.10(top): © Vol. 44/PhotoDisc/Getty Images; 16.10(bottom): © Vol. 19/Photo Disc/Getty Images; 16.11: © Edward R. Degginger/Bruce Coleman; 16.12a: © Douglas Faulkner/Sally Faulkner Collection; 16.12b: © CORBIS royalty-free; 16.13a: © Vol. 6/PhotoDisc/Getty Images; 16.13b: © CORBIS royalty-free; 16.13c: © Eldon Enger; 16.16: © Vol. 16/ PhotoDisc/Getty Images; 16.17: © CORBIS royalty-free; 16.18: © Vol. 44/PhotoDisc/ Getty Images; 16.19: © Vol. 36/PhotoDisc/ Getty Images; 16.20(left, right), 6.21(left): © CORBIS royalty-free; 16.21(right): © Vol. 44/PhotoDisc/Getty Images; 16.22(left, right): © CORBIS royalty-free; 16.23(left): © Getty RF; 16.23 (right), 16.25: © Eldon Enger; 16.29(dandelions): © Vol. 26/PhotoDisc/Getty Images; 16.29(mute swans): © Stockbyte RF; 16.30a: © Hulton-Deutsch Collection/ CORBIS; 16.30b: © Vol. 31/PhotoDisc/ Getty Images; p. 347 (top): © Judy Enger; 16.31(left): Courtesy U.S. Fish & Wildlife Service; 16.31(right): Courtesy U.S. Fish & Wildlife Service, Tracy Brooks; 16.32(top): © Vol. 19/PhotoDisc/Getty Images; 16.32 (middle): © Vol. 16/PhotoDisc/Getty Images; 16.32(bottom): © Vol. 19/Photo Disc/Getty Images; 16.33(top): Courtesy USDA; 16.33(bottom): © Vol. 19/Photo Disc/Getty Images.

Chapter 17

Opener: © CORBIS royalty-free; 17.4: © Vol. 80/PhotoDisc/Getty Images; 17.5: © Peter Johnson/CORBIS; 17.6(top): © Vol. 6/PhotoDisc/Getty Images; 17.6(middle): © CORBIS royalty-free; 17.6(bottom): © Vol. EP073/Getty Images; 17.7a: © Stephen J. Krasemann/Photo Researchers; 17.7b: © Walt Anderson/ Visuals Unlimited; p. 371: © The Hudson Bay Project. Photo provided by Robert L. Jefferies; 17.15: © Vol. 67/PhotoDisc/Getty Images; p. 373: © Bettmann/CORBIS; 17.16: © CORBIS royalty-free; 17.17: © Vol. 25/PhotoDisc/Getty Images; 17.18: © Vol. 6/PhotoDisc/Getty Images.

Chapter 18

Opener: © Eldon Enger; 18.2: © Vol. 61/ PhotoDisc/Getty Images; p. 380: © Leonard Choi; 18.4a: © Eldon Enger; 18.4b: © Getty RF; 18.5: © Paul A. Souders/CORBIS; 18.6(top, bottom): © Lincoln P. Brower Nature Photography; 18.7: © Vol. EP004/Getty Images; 18.8a: Courtesy of Sybille Kalas; 18.8b: © Vol. 6/ PhotoDisc/Getty Images; 18.10: © CORBIS royalty-free; 18.11: © Punchstock; 18.14: © Vol. 44/PhotoDisc/Getty Images; 18.15: © CORBIS royalty-free; 18.16: © Vol. 112/PhotoDisc/Getty Images; 18.17: © Eldon Enger; 18.18: © CORBIS royalty-free; 18.19: © Harry Rogers/Photo Researchers; 18.20: © Vol. 44/PhotoDisc/Getty Images; 18.22: © Gallo Images/CORBIS; 18.23: © Eldon Enger; 18.24(left, right): © Vol. EP073/Getty Images.

Chapter 19

Opener: © Dr. Tim Baker/Visuals Unlimited; 19.7: Courtesy of Abe Tetsuya; 19.11a: © Clouds Hill Imaging Ltd./ CORBIS; 19.11b: © Niall Benvie/CORBIS; 19.13: © Fred Ross.

Chapter 20

Opener(American robin): © Arthur Morris/CORBIS; (blackbird, European robin): © Stockbyte; 20.1: © Vol. 1/ PhotoDisc/Getty Images; 20.2: © Stock Montage; 20.4a: © CORBIS royalty-free; 20.4b: © A.M. Siegelman/Visuals Unlimited; 20.4c: © Eldon Enger; 20.4d: © Steven P. Lynch; 20.5a: © 2006 W.B. Saunders/Biological Photo Service; 20.5b: © Eldon Enger; 20.5c: US Geological Survey; 20.6: © CORBIS royalty-free; 20.7a: © David Denning Photography/BioMEDIA Associates; 20.7b: © Don & Pat Valenti/Tom Stack & Associates; 20.7c: © David Denning Photography/BioMEDIA Associates; 20.10: © The McGraw-Hill Companies, Inc./John A. Karachewski, photographer; 20.11(algae): © Steven P. Lynch; 20.11(red Algae): © Steven P. Lynch; 20.11(protozoan): © M. I. Walker/Photo Researchers; 20.11(slime mold): © Bill Beatty/ Visuals Unlimited; 20.12(top left): © Vol. 18/ PhotoDisc/Getty Images; 20.12(bottom left): © Judy Enger; 20.12(right): © CORBIS royalty-free; 20.13(ferns, plants, conicers): © Eldon Enger; 20.13(mosses, algae): © Steven P. Lynch; 20.14(Sea Anemone): © Vol. 44/Photo Disc/Getty Images; 20.14(sponge): © Vol. 112/PhotoDisc/Getty Images; 20.14(earthworm): © Vol. OS18/Getty Images; 20.14(insect): © PunchStock; 20.14(mammal, bird): © Digital Vision.

Chapter 21

Opener: © D. Hurst/Alamy Images; 21.2: © PunchStock RF; p. 444(Streptoccocus): © David Phillips/ Visuals Unlimited; p. 444(E. coli): © R. Kessel/G. Shih/Visuals Unlimited; p. 444(Lactobacillus): © Moredum Animal Health/SPL/Photo Researchers; p. 444(Corynebacterium): © CNRI/SPL/ Photo Researchers; 21.3: © Jerome Wexler/Photo Researchers; 21.4: USDA; 21.5a: © 2006 T.J. Beveridge/Biological Photo Service; 21.5b: © Steven P. Lynch; 21.6: NOAA; 21.8c: © Vol. 4/PhotoDisc/ Getty Images; 21.9a,b: © Steven P. Lynch; 21.10(left): Courtesy of Leland Johnson; 21.10(middle, right): © Steven P. Lynch; 21.11(top,bottom): © Lester V. Bergman/ CORBIS; 21.12(top): © Michael Abbey/ Photo Researchers; 21.12(bottom): © Science VU/Visuals Unlimited; 21.15: © Edward S. Ross; p. 456: © Brand X/PunchStock; 21.16: © Cabisco/Visuals Unlimited; 21.17c: © Steven P. Lynch; 21.18: © Michael & Patricia Fogden/ CORBIS; 21.19(chytridiomycota): Courtesy of Tom Volk, University of Wisconsin-La Crosse; 21.19(zygomycota): © D. Yeske/Visuals Unlimited; 21.19(ascomycota): © CORBIS royalty-free; 21.19(basidiomycota): © Edward S. Ross; p. 459 (bottom): © Nancy M. Wells/Visuals Unlimited; p. 460(top): © 2006 Richard Humbert/Biological Photo Service; 21.21a: © Steven P. Lynch; 21.21b: © Vol. BS29/Getty Images; 21.22: © David M. Dennis/Tom Stack & Associates; 21.23: © CORBIS royalty-free.

Chapter 22

Opener: © Steven P. Lynch; 22.1(left): © Vol. 24/PhotoDisc/Getty Images; 22.1(middle): © Digital Vision/PunchStock; 22.1(right): © Vol. 36/PhotoDisc/Getty Images; 22.5(all): © Steven P. Lynch; 22.6(right): Courtesy of Wilfred A. Cote, SUNY College of Environmental Science and Forestry; 22.7(left): © Randy Moore/Visuals Unlimited; 22.10: © Steven P. Lynch; 22.14(left): © Vol. DV312/Getty Images; 22.14(right): © Digital Vision; 22.15(left, right): © Steven P. Lynch; 22.15(middle): © Vol. 18/Getty Images; 22.18 (left): © The McGraw-Hill Companies, Inc./Lars A. Niki, photographer; 22.18 (middle): © Steven P. Lynch; 22.18(right): © Bill Beatty/Visuals Unlimited; 22.21(left): © Steven P. Lynch; 22.21(right): © Vol. 1/Getty Images; 22.22burr, burr on shirt, Milkweed): © Eldon Enger; 22.22(raspberries): © Judy Enger; 22.25: © Fred Ross; 22.27 © Judy Enger; 22.28: © Fred Ross; p. 487: © Eldon Enger.

Chapter 23

Opener: © CORBIS royalty-free; 23.1(frog): © Fred Ross; 23.1(snail): © Vol. OS50/Getty Images; 23.1(jellyfish): © Vol. 112/PhotoDisc/Getty Images; 23.1(mantis): © Getty RF; 23.1(starfish): © Brand X/PunchStock; p. 494: © Charles Cole; 23.9b: © Brand X/PunchStock; 23.9c: © CORBIS royalty-free; 23.11(jellyfish): © CORBIS royalty-free; 23.11(anemone): © Vol. 112/Photo Disc/Getty Images; 23.11(brain coral): © PunchStock; 23.12a,b: © Bruce Russell/ BioMedia Associates; 23.12c: © NIBSC/ Science Photo Library/Photo Researchers; 23.18(top): © Robert DeGoursey/Visuals Unlimited; 23.18(middle): © Michael DiSpezio; 23.18(bottom): © E. R. Degginger/ Photo Researchers; 23.18(right): © Gary Meszaros/Visuals Unlimited; 23.19(chiton): © Robert A. Ross/RARE Photography; 23.19(snail): © Vol. DV13/Getty Images; 23.19(clam): © age fotostock/SuperStock; 23.19(octopus): © Brand X/PunchStock; 23.20(grasshopper): © CORBIS royalty-free; 23.20(millipede, centipede): © Brand X/Getty Images; 23.20(shrimp): © Brand X/PunchStock; 23.20(scorpion): © Digital Vision/Getty Images; 23.22(starfish): © CORBIS royalty-free; 23.22(brittle Star): © 2002 Jim Greenfield/Image Quest 3-D; 23.22(urchin): © David Wrobel /Visuals Unlimited; 23.22(cucumber): © Michael DiSpezio; 23.22(feather star): © Brand X/PunchStock; 23.24a: © Russ Kinne/ Photo Researchers; 23.24b: © Tom Stack/ Tom Stack & Associates; 23.25a,b: © Brand X/PunchStock; 23.26(top): © Brand X/PunchStock; 23.26(bottom): © Vol. 112/PhotoDisc/Getty Images; 23.28(egg): © Kent T. Breck/Animals Animals/Earth Scenes; 23.28(larva, pupa): © PunchStock; 23.28(pupa w/butterfly): © Raymond Coleman/Visuals Unlimited; 23.28(adult butterfly): © PunchStock; p. 512: © Raymond Tercafs - Productions Services/Bruce Coleman; 23.19(tadpole): © DK Limited/CORBIS; 23.19(salamander): © Brand X/PunchStock; 23.19(painted frog): © Digital Vision/Getty Images; 23.19(Inflated frog: © Vol. 44/PhotoDisc/ Getty Images; 23.31(turtle): © Vol. 08/ PhotoDisc/Getty Images; 23.31(alligators): © PunchStock; 23.31(garden snake): © Vol. 1/PhotoDisc/Getty Images; 23.31(lizard): © CORBIS royalty-free; 23.32(hummingbird): © CORBIS royalty-free; 23.32(ostrich): © Digital Vision; 23.32(eagle): © Vol. 44/PhotoDisc/Getty Images; 23.33(echidna): © Kevin Schafer/ CORBIS; 23.22(kangaroo): © CORBIS royalty-free; 23.22(zebra): © Brand X Pictures.

Chapter 24

Opener: © Fred Ross; p. 526: © Vol. 103/ PhotoDisc/Getty Images; p. 522: © Vol. 18/ PhotoLink/Getty Images; p. 526: © Vol. EP046/PhotoDisc/Getty Images; p. 533(house, plaque): © Fred Ross; 24.11(inset): Photo by Susuma Ito, from Flickinger, et al., *Medical Cellular Biology*, © 1979 Elsevier Inc.; 24.11(background): © Manfred Kage/Peter Arnold.

Chapter 25

Opener: © CORBIS royalty-free; p. 542: Copyright © LECO Corporation, St. Louis, Michigan. Reprinted by permission.; 25.1(bread, cakes): © Vol. 20/ PhotoDisc/Getty Images; 25.1(healthy food): © Digital Vision/GettyImages; 25.1(junk food): © Vol. 93/PhotoDisc/ Getty Images; 25.1(rice): © Vol. 19/Photo Disc/Getty Images; p. 544: © Kate Brooks/ CORBIS; p. 547: © Bettmann/CORBIS; p. 548 (top & bottom): © The McGraw-Hill Companies, Inc./Bob Coyle, photographer; p. 549: © The McGraw-Hill Companies, Inc./Bob Coyle, photographer; p. 551(top left and bottom right): © The McGraw-Hill Companies, Inc./Bob Coyle, photographer; p. 551 (right): © Fred Ross; p. 554: © Image 100/RF; p. 555(left): © Tom & Dee Ann McCarthy/CORBIS; p. 555(right): © Vol. 48/PhotoDisc/Getty Images; p. 556: © Randy Faris/CORBIS; 25.3(both) Courtesy of Dr. Randy Sansone; 25.4: © Paul A. Souders; 25.6a,b: Courtesy of David Dempster; 25.6c: © Vol. 40/PhotoDisc/Getty Images; p. 561(bottom): © David H. Wells/CORBIS; p. 562(left): © CORBIS royalty-free; p. 563 (top): © Henry Ray Abrams/ Reuters/CORBIS; p. 563(bottom): © Vol. 20/PhotoDisc/Getty Images;

Chapter 26

Opener: © Vol. 18/PhotoDisc/Getty Images; p. 574: © Vol. 54/PhotoDisc/Getty Images; 26.16b: © Victor B. Eichler, Ph.D.; p. 586: Centers for Disease Control; 26.19b: Courtesy Dr. Arthur J. Olson, Scripps Institute; 26.20a: © Matt Meadows/ Peter Arnold; p. 590: © Fred Ross; p. 592: National Library of Medicine, Historical Images Collection.

Chapter 27

Opener: © CORBIS royalty-free; 27.1: Margaret Mead Collection, Library of Congress; p. 597: © Alen MacWeeney/ CORBIS; p. 598: (man, girl): © Fred Ross; p. 599 (top): Courtesy of Thomas G. Brewster, Foundation for Blood Research, Scarborough, ME. From T.G. Brewster and S. Gerald Park, "Chromosome Disorders Associated with Mental Retardation" in Pediatric Annals, 7(2), 1978; 27.3: From M. Bartalos and T.A. Baramski, *Medical Cytogentetics*. Williams & Wilkins, 1967, p. 133, f. 9-2; 27.4 (left) and (right): Courtesy of Dr. Kenneth L. Becker from Fertility and Sterility, 23:5668-78, Williams & Wilkins, 1972; 27.11: © SIU/ Peter Arnold, Inc.; p. 607: © McGraw-Hill Digital Library; p. 609: © Vol. 29/Photo Disc/Getty Images; p. 610: © Digital Vision/PunchStock; p. 611(top): © Vol. SS17/ Getty Images; p. 611(bottom): © Vol. SS16/ Getty Images; 27.17 Courtesy Kelly Kurchak; p. 612 (suppositories): © Fred Ross; p. 613(pills): © The McGraw-Hill Companies, Inc./Bob Coyle, photographer; p. 613(implants): © Hank Morgan/Photo Researchers, Inc.; p. 613 (diaphragm): © The McGraw-Hill Companies, Inc./Bob Coyle, photographer; p. 614: Centers for Disease Control, James Gathany; p. 615 (male condom, female condom): © The McGraw-Hill Companies, Inc./Bob Coyle, photographer; p. 615(IUD): © The McGraw-Hill Companies, Inc./Vincent Ho, photographer; p. 627: © CORBIS royalty-free.

Index

B

S